# Applied Geometry and Discrete Mathematics

## The Victor Klee Festschrift

# DIMACS
Series in Discrete Mathematics
and Theoretical Computer Science

Volume 4


# Applied Geometry and Discrete Mathematics

## The Victor Klee Festschrift


Peter Gritzmann
Bernd Sturmfels
Editors



NSF Science and Technology Center
in Discrete Mathematics and Theoretical Computer Science
A consortium of Rutgers University, Princeton University,
AT&T Bell Labs, Bellcore

**American Mathematical Society**
**Association for Computing Machinery**

1980 *Mathematics Subject Classification* (1985 *Revision*). Primary 05, 15, 28, 46, 51, 52, 57, 65, 68, 90.

---

**Library of Congress Cataloging-in-Publication Data**

Applied geometry and discrete mathematics: The Victor Klee festschrift/Peter Gritzmann, Bernd Sturmfels, editors.

p. cm.—(DIMACS series in discrete mathematics and theoretical computer science, ISSN 1052-1798; v. 4)

Includes bibliographical references.

AMS: ISBN 0-8218-6593-5 (acid-free paper)

ACM: ISBN 0-89791-385-X (acid-free paper)

1. Mathematics. 2. Klee, Victor. I. Gritzmann, Peter, 1954- . II. Sturmfels, Bernd, 1962- . III. Klee, Victor. IV. Series.

QA7.A664 1991 90-26934

510—dc20 CIP

---

To order through AMS contact the AMS Customer Services Department, P.O. Box 6248, Providence, Rhode Island 02940-6248 USA. For VISA or MASTERCARD orders call 1-800-321-4AMS. Order code DIMACS/4.

To order through ACM contact ACM Order Department, P.O. Box 64145, Baltimore, Maryland 21264. Phone 301-528-4261. Order number 222912.



Printed in the United States of America.

The paper used in this book is acid-free and falls within the guidelines established to ensure permanence and durability. ♾

This publication was typeset using AMS-TEX, the American Mathematical Society's TEX macro system.

10 9 8 7 6 5 4 3 2 1 95 94 93 92 91

# Foreword

This DIMACS volume, The "Victor Klee Festschrift," is a collection of research and survey papers that are related to the work of Victor Klee. The publication of this book on the occasion of Professor Klee's 65th birthday mirrors the breadth of his mathematical contributions. We especially thank the editors, Peter Gritzmann and Bernd Sturmfels, for preparing a volume that contains articles on such a variety of subjects, and that is a suitable tribute to a leader in the field of discrete mathematics.

Daniel Gorenstein, Director
Robert Tarjan, Co-Director
Fred S. Roberts, Associate Director

# Brief Contents


| | |
|---|---|
| Preface | ix |
| Biography of Victor Klee | xi |
| Bibliography of Victor Klee | xvii |
| Contents (in alphabetical order by author) | xxxi |
| List of Papers (by subjects) | xxxv |
| Contributed Articles | 1–608 |

# Preface

This volume comprises a collection of research articles dedicated to Victor Klee on the occasion of his 65th birthday in September 1990. All papers are related to Victor Klee's research work, and, in view of his broad interests, a wide range of areas in mathematics and its applications are touched upon here. These areas include

- Discrete and Computational Geometry,
- Classical and Computational Convexity,
- Convex Polytopes and their Relatives,
- Combinatorics, Polyhedral Combinatorics, and Graph Theory,
- Functional Analysis,
- Mathematical Programming and Optimization, and
- Theoretical Computer Science.

Victor Klee has made significant contributions not only to all of the above fields, but also to mathematics education, mathematical methods in economics and the decision sciences, applications of discrete mathematics in the biological and social sciences, and information linkage between applied mathematics and industry. Rather than attempting to summarize or comment on Victor Klee's numerous professional achievements, we let his vita and bibliography speak for themselves.

Following the spirit of Victor Klee's holistic view of mathematics, the present collection is not divided into mathematical subcategories, but the articles appear in alphabetical order by first author. In order to facilitate browsing through this volume and to give easy access to papers belonging to the same area, we include a list of papers by subject area.

We are indebted to the Center for Discrete Mathematics and Theoretical Computer Science, in particular to its director Daniel Gorenstein, and to the American Mathematical Society for their help in arranging the publication of this volume. We wish to thank the referees for their invaluable help and the authors for their enthusiastic support throughout this project. But, above all, we join all contributors in their birthday wishes expressing the deepest gratitude to Victor Klee for all that he has given to us.

Peter Gritzmann and Bernd Sturmfels
September 1990

*photograph by Lisette Klee*

— Victor Klee —

# Biography of Victor Klee

## Personal

Born in San Francisco, 1925

## Education

Ph.D., University of Virginia, 1949

B.A., Pomona College, 1945

## Honorary Degrees

D.Sc., Université de Liège, 1984

D.Sc., Pomona College, 1965

## Awards

Pomona College

- David Prescott Barrows Award for Distinguished Achievement, 1988

Reed College

- Vollum Award for Distinguished Accomplishment in Science and Technology, 1982

Alexander von Humboldt Stiftung

- Preisträger, 1980–1981

Mathematical Association of America

- C. B. Allendoerfer Award, 1980
- Annual Award for Distinguished Service to Mathematics, 1977
- L. R. Ford Award, 1972

University of Virginia

- President's and Vistor's Research Prize, 1952

## Full-time Employment

University of Washington

- Professor of Mathematics, 1957–present
- Associate Professor, 1954–1957
- Assistant Professor, 1953–1954

Adjunct Professor of Computer Science, 1974–present
Professor or Adjunct Professor of Applied Mathematics, 1976–1984

University of Western Australia
Visiting Professor, 1979

University of Victoria
Visiting Professor, 1975

T. J. Watson Research Center, IBM
Full-time Consultant, 1972

University of Colorado
Visiting Professor, 1971

University of California, Los Angeles
Visiting Associate Professor, 1955–1956

University of Virginia
Assistant Professor, 1949–1953
Instructor, 1947–1948

## Fellowships

Senior Fellow, Institute for Mathematics and its Applications, Minneapolis, 1987

Mathematical Sciences Research Institute, Berkeley, 1985–1986

Guggenheim Fellow, University of Erlangen-Nürnberg, 1980–1981

Center for Advanced Study in the Behavioral Sciences, Stanford, 1975–1976

Sloan Foundation Fellow, University of Copenhagen, 1959–1960

National Science Foundation Senior Postdoctoral Fellow, University of Copenhagen, 1958–1959

Research Fellow of the Alfred P. Sloan Foundation, 1956–1958

National Research Council, Institute for Advanced Study, 1951–1952

A. E. C. Predoctoral, University of Virginia, 1948–1949

Du Pont Predoctoral, University of Virginia, 1945–1947

## Part-time Consultant

W. H. Freeman and Company, 1976–present

Holt, Rinehart and Winston, Inc., 1966–1976

E. I. du Pont de Nemours, Inc., 1968–1972

The RAND Corporation, 1966–1970

Boeing Scientific Research Laboratories, 1963–1969

## Professional Societies

American Mathematical Society
Associate Secretary, 1955–1958

Symposium on Convexity, Chairman, Organizing Committee, 1961
Council, 1964–1966, 1969–1971
Executive Committee, 1969–1970

American Association for the Advancement of Science
Chairman of Section A, 1975
Fellow, 1976–present

Mathematical Association of America
Board of Governors, 1967–1978
First Vice-President, 1968–1970
President-Elect, 1970–1971
President 1971–1973

Sigma Xi
National Lecturer, 1969

Society for Industrial and Applied Mathematics
Council, 1966–1968

Also member of Association for Computing Machinery, International Linear Algebra Society, Mathematical Programming Society, Operations Research Society, and Phi Beta Kappa.

**Invited Lectures**

International Congress of Mathematicians, Vancouver, 1974

Eighth International Symposium on Mathematical Programming, Stanford, 1973 (Plenary Speaker)

Invited lectures at various annual meetings of American Mathematical Society, Mathematical Association of America, Society for Industrial and Applied Mathematics, American Association for the Advancement of Science, Canadian Mathematical Congress, Deutsche Mathematiker Vereinigung, Deutsche Gesellschaft für Mathematik, Ökonomie und Operations Research

Invited hour addresses at national or international conferences devoted to the following subjects:

Applications of Combinatorics in the Biological and Social Sciences (1988)
Applied Linear Algebra (1988)
Combinatorics and Geometry (1989)
Discrete Optimization (1981, 1980, 1977)
Discrete Geometry (1981, 1966, 1962)
Convexity (1980, 1975, 1965, 1961)
Information Linkage between Applied Mathematics and Industry (1978)

Combinatorial Mathematics (1990, 1978, 1969, 1969, 1968, 1968, 1963)

Computing in Algebra and Number Theory (1975)

Mathematical Programming (1988, 1982, 1973, 1967)

Set-Theoretic Topology (1973, 1961, 1955)

Mathematical Methods of Economics (1972)

Graph Theory and its Applications (1983, 1972, 1969)

Algorithmic and Applied Combinatorics (1986, 1983, 1971, 1969)

Teaching of Geometry (1990, 1988, 1969, 1967)

Calculus of Variations and Control Theory (1968)

Mathematics of the Decision Sciences (1967)

Scientific Computing (1964)

Functional Analysis (1964, 1960, 1960)

Polytopes and Convex Sets (1990)

Applied and Computational Convexity (1990)

Operations Research (1990, 1989, 1987, 1983, 1983, 1980)

## Current Editorships

Discrete Mathematics

Discrete Applied Mathematics

Journal of Combinatorial Theory

Linear Algebra and Applications

Mathematics of Operations Research

Discrete and Computational Geometry

## Ph.D. Students

29 in Mathematics

1 in Applied Mathematics

1 in Computer Science

## Some other activities of the last six years

Institute for Mathematics and its Applications

Chairman of Organizing Committee for Year-long Program in Applied Combinatorics

Coordinator, Program in Discrete and Computational Geometry

Advisory Committee, Program in Applied Linear Algebra

Board of Governors

Mathematical Sciences Research Institute

Board of Trustees

Centre de Recherche de Mathématiques Appliquées, Université de Montréal

Steering Committee

Cornell University

Advisory Panel, Center for Research in Discrete Optimization

University of Florida

Scientific Board, Center for Research in Discrete Optimization

American Mathematical Society

Nominating Committee

Centennial Fellowship Committee

Organizing Committee, Summer Workshop on Mathematical Developments Related to Linear Programming

Invited speaker at special sessions on differential equations, functional analysis, convex sets, combinatorics, discrete geometry

Mathematical Association of America

Committee on the Annual Award for Distinguished Service

Ad Hoc Committee on Awards

Mathematical Programming Society

International Program Committee

Association for Computing Machinery

Program Committee for a symposium on computational geometry

Operation Research Society of America

Arranged special session of papers on mathematical aspects of linear programming

# Bibliography of Victor Klee


**1946**

1. *On the equation,* $\phi(x) = 2m$, Amer. Math. Monthly **53**, 327–328.

**1947**

2. *On a conjecture of Carmichael,* Bull. Amer. Math. Soc. **53**, 1183–1186.
3. *On completing a determinant,* Amer. Math. Monthly **54**, 96–97.
4. *Some remarks on Euler's totient,* Amer. Math. Monthly **54**, 332.

**1948**

5. *A generalization of Euler's* $\phi$*-function,* Amer. Math. Monthly **55**, 358–359.
6. *The support property of a convex set in a linear normed space,* Duke Math. J. **15**, 767–772.

**1949**

7. *On a problem of Erdös,* Amer. Math. Monthly **56**, 21–22.
8. *A note on Fermat's congruence,* Amer. Math. Monthly **56**, 626–628.
9. *A characterization of convex sets,* Amer. Math. Monthly **56**, 247–249.
10. *Dense convex sets,* Duke Math. J. **16**, 351–354.

**1950**

11. *Some characterization of reflexivity,* Rev. Cienc. (Lima) **52**, 15–23.
12. *Decomposition of an infinite-dimensional linear system into ubiquitous convex sets,* Amer. Math. Monthly **57**, 540–541.

**1951**

13. *Some characterizations of compactness,* Amer. Math. Monthly **58**, 389–393.

14. *On certain intersection properties of convex sets*, Canad. J. Math. **3**, 272–275.
15. *Convex sets in linear spaces*, Duke Math. J. **18**, 443–466.
16. *Convex sets in linear spaces.* II, Duke Math. J. **18**, 877–883.

**1952**

17. *Invariant metrics in groups (solution of a problem of Banach)*, Proc. Amer. Math. Soc. **3**, 484–487.
18. *Convex functions and upper semi-continuous collections*, (with R. D. Anderson), Duke Math. J. **19**, 349–357.

**1953**

19. *The critical set of a convex body*, Amer. J. Math. **75**, 178–188.
20. *Convex sets in linear spaces.* III, Duke Math. J. **20**, 105–112.
21. *Convex bodies and periodic homeomorphisms in Hilbert space*, Trans. Amer. Math. Soc. **74**, 10–43.
22. *On a theorem of Béla Sz.-Nagy*, Amer. Math. Monthly **60**, 618–619.

**1954**

23. *Invariant extension of linear functionals*, Pacific J. Math. **4**, 37–46.
24. *Some remarks on continuous transformations*, (with W. R. Utz), Proc. Amer. Math. Soc. **5**, 182–184.
25. *A characterization of reflexivity by the lattice of closed subspaces*, (with E. E. Floyd), Proc. Amer. Math. Soc. **5**, 655–661.
26. *Common secants for plane convex sets*, Proc. Amer. Math. Soc. **5**, 639–641.

**1955**

27. *A note on extreme points*, Amer. Math. Monthly **62**, 30–32.
28. *Some topological properties of convex sets*, Trans. Amer. Math. Soc. **78**, 30–45.
29. *Separation properties of convex cones*, Proc. Amer. Math. Soc. **6**, 313–318.
30. *Boundedness and continuity of linear functionals*, Duke Math. J. **22**, 263–270.
31. *Some finite-dimensional affine topological spaces*, Portugal. Math. **14**, 27–30.
32. *On metric independence and linear independence*, (with L. M. Blumenthal), Proc. Amer. Math. Soc. **6**, 732–734.
33. *Topological structure of normed linear spaces*, Summary of Lectures and Seminars, Summer Institute on Set Theoretic Topology, Madison, Wisconsin, Amer. Math. Soc., Providence, pp. 132–134.

**1956**

34. *Solution of a problem of E. M. Wright on convex functions*, Amer. Math. Monthly **63**, 106–107.
35. *A note on topological properties of normed linear spaces*, Proc. Amer. Math. Soc. 7, 673–674.
36. *Strict Separation of Convex Sets*, Proc. Amer. Math. Soc. 7, 735–737.
37. *The structure of semispaces*, Math. Scand. **4**, 54–64.
38. *An example in the theory of topological linear spaces*, Arch. Math. **3**, 362–366.
39. *Iteration of the 'Lin' operation for convex sets*, Math. Scand. **4**, 231–238.
40. *A note on certain function spaces*, (with M. E. Rudin), Arch. Math. **7**, 469–470.
41. *Fixed-point sets of periodic homeomorphisms of Hilbert space*, Ann. of Math. **64**, 393–395.

**1957**

42. *Extremal structure of convex sets*, Arch. Math. **8**, 234–240.
43. *On a method of mapping due to Kadeč and Bernstein*, (with R. G. Long), Arch. Math. **8**, 280–285.
44. *On a problem of Banach*, Colloq. Math. **5**, 78.
45. *Homogeneity of infinite-dimensional parallelotopes*, Ann. of Math. **66**, 454–460.

**1958**

46. *Extremal structure of convex sets.* II, Math. Z. **69**, 90–104.
47. *On the Borelian and projective types of linear subspaces*, Math. Scand. **6**, 189–199.

**1959**

48. *Some characterizations of convex polyhedra*, Acta Math. **102**, 79–107.
49. *Some new results on smoothness and rotundity in normed linear spaces*, Math. Ann. **139**, 51–63.
50. *Continuous convex sets*, (with D. Gale), Math. Scand. **7**, 379–391.

**1960**

51. *Polyhedral sections of convex bodies*, Acta Math. **103**, 243–267.
52. *An example related to the fixed-point property*, Nieuw Arch. Wisk. **8**, 81–82.
53. *Shrinkable neighborhoods in Hausdorff linear spaces*, Math. Ann. **141**, 281–285.
54. *Leray-Schauder theory without local convexity*, Math. Ann. **141**, 286–296. (Corrections Math. Ann. **145** (1962), 464–465.)

55. *Asymptotes and projections of convex sets*, Math. Scand. **8**, 356–362.
56. *Mappings into normed linear spaces*, Fund. Math. **49**, 25–34.
57. *Circumspheres and inner products*, Math. Scand. **8**, 363–370.

**1961**

58. *Stability of the fixed-point property*, Colloq. Math. **8**, 43–46.
59. *Convexity of Chebyshev sets*, Math. Ann. **142**, 292–304.
60. *Topological equivalence of a Banach space with its unit cell*, Bull. Amer. Math. Soc. **67**, 286–290.
61. *Relative extreme points*, Proceedings of the International Symposium on Linear Spaces, Jerusalem, Israel, 1960, pp. 286–290.
62. *A question of Katětov concerning the Hilbert parallelotope*, Proc. Amer. Math. Soc. **12**, 900–903.

**1962**

63. *A conjecture on weak compactness*, Trans. Amer. Math. Soc. **104**, 394–402.
64. *Exotic topologies for linear spaces*, Proceedings of the International Symposium on Topology, Prague, Czechoslovakia, 1961, pp. 238–249.

**1963**

65. *Topological structure of infinite-dimensional linear spaces: the classification problem*, (special volume devoted to the International Conference on Functional Analysis, Warsaw, Poland, 1960), Studia Math. 69–71.
66. *Barycentric calculus*, Encyclopedia Brittanica, vol. 1, p. 211.
67. *The finite topology of a linear space*, (with S. Kakutani), Arch. Math. **14**, 55–58.
68. *Idempotency of the hull-formation* $H^{\gamma}$, Z. Wahrscheinlichkeitstheorie **1**, 258–262.
69. *The Euler characteristic in combinatorial geometry*, Amer. Math. Monthly **70**, 119–127.
70. *On a problem of Hirschfeld*, Nieuw Arch. Wisk. **11**, 22–26.
71. *Rearrangements of series of vectors*, Math. Z. **81**, 46–51.
72. *On a question of Bishop and Phelps*, Amer. J. Math. **85**, 95–98.
73. *On a conjecture of Lindenstrauss*, Israel J. Math. **1**, 1–4.
74. *Helly's theorem and its relatives*, (with L. Danzer and B. Grünbaum), in Convexity (V. Klee, ed.) Proc. Sympos. Pure Math., vol. 7, Amer. Math. Soc., Providence, pp. 101–180.
75. *Topological classification of convex sets*, (with H. Corson), Proc. Sympos. Pure Math. vol. 7, Amer. Math. Soc., Providence, pp. 37–51.
76. *Infinite-dimensional intersection theorems*, in Convexity (V. Klee, ed.) Proc. Sympos. Pure Math., vol. 7, Amer. Math. Soc., Providence, pp. 349–360.
77. *The generation of affine hulls*, Acta Sci. Math. (Szeged) **24**, 60–81.

78. *The generation of convex hulls*, (with W. Bonnice), Math. Ann. **152**, 1–29.
79. *On a theorem of Dubins*, J. Math. Anal. Appl. **7**, 425–427.
80. *Convexity*, Proc. Sympos. Pure Math. vol. 7, Amer. Math. Soc., Providence, 516 + xv pages, (Editor).

## 1964

81. *Combinatorial geometry in the plane* (with H. Hadwiger and H. Debrunner), Holt, New York, 113 + v pages, (Translator and author of one chapter).
82. *Extreme points of convex sets without completeness of the scalar field*, Mathematika **10**, 59–63.
83. *Every simple closed curve in $E^3$ is unknotted in $E^4$*, (with R. H. Bing), J. London Math. Soc. **39**, 86–94.
84. *On the angle between two lines in a Minkowski plane*, (with P. Katz), Nieuw Arch. Wisk. **12**, 102–105.
85. *Connectedness in topological linear spaces*, Israel J. Math. **2**, 127–131.
86. *Some semicontinuity theorems for convex polytopes and cell-complexes*, (with H. G. Eggleston and B. Grünbaum), Comment. Math. Helv. **39**, 165–188.
87. *A 'string algorithm' for shortest paths in directed networks*, Oper. Res. **12**, 428–432.
88. *A combinatorial analogue of Poincaré's duality theorem*, Canad. J. Math. **16**, 517–531.
89. *Diameters of polyhedral graphs*, Canad. J. Math. **16**, 602–614.
90. *The number of vertices of a convex polytope*, Canad. J. Math. **16**, 701–720.
91. *A property of polyhedral graphs*, J. Math. Mech. **13**, 1039–1042.
92. *Utility functions and the 'lin' operation for convex sets*, Israel J. Math. **2**, 191–197.
93. *Two topological properties of topological linear spaces*, (with C. Bessaga), Israel J. Math. **2**, 211–220.

## 1965

94. *A theorem on convex kernels*, Mathematika **12**, 89–93.
95. *Summability in $l(p_1, p_2, \ldots)$ spaces*, Studia Math. **25**, 277–280.
96. *Two examples in the theory of topological linear spaces*, Studia Math. **25**, 385–390.
97. *Problem in barycentric coordinates*, J. Appl. Phys. **36**, 1854–1856.
98. *A class of linear programming problems requiring a large number of iterations*, Numer. Math. **7**, 313–321.
99. *Heights of convex polytopes*, J. Math. Anal. Appl. **11**, 176–190.

100. *Paths on polyhedra*, I, J. Soc. Indust. Appl. Math. **13**, 946–956.

____ Translation: Russian translation of Hadwiger-Debrunner (Izv. "Nauka," Moscow, 1965) includes translation of part of my added chapter. (See No. 81.)

## 1966

101. *Every non-normable F-space is homeomorphic with its closed convex bodies*, (with C. Bessaga), Math. Ann. **163**, 161–166.

102. *Convex polytopes and linear programming*, Proceedings of the IBM Scientific Computing Symposium on Combinatorial Problems, March 16–18, 1964, IBM Data Processing Division, pp. 123–158.

103. *Exposed points of convex sets*, (with G. Choquet and H. Corson), Pacific J. Math. **16**, 33–43.

104. *Paths on polyhedra*. II, Pacific J. Math. **16**, 249–262.

105. *A comparison of primal and dual methods for linear programming*, Numer. Math. **9**, 227–235;

____ Reprinted: No. 101 in Contributions to functional analysis, Springer, New York.

## 1967

106. *Remarks on nearest points in normed linear spaces*, Proceedings of the Colloquium on Convexity, Copenhagen, Denmark 1965, pp. 168–176.

107. *Problem size in linear programming*, Proceedings of the Colloquium on Convexity, Copenhagen, Denmark, 1965, pp. 177–184.

108. *The d-step conjecture for polyhedra of dimension $d < 6$*, (with D. Walkup), Acta Math. **117**, 53–78.

109. *Lengths of snakes in boxes*, (with L. Danzer), J. Combin. Theory **2**, 258–265.

110. *A method for constructing circuit codes*, J. Assoc. Comput. Mach. **14**, 520–529.

111. *Diameters of polytopes*, Chapter 16 (pp. 341–355) of Convex Polytopes, B. Grünbaum, Wiley, New York.

112. *Long paths and circuits on polytopes*, Chapter 17 (pp. 356–389) of Convex Polytopes, B. Grünbaum, Wiley, New York.

113. *Asymptotes of convex bodies*, Math. Scand. **20**, 89–90.

114. *Characterizations of a class of convex sets*, (with C. Olech), Math. Scand. **20**, 290–296.

115. *Applications of geometry*, Proceedings of the CUPM Geometry Conference, Santa Barbara, 1967. Part I: Convexity and Applications (L. Durst, ed.), Mathematical Association of America, Committee on the Undergraduate Program in Mathematics, Berkeley, pp. 7–42.

## 1968

116. *Convex functions on convex polytopes*, (with D. Gale and R. T. Rockafellar), Proc. Amer. Math. Soc. **19**, 867–873.
117. *Behavior of linear forms on extreme points*, Illinois J. Math. **12**, 254–263.
118. *Maximal separation theorems for convex sets*, Trans. Amer. Math. Soc. **134**, 133–147.
119. *Facets and vertices of transportation polytopes*, (with C. Witzgall), Lectures in Appl. Math. (G. Dantzig and A. Veinott, eds.), vol. 11, Amer. Math. Soc., Providence, pp. 257–282.
120. *Helly's theorem and its applications*, (with L. Danzer and B. Grünbaum), (Expanded version of No. 74, translated into Russian by S. Zalgaller), "Mir", Moscow, 160 pages.

____ Reprinted: No. 98 in Lectures in Appl. Math., vol. 11, Amer. Math. Soc., Providence, pp. 65–76.

## 1969

121. *Convexity*, Encyclopedia Brittanica, pp. 436–437.
122. *Can a plane convex body have two equichordal points?*, Amer. Math. Monthly **76**, 54–55.
123. *Can nine tetrahedra form a neighboring family?*, Amer. Math. Monthly **76**, 178–179.
124. *Is every polygonal plane region illuminable from some point?*, Amer. Math. Monthly **76**, 180.
125. *What is the expected volume of a simplex whose vertices are chosen at random from a convex body?*, Amer. Math. Monthly **76**, 286–288.
126. *Is there an $n$ for which $\phi(x) = n$ has a unique solution?*, Amer. Math. Monthly **76**, 288–289.
127. *Can the boundary of a $d$-dimensional convex body contain segments in all directions?*, Amer. Math. Monthly **76**, 408–410.
128. *Is a body spherical if its $HA$-measurements are constant?*, Amer. Math. Monthly **76**, 539–542.
129. *Can all convex Borel sets be generated in a Borelian manner within the realm of convexity?*, Amer. Math. Monthly **76**, 678–679.
130. *Intersection theorems for positive sets*, (with W. Hansen), Proc. Amer. Math. Soc. **22**, 450–457.
131. *What are the intersection graphs of arcs in a circle?*, Amer. Math. Monthly **76**, 810–813.
132. *On a lemma of Fullerton and Braunschweiger*, Math. Ann. **182**, 249–250.
133. *Invertibly positive linear operators on spaces of continuous functions*, (with T. A. Brown and M. Juncosa), Math. Ann. **183**, 105–114.

134. *Two renorming constructions related to a question of Anselone*, Studia Math. **33**, 231–242.
135. *Separation and support properties of convex sets—a survey*, Control Theory and the Calculus of Variations (A. Balakrishnan, ed.), Academic Press, New York, pp. 235–303.

**1970**

136. *What is the maximum length of a d-dimensional snake?*, Amer. Math. Monthly **77**, 63–65.
137. *Which isoperimetric ratios are bounded?*, Amer. Math. Monthly **77**, 288–289.
138. *Must a compact endset have zero measure?*, (with M. Martin), Amer. Math. Monthly **77**, 616–618.
139. *Convexité*, Encyclopedia Universalis, **4**, 982–985.
140. *The use of circuit codes in analog-to-digital conversion*, Graph Theory and its Applications (B. Harris, ed), Academic Press, New York, pp. 121–131.
141. *Shapes of the future—unsolved problems in geometry. Part* I: *two dimensions*, 25-minute color film and 20-page viewer's manual, Individual Lecture Film Project of the Mathematical Association of America.

**1971**

142. *Semicontinuity of the face-function of a convex set*, (with M. Martin), Comment. Math. Helv. **46**, 1–12.
143. *The use of research problems in high school geometry*, Educational Studies in Mathematics, vol. 3, pp. 482–489. 1971–72 (Reprinted in The Teaching of Geometry at the Pre-College Level (H.-G. Steiner, ed.), Reidel, Dordrecht, pp. 206–213.)
144. *Shapes of the future*, Amer. Sci. **59**, 84–91 (Reprinted in Science Today (India) **5**, no. 12, June 1971, 41–49, and in The Two-Year College Math. J. **2**, 14–27).
145. *The greedy algorithm for finitary and cofinitary matroids*, Combinatorics (T. Motzkin, ed.), Proc. Sympos. Pure Math., vol. 19, Amer. Math. Soc., Providence, pp. 137–152.
146. *What is a convex set?*, Amer. Math. Monthly **78**, 616–631.
147. *Shapes of the future—unsolved problems in geometry. Part* II: *three dimensions*, 40-minute color film and 27-page viewer's manual, Individual Lecture Film Project of the Mathematical Association of America.
148. *Monthly research problems*, 1969–1971, (with R. Guy), Amer. Math. Monthly **78**, 1113–1122.

**1972**

149. *How good is the simplex algorithm*?, (with G. Minty), Inequalities III (O. Shisha, ed), Academic Press, New York, pp. 159–175.
150. *Experimental designs by level reduction of the d-dimensional cuboctahedron*, (with D. Doehlert), Discrete Math., **2**, 309–334.

151. *Unions of increasing and intersections of decreasing sequences of convex sets*, Israel J. Math., **12**, 70–78.

152. *Which generalized prisms admit H-circuits?*, Graph Theory and Applications (Y. Alavi, D. R. Lick, and A. T. White, eds.), Springer, Berlin, pp. 173–178.

153. *On a question of Colin Clark concerning three properties of convex sets*, Canad. Math. Bull., **15**, 535–537.

**1973**

154. *A remark on 'Some properties of ordered finite-dimensional spaces'*, Mathematical Models in Economics (J. Los and M. W. Los, eds.), Polish Scientific Publishers, Warsaw; and North-Holland, Amsterdam, pp. 329–331.

**1974**

155. *Polytope pairs and their relationship to linear programming*, Acta Math. **133**, 1–25.

156. *Shellings of spheres and polytopes*, (with G. Danaraj), Duke Math. J. **41**, 443–451.

**1975**

157. *Some proximate concepts in topology*, (with A. Yandl), Sympos. Math. **16**, 21–39.

158. *Convex polyhedra and mathematical programming*, Proceedings of the 1974 International Congress of Mathematicians in Vancouver, vol. 1, pp. 485–490.

159. *Unique reducibility of subsets of commutative topological groups and semigroups*, (with D. Gale), Math. Scand. **36**, 174–198.

160. *Spira's theorems on complete linear proofs of systems of linear inequalities*, Mathematatika **22**, 112–121.

161. *Ratio-sequences of chains in connected metric spaces*, The Geometry of Metric and Linear Spaces (L. M. Kelly, ed), Springer, Berlin, pp. 134–146.

162. *A d-pseudomanifold with $f_0$ vertices has at least $df_0 - (d-1)(d+2)$ d-simplices*, Houston J. Math. **1**, 81–86.

**1976**

163. *Minimum graphs of specified diameter, connectivity and valence*. I, (with H. Quaife), Math. Oper. Res. **1**, 28–31.

**1977**

164. *A linearly compact convex set dense in every vector topology*, Arch. Math. **28**, 80–81.

165. *The connectedness game and the c-complexity of certain graphs*, (with G. Danaraj), SIAM J. Appl. Math. **32**, 431–442.

166. *When is a matrix sign stable?*, (with C. Jeffries and P. van den Driessche), Canad. J. Math. **29**, 315–326.

167. *Can the measure of* $\bigcup_1^n [a_j, b_j]$ *be computed in less than* $0(n \log n)$ *steps?*, Amer. Math. Monthly **84**, 284–285.

168. *Classification and enumeration of minimum* $(d, 1, 3)$*-graphs and minimum* $(d, 2, 3)$*-graphs*, (with H. Quaife), J. Combin. Theory B **23**, 83–93.

169. *Impressions of mathematical education in the People's Republic of China*, Amer. Math. Monthly **84**, 509–517.

170. *Pure and applied mathematics in the People's Republic of China* (with E. Brown, G. Carrier, W. Feit, A. Fitzgerald, J. Keller, J. Kohn, C. Leban, S. MacLane, H. Pollak, and H. Wu) (S. MacLane and A. Fitzgerald, eds.), National Academy of Sciences, Washington, D.C., 116 + ix pages.

171. *Linear algorithms for testing the sign stability of a matrix and for finding Z-maximum matchings in acyclic graphs*, (with P. van den Driessche), Numer. Math. **28**, 273–285.

**1978**

172. *Which spheres are shellable?*, (with G. Danaraj), Algorithmic Aspects of Combinatorics (B. Alspach et al., eds), Ann. Discrete Math. **2**, 33–52.

173. *A representation of two-dimensional pseudomanifolds and its use in the design of a linear-time shelling algorithm*, (with G. Danaraj), Algorithmic Aspects of Combinatorics (B. Alspach et al., eds), Ann. Discrete Math. **2**, 53–63.

174. *Adjoints of projective transformations and face-figures of convex polytopes*, Polyhedral Combinatorics (M. Balinski and A. Hoffman, eds.), Mathematical Programming Studies **8**, 208–216.

**1979**

175. *Use of Floyd's algorithm to find shortest restricted paths*, (with D. Larman), Discrete Optimization (P. Hammer et al., eds.), Ann. Discrete Math. **4**, 237–249.

176. *Combinatorial optimization: what is the state of the art?*, Information Linkage Between Applied Mathematics and Industry (P. Wang, ed.), Academic Press, New York, pp. 71–136.

177. *Some unsolved problems in plane geometry*, Math. Mag. **52**, 131–145.

**1980**

178. *On the complexity of d-dimensional Voronoi diagrams*, Arch. Math. **34**, 75–80.

179. *Another generalization of Carathéodory's theorem*, Arch. Math. **34**, 560–562.

180. *Classification and enumeration of minimum* $(d, 3, 3)$*-graphs for odd* $d$, J. Combin. Theory B **28**, 184–207.

181. *The diameter of almost all bipartite graphs*, (with D. Larman and E. Wright), Studia Sci. Math. Hungar. **15**, 39–43.

____ Updated version of paper 176 above is in Math. Oper. Res. **5**, 1–26.

**1981**

182. *Qualitative matrices: strong sign-solvability and weak-satisfiability*, (with R. Ladner), Computer-Assisted Analysis and Model Simplification (H. Greenberg and J. Maybee, eds.), Academic Press, New York, pp. 293–320.

183. *Diameters of random graphs*, (with D. Larman), Canad. J. Math. **33**, 618–640.

184. *Dispersed Chebyshev sets and coverings by balls*, Math. Ann. **257**, 251–260.

185. *The proportion of labelled bipartite graphs which are connected*, (with D. Larman and E. Wright), J. London Math. Soc. **24**, 397–404.

**1982**

186. *How many steps?*, Convexity and Related Combinatorial Geometry (M. Breen and D. Kay, eds.), Marcel Dekker, New York, pp. 1–6.

187. *A note on convex cones and constraint qualifications in infinite-dimensional vector spaces*, J. Optim. Theory and Appl. **37**, 277–284.

**1983**

____ Chinese translation of paper 176 above appears in Applied Mathematics and Mathematics of Computation (People's Republic of China) **6**, 49–65.

**1984**

188. *Sign-solvability revisited*, (with R. Ladner and R. Manber), Linear Algebra Appl. **59**, 131–157.

189. *Diameters of random bipartite graphs*, (with B. Bollobás), Combinatorica **4**, 7–19.

**1985**

190. *Finding the smallest triangles containing a given convex polygon*, (with M. C. Laskowski), J. Algorithms **6**, 359–375.

**1986**

191. *Inspheres and inner products*, (with E. Maluta and C. Zanco), Israel J. Math. **55**, 1–14.

192. *Tiling with smooth and rotund tiles*, (with E. Maluta and C. Zanco), Fund. Math. **126**, 269–290.

193. *Limits of starshaped sets*, (with G. Beer), Arch. Math. **46**, 241–249.
194. *Facet centroids and volume minimization*, Studia Sci. Math. Hungar. **21**, 143–147.
195. *Do infinite-dimensional Banach spaces admit nice tilings?*, Studia Sci. Math. Hungar. **21**, 415–427.

**1987**

196. *Qualitative stability of linear systems*, (with C. Jeffries and P. van den Driessche), Linear Algebra Appl. **87**, 1–48.
197. *Locally countable plump tilings are flat*, (with C. Tricot), Math. Ann. **277**, 315–325.
198. *Recursive structure of $S^*$-Matrices, and an $0(m^2)$ algorithm for recognizing strong sign-solvability*, Linear Algebra Appl. **96**, 233–247.
199. *The $d$-step conjecture and its relatives*, (with P. Kleinschmidt), Math. Oper. Res. **12**, 718–755.

**1988**

200. *A qualitative analysis of $\dot{x} = Ax + b$*, (with T. Bone and C. Jeffries), Discrete Appl. Math. **20**, 9–30.

**1989**

201. *Sign-patterns and stability*, Applications of Combinatorics and Graph Theory to the Biological and Social Sciences (F. Roberts, ed.), IMA Volumes in Mathematics and Its Applications, vol. 17, Springer, New York, pp. 203–219.
202. *On the 0-1 maximization of positive definite quadratic forms*, (with P. Gritzmann), Operations Research Proceedings 1988, Springer, Berlin, pp. 222–227.
203. *Optimization of globally convex functions*, (with T. C. Hu and D. G. Larman), SIAM J. Control Optim. **27**, 1026–1047.
204. *Edward James McShane, 1904–1989*, (with W. Fleming), Notices Amer. Math. Soc. **36**, 828–830.
205. *Minimum graphs of specified diameter, connectivity and valence.* II, (with E. Engelhardt, K. Li, and H. Quaife), Discrete Math. **78**, 257–266.

**1990**

206. *Geometry of the Gass-Saaty parametric cost LP algorithm*, (with P. Kleinschmidt), Discrete Comput. Geom. **5**, 13–26.
207. *Computational complexity of norm-maximization*, (with H. L. Bodlaender, P. Gritzmann, and J. van Leeuwen), Combinatorica **10**, 203–225.

208. *On the limited power of linear probes and other optimization oracles*, (with P. Gritzmann and J. Westwater), Proceedings of the Sixth Annual Symposium on Computational Geometry, Assoc. for Comput. Mach., pp. 92–101.

**1991-**

209. *Good and bad radii of convex polygons*, (with P. Gritzmann and L. Habsieger), SIAM J. Comput., **20**, 395–405.
210. *Convex polytopes and related complexes*, (with P. Kleinschmidt), Handbook of Combinatorics (R. Graham, M. Grötschel and L. Lovasz, eds.), North-Holland, Amsterdam, accepted for publication, 1991.
211. *Convex geometry in undergraduate instruction*, (to appear).
212. *Old and new unsolved problems in plane geometry and number theory*, (with S. Wagon), Mathematical Association of America, (to appear).
213. *Inner and outer j-radii of convex bodies in finite-dimensional normed spaces*, (with P. Gritzmann), Discrete Comput. Geom., (to appear).
214. *Computational complexity of inner and outer j-radii of convex polytopes*, (with P. Gritzmann), Math. Prog., (to appear).
215. *Uniform properties of collections of convex bodies*, (with E. Maluta and C. Zanco) (to appear).

*E-mail address*: KLEE@MATH.WASINGTON.EDU

# Contents

(in alphabetical order by author)

A Dual Forest Algorithm for the Assignment Problem
HANS ACHATZ, PETER KLEINSCHMIDT, AND KONSTANTINOS PAPARRIZOS 1

Self-duality Groups and Ranks of Self-dualities
JONATHAN ASHLEY, BRANKO GRÜNBAUM, G. C. SHEPHARD, AND WALTER STROMQUIST 11

Do Projections Go to Infinity?
IMRE BÁRÁNY, JACOB E. GOODMAN, AND RICHARD POLLACK 51

The Minimal Projective Plane Polyhedral Maps
D. W. BARNETTE 63

Packing Euclidean Space with Congruent Cylinders and with Congruent Ellipsoids
A. BEZDEK AND W. KUPERBERG 71

Extended Euler-Poincaré Relations for Cell Complexes
ANDERS BJÖRNER AND GIL KALAI 81

Computing the Convex Hull in the Euclidean Plane in Linear Expected Time
KARL HEINZ BORGWARDT, NORBERT GAFFKE, MICHAEL JÜNGER, AND GERHARD REINELT 91

Measures of $F$-Stars in Finitely Starlike Sets
MARILYN BREEN 109

On Sign-Nonsingular Matrices and the Conversion of the Permanent into the Determinant
RICHARD A. BRUALDI AND BRYAN L. SHADER 117

Recognizing Properties of Periodic Graphs
EDITH COHEN AND NIMROD MEGIDDO 135

On Generic Global Rigidity
ROBERT CONNELLY 147

Some Regular Maps and Their Polyhedral Realizations
H. S. M. COXETER AND G. C. SHEPHARD 157

Volumes of a Random Polytope in a Convex Set
L. DALLA AND D. G. LARMAN 175

Bodies of Constant Width in Riemannian Manifolds and Spaces of Constant Curvature
B. V. DEKSTER 181

Uniquely Remotal Hulls
DUANE DETEMPLE, JACK ROBERTSON, AND GRAHAM WOOD 193

The Symmetries of the Cut Polytope and of Some Relatives
M. DEZA, V. P. GRISHUKHIN, AND M. LAURENT 205

Complete Descriptions of Small Multicut Polytopes
M. DEZA, M. GRÖTSCHEL, AND M. LAURENT 221

A Hyperplane Incidence Problem with Applications to Counting Distances
HERBERT EDELSBRUNNER AND MICHA SHARIR 253

Gaps in Difference Sets, and the Graph of Nearly Equal Distances
PAUL ERDŐS, ENDRE MAKAI, JÁNOS PACH, AND JOEL SPENCER 265

Remarks on 5-Neighbor Packings and Coverings with Circles
G. FEJES TÓTH AND L. FEJES TÓTH 275

Symmetric Solutions to Isoperimetric Problems for Polytopes
P. FILLIMAN 289

A Global Newton Method
A. A. GOLDSTEIN 301

Volume Approximation of Convex Bodies by Circumscribed Polytopes
PETER M. GRUBER 309

Points Sets with Small Integral Distances
HEIKO HARBORTH AND LOTHAR PIEPMEYER 319

Convex Minimizers of Variational Problems
ERHARD HEIL 325

Flattening a Rooted Tree
PAUL HILFINGER, EUGENE L. LAWLER, AND GÜNTER ROTE 335

The Geometric Complementarity Problem and Transcending Stationarity in Global Optimization
REINER HORST AND HOANG TUY 341

Every Tree is Graceful (But Some are More Graceful than Others)
T. C. HU AND A. B. KAHNG 355

Qualitative Analysis of Schur Complements
CHARLES R. JOHNSON AND JOHN MAYBEE 359

Centers and Invariant Points of Convex Bodies
M. J. KAISER, T. L. MORIN, AND T. B. TRAFALIS 367

The Diameter of Graphs of Convex Polytopes and $f$-Vector Theory
GIL KALAI 387

Multiply Perspective Simplices, Desmic Triads and the Edelstein Theorems
L. M. KELLY 413

Submanifolds of the Cube
W. KÜHNEL AND CH. SCHULZ 423

Finite Unions of Closed Subgroups of the $n$-Dimensional Torus
JIM LAWRENCE 433

Regular Triangulations of Convex Polytopes
CARL W. LEE 443

On the Number of Antipodal or Strictly Antipodal Pairs of Points in Finite Subsets of $\mathbb{R}^d$
E. MAKAI, JR. AND H. MARTINI 457

Multi-Order Convexity
JUAN-ENRIQUE MARTÍNEZ-LEGAZ AND IVAN SINGER 471

Almost Orthogonal Lines in $E^d$
MOSHE ROSENFELD 489

Chiral Polytopes
EGON SCHULTE AND ASIA IVIĆ WEISS 493

Exact Upper Bounds for the Number of Faces in $d$-Dimensional Voronoi Diagrams
RAIMUND SEIDEL 517

Stretchability of Pseudolines is NP-Hard
PETER W. SHOR 531

A Zonotope Associated with Graphical Degree Sequences
RICHARD P. STANLEY 555

Geometry of Spaces of Homogeneous Polynomials
on Banach Lattices
K. SUNDARESAN 571

The Combinatorics of Bivariate Splines
WALTER WHITELEY 587

# List of Papers

(by subjects)

## Algebraic Combinatorics Geometry

Extended Euler-Poincaré Relations for Cell Complexes
A. Björner and G. Kalai 81

The Diameter of Graphs of Convex Polytopes and $f$-Vector Theory
G. Kalai 387

Chiral Polytopes
E. Schulte and A. Ivić Weiss 493

A Zonotope Associated with Graphical Degree Sequences
R. P. Stanley 555

## Combinatorial Geometry

Self-duality Groups and Ranks of Self-dualities
J. Ashley, B. Grünbaum, G. C. Shephard, and W. Stromquist 11

The Minimal Projective Plane Polyhedral Maps
D. W. Barnette 63

Measures of $F$-stars in Finitely Starlike Sets
M. Breen 109

Some Regular Maps and Their Polyhedral Realizations
H. S. M. Coxeter and G. C. Shephard 157

A Hyperplane Incidence Problem with Applications to Counting Distances
H. Edelsbrunner and M. Sharir 253

Point Sets with Small Integral Distances
H. Harborth and L. Piepmeyer 319

Multiply Perspective Simplices, Desmic Triads
and the Edelstein Theorems
L. M. KELLY 413

On the Number of Antipodal or Strictly Antipodal Pairs of Points in
Finite Subsets in $\mathbb{R}^d$
E. MAKAI, JR. AND H. MARTINI 457

Almost Orthogonal Lines in $E^d$
M. ROSENFELD 489

**Combinatorial Optimization**

A Dual Forest Algorithm for the Assignment Problem
H. ACHATZ, P. KLEINSCHMIDT, AND K. PAPARRIZOS 1

The Symmetries of the Cut Polytope and of Some Relatives
M. DEZA, V. P. GRISHUKHIN, AND M. LAURENT 205

Complete Descriptions of Small Multicut Polytopes
M. DEZA, M. GRÖTSCHEL, AND M. LAURENT 221

**Combinatorial Topology**

Extended Euler-Poincaré Relations for Cell Complexes
A. BJÖRNER AND G. KALAI 81

Submanifolds of the Cube
W. KÜHNEL AND CH. SCHULZ 423

Finite Unions of Closed Subgroups of the $n$-Dimensional Torus
J. LAWRENCE 433

**Computational Geometry**

Computing the Convex Hull in the Euclidean Plane in
Linear Expected Time
K. H. BORGWARDT, N. GAFFKE, M. JÜNGER, AND G. REINELT 91

A Hyperplane Incidence Problem with Applications to
Counting Distances
H. EDELSBRUNNER AND M. SHARIR 253

Exact Upper Bounds for the Number of Faces in $d$-Dimensional
Voronoi Diagrams
R. SEIDEL 517

Stretchability of Pseudolines Is NP-Hard
P. W. SHOR 531

## Computer Science

Recognizing Properties of Periodic Graphs
E. Cohen and N. Megiddo 135

Flattening a Rooted Tree
P. Hilfinger, E. L. Lawler, and G. Rote 335

## Convex Geometry

Volumes of a Random Polytope in a Convex Set
L. Dalla and D. G. Larman 175

Uniquely Remotal Hulls
D. DeTemple, J. M. Robertson, and G. Wood 193

Symmetric Solutions to Isoperimetric Problems for Polytopes
P. Filliman 289

Volume Approximation of Convex Bodies by Circumscribed Polytopes
P. M. Gruber 309

Convex Minimizers of Variational Problems
E. Heil 325

Centers and Invariant Points of Convex Bodies
M. J. Kaiser, T. L. Morin, and T. Trafalis 367

On the Number of Antipodal or Strictly Antipodal Pairs of Points in Finite Subsets in $\mathbb{R}^d$
E. Makai, Jr. and H. Martini 457

## Convexity in General Spaces, Generalized Convexity

Bodies of Constant Width in Riemannian Manifolds and Spaces of Constant Curvature
B. V. Dekster 181

Multi-Order Convexity
J-E Martínez-Legaz and I. Singer 471

Geometry of Spaces of Homogeneous Polynomials on Banach Lattices
K. Sundaresan 571

## Discrete Geometry

Do Projections Go to Infinity?
I. Bárány, J. E. Goodman, and R. Pollack 51

Packing Euclidean Space with Congruent Cylinders and with Congruent Ellipsoids
A. Bezdek and W. Kuperberg 71

On Generic Global Rigidity
R. Connelly 147

Gaps in Difference Sets, and the Graph of Nearly Equal Distances
P. Erdös, E. Makai, J. Pach, and J. Spencer 265

Remarks on 5-Neighbor Packings and Coverings
G. Fejes Tóth and L. Fejes Tóth 275

**Functional Analysis**

Geometry of Spaces of Homogeneous Polynomials on Banach Lattices
K. Sundaresan 571

**Graph Theory**

Recognizing Properties of Periodic Graphs
E. Cohen and N. Megiddo 135

Gaps in Difference Sets, and the Graph of Nearly Equal Distances
P. Erdös, E. Makai, J. Pach, and J. Spencer 265

Every Tree is Graceful (But Some Are More Graceful than Others)
T. C. Hu and A. B. Kahng 355

A Zonotope Associated with Graphical Degree Sequences
R. P. Stanley 555

**Mathematical Programming**

A Dual Forest Algorithm for the Assignment Problem
H. Achatz, P. Kleinschmidt, and K. Paparrizos 1

The Symmetries of the Cut Polytope and of Some Relatives
M. Deza, V. P. Grishukhin, and M. Laurent 205

Complete Descriptions of Small Multicut Polytopes
M. Deza, M. Grötschel, and M. Laurent 221

A Global Newton Method
A. A. Goldstein 301

The Geometric Complementarity Problem and Transcending Stationarity in Global Optimization
R. Horst and H. Tuy 341

Centers and Invariant Points of Convex Bodies
M. J. Kaiser, T. L. Morin, and T. Trafalis 367

## Matrix Theory

On Sign-Nonsingular Matrices and the Conversion of the Permanent into the Determinant
R. A. Brualdi and B. L. Shader 117

Qualitative Analysis of Schur Complements
C. R. Johnson and J. Maybee 359

## Nonlinear and Global Optimization

A Global Newton Method
A. A. Goldstein 301

The Geometric Complementarity Problem and Transcending Stationarity in Global Optimization
R. Horst and H. Tuy 341

## Packing and Covering

Packing Euclidean Space with Congruent Cylinders and with Congruent Ellipsoids
A. Bezdek and W. Kuperberg 71

Remarks on 5-Neighbor Packings and Coverings
G. Fejes Tóth and L. Fejes Tóth 275

## Polyhedra

Self-duality Groups and Ranks of Self-dualities
J. Ashley, B. Grünbaum, G. C. Shephard, and W. Stromquist 11

The Minimal Projective Plane Polyhedral Maps
D. W. Barnette 63

Some Regular Maps and Their Polyhedral Realizations
H. S. M. Coxeter and G. C. Shephard 157

Symmetric Solutions to Isoperimetric Problems for Polytopes
P. Filliman 289

The Diameter of Graphs of Convex Polytopes and $f$-Vector Theory
G. Kalai 387

Finite Unions of Closed Subgroups of the $n$-Dimensional Torus
J. Lawrence 433

Regular Triangulations of Convex Polytopes
C. W. LEE 443

Chiral Polytopes
E. SCHULTE AND A. IVIĆ WEISS 493

Exact Upper Bounds for the Number of Faces in $d$-Dimensional Voronoi Diagrams
R. SEIDEL 517

## Rigidity Theory and Splines

On Generic Global Rigidity
R. CONNELLY 147

The Combinatorics of Bivariate Splines
W. WHITELEY 587

## Stochastic Geometry

Computing the Convex Hull in the Euclidean Plane in Linear Expected Time
K. H. BORGWARDT, N. GAFFKE, M. JÜNGER, AND G. REINELT 91

Volumes of a Random Polytope in a Convex Set
L. DALLA AND D. G. LARMAN 175

## Theory of Variations

Convex Minimizers of Variational Problems
E. HEIL 325

DIMACS Series in Discrete Mathematics
and Theoretical Computer Science
Volume 4, 1991

Applied Geometry
and Discrete Mathematics
THE VICTOR KLEE FESTSCHRIFT

# A Dual Forest Algorithm for the Assignment Problem

HANS ACHATZ, PETER KLEINSCHMIDT,
AND KONSTANTINOS PAPARRIZOS

ABSTRACT. A new algorithm for the (linear) assignment problem is presented. The algorithm is of dual nature but dual feasibility is relaxed. The iterations are guided by the valency structure of forests which represent the dual solutions. Some computational experience is presented.

## 1. Introduction

Recently, there has been some revised interest in algorithms for the assignment problem. (For a recent survey see [**AMO**].) Part of this interest comes from theoretical considerations: Balinski [**B2, B3**] described an $O(n^3)$ simplex algorithm for the dual assignment problem using his proof of the Hirsch-conjecture for dual transportation polyhedra [**B1**]. Goldfarb [**G**] described an $O(nk+n^2 \log n)$-algorithm for the sparse assignment problem with $k$ admissible arcs using "Fibonacci-heaps", a data structure recently proposed by Fredman and Tarjan.

All these algorithms use the valency structure of the trees which represent dual feasible solutions. A similar dual algorithm by Hung and Rom [**HR**] is shown to be equivalent to Balinski's algorithm in [**KLS**]. A primal "valency guided" approach to the assignment problem was proposed in [**Cu**] and independently in [**BGK**].

While there have been several recent papers on efficient implementations of primal-dual algorithms for the assignment problem [**CMT, CT, DM, JV**], the only papers which report on some computational experience with the mentioned dual algorithms seem to be [**KLS, HR**].

In this paper we will present a new valency related algorithm which is neither primal-dual nor does it preserve dual or primal feasibility throughout.

---

1980 *Mathematics Subject Classification* (1985 *Revision*). Primary 90C08.
The second author's research was supported in part by grant no. I-84-095.6/88 of the German-Israeli Foundation for Scientific Research & Development.

Our algorithm is a more efficient variant of an approach of Paparrizos **[P2]**.

After presenting the notation in this section, we will describe our algorithm and prove its correctness and complexity bounds in §2, and present some computational experience in §3.

The assignment problem with an $(n \times n)$ cost matrix $(c_{ij})$ can be stated (in its LP form) as follows:

$$(P) \qquad \min \sum_{i,j} c_{ij} x_{ij},$$

$$\sum_{j=1}^{n} x_{ij} = 1, \qquad 1 \le i \le n,$$

$$\sum_{i=1}^{n} x_{ij} = 1, \qquad 1 \le j \le n,$$

$$x_{ij} \ge 0, \qquad 1 \le i, j \le n.$$

The dual problem is

$$(D) \qquad \max \sum_{i=1}^{n} u_i + \sum_{j=1}^{n} v_j,$$

$$u_i + v_j \le c_{ij}, \qquad 1 \le i, j \le n.$$

Throughout we set $w_{ij} := c_{ij} - u_i - v_j$.

With an assignment problem we associate the complete bipartite graph $G(I, J)$, where $I = \{1, \dots, n\}$ consists of the "row nodes" and $J = \{1, \dots, n\}$ consists of "column nodes". Each number $i$ of $I$ or $J$ is associated with a dual variable $u_i$ or $v_i$, respectively.

A subgraph $F$ of $G(I, J)$ is called *dual feasible* iff

(i) there is a vector $(u, v) = (u_1, \dots, u_n, v_1, \dots, v_n)$ which is feasible for $(D)$, and
(ii) if $(i, j)$ is an edge of $F$, $i \in I$, $j \in J$, then $w_{ij} = 0$.

It is well known that the spanning trees of $G(I, J)$ which are dual feasible describe the basic feasible solutions of $(D)$.

A subgraph $F$ of $G(I, J)$ which contains all nodes and does not contain a cycle is called a spanning forest. Note that a spanning forest may have isolated nodes.

$F$ is called *primal feasible* if there is a solution $x = (x_{ij})$ of $(P)$ such that $x_{ij} = 0$ for $(i, j) \notin F$.

If $F$ is both primal and dual feasible then it follows from complementary slackness that $(u, v)$ and $x$ are optimal solutions for $(D)$ and $(P)$, respectively.

Let a set of certain nodes in $J$ be called *roots*. Then a spanning forest $F$ of $G(I, J)$ is called a *superforest* provided the following two conditions hold:

(i) Every connected component of $F$ contains precisely one root.
(ii) Every node in $J$ which is not a root is 2-valent.

Let $F$ be a superforest. Then we partition $F$ into two components $F^S$ (the surplus forest) and $F^D$ (the deficit forest), where $F^S$ is the union of those components of $F$ whose root is at least 2-valent and $F^D := F \setminus F^S$. Then we have the following

LEMMA 1. *If $F$ is a dual feasible superforest and $F^S = \emptyset$ then the dual vector $(u, v)$ corresponding to $F$ is an optimal solution for the dual assignment problem.*

PROOF. $F$ consists only of the deficit forest $F^D$ which has only 1-valent or isolated roots. Let $K$ be a connected component of $F$ with $k$ column nodes, one of which is a 1-valent root. From property (ii) of a superforest it follows that $K$ has $2k-1$ edges and hence also $k$ row nodes. From property (i) it follows that in $F$ there are no isolated row nodes and hence every row node belongs to one component with a 1-valent root. This shows that there are no isolated roots. Hence, $F$ is partitioned into tree components $K_i$ with $k_i$ column nodes and $k_i$ row nodes. One column node (the root of $K_i$) has valency 1 and the other column nodes have valency 2. The $k_i$ row and column nodes of $K_i$ correspond to a square $k_i \times k_i$ submatrix of $(c_{ij})$. For the $k_i \times k_i$ assignment problem corresponding to this submatrix the dual variables of $K_i$ are dual feasible as they are dual feasible for the whole problem. As the tree $K_i$ has the special valency structure (as in **[B2]**) every column node of $k_i$ has a unique adjacent row node as a child (with respect to the root). Hence, $K_i$ is primal feasible for the $k_i \times k_i$ subproblem. As $F$ is partitioned into such $K_i$, $F$ is primal feasible and hence $(u, v)$ is optimal. □

The idea of our algorithm is as follows:

(i) Start with a dual feasible superforest $F_0$.
(ii) Construct a sequence $F_0, \dots, F_k$ of superforests by a special pivoting strategy such that $F_k$ is dual feasible and $F_k^S = \emptyset$.

By Lemma 1 $F_k$ contains an optimal assignment.

## 2. Description of the algorithm

The initial forest $F_0$ can be constructed from any feasible basic dual solution $(u, v)$ for $(D)$. So any good dual heuristic can be used for the first forest.

Let $(u, v)$ be a basic dual feasible solution for $(D)$ and $T$ a corresponding tree. Choose any node $r$ in $T$ as a source and delete all edges $(i, j)$ in $T$ such that $j$ is a column node and $i$ is the father of $j$ with respect to $r$. This, of course, yields a spanning forest $F_0$ which is a superforest if we let all column nodes be roots (condition (ii) for a superforest becomes void). We start the algorithm with $F_0$ and its dual variables. Of course, $F_0$ is dual feasible as $(u, v)$ is dual feasible.

We now describe how to obtain a superforest $F_{m+1}$ from a superforest $F_m$:

First we determine an edge $(g, h)$ which will be added to $F_m$.

Let $\delta = w_{gh} = \min\{w_{ij} \mid i \in F_m^S,\ i \in I,\ j \in F_m^D,\ j \in J\}$. Let $\varphi(g, h)$ be the unique path in $F_m \cup (g, h)$ which joins two roots.

We update the forest and its roots according to the following cases:

(1) $h$ is not an isolated root in $F_m$
 (a) $h$ is a 1-valent root,
 (b) or $h$ is not a root (hence 2-valent).

(2) $h$ is an isolated root in $F_m$ and
 (a) the unique root $j$ in $F_m^S \cap \varphi(g, h)$ is 2-valent,
 (b) $j$ is not 2-valent.

CASE (1a). Let $F_m^*$ be the component of $F_m$ which contains $h$. Set $F_{m+1}^D := F_m^D \backslash F_m^*$ and $F_{m+1}^S := F_m^S \cup F_m^* \cup (g, h)$. The node $h$ is no longer a root.

CASE (1b). Let $(k, h)$ be the edge of $F_m$, where $k$ is the father of $h$ with respect to its root. Let $F_m^*$ be the component of $F_m^D \backslash (k, h)$ which contains $h$. Set $F_{m+1}^D := F_m^D \backslash (F_m^* \cup (k, h))$ and $F_{m+1}^S := F_m^S \cup F_m^* \cup (g, h)$. The roots do not change.

CASE (2a). Let $F_m^*$ be the component of $F_m^S$ which contains $g$. Set $F_{m+1}^S := F_m^S \backslash F_m^*$ and $F_{m+1}^D := F_m^D \cup F_m^* \cup (g, h)$. The node $j$ is no longer a root.

CASE (2b). Let $i$ be the son of $j$ which is in $\varphi(g, h)$. Let $F_m^*$ be the component of $F_m^S \backslash (j, i)$ which contains $g$. Set $F_{m+1}^D := F_m^D \cup F_m^* \cup (g, h)$ and $F_{m+1}^S := F_m^S \backslash (F_m^* \cup (j, i))$. The roots do not change.

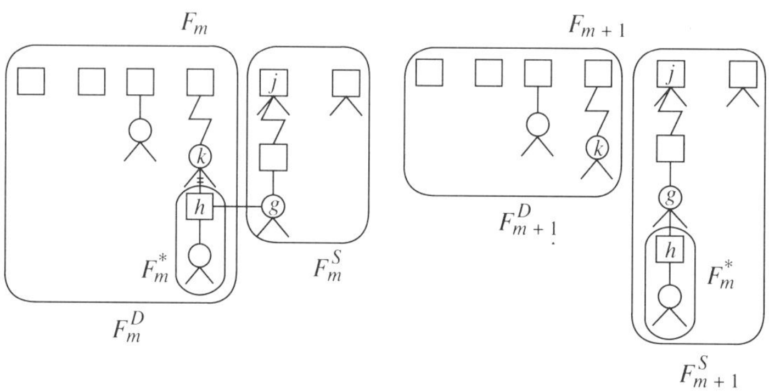


FIGURE 1. Iteration of type (1b)

An iteration of type (1b) is illustrated in Figure 1. It is trivial that all $F_m$ obtained by the described changes are superforests.

The dual variables are updated as follows: Let $u_i$, $v_j$ and $u'_i$, $v'_j$ be the dual variables associated with $F_m$ and $F_{m+1}$, respectively.

Dual updates for iterations for type (1):

$$\begin{array}{ll} u_i' := u_i - \delta & \text{if } i \in F_m^*, \\ v_j' := v_j + \delta & \text{if } j \in F_m^*, \\ u_i' := u_i \text{ and } v_j' := v_j & \text{otherwise.} \end{array}$$

Dual updates for iterations for type (2):

$$\begin{array}{ll} u_i' := u_i + \delta & \text{if } i \in F_m^*, \\ v_j' := v_j - \delta & \text{if } j \in F_m^*, \\ u_i' := u_i \text{ and } v_j' := v_j & \text{otherwise.} \end{array}$$

Observe that the updates take place only in the "cut off" subtree $F_m^*$. The algorithm stops when $F_m^S = \varnothing$.

THEOREM. *The algorithm solves the assignment problem correctly in time* $O(n^3)$.

For the proof of the Theorem we need some lemmas which have already been proved for a similar algorithm in **[P2]**. However, because of the more restricted structure of superforests, the proof below is more compact. As the specification of the dual variables is dependent on the current forests $F_m$, we will write them as $u_i(F_m)$ and $v_j(F_m)$. Also, we write

$$w_{ij}(F_m) = c_{ij} - u_i(F_m) - v_j(F_m),$$
$$\delta_m = \min\{w_{ij}(F_m) \mid i \in F_m^S,\ j \in F_m^D,\ i \in I,\ j \in J\},$$
$$\delta_m' = \min\{w_{ij}(F_m) \mid i \in F_m^D,\ j \in F_m^S,\ i \in I,\ j \in J\}.$$

LEMMA 2. *If $F_m^D$ and $F_m^S$ are dual feasible, then $\delta_{m+1} \ge \delta_m$ and $\delta_{m+1}' \ge -\delta_m$.*

PROOF. As the argument for iterations of type (2) is completely analogous we only prove Lemma 2 for iterations of type (1). Assume $F_{m+1}$ is obtained from $F_m$ by an iteration of type (1). Then $\delta_{m+1} = \min\{\varepsilon_1, \varepsilon_2\}$, where

$$\varepsilon_1 = \min\{w_{ij}(F_{m+1}) \mid i \in F_m^S,\ j \in F_{m+1}^D\},$$
$$\varepsilon_2 = \min\{w_{ij}(F_{m+1}) \mid i \in F_m^*,\ j \in F_{m+1}^D\}.$$

Clearly, $\varepsilon_1 \ge \delta_m$ as $F_m^S \subset F_{m+1}^S$ and $F_{m+1}^D \subset F_m^D$. For $i \in F_m^*$ and $j \in F_{m+1}^D$ we have

$$w_{ij}(F_{m+1}) = c_{ij} - (u_i(F_m) - \delta_m) - v_j(F_m) = w_{ij}(F_m) + \delta_m.$$

As $(i, j) \in F_m^D$ and $F_m^D$ is dual feasible, we have $w_{ij}(F_m) \ge 0$ and hence $w_{ij}(F_{m+1}) \ge \delta_m$ and $\varepsilon_2 \ge \delta_m$. This proves the first part of Lemma 2. Let

$$\varepsilon_1' = \min\{w_{ij}(F_{m+1}) \mid i \in F_{m+1}^D,\ j \in F_m^S\},$$
$$\varepsilon_2' = \min\{w_{ij}(F_{m+1}) \mid i \in F_{m+1}^D,\ j \in F_m^*\}.$$

Then $\delta'_{m+1} = \min\{\varepsilon'_1, \varepsilon'_2\}$. Clearly, $\varepsilon'_1 \geq \delta'_m$ as $F^D_{m+1} \subset F^D_m$ and $F^S_m \subset F^S_{m+1}$.

For $i \in F^D_{m+1}$ and $j \in F^*_m$ we have

$$w_{ij}(F_{m+1}) = c_{ij} - u_i(F_m) - (v_j(F_m) + \delta_m) = w_{ij}(F_m) - \delta_m.$$

As $(i, j) \in F^D_m$ and $F^D_m$ is dual feasible, we have $w_{ij}(F_m) \geq 0$ and hence $w_{ij}(F_{m+1}) \geq -\delta_m$ and $\varepsilon'_2 \geq -\delta_m$. So, $\delta'_{m+1} \geq \min\{\delta'_m, -\delta_m\}$. To show $\delta'_{m+1} \geq -\delta_m$ we proceed by induction on $m$. As $F_0$ is dual feasible we have $\delta'_0, \delta_0 \geq 0$ and hence $\delta'_1 \geq \min\{\delta'_0, -\delta_0\} = -\delta_0$.

Furthermore, assuming by induction that $\delta'_m \geq -\delta_{m-1}$ $(\geq -\delta_m)$ we have $\delta'_{m+1} \geq \min\{\delta'_m, -\delta_m\} = -\delta_m$. This completes the proof of Lemma 2. □

LEMMA 3. *If $F^D_m$ and $F^S_m$ are dual feasible then so are $F^D_{m+1}$ and $F^S_{m+1}$.*

PROOF. As in Lemma 2 we concentrate on iterations of type (1). We have to show that $w_{ij}(F^D_{m+1})$ and $w_{ij}(F^S_{m+1})$ are nonnegative. The only nontrivial case is $i \in F^*_m$ and $j \in F^S_m$. Then $w_{ij}(F_{m+1}) = c_{ij} - (u_i(F_m) - \delta_m) - v_j(F_m) = w_{ij}(F_m) + \delta_m \geq \delta'_m + \delta_m$. From Lemma 2 it follows that $\delta'_m + \delta_m \geq -\delta_{m-1} + \delta_m \geq 0$ and hence $w_{ij}(F_{m+1}) \geq 0$. □

PROOF OF THE THEOREM. If $F^D_m$ contains $k$ isolated roots we say that $F_m$ has *level* $k$.

Let $F_i$ be the first forest of level $k$ and $F_j$ the first forest of level $k-1$. In $F^D_i$ there can be at most $n-k-1$ column nodes which are different from the $k$ isolated roots. Therefore, in level $k$, there can be at most $n-k-1$ successive iterations of type (1) before a level reducing iteration of type (2) occurs. Hence, it takes at most $n-k$ iterations to get from $F_i$ to $F_j$.

After at most

$$\sum_{k=1}^{n-1}(n-k) = n(n-1)/2$$

iterations we have level 0, i.e. all isolated roots have disappeared.

A simple counting argument as in the proof of Lemma 1 shows that in this case the surplus forest is empty. By Lemma 1 the final forest is primal feasible and by Lemma 3 it is dual feasible as the whole forest consists of the dual feasible deficit forest only. Hence, after at most $n(n-1)/2$ iterations we have reached an optimal solution for the dual problem.

The optimal primal assignments correspond to the arcs $(i, j)$, where the column node $j$ is the father of the row node $i$.

It remains to show the complexity bound.

Except for the determination of the incoming arc $(g, h)$ all operations during one iteration (dual updates, tracing paths to the root) take at most time $O(n)$.

The arc $(g, h)$ is determined by $\delta_m = w_{gh} = \min\{w_{ij} \mid i \in F^S_m, j \in F^D_m\}$. This computation may involve $O(n^2)$ comparisons during one iteration.

However, as during one level the set $F_m^D \cap J$ is decreasing and the set $F_m^S \cap I$ is increasing, we can, for each fixed $j \in F_m^D \cap J$, store the node $i \in F_m^S \cap I$, where $w_{ij}$ is minimal. After an iteration we only have to compare this value to those $w_{ij}$ where $i$ has been added to the surplus forest. Hence, it takes at most $O(n^2)$ work per level to compute the $\delta_m$ and the incoming arcs. As we have only $O(n)$ levels, the complexity bound follows. □

## 3. Computational experience

We have implemented the algorithm of §2 in standard Fortran 77 to make it comparable to other codes for the assignment problem.

To get our initial forest we choose essentially the same procedure as in **[CMT]**. To each row node $i$ we associate the column node where $\min\{c_{ij} \mid j \in J\}$ is achieved. Furthermore, we increased the number of assignments in that graph by appe ding isolated column nodes in a dual feasible way to components with a root of valency exceeding 1. This way we reduce the level of the first forest.

No sophisticated data structuring was used. Our test setup follows more or less the one chosen in **[CMT]**.

Three classes of cost matrices have been tested:

(i) Costs $c_{ij}$ uniformly random between 1 and $10^2$.
(ii) Costs $c_{ij}$ uniformly random between 1 and $10^6$.
(iii) Costs $c_{ij}$ uniformly random between 1 and $i \cdot j$.

For each class we produced 20 problems as the average CPU-time is quite stable in that order of magnitude. We compared our code to three codes that were available to us:

(i) code MHR (the dual relaxation algorithm of **[HR]**, implemented by Schannath (see **[KLS]**),
(ii) code LSAP (primal-dual code from **[BD]**),
(iii) code APC (primal-dual code from **[CMT]**).

The tests were run on a PC-AT and other machines. However, there were no significant differences in the relative CPU-time on those machines. The tables below report the relative CPU-time on the AT (the time of our code "FA" being normed to 1). CPU-time was measured after the input was performed and prior to the output of the solution. The initialization time was included.

It turns out that on all three problem sets our code is superior. For small costs $1 \leq c_{ij} \leq 100$ our results (compared to APC) are not as good as for the remaining problems. However, a special strategy to handle small costs could be incorporated into our code. On the "hard" problems for $1 \leq c_{ij} \leq i \cdot j$ FA is performing especially well.

From our results we may deduce that FA has a similar performance as two other codes described in **[CT]** and **[JV]**. However, those codes were not available to us.

TABLE 1
Average relative CPU-time for twenty runs and $1 \le c_{ij} \le 100$

| $n$ | FA | MHR | APC | LSAP |
|---|---|---|---|---|
| 50 | 1 | 1.69 | 1.48 | 1.79 |
| 100 | 1 | 2.08 | 1.59 | 2.35 |
| 150 | 1 | 2.07 | 1.56 | 2.71 |
| 200 | 1 | 2.28 | 1.80 | 3.20 |
| 250 | 1 | 2.45 | 1.89 | 4.32 |
| 300 | 1 | 2.26 | 1.89 | 5.88 |

TABLE 2
Average relative CPU-time for twenty runs and $1 \le c_{ij} \le 10^6$

| $n$ | FA | MHR | APC | LSAP |
|---|---|---|---|---|
| 50 | 1 | 1.63 | 1.77 | 1.80 |
| 100 | 1 | 1.81 | 2.02 | 2.01 |
| 150 | 1 | 1.89 | 2.13 | 2.19 |
| 200 | 1 | 2.08 | 2.14 | 2.25 |
| 250 | 1 | 2.01 | 2.13 | 2.22 |
| 300 | 1 | 1.97 | 2.20 | 2.50 |

TABLE 3
Average relative CPU-time for twenty runs and $1 \le c_{ij} \le i \cdot j$

| $n$ | FA | MHR | APC | LSAP |
|---|---|---|---|---|
| 50 | 1 | 2.09 | 1.86 | 2.14 |
| 100 | 1 | 2.34 | 2.50 | 2.59 |
| 150 | 1 | 2.67 | 2.74 | 2.59 |
| 200 | 1 | 2.72 | 2.85 | 2.93 |
| 250 | 1 | 2.76 | 2.95 | 2.93 |
| 300 | 1 | 2.81 | 3.04 | 3.03 |

We may conclude that our algorithm, which has a dual nature but which relaxes dual feasibility, is at least competitive with primal-dual approaches.

It may be of interest that the average number of iterations of FA was just slightly higher than $n$ for problems of the first two classes. For $1 \le c_{ij} \le i \cdot j$ the number of iterations was about $1.5n$. This contrasts the bound of $2n$ dual iterations observed in [**APT**]. Our gain may come from the fact that we relax dual feasibility. Our theoretical bound for the number of iterations is

$n \cdot (n-1)/2$ which is smaller than the upper bound for the diameter of the edge-graph of the dual polyhedron. This bound is $(n-1)^2$ and corresponds to the bound stated in the Hirsch-conjecture **[KK]** for general polyhedra. The Hirsch-conjecture for dual transportation polyhedra was proved in **[B1]** and this was the basis for Balinski's signature algorithm **[B3]** which is closely related to ours.

It may be guessed that our approach of passing through the exterior of the dual feasible region can be extended to more general LP problems. In fact, such generalizations have been found in **[P1, P3]**. "Exterior" paths of this sort may also yield some good worst case bounds on the number of iterations in the spirit of the Hirsch-conjecture.

In concluding we want to remark that our algorithm can handle sparsity very well. If no incoming arc $(g, h)$ exists it follows easily from the structure of the forest that the problem is infeasible. Hence no artificial arcs have to be introduced. Following the ideas of **[G]**, the second author and A. Hefner have shown that if there are $k$ admissible arcs the algorithm has complexity $O(nk + n^2 \log n)$ if Fibonacci heaps or other priority queues are used for determining the $\delta_m$.

A first implementation of $A$. Hefner using priority queues shows that our sparse version is about 4.5 times faster on problems with 5% density and $n = 300$ than the code APS from **[CMT]**. However, for larger problems, variants of Fibonacci heaps may yield even better results. Similar to the algorithm of **[KLS]** the new algorithm also solves $(m \times n)$-transportation problems with $m$ unit demands on the $m$ source nodes in time $O(mn^2)$. A variant of our code for this problem has been successfully used in a system for manpower scheduling **[BK]** and a system for automated classification of human chromosomes **[GKMT]**.

## References


**[AMO]** R. K. Ahuja, T. L. Magnanti, and J. B. Orlin, *Network flows*, in Optimization (Handbook in OR & MS, Vol. 1), Elsevier, 1989, pp. 211–369.

**[APT]** I. A. Ali, R. Padman, and H. Thiagarajan, *Dual algorithms for pure network problems*, Oper. Res. **37** (1989), 159–171.

**[B1]** M. L. Balinski, *The Hirsch conjecture for dual transportation polyhedra*, Math. Oper. Res. **9** (1984), 629–633.

**[B2]** ——, *Signature methods for the assignment problem*, Oper Res. **33** (1985), 527–537.

**[B3]** ——, *A competitive (dual) simplex method for the assignment problem*, Math. Programming **34** (1986), 125–141.

**[BGK]** R. Barr, F. Glover, and D. Klingman, *An alternating path basis for assignment problems*, Math. Programming **13** (1977), 1–13.

**[BD]** R. E. Burkard and U. Derigs, *Assignment and matching problems: Solution methods with FORTRAN-programs*, Springer, Berlin, Heidelberg, and New York, 1980.

**[BK]** R. Bühner and P. Kleinschmidt, *Rechnerunterstützter Personaleinsatz—dargestellt am Beispiel einer Fertigungsinsel-Organisation, Die Betriebswirtschafts* 6/89 (1989), 761–768.

**[CMT]** G. Carpaneto, S. Martello, and P. Toth, *Algorithms and codes for the assignment problem*, Ann. Oper. Res. **13** (1988), 193–223.

[CT] G. Carpaneto and P. Toth, *Primal dual algorithms for the assignment problem*, Discrete Appl. Math. **18** (1987), 137–153.

[Cu] W. H. Cunningham, *A network simplex method*, Math. Programming **11** (1976), 105–116.

[DM] U. Derigs and A. Metz, *An efficient labeling technique for solving sparse assignment problems*, Computing **36** (1986), 301-311.

[G] D. Goldfarb, *Efficient dual simplex algorithms for the assignment problem*, Math. Programming **33** (1985), 969–982.

[GKMT] J. Graham, P. Kleinschmidt, I. Mitterreiter, and M. Tso, *An efficient transportation procedure for automatic chromosome karyotyping, Pattern recognition letters* (submitted).

[HR] M. S. Hung and W. D. Rom, *Solving the assignment problem by relaxation*, Oper. Res. **28** (1980), 969–982.

[JV] R. Jonker and A. Volgenant, *A shortest path algorithm for dense and sparse linear assignment problems*, Computing **38** (1987), 325–340.

[KK] V. L. Klee and P. Kleinschmidt, *The d-step conjecture and its relatives*, Math. Oper. Res. **12** (1987), 718–755.

[KLS] P. Kleinschmidt, C. W. Lee, and H. Schannath, *Transportation problems which can be solved by the use of Hirsch-paths for the dual problems*, Math. Programming **37** (1987), 153–168.

[P1] K. Paparrizos, *A non-dual signature method for the assignment problem and a generalization of the dual simplex method for the transportation problem*, RAIRO Rech. Opér. **22** (1988), 269–289.

[P2] ——, *An infeasible (exterior point) simplex algorithm for assignment problems*, Math. Programming (to appear).

[P3] ——, *A dual simplex algorithm with a new monotonic property*, manuscript, 1989.

(H. Achatz and P. Kleinschmidt) Fakultät für Mathematik und Informatik, Universität Passau, Innstrasse 33, 8390 Passau, Federal Republic of Germany
*E-mail address*: kleinsch@unipas.fmi.uni-passau.de

(K. Paparrizos) Department of Mathematics, Democritus University of Thrace, 67100 Xanthi, Greece

DIMACS Series in Discrete Mathematics
and Theoretical Computer Science
Volume 4, 1991

Applied Geometry
and Discrete Mathematics
THE VICTOR KLEE FESTSCHRIFT

# Self-duality Groups and Ranks of Self-dualities

JONATHAN ASHLEY, BRANKO GRÜNBAUM, G. C. SHEPHARD,
AND WALTER STROMQUIST

ABSTRACT. In many contexts, self-dual objects (which are both isomorphic and dual to themselves) are of special interest. The concept of *rank* of a self-dual object has recently been introduced, and the question was raised whether there exist self-dual objects of rank greater than 2. Extending results obtained by other workers we show that the answer is affirmative for convex polyhedra, for planar tilings, and for configurations of points and lines in the projective plane. However, the detailed answers are different in the three cases, and also depend on various natural conditions that may be imposed.

## 1. Introduction

The concept of *duality* occurs in almost every branch of mathematics. While *isomorphism* preserves certain relations between the mathematical entities under consideration, duality reverses them. In this paper we shall be concerned with duality in three particular contexts:

(i) tilings of the Euclidean plane;
(ii) polyhedra in the Euclidean space of three dimensions; and
(iii) configurations of points and lines in the projective plane.

Other areas in which duality occurs abound—for example, vector spaces and topological graph theory—but these will not be discussed here.

In every situation it is interesting to look for those special objects in which isomorphism and duality occur simultaneously. These are said to be *self-dual.* Again it is possible to find many familiar examples with this property: the regular tiling of the Euclidean plane by squares; three-dimensional convex pyramids; the configurations of Pappus and Desargues in projective

1980 *Mathematics Subject Classification* (1985 *Revision*). Primary 51M20; Secondary 06B99, 51A10, 51F99.

The first author's research was supported by an NSF Postdoctoral Fellowship in the Mathematical Sciences.

The second author's research was supported in part by NSF grant DMS-8620181. Essential for the discovery of some of the results of this paper was a transparent sphere, a gift from Professor Robert Connelly.

geometry. Also, Euclidean spaces and, more generally, all complete normed inner-product spaces are self-dual.

In spite of the fact that self-duality is so familiar and has been investigated for well over a century, one aspect of it escaped attention until very recently. In a short note published in 1988 [**17**] the *rank* of a self-duality was defined, and some questions about its possible values were raised. (The rank, which will be formally defined in the next section, is simply the order of the self-duality regarded as a mapping.) This note spurred interest in the area and led to a number of new results. The only publication so far is by Jendrol [**19**], which described a polyhedron with 14 vertices and rank 4. This example is shown in Figure 25, and its significance will be explained below. Other examples and related results have been circulated informally, or are in the process of being published (see Archdeacon and Richter [**2**], McCanna [**26**], McKee [**27**], Servatius et al. [**31**]). It is the aim of this note to present some new results and to point out some of the intriguing problems that are still open.

The paper is organized as follows. In §2 we introduce the necessary terminology in a general context, illustrated by reference to a familiar polyhedron. In particular, we introduce the concept of the self-duality group of a given geometric structure; this is the group whose elements are the automorphisms and self-dualities admitted by the structure being investigated. §3 is devoted to self-dual tilings, and in §4 we are concerned with self-dual polyhedra and show how self-dual tilings may be used to construct them. In each case the possible symmetry groups will be investigated. In §5 we discuss self-dual configurations in the projective plane, while the final section is devoted to remarks and open problems. Throughout we find it interesting to consider not only the full self-duality groups, but also the subgroups that arise by limiting attention to isomorphisms and self-dualities that are induced by geometrically interesting mappings.

## 2. Terminology

Although our interest is in geometric applications we begin by presenting the fundamental definitions in the more abstract context of partially ordered sets. If $F$ and $F'$ are partially ordered sets, then an *isomorphism* $\alpha: F \to F'$ is a bijection which preserves the ordering; that is, $x < y$ (in $F$) if and only if $\alpha(x) < \alpha(y)$ (in $F'$). A *duality* $\delta: F \to F'$ is a bijection which reverses the ordering; that is, $x < y$ if and only if $\delta(x) > \delta(y)$. A duality or isomorphism from $F$ to $F$ is called a *self-duality* or *automorphism* of $F$, respectively. If $\delta$ is a self-duality of $F$ then $\delta^2 = \delta \circ \delta$ is an automorphism.

The *rank* of a self-duality $\delta$, denoted by $r(\delta)$, is its order or period—in other words, $r(\delta)$ is the least positive integer $r$ for which $\delta^r$ is the identity; if there is no such integer then we define $r(\delta) = \infty$. We say that $F$ is *self-dual* if $F$ admits at least one self-duality, and in that case we define the *rank* of $F$, denoted by $r(F)$, as the smallest $r$ such that $F$ has a self-duality of rank $r$. If $r(F)$ is finite then it must be a power of 2; for if $F$ admits a self-duality $\delta$ of rank $2^k n$ for some odd $n > 1$, then $\delta^n$ is a self-duality of smaller rank $2^k$.

The automorphisms of $F$ form a group called the *automorphism group*, $A(F)$. The self-dualities and automorphisms of $F$ together form a group which we call the *self-duality group* of $F$ and denote by $D(F)$. If $F$ is not self-dual then $D(F) = A(F)$, but if $F$ is self-dual then $A(F)$ is a subgroup of index 2 in $D(F)$. For self-dual $F$, the rank $r(F)$ equals the minimal order of the elements of $D(F)$ that are not in $A(F)$. It is easily verified that $D(F)$ is isomorphic to a well-determined subgroup of $A(F \oplus F^*)$, where $F \oplus F^*$ is the disjoint unordered union of $F$ and a partially ordered set $F^*$, which is isomorphic to the set obtained from $F$ by reversing the order. We may think of $F$ and of $F^*$ as of two different "colors"—then elements of $A(F)$ preserve the colors, whereas the elements of $D(F)$ not in $A(F)$ reverse them. In this way $D(F)$ can be regarded as a "2-color group". In the case of the Euclidean plane, the discrete 2-color groups are well known (see Grünbaum and Shephard [**16, Chapter 8**]), and it would be interesting to determine which of them can actually occur as representing self-duality groups.

In a geometric context, let us first consider convex polyhedra in ordinary Euclidean space $\mathbb{E}^3$. If $P$ is such a polyhedron, then its proper faces (of dimension 0, 1, or 2) form a set which is partially ordered by inclusion; we call it the *face lattice* of $P$, and denote it by $F_P$. All of the definitions given above for partially ordered sets can be applied to $F_P$, and we may, without possibility of confusion, use the same terms for the polyhedra themselves. For example, an isomorphism from $F_P$ to $F_Q$ can also be called an isomorphism (or "combinatorial isomorphism", or "combinatorial equivalence") from $P$ to $Q$. A duality $\delta$ of $P$ maps 0-dimensional faces (that is, vertices) to 2-dimensional faces, and vice versa. (The action of $\delta$ on the 1-dimensional faces (edges) is determined by its action on the 0- and 2-dimensional faces, so we do not need to specify it separately.) It will be convenient and should cause no confusion if the automorphism and self-duality groups $A(F_P)$ and $D(F_P)$ of $F_P$ are denoted $A(P)$ and $D(P)$, respectively.

As an illustration of these concepts, consider the six-sided pyramid $P$ with apex $A$, base $B$, base-vertices $F_1, F_3, \dots, F_{11}$, and mantle-faces $F_2$, $F_4, \dots, F_{12}$, shown in Figure 1. A self-duality $\delta$ of $P$ is defined by $\delta(A) = B$, $\delta(B) = A$, $\delta(F_j) = F_{j+1}$ (for $j = 1, \dots, 12$). It is clear that $r(\delta) = 12$ and that $\delta^3$ is also a self-duality of $P$, with $r(\delta^3) = 4$. We note that $P$ has other self-dualities; for example, the self-duality $\varepsilon$ defined by $\varepsilon(A) = B$, $\varepsilon(B) = A$, $\varepsilon(F_i) = F_j$ if and only if $i + j = 13$ has rank 2, hence $r(P) = 2$. Some additional checking shows that $\delta$ and $\varepsilon$ generate the self-duality group $D(P)$ of $P$, which is (isomorphic to) the dihedral group $\mathbf{D}_{12}$ of order 24. Also, $A(P)$ is (isomorphic to) the dihedral group $\mathbf{D}_6$ of order 12. More generally, if $P_n$ denotes the $n$-sided pyramid then $A(P_n)$ is (isomorphic to) the dihedral group $\mathbf{D}_n$ and $D(P_n)$ to $\mathbf{D}_{2n}$. (In §4 we shall give a more instructive description of $D(P_n)$.) It can be verified easily that every pyramid has rank 2; in fact, *all* self-dual polyhedra described in the literature prior to 1988 are of rank 2.

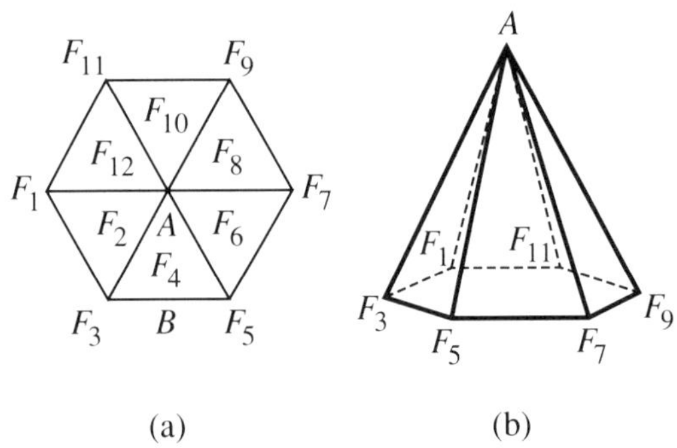


FIGURE 1. A six-sided pyramid $P$, illustrating the concept of self-duality. (a) A Schlegel diagram of $P$, with vertices and 2-dimensional faces labelled so as to indicate the self-duality $\delta$ of $P$ discussed in the text. (b) A perspective view of the same pyramid $P$. To simplify the diagram only the vertices are labelled.

If a polyhedron $P$ has a self-duality of rank 2, then a simple way of indicating this self-duality is by *balanced labelling*, that is by assigning the same symbol to any vertex and to the face corresponding to it by the self-duality. (Similar considerations apply to other partially ordered sets.) For the self-duality $\varepsilon$ of the six-sided pyramid $P$ discussed above a balanced labelling is indicated in Figure 2. It is remarkable—from mathematical as well as from psychological and sociological points of view—that due to a confusion of concepts, many of the authors who defined self-duality in the same way as it is defined here nevertheless treated it as if self-duality were equivalent to the existence of a balanced labelling (see Grünbaum and Shephard [**17**] for references to instances of this phenomenon); hence they could not even consider the possibility that self-dualities of rank higher than 2 might exist. It is true that some writers did not make this logical error; instead, they *defined* self-duality as meaning the existence of a balanced labelling. Besides being inappropriate terminology, such an approach only avoids the difficulties and does not lead to any insights into the nature of the problem.

In the sequel, we shall adopt the very convenient balanced labelling whenever it is appropriate, that is, for indicating a self-duality of rank 2. For self-dualities $\delta$ of rank $n$ it is convenient to use a letter with subscripts for

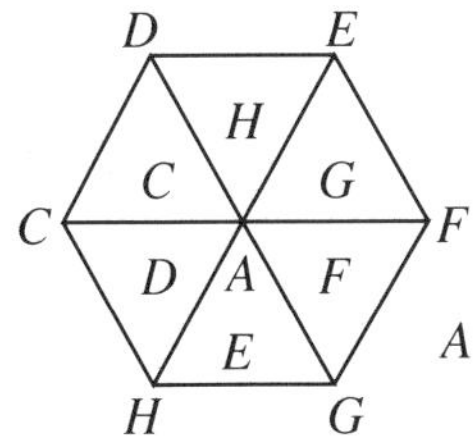


FIGURE 2. A balanced labelling of the six-sided pyramid $P$, which shows a self-duality of $P$ of rank 2.

elements which arise as images under $\delta$ and its powers, thus

$$\to A_{n-1} \to A_0 \to A_1 \to A_2 \to \cdots ,$$

where the odd subscripts relate to points, and the even subscripts to other elements (tiles, faces of a polyhedron, lines in the projective plane, and so on). In this notation, if $m$ is any odd integer then

$$\delta^m : A_j \to A_{j+m} \pmod n$$

is also a self-duality, but—as can be seen from the example of the hexagonal pyramid given above—in general not all self-dualities can be expressed in this way.

To avoid confusion we should stress that when treating polyhedra, tilings, and other objects of our consideration as combinatorial entities, then the various groups defined above will be regarded as abstract groups. In contrast, when considering the geometry of these objects, the groups will be groups of isometries or other geometric transformations. This will be reflected in the terminology as well as in the notation.

## 3. Self-dual tilings

Throughout, and without any further explicit statement of the fact, we shall limit attention to tilings $\mathcal{T}$ of the plane that are well behaved in the sense that they are *normal.* This means that each tile is a closed topological disk, the diameters of the tiles are uniformly bounded from above, their inradii are uniformly bounded from below, and the intersection of any two tiles is either empty, or a single point, or a single arc (see Grünbaum and Shephard **[16, Chapters 3 and 4]** for a discussion of these requirements and their consequences; throughout, we shall use the terminology of this book).

Since the vertices, edges and tiles of a tiling $\mathcal{T}$ from a set $F_{\mathcal{T}}$ partially ordered by inclusion, the definitions of automorphisms, self-duality, and the other concepts introduced above apply to $F_{\mathcal{T}}$. However, the geometric context suggests that besides the groups $A(\mathcal{T})$ and $D(\mathcal{T})$, various subgroups are of interest. This leads to other meanings of "self-duality", for which we have to develop an appropriate terminology. To begin with, if $F_{\mathcal{T}}$ is self-dual as defined above, then we shall say that $\mathcal{T}$ is *combinatorially self-dual.* Using well-known techniques of map-extension (see §§4.1 and 4.2 of Grünbaum and Shephard **[16]**), it can be shown that $\mathcal{T}$ is combinatorially self-dual if and only if there exists a tiling $\mathcal{T}^*$ of the plane such that $\mathcal{T}^*$ is a homeomorphic image of $\mathcal{T}$, and each tile of $\mathcal{T}$ or $\mathcal{T}^*$ contains precisely one vertex of the other tiling in its interior, and that each edge of each of the tilings crosses at a single point precisely one edge of the other tiling. Two tilings $\mathcal{T}$ and $\mathcal{T}^*$ related in this way are said to be *dually situated* or *in dual position*, and we may say that $\mathcal{T}$ and $\mathcal{T}^*$ are *topologically dual.* (The tilings in Figures 3 and 4 can serve as illustrations.) However, as the concepts of self-duality in the discrete (combinatorial) sense and in the continuous (topological) sense coincide, we do not need new terminology for this situation. On the other hand, given a self-dual tiling $\mathcal{T}$, it is not always possible to find an isometric image $\mathcal{T}^*$ and $\mathcal{T}$ such that $\mathcal{T}$ and $\mathcal{T}^*$ are in dual position (see, for example, Figure 5). If this can be done (as it can, for example, in the case of the regular tiling by squares, or the tilings in Figures 3 and 4) then we say that $\mathcal{T}$ and $\mathcal{T}^*$ are *metrically self-dual.*

Since tilings have infinite numbers of faces and vertices, the rank of a self-duality of a tiling is not necessarily finite; in fact, the rank $r(\mathcal{T})$ of a tiling can be infinite. As an example, consider the periodic tiling $\mathcal{T}$ shown in Figure 3(a), obtained by a systematic modification of the square tiling. It is easy to verify that $\mathcal{T}$ admits a metric self-duality (illustrated by the tilings $\mathcal{T}$ and $\mathcal{T}^*$ in dual position shown in Figure 3(b); moreover, $\mathcal{T}$ has the property that if $\delta$ is any combinatorial self-duality of $\mathcal{T}$ then $\delta$ can be extended to a metric self-duality of $\mathcal{T}$, such that $\delta^2$ is a nontrivial translational symmetry of $\mathcal{T}$. Hence $r(\mathcal{T}) = \infty$. As demonstrated by its labelling, the tiling $\mathcal{T}$ shown in Figure 4(a) has a combinatorial self-duality of rank 4; in fact, as shown by Figure 4(b) this is a metric self-duality of rank 4. Moreover, it is easy to show that every combinatorial self-duality of $\mathcal{T}$ is of rank at least 4, and hence $r(T) = 4$.

THEOREM 1. *The rank of any combinatorially self-dual tiling of the plane is either* 2, *or* 4, *or* $\infty$. *For each of these ranks there exist tilings which are metrically self-dual.*

PROOF. The tilings shown in Figures 3 and 4 and the square tiling are metrically self-dual of rank $\infty$, 4, and 2, respectively, and hence establish the second part of the theorem. For the first assertion we assume that the rank $r = r(\delta)$ of a combinatorial self-duality $\delta$ of a tiling $\mathcal{T}$ is finite. We shall then show that $r$ must be either 2 or 4.

To see this, let $\mathcal{T}$ and $\mathcal{T}^*$ be in dual position, and let $\mathcal{D}$ be the tiling formed by superposition of $\mathcal{T}$ and $\mathcal{T}^*$ (this is illustrated in Figures 3(b) and 4(b)). This means that each tile of $\mathcal{D}$ is a quadrilateral (that is, has four adjacent tiles), its edges are the "halfedges" of $\mathcal{T}$ and the "halfedges" of $\mathcal{T}^*$, and its vertices comprise the vertices of $\mathcal{T}$, the vertices of $T^*$, and

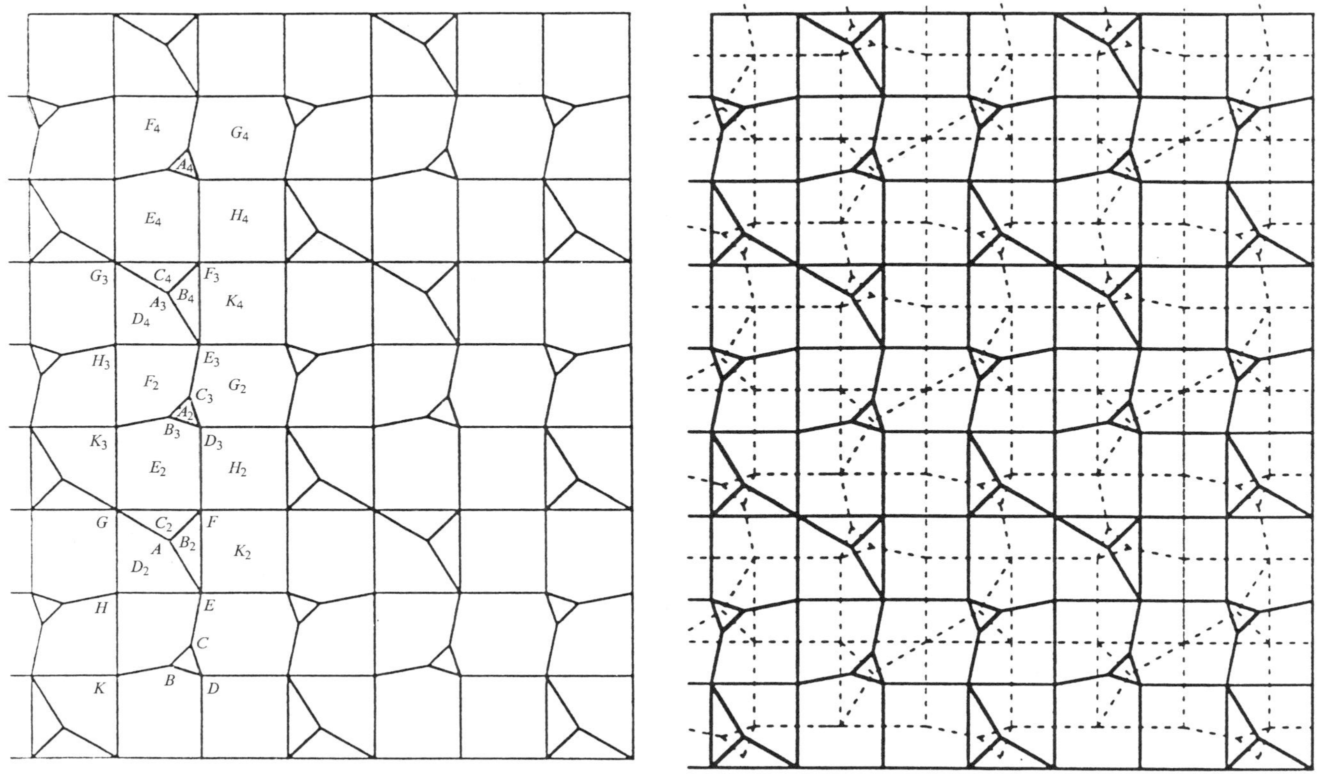


(a)

(b)

FIGURE 3. A self-dual tiling $T$ of infinite rank. The labels in (a) indicate one self-duality $\delta$ of $T$ which is of infinite rank; the application of $\delta$ increases the subscripts by 1. (b) The metric self-duality of $T$; two copies of $T$ (that is, $T$ and its dual $T^*$) are shown in dual position. Analogous illustrations will be used in several other figures.

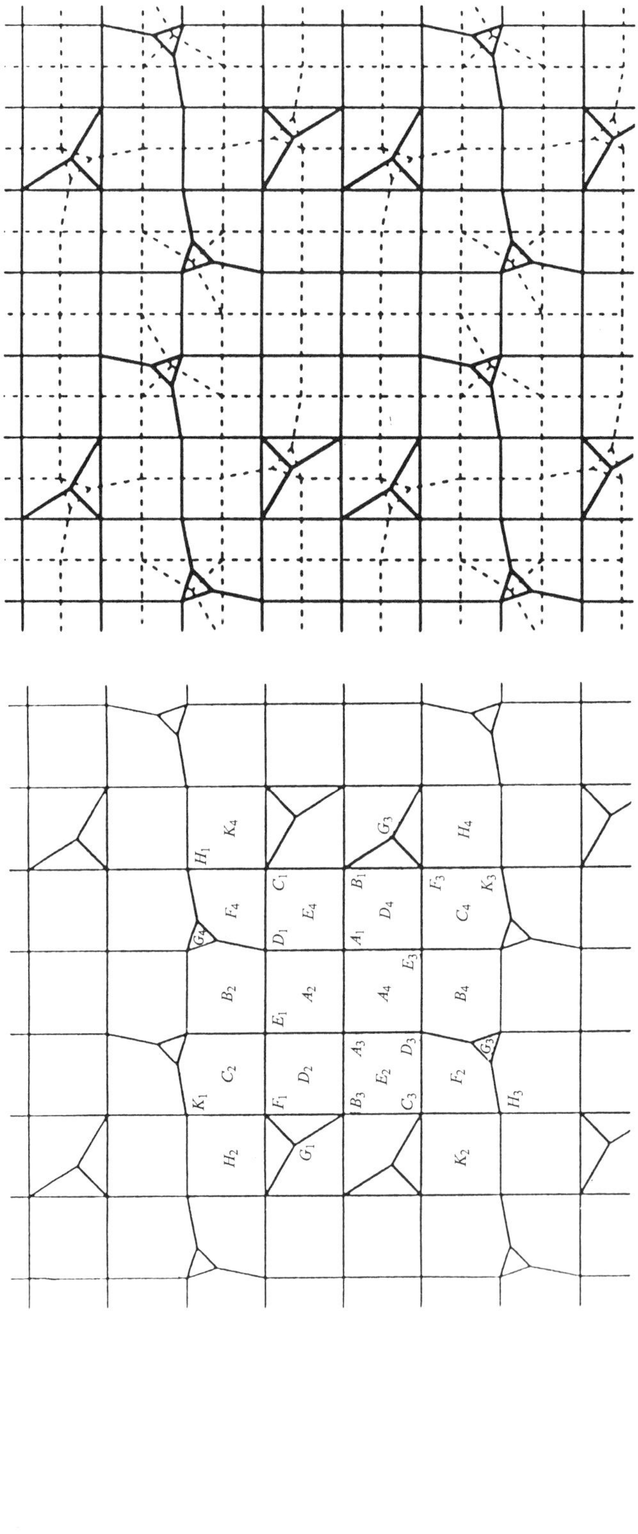


FIGURE 4. A self-dual tiling $T$ of rank 4. The interpretation of the two parts of this figure is as in Figure 3, except that the subscripts are to be taken mod 4.

the intersection-points of pairs of edges, one from $\mathscr{T}$, the other from $\mathscr{T}^*$. We shall refer to $\mathscr{D}$ as the *dual superposition tiling* of $\mathscr{T}$ (or of $\mathscr{T}^*$).

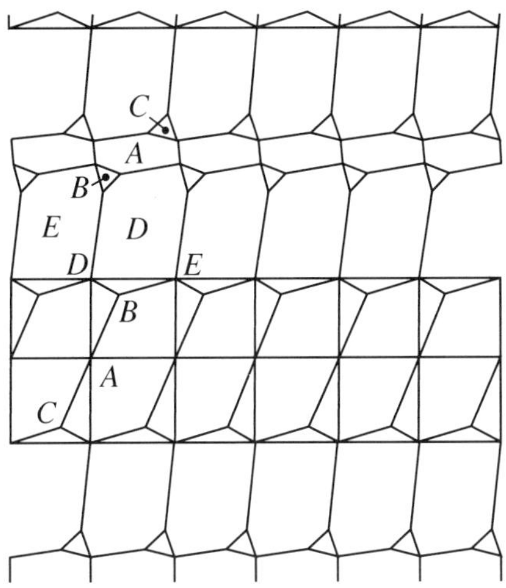


FIGURE 5. A self-dual tiling of rank 2. A self-duality is indicated by the balanced labelling of a few vertices and tiles. This tiling is not metrically self-dual since the small triangles (such as $B$ and $C$) are too close to each other to be able to contain an isometric image of the two corresponding trivalent vertices.

For each self-duality $\delta$ of $\mathscr{T}$ we shall define a homeomorphism $\phi$ of the plane which maps the tiling $\mathscr{D}$ onto itself (see Figure 6(a)). Let $P$ be any vertex of $\mathscr{T}$. Then $\delta(P)$ is a tile of $\mathscr{T}$ which contains in its interior a unique vertex $P^*$ of $\mathscr{T}^*$, so we define $\phi(P) = P^*$. A vertex $Q^*$ of $\mathscr{T}^*$ is contained in a unique tile of $\mathscr{T}$ which, by $\delta$, maps into a vertex $Q$ of $\mathscr{T}$, so we define $\phi(Q^*) = Q$. If $P$ and $Q^*$ are "opposite" vertices of a tile $F_1$ of $\mathscr{D}$ then $P^*$ and $Q$ are "opposite" vertices of a tile $F_2$ of $\mathscr{D}$. Since $\delta$ maps the edges of $\mathscr{T}$ incident with $P$ onto the edges of the tile of $\mathscr{T}$ containing $P^*$, and these correspond to the edges of $\mathscr{T}^*$ incident with $P^*$, we may define the image under $\phi$ of an edge $PR$ of $F_1$ (a "halfedge" of $\mathscr{T}$) to be the corresponding edge $P^*R^*$ of $F_2$. In a similar way, each of the other edges $RQ^*$, $Q^*S$, and $SP$ of $F_1$ are mapped by $\phi$ onto the edges $R^*Q$, $QS^*$, and $S^*P^*$ of $F_2$. By map-extensions (Grünbaum and Shephard [**16**, §§**4.1, 4.2**]) $\phi$ can be extended to a homeomorphism of the whole plane onto itself, and by construction $\phi(\mathscr{D}) = \mathscr{D}$. Moreover, it is clear that $\phi^r$ is the identity if and only if $\delta^r$ is the identity—they are necessarily maps of the same order.

For any tile $F$ of $\mathscr{D}$ consider the union of the images of $F$ under $\phi$ and its powers. If $\delta$ is of finite order, then so is $\phi$ and this set may be written $G = \bigcup_{0\leq n<r} \phi^n(F)$. If $C$ is any finite union of tiles of $\mathscr{D}$ which is

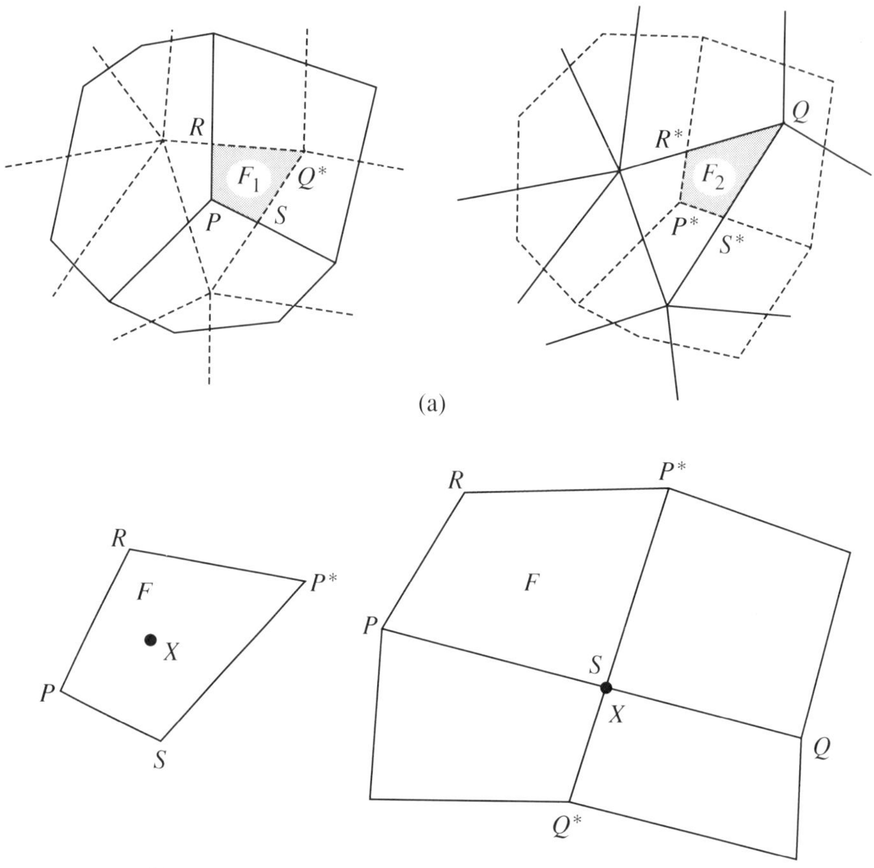


FIGURE 6. An illustration of the steps in the proof of Theorem 1; details are explained in the text.

connected and contains $G$, then $H = \bigcup_{0 \leq n < r} \phi^n(C)$ is also a finite connected set of tiles of $\mathscr{D}$, which is invariant under $\phi$. It is possible that $H$ is not simply connected, in which case we augment $H$ by adjoining a finite number of tiles (to fill up the "holes") to form a set $J$ which is connected and simply connected. Clearly $J$ is also invariant under $\phi$. By the Brouwer Fixed Point Theorem (see, for example, Alexandroff-Hopf [**1, Satz IIa on p. 480**], or Eilenberg-Steenrod [**12, Theorem 3.3 on p. 301**]), $J$ contains a point $X$ which is fixed under $\phi$. The remaining part of the proof is divided into two cases, depending on whether $X$ belongs to the interior or to the boundary of a tile $F$ of $\mathscr{D}$. Let $F$ have vertices $PRP^*S$ in order, where $P$ and $P^*$ are vertices of $\mathscr{T}$ and $\mathscr{T}^*$ respectively, and each of $R$, $S$ is a point of intersection of an edge of $\mathscr{T}$ with an edge of $\mathscr{T}^*$.

CASE 1. $X$ is an interior point of the tile $F$ of $\mathscr{D}$ (see Figure 6(b)). Since $\phi$ must map $F$ onto itself, it follows that $\phi(P) = P^*$ and $\phi(P^*) = P$. Also, $\phi$ either interchanges $R$ and $S$, or else leaves them invariant. In either case, $\phi^2(R) = R$ and $\phi^2(S) = S$. Therefore $\phi^2$, and hence also $\delta^2$, is the identity on $\mathscr{D}$, and hence also on $\mathscr{T}$; thus, in Case 1 we have $r(\mathscr{T}) = r(\delta) = 2$.

CASE 2. $X$ lies on the boundary of the tile $F$ of $D$. Since $\phi$ maps $PR$ onto $P^*R$ and $PS$ onto $P^*S$, it is clear that $X$ cannot be an interior point of an edge of $F$; hence let us suppose it coincides with $S$ (see Figure 6(c)) and so is the intersection of the edge $PQ$ of $\mathscr{T}$ with the edge $P^*Q^*$ of $\mathscr{T}^*$. Since the edges $PS$, $P^*S$, $QS$, $Q^*S$ of $\mathscr{D}$ (halfedges of $\mathscr{T}$) must be permuted by $\phi$ and since $\phi(P) = P^*$, there are only two possibilities: either $\phi(P^*) = P$, $\phi(Q) = Q^*$, and $\phi(Q^*) = Q$, or else $\phi(P^*) = Q$, $\phi(Q) = Q^*$, and $\phi(Q^*) = P$. In the first case, $\phi^2$ is the identity, and therefore so is $\delta^2$, hence $r(\delta) = 2$ and $r(\mathscr{T}) = 2$. In the second case, $\phi^4$ is the identity, hence so is $\delta^4$, and $r(\delta) = r(\mathscr{T}) = 4$.

Since these cases cover all possibilities, we deduce that $r(\mathscr{T})$ is 2 or 4, and so Theorem 1 is proved. □

A tiling is called *periodic* if it admits as symmetries translations in two independent directions. Although Theorem 1 applies to any self-dual normal tiling of the plane, the examples cited to establish the second assertion of the theorem show that the three possible ranks $(2, 4, \infty)$ can each be realized even with periodic tilings.

It is well known that the symmetry groups of periodic tilings fall into 17 classes. These are often called "wallpaper groups" (see, for example, Grünbaum and Shephard **[16, §1.4]**) and it is convenient to denote them by their "international crystallographic symbols". In the following, some familiarity with these groups and their notation is assumed.

Our next objective is to determine the possible symmetry groups of self-dual tilings.

THEOREM 2. *A periodic self-dual tiling $\mathscr{T}$ can have symmetry group $S(\mathscr{T})$ of one of the classes* **p1, p2, p4, pm, pg, cm, pmm, pmg, pgg, cmm, p4m**, *or* **p4g**. *It cannot have a symmetry group of any of the remaining classes* **p3, p31m, p3m1, p6**, *or* **p6m**.

PROOF. The first part of the theorem is established by the existence of the tilings in Figure 7(a)–(l). Each of these is obtained by modifying the square tiling (which has symmetry group **p4m**) using the construction described (for polyhedra) by Kirkman **[23]** (see also Brückner **[4, p. 93]**). This states that if a self-dual object is changed in a way that is compatible with any particular duality (that is, by making dual changes at dually corresponding elements), then the resulting object will also be self-dual.

For the second part of the proof we need to deal with each of the five classes of groups separately. We give full details in two cases.

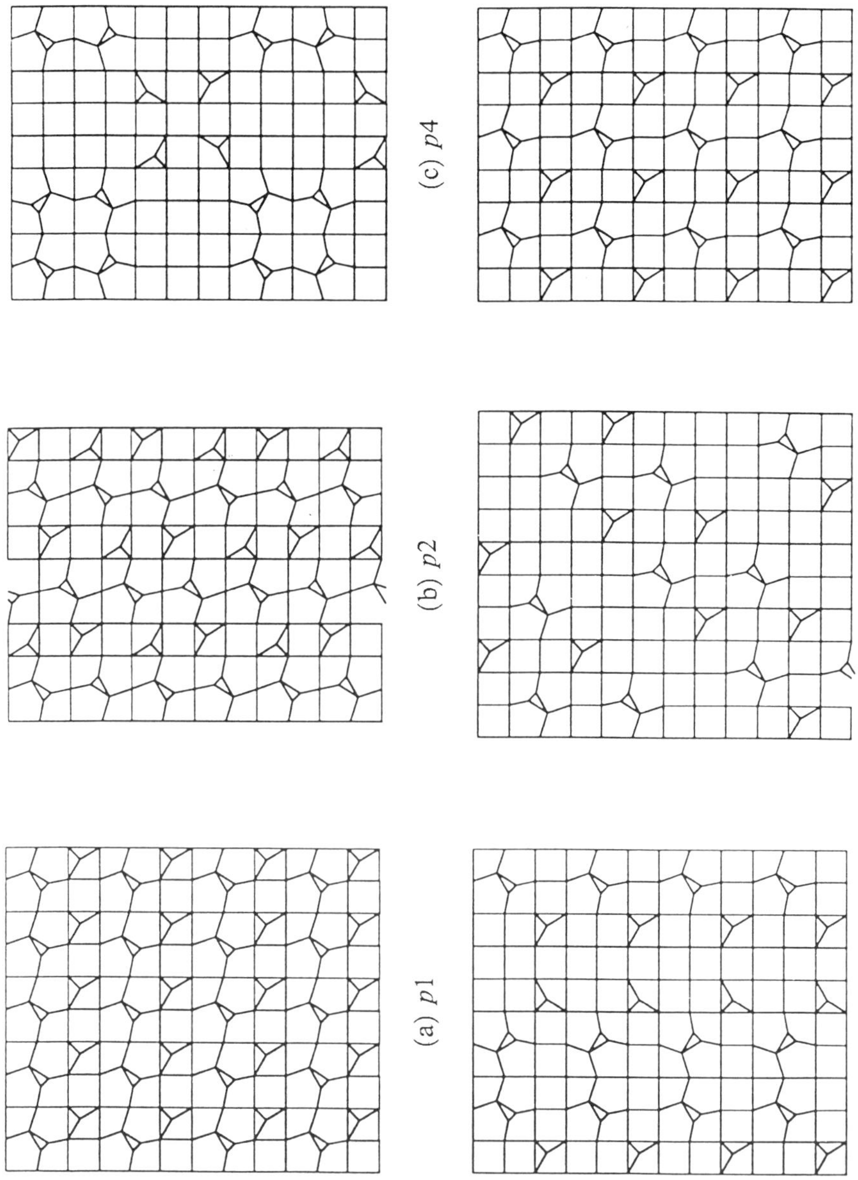


FIGURE 7, first part. Periodic metrically self-dual tilings of rank 2 with various symmetry groups. Except for (l), all tilings are harmonious.

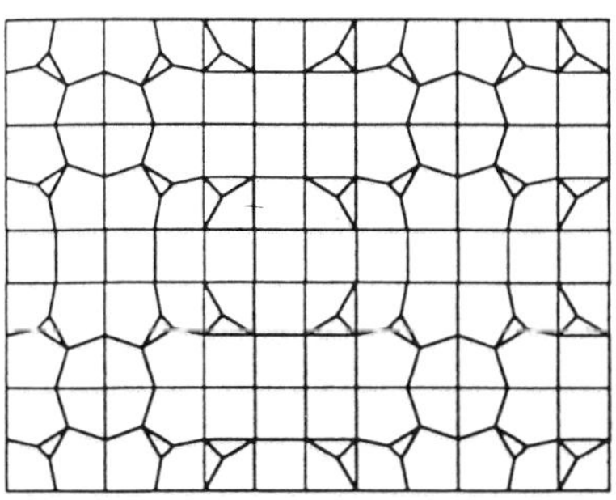

(g) *pmm*

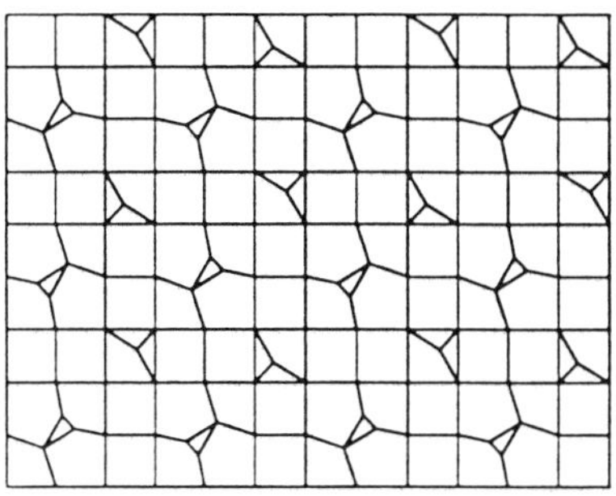

(h) *pmg*

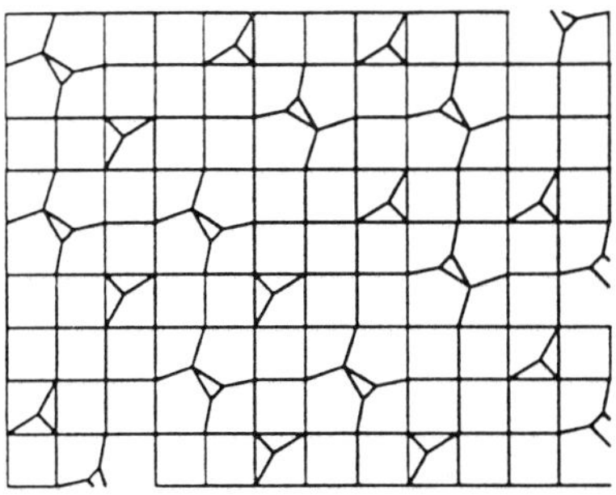

(i) *pgg*

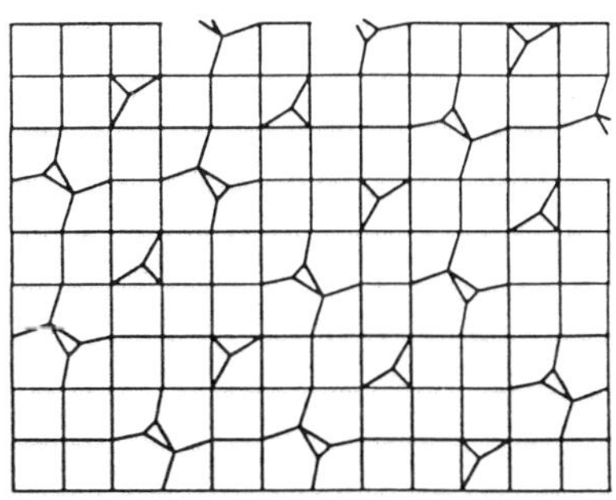

(j) *cmm*

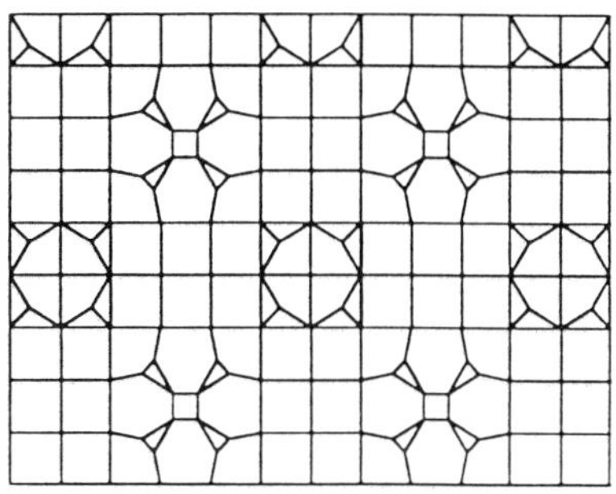

(k) *p4m*

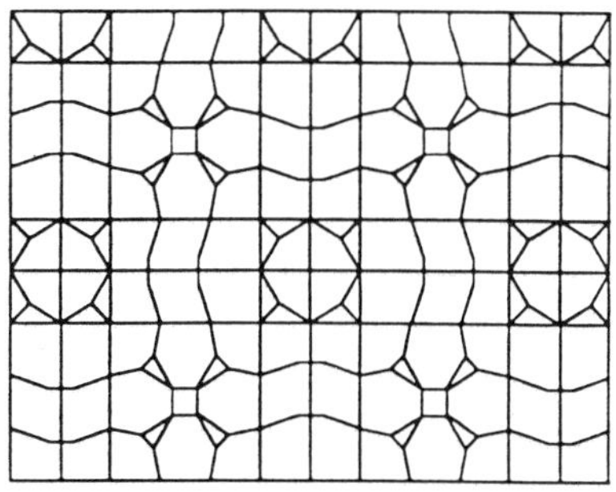

(l) *p4g*

FIGURE 7, second part.

CLASS **p6**. Here the centers of 6-fold rotational symmetry (or 6-centers, as we shall call them) are arranged in the form of an "equilateral triangle lattice" (or "dot pattern of type DPP51" in the notation of Grünbaum and Shephard [**16, §5.3**]). See Figure 8(a) where these centers are denoted by small hexagons. Construct the regular hexagonal tiling $\mathscr{H}$ (which is a Dirichlet tiling for the 6-centers) as shown in the diagram; the 6-centers are the centers of the hexagonal tiles. Now suppose $\mathscr{T}$ is a self-dual tiling whose symmetry group of class **p6** has this set of 6-centers in its group diagram (see [**16, §1.4**]); we shall show that this assumption leads to a contradiction. Since the 6-centers are all of "one kind", that is, they form one transitivity class under the group, there are two possibilities:

(i) each 6-center is a vertex of $\mathscr{T}$ of valence $6n$ $(n \geq 1)$; or
(ii) each 6-center is the center of a $6m$-gonal tile of $T$ (with $m > 1$).

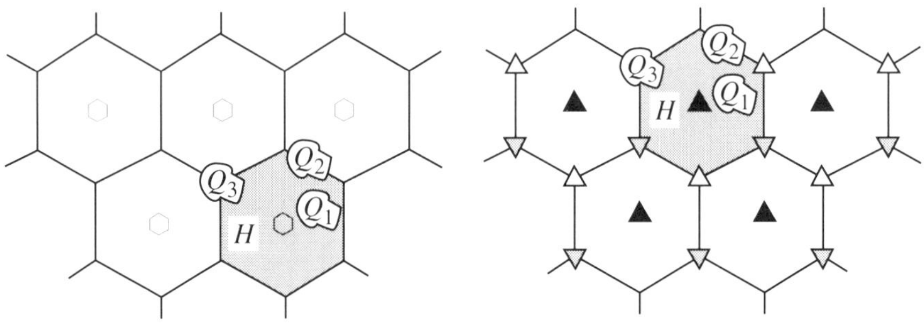


FIGURE 8. Illustrations of the steps in the proof of Theorem 2; details are explained in the text.

For case (i), the self-duality $\delta$ of $T$ whose existence is assumed must map the vertex $V$ at a 6-center into a $6n$-gon $Q$, and we consider the possible positions of $Q$ relative to a hexagon $H$ of $\mathscr{H}$. If $Q$ is interior to $H$ (as $Q_1$ in Figure 8(a)) then the operations of the group **p6** will map this into five other tiles equivalent to $Q$; hence $H$ will contain at least six copies of tiles equivalent to $Q$. If $Q$ straddles an edge of $H$ (as in $Q_2$, say) then the operations of **p6** (the 6-fold rotations about the center of $H$, and the halfturns about the midpoints of the edges of $H$) will map the parts of $Q_2$ in $H$ and outside $H$ into parts of other tiles which together amount to a total of either three or six tiles of the transitivity class of $Q$ contained in $H$. The former case arises if $Q$ contains the midpoint of the edge of $H$, the latter if it does not contain the midpoint. If $Q$ contains a vertex of $H$ (as $Q_3$, say), then $H$ contains one-third of $Q$ and, by symmetry, five other one-third parts of tiles equivalent to $Q$, making a total of two copies of $Q$ (actually six one-third copies) in $H$. Now $\delta$ maps each vertex $V$ into just *one* tile $Q$, and the hexagons are in one-to-one correspondence with the vertices. It follows that each hexagon must contain precisely one copy of $Q$. As we have seen, this does not happen; whatever the position of $Q$ at least two copies lie in $H$. This contradiction shows that case (i) cannot occur.

Case (ii) is dealt with similarly. We consider the possible positions of the $6m$-valent vertex which is the image of the $6m$-gon $Q$ under $\delta$. According to the position of that vertex relative to $H$ there are either six, three, or two copies of it in $H$. Again a contradiction is reached, since an argument analogous to the one given above shows that each $H$ must contain precisely one such vertex.

CLASS **p6m**. The argument here is exactly similar except that in this case the number of elements (vertices of valence $6n$ or $6m$-gons) in $H$ is either 2, 3, 6, or 12.

CLASS **p3**. The centers of 3-fold rotational symmetry (the 3-centers) form a triangular lattice; see Figure 8(b) where they are indicated by triangles. Suppose $\mathscr{T}$ is a self-dual tiling whose symmetry group has this group diagram; we shall show this assumption leads to a contradiction. The 3-centers are of three kinds (belong to three transitivity classes under the group) as indicated by the symbols in the diagram. For the centers indicated by solid triangles we construct the hexagonal tiling $\mathscr{H}$ whose vertices are the 3-centers of the other two kinds. This tiling is indicated by the lines in the diagram. Consider the possibilities:

(i) each 3-center indicated by a solid triangle is a vertex of $\mathscr{T}$ of valence $3n$ $(n \geq 1)$; or
(ii) each 3-center indicated by a solid triangle is the center of a $3m$-gonal tile of $\mathscr{T}$ (where $m \geq 1$).

For case (i) we proceed as described above for case (i) of the group **p6**. Let $V$ be a vertex as described, with $\delta(V) = Q$, and consider possible positions of $Q$ relative to a tile $H$ of $\mathscr{H}$. If $Q$ is interior to $H$ (at $Q_1$, say) then the operations of the group lead to three copies of $Q$ in $H$. If $Q$ straddles an edge of $H$ (at $Q_2$, say), then the operations of the group lead to three half-copies of $Q$ in $H$. Both cases lead to a contradiction as before. The only remaining possibility is that $Q$ is at a vertex of $H$ (at $Q_3$, say) represented by the open (hollow) triangles. In this case one copy (actually, three one-third copies) of $Q$ lies in $H$, and there is no immediate contradiction. However, we now proceed with an exactly similar argument for the remaining 3-centers. Let us denote such a 3-center by $V'$. If $V'$ is a $3p$-valent vertex, then we count the number of $3p$-gonal tiles $Q'$ in a hexagon $H'$. Again there are three copies (if $Q'$ lies interior to $H'$) or three half-copies (if $Q'$ straddles an edge of $H'$); however, the exceptional case of three one-third copies cannot arise now, since the vertices of $H'$ are already allocated to vertices or tiles. Hence a contradiction is reached. An exactly similar argument can be used in case (ii), and so group **p3** cannot occur.

CLASSES **p31m** AND **p3m1**. The argument for **p3m1** is identical to that for **p3**. For **p31m** we employ an argument similar to that for **p6**. There are two sorts of 3-centers: each center of the first sort has three lines of reflection passing through it, and all such centers belong to one transitivity class under the group. Take the corresponding Dirichlet tiling and consider how many 3-centers of the second sort (which do not lie on any lines of reflection) are contained in each tile. A contradiction is reached as before, showing that a group of this class cannot arise as a symmetry group of a self-dual tiling.

This completes the proof of the theorem. □

We conclude this section with some remarks about Theorem 2. Let us say that a tiling $\mathcal{T}$ is *harmonious* if every automorphism of $\mathcal{T}$ is induced by an (isometric) symmetry of the tiling. In a similar way, we may say that $\mathcal{T}$ is *harmoniously self-dual* if $\mathcal{T}$ is harmonious and there exists an isometric copy $\mathcal{T}^*$ of $\mathcal{T}$ which establishes that $\mathcal{T}$ is metrically self-dual, and, moreover, every symmetry of $\mathcal{T}$ is also a symmetry of $\mathcal{T}^*$. If $\mathcal{T}$ is harmoniously self-dual, then we refer to this special superposition of $\mathcal{T}$ and $\mathcal{T}^*$ as a *harmoniously dual superposition*, and denote it by $\mathcal{T}^H$. Clearly, for harmoniously self-dual tilings $\mathcal{T}$ the self-duality group $D(\mathcal{T})$ is isomorphic to the symmetry group $S(\mathcal{T}^H)$, and the latter, as we have remarked, may be regarded as a 2-color group.

These concepts may appear very special. However, no periodic tiling is known which is not homeomorphic to a harmonious tiling, and no self-dual tiling is known which is not homeomorphic to a harmoniously self-dual tiling. We conjecture that, in fact, no exceptions exist, and all periodic tilings are homeomorphic to harmonious ones, and the same is true for self-dual tilings.

The theorem shows that a group of the class **p4g** can occur as the symmetry group of a self-dual tiling. However, it is of interest to note that this is not so if we restrict attention to harmoniously self-dual tilings. (It will be observed that the tiling in Figure 7(l) with this group is *not* harmoniously self-dual.) To see this, suppose $\mathcal{T}$ is harmoniously self-dual, with symmetry group of class **p4g**. As in the proof of the second part of Theorem 2, each 4-center must be either a vertex (of valence $4n$, $n \geq 1$) or the center of a tile (with $4m$ edges, $m \geq 1$). Since $\mathcal{T}$ is harmoniously self-dual, centers of both kinds (with $n = m$) must exist, and so the set of 4-centers of the first kind must be congruent to the set of 4-centers of the second kind. But this is impossible since, in group **p4g**, all 4-centers are of the same kind, that is, belong to the same transitivity class (see, for example, Figure 1.4.2 of **[16]**). This is a contradiction showing that such a tiling cannot exist.

Another remark concerns the relationship between the symmetry group of a self-dual tiling, and its rank. This has not been fully investigated, and in Table 1 we show the information available to us. A plus sign $(+)$ indicates that there exists a tiling of the rank in question $(2, 4, \text{ or } \infty)$; a cross $(\times)$ indicates that no such tiling exists; and a minus sign $(-)$ indicates that no tiling is known, although we are not sure that such a tiling cannot exist.

TABLE 1

| Group | p1 | pg | pm | cm | p2 | pgg | pmg | pmm | cmm | p4 | p4g | p4m |
|---|---|---|---|---|---|---|---|---|---|---|---|---|
| Rank 2 | + | + | + | + | + | + | + | + | + | + | + | + |
| 4 | × | × | × | × | + | − | − | − | + | − | − | − |
| ∞ | + | + | + | − | + | − | − | − | − | − | − | − |

In Figures 7, 9, and 10 we show examples of self-dual tilings with all combinations of symmetry group and rank for which we assert existence in Table 1.

We remarked in §2 that self-duality groups are best described in terms of 2-color groups. A 2-color group is the group of automorphisms of a 2-colored tiling (that is, a tiling in which each tile has one of the two given colors), where the allowable transformations are those that map the tiling isometrically onto itself and either preserve the colors of all tiles, or else reverse them all. It is well known that there are 46 classes of such groups of periodic tilings of the plane, and they are particularly suitable for specifying the self-duality groups of harmonious tilings. (For details of the theory of color groups see [**16, Chapter 8**], or [**30**], where references to the ample literature can be found as well.) There exist no generally accepted symbols for these groups, but the symbols recently proposed by Coxeter [**9**] are very convenient and may well become standard. (For a comparison of the different systems of notation for 2-color groups see Washburn and Crowe [**33**].) The Coxeter notation for the 2-color symmetry group of a 2-colored periodic figure $F$ (or, more generally, for any figure with a binary property) is a symbol of the type $\mathbf{G}_1/\mathbf{G}_2$, in which $\mathbf{G}_1$ is the symmetry group of $F$ without regard to the color, and $\mathbf{G}_2$ is the subgroup (of index 2) consisting of those isometries that map $F$ onto itself while conserving the colors. (It should be stressed that the symbol $\mathbf{G}_1/\mathbf{G}_2$ is meant to indicate that the 2-color group is related to two other groups which express symmetry properties of a 2-colored figure; it is not intended as an identification of the 2-color group with a quotient group.) In the case of 2-color symmetry groups of tilings, one symbol $\mathbf{G}_1/\mathbf{G}_2$ fails to determine the class of the 2-color group uniquely, and an addition to the notation is necessary to distinguish between the two possibilities. Using the notation of 2-color groups, in Table 2 we list the self-duality groups of the harmonious tilings shown in the diagrams of this paper; for convenience of reference, we also give the symbols for 2-color groups used in [**16**]. It is clear that these examples do not exhaust the possibilities; in fact, we know of several other self-duality groups, which are listed below. It would be desirable for a complete list of self-duality groups of harmonious tilings to be worked out. Since the rank of a harmonious self-dual tiling $\mathscr{T}$ equals the minimum of the orders of the "color-reversing" symmetries of $\mathscr{T}^H$, this would lead to an independent proof of Theorem 1 in the harmonious case. We must stress that for nonharmonious tilings there is no such connection between the color symmetries of $\mathscr{T}^H$ and the self-duality group of $\mathscr{T}$. This can be easily seen from the example of the tiling in Figure 7(l), which has symmetry group **p4g**,

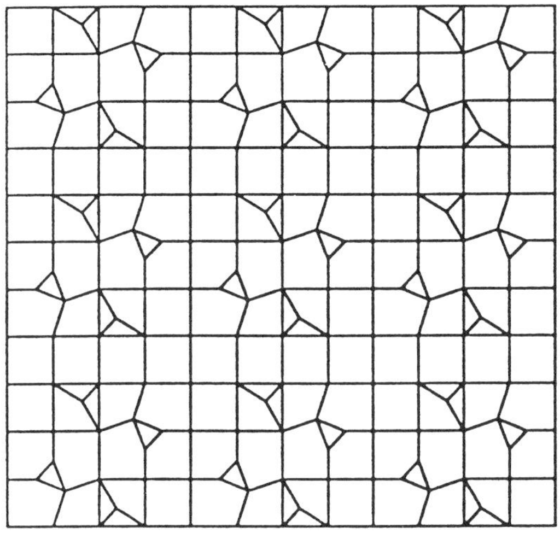

*p*2

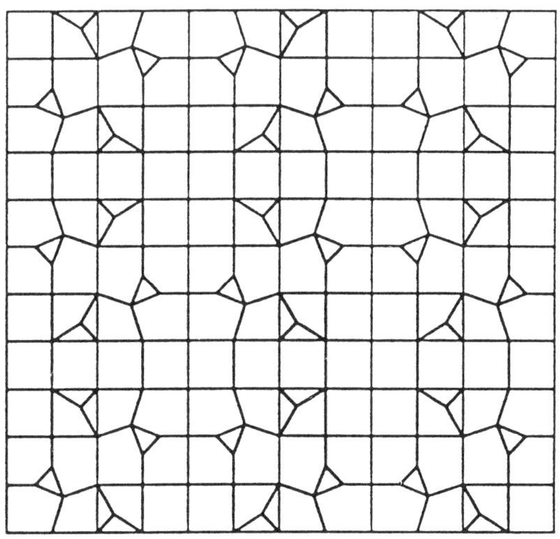

*cmm*

FIGURE 9. Periodic metrically self-dual and harmonious tilings of rank 4, with various symmetry groups.

automorphisms group **p4m**, self-duality group $\mathbf{p4m[2]_5} = \mathbf{p4m/p4m}$, and for which $\mathscr{T}^H$ has color symmetry group $\mathbf{pmg[2]_5} = \mathbf{pmg/p2}$.

In addition to the the listings in Table 2, we have examples of harmonious tilings of rank 2 with self-duality groups $\mathbf{cm[2]_1} = \mathbf{cm/p1}$, $\mathbf{pm[2]_2} = \mathbf{pm/cm}$, $\mathbf{pm[2]_5} = \mathbf{pm/pm(m')}$, $\mathbf{pmg[2]_3} = \mathbf{pmg/pgg}$, $\mathbf{pmg[2]_1} = \mathbf{pmg/pmg}$, $\mathbf{pmm[2]_4} = \mathbf{pmm/pmg}$, $\mathbf{pmm[2]_1} = \mathbf{pmm/pmm}$, $\mathbf{pgg[2]_1} = \mathbf{pgg/pg}$, $\mathbf{cmm[2]_4} =$

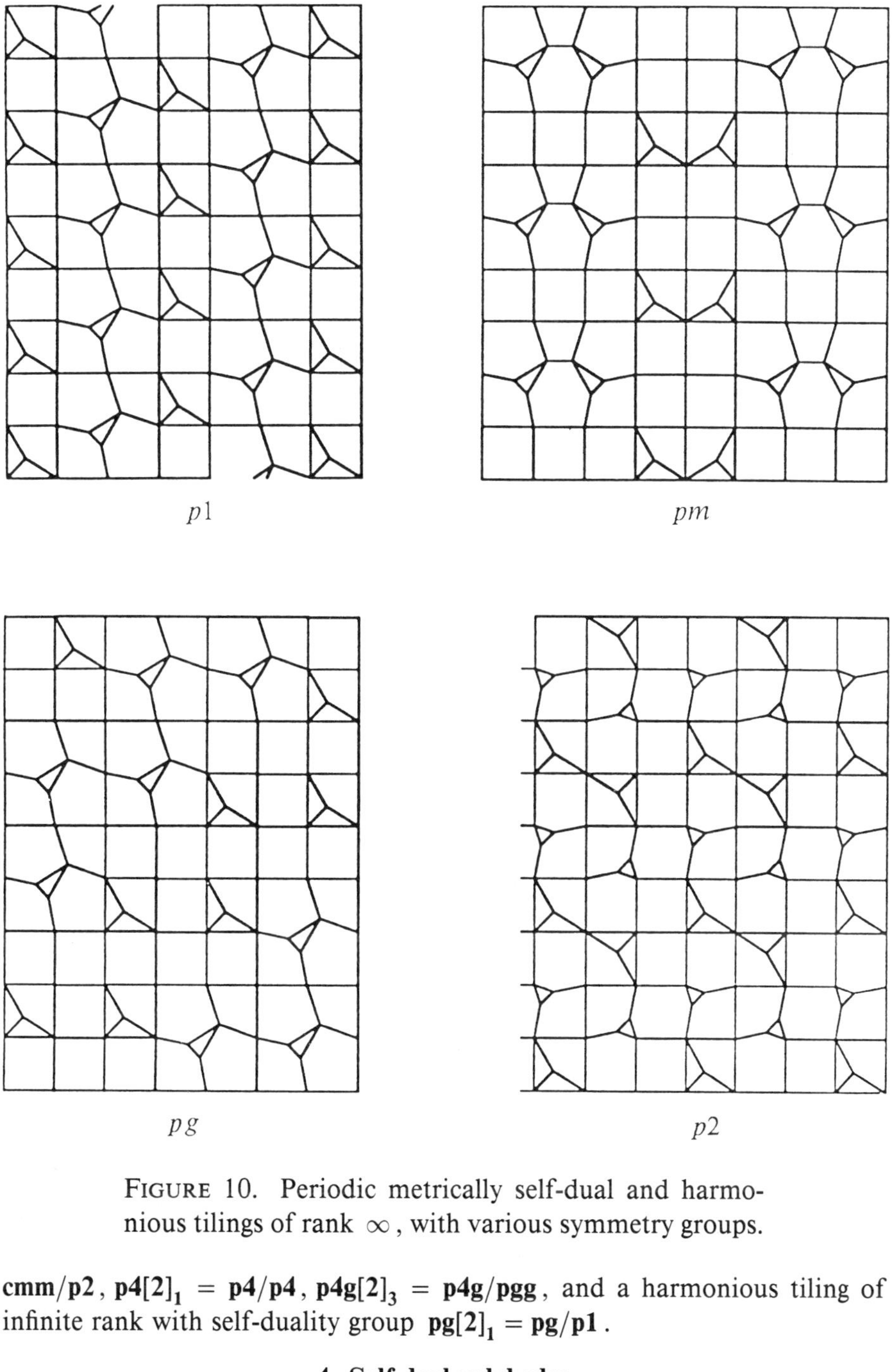

FIGURE 10. Periodic metrically self-dual and harmonious tilings of rank $\infty$, with various symmetry groups.

**cmm/p2**, $\mathbf{p4[2]_1}$ = **p4/p4**, $\mathbf{p4g[2]_3}$ = **p4g/pgg**, and a harmonious tiling of infinite rank with self-duality group $\mathbf{pg[2]_1}$ = **pg/p1**.

## 4. Self-dual polyhedra

Athough the theory of self-duality for polyhedra differs in many respects from that of tilings, the latter provide a very convenient approach to the construction of self-dual polyhedra. (For other constructions, see Archdea-

TABLE 2

| Tiling | Self-duality group | Tiling | Self-duality group |
|---|---|---|---|
| Fig. 3 | $\mathbf{pgg}[2]_2 = \mathbf{pgg}/\mathbf{p2}$ | Fig. 4 | $\mathbf{p4}[2]_2 = \mathbf{p4}/\mathbf{p2}$ |
| Fig. 7(a) | $\mathbf{p2}[2]_1 = \mathbf{p2}/\mathbf{p1}$ | Fig. 7(b) | $\mathbf{p2}[2]_2 = \mathbf{p2}/\mathbf{p2}$ |
| (c) | $\mathbf{p4g}[2]_1 = \mathbf{p4g}/\mathbf{p4}$ | (d) | $\mathbf{pmg}[2]_4 = \mathbf{pmg}/\mathbf{pm}$ |
| (e) | $\mathbf{pm}[2]_1 = \mathbf{pm}/\mathbf{pg}$ | (f) | $\mathbf{cmm}[2]_2 = \mathbf{cmm}/\mathbf{cm}$ |
| (g) | $\mathbf{cmm}[2]_5 = \mathbf{cmm}/\mathbf{pmm}$ | (h) | $\mathbf{pmg}[2]_1 = \mathbf{pmg}/\mathbf{pmg}$ |
| (i) | $\mathbf{cmm}[2]_1 = \mathbf{cmm}/\mathbf{pgg}$ | (j) | $\mathbf{pmm}[2]_3 = \mathbf{pmm}/\mathbf{cmm}$ |
| (k) | $\mathbf{p4m}[2]_5 = \mathbf{p4m}/\mathbf{p4m}$ | | |
| Fig. 8(a) | $\mathbf{p4}[2]_2 = \mathbf{p4}/\mathbf{p2}$ | Fig. 8(b) | $\mathbf{p4g}[2]_2 = \mathbf{p4g}/\mathbf{cmm}$ |
| Fig. 9(a) | $\mathbf{p1}[2] = \mathbf{p1}/\mathbf{p1}$ | Fig. 9(b) | $\mathbf{cm}[2]_3 = \mathbf{cm}/\mathbf{pm}$ |
| (c) | $\mathbf{cm}[2]_2 = \mathbf{cm}/\mathbf{pg}$ | (d) | $\mathbf{pgg}[2]_2 = \mathbf{pgg}/\mathbf{p2}$ |
| Fig. 11 | $\mathbf{pg}[2]_1 = \mathbf{pg}/\mathbf{p1}$ | | |

con and Richter [**2**]. The related topic of self-dual graphs is discussed by Servatius et al. [**31**] and McKee [**27**].) In particular, we shall show how to construct polyhedra with rank $2^k$ for any $k \geq 1$ (Corollary to Theorem 3). If, however, we also impose the condition of central symmetry, the situation changes drastically and only $k = 1$ and $k = 2$ are possible (Theorem 5). As for tilings so for polyhedra, the automorphism group $A(P)$ limits the possibilities, but does not in any sense determine the self-duality group $D(P)$; the possible pairs of groups $\{A(P), D(P)\}$ have yet to be determined. We explained in the previous section that $D(P)$ is best regarded as a 2-color group; in an obvious modification of the Coxeter notation, we use for 2-color groups on the sphere symbols of the type $\mathbf{G_1}/\mathbf{G_2}$, where $\mathbf{G_1}$ is a group and $\mathbf{G_2}$ is a suitable subgroup of index 2. We use $v(P)$ to denote the number of vertices of the polyhedron $P$, and $\mathbf{C_k}$ for the cyclic group of order $k$.

THEOREM 3. *For every positive integer $n$, there exists a convex polyhedron $P_n$ such that $A(P_n) = \mathbf{C_n}$ and $D(P_n) = \mathbf{C_{2n}}/\mathbf{C_n}$. Moreover, these polyhedra can be chosen so that $v(P_1) = 7$, $v(P_2) = 13$, and $v(P_n) = 5n+1$ for $n \geq 3$.*

PROOF OF THEOREM 3. The polyhedra $P_1$ and $P_2$ in Figures 11 and 12 establish the theorem for $n = 1$ and $n = 2$.

For $n \geq 3$, we consider the periodic tiling in Figure 13(a), which is harmoniously self-dual and of infinite rank, as illustrated in Figure 13(b). We take a "strip" consisting of $n \geq 3$ rows (in Figure 13(a) such a strip with $n = 8$ is shaded), identify the top and bottom ends of the strip, and identify as one vertex $V$ all the vertices along the right edge (see Figure 13(c)). This leads to a 3-connected planar graph $G_n$ with $5n + 1$ vertices, whose unbounded region we denote by $W$ (see also Figure 13(d)). By Steinitz's Theorem (see Grünbaum [**13, Chapter 13**]) this implies that there exists a convex polyhedron $P_n$, of a unique combinatorial equivalence class, such that the graph of vertices and edges of $P_n$ is isomorphic with $G_n$; in fact, Figure 13(d) can serve as a Schlegel diagram of $P_n$. The self-duality group of $P_n$ is generated by the self-duality $\delta$ of rank $r(\delta) = 2n$ which is defined by $\delta(V) = W$, $\delta(W) = V$, and $\delta(X_j) = X_{j+1}$ (subscripts $\bmod 2n$), where $X$ represents

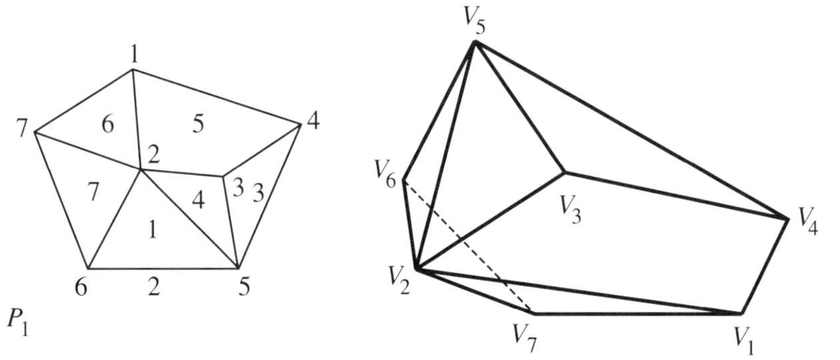


(a) (b)

FIGURE 11. A Schlegel diagram (a), and a view from above (b), of a combinatorially self-dual polyhedron $P_1$ of rank 2, with seven vertices and such that $A(P_1) = \mathbf{C_1}$ and $D(P_1) = \mathbf{C_2}$. In (a), a balanced labelling is indicated; the vertices in (b) have the coordinates given in the text.

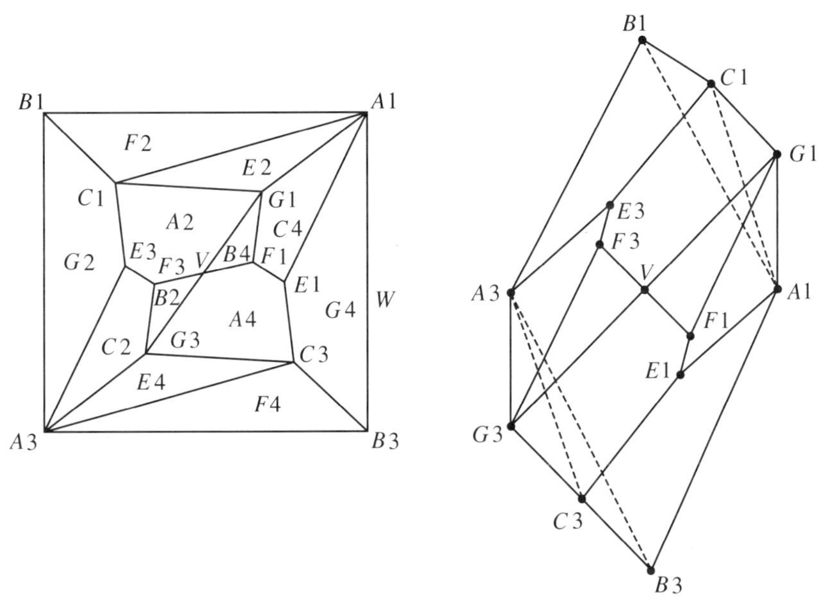


FIGURE 12. A Schlegel diagram (a), and a projection (b), of a combinatorially self-dual polyhedron $P_2$ of rank 4, with 13 vertices and such that $A(P_2) = \mathbf{C_2}$ and $D(P_2) = \mathbf{C_4}$.

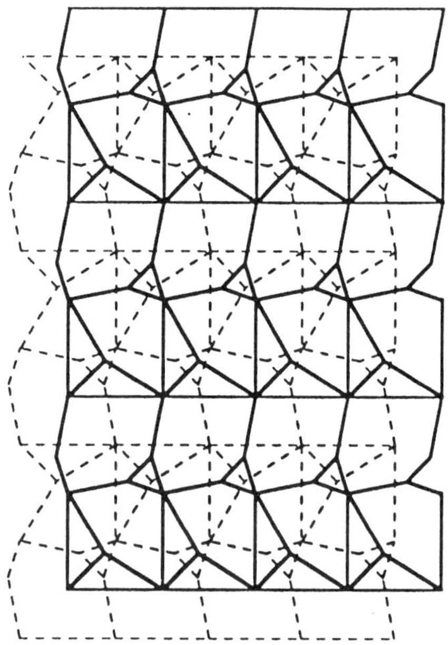

(b)

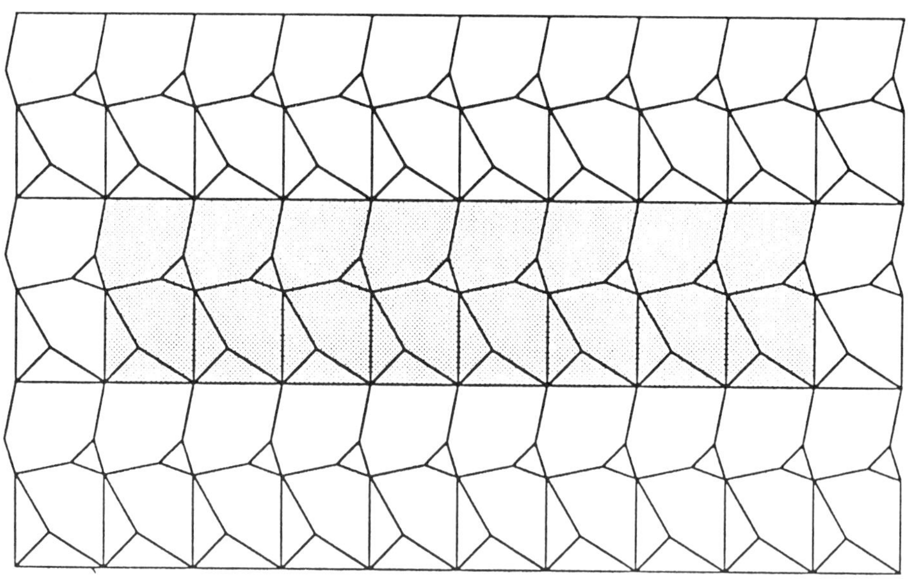

(a)

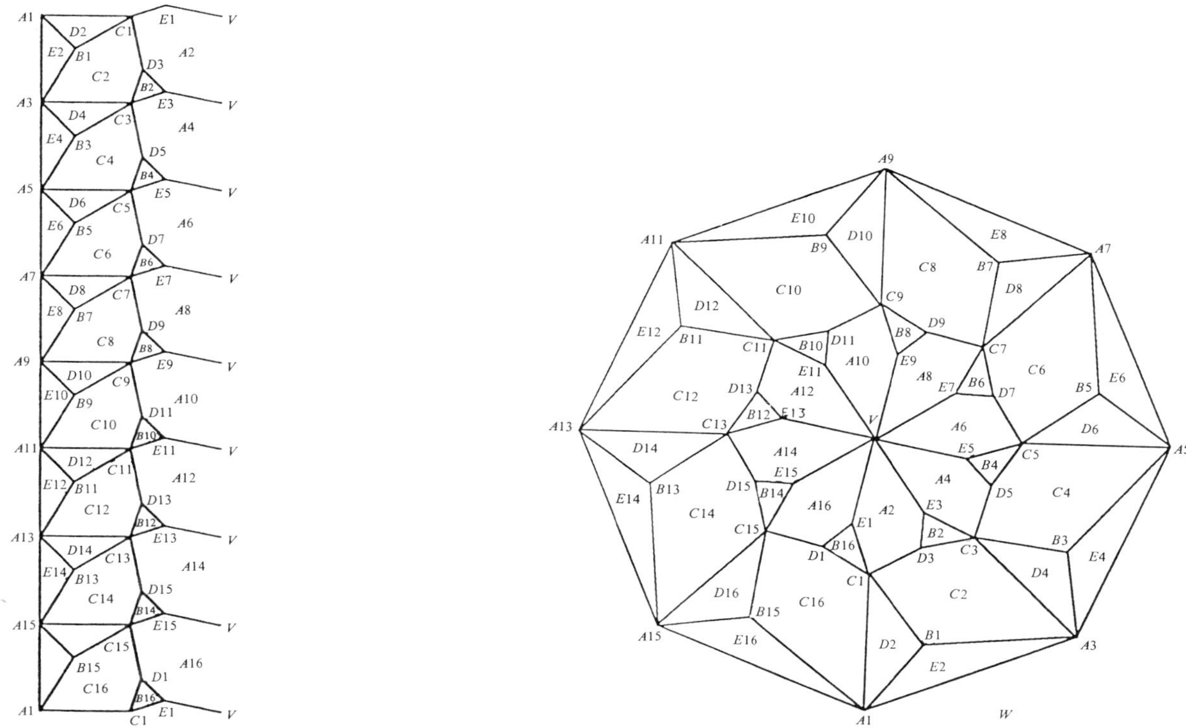

(c) (d)

FIGURE 13. An illustration of the proof of Theorem 3. (a) A self-dual tiling $T$ (of infinite rank), and a "strip" of $T$ (indicated by shading) which is used in part (c) to construct a self-dual polyhedron. (b) The tiling $T$ from part (a) is metrically self-dual. (c) The strip from part (a), with labels to indicate the identifications that are needed for the construction. (d) The graph formed by the identifications indicated in part (c); it can also serve as a Schlegel diagram of the polyhedron constructed in the text. The subscripts of the labels taken mod 16 indicate a self-duality of the polyhedron, which has rank 16, has automorphism group $\mathbf{C}_8$, and self-duality group $\mathbf{C}_{16}$.

any of the letters $A$, $B$, $C$, $D$, $E$ used for faces and vertices in the figure. The odd powers of $\delta$ are the only self-dualities of $P_n$, and the even powers are the only automorphisms. It follows that $P_n$ satisfies the requirements of Theorem 3. In particular, Figure 13(d) illustrates $P_8$, which is of rank 16. □

COROLLARY. *For every positive integer* $k$, *there exists a convex polyhedron* $Q_k$ *such that* $r(Q_k) = 2^k$. *Moreover,* $Q_k$ *can be chosen so that* $v(Q_1) = 4$, $v(Q_2) = 13$, *and* $v(Q_k) = 5 \cdot 2^{k-1} + 1$ *for* $k \geq 3$.

For $k \geq 2$, the corollary follows from the theorem by taking $Q_k = P_n$ for $n = 2^{k-1}$. In general, the rank of $P_n$ is the highest power of 2 that divides $2n$. For $k = 1$ the situation is special, since the regular tetrahedron satisfies the requirements of the corollary for $k = 1$, although it has too many automorphisms to play the role of $P_1$ in Theorem 3.

To formulate our next result, we need the concept of polarity for convex polyhedra in $\mathbb{E}^3$. If $P$ is a convex polyhedron $P$ then its *polar* polyhedron $P^*$ is defined by

$$P^* = \{Y \in \mathbb{E}^3 : \langle X, Y \rangle \leq 1 \text{ for all } X \in P\}.$$

This relationship is reciprocal; that is, $P^{**}$ and $P$ are the same subset of $\mathbb{E}^3$ (see, for example, [**13**, §**3.4**]). There is a natural duality (in the sense defined in §1) $\delta$ from $P$ to $P^*$, which maps each face $F$ of $P$ to the face $F^*$ of $P^*$ defined by

$$\delta(F) = F^* = \{Y \in P^* : \langle X, Y \rangle \leq 1 \text{ for all } X \in F\}.$$

(Another way of describing the polarity between convex polyhedra is to say that $P^*$ is obtained from $P$ by *reciprocation* in a unit sphere $U$.) For convex polyhedra $P$ and $Q$, we define a *metric isomorphism* from $P$ to $Q$ as an isometry of $\mathbb{E}^3$ which maps $P$ onto $Q$, and a *metric duality* from $P$ to $Q$ as an isometry of $\mathbb{E}^3$ which maps $P$ onto its polar $Q^*$. Similarly, we can define the *metric automorphism group* $S(P)$ and the *metric self-duality group* of $P$. It is clear that every metric automorphism or self-duality of $P$ induces a combinatorial automorphism or self-duality (so that the groups just defined can be considered as subgroups of $A(P)$ and $D(P)$, respectively), but the reverse is generally not the case. A polyhedron $P$ is *harmonious* if $S(P)$ is isomorphic with $A(P)$. A polyhedron $P$ is *harmoniously self-dual* if it is harmonious and there is a unit sphere $U$ such that the polyhedron $P^*$ obtained from $P$ by reciprocation in $U$ is congruent with $P$ and so situated that every symmetry of $P$ is also a symmetry of $P^H = P \cup P^*$. Then the self-dualities of $P$ can be realized by isometries of $\mathbb{E}^3$ which map $P$ onto $P^*$ and vice versa.

It is well known (Mani [**25**]) that for each convex polyhedron $P$ in $\mathbb{E}^3$ there exists an isomorphic polyhedron $Q$ for which each automorphism is induced by a symmetry. However, it is an open question whether each self-dual convex polyhedron in $\mathbb{E}^3$ is isomorphic to a harmoniously self-dual polyhedron. The next theorem shows that the polyhedra in Theorem 3 are isomorphic to such polyhedra, with the consequence that Theorem 3 is true in a metric as well as in a combinatorial sense. In Theorem 4 we use the usual notation for symmetry groups of a 2-sphere; see, for example, Coxeter and Moser [**10**] or Grünbaum and Shephard [**15**]. We note that the self-duality group of a harmoniously self-dual polyhedron $P$, that is, the symmetry group of $P^H$, can be described by a 2-color group.

THEOREM 4. *For all positive integers $n$ and $k$, the polyhedra $P_n$ and $Q_k$ described in the proofs of Theorem 3 and its corollary can be chosen to be harmoniously self-dual; that is, their automorphisms and self-dualities are all induced by isometries of $\mathbb{E}^3$ and reciprocations. In particular, the choice can be made so that $A(P_n) = S(P_n) = [\mathbf{n}]^+$ and $D(P_n) = S(P_n^H) = [\mathbf{2}^+, \mathbf{2n}^+]/[\mathbf{n}]^+$ for all $n \geq 1$.*

PROOF OF THEOREM 4. In the proofs of Theorem 3 and its corollary, the polyhedra $P_n$ and $Q_k$ were defined only up to isomorphism. To complete the proof of Theorem 4, we need to specify points in $\mathbb{E}^3$ for the vertices of each $P_n$ and $Q_k$.

In the case $P_1 = Q_2$ we locate the vertices as follows (see Figure 11(b)):

$$\begin{aligned} V_1 &= (\sqrt{2}, 1, 0), & V_2 &= (0, -1, 0), \\ V_3 &= ((w/\sqrt{2}), w-1, 2w), & V_4 &= (\sqrt{2}, 1, 1), \\ V_5 &= (-\sqrt{2}, 1, 0), & V_6 &= (-\sqrt{2}, 1, -\sqrt{2}), \\ V_7 &= (0, 1, -\sqrt{2}), \end{aligned}$$

where $w = \frac{2}{5}(\sqrt{6} - 1)$. Now the linear transformation $(x, y, z) \to (-x, -y, z)$ maps $P_1$ onto its polar, and so induces the unique self-duality of $P_1$.

For $P_2 = Q_2$ we choose the vertices as follows (see Figure 12(b)):

$$\begin{aligned} V &= (0, 0, +1), \\ A_1 &= (+1, 0, -1), & A_3 &= (-1, 0, -1), \\ B_1 &= (0, +1, -1), & B_3 &= (0, -1, -1), \\ C_1 &= (+\tfrac{1}{2}, +\tfrac{3}{2}, -\tfrac{1}{2}), & C_3 &= (-\tfrac{1}{2}, -\tfrac{3}{2}, -\tfrac{1}{2}), \\ E_1 &= (+\tfrac{1}{3}, -\tfrac{2}{3}, +\tfrac{1}{3}), & E_3 &= (-\tfrac{1}{3}, +\tfrac{2}{3}, +\tfrac{1}{3}), \\ F_1 &= (+\tfrac{1}{3}, -\tfrac{1}{3}, +\tfrac{2}{3}), & F_3 &= (-\tfrac{1}{3}, +\tfrac{1}{3}, +\tfrac{2}{3}), \\ G_1 &= (+1, +1, 0), & G_3 &= (-1, -1, 0). \end{aligned}$$

Regarding the points of $\mathbb{E}^3$ as column vectors, let $\delta$ be the isometry of $\mathbb{E}^3$ defined by left multipliation by the matrix

$$\begin{bmatrix} 0 & -1 & 0 \\ 1 & 0 & 0 \\ 0 & 0 & -1 \end{bmatrix}.$$

Then $\delta$ maps $Q_2$ onto $Q_2^*$, and thus induces a self-duality of $Q_2$. The isometry, and hence the self-duality, has rank 4, and it is easy to see on combinatorial grounds that no self-duality of $Q_2$ has smaller rank. It is also easy to check that $Q_2$ is harmoniously self-dual and has the stated groups of automorphisms and self-dualities. (To verify that $\delta$ maps $Q_2$ onto $Q_2^*$ it is only necessary to show that $\langle \delta(X), Y \rangle \leq 1$ for any pair $X$, $Y$ of vertices of $Q_2$, with equality in the appropriate cases. We omit the details.)

There remains the case of $P_n$ with $n \geq 3$, which is somewhat more difficult to calculate. We begin by constructing a simpler polyhedron $T_n$, whose underlying graph is illustrated in Figure 14(a) (for the case $n = 8$). Then $T_n$ satisfies $A(T_n) = \mathbf{D_n}$ and $D(T_n) = \mathbf{D_{2n}}$. To realize $T_n$ as a polyhedron in $\mathbb{E}^3$ we locate the vertices as follows:

$$V = (0, 0, 1),$$
$$C_j = \left(\frac{1}{b} \cdot \cos \frac{\pi(j+1)}{n}, \frac{1}{b} \cdot \sin \frac{\pi(j+1)}{n}, 0\right),$$
$$A_j = \left(b \cdot \cos \frac{\pi j}{n}, b \cdot \sin \frac{\pi j}{n}, -1\right),$$
$$F_j = \left(b \cdot \cos \frac{\pi j}{n}, b \cdot \sin \frac{\pi j}{n}, 1 - b^4\right)$$

for $j = 1, 3, \ldots, 2n-1$, where $b = \sqrt{\cos(\pi/n)}$. It can be verified that $S(T_n) = [\mathbf{n}]$ and $S(T_n^H) = [\mathbf{2}^+, \mathbf{2n}]/[\mathbf{n}]$. (We note that a right $n$-sided pyramid with a regular basis has the same groups as $T_n$; this is the description of the self-duality group of pyramids alluded to in §2.)

Now, again understanding the vertices as column vectors, let the isometry $\delta$ be defined by left multiplication by the matrix

$$\begin{bmatrix} +\cos\frac{\pi}{n} & -\sin\frac{\pi}{n} & 0 \\ +\sin\frac{\pi}{n} & +\cos\frac{\pi}{n} & 0 \\ 0 & 0 & -1 \end{bmatrix}$$

which represents a rotary reflection about the $Z$-axis through $1/2n$ of the full turn. Then it can be verified (we omit the calculations) that $\delta$ maps $T_n$ onto $T_n^*$ and that $\delta$ has order $2n$. To construct $P_n$, we must modify $T_n$ in such a way as to destroy the symmetries which are not induced by powers of $\delta$. We do this by cutting off each of the vertices $F_j$ $(j = 1, 3, \ldots)$, adding in its place a new face $B_{i-1}$ which passes through the vertex $C_j$ and which intersects the edges through $F_j$ at points we call $D_j$ and $E_j$. To preserve the self-duality of the polyhedron under $\delta$, we also need to add a vertex $B_j$ $(j = 1, 3, \ldots)$ which extends the face $C_{j+1}$ and to replace the face $F_{j+1}$ by new faces $D_{j+1}$ and $E_{j+1}$, as shown in Figure 14(b); the position of $B_j$ is determined by polarity from the position of the plane through $C_j$, $D_j$, and $E_j$. (To avoid cluttering up the diagram, the modifications are shown for one value of $j$ only; they need to be carried out at all vertices $F_j$.) The calculations of the coordinates of the vertices of the modified polyhedron

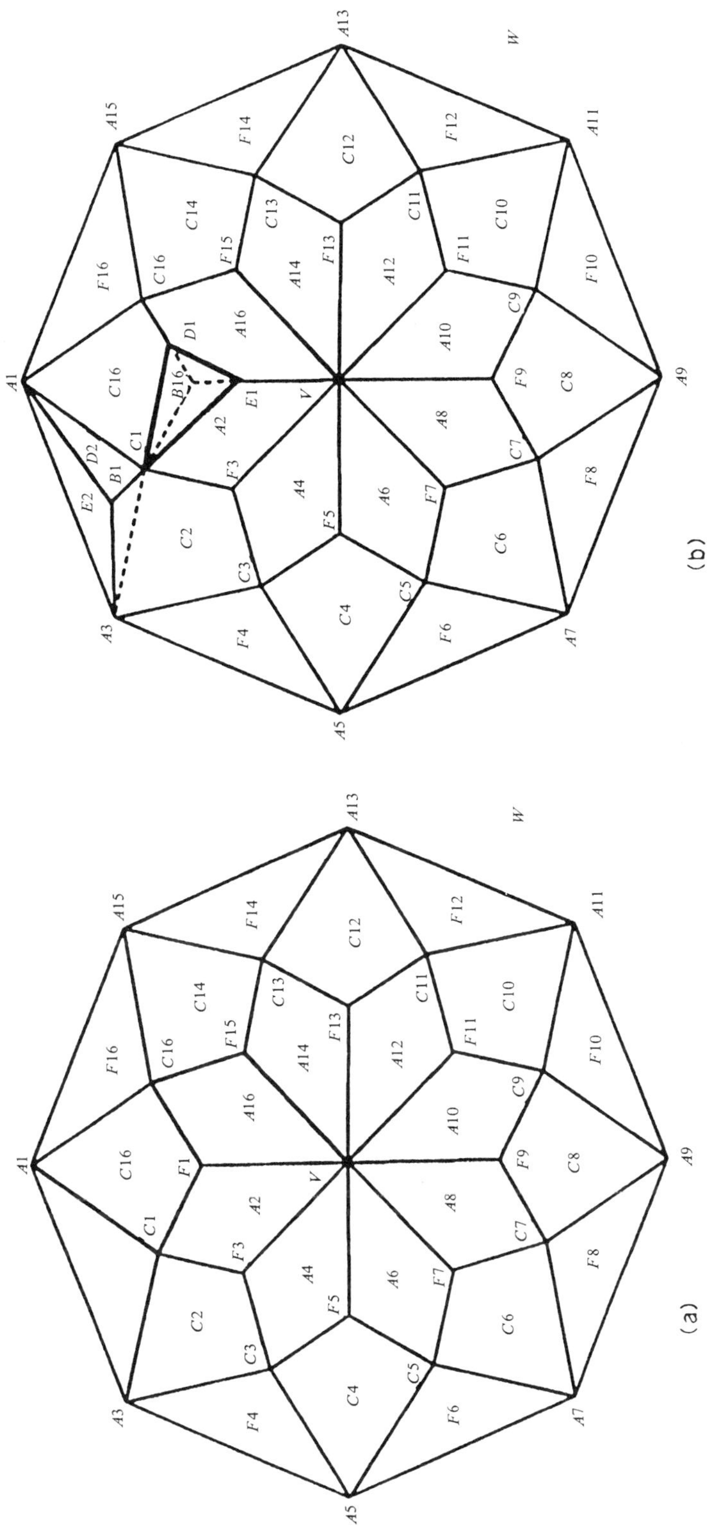


FIGURE 14. (a) A Schlegel diagram of the polyhedron $T_n$ used in the proof of Theorem 4. (b) An illustration of the process of cutting of the vertices $F_j$, for the proof of Theorem 4. For simplicity, only one vertex $F_j$ has been cut off.

are essentially elementary, and we do not give details here. It is clear that duality of the new polyhedron $P_n$ and its image under $\delta$ is maintained, that the polyhedron obtained is isomorphic to $P_n$ in Theorem 3, and that the groups $S(P_n)$ and $S(P_n^H)$ are as claimed. □

We remark that a slight modification of the above construction leads to another family of harmoniously self-dual polyhedra: convex polyhedra $R_n$ such that $A(R_n) = S(R_n) = [\mathbf{n}]^+$ and $D(R_n) = S(R_n^H) = [\mathbf{2}, \mathbf{n}]^+/[\mathbf{n}]^+$, for all $n \geq 2$. For $R_2$ the polyhedron of Figure 15 can be chosen, and for $n \geq 3$ a typical polyhedron is shown in Figure 16 (where we have taken $n = 8$). We note that these polyhedra have rank 2; in fact, all their self-dualities have rank 2.

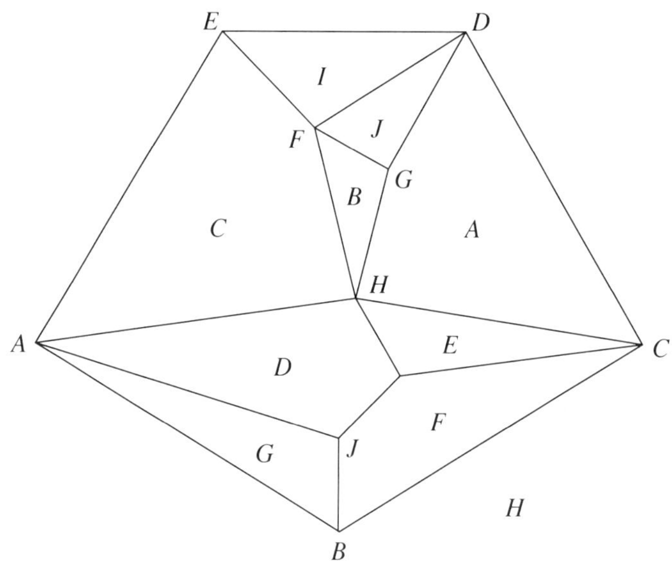


FIGURE 15. A Schlegel diagram of a polyhedron $R_2$.

Concerning the possibilities present for centrally symmetric polyhedra we have:

THEOREM 5. *If $P$ is a combinatorially self-dual polyhedron with a center of symmetry, then its rank $r(P)$ is either* 2 *or* 4. *Moreover, there exist centrally symmetric convex polyhedra $C_2$ and $C_4$, with center at the origin, which are harmoniously self-dual and such that* $r(C_2) = 2$, $S(C_2) = [\mathbf{2}, \mathbf{2}]$, $S(C_2^H) = [\mathbf{2}, \mathbf{4}]/[\mathbf{2}, \mathbf{2}]$, *and* $r(C_4) = 4$, $S(C_4) = [\mathbf{2}, \mathbf{2}^+]$, $S(C_4^H) = [\mathbf{2}, \mathbf{4}^+]/[\mathbf{2}, \mathbf{2}^+]$.

PROOF. The fact that 2 and 4 are the only possible values for the rank of a combinatorially self-dual, centrally symmetric polyhedron follows almost word for word as in the analogous part of the proof of Theorem 1, with three

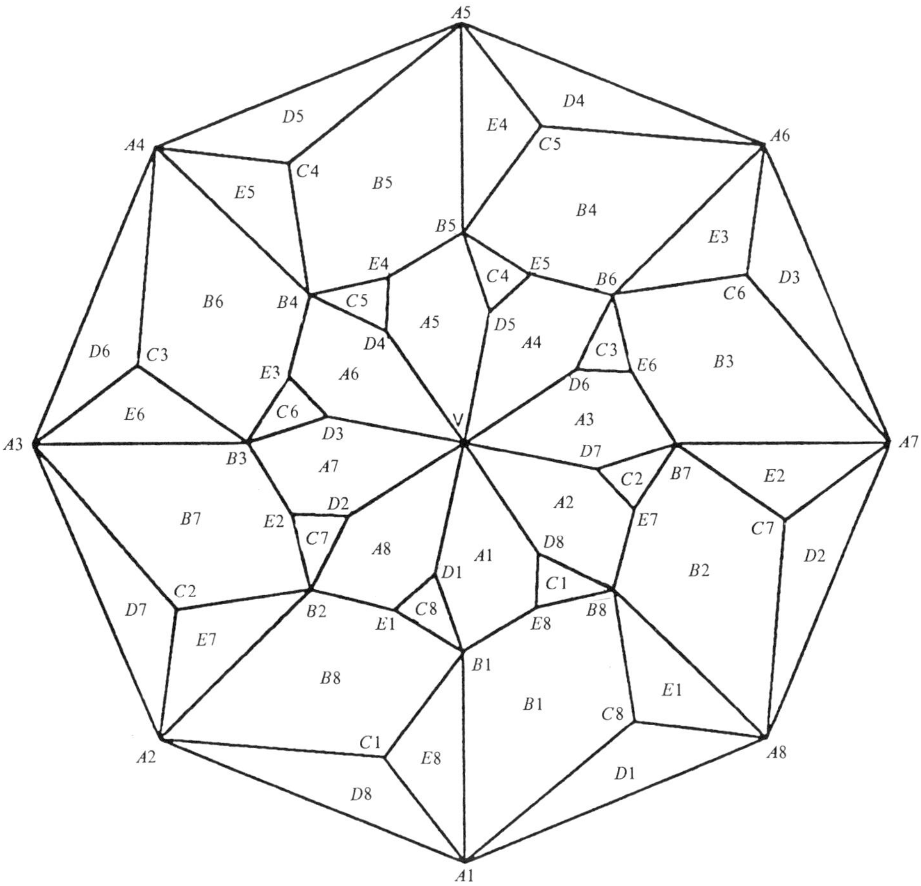


FIGURE 16. A Schlegel diagram of a polyhedron $R_n$ (for $n = 8$) with $5n + 1$ vertices, such that $S(R_n) = [\mathbf{n}]^+$ and $S(R_n^H) = [\mathbf{2}, \mathbf{n}]^+/[\mathbf{n}]^+$. In this diagram balanced labelling is used.

points of difference: (i) Instead of working with the polyhedra themselves, we consider their projections onto a fixed sphere on which they induce tilings; thus we discuss tilings of the sphere instead of the plane. (ii) Since the mapping $\phi$ may be assumed to commute with the antipodal map of the sphere, it has a fixed point (see, for example, Alexandroff-Hopf [**1, Theorem IV on p. 48**]). (iii) Since the number of faces of a polyhedron is finite, the rank of a polyhedron cannot be infinite. Thus we only need to prove the existential part of the theorem.

The smallest centrally symmetric self-dual convex polyhedron that can be taken as $C_2$ has eight vertices; a Schlegel diagram of $C_2$ is shown in Figure 17(a). This polyhedron was first mentioned as being (combinatorially) self-dual by Kirkman [**23**], and later also by Hermes [**18**] and Brückner [**4**]; however, none of these authors noticed that $C_2$ could be realized as a centrally symmetric polyhedron. This fact was first stated by Jucovic [**21**], while the realization of $C_2$ as a harmoniously self-polar polyhedron (as shown in Figure 17(b)) is due to Leichtweiss [**24**]. The construction of $C_4$, which appears to be new, is indicated in Figure 18. In Figure 18(a) a Schlegel diagram of $C_4$ is given, while a harmoniously self-dual realization is shown in Figure 18(b); the coordinates of the vertices of the polyhedra in Figures 17(b) and 18(b) are listed in the captions. The verification of the claims concerning groups of symmetries and the harmonious self-duality of these polyhedra is routine, and is omitted. □

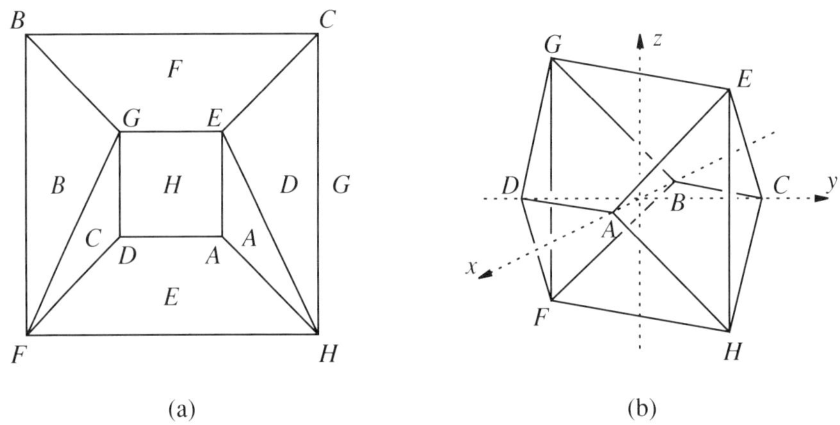


FIGURE 17. The self-dual centrally symmetric polyhedron $C_2$ of rank 2. (a) A Schlegel diagram of the polyhedron, with balanced labelling. (b) A projection of a harmoniously self-dual, centrally symmetric realization of $C_2$, with vertices $A = (1, 0, 0)$, $B = (-1, 0, 0)$, $C = (0, 1, 0)$, $D = (0, -1, 0)$, $E = (1, 1, 1)$, $F = (-1, -1, -1)$, $G = (-1, -1, 1)$, $H = (1, 1, -1)$.

We conjecture that each of the polyhedra described in this section has the smallest possible numbers of vertices among all self-dual polyhedra with the properties stated in each case.

## 5. Self-dual configurations in the projective plane

For our third, and final, example of self-dual objects whose rank exceeds 2 we turn to configurations of points and lines in the real projective plane. We shall be exclusively concerned with $(n_m)$-configurations, that is, sets of $n$ points $P_i$ and $n$ lines $L_i$, $i = 1, 2, \ldots, n$, such that every $P_i$ lies on exactly $m$ of the lines $L_j$, and conversely. Duality is defined as in §1, and it is worth remarking that many of the configurations that occur in elementary projective geometry are self-dual—for example, the configurations of Pappus and Desargues are self-dual $(9_3)$- and $(10_3)$-configurations.

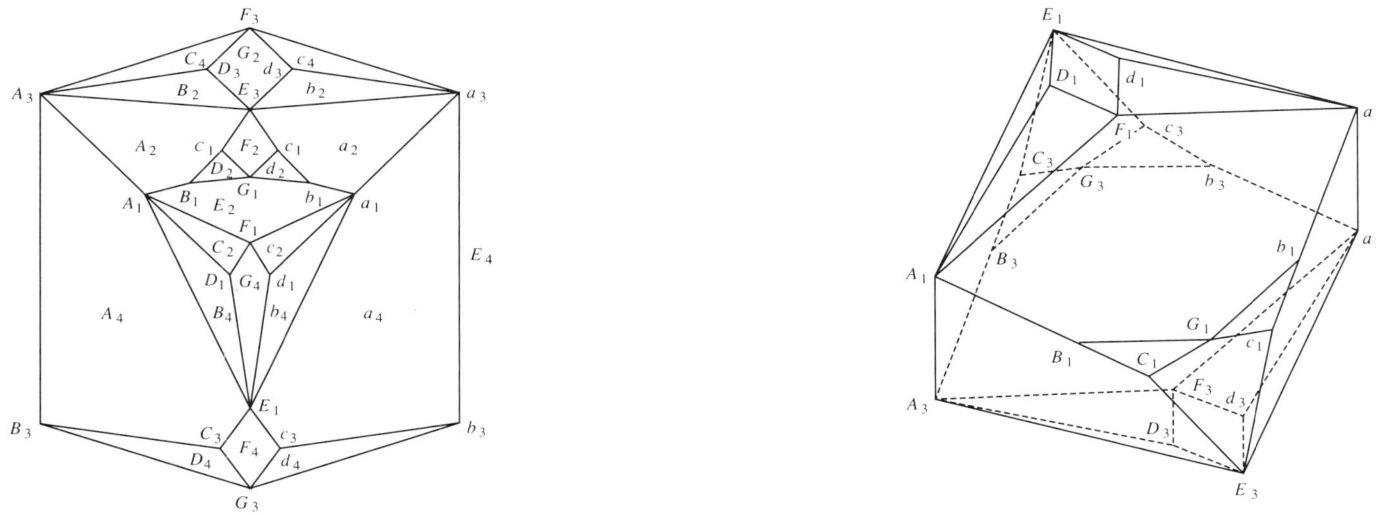


FIGURE 18. The self-dual centrally symmetric polyhedron $C_4$ of rank 4. (a) A Schlegel diagram of the polyhedron, with labels indicating a self-duality of rank 4. A capital letter with a subscript is used for each vertex, and the self-duality map increases the subscript by 1( mod 4). (b) A projection of a centrally symmetric metrically self-dual realization, with vertices $A_1 = (60, 0, 60)$, $A_3 = (60, 0, -60)$, $B_1 = (30, 30, 45)$, $B_3 = (30, -30, -45)$, $C_1 = (18, 42, 33)$, $C_3 = (18, -42, -33)$, $D_1 = (10, -40, 50)$, $D_3 = (10, 40, -50)$, $E_1 = (0, -60, 30)$, $E_3 = (0, 60, -30)$, $F_1 = (0, -20, 70)$, $F_3 = (0, 20, -70)$, $G_1 = (0, 40, 40)$, $G_3 = (0, -40, -40)$. The vertices labelled by lower case letters differ from the ones marked with the same upper case letter by the sign of the $x$-coordinate, and all coordinates should be divided by 60. The isometry mapping of $C_4$ onto its polar $C_4^*$, which is needed for the self-duality of rank 4, is a rotation about the $x$-axis that carries the $z$-axis to the $y$-axis.

As with tilings, several distinct meanings can be attached to self-duality of configurations. The first, alluded to in the preceding paragraph, is the *combinatorial* sense, with isomorphisms and dualities being understood in the sense of the incidence lattice. Of independent interest are *projective* self-dualities, in which isomorphisms are induced by collineations, and dualities by correlations, of the projective plane in which the configuration has been embedded. (In fact, an even more detailed classification arises from the possibility of using projective planes over various fields; here, however, we shall restrict attention to the *real* projective plane.) For a self-dual configuration $C$ embedded in the projective plane we shall use $A(C)$ and $D(C)$ to denote the automorphism and self-duality groups, while $A_L(C)$ and $D_L(C)$ will indicate the groups of collineations, or of collineations and correlations, of $C$, respectively.

As an illustration we consider the $(21_4)$-configuration $C$ shown in Figure 19. This self-dual configuration has been extensively studied by many authors, but only recently has it been observed that it admits an embedding in the *real* Euclidean plane $\mathbb{E}^2$ (embeddings in the complex plane and in finite planes were found long ago); see Grünbaum and Rigby **[14]** for the properties of this configuration, and for references to earlier works. In can be shown that $A(C) = \mathrm{PGL}(2, 7)$ of order 336, and that $D(C)$ has order 672; in contrast, $A_L(C) = \mathbf{D}_7$, $D_L(C) = \mathbf{D}_7 \times \mathbf{C}_2$, and no embedding of $C$ in $\mathbb{E}^2$ has a larger group of collineations and correlations.

From now on we shall consider only $(n_3)$-configurations. For some special cases, including the configurations of Pappus and Desargues, the various groups have been studied; see, for example, Kagno **[22]** and Coxeter **[7, 8]**. However, the question of rank of the configurations has not been considered before. (We may note that Coxeter **[6, p. 67]** mentions that a "polarity" is an involutory correlation; he thus comes close to the question of rank of a self-duality—but he does not pursue the topic.) The following is analogous to the result on polyhedra.

THEOREM 6. *There exist self-dual $(n_3)$-configurations $C_k$ in the projective plane ($k \geq 1$) with $r(C_k) = 2^k$. We may choose $C_1$ to be a $(9_3)$-configuration, and $C_k$, for $k \geq 2$, to be a $(64q_3)$-configuration with $A(C_k) = A_L(C_k) = \mathbf{C}_q$, and $D(C_k) = D_L(C_k) = \mathbf{C}_{2q}$, where $q = 2^{k-1}$.*

PROOF. For $C_1$ we may take the Pappus configuration. For $k \geq 2$ we construct $C_k$ using the fragments $\mathbf{F}$ and $\mathbf{G}$ shown in Figure 20. Fragment $\mathbf{F}$ is obtained from a configuration usually denoted $(10_3)_5$ (see, for example, Schroeter **[29]**, Dorwart **[11]**, and Bokowski and Sturmfels **[3, p. 44]**), by breaking one of its incidences. Fragment $\mathbf{G}$ is dual to $\mathbf{F}$, and the points and lines of both fragments have been labelled following the convention made in §2. Using three copies of $\mathbf{F}$ and three copies of $\mathbf{G}$, together with additional points and lines labelled by $W$ and $X$ (with appropriate primes and/or sub-

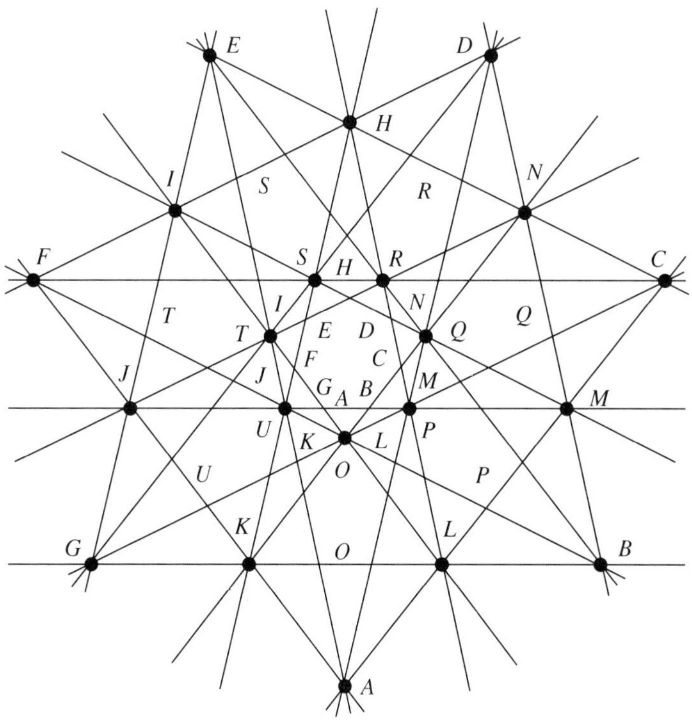


FIGURE 19. A self-dual configuration $(21_4)$; a balanced labelling is shown.

scripts), we construct a larger fragment **H**, which is schematically indicated in Figure 21. Using $q = 2^{k-1}$ copies of **H**, placed in a necklace-like circuit, yields the configuration $C_k$. A self-duality $\delta$ of this configuration maps $A_1, B_1, C_1, \ldots$ onto $A_2, B_2, C_2, \ldots$; these map into $A_3, B_3, C_3, \ldots$ in the next fragment, and so on. (Subscripts are taken modulo $2q$.) To see that $C_k$ satisfies the requirements of Theorem 6, it is convenient to consider its Levi-diagram (Figure 23, see **[5]** for more detailed explanations of Levi graphs); this is a graph in which the points of the configuration are represented by black vertices (solid dots), the lines of the configuration by white vertices (hollow dots), and vertices corresponding to incident points and lines determine edges of the diagram. The parts of the Levi-diagram corresponding to **F** and **G** are shown in Figure 22; from these it is clear that **G** is dual to **F**, and from the different roles $A$ and $V$ play in the diagrams it is obvious that the only way a duality can operate in $C_k$ is by mapping a copy of **F** onto a copy of **G**. The points and lines added in the formation of **H** also map appropriately under dualities that carry copies of **F** onto copies of **G** (and vice versa), as is visible from the Levi-diagrams of two consecutive copies of **H**. It follows therefore that all possible dualities are odd powers

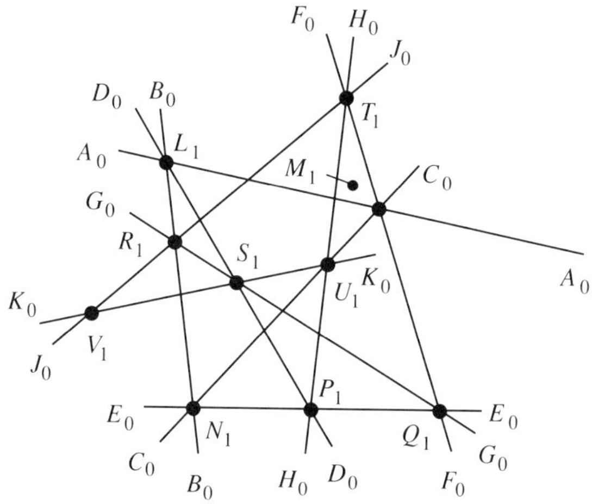


(a) $F$

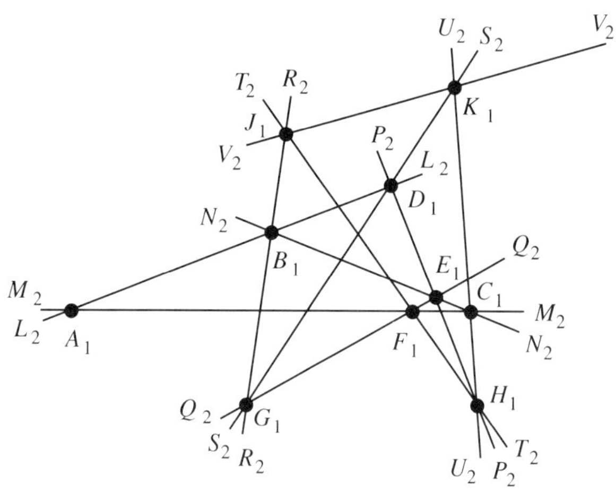


(b) $G$

FIGURE 20. Fragments **F** and **G** of configurations used in the construction of the self-dual configuration described in the text.

of the duality $\delta$ described above. Each of these is of rank $2q = 2^k$, which completes the proof of Theorem 6. □

We do not assert that $C_k$ is the smallest configuration of rank $2^k$, but we believe that the numbers of points and lines cannot be decreased greatly unless a radically different method of construction is used. Also, it would be of

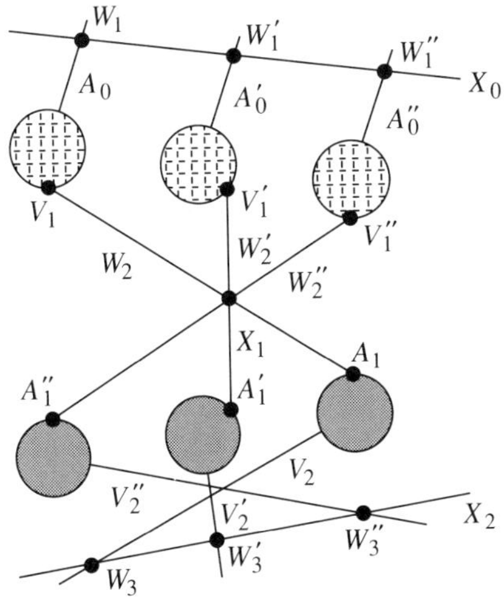


FIGURE 21. A larger fragment, **H**, of a configuration, used in the proof of Theorem 6.

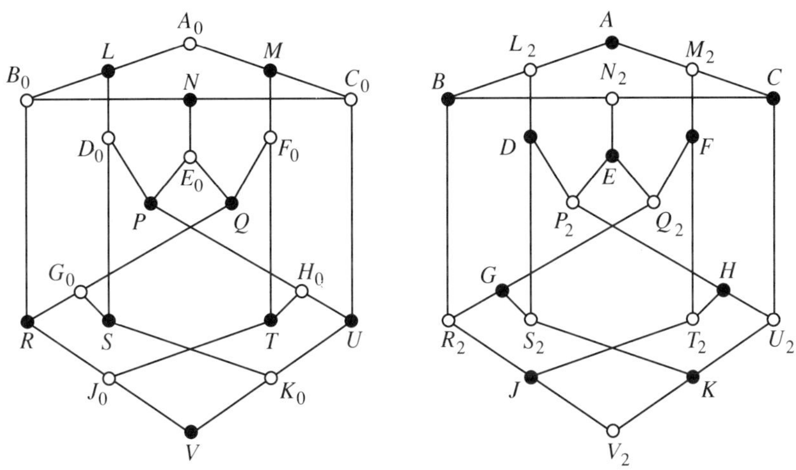


(a) Levi-diagram of $F$. (b) Levi-diagram of $G$.

FIGURE 22. Levi diagrams of the fragments **F** and **G**. The solid dots represent points, the hollow ones represent lines.

interest to decide whether the configurations we described have embeddings for which $A_L(C_k) = \mathbf{C}_q$, and $D_L(C_k) = \mathbf{C}_{2q}$.

It seems likely that there exist self-dual $(n_m)$-configurations with $m > 3$ and of arbitrarily large rank; however, this has not been established so far.

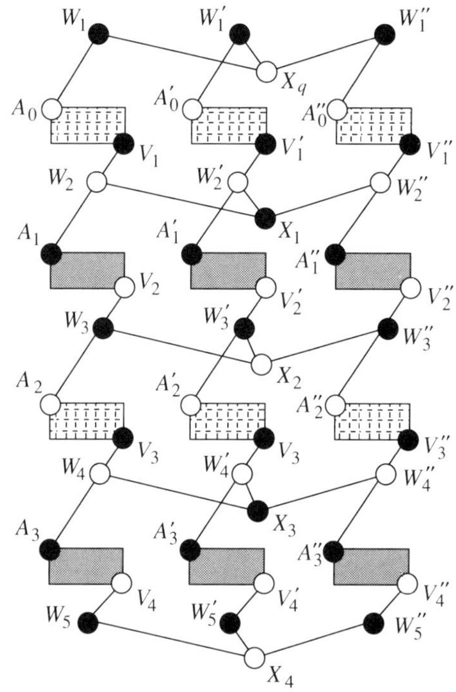


FIGURE 23. Levi diagram of the fragment **H**.

## 6. Remarks and problems

(1) Plane self-dual tilings of infinite rank can be used to construct self-dual tilings of the torus of arbitrary rank $r = 2^k$. The method is illustrated in Figure 24 for $r = 8$. No results seem to be available concerning the possible automorphism and self-duality groups of such tilings. Also, it appears that self-dual tilings on manifolds of higher genus have not been investigated so far.

(2) The construction of self-dual polyhedra or tilings by performing "dual changes" on "dually corresponding" elements has been used to obtain almost all the examples in this paper, even if the presentation does not indicate this. Although this technique is as old as the investigation of self-dual polyhedra (Kirkman [**23**]), it applies here with an essential modification: if the self-duality is of rank $r$ greater than 2, then the changes must be made at all $r$ elements involved. Thus it is possible that, from a self-dual polyhedron $P$ of rank 2, the use of a self-duality of $P$ with rank $r > 2$ can lead to a self-dual polyhedron of rank $r$. This happens, for example, in the construction of the Jendrol polyhedron (see Figure 25). Similarly, the use of self-dualities of infinite rank of the square tiling (which is of rank 2) leads to tilings with infinite rank, such as those in Figures 3, 5, and 10.

(3) In the hyperbolic plane there exist infinitely many self-dual tilings with a high degree of regularity. However, no characterization of the possible automorphism groups or self-duality groups is known.

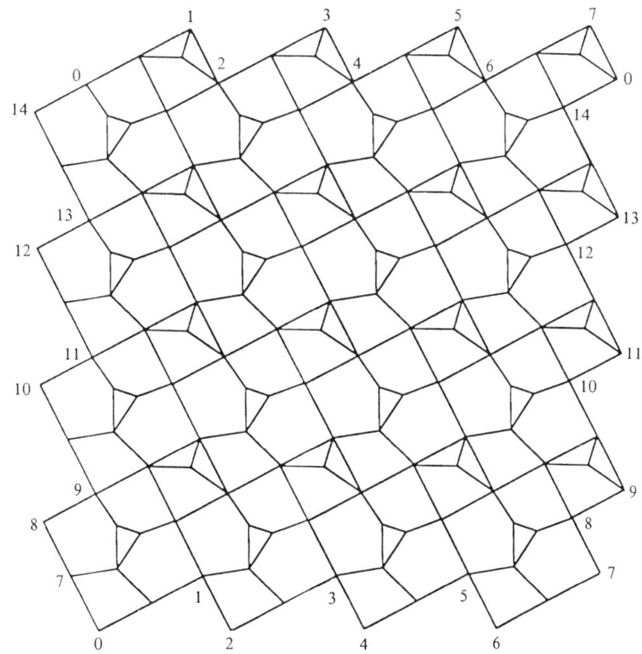


FIGURE 24. Suitable period parallelograms of a self-dual tiling of the plane can be used to construct self-dual tilings of the torus. Here a tiling with symmetry group **p1** and infinite rank is used to define a self-dual tiling of the torus of rank 8.

(4) As observed by Leichtweiss [**24**] and Sztencel and Zaremba [**32**], the polyhedron $C_2$ described in §4 can serve as the unit sphere of a 3-dimensional normed space $X$ with the property that the dual (adjoint) space $X^*$ is isometric to $X$ (thus $X$ is self-dual) but only the second dual $X^{**}$ is canonically isometric to $X$. Analogously, the polyhedron $C_4$ is the unit sphere of a 3-dimensional normed space $Y$ which is isometric to its dual $Y^*$, but the fourth dual $Y^{****}$ is the first which is canonically isometric to $Y$. Moreover, Theorem 6 implies that for no 3-dimensional normed space isometric to its dual does one have to go beyond the fourth dual to find a canonical isometry. It is known that there exist self-dual centrally symmetric polytopes of all dimensions and of rank 2; these are of interest as unit balls of self-dual normed spaces without inner product norms (see Partingon [**28**]). However, it is not known whether ranks greater than 2 are possible for centrally symmetric convex polytopes of dimension exceeding 3. If central symmetry is not required, the polyhedra constructed in §4 can be used to obtain self-dual convex polytopes of dimension 4 or higher, with arbitrarily high rank.

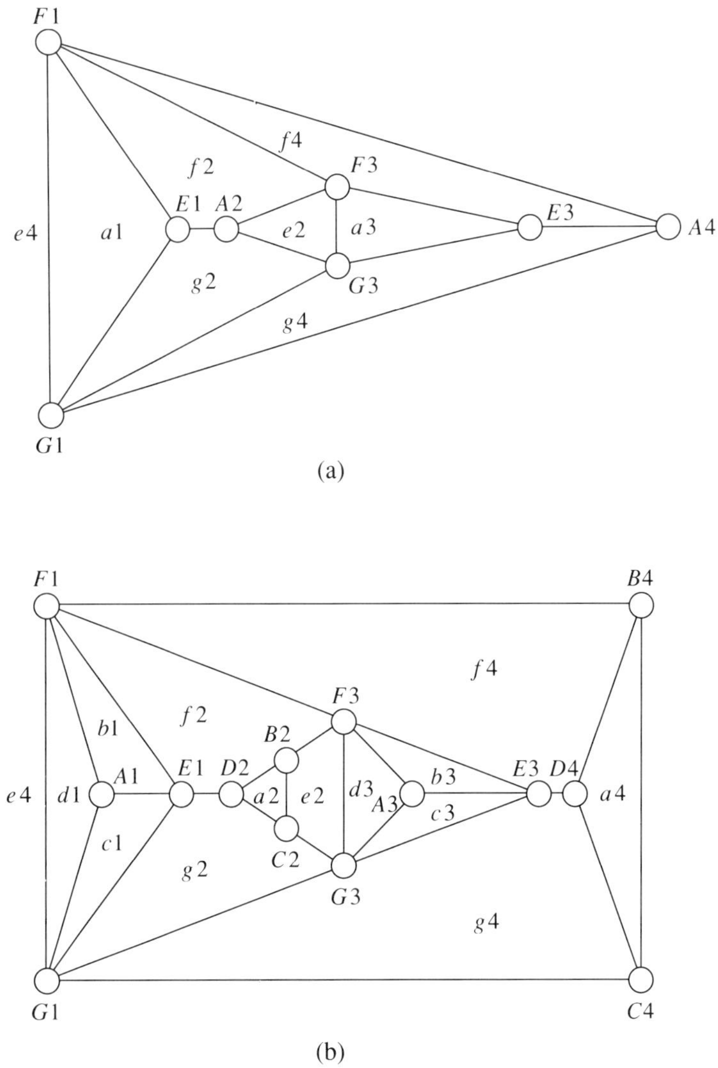


FIGURE 25. (a) Another Schlegel diagram of the self-dual polyhedron $C_2$ of Theorem 4, and a self-duality of $C_2$ rank 4, which increases all subscripts by 1 (mod 4). (b) The self-dual polyhedron of rank 4 described by Jendrol [**19**] can be obtained from the self-dual polyhedron in part (a) by subdividing the faces $a_1$ and $a_3$, and simultaneously truncating the vertices $A_2$ and $A_4$.

(5) S. Jendrol [**20**] has determined all symmetry groups that are possible for 3-dimensional self-dual convex polyhedra.

## References


1. P. Alexandroff and H. Hopf, *Topologie*, Springer, Berlin, 1935.
2. D. Archdeacon and R. B. Richter, *The construction and classification of self-dual polyhedra*, J. Combinat. Theory, Ser. B (to appear).
3. J. Bokowski and B. Sturmfels, *Computational synthetic geometry*, Lecture Notes in Math., vol. 1355, Springer-Verlag, New York, 1989.
4. M. Brückner, *Vielecke und Vielflache*, Teubner, Leipzig, 1900.
5. H. S. M. Coxeter, *Self-dual configurations and regular graphs*, Bull. Amer. Math. Soc. **56** (1950), 413–455 = *Twelve geometric essays*, Southern Illinois Univ. Press, Carbondale, Ill. 1968, pp. 106–149.
6. ——, *The real projective plane*, 2nd ed., Cambridge Univ. Press, 1955.
7. ——, *Desargues configurations and their collineation groups*, Math. Proc. Cambridge Philos. Soc. **78** (1975), 227–246.
8. ——, *The Pappus configuration and the self-inscribed octagon*. I, II, III, Nederl. Akad. Wetensch. Proc. Ser. A **80** (1977), 256–300.
9. ——, *Coloured symmetry*, in M. C. Escher: Art and Science (H. S. M. Coxeter et al., eds.), Elsevier, Amsterdam, 1986, pp. 15–33.
10. H. S. M. Coxeter and W. O. J. Moser, *Generators and relations for discrete groups*, 4th ed., Springer-Verlag, Berlin, 1980.
11. H. L. Dorwart, *Configurations*, WFF'N PROOF, New Haven, Conn., 1967.
12. S. Eilenberg and N. Steenrod, *Foundations of algebraic topology*, Princeton Univ. Press, Princeton, N.J., 1952.
13. B. Grünbaum, *Convex polytopes*, Interscience, London, 1967.
14. B. Grünbaum and J. F. Rigby, *The real configuration* $(21_4)$, J. London Math. Soc. (2) **41** (1990), 336–346.
15. B. Grünbaum and G. C. Shephard, *Patterns on the 2-sphere*, Mathematika **28** (1981), 1–35.
16. ——, *Tilings and patterns*, Freeman, New York, 1987.
17. ——, *Is selfduality involutory*? Amer. Math. Monthly **95** (1988), 729–733.
18. O. Hermes, *Die Formen der Vielflache*, J. Reine Angew. Math. **122** (1900), 124–154.
19. S. Jendrol, *A non-involutory selfduality*, Discrete Math. **74** (1989), 325–326.
20. ——, *On symmetry groups of selfdual polyhedra* (to appear).
21. E. Jucovič, *Characterization of the p-vector of a self-dual 3-polytope*, Combinatorial Structures and their Applications (R. Guy et al., eds.), Gordon and Breach, New York, 1970, pp. 185–187.
22. I. N. Kagno, *Desargues' and Pappus' graphs and their groups*, Amer. J. Math. **69** (1947), 859–862.
23. T. P. Kirkman, *On autopolar polyhedra*, Philos. Trans. Roy. Soc. London **147** (1857), 183–215.
24. K. Leichtweiss, *Zur expliziten Bestimmung der Norm der selbstadjungierten Minkowski-Räume*, Resultate Math. **1** (1978), 61–87.
25. P. Mani, *Automorphismen von polyedrischen Graphen*, Math. Ann. **192** (1971), 279–303.
26. J. McCanna, *Is self-duality always involutory* ? Proc. Twentieth Southeastern Conference on Combinatorics, Graph Theory and Computing (Boca Raton, FL, 1989). Congr. Numer. 72 (1990), 175–178.
27. T. A. McKee, *Neighborhood and self-dual multigraphs* (to appear).
28. J. R. Partington, *Self-conjugate polyhedral Banach spaces*, Bull. London Math. Soc. **18** (1986), 284–286.
29. H. Schroeter, *Ueber die Bildungsweise und geometrische Konstruction der Konfigurationen* $(10_3)$, Nachr. Akad. Wiss. Göttingen **8** (1889), 193–236.
30. R. L. E. Schwarzenberger, *Colour symmetry*, Bull. London Math. Soc. **16** (1984), 209–240.
31. B. Servatius, P. R. Christopher, A Juels, A. Oudaise, and P. S. Berkebile, *Construction of self-dual graphs* (to appear).
32. R. Sztencel and P. Zaremba, *On self-conjugate Banach spaces*, Colloq. Math. **44** (1981), 112–115.

33. D. K. Washburn and D. W. Crowe, *Symmetries of culture: Theory and practice of plane pattern analysis*, Univ. of Washington Press, Seattle, Wash., 1988.

DEPARTMENT OF MATHEMATICS, UNIVERSITY OF WASHINGTON, SEATTLE, WASHINGTON 98195

DEPARTMENT OF MATHEMATICS, UNIVERSITY OF WASHINGTON, SEATTLE, WASHINGTON 98195
*E-mail address*: grunbaum@math.washington.edu

DEPARTMENT OF MATHEMATICS, UNIVERSITY OF EAST ANGLIA, NORWICH, ENGLAND

DANIEL H. WAGNER ASSOCIATES, PAOLI, PENNSYLVANIA 19301

DIMACS Series in Discrete Mathematics
and Theoretical Computer Science
Volume **4**, 1991

Applied Geometry
and Discrete Mathematics
THE VICTOR KLEE FESTSCHRIFT

# Do Projections Go to Infinity?

IMRE BÁRÁNY, JACOB E. GOODMAN, AND RICHARD POLLACK

ABSTRACT. Generalizing a theorem of Aharoni, Duchet, and Wajnryb, we prove their conjecture that the orbit of a point under any sequence of projections on a family of lines in the plane, any two of which intersect in the unit disk, is bounded. Moreover, we show that while the natural extension of this theorem is false in dimension greater than 2, with appropriate additional hypotheses it remains true.

## 1. Introduction

In a 1984 paper by R. Aharoni, P. Duchet, and B. Wajnryb [**1**], the following theorem appears:

THEOREM. *If* $\mathscr{F}$ *is a finite family of flats in* $\mathbb{R}^d$ *then any sequence of projections on* $\mathscr{F}$ *is bounded.*

At the end of the paper, the authors pose the problem of extending the result to infinite families, and offer the

CONJECTURE. Let $\mathscr{F}$ be a family of flats in $\mathbb{R}^d$ and let $K = \bigcap\{F^{\perp\perp} : F \in \mathscr{F}\}$ (i.e., the maximal subspace parallel to all members of $\mathscr{F}$). Then every sequence of projections on $\mathscr{F}$ is bounded if and only if the set

$$\{x \in K^{\perp} : \{x\} = \bigcap \mathscr{G} \cap K^{\perp} \text{ for some } \mathscr{G} \subset \mathscr{F}\}$$

is bounded.

In the plane, in particular, this would say that if a family $\mathscr{F}$ of lines has the property that every two members of $\mathscr{F}$ meet in the unit disk, then the

1980 *Mathematics Subject Classification* (1985 *Revision*). Primary 51M99; Secondary 52A30, 05B30.

The authors were supported in part by the Cowles Foundation of Yale University and the Center for Discrete Mathematics and Theoretical Computer Science (DIMACS), a National Science Foundation Science and Technology Center, under NSF grant STC88-09648 (Bárány); by NSF grant DMS-8501492, NSA grant MDA904-89-H-2038, PSC-CUNY grant 666426, and DIMACS (Goodman); and by NSF grants DMS-8501947 and CCR-8901484, NSA grant MDA904-89-H-2030, and DIMACS (Pollack).


1052-1798/91 $1.00 + $.25 per page

orbit of a point under any sequence of projections onto members of $\mathcal{F}$ is bounded.

The intuition behind this is that when seen from "far away" such a family $\mathcal{F}$ appears to be a pencil, and the conclusion is obvious for a pencil since successive projections always move closer to the vertex of the pencil. Thus, if the orbit of a point moves far from the origin, the orbit should then move in toward the origin until it is reasonably close and hence the orbit should stay in a bounded region of the plane.

The purpose of this paper is to prove this conjecture for $d = 2$, and to point out that—unlike the original theorem—it is false for $d > 2$ without additional hypotheses.

In §2 the proof is given for line families in the plane. (For simplicity we assume that no two lines are parallel, but this restriction could be eliminated from the theorem at the cost of a slightly more complicated argument.) §3 contains a discussion of the situation for families of lines in higher dimensions, including some counterexamples to the conjecture and both necessary and sufficient conditions for the validity of the corresponding theorem in dimension 3 or more. The idea common to §§2 and 3 is the construction of a large hypersurface that is orthogonal to the given line family. Such a hypersurface will bound a set invariant under projection.

We thank the referees for some helpful suggestions.

We begin with an observation which holds in any dimension.

PROPOSITION 1. *Given a family $\mathcal{F}$ of flats and a point $x$ in $\mathbb{R}^d$, let $\pi_i \colon \mathbb{R}^d \to F_i$ denote orthogonal projection onto $F_i \in \mathcal{F}$. The following are equivalent*:

(i) *Each sequence of projections $\pi_{i_1}, \ldots, \pi_{i_k}, \ldots$ onto members of $\mathcal{F}$, beginning with $x$, is bounded.*

(ii) *All sequences of projections $\pi_{i_1}, \ldots, \pi_{i_k}, \ldots$ onto members of $\mathcal{F}$, beginning with $x$, are uniformly bounded.*

(iii) *$x$ is contained in a bounded set, invariant under projection onto members of $\mathcal{F}$.*

PROOF. The only implication that is not obvious is (i) $\Rightarrow$ (ii), which we proceed to show.

Suppose (ii) were false. We could then inductively choose points $x = x_0, y_1, x_1, y_2, x_2, \ldots$ and finite sequences $(F_{i_{1,1}}, \ldots, F_{i_{1,k_1}})$, $(F_{i_{2,1}}, \ldots, F_{i_{2,k_2}}), \ldots$ of members of $\mathcal{F}$ satisfying

$$(1) \qquad y_j = \pi_{i_{j,k_j}} \cdot \cdots \cdot \pi_{i_{j,1}}(x), \qquad x_j = \pi_{i_j,k_j} \cdot \cdots \cdot \pi_{i_{j,k_j}}(x_{j-1}),$$

$$(2) \qquad |y_j| \geq |x_{j-1}| + |x_{j-1} - x| + 1.$$

Since every composition of projection mappings belongs to Lip(1), it follows from (1) that $|x_j - y_j| \leq |x_{j-1} - x|$. Hence, $|x_j| \geq |y_j| - |x_{j-1} - x| \geq$

$|x_{j-1}| + 1$ by (2), so that by induction we have $|x_j| \geq |x| + j$, which shows that the sequence of projections of $x$ onto the flats

$$F_{i_{1,1}}, \dots, F_{i_{1,k_1}}, F_{i_{2,1}}, \dots, F_{i_{2,k_2}}, \dots$$

is unbounded. □

(Notice that the implication (i) ⇒ (ii) does not depend on the fact that the mappings $\pi_i$ are projections nor on the fact that they and their compositions belong to Lip(1). It is easily seen, in fact, that it is sufficient to assume the existence of a real-valued function $g$ such that for any $f$ which is a composition of the functions $\pi_i$ we have $|f(x) - f(y)| \leq g(x, y)$.)

COROLLARY 1. *If $\mathcal{F}$ is as above and $x, y \in \mathbb{R}^d$, then $x$ belongs to a bounded set invariant under projection onto members of $\mathcal{F}$ if and only if $y$ does.*

PROOF. Since $\pi_F: \mathbb{R}^d \to F$ belongs to Lip(1), $x$ satisfies condition (i) if and only if $y$ does. The conclusion therefore follows from Proposition 1. □

We will say that a family $\mathcal{F}$ of flats which satisfies the conditions of Proposition 1 has the *bounded projection property*; notice that these conditions are independent of the point $x$, by Corollary 1.

## 2. Line families in the plane

THEOREM 1. *Suppose $\mathcal{L}$ is a family of lines in $\mathbb{R}^2$, no two of which are parallel. If the intersection of any two members of $\mathcal{L}$ is contained in the unit disk $B^2$, then $\mathcal{L}$ has the bounded projection property. Conversely, if $\mathcal{L}$ has the bounded projection property, then the pairwise intersections of all members of $\mathcal{L}$ are bounded.*

PROOF. The converse follows immediately from the observation that the point of intersection of two lines can be approximated arbitrarily closely by a sequence of projections alternating between the two lines. If such points of intersection formed an unbounded set, we could therefore find sequences of projections receding arbitrarily far from the origin, thereby violating condition (ii) of Proposition 1.

For the direct implication it is enough to show, by Corollary 1 above, that for some $x \in \mathbb{R}^2$ each sequence of projections onto members of $\mathcal{L}$, beginning with $x$, is bounded.

For each line $L \in \mathcal{L}$, let $\theta = \theta(L)$ be the direction of $L$ $(0 \leq \theta \leq \pi)$ and $r(\theta)$ the vector from $O$ to the closest point of $L$. We have $|r(\theta)| \leq 1$. We claim first that $r(\theta) \in \text{Lip}(1)$.

To see this, notice that since two lines, in directions $\theta_1$ and $\theta_2$, say, meet in $B^2$, the farthest apart the values of $r(\theta_1)$ and $r(\theta_2)$ can be occurs when the lines meet on the boundary of $B^2$, as in Figure 1. But then it is easily

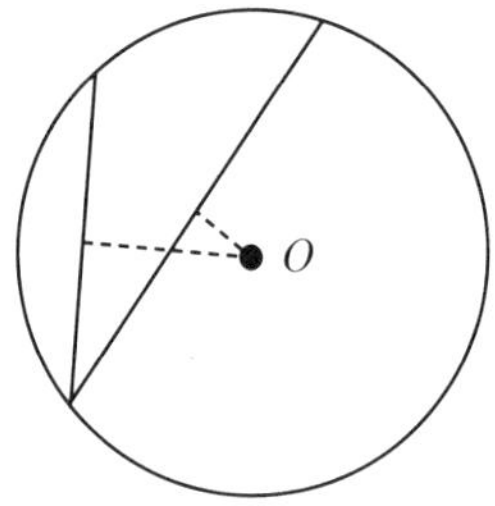


FIGURE 1

seen that $|r(\theta_2)-r(\theta_1)| = |\sin(\theta_2-\theta_1)|$, so that in particular $|r(\theta_2)-r(\theta_1)| \leq |\theta_2-\theta_1|$.

We can therefore extend the function $r(\theta)$ from its domain, which may be an arbitrary subset of $[0, \pi]$, to the entire interval, keeping it in $\mathrm{Lip}(1)$, and preserving the property that $|r(\theta)| \leq 1$; let us call the extended function $r(\theta)$ as well. This amounts to extending the original family of lines to a new family $\widetilde{\mathscr{L}}$, one for each direction.

A calculation similar to the one above shows that the new family $\widetilde{\mathscr{L}}$ satisfies nearly the same conditions that $\mathscr{L}$ did: the intersection of any two members of $\widetilde{\mathscr{L}}$ is contained in the disk $\rho B^2$, where $\rho$ $(\approx 1.48\ldots) = \csc\alpha$ for $\alpha$ the solution of $\cos\alpha = \alpha$. In particular, $\widetilde{\mathscr{L}}$ is a so-called *spread* **[2]**, and every point $x$ of $\mathbb{R}^2$ belongs to at least one line of $\widetilde{\mathscr{L}}$, and in fact to a unique such line if $x \notin \rho B^2$. Moreover, it follows from the fact that all members of $\widetilde{\mathscr{L}}$ meet inside $\rho B^2$ that as a point $x$ runs around the circle centered at the origin of radius $R > \rho$, the direction of the line $L \in \widetilde{\mathscr{L}}$ passing through $x$ varies monotonically.

Now suppose $\theta(L_i) < \theta(L_j) < \theta(L_k)$ are sufficiently close, and a point $x$ on $L_i$ is sufficiently far from $O$. Then it follows that $\pi_k(\pi_j(x))$ "dominates" $\pi_k(x)$, i.e., lies further from $O$ (see Figure 2(a)), and—passing to the limit while remembering the monotonicity of $\theta(L)$—that the orthogonal trajectory through $x$ to the family $\widetilde{\mathscr{L}}$ also dominates $\pi_k(x)$, as shown in Figure 2(b). (The simplest way to see that this orthogonal trajectory in fact exists is to construct it by considering, for each line $L_k$ as in Figure 2(a), the supremum of all the points on $L_k$ obtained by a sequence of projections, starting with $x$, along a finite system of rays such as $L_j$ between $L_i$ and $L_k$, and to show that the resulting curve, which is easily seen to be continuous, is in fact orthogonal to each ray.)

For each $\theta = \theta(L)$, let $t$ be the directed distance along $L$ in the direction $\theta$, measured from the midpoint between the two points at which $L$ meets the unit circle. (Thus each line $L \in \widetilde{\mathscr{L}}$ has an "origin" $O_L$, as shown in Figure 3.) Assume also that $x$ lies on a line $L_0$ with $\theta(L_0) = 0$.

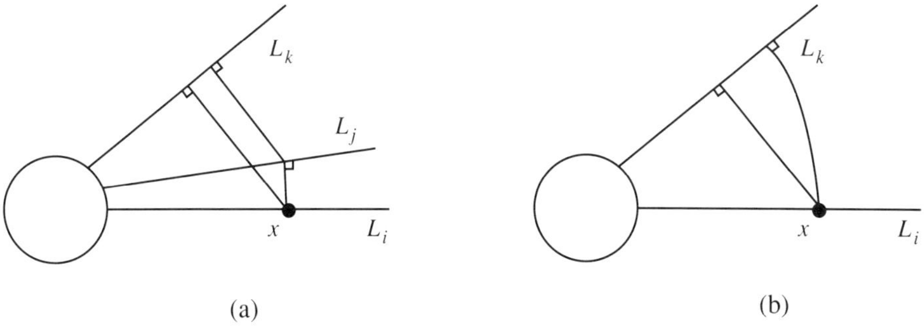


FIGURE 2

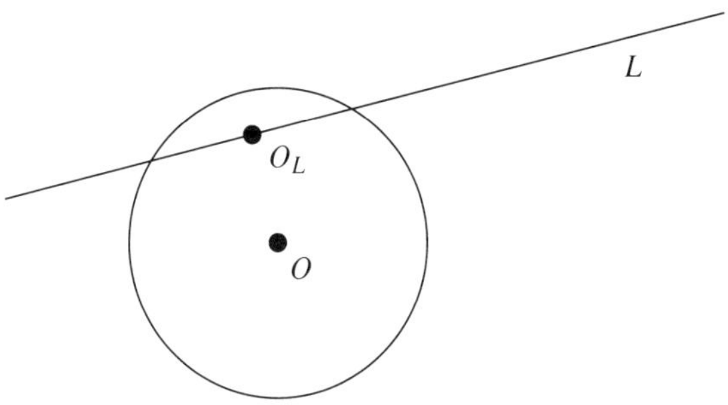


FIGURE 3

In terms of this "local" coordinate system, we are therefore reduced to showing that as $\theta$ varies through $2\pi$, the orthogonal trajectory $t = t(\theta)$ to $\widetilde{\mathscr{L}}$ which begins at $x$ both returns to $x$ and stays outside $\rho B^2$ all the way: the first assertion is equivalent to $t(2\pi) = t(0)$, and the second is implied by $t(\theta) > \rho$ for all $\theta$ from 0 to $2\pi$.

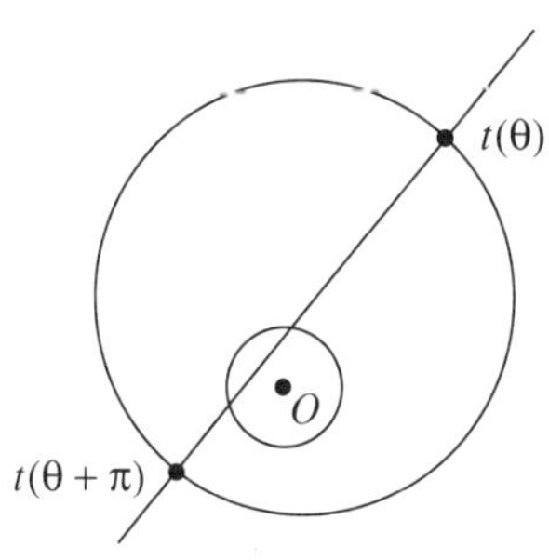


FIGURE 4

Ignoring the second question for the moment, for each $\theta = \theta(L)$ let $f(\theta)$ denote the length of the portion of $L$ between $t(\theta)$ and $t(\theta+\pi)$ (see Figure 4). It follows by looking at the discrete situation (see Figure 5) and passing to the limit that $f'(\theta) = 0$ for all $\theta$, which proves the first assertion.

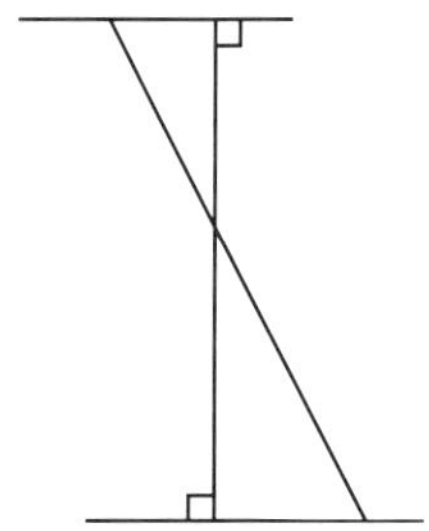

FIGURE 5

As for the second, one can argue as follows: If two lines pass through $B^2$ and form an angle of $\alpha$, the decrease in the value of $t$ from one to the other along the orthogonal trajectory is no worse than the decrease in $t$ by direct projection (see Figure 2). Moreover, this decrease is maximized if the first line is tangent to the disk (See Figure 6), in which case the map is given by $t \to t\cos\alpha - \sin\alpha$, or

$$t + \frac{\sin\alpha}{1-\cos\alpha} \to \cos\alpha\left(t + \frac{\sin\alpha}{1-\cos\alpha}\right).$$

Iterating $2\pi/\alpha$ times, solving for $t$, and passing to the limit as $\alpha \to 0$, we get $t \to t - 2\pi$; in particular, if $t > 2\pi + \rho$ to begin with, a complete rotation will end up with $t > \rho$.

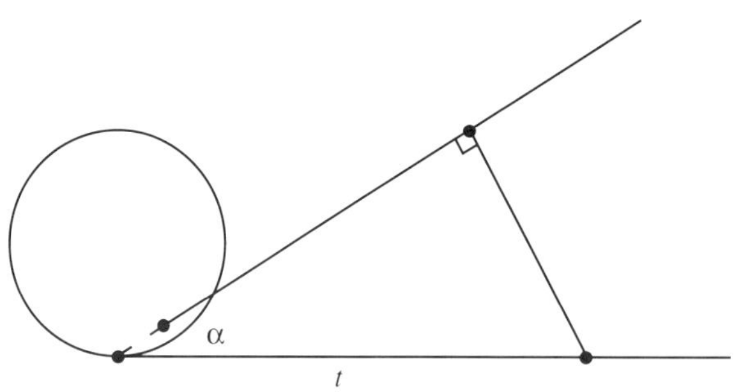


FIGURE 6

Thus if we start with any point $x$ at least $2\pi + \rho$ from the "origin" $O_L$ on the line $L \in \widetilde{\mathscr{L}}$ containing $x$, the orthogonal trajectory to $\widetilde{\mathscr{L}}$ beginning at $x$ will remain outside $\rho B^2$ and will close up, which shows that $\widetilde{\mathscr{L}}$, and hence $\mathscr{L}$, has the bounded projection property. □

## 3. Higher dimensions

Let $\mathscr{L}$ be a family of lines in $\mathbb{R}^d$ $(d \geq 2)$. As in the planar case, a line $l \in \mathscr{L}$ is given by its direction $u \in \mathbf{S}^{d-1}$ (or $-u \in \mathbf{S}^{d-1}$) and the vector $F(u) \in l$ nearest to the origin. $F$ is defined on a symmetric subset $T \in \mathbf{S}^{d-1}$ which we assume to be closed. (This is somewhat more than assuming that $\mathscr{L}$ contains no parallel lines.) Clearly, $F$ has the properties

$$F(u) = F(-u) \quad \text{for all } u \in T, \tag{3}$$

$$u \cdot F(u) = 0 \quad \text{for all } u \in T. \tag{4}$$

If $\mathscr{L}$ is to have the bounded projection property then $F$ itself must certainly be bounded. In addition, we will need some further conditions below on the function $F$:

$$|(F(u) - F(v)) \cdot (u - v)| \leq c|u - v|^2 \quad \text{for all } u, v \in T, \tag{5}$$

$$|F(u) - F(v)| \leq c|u - v| \quad \text{for all } u, v \in T, \tag{6}$$

$$\int_\Gamma F(u)\, du = 0 \quad \text{for all smooth closed curves } \Gamma \subset T. \tag{7}$$

We begin by giving several conditions which are necessary in order for $\mathscr{L}$ to have the bounded projection property.

THEOREM 2. *Suppose $\mathscr{L}$ has the bounded projection property. Then $F$ is bounded and satisfies* (5) *with a constant $c$ which depends only on $\mathscr{L}$. Moreover, if $F \in \mathscr{C}^1$ on $T$, then $F$ satisfies* (7) *as well.*

PROOF. Let $l_u$, $l_v \in \mathscr{L}$ be two lines with directions $u$, $v \in \mathbf{S}^{d-1}$ chosen so that $u \cdot v \geq 0$. The sequence of projections alternating between $l_u$ and $l_v$ will approach oscillation between $p = p(u, v) \in l_u$ and $q = p(v, u) \in l_v$, where

$$|p - q| = \operatorname{dist}(l_u, l_v).$$

A simple calculation gives

$$\begin{aligned} p(u, v) &= F(u) + \frac{F(v) \cdot u + F(u) \cdot v(u \cdot v)}{1 - (u \cdot v)^2} u \\ &= F(u) - \frac{F(u) \cdot v}{1 + u \cdot v} u + \frac{F(v) \cdot u + F(u) \cdot v}{1 - (u \cdot v)^2} u. \end{aligned}$$

The first two terms here are bounded, while the last is bounded if and only if

$$|F(v) \cdot u + F(u) \cdot v| \leq c(1 - u \cdot v)$$

for some constant $c$. Since $1 - u \cdot v = \frac{1}{2}|u - v|^2$ and $F(u) \cdot v + F(v) \cdot u = -(F(u) - F(v)) \cdot (u - v)$ by (4), condition (5) expresses the boundedness of the set $\{p(u, v): u, v \in T\}$ in terms of $F$, which is clearly a necessary condition.

To establish condition (7), let $u(t)$ be a smooth parametrization of $\Gamma$, with $u(0) = u(1)$ and $0 \le t \le 1$, say. Consider the curve

$$z(t) = F(u(t)) + x(t)u(t) \in l_{u(t)}$$

starting from $z(0)$ with $x(0) > 0$ large enough. $z(t)$ will be a trajectory of infinitesimal projections along the one-parameter family of lines $l_{u(t)}$ if the direction of $z(t)$ is orthogonal to $u(t)$ for all $t \in [0, 1]$. That is,

$$0 = u(t) \cdot \dot{z}(t) = u(t) \cdot \left[\frac{dF}{du} \cdot \dot{u}(t) + \dot{x}(t)u(t) + x(t)\dot{u}(t)\right].$$

Taking the derivative of (4) after substituting $u = u(t)$, we get $u \cdot (\frac{dF}{du} \cdot \dot{u}) + \dot{u} \cdot F = 0$, so that

$$\dot{u} \cdot F = u \cdot [\dot{x}u + x\dot{u}] = \dot{x}$$

because $u^2(t) \equiv 1$. Then

$$x(t) = x(0) + \int_{\tau=0}^{t} F(u(\tau)) \cdot \dot{u}(\tau)\, d\tau = x(0) + \int_{u(0)}^{u(t)} F(u)\, du.$$

In particular, $x(1) = x(0) + \int_\Gamma F(u)\, du$. Now if $\int_\Gamma F(u)\, du \neq 0$, then—changing the direction of $\Gamma$ if necessary—we may assume that $\int_\Gamma F(u)\, du > 0$. But $z(1) \in l_{u(0)}$, hence we are back on the same line we started from except that $z(1)$ is farther away from $z(0)$ by a distance of $\int_\Gamma F(u)\, du$. Repeating this sequence of infinitesimal projections along $\Gamma$, we see that $z(t)$ will go to infinity. □

THEOREM 3. *Assume $T = \mathbf{S}^{d-1}$ and $F \in \mathscr{C}^1$. If $F$ satisfies condition (7), then $\mathscr{L}$ has the bounded projection property.*

PROOF. First we claim that the lines in $\mathscr{L}$ cover $\mathbb{R}^d$, except for a bounded set, exactly once. Notice that since $F \in \mathscr{C}^1$, $F$ satisfies condition (5). As we have seen in the proof of Theorem 2, (5) is equivalent to the boundedness of the set $\{p(u, v): u, v \in \mathbf{S}^{d-1}, u \neq v\}$, which implies, in particular, that the set of intersections $\{l_u \cap l_v: u, v \in \mathbf{S}^{d-1}, u \neq v\}$ is bounded. Hence the lines in $\mathscr{L}$ cover $\mathbb{R}^d$, except for a bounded set, at most once. Then clearly the function

$$H(u, x) = F(u) + xu$$

defined on $\mathbf{S}^{d-1} \times (x_0, \infty)$ for some large $x_0$ will be injective. It is continuous, and hence, by a theorem of Brouwer [3], open. But since $H$ is also closed, the boundary of its image will be contained in $H(\mathbf{S}^{d-1}, x_0)$. This proves that the equation $H(u, x) = z$ is uniquely solvable for $|z| > c$ provided $c$ is large enough, proving the claim.

Now define $f\colon \mathbf{S}^{d-1} \times (x_0, \infty) \to \mathbb{R}$ by

$$f(u, x) = x + \int_{u_0}^{u} F(w)\,dw,$$

where the integral is taken along any smooth curve in $\mathbf{S}^{d-1}$ connecting $u_0$ to $u$. Here $u_0 \in \mathbf{S}^{d-1}$ is fixed. This definition is legitimate by (7). For fixed $x \in (x_0, \infty)$ we have

$$\operatorname{grad}_u f(u, x) = F(u) \quad \text{for all } u \in \mathbf{S}^{d-1}.$$

Consider the map $G\colon \mathbf{S}^{d-1} \times (x_0, \infty) \to \mathbb{R}^d$ defined by

$$G(u, x) = H(u, f(u, x)) = F(u) + u\left(x + \int_{u_0}^{u} F(w)\,dw\right).$$

It follows from what we have proved about $H$ that the equation $G(u, x) = z$ has a unique solution for every $z \in \mathbb{R}^d$ provided $|z| > c$.

Now fix $x \in (x_0, \infty)$ and set

$$Z_x = \{z \in \mathbb{R}^d : z = G(u, x) \text{ for some } u \in \mathbf{S}^{d-1}\}.$$

We claim that $Z_x$ is the boundary of a compact convex set if $x$ is large enough. We show first that

$$u \cdot (G(v, x) - G(u, x)) \le 0 \tag{8}$$

for all $u$, $v \in \mathbf{S}^{d-1}$. This inequality is equivalent to

$$u \cdot \left(F(v) - F(u) + v\int_{u_0}^{v} F(w)\,dw - u\int_{u_0}^{u} F(w)\,dw\right) \le \frac{x}{2}(u - v)^2.$$

A straightforward calculation (which we omit) shows that the left-hand side belongs to $O((u - v)^2)$ as $u \to v$, and this is enough since we may choose $x$ as large as we wish. Now let $K_x$ be the convex hull of $Z_x$. $K_x$ is convex and compact. We prove that its boundary is $Z_x$. Take any point $z \in Z_x$. Then $z = G(u, x)$ for some $u \in \mathbf{S}^{d-1}$, and (8) shows that $z$ is on the boundary of $K_x$ (with outer normal $u$). On the other hand, if $z$ is on the boundary of $K_x$ with outer normal $u \in \mathbf{S}^{d-1}$, then consider the set $Y = \{y \in K_x : u \cdot (y - z) = 0\}$. $Y$ is the convex hull of the set $\{y \in Z_x : u \cdot (y - z) = 0\}$. But since every point $y \in Z_x$ has a unique outer normal to it, this latter set consists of a single point. Then so does $Y$, and we have $Y = \{z\}$ with $z = G(u, x) \in Z_x$.

Finally, since the outer normals to $K_x$ at $z_1 = G(u, x)$ and $z_2 = G(-u, x)$ determine the same line $l_u = l_{-u} \in \mathscr{L}$, we have $z_1$, $z_2 \in l_u$, and it follows that the projection of $K_x$ onto $l_u$ coincides with $K_x \cap l_u$. This and the convexity of $K_x$ imply that projections do not leave $K_x$, i.e., that $\mathscr{L}$ has the bounded projection property. $\square$

We remark that if $F \in \mathscr{C}^1$, then $F$ satisfies (6) which, in turn, implies (5). Hence, for $F \in \mathscr{C}^1$ and $T = \mathbf{S}^{d-1}$ the necessary condition of Theorem 2 and the sufficient condition of Theorem 3 coincide and are equivalent to (7), i.e., to the statement that the vector field $F$ is conservative. When $F \in \mathscr{C}^1$ this, in turn, is equivalent to the condition that the derivative tensor of $F$ is symmetric.

What happens when $F \in \mathscr{C}^1$ on $T$, but $T \neq \mathbf{S}^{d-1}$? In that case one tries to extend $F$ from $T$ to $\mathbf{S}^{d-1}$, as in Theorem 1, preserving properties (5) and (7), i.e., keeping $F$ conservative. There seem to be no natural geometric conditions which will insure that such an extension is possible. There are, however, a variety of analytic conditions which with the help of the Whitney Extension Theorem (see **[4, pp. 176–177]**) will guarantee such an extension. In the absence of any nice condition, however, we do not pursue this line any further.

We remark also that in the plane condition (7) is trivially satisfied. Indeed, if $T \neq \mathbf{S}^1$, then $\int_\Gamma F(u)\,du = 0$ for every closed curve $\Gamma$, and if $T = \mathbf{S}^1$ then there is essentially one nontrivial closed curve, $\Gamma = \mathbf{S}^1$, but for that curve the integral vanishes because of (3). Hence (7) is always satisfied. Moreover, in the plane the shortest distance between two lines in $\mathscr{L}$ is zero (there are no parallel lines by assumption). So in this case (5) reduces to (6), i.e., to the condition that $r(\theta) \in \mathrm{Lip}(1)$.

As for the original conjecture, it is false for several reasons. First, for a line family $\mathscr{L}$ in $\mathbb{R}^3$, say, the corresponding vector field $F\colon T \to \mathbb{R}^3$ may not be bounded, while $l_1 \cap l_2 = \varnothing$ for every pair $l_1, l_2 \in \mathscr{L}$. But even if $F$ is bounded and $l_1 \cap l_2 = \varnothing$ for every pair in $\mathscr{L}$, $F$ may not satisfy condition (5). An example which does satisfy (5) and still does not have the bounded projection property is $F\colon \mathbf{S}^2 \to \mathbb{R}^3$ defined by

$$F(x_1, x_2, x_3) = (-x_3x_2, x_3x_1, 0).$$

$F$ clearly satisfies (3), (4), and (5), but does not satisfy (7). Hence this provides another counterexample to the conjecture. (A similar example was found by Robert Connelly.)

Actually, $F$ can be restricted to a set $T_\alpha = \{x \in \mathbf{S}^2, x_3 = \pm\alpha\}$ for any $0 < \alpha < 1$, and the resulting line family will also not have the bounded projection property. As a matter of fact, this same vector field $F$ can be restricted to another set $T \subset \mathbf{S}^2$ that has only two accumulation points, $(0, 0, 1)$ and $(0, 0, -1)$, in such a way that the bounded projection property still fails.

We mention finally that there is an example of a family $\mathscr{F}$ of planes in $\mathbb{R}^3$ such that any three of the planes meet in the unit ball, yet $\mathscr{F}$ does not have the bounded projection property. This example can be constructed with the help of the fact that if $\mathscr{F}$ has the bounded projection property, then the line family $\mathscr{L} = \{F_1 \cap F_2 : F_1, F_2 \in \mathscr{F},\ F_1 \neq F_2\}$ will have it as well.

## References

1. R. Aharoni, P. Duchet, and B. Wajnryb, *Successive projections on hyperplanes*, J. Math. Anal. Appl. **103** (1984), 134–138.
2. B. Grünbaum, *Arrangements and spreads*, Amer. Math. Soc., Providence, R. I., 1972.
3. W. Hurewicz and H. Wallman, *Dimension theory*, Princeton Univ. Press, Princeton, N. J., 1948.
4. E. M. Stein, *Singular integrals and differentiability properties of functions*, Princeton Univ. Press, Princeton, N. J., 1970.

Mathematical Institute, Hungarian Academy of Sciences, Pf. 127, H-1364, Budapest, Hungary
*E-mail address*: ibarany@yalevm.edu

Department of Mathematics, City College, City University of New York, New York, New York 10031
*E-mail address*: jegcc@cunyvm.cuny.edu

Courant Institute of Mathematical Sciences, New York University, New York, New York 10012
*E-mail address*: pollack@geometry.nyu.edu

DIMACS Series in Discrete Mathematics
and Theoretical Computer Science
Volume 4, 1991

Applied Geometry
and Discrete Mathematics
THE VICTOR KLEE FESTSCHRIFT

# The Minimal Projective Plane Polyhedral Maps

D. W. BARNETTE

ABSTRACT. A well-known theorem of Steinitz [**3**] states that the graphs of convex 3-dimensional polytopes can be generated from the graph of the tetrahedron by a process called facet splitting.

One generalization of convex 3-polytopes is polyhedral maps in 2-manifolds. In [**2**] we show how to generate the polyhedral maps in the projective plane from seven minimal maps using face splitting and its dual, vertex splitting. In this paper we show how to generate the polyhedral maps in the projective plane from 16 minimal maps using just vertex splitting. By duality, this is equivalent to giving a generating procedure using just face splitting.

## 1. Definitions and notation

In this paper all manifolds are compact 2-dimensional manifolds. If a graph $G$ is embedded in a manifold $M$ then the closures of the connected components of $M - G$ are called the *faces* of $G$. If each face is a closed 2-cell and each vertex has valence at least three then $G$ is called a *map* in $M$. If, in addition, no two fases have a multiply connected union then $G$ is called a *polyhedral map* in $M$. This condition on the union of two faces is equivalent to saying that any two faces that meet, meet on a single vertex or a single edge. When two faces in a map meet in one of these two ways we say that they *meet properly*.

If $G$ is a polyhedral map we can create a new polyhedral map $G'$ by adding a new edge $xy$ across a face of $G$ such that $x$ and $y$ do not lie on the same edge of $G$. We say that $G'$ is obtained from $G$ by *face splitting*. The vertices $x$ and $y$ may be vertices of $G$, or one or both of them may lie in the relative interior of edges of $G$. If $x$ and $y$ are vertices of $G$ we say that the splitting is *vertex-to-vertex*.

1980 *Mathematics Subject Classification* (1985 *Revision*). Primary 52A37.

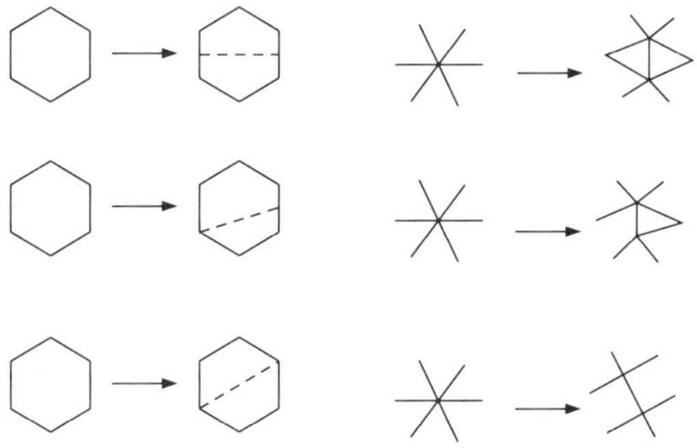

FIGURE 1

The dual of face splitting is *vertex splitting* (in Figure 1 we show three types of face splittings and their corresponding vertex splittings in the dual). The inverse of vertex splitting is *edge shrinking*. When we shrink an edge, any resulting double edges bounding 2-sided faces are coalesced into single edges.

The inverse of face splitting is *removing edges*. When an edge is removed, any two remaining edges meeting at a resulting 2-valent vertex are merged into single edges.

We say that a map $G$ is *minimal* with respect to vertex splitting (face splitting, vertex splitting, and face splitting) provided shrinking an edge (removing an edge, removing an edge, or shrinking an edge) always produces a graph which is not a polyhedral map.

In [2] we prove that the polyhedral maps in the projective plane (hereafter called PPPMs) can be generated from a set $\mathcal{G}_0$ of seven polyhedral maps (Figure 2) by vertex and face splittings. That is, given any PPPM $G$, there is a map $G_0 \in \mathcal{G}_0$ and a sequence of PPPMs $G_0, G_1, \dots, G_n = G$ such that for $1 \leq i \leq n$, $G_i$ is obtained from $G_{i-1}$ by either vertex or face splitting. The set of seven maps are in fact minimal with respect to vertex and face splitting.

If $P$ is a path in $G$ and if $a$ and $b$ are vertices of $P$ then $P[a, b]$ denotes the portion of $P$ that joins $a$ to $b$. We use a parenthesis instead of a bracket when we wish to exclude an endpoint of the path.

We represent our maps by graphs drawn on a disk. The map in the projective plane is obtained by identifying antipodal points on the boundary of the disk. No edges lie on the boundary of the disk in our representations and, with two exceptions, no vertices lie on the boundary. The exceptions are maps $N_6$ and $N_7$ in Figure 3, which have one and two vertices on the boundary, respectively. This was done to enable us to show the symmetry of these two maps.

If shrinking an edge $e$ of a polyhedral map produces another polyhedral map we say that $e$ is *shrinkable*.

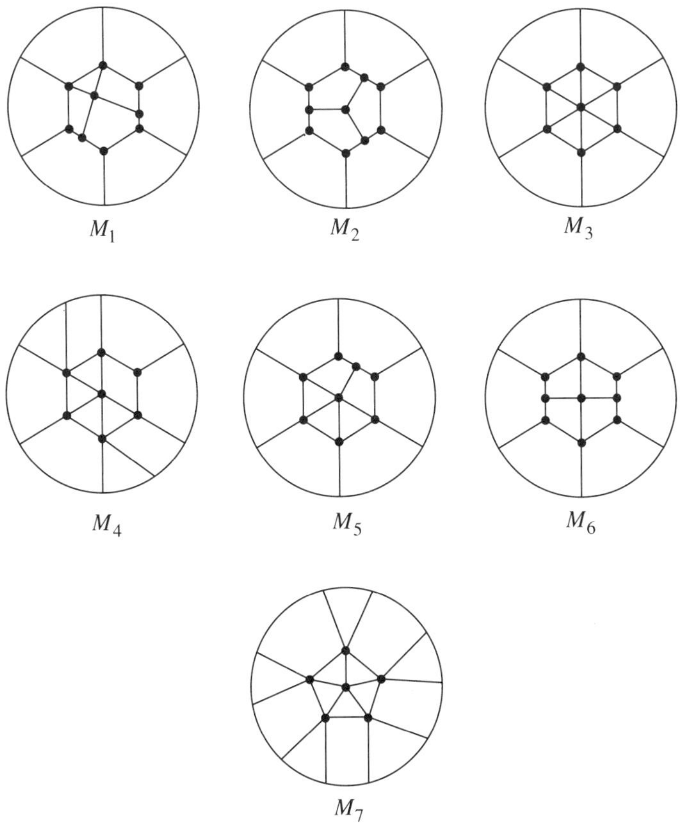


FIGURE 2

If an edge $e$ meets two faces $F_1$ and $F_2$ and $F_1 \cap F_2 \neq \varnothing$ but $F_1 \cap F_2 \cap e = \varnothing$, and if $e$ does not belong to a triangular face that meets $F_1$ and $F_2$ on edges, then we say that $e$, $F_1$, and $F_2$ form a 3-*chain*.

## 2. Preliminary lemmas

LEMMA 1. *If $e$ is an edge of a polyhedral map in a manifold (other than the sphere) then $e$ is not shrinkable if and only if it belongs to a 3-chain.*

The proof of Lemma 1 is found in [**2**] (where the 3-chain is referred to as a "3-chain of type II").

If the complement in $M$ of a 3-chain contains a connected component that is an open cell then we say that the chain is *planar*, otherwise it is *nonplanar*.

LEMMA 2. *If $a$ is a 3-valent vertex of a triangular face $F$ of a polyhedral map $G$ in a manifold other than the sphere, then the edge of $G$ not belonging to $F$ and meeting $F$ at $a$ is shrinkable.*

PROOF. This is essentially the dual of Lemma 2 in [**1**] where it is proved for any polyhedral map in the sphere, other than the graph of the tetrahedron. The same proof proves the lemma in all other manifolds. □

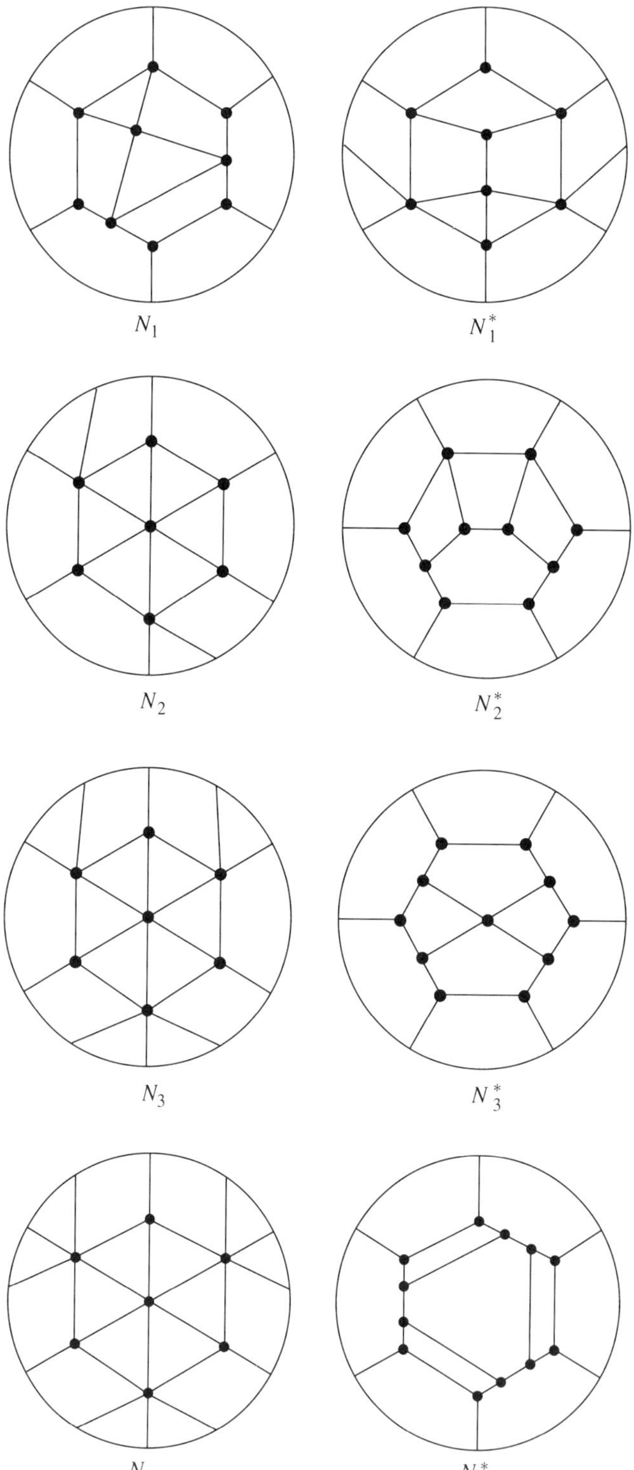


FIGURE 3

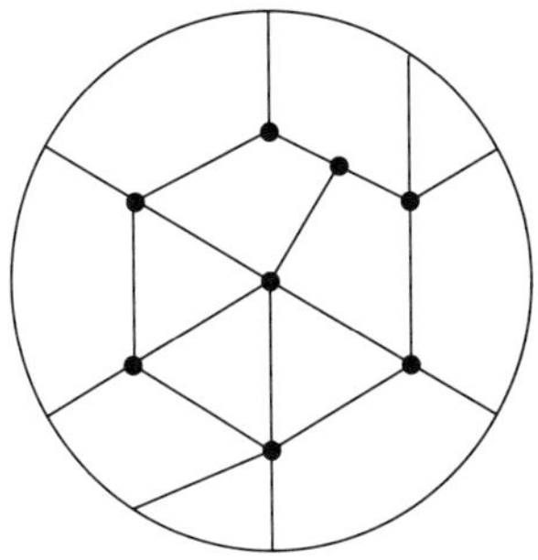

$N_5$

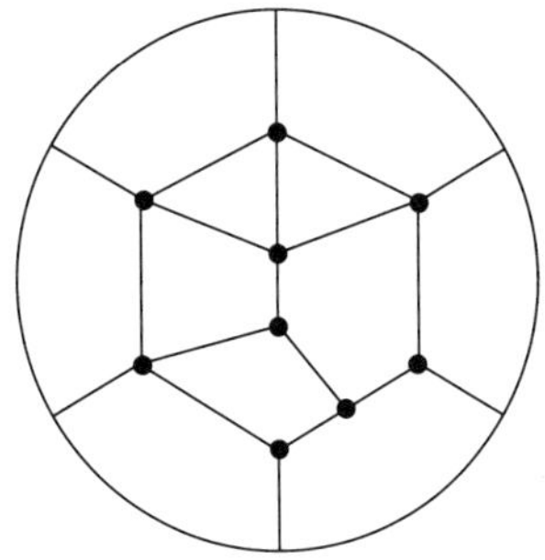

$N_5^*$

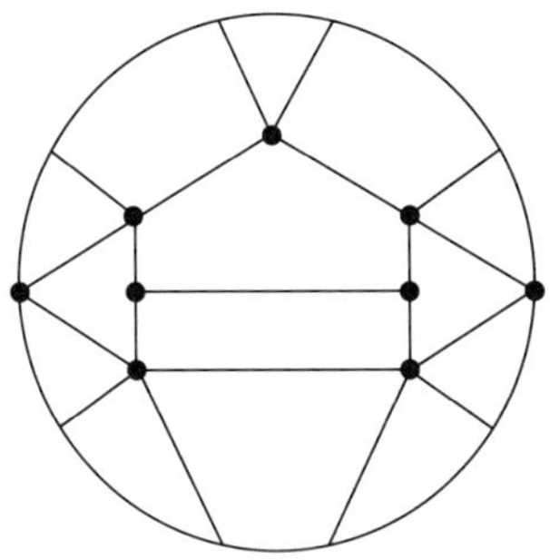

$N_6$

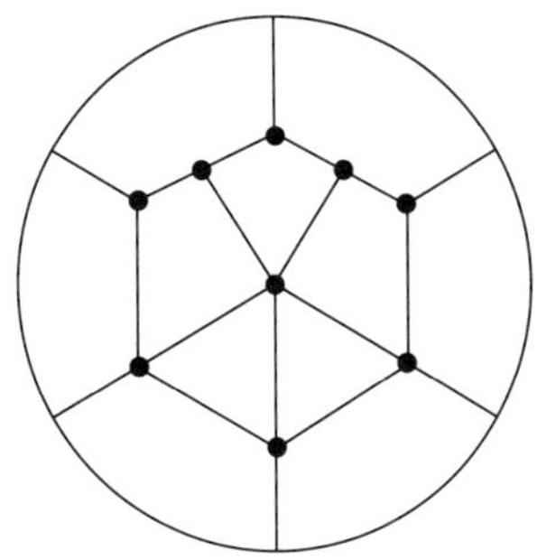

$N_6^*$

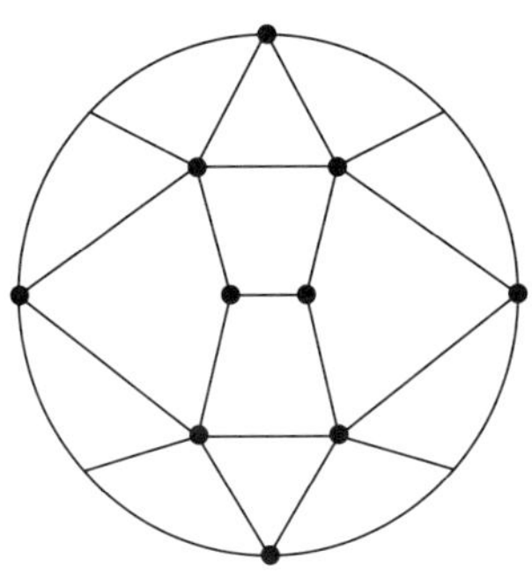

$N_7$

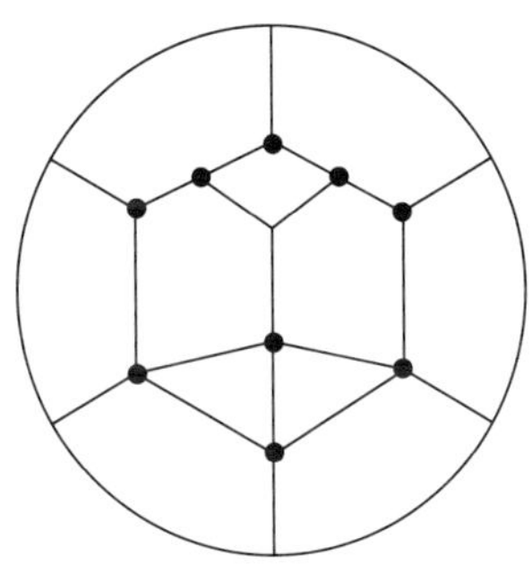

$N_7^*$

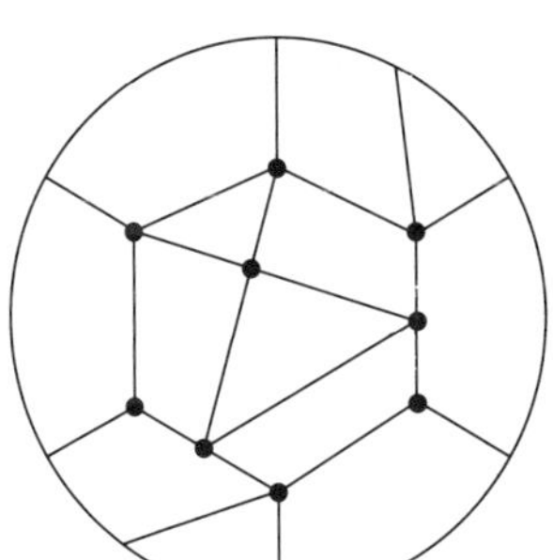

$N_8$

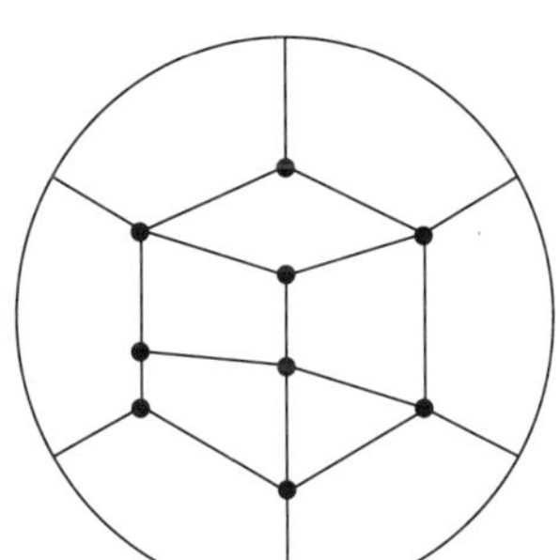

$N_8^*$

FIGURE 3 (continued)

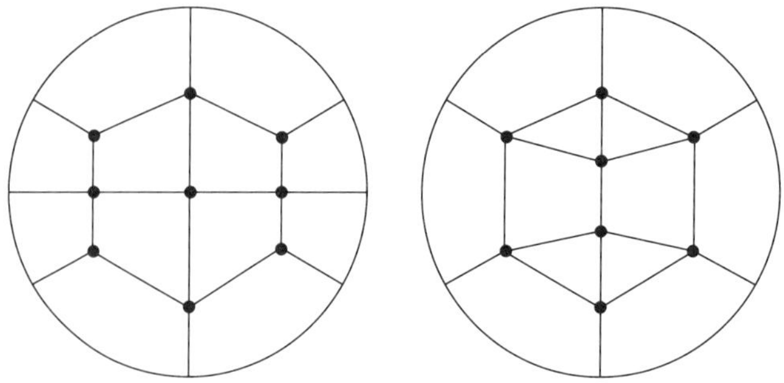


FIGURE 3 (continued)

## 3. The minimal maps

The following theorem will allow us to construct the minimal maps with respect to vertex splitting from the seven maps in Figure 2.

THEOREM 1. *Let $M$ be a 2-manifold other than the sphere and let $G$ be a polyhedral map of $M$ minimal with respect to vertex splitting. Then $G$ can be obtained from a polyhedral map $G_0$ which is minimal with respect to vertex and face splitting by a sequence of vertex-to-vertex face splittings.*

PROOF. Let $\mathscr{G}_0$ be the set of polyhedral maps of $M$ that are minimal with respect to vertex and face splittings, and let $\mathscr{G}_1$ be the set of polyhedral maps minimal with respect to vertex splitting.

If $G$ is in $\mathscr{G}_1$ and not in $\mathscr{G}_0$ then there is an edge $e = xy$ of $G$ that is removable. First we show that $x$ and $y$ have valence at least four. Suppose $x$ is 3-valent. Let $xz$ and $xw$ be the other two edges meeting $x$. If $yw$ is an edge of $G$ then shrinking $xw$ produces the same graph as does removing $xy$. Thus $G \notin \mathscr{G}_1$, a contradiction.

If $yw$ is not an edge then shrinking edge $xw$ produces the same map as does removing $e$ and then adding the edge $yw$. Thus $xw$ is shrinkable, a contradiction.

Next we show that the map $G'$ produced by removing $e$ from $G$ is minimal with respect to vertex splitting.

Suppose an edge $e_1 = ab$ in $G'$ is shrinkable. Since $x$ and $y$ are at least 4-valent in $G$, $e_1$ is an edge of $G$. In $G$, $e_1$ is not shrinkable, thus, there is a 3-chain consisting of $e_1$ and faces $F_1$ and $F_2$ such that $a \in F_1$, $b \in F_2$, $F_1 \cap F_2 \neq \emptyset$, and $e_1$ does not lie on a triangular face meeting $F_1$ and $F_2$ on edges. We now treat three cases.

*Case* I. $F_1 \cup F_2 \cup e_1$ is nonplanar. In this case, if $F_1$ and $F_2$ exist in $G'$ then $F_1 \cup F_2 \cup e_1$ is nonplanar in $G'$ and $e_1$ is not shrinkable.

If one face, say $F_1$, does not exist in $G'$ then $F_1$ was merged with a face when $e$ was removed, producing a new face $F_3$. Now $F_3 \cup F_2 \cup e_1$ is nonplanar and $e_1$ is not shrinkable.

*Case* II. $F_1 \cup F_2 \cup e_1$ is planar and $F_1$ is not a face in $G'$. The face $F_1$ was thus merged with a face $F_3$ producing a face $F_4$ when $e$ was removed.

Let the cell enclosed by $F_1 \cup F_2 \cup e_1$ be $A$. Then $A$ is bounded by $e$, a path $P_1[a, c]$ along $F_1$, and a path $P_2[b, c]$ along $F_2$.

*Case* IIa. $e_1 \subseteq F_4$. Now we have $a \in F_4$, $b \in F_2$, $F_4 \cap F_2 \neq \emptyset$ in $G'$, and $F_4 \cap F_2 \cap e_1 = \emptyset$. Since $e_1$ is shrinkable in $G'$, there is a triangular face $F_5$ in $G'$ with $e_1 \subseteq F_5$, and $F_4$ and $F_2$ meeting $F_5$ on edges. Let the third vertex of $F_5$ be $d$. If $F_5$ does not lie in $A$ then $A \cup F_1 \cup F_2 \cup F_5$ is a sphere, which is a contradiction. Thus $F_5$ lies in $A$ and $d \in P_2$.

We have $bd \subseteq F_2$ and $ad \subseteq F_4$; thus, either $ad \subseteq F_1$ or $ad \subseteq F_3$. But by our assumption that $e_1$, $F_1$, $F_2$ form a 3-chain we have $ad \not\subseteq F_1$; thus, $ad \subseteq F_3$. Furthermore, since $d \in F_1 \cup F_3$ and $c \in F_1 \cup F_3$ we have that $P_2[d, c] \subseteq F_1 \cup F_3$ and, since $P_2 \cap F_1 = c$, we have $P_2[d, c] \subseteq F_3$.

Now, $d$ is 3-valent in $G$ and, by Lemma 2, $G$ contains a shrinkable edge, a contradiction.

*Case* IIb. $e_1 \subseteq F_4$. Since $b \in F_4$ and $c \in F_4$ we have $b$ and $c$ lying on $F_2$ and $F_4$ in $G'$ and thus $bc$ is an edge in $G'$. Since $bc$ is not an edge of $F_1$, it is an edge of $F_3$. Now $a$ and $c$ lie on $F_3$ and $F_1$; thus, $ac$ is an edge of $G$, and $F_3$ is a triangular face with $A = \operatorname{int} F_3$. Thus, $F_1$, $F_2$, and $e_1$ do not form a 3-chain, a contradiction.

*Case* III. $F_1 \cup F_2 \cup e_1$ is planar and $F_1$ and $F_2$ are faces in $G'$. Let $A$ be defined as in Case II. Since $e_1$ is shrinkable in $G'$, $A$ is the interior of a triangular face $H$ in $G'$. Since $e_1$ is not shrinkable in $G$, $H$ is not a face in $G$. We must have that $H$ is the new face formed when $e$ is removed. Let the two faces merged to form $H$ be $H_1$ and $H_2$. Then the closure of $A$ is $H_1 \cup H_2$; thus, $e$ is an edge across $A$.

If $e$ meets $P_1(a, c)$ and $P_2(b, c)$ then clearly $e$ is shrinkable. If $e$ meets $a$ and $P_2(b, c)$ at a vertex $d$ then $P_2[b, d]$ is a shrinkable edge. A similar argument holds if $e$ meets $b$ and $P_1(a, c)$. Since $e_1$ is an edge in $G$, these are the only possible locations for $e$ and we have exhausted the cases.

We now have that $G'$ is minimal with respect to vertex shrinking. We now repeat this process, removing an edge from $G'$ which meets vertices of valence at least four, producing another polyhedral map that is minimal with respect to vertex splitting. The process stops when we reach a polyhedral map $G_0 \in \mathcal{G}_0$. □

We note that according to the proof of Theorem 1 any map in $\mathcal{G}_1$ can be constructed from a map in $\mathcal{G}_0$ by face splitting, vertex-to-vertex such that all intermediate maps are minimal with respect to vertex splitting.

The following two lemmas will greatly simplify the task of constructing the maps in $\mathcal{G}_1$.

LEMMA 3. *In constructing any map in $\mathcal{G}_1$ from a map in $\mathcal{G}_0$ we need never do a face splitting that creates a triangular face with a 3-valent vertex.*

PROOF. By Lemma 2 the map resulting from such a splitting would have a shrinkable edge. □

LEMMA 4. *If* $G \in \mathcal{G}_1$ *is constructed from* $G_0 \in \mathcal{G}_0$ *by vertex-to-vertex face splittings and if adding an edge* $e$ *to* $G_0$ *creates a triangular face with a* 3-*valent vertex, then* $e$ *need never be added in the sequence of face splittings that produces* $G$.

PROOF. In the proof of Theorem 1 we show that if an edge $e_1$ is not shrinkable in a map $G$ then removing an edge whose vertices are at least 4-valent leaves $e_1$ not shrinkable in the resulting map. The contrapositive of this is that if an edge $e_1$ in a map $G$ is shrinkable then splitting a face vertex-to-vertex produces a map in which $e_1$ is shrinkable.

Suppose the sequence of maps produced in constructing $G$ is $G_0, G_1, \ldots, G_n = G$. If $G_i$ is produced from $G_{i-1}$ by adding edge $e$ then $G_i$ can be constructed by adding $e$ to $G_0$ and then adding the edges that produced $G_1, \ldots, G_{i-1}$. Adding $e$ to $G_0$ produces a shrinkable edge by Lemma 2, and, by the above observation, $G_1, \ldots, G_n$ will all have a shrinkable edge, which is a contradiction. □

Using the above lemmas we have constructed the set $\mathcal{G}_1$ of minimal maps with respect to vertex splitting. In addition to the maps in $\mathcal{G}_0$ there are nine other maps in $\mathcal{G}_1$. These maps $(N_1, \ldots, N_9)$ are given in Figure 3 along with their duals $(N_1^*, \ldots, N_9^*)$.

From $M_1$ one gets $N_1, N_2$;
from $M_2$ and $M_7$ one gets no new maps;
from $M_3$ one gets $N_2, N_3, N_4$;
from $M_4$ one again gets $N_2, N_3, N_4$
from $M_5$ one gets $N_5, N_6, N_7$
from $M_6$ one gets $N_9$.

Thus we have

THEOREM 2. *The PPPMs can be generated from* $\mathcal{G}_0 \cup \{N_1, \ldots, N_9\}$ *by vertex splitting.*

THEOREM 3. *The PPPMs can be generated from* $\mathcal{G}_0 \cup \{N_1^*, \ldots, N_9^*\}$ *by face splitting.*

(We need not list the duals of the maps in $\mathcal{G}_0$ because $M_1^* = M_4$, $M_2^* = M_7$, $M_3^* = M_6$, and $M_5^* = M_5$.

## REFERENCES

1. D. Barnette and E. Jucovič, *Hamiltonian circuits on* 3-*polytopes*, J. Combin. Theory **9** (1970), 54–59.
2. D. Barnette, *Generating projective plane polyhedral maps*, J. Combin. Theory Ser. B (to appear).
3. E. Steinitz and H. Rademacher, *Vorlesungen über die Theorie der Polyeder*, Springer, Berlin, 1934.

DEPARTMENT OF MATHEMATICS, UNIVERSITY OF CALIFORNIA, DAVIS, CALIFORNIA 95616

DIMACS Series in Discrete Mathematics
and Theoretical Computer Science
Volume **4**, 1991

Applied Geometry
and Discrete Mathematics
THE VICTOR KLEE FESTSCHRIFT

# Packing Euclidean Space with Congruent Cylinders and with Congruent Ellipsoids

A. BEZDEK AND W. KUPERBERG

ABSTRACT. In a previous paper we proved that the maximum density of a packing of the three-dimensional Euclidean space with congruent circular cylinders of infinite height (in both directions) is attained when all cylinders are arranged parallel to one direction, thus the maximum density is $\pi/\sqrt{12}$, the same as that for congruent circle packings in the plane. Here, we turn to cylinders (prisms) with noncircular bases and of finite height in the three-dimensional Euclidean space, and we show that a nonparallel arrangement of the cylinders can be denser than any parallel one, even under the assumption of compactness, convexity, and central symmetry of the cylinder's base. In particular, certain congruent elliptical cylinders of a finite height admit a packing of density greater than $\pi/\sqrt{12}$, which is the maximum density for parallel packings with such cylinders. A similar construction applied to certain ellipsoids yields a packing of density greater than that of any lattice packing with congruent ellipsoids, in any dimension $n \geq 3$, which is not possible for $n = 2$. The construction is partially generalized to certain affine images of bodies of constant width.

## 1. Introduction and preliminaries

A family $\mathscr{P}$ of subsets of the Euclidean $n$-dimensional space $E^n$ is called a *packing* if the members of $\mathscr{P}$ have mutually disjoint interiors. If each member of $\mathscr{P}$ is congruent to the same set $S$ then we say that $\mathscr{P}$ is a *packing of $E^n$ with congruent copies of $S$*. If $\mathscr{F}$ is a family of sets, then $\mathscr{F}^*$ will denote the union of all members of $\mathscr{F}$. A packing $\mathscr{T}$ of $E^n$ with congruent copies of a set $T$ is called a *tiling* provided that $\mathscr{T}^* = E^n$, and any set $T$ for which such a tiling exists is called a *tile.* A *body* is a compact connected set with an interior point. Among all packings with congruent

1980 *Mathematics Subject Classification* (1985 *Revision*). Primary 52A45, 52A43; Secondary 10E30, 05B40.

*Key words and phrases.* Packing, convex body, lattice packing, tiling, packing density, cylinder, ellipsoid, affinity.

The first author was partially supported by the Hungarian National Foundation for Scientific Research, grant number 1238.

copies of a body $K$ we distinguish two special types, namely *packings with translates*, in which all copies of $K$ are translates of each other, and *lattice packings*, in which, in addition to the translation condition, it is required that the vectors corresponding to the translations form a *lattice*, defined as the set of all integer-linear combinations of some vector base for the space

The (*upper*) *density* of a packing $\mathscr{P}$, denoted by $d(\mathscr{P})$, is defined by

$$d(\mathscr{P}) = \limsup_{r\to\infty}\{\mathrm{Vol}[B(0, r) \cap \mathscr{P}^*]/\,\mathrm{Vol}[B(0, r)]\},$$

where $\mathrm{Vol}(S)$ denotes the volume of the set $S$ and $B(0, r)$ denotes the ball of radius $r$, centered at the origin. Intuitively speaking, the number $d(\mathscr{P})$ represents the relative size (by volume) of the part of the space occupied by $\mathscr{P}$ as compared to the whole space.

For a given body $K$, consider the collection of all packings of $E^n$ with congruent copies of $K$, the collection of all packings with translates of $K$, and the collection of all lattice packings with $K$. Each of these collections contains a densest packing, which follows from a general result of H. Groemer [7]. These maximum densities are called the *packing density* of $K$, $\delta(K)$, the *translation packing density* of $K$, $\delta_T(K)$, and the *lattice packing density* of $K$, $\delta_L(K)$, respectively. Obviously, $0 < \delta_L(K) \le \delta_T(K) \le \delta(K) \le 1$ for every body $K$, and it can be easily shown by a compactness-type argument that the equality $\delta(K) = 1$ holds only if $K$ is a tile. The survey paper [**4**] by G. Fejes Tóth gives more detailed definitions of these and related concepts and a review of recent results concerning them. Although for $n = 2$ many specific and general results concerning the values of $\delta(K)$ have been obtained, and for higher dimensions some estimates exist, mostly of an asymptotic nature (as $n \to \infty$), the exact value of $\delta(K)$ has not yet been found for any convex body in any dimension $n \ge 3$, with the obvious exception of the tiles. In particular, the famous sphere packing conjecture concerning the packing density of the ball in dimension three still remains open. Only very recently, a first noncompact convex set $C$ in $E^3$ with $\delta(C) \ne 1$ had its packing density $\delta(C)$ explicitly computed. Namely, it was shown in [**1**] that the maximum density of any packing of $E^3$ with congruent circular cylinders of infinite height (in both directions) is $\pi/\sqrt{12}$, the same as that for congruent circle packings in the plane, attained in the natural parallel arrangement of the cylinders, corresponding to the well-known maximum density packing of the plane with congruent circles (see A. Thue [**13**]). J. B. Wilker [**14**] conjectures that the same holds true in $E^3$ for congruent circular cylinders of finite height, and even for circular disks (i.e., "cylinders of height zero") packed in $E^3$ so that their relative interiors are disjoint, but neither part of this conjecture has been settled yet.

Given a body $K$ in $E^{n-1}$ $(n \ge 3)$, and a number $h > 0$, by a (*straight*) *cylinder* (or *prism*) of height $h$ over $K$, or a *cylinder with base $K$ and height $h$*, we understand a set in $E^n$ congruent to the Cartesian product $K \times [0, h] \subset E^n$. Observe that every such cylinder is the union of parallel line segments of length $h$ (*generating lines*) with one end-point on the base, all perpendicular to the hyperplane of the base, and all lying on the same side of that hyperplane. The generic cylinder with base $K$ and height $h$ will be denoted by $C(K, h)$. Obviously, $\delta(C(K, h)) \geq \delta(K)$, the corresponding packing densities taken in $E^n$ and in $E^{n-1}$, respectively. However, generalizing Wilker's conjecture, one might expect that the most economical way to pack cylinders over $K$ in $E^n$ is to stack them in infinitely long beams and then to arrange the beams in a pattern following a maximum density packing of copies of $K$ in $E^{n-1}$. In other words, one might conjecture that $\delta(C(K, h)) = \delta(K)$ for all $K$ and all $h > 0$, or at least for all convex bodies $K$, perhaps with the additional assumption of central symmetry of $K$.

In the following sections we consider packing of $E^3$ with congruent noncircular cylinders of finite height, elliptic cylinders among them, and we produce counter-examples to the conjectures discussed above. Some of the examples are generalized to higher dimensions. The case of the elliptic cylinders leads to the discovery of nonlattice packings of $E^n$ with suitably chosen congruent ellipsoids, packings denser than any lattice packing of $E^n$ with congruent spheres (for every $n \geq 3$). Thus, if the sphere packing conjecture in $E^3$ is true, then certain congruent ellipsoids can be packed more densely in $E^3$ than congruent spheres can, which is in stark contrast with the corresponding situation in the plane. By a well-known result of L. Fejes Tóth [**5**] (see also [**6, p. 88**]), congruent ellipses, and even ellipses of equal areas, can be packed in the plane only as densely as congruent circles can, and the maximum density is attained in a lattice packing.

The authors express their gratitude to Gábor Fejes Tóth, Greg Kuperberg, and Jörg Wills for their valuable comments and suggestions.

## 2. A compact nonconvex cylindrical tile whose base is not a tile

Let $K$ be a $3 \times 4$ rectangle with a rectangular $1 \times 2$ hole cut in it, as shown in Figure 1(a). Obviously, $K$ is not a plane tile, in fact, $\delta(K) = 5/6$. However, the cylinder of height 1 over $K$, $C(K, 1)$, shown in Figure 1(b), is a tile in $E^3$, thus $\delta(C(K, 1)) = 1$. The tiling pattern is shown in Figure 2: first a chain of linked congruent copies of $C(K, 1)$ is arranged (Figure 2(a)), infinitely long in both directions, to form a solid beam whose perpendicular cross-section resembles the cross on the Swiss flag, and then translated copies of the beam are arranged to tile $E^3$ in the same manner as the "Swiss crosses" tile the plane (Figure 2(b)).

(2.1) Remark. Notice the similarity between the above example and the example of K. Reinhardt which he produced in [**12**] for a different purpose, namely to solve a part of Hilbert's Problem 18. Other examples of polyhedral tiles in $E^3$, of every possible genus, have been constructed by H. Debrunner in [**3**].

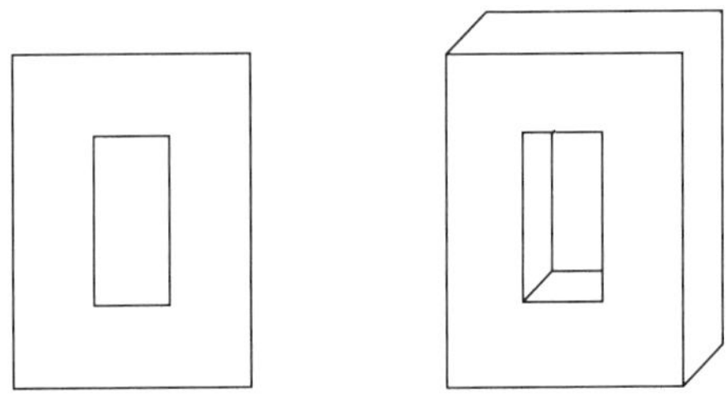

(a) (b)

FIGURE 1

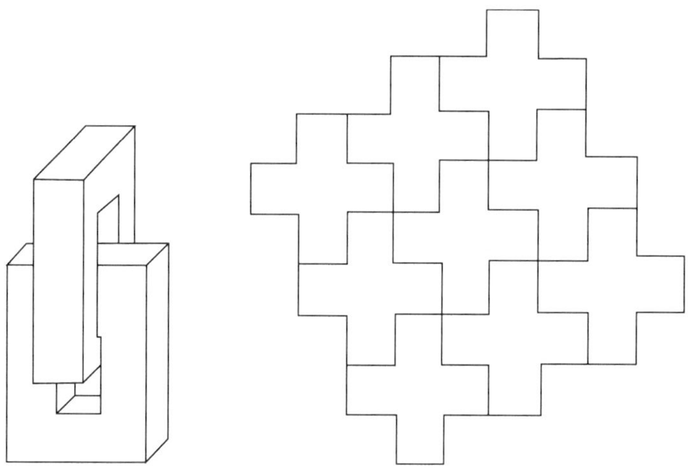

(a) (b)

FIGURE 2

(2.2) REMARK. The lack of simple connectedness of $K$ can be remedied by cutting $K$ in two congruent nontiling pieces as shown in Figure 3. The modified example is not just simply connected, but even star-shaped. We do not know whether a convex example exists in $E^3$, we believe not. B. Grünbaum [**8**] conjectures that in dimensions higher than 3 such examples might exist.

## 3. A convex nontiling plane pentagon over which short cylinders almost tile space

Start with cutting a regular hexagon of unit edge in three congruent pentagons as shown in Figure 4. The resulting pentagonal tile has two right

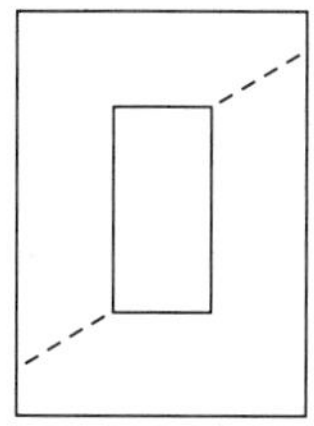

FIGURE 3

angles and three angles measuring $120^\circ$ each. Take one of these pentagons and stretch it along the line marked dashed in Figure 5 by a factor of $\sqrt{2}$ (certain other factors would do as well). The stretching causes a distortion of angles so that each angle of the resulting pentagon $P$ measures more than $90^\circ$ but less than $120^\circ$, thus $P$ is not a tile. Consequently, $\delta(P) < 1$.

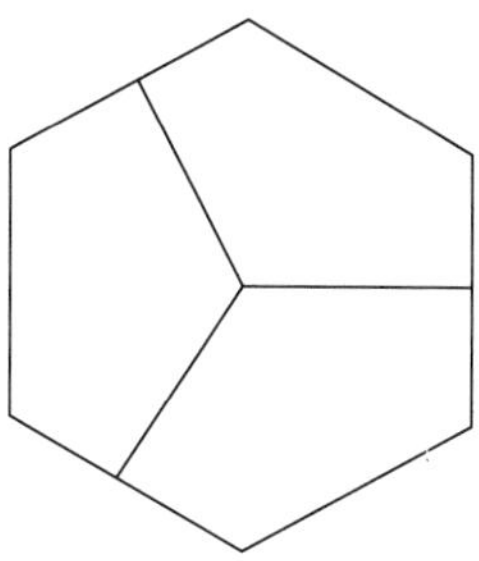

FIGURE 4

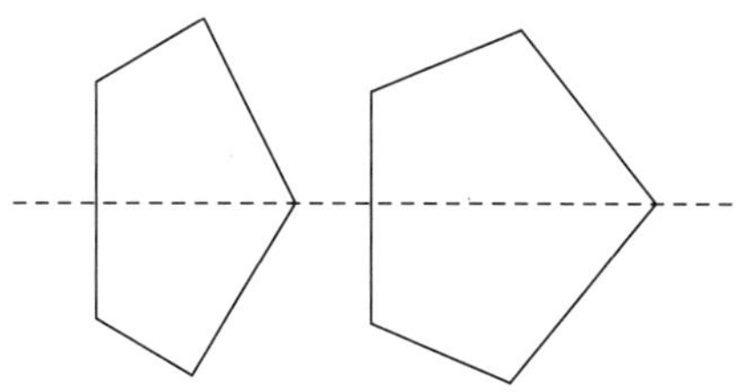

FIGURE 5

We assert now that congruent, sufficiently short cylinders over $P$ can pack $E^3$ with density arbitrarily close to 1. Specifically, we assert that $\lim_{h\to 0} \delta(C(P, h)) = 1$, from which it will follow directly that $\delta(C(P, h)) > \delta(P)$ for all sufficiently small $h$.

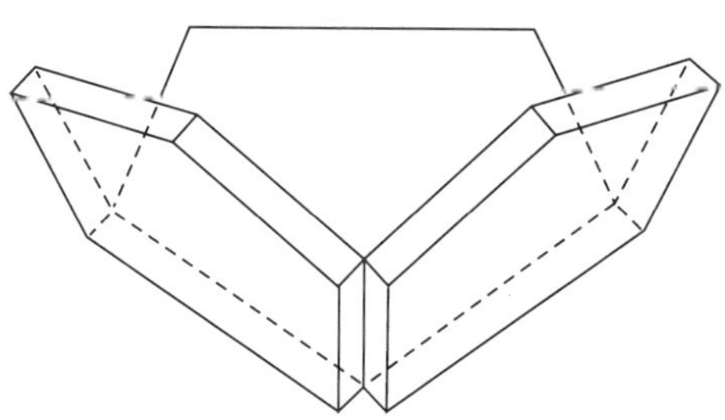

FIGURE 6

PROOF. Position three congruent copies of $C(P, h)$ so as to form a three-sided "cup" (see Figure 6) whose flat inside faces, copies of $P$, are inclined to the horizontal plane $H$ at a $45^\circ$ angle, and these three inside faces of the cup, when projected perpendicularly down to $H$, reproduce the regular hexagon from the beginning of this section. Next, congruent copies of the cup can be stacked, each inside and touching the previous one, by parallel translations perpendicularly to $H$, forming a vertical stack, infinitely long over and under $H$. The perpendicular projection of the stack on $H$ is contained in a regular hexagon of side $1 + h/\sqrt{2}$. Now, tile the plane $H$ with such hexagons and position a stack of cups corresponding to each of the tiles, so that each stack projects into its assigned tile. The result is a packing of $E^3$ with congruent copies of $C(P, h)$, and the density of this packing, easy to compute, equals $(1 + h/\sqrt{2})^{-2}$, which proves the assertion. □

(3.1) REMARK. Analogous examples exist in dimensions higher than 3. Begin with a solid in the shape of a pyramid, obtained by joining the center of an $(n-1)$-dimensional cube with one of its faces, then stretch it somewhat along the line perpendicular to its base so as to obtain a nontiling convex polyhedron which can play the role of the pentagon $P$. Proceed by building "cups" and "stacks of cups" in virtually the same way as before.

## 4. Packing $E^3$ with congruent elliptical cylinders

Let $K$ be an ellipse of axial ratio $k > \sqrt{3} + 1$. We will show that there exists a number $h_0 > 0$ such that, for each positive number $h \leq h_0$, the inequality $\delta(C(K, h)) > \delta(K) = \pi/\sqrt{12}$ holds. We begin with the familiar densest lattice packing of the plane with unit circles (Figure 7(a)) to which we apply the transformation of stretching by a factor of $k$ in the direction of the line drawn dashed in Figure 7. The result is a lattice packing of the plane with ellipses of axial ratio $k$ (Figure 7(b)). The density of this packing is $\pi/\sqrt{12}$, the same as that of the circle packing. Observe that the minor axes of the ellipses are of length 2, while the gaps between the ellipses are large enough to contain a segment of length greater than 2. In fact, there exists a number $h_0 > 0$ such that each gap contains a rectangle with sides 2 and $h_0$. This is all we need, because now the elliptical cylinders of height $h$ (for any $h < h_0$) can be stacked in beams over and under the ellipses of the plane packing, forming a space packing of density $\pi/\sqrt{12}$, after which additional cylinders can be inserted vertically in the gaps between the beams (see Figure 8), resulting in a nonlattice packing whose density is greater than $\pi/\sqrt{12}$.

(4.1) REMARK. A similar construction can be applied to each centro-symmetric convex plane body $K$ which is not a tile, i.e., other than a hexagon or a parallelogram. By the theorem of L. Fejes Tóth [**5**], the maximum density packing of the plane with congruent copies of $K$ is obtained by circumscribing a centro-symmetric hexagon of minimum area about $K$ and by tiling the

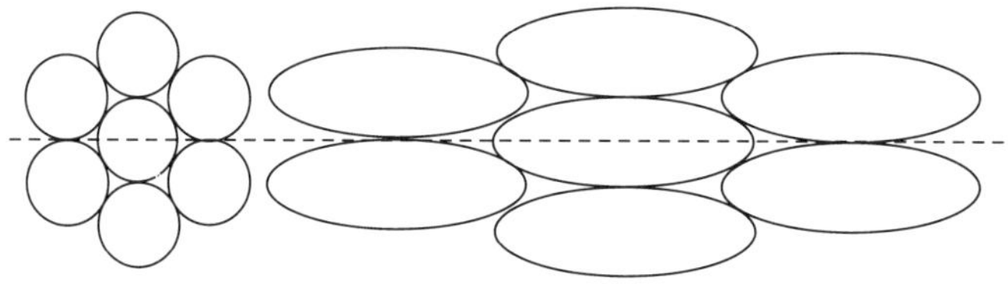


FIGURE 7

FIGURE 8

plane with copies of the hexagon. Since $K$ does not fill the hexagon, there is a gap at one of the hexagon's vertices. A suitable stretching transformation in the direction of one line can be applied to lengthen the gap so that it can contain a line segment longer than the width of the transformed $K$.

## 5. Packing $E^n$ with congruent ellipsoids

Two facts are crucial for our construction of nonlattice ellipsoid packing denser than any lattice sphere packing. The first one is a theorem of A. Heppes [**9**], generalized to higher dimensions by J. Horváth [**11**]: For each lattice sphere packing in dimension $n \geq 3$ there exist $n$ linearly independent directions in which a line can be drawn not intersecting any of the spheres. For our purpose we need just one such line. The second fact is the following.

(5.1) LEMMA. *For every pair of ellipsoids $K_1$ and $K_2$, in $E^n$ with $\mathrm{Vol}(K_1) = \mathrm{Vol}(K_2)$ there exists an affinity $A: E^n \to E^n$ such that $A(K_1)$ and $A(K_2)$ are congruent.*

PROOF. Without loss of generality we may assume that $K_1$ is a unit sphere and that $K_2$ is in a standard position with respect to a coordinate system for $E^n$. Denote the half-axes of $K_2$ by $a_1, a_2, \ldots, a_n$. By the assumption of equality of volumes, $a_1 a_2 \cdots a_n = 1$. The affinity $A$ defined by a diagonal $n \times n$ matrix whose diagonal consists of the numbers $1, a_1, a_1 a_2, \ldots,$ $a_1 a_2 \cdots a_{n-1}$ turns $K_1$ and $K_2$ into congruent ellipsoids. Indeed, the half-axes of the ellipsoid $A(K_1)$ are $b_1 = 1$, $b_2 = a_1$, $b_3 = a_1 a_2, \ldots, b_n =$ $a_1 a_2 \cdots a_{n-1}$, and the half-axes of $A(K_2)$, namely $c_1 = a_2$, $c_2 = a_1 a_2, \ldots,$ $c_n = a_1 a_2 \cdots a_n = 1$, form a permutation of those of $A(K_1)$. □

Now, given an arbitrary lattice packing of $E^n$ with spheres, we start with a line disjoint with all of the spheres. There exists a number $r > 0$ such that a cylinder of radius $r$, infinitely long in both directions, is disjoint with all of the spheres as well. The translates of the cylinder by the vectors of the lattice (except for the vectors parallel to the cylinder) produce a lattice of such cylinders (based on a lattice in $E^{n-1}$), which we may assume are disjoint not only with the spheres, but with each other as well, reducing the size of the radius $r$, if necessary. In each of these cylinders we inscribe a string of congruent ellipsoids, each ellipsoid long enough to match the sphere's volume. We have now a packing in which some members are spheres and some are ellipsoids; the density of this "mixed" packing exceeds that of the original lattice sphere packing by the density of the family of ellipsoids. The spheres are translates of each other and so are the ellipsoids. Let $A$ be an affinity which turns one of the spheres and one of the ellipsoids into congruent ellipsoids, given by Lemma (5.1). The transformation $A$ turns all members of our packing into congruent ellipsoids while preserving the density.

(5.2) REMARK. A line disjoint with all spheres in the maximum density lattice packing of $E^3$ with unit spheres can be found explicitly in the following way. The packing consists of parallel strings of spheres (by a *string of spheres* we understand a collection of nonoverlapping unit spheres whose centers are collinear and in which each sphere is tangent to two other spheres, and two strings are *parallel* if their center lines are parallel, see [**2**]). Let $H$ be a plane perpendicular to the strings. The intersection points of the center lines of the strings with $H$ are the vertices of congruent triangles with sides $\sqrt{3}$, $\sqrt{3}$, and 2, tiling $H$ in a regular, side-to-side manner. The perpendicular projection of each string on $H$ is a unit circular disk centered at a vertex of such a triangle. Three unit disks centered at the vertices of the $(\sqrt{3}, \sqrt{3}, 2)$-triangle do not cover it, leaving a small gap in it (see Figure 9). Every line parallel to the strings and intersecting $H$ in a point inside the gap is disjoint from the entire packing.

## 6. Bodies of constant width and their affine images

Since the theorem of Heppes used in §5 has been generalized by I. Hortobágyi [**10**] to bodies of constant width in $E^3$, our construction from that section also can be generalized to such bodies. However, for this purpose we need to modify Lemma (5.1) as follows.

(6.1) LEMMA. *For every pair of affinities, $A_1$ and $A_2$, of $E^n$ with*

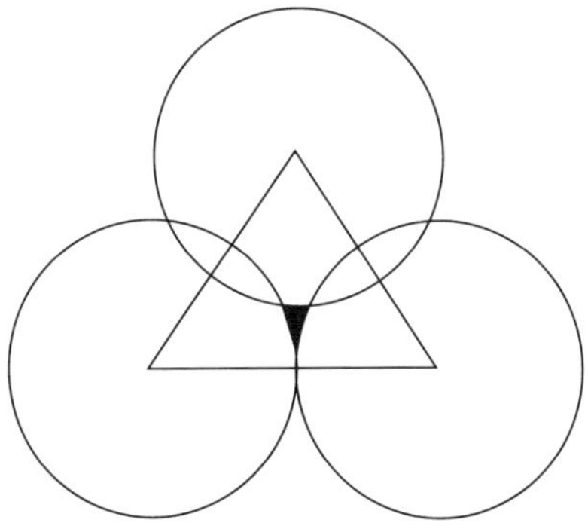

FIGURE 9

$|\det(A_1)| = |\det(A_2)|$, *there exist an affinity* $A$ *and isometries* $B$ *and* $C$ *such that* $BAA_1 = AA_2C$.

PROOF. Let $S$ be the unit sphere in $E^n$. Since $|\det(A_1)| = |\det(A_2)|$, the ellipsoids $A_1(S)$ and $A_2(S)$ are of the same volume. By Lemma (5.1), there exists an affinity $A$ such that the ellipsoids $AA_1(S)$ and $AA_2(S)$ are congruent; let $B$ be an isometry sending $AA_1(S)$ to $AA_2(S)$. Then the affinity $A_2^{-1}A^{-1}BAA_1$ sends the unit sphere $S$ onto itself, therefore it is an isometry; denote it by $C$. The equality $BAA_1 = AA_2C$ follows directly. □

Now, given a lattice packing of $E^3$ with copies of a convex body $K$ of constant width $w$, by the theorem of Hortobágyi, there exists a cylinder, disjoint with the packing. Therefore there exists a lattice of mutually disjoint cylinders, each disjoint with the packing. Let $r$ denote the radius of the cylinders. Let $A_1$ be the identity and let $A_2$ be the composition of a homothety with the coefficient $r/2w$ followed by a stretching in the direction of the cylinders by a factor of $(2w/r)^3$. Notice that $|\det(A_1)| = |\det(A_2)|$. Let $A$, $B$, and $C$ be the transformations given by the conclusion of Lemma (6.1). Observe now that a copy of $A_2C(K)$ can be placed in the cylinder by a parallel translation, thus a string of such copies can be placed in each of the cylinders, increasing the density of the original packing. The affinity $A$ turns all members of the obtained packing into congruent copies of an affinely distorted $K$, preserving the increased density.

## 7. Additional remarks and problems

The following improvement of the ellipsoid packing scheme has been suggested by G. Kuperberg. The entire "tunnel" (prism) over the gap shown in Figure 9 can be packed with translates of a very long, thin ellipsoid so that the density in the tunnel is arbitrarily close to $\pi/\sqrt{18} = 0.74048049\ldots$ (the lattice packing density of the sphere). Still, the ellipsoid's volume can match that of the sphere, so that Lemma (5.1) can be applied. So modified packing of $E^3$ with congruent ellipsoids is of density higher than that which we described in §5. An easy computation shows that such a packing can have density arbitrarily close to $(24\sqrt{2}-6\sqrt{3}-2\pi)\pi/72 = 0.753355461\ldots$. J. Wills informs us that an additional modification of the packing scheme can produce even higher density, namely $0.7585\ldots$.

Concerning packings with elliptical cylinders which we considered in §4, a similar modification of the packing scheme leads to a surprisingly high density of over 99%. To be precise, certain congruent elliptical cylinders can be packed in $E^3$ with density arbitrarily close to $(4\sqrt{3}-\pi)\pi/12 = 0.991332331\ldots$.

It seems that there should exist a common upper bound smaller than 1 for the density of all packings of $E^3$ with congruent ellipsoids and with congruent elliptical cylinders as well, but we know of no proof to that effect.

## REFERENCES


1. A. Bezdek, and W. Kuperberg, *Maximum density space packing with congruent circular cylinders of infinite length*, Mathematika **37** (1990), 74–80.
2. A. Bezdek, W. Kuperberg, and E. Makai, *Maximum density space packing with parallel strings of spheres*, J. Discrete Comput. Geom. (to appear).
3. H. E. Debrunner, *Tiling three-space with handlebodies*, Studia Sci. Math. Hungar. **21** (1986), 201–202.
4. G. Fejes Tóth, *New results in the theory of packing and covering*, in Convexity and its Applications (P. M. Gruber and J. M. Wills, eds.), Birkhäuser-Verlag, Basel/Boston/Stuttgart, 1983, pp. 318–359.
5. L. Fejes Tóth*Some packing and covering theorems*, Acta Sci. Math. (Szeged) **12** (1952), 62–67.
6. ——, *Lagerungen in de Ebene, auf der Kugel und im Raum*, Springer-Verlag, Berlin/Heidelberg/New York, 1972.
7. H. Groemer, *Existenzsätze für Lagerungen im Euklidischen Raum*, Math. Z. **81** (1963), 260–278.
8. B. Grünbaum, Private communication (letter), 1989.
9. A. Heppes, *Ein Satz über gitterförmige Kugelpackungen*, Ann. Univ. Sci. Budapest. Eötvös Sect. Math. **3–4** (1960/61), 89–90.
10. I. Hortobágyi, *Extreme problems connected with packings of sets of constant width*, Thesis, Hungarian Acad. Sci. Math. Inst., Budapest, 1977.
11. J. Horváth, *Über die Durchsichtigkeit gitterförmiger Kugelpackungen*, Studia Sci. Math. Hungar. **5** (1970), 421–426.
12. K. Reinhardt, *Zur Zerlegung der euklidischen Räume in kongruente Polytope*, S.-ber. Preuss. Akad. Wiss., 1928, pp. 150-155.
13. A. Thue, *Über die dichteste Zusammenstellung von kongruenten Kreisen in der Ebene*, Christiania Vid. Selsk. Skr. **1** (1910), 3–9.
14. J. B. Wilker, *Problem* II, in Intuitive Geometry, Colloq. Math. Soc. János Bolyai **48** (1987), 700.


MATHEMATICAL INSTITUTE OF THE HUNGARIAN ACADEMY OF SCIENCES, P.O.B. 127, H-1364 BUDAPEST, HUNGARY

*E-mail address*: h1159bez@ella.uucp

DIVISION OF MATHEMATICS, F.A.T., AUBURN UNIVERSITY, AUBURN, ALABAMA 36849–5310

*E-mail address*: wkuperb@ducvax.auburn.edu

DIMACS Series in Discrete Mathematics
and Theoretical Computer Science
Volume 4, 1991

Applied Geometry
and Discrete Mathematics
THE VICTOR KLEE FESTSCHRIFT

# Extended Euler-Poincaré Relations for Cell Complexes

ANDERS BJÖRNER AND GIL KALAI

ABSTRACT. The pairs $(f, \beta)$ of $f$-vectors and Betti number sequences arising from regular cell complexes with the intersection property are characterized. This extends a characterization previously known for simplicial complexes to a much more general class that includes all polyhedral complexes.

## 1. Introduction

Let $\Delta$ be a finite CW complex with $f_i$ $i$-dimensional cells and Betti numbers $\beta_i$ (in cellular or singular reduced homology over a field $F$). Define

$$\chi_{k-1} = \sum_{j \geq k} (-1)^{j-k} (f_j - \beta_j)$$

for $k \geq 0$. The well-known Euler-Poincaré formula states that

(i) $$\chi_{-1} = 1.$$

For the case of simplicial complexes $\Delta$, other nontrivial relations between the $f$-vector $(f_0, f_1, \ldots)$ and the Betti sequence $(\beta_0, \beta_1, \ldots)$ were given in **[BK]**, leading to a complete characterization of $(f, \beta)$-pairs. It is the purpose of this paper to extend these additional Euler-Poincaré relations, and hence the entire characterization, to a much wider class of cell complexes, including all polyhedral complexes.

For $n, k \geq 1$ there is a unique expansion

$$n = \binom{a_k}{k} + \binom{a_{k-1}}{k-1} + \cdots + \binom{a_i}{i},$$

such that $a_k > a_{k-1} > \cdots > a_i \geq i \geq 1$. This given, define

$$\partial_{k-1}(n) = \binom{a_k}{k-1} + \binom{a_{k-1}}{k-2} + \cdots + \binom{a_i}{i-1}.$$

Also, let $\partial_{k-1}(0) = 0$.

1980 *Mathematics Subject Classification* (1985 *Révision*). Primary 55M99, 05A99; Secondary 55N99, 57Q05.

This work was supported by a U.S.A.-Israel BSF grant and by the Swedish National Board for Technical Development.

THEOREM 1. *Let $\Delta$ be a finite regular CW complex. Assume that every nonempty intersection of two closed cells is a closed cell. Define $(\chi_0, \chi_1, \dots)$ as before in terms of the $f$-vector $(f_0, f_1, \dots)$ and the Betti sequence $(\beta_0, \beta_1, \dots)$ of $\Delta$. Then, for all $k \geq 1$,*

$$\text{(ii)} \qquad \partial_k(\chi_k + \beta_k) \leq \chi_{k-1}.$$

In **[BK]** it was shown that any pair of finite nonnegative integer sequences $f = (f_0, f_1, \dots)$ and $\beta = (\beta_0, \beta_1, \dots)$ satisfying the complete set of Euler-Poincaré relations (i) and (ii) arises as the $f$-vector and Betti sequence (over $F$) of some simplicial complex. Thus, the family of $(f, \beta)$-pairs realized by simplicial complexes coincides with that realized by the more general class of regular CW complexes with the stated intersection property. The situation for more general classes of regular and nonregular CW complexes is described in **[BK, Appendix]**.

## 2. Preliminaries

A CW complex is called *regular* if the closure of each cell is homeomorphic to a closed ball via an attaching map. Familiarity with some basic properties of regular CW complexes will be assumed. **[CF, LW]** are good references for this.

The closed cells of a regular CW complex ordered by inclusion and augmented with a new least element $\hat{0}$ form a poset (partially ordered set), called the *face poset*. The exact relationship between regular CW complexes and posets is discussed in **[B]**. The class of face posets of regular CW complexes is too restrictive for our recursive arguments in §3 (see Remark 2 at the end of the paper), and will be enlarged to a somewhat wider class of combinatorially defined posets. In this connection some poset terminology is used that we now review.

Let $P = (P, \leq)$ be a finite poset. A subset $A \subseteq P$ is called an *ideal* if $x \leq y \in A$ implies $x \in A$, and a *filter* if $x \geq y \in A$ implies $x \in A$, for all $x, y \in P$. Subsets of the form $[x, y] = \{z \in P \mid x \leq z \leq y\}$ are *closed intervals*, and closed intervals of cardinality 2 (i.e., $[x, y] = \{x, y\}$) are *coverings*. If $[x, y]$ is a covering we say that $y$ *covers* $x$, written $x \prec y$. A totally ordered subset (all pairs comparable) is called a *chain* and a totally unordered subset (no comparable pairs) an *antichain*.

A poset is a *meet-semilattice* if every pair of elements has a greatest lower bound (called a *meet*). Every interval $[x, y]$ in a meet-semilattice is a *lattice* (i.e., every pair of elements has a l.u.b. as well as a g.l.b.). Suppose that $P$ is a finite meet-semilattice such that every closed interval of length 2 has at least two elements in the middle. Then every interval $[x, y]$ has the property that if $x \prec a \leq y$ there exists $x \leq c \prec y$ such that $a \nleq c$ (more generally, as a lattice $[x, y]$ is *complemented*).

## 3. Proofs

Let $P$ be a finite poset with the following three properties:

(I) $P$ has a least element $\widehat{0}$ and is graded: there exists a dimension function $\dim: P \to \mathbf{Z}$ such that $\dim(\widehat{0}) = -1$ and if $x \prec y$ ($y$ covers $x$), then $\dim(y) = \dim(x) + 1$.

(II) $P$ is a meet-semilattice.

(III) All closed intervals $[x, y]$ in $P$ of length 2 have exactly two elements in the middle (i.e., are isomorphic to 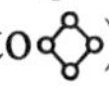).

Suppose that $\lambda$: $\mathrm{cov}(P) \to F^*$ is a mapping from the set of coverings in $P$ to nonzero elements in some field $F$. Require that:

(IV) For every interval $[x, y]$ in $P$ of length 2

$$\sum_{x \prec z \prec y} \lambda(x \prec z) \cdot \lambda(z \prec y) = 0.$$

We define homology groups of $(P, \lambda)$ with coefficients in the field $F$ as follows. Let $P_k = \{x \in P \mid \dim(x) = k\}$ and let $C_k = C_k(P)$ be the $F$-vector space freely generated by $P_k$. For $x \in P_k$ let

$$d_k(x) = \sum_{y \prec x} \lambda(y \prec x) \cdot y \in C_{k-1},$$

and extend linearly to a mapping $d_k\colon C_k \to C_{k-1}$. Put $d_i = 0$ for $i < 0$ and $i > \dim(P)$. Condition (IV) implies that $d_{k-1} \circ d_k = 0$ for all $k \in \mathbf{Z}$, so we can define (reduced) homology

$$\widetilde{H}_k(P, \lambda; F) = \mathrm{Ker}\, d_k / \mathrm{Im}\, d_{k+1}.$$

This leads to our last requirement.

(V) All closed intervals $[x, y]$, $x < y$, in $P$ are acyclic.

This means that $\widetilde{H}_k([x, y], \lambda; F) = 0$ for all $k \in \mathbf{Z}$, where homology is computed on the subposet $[x, y]$ with respect to the induced edge-labeling $\lambda$.

For $P$ and $\lambda$ as above define face numbers

$$f_k = \mathrm{card}\, P_k = \dim_F C_k,$$

and Betti numbers

$$\beta_k = \dim_F \widetilde{H}_k(P, \lambda; F).$$

Also, let

$$\chi_{k-1} = \sum_{j \geq k} (-1)^{j-k} (f_j - \beta_j) \quad \text{for } k \geq 0.$$

THEOREM 2. *For any poset $P$ and edge-labeling $\lambda$ satisfying conditions* (I)–(V) *the following relations hold*:

(i) $$\chi_{-1} = 1,$$

(ii) $$\partial_k(\chi_k + \beta_k) \leq \chi_{k-1} \quad \textit{for all } k \geq 1.$$

Following the proof of Theorem 2 we will show that it implies Theorem 1.

PROOF OF THEOREM 2. In what follows it will be important that the class of all edge-labeled posets $(P, \lambda)$ satisfying (I)–(V) is *hereditary*, meaning that it is closed under taking principal filters $P^{\geq v} = \{x \in P \mid x \geq v\}$ and arbitrary ideals. Let $\mathscr{C}$ denote this class.

First recall the elementary fact that if

$$0 \to E_n \to E_{n-1} \xrightarrow{e_{n-1}} \cdots \to E_k \xrightarrow{e_k} \cdots \xrightarrow{e_0} E_{-1} \to 0$$

is any semi-exact ($e_k \circ e_{k+1} = 0$ for all $k$) sequence of finite-dimensional $F$-vector spaces and linear maps, then

$$\dim \operatorname{Im}(e_k) = \sum_{j \geq k} (-1)^{j-k} (\dim E_j - \dim H_j) \quad \text{for } k \geq 0, \tag{1}$$

where $H_j = \operatorname{Ker}(e_j)/\operatorname{Im}(e_{j+1})$. Hence

$$\chi_{-1} = \dim_F \operatorname{Im}(d_0) = \dim_F C_{-1} = \operatorname{card} P_{-1} = 1,$$

which is relation (i).

Define

$$\begin{aligned} \varphi_k &= f_k - f_{k+1} + f_{k+2} - f_{k+3} + \cdots, \\ \gamma_k &= \beta_k - \beta_{k+1} + \beta_{k+2} - \beta_{k+3} + \cdots. \end{aligned}$$

Then $\chi_{k-1} = \varphi_k - \gamma_k$, so relations (ii) can be rewritten as

$$\partial_k(\varphi_{k+1} + \gamma_k) \leq \varphi_k - \gamma_k \quad \text{for } k \geq 1. \tag{$*$}$$

We will prove $(*)$ by induction on the size of $P$. If $f_1 = 0$ (in particular, if $|P| \leq 3$), then $\varphi_k = \gamma_k = 0$ for all $k \geq 1$ and there is nothing to prove. In the following argument we use that $(*)$ is satisfied also for $k = 0$ (since $\partial_0 \leq 1 = \chi_{-1}$) irrespective of the size of $P$.

Assume that $f_1 \neq 0$, and for the induction step select an atom $v \in P_0$ which is not maximal in $P$. Since $\mathscr{C}$ is hereditary, both the *star* $\operatorname{st}(v) = P^{\geq v}$ and the *antistar* $\operatorname{ast}(v) = P - P^{\geq v}$ belong to $\mathscr{C}$. Hence, by the induction assumption:

$$\partial_k(\overset{\text{a}}{\varphi}_{k+1} + \overset{\text{a}}{\gamma}_k) \leq \overset{\text{a}}{\varphi}_k - \overset{\text{a}}{\gamma}_k, \tag{2}$$

$$\partial_{k-1}(\overset{\text{s}}{\varphi}_k + \overset{\text{s}}{\gamma}_{k-1}) \leq \overset{\text{s}}{\varphi}_{k-1} - \overset{\text{s}}{\gamma}_{k-1}, \tag{3}$$

where superscript “a” refers to the antistar and “s” to the star at $v$.

The partitioning $P = \operatorname{ast}(v) \cup \operatorname{st}(v)$ shows at once that $f_k = \overset{\text{a}}{f}_k + \overset{\text{s}}{f}_{k-1}$, hence

$$\varphi_k = \overset{\text{a}}{\varphi}_k + \overset{\text{s}}{\varphi}_{k-1}. \tag{4}$$

The proof of the next two relations, which involves Betti numbers, will be postponed until later:

$$(5) \qquad \gamma_k \le \overset{\mathrm{a}}{\gamma}_k + \overset{\mathrm{s}}{\gamma}_{k-1},$$

$$(6) \qquad \overset{\mathrm{s}}{f}_{k-1} \le \varphi_k - \gamma_k .$$

Finally we need the following relation, due to Eckhoff and Wegner [**EW**]:

$$(7) \quad \partial_k(m+n) \le \max\{\partial_k(m), n\} + \partial_{k-1}(n), \qquad m, n \ge 0,\ k \ge 1.$$

These inequalities provide the induction step:

$$\begin{aligned}\partial_k(\varphi_{k+1}+\gamma_k) &\le_{(4,5)} \partial_k(\overset{\mathrm{a}}{\varphi}_{k+1}+\overset{\mathrm{s}}{\varphi}_k+\overset{\mathrm{a}}{\gamma}_k+\overset{\mathrm{s}}{\gamma}_{k-1})\\ &\le_{(7)} \max\{\partial_k(\overset{\mathrm{a}}{\varphi}_{k+1}+\overset{\mathrm{a}}{\gamma}_k), \overset{\mathrm{s}}{\varphi}_k+\overset{\mathrm{s}}{\gamma}_{k-1}\} + \partial_{k-1}(\overset{\mathrm{s}}{\varphi}_k+\overset{\mathrm{s}}{\gamma}_{k-1}).\end{aligned}$$

Furthermore,

$$\begin{aligned}&\partial_k(\overset{\mathrm{a}}{\varphi}_{k+1}+\overset{\mathrm{a}}{\gamma}_k)+\partial_{k-1}(\overset{\mathrm{s}}{\varphi}_k+\overset{\mathrm{s}}{\gamma}_{k-1})\\ &\quad\le_{(2,3)} \overset{\mathrm{a}}{\varphi}_k-\overset{\mathrm{a}}{\gamma}_k+\overset{\mathrm{s}}{\varphi}_{k-1}-\overset{\mathrm{s}}{\gamma}_{k-1}\\ &\quad= \overset{\mathrm{a}}{\varphi}_k+\overset{\mathrm{s}}{\varphi}_{k-1}-(\overset{\mathrm{a}}{\gamma}_k+\overset{\mathrm{s}}{\gamma}_{k-1}) \le_{(4,5)} \varphi_k-\gamma_k,\end{aligned}$$

and

$$\overset{\mathrm{s}}{\varphi}_k+\overset{\mathrm{s}}{\gamma}_{k-1}+\partial_{k-1}(\overset{\mathrm{s}}{\varphi}_k+\overset{\mathrm{s}}{\gamma}_{k-1}) \le_{(3)} \overset{\mathrm{s}}{\varphi}_k+\overset{\mathrm{s}}{\varphi}_{k-1} = \overset{\mathrm{s}}{f}_{k-1} \le_{(6)} \varphi_k-\gamma_k .$$

Hence, in either case $(*)$ follows.

It remains to prove (5) and (6).

In addition to the subposets star $\mathrm{st}(v)$ and antistar $\mathrm{ast}(v)$ previously defined, we now also need the *cone* $\mathrm{cn}(v)$ and *link* $\mathrm{lk}(v)$ at $v$:

$$\begin{aligned}\mathrm{cn}(v) &= \{x \in P \mid x \le y \text{ for some } y \ge v\},\\ \mathrm{lk}(v) &= \{x \in P \mid x \le y \text{ for some } y \ge v \text{ and } x \not\ge v\}.\end{aligned}$$

Both the cone and link with their induced edge-labeling belong to $\mathscr{C}$, since $\mathscr{C}$ is hereditary. Clearly,

$$(8) \qquad P = \mathrm{ast}(v) \cup \mathrm{cn}(v), \qquad \mathrm{lk}(v) = \mathrm{ast}(v) \cap \mathrm{cn}(v).$$

Consider for a moment two arbitrary ideals $I$ and $J$ in $P$. For each $k$ there is a short exact sequence

$$0 \to C_k(I \cap J) \xrightarrow{(i,-i)} C_k(I) \oplus C_k(J) \xrightarrow{p+q} C_k(I \cup J) \to 0,$$

where $i$, $p$, and $q$ are the natural injections. These maps commute with the respective boundary maps $d_k$, so we obtain a long exact homology sequence (Mayer-Vietoris)

$$(9) \qquad \begin{aligned}\cdots &\to \widetilde{H}_{k+1}(I \cup J) \to \widetilde{H}_k(I \cap J) \to \widetilde{H}_k(I) \oplus \widetilde{H}_k(J)\\ &\to \widetilde{H}_k(I \cup J) \to \widetilde{H}_{k-1}(I \cap J) \to \cdots .\end{aligned}$$

This sequence will be used to prove the following lemma.

(10) *Let $A$ be an antichain in $P$ such that $\wedge A \neq \widehat{0}$. Then the ideal $\overline{A} = \{x \in P \mid x \leq a \in A\}$ is acyclic.*

Argue by induction on $|A| = t$. If $t = 1$, then $\overline{A}$ is an interval and hence acyclic by condition (V). Let $t > 1$, $x \in A$, and assume by induction that $\overline{A - \{x\}}$ is acyclic. The ideal $\overline{A - \{x\}} \cap \overline{\{x\}}$ has at most $t-1$ maximal elements (since $P$ is a meet-semilattice), hence by induction it is also acyclic. Since the ideals $\overline{A - \{x\}}$ and $\overline{\{x\}} = [\widehat{0}, x]$ and their intersection are acyclic, it follows from (9) that their union $\overline{A}$ is acyclic, which proves (10).

Applying (10) to the antichain of maximal elements of $\mathrm{cn}(v)$ we conclude that

(11) $$\mathrm{cn}(v) \text{ is acyclic.}$$

In view of (8) and (11), relation (1) applied to (9) gives

$$0 \leq \beta^{\mathrm{l}}_{k-1} - \beta_k + \beta^{\mathrm{a}}_k - \beta^{\mathrm{l}}_k + \beta_{k+1} - \cdots \quad (\text{where } \beta^{\mathrm{l}}_i = \beta_i(\mathrm{lk}), \text{ etc.})$$

or, which is the same,

(12) $$\gamma_k \leq \gamma^{\mathrm{a}}_k + \gamma^{\mathrm{l}}_{k-1}.$$

Next we show that $\beta^{\mathrm{s}}_k = \beta^{\mathrm{l}}_k$, and hence $\gamma^{\mathrm{s}}_k = \gamma^{\mathrm{l}}_k$ for all $k$. Together with (12) this gives relation (5).

For a poset $P \in \mathscr{C}$ and an ideal $J$ of $P$ define $C_k(P, J) = C_k(P)/C_k(J)$. The boundary map $d_k \colon C_k(P) \to C_{k-1}(P)$ induces a boundary map $d^J_k \colon C_k(P, J) \to C_{k-1}(P, J)$, and we obtain the relative homology group

$$H_k(P, J) = H_k(P, J, \lambda; F) = \mathrm{Ker}\, d^J_k / \mathrm{Im}\, d^J_{k+1}.$$

For each $k$ there is a short exact sequence $0 \to C_k(J) \to C_k(P) \to C_k(P, J) \to 0$, whose maps are the natural injection and projection. These maps commute with the respective boundary maps, so we obtain a long exact sequence

(13) $$\cdots \to \widetilde{H}_{k+1}(P) \to H_{k+1}(P, J) \to \widetilde{H}_k(J) \to \widetilde{H}_k(P) \to \cdots.$$

Applying this to the case $P = \mathrm{cn}(v)$ and $J = \mathrm{lk}(v)$ we deduce using (11) that $H_{k+1}(\mathrm{cn}(v), \mathrm{lk}(v)) \cong \widetilde{H}_k(\mathrm{lk}(v))$. On the other hand, it follows directly from the definition that $\widetilde{H}_k(\mathrm{st}(v)) \cong H_{k+1}(\mathrm{cn}(v), \mathrm{lk}(v))$. Hence,

(14) $$\widetilde{H}_k(\mathrm{st}(v)) \cong \widetilde{H}_k(\mathrm{lk}(v)) \quad \text{for all } k.$$

It is left to prove relation (6). Put

$$(\mathrm{st})_k = \{x \in \mathrm{st}(v) \mid \dim_{\mathrm{st}(v)}(x) = k = \dim_P(x) - 1\}.$$

For every $x \in (\mathrm{st})_{k-1}$ let $u(x)$ be a maximal complement of $v$ in the interval $[\widehat{0}, x]$. Consider the set $\{d_k(x) \mid x \in (\mathrm{st})_{k-1}\}$. As a subset of $B_{k-1}(P) = \mathrm{Im}\, d_k$ this set is independent, because $u(x)$ has nonzero coefficient in $d_k(x)$, but from the semilattice property it follows that $u(x) \not< y$, and hence $u(x)$ has a zero coefficient in $d_k(y)$ for $y \in (\mathrm{st})_{k-1}$, $y \neq x$. Hence,

$$f^{s}_{k-1} = \mathrm{card}(\mathrm{st})_{k-1} \leq \dim B_{k-1} \overset{(1)}{=} \varphi_k - \gamma_k,$$

which is relation (6). □

PROOF OF THEOREM 1. Let $P = P(\Delta)$ be the face poset of a finite regular CW complex with the intersection property (the family of closed cells is closed under nonempty intersections). Let $\lambda$ be the labeling of coverings in $P$ by the incidence numbers over $F$ induced by an orientation. (See **[CF]** or **[LW]** for this construction in cellular topology.) It follows from the standard facts proven in these sources (e.g., **[LW, pp. 168–169]**) that conditions (I), (III), and (IV) are satisfied and that the combinatorially defined homology groups $\widetilde{H}_k(P(\Delta), \lambda; F)$ coincide with the singular homology groups of $\Delta$. Condition (II) is implied by (in fact, equivalent to) the intersection property. Hence, to deduce Theorem 1 from Theorem 2 it remains only to verify condition (V) for $P(\Delta)$.

Let $x < y$ in $P = P(\Delta)$. If $x = \hat{0}$ we are done, since the interval $[\hat{0}, y]$ is the face lattice of the closed cell $y$ which is acyclic. So, assume that $x \neq \hat{0}$. We claim that

(15) the poset $[\hat{0}, y] \backslash [x, y]$ is acyclic.

To prove this we use that the poset $[\hat{0}, y] \backslash [x, y]$ is the face poset of a subcomplex $\hat{\Delta}$ of $\Delta$. This subcomplex $\hat{\Delta}$ lies in the boundary $\dot{y}$ of the cell $y$, which is a sphere (since $\Delta$ is regular). The complement $\dot{y} \backslash \hat{\Delta}$ is the open star of the cell $x$, i.e., the union of all open cells in $\dot{y}$ whose closure contains $x$. This open star is acyclic, and by Alexander duality so is then also $\hat{\Delta}$, which proves (15). [To see this last argument with particular ease, it is helpful to pass to barycentric subdivisions. The simplicial complex of chains (totally ordered subsets) of the open interval $(\hat{0}, y)$ triangulates a sphere, and the subcomplex determined by $[x, y)$ is a cone (with cone-point $x$), and hence acyclic. By Alexander duality the chain complex of $(\hat{0}, y) \backslash [x, y)$ is therefore also acyclic, and this complex is the barycentric subdivision of $\hat{\Delta}$.]

From (15) we deduce that $[x, y]$ is acyclic as follows. Note that in the poset $[\hat{0}, y]$ we have $\mathrm{cn}(x) = [\hat{0}, y]$, $\mathrm{st}(x) = [x, y]$, and $\mathrm{lk}(x) = [\hat{0}, y] \backslash [x, y]$. Since $[\hat{0}, y]$ is acyclic we can apply the long exact sequence (13), as in (14), to deduce from (15) that $[x, y]$ is acyclic.

Condition (V) has now been verified and the proof is complete. □

## 4. Final remarks

1. The only relations beyond the Euler-Poincaré formula which hold for arbitrary regular cell complexes are $\chi_k \geq 1$, $k \geq 0$ (see **[BK, Appendix]**). The following example shows that the conclusion of Theorem 1 fails to hold

if the intersection condition on cells is relaxed and it is only assumed that every nonempty intersection of cells is homeomorphic to a ball. Let $\Delta$ be the regular complex obtained from the boundary of the tetrahedron by subdividing two opposite edges and gluing a 2-cell to the remaining four edges. We have $f = (6, 8, 5)$, $\beta = (0, 0, 2)$, and inequalities (ii) of Theorem 1 are violated for $k = 2$.

2. If $P$ is the face poset of a regular CW complex and $x \in P$, $x \neq \widehat{0}$, then $P^{\geq x}$ is not necessarily isomorphic to the face poset of some regular CW complex. This follows, e.g., from R. D. Edwards' result (see **[D]**) that the double suspension of any homology $d$-sphere is homeomorphic to the $(d+2)$-sphere. Therefore the class of face posets of regular CW complexes is not hereditary in the sense needed for our proof.

3. The proof of Theorem 1 is new even for the special case of simplicial complexes. Several parts of the proof are considerably simpler in this case. For instance, $\mathrm{lk}(v)$ and $\mathrm{st}(v)$ are combinatorially isomorphic so that the isomorphism of their homology is immediate. Also, the class of face posets of simplicial complexes is hereditary.

4. Inequality (7) was proved by Eckhoff and Wegner as part of their simple proof of the Kruskal-Katona theorem **[EW]**. This inequality also follows at once from the Kruskal-Katona theorem itself. Wegner **[W]** used this inequality to extend the Kruskal-Katona theorem to a large class of graded meet-semilattices.

5. A basic idea behind our proof is to derive complicated nonlinear inequalities (inequalities (ii) of Theorem 1) for face numbers and Betti numbers of a complex, from simpler linear inequalities (inequalities (5) and (6)) which involve face numbers and Betti numbers of stars and antistars of faces. This is a central idea (just for face numbers) in the papers of Eckhoff-Wegner **[EW]** and Wegner **[W]** mentioned above. A similar idea (applied to $h$-numbers) played a role already in McMullen's proof of the upper bound theorem **[Mc]**.

6. We thank Ron Adin for helpful comments.

## References

**[B]** A. Björner, *Posets, regular CW complexes and Bruhat order*, European J. Combin. **5** (1984), 7–16.

**[BK]** A. Björner and G. Kalai, *An extended Euler-Poincaré theorem*, Acta Math. **161** (1988), 279–303.

**[CF]** G. E. Cooke and R. L. Finney, *Homology of cell complexes*, Princeton Univ. Press, Princeton, N.J., 1967.

**[D]** R. J. Daverman, *Decompositions of manifolds*, Academic Press, Orlando, Fla., 1986.

**[EW]** J. Eckhoff and G. Wegner, *Über einen Satz von Kruskal*, Period. Math. Hungar. **6** (1975), 137–142.

**[LW]** A. T. Lundell and S. Weingram, *The topology of CW complexes*, Van Nostrand, New York, 1969.

[Mc] P. McMullen, *The maximum number of faces of a convex polytope*, Mathematika **17** (1970), 179–184.

[W] G. Wegner, *Kruskal-Katona's theorem in generalized complexes*, in Finite and Infinite Sets, Vol. 2, Colloq. Math. Soc. János Bolyai, vol. 37, North-Holland, Amsterdam, 1984, pp. 821–827.

Department of Mathematics, Royal Institute of Technology, S-10044 Stockholm, Sweden

*E-mail address*: andersb@nada.kth.se

Edmund Landau Center for Research in Mathematical Analysis, Institute of Mathematics and Computer Science, Hebrew University, Jerusalem, Israel

DIMACS Series in Discrete Mathematics
and Theoretical Computer Science
Volume **4**, 1991

Applied Geometry
and Discrete Mathematics
THE VICTOR KLEE FESTSCHRIFT

# Computing the Convex Hull in the Euclidean Plane in Linear Expected Time

KARL HEINZ BORGWARDT, NORBERT GAFFKE,
MICHAEL JÜNGER, AND GERHARD REINELT

ABSTRACT. There are many algorithms for computing the convex hull of a set of $n$ points in the Euclidean plane in worst case time $O(n \ln n)$. It is also known that $O(n)$ average time behavior can be achieved under the assumption that the $n$ points are independently and uniformly distributed over the unit square or the unit disk. We will show how any $O(n \ln n)$ worst case time algorithm for the convex hull can be converted into an $O(n)$ average time algorithm under this assumption. Also, it turns out that $O(n)$ average time is still obtained if algorithms of higher worst case time complexity are used. Finally, we will give a computational comparison of our method with other linear expected time algorithms for problem instances in the unit square.

## 1. Introduction

The problem of computing the convex hull for $n$ points in the Euclidean plane has received considerable attention in computational geometry. Several algorithms with worst case time complexity $O(n \ln n)$ (resp. $O(n \ln H)$, where $H$ is the number of vertices of the convex hull) are presented in the literature. This complexity is best possible if the output of the algorithm consists of a list of the vertices of the convex hull. Assuming a uniform and independent distribution of the $n$ points over the unit square or the unit disk, linear expected time algorithms have been given by Bentley and Shamos [**3**], Akl and Toussaint [**1**], Devroye [**4**], Devroye and Toussaint [**5**], Kirkpatrick and Seidel [**9**], and Golin and Sedgewick [**7**]. For a survey on these and related subjects cf. Lee and Preparata [**10**].

We will show how any $O(n \ln n)$ worst case time algorithm for the convex hull can be converted into an $O(n)$ average case time algorithm under this assumption.

1980 *Mathematics Subject Classification* (1985 *Revision*). Primary 52-04; Secondary 60D05, 68C25.

*Key words and phrases.* Computational geometry, computational complexity, convex hull.

For the unit square, a similar result follows from the work of Devroye [**4**] who uses the fact that the set of the maximal vectors is of expected cardinality $O(\ln n)$ and can be computed in $O(n)$ expected time by a divide and conquer technique described in Bentley et al. [**2**]. Any polynomial worst case time algorithm applied to the set of maximal vectors yields a linear expected time method. On the other hand, Devroye and Toussaint [**5**] show that applying an $O(n^2)$ worst case time algorithm to all points outside a certain polygon of up to eight vertices also yields an $O(n)$ expected time algorithm. We refine their elegant "throw-away" principle so that for the unit square we can strengthen their result to the one of Devroye [**4**]. In addition, for the unit disk, we can apply any algorithm of worst case time complexity $O(n^r)$, where $r < 3$, to obtain a linear expected time method. In a way, this is best possible because Raynaud [**13**] has shown that the convex hull of $n$ independently and uniformly distributed points in the unit disk is expected to have $O(n^{1/3})$ vertices.

Finally we will give a computational comparison of the new algorithm with other linear expected time algorithms for problem instances in the unit square.

## 2. Analysis for the unit square

Let $A = \{a_1, a_2, \dots, a_n\}$ be $n$ random points generated independently from the uniform distribution on the unit square $[0, 1] \times [0, 1]$. We wish to compute the convex hull $\mathrm{conv}(A)$. In the following we assume that we have an algorithm CH which computes the convex hull of a set of $n$ points in the unit square in worst case time $O(n \ln n)$. Below we will also allow CH to have greater (but polynomial) worst case running time. The basic idea to obtain still a linear running time on the average is to compute the convex hull for a small subset $S$ of $A$ such that $\mathrm{conv}(S) = \mathrm{conv}(A)$ with high probability.

To this end we classify the $n$ points with a function $h\colon [0, 1] \times [0, 1] \to [0, 1]$ such that $h(x)$ will express whether $x$ is likely to be an interior point of $\mathrm{conv}(A)$. With increasing $h(x)$ the probability that $x$ can be discarded as a candidate for a vertex of the convex hull increases.

The algorithm can be stated as follows:

PHASE 1

(1) Determine a parameter $\alpha = \alpha(n) > 0$ depending on $n$.
(2) Let $S = \{a_i \in A | h(a_i) \leq \alpha\}$.
(3) Apply algorithm CH to the subset $S$.
(4) Check if $\mathrm{conv}(S) \supseteq \{x \in [0, 1] \times [0, 1] | h(x) > \alpha\}$. If this is true then stop: $\mathrm{conv}(S) = \mathrm{conv}(A)$. Otherwise, go to Phase 2.

PHASE 2

(5) Apply algorithm CH to the complete set $A$.

Let $|S| = k$, $F_\alpha = \{x \in [0, 1] \times [0, 1] | h(x) \leq \alpha\}$, and $G_\alpha = \{x \in$

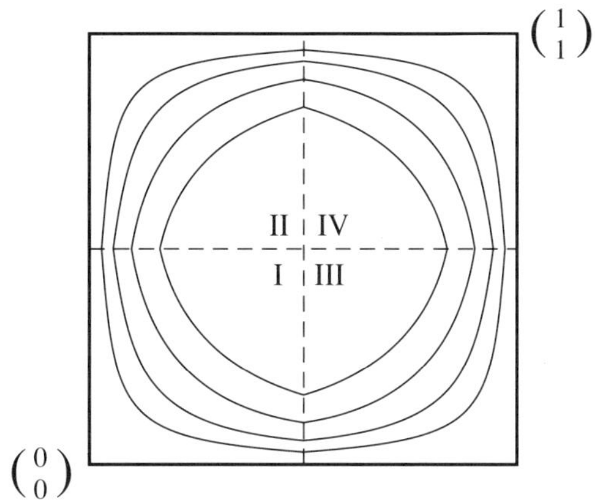


FIGURE 1. The isoclinals of $h$

$[0, 1] \times [0, 1] | h(x) > \alpha\}$. Obviously, step (2) takes linear time in $n$. Step (3) takes time $O(k \ln k)$ by our assumption on CH, and step (4) can be done in $O(k)$ time (see below). The overall computation time depends mainly on the result of step (4). If $\operatorname{conv}(S) \not\supseteq G_\alpha$ we have to apply CH again (this time to all $n$ points) which takes time $O(n \ln n)$. In the following we will estimate the expectation of the time $T_{\text{CH}}$ which is spent in running algorithm CH. If this is linear (in $n$), then the overall algorithm has linear expected running time. We have

$$\begin{aligned} E[T_{\text{CH}}] &\leq \sum_{k=1}^{n} k \ln k \cdot P(|S| = k) + n \ln n \cdot P(\operatorname{conv}(S) \not\supseteq G_\alpha) \\ &\leq \ln n \cdot \sum_{k=1}^{n} k \cdot P(|S| = k) + n \ln n \cdot P(\operatorname{conv}(S) \not\supseteq G_\alpha) \\ &= \ln n \cdot E[|S|] + n \ln n \cdot P(\operatorname{conv}(S) \not\supseteq G_\alpha). \end{aligned} \tag{1}$$

Note that in the above formula we have assumed running time $n \ln n$ for algorithm CH rather than $Cn \ln n$ for some constant $C$. This simplifies our computations but does not affect the results. Before we evaluate $E[T_{\text{CH}}]$ asymptotically for $n \to \infty$ we specify the function $h(x)$ and the criterion in step (4) which allows us to state success or failure of Phase 1. We define $h: [0, 1] \times [0, 1] \to [0, 1]$ by

$$h(x) = h(x_1, x_2) = \min\{x_1, 1 - x_1\} \cdot \min\{x_2, 1 - x_2\}.$$

For two points $a$ and $b$ in $[0, 1] \times [0, 1]$ we write $[a, b] := \{\lambda a + (1-\lambda) b | 0 \leq \lambda \leq 1\}$ for the line segment connecting $a$ and $b$.

The isoclinals of $h$ have the shape shown in Figure 1. The function $h$ is symmetric with respect to the lines $x_1 = \frac{1}{2}$ and $x_2 = \frac{1}{2}$, as well as to the diagonals $x_1 = x_2$ and $x_1 = 1 - x_2$. This will enable us to restrict our considerations to quadrant $\mathrm{I} = [0, \frac{1}{2}] \times [0, \frac{1}{2}]$ of Figure 1. In this quadrant the function $h$ simplifies to $h(x) = x_1 \cdot x_2$. The parameter $\alpha = \alpha(n)$ in our algorithm will be chosen as

$$\alpha = \sqrt{e}\frac{(2+\sigma)\ln n}{n}, \tag{2}$$

with some fixed $\sigma > 0$ (not depending on $n$). Throughout this section we will assume that $n$ is large enough to satisfy $0 < \alpha < \frac{1}{4}$.

Suppose we have constructed $\mathrm{conv}(S)$, where $|S| = k$ in Phase 1 of the algorithm. We know the sequence of vertices $a_{i_1}, a_{i_2}, \ldots, a_{i_l}$, where $l \leq k$, of $\mathrm{conv}(S)$ sorted by increasing angles (assuming polar coordinates originating from an interior point of the convex hull). We can assume the availability of the sorted sequence, since CH may take time $O(k \ln k)$ for constructing $\mathrm{conv}(S)$. It is easy to decide in time $O(k)$ whether the point $(\frac{1}{2}, \frac{1}{2})$ is an interior point of $\mathrm{conv}(S)$. If this is not the case, we declare the failure of Phase 1. Otherwise, we check if each of the line segments $[a_{i_r}, a_{i_{r+1}}]$, $r = 1, 2, \ldots, l-1$, and the segment $[a_{i_l}, a_{i_1}]$ are contained in $F_\alpha$, which also takes time $O(k)$. Due to the convexity of $G_\alpha$, in this case $G_\alpha \subseteq \mathrm{conv}(S)$ and we declare the success of Phase 1. Otherwise we declare the failure of Phase 1. We are going to give a sufficient condition for the success of Phase 1, which will enable us to give an estimate for $P(\mathrm{conv}(S) \not\supseteq G_\alpha)$. To this end let

$$\beta = \beta(n) = \frac{(2+\sigma)\ln n}{n}.$$

In addition to $F_\alpha$ we also consider $F_\beta$, for which, due to $\beta < \alpha$, we have $F_\beta \subset F_\alpha$. The limiting curve between $F_\beta$ and $G_\beta$ in quadrant I is $x_1 x_2 = \beta$, or $x_2 = \beta / x_1$ for $x_1 \in [2\beta, \frac{1}{2}]$. We partition $F_\beta \cap \mathrm{I}$ into cells $T_1, T_2, \ldots, T_{2\lfloor \ln n \rfloor + 1}$ as follows. We define numbers $t_i = \sqrt{\beta} \cdot \lfloor 2\sqrt{\beta} \rfloor^{-i/\lfloor \ln n \rfloor}$ for $i = 0, 1, \ldots, \lfloor \ln n \rfloor$ so that $t_0 = \sqrt{\beta}$ and $t_{\sqrt{\lfloor \ln n \rfloor}} = \frac{1}{2}$. For $i = 1, 2, \ldots, \lfloor \ln n \rfloor$, we set $T_i = \{x \in \mathrm{I} \cap F_\beta | t_{i-1} < x_1 \leq t_i\}$ and $T_{i+\lfloor \ln n \rfloor} = \{x \in \mathrm{I} \cap F_\beta | t_{i-1} < x_2 \leq t_i\}$. In addition we define the corner cell $T_{2\lfloor \ln n \rfloor + 1} = \{x | 0 \leq x_1 \leq \sqrt{\beta}, 0 \leq x_2 \leq \sqrt{\beta}\}$ which will be ignored whenever we refer to the set of cells. An illustration for this partition is given in Figure 2.

Let $\Lambda(M)$ denote the Lebesgue measure of a subset $M$ of the plane. We have

$$\Lambda(T_{2\lfloor \ln n \rfloor + 1}) = \beta$$

and

$$\begin{aligned}\Lambda(T_i) &= \int_{t_{i-1}}^{t_i} \frac{\beta}{x}\,dx = \beta[\ln t_i - \ln t_{i-1}] = \beta \ln \frac{t_i}{t_{i-1}} \\ &= \beta \ln[[2\sqrt{\beta}]^{-1/\lfloor \ln n \rfloor}] = \frac{\beta}{\lfloor \ln n \rfloor} \ln \left[\frac{1}{2\sqrt{\beta}}\right]\end{aligned}$$

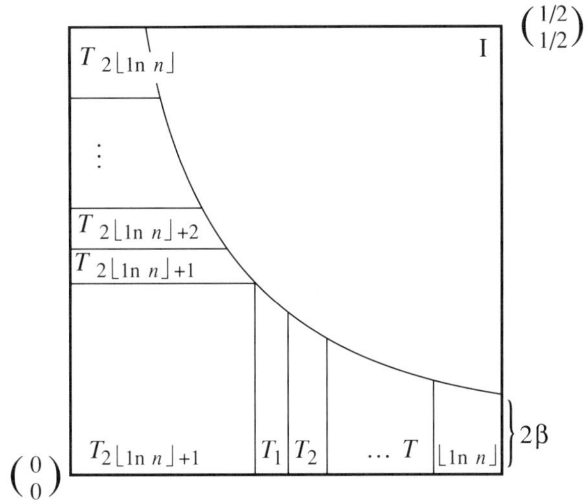


FIGURE 2. The partition $F_\beta \cap \mathrm{I}$

for all $i = 1, 2, \ldots, \lfloor \ln n \rfloor$, and, due to symmetry, for all $i = \lfloor \ln n \rfloor + 1, \ldots, 2\lfloor \ln n \rfloor$.

LEMMA 2.1. *If every* $T_i$, $i = 1, 2, \ldots, 2\lfloor \ln n \rfloor$, *and every corresponding set in the quadrants* II, III, IV *contains at least one point of* $A$, *then* $\operatorname{conv}(A) \supseteq G_\alpha$.

PROOF. Without loss of generality we assume $a_i \in T_i$ for $i = 1, 2, \ldots, 2\lfloor \ln n \rfloor$. Analogous assumptions are made for the quadrants II, III, IV. The (not necessarily convex) closed polygon with vertices $a_{2\lfloor \ln n \rfloor}, a_{2\lfloor \ln n \rfloor - 1}, \ldots, a_{\lfloor \ln n \rfloor + 2}, a_{\lfloor \ln n \rfloor + 1}, a_1, a_2, \ldots, a_{\lfloor \ln n \rfloor - 1}, a_{\lfloor \ln n \rfloor}$ is contained in $\operatorname{conv}(A)$. So it is sufficient to prove that all line segments $[a_i, a_{i+1}]$, $i = 1, 2, \ldots, \lfloor \ln n \rfloor - 1, \lfloor \ln n \rfloor + 1, \lfloor \ln n \rfloor + 2, \ldots, 2\lfloor \ln n \rfloor - 1$, and $[a_1, a_{\lfloor \ln n \rfloor + 1}]$, and a representative line segment between adjacent cells of adjacent quadrants are contained in $F_\alpha$.

CASE 1. We consider a line segment $[a_i, a_{i+1}]$ for which $1 \le i < \lfloor \ln n \rfloor$. $[a_i, a_{i+1}]$ belongs to the convex hull of $T_i \cup T_{i+1}$, and so we have to show that $\max\{h(x) | x \in \operatorname{conv}(T_i \cup T_{i+1})\} \le \alpha$. Obviously, the maximum is attained at some point from the line segment $\{x(\lambda) = \lambda(t_{i-1}, \beta/(t_{i-1})) + (1-\lambda)(t_{i+1}, \beta/(t_{i+1})) | 0 \le \lambda \le 1\}$. We obtain

$$\begin{aligned} h(x(\lambda)) &= [\lambda t_{i-1} + (1-\lambda)t_{i+1}]\left[\lambda \frac{\beta}{t_{i-1}} + (1-\lambda)\frac{\beta}{t_{i+1}}\right] \\ &= \lambda^2 \beta + (1-\lambda)^2 \beta + \lambda(1-\lambda)\beta c, \end{aligned}$$

where $c = (t_{i-1})/(t_{i+1}) + (t_{i+1})/(t_{i-1}) \ge 2$, since the function $1/x + x$ is strictly increasing for $x \geq 1$. Thus

$$\begin{aligned} h(x(\lambda)) &= \beta[(2-c)\lambda^2 + (c-2)\lambda + 1] \\ &= \beta(c-2)\lambda(1-\lambda) + \beta \\ &\leq \frac{1}{4}\beta(c-2) + \beta \\ &= \beta\left[\frac{1}{2} + \frac{1}{4}\left(\frac{t_{i-1}}{t_{i+1}} + \frac{t_{i+1}}{t_{i-1}}\right)\right]. \end{aligned}$$

By our choice of the $t_i$

$$\begin{aligned} \frac{t_{i+1}}{t_{i-1}} &= \left[\frac{1}{4\beta}\right]^{1/\lfloor \ln n \rfloor} = \left[\frac{n}{4(2+\sigma)\ln n}\right]^{1/\lfloor \ln n \rfloor} \leq \left[\frac{n}{4}\right]^{1/\lfloor \ln n \rfloor} \\ &= \left[\frac{1}{4}\right]^{1/\lfloor \ln n \rfloor} e^{\ln n/\lfloor \ln n \rfloor} \leq e\left[\frac{1}{4}e\right]^{1/\lfloor \ln n \rfloor} \leq e \end{aligned}$$

which implies

$$\frac{t_{i-1}}{t_{i+1}} + \frac{t_{i+1}}{t_{i-1}} \leq \frac{1}{e} + e \leq \frac{1}{3} + 3 = \frac{10}{3}.$$

Hence

$$h(x(\lambda)) \leq \beta \cdot \tfrac{4}{3} \leq \beta \cdot \sqrt{e} = \alpha.$$

CASE 2. We consider the segment $[a_1, a_{\lfloor \ln n \rfloor + 1}]$. The function $h$ restricted to the convex hull $\operatorname{conv}(T_1 \cup T_{\lfloor \ln n \rfloor + 1})$ attains its maximum at some point $x(\lambda)$, $0 \leq \lambda \leq 1$, from the line segment joining $(t_1, \beta/t_1)$ and $(\beta/t_1, t_1)$. We have

$$\begin{aligned} h(x(\lambda)) &= \left[\lambda t_1 + (1-\lambda)\frac{\beta}{t_1}\right]\left[\lambda\frac{\beta}{t_1} + (1-\lambda)t_1\right] \\ &= \lambda^2\beta + (1-\lambda)^2\beta + \lambda(1-\lambda)\left[\frac{t_1^2}{\beta} + \frac{\beta}{t_1^2}\right] \cdot \beta, \end{aligned}$$

and

$$\frac{t_1^2}{\beta} = \left[\frac{1}{4\beta}\right]^{1/\lfloor \ln n \rfloor}$$

which is equal to $(t_{i+1})/(t_{i-1})$ from Case 1. So $h(x(\lambda))$, as a function of $\lambda \in [0, 1]$, is the same as in Case 1.

CASE 3. We consider a segment between $a_{\lfloor \ln n \rfloor}$ and a point in the cell of quadrant III which is adjacent to $T_{\lfloor \ln n \rfloor}$. On the convex hull of these two cells the function $h$ attains its maximum at some point from the line segment joining $(t_{\lfloor \ln n \rfloor - 1}, \beta/t_{\lfloor \ln n \rfloor - 1})$ and $(1 - t_{\lfloor \ln n \rfloor - 1}, \beta/t_{\lfloor \ln n \rfloor - 1})$. Obviously the maximizing point is given by $x_1 = \frac{1}{2}$ and $x_2 = \beta / t_{\lfloor \ln n \rfloor - 1}$, and

$$\begin{aligned} h(x_1, x_2) &= \frac{1}{2}\frac{\beta}{t_{\lfloor \ln n \rfloor - 1}} = \frac{1}{2}\beta\left[2\left[\frac{1}{2\sqrt{\beta}}\right]^{1/\lfloor \ln n \rfloor}\right] \\ &= \beta\left[\frac{1}{4\beta}\right]^{1/2\lfloor \ln n \rfloor} \leq \beta \cdot \sqrt{e} = \alpha, \end{aligned}$$

where we have used $[1/4\beta]^{1/\lfloor \ln n \rfloor} \leq e$ from Case 1. □

LEMMA 2.2. $P(\text{conv}(S) \not\supseteq G_\alpha) \leq 8[4(2+\sigma)]^{1+\sigma/2} \cdot (\ln n)^{2+\sigma/2} \cdot n^{-(1+\sigma/2)}$.

PROOF. Consider the cells $T_i$, $i = 1, 2, \ldots, 8\lfloor \ln n \rfloor$, from Lemma 2.1. By Lemma 2.1 $P(\text{conv}(S) \not\supseteq G_\alpha) \leq P(A \cap T_i = \emptyset)$ for at least one $i \in \{1, 2, \ldots, 8\lfloor \ln n \rfloor\}) \leq \sum_{i=1}^{8\lfloor \ln n \rfloor} P(A \cap T_i = \emptyset)$.

Now, for a given $i \in \{1, 2, \ldots, 8\lfloor \ln n \rfloor\}$, the event that $A \cap T_i = \emptyset$ means that all the $n$ points $a_1, \ldots, a_n$ fall into $([0, 1] \times [0, 1]) \setminus T_i$. Since we have assumed $a_1, \ldots, a_n$ to be independently and uniformly distributed on $[0, 1] \times [0, 1]$, we obtain

$$\begin{aligned} P(A \cap T_i = \emptyset) &= (1 - \Lambda(T_i))^n = \left(1 - \frac{n\Lambda(T_i)}{n}\right) \\ &\leq e^{-n\Lambda(T_i)} = e^{-n\beta \ln[1/2\sqrt{\beta}]/\lfloor \ln n \rfloor}. \end{aligned}$$

Inserting $\beta = (2+\sigma)\ln n/n$ and observing $\ln[1/2\sqrt{\beta}] > 0$, we get

$$\begin{aligned} P(A \cap T_i = \emptyset) &\leq e^{-(2+\sigma)\cdot[-\ln 2 - \ln(2+\sigma)/2 - \ln(\ln n)/2 + (\ln n)/2]} \\ &= 2^{2+\sigma} \cdot (2+\sigma)^{1+\sigma/2} \cdot (\ln n)^{1+\sigma/2} \cdot n^{-(1+\sigma/2)} \\ &= [4(2+\sigma)]^{1+\sigma/2} \cdot (\ln n)^{1+\sigma/2} \cdot n^{-(1+\sigma/2)}, \end{aligned}$$

and hence

$$\begin{aligned} P(\text{conv}(S) \not\supseteq G_\alpha) &\leq 8\lfloor \ln n \rfloor \cdot P(A \cap T_i = \emptyset) \\ &\leq 8 \cdot [4(2+\sigma)]^{1+\sigma/2} \cdot (\ln n)^{2+\sigma/2} \cdot n^{-(1+\sigma/2)} \quad \square \end{aligned}$$

LEMMA 2.3. $E[|S|] \leq 4\sqrt{e}(2+\sigma)(\ln n)^2$.

PROOF. Since $|S|$ is binomially distributed with parameters $n$ and $p = \Lambda(F_\alpha)$, we have

$$\begin{aligned} E[|S|] &= n \cdot p = n\left[4\alpha + 4\int_{2\alpha}^{1/2} \frac{\alpha}{x}\,dx\right] = 4\alpha n\left[1 + \ln\frac{1}{2} - \ln 2\alpha\right] \\ &= 4\sqrt{e}(2+\sigma)\ln n\left[1 + \ln\frac{1}{2} - \ln\left[2\sqrt{e}\frac{(2+\sigma)\ln n}{n}\right]\right] \\ &= 4\sqrt{e}(2+\sigma)\ln n[1 + \ln n - \ln 2 - \ln(2\sqrt{e}) - \ln(2+\sigma) - \ln\ln n] \\ &\leq 4\sqrt{e}(2+\sigma)(\ln n)^2. \quad \square \end{aligned}$$

LEMMA 2.4

$$E[T_{\mathrm{CH}}] \le 4\sqrt{e}(2+\sigma)(\ln n)^3 + 8[4(2+\sigma)]^{1+\sigma/2} \cdot (\ln n)^{3+\sigma/2} \cdot n^{-\sigma/2}.$$

PROOF. Immediate by (1), Lemma 2.2, and Lemma 2.3. □

So the overall complexity of our algorithm is dominated by the selection of the points to be considered in step (2) of Phase 1. We have thus proved

THEOREM 2.5. *The expected running time of the two-phase algorithm is linear if the given points are generated independently from a uniform distribution on the unit square.* □

Obviously there can be no algorithm for the convex hull with sublinear expected time complexity. In this sense our results cannot be improved. However, we can strengthen our result in another sense. The number of points considered in Phase 1 is so small that the two-phase method will run in expected linear time even if algorithm CH takes worst case time more than $O(n \ln n)$.

THEOREM 2.6. *For any algorithm CH with polynomial worst case running time there is a version of the two-phase method of linear expected running time.*

PROOF. Let CH be of worst case time complexity $O(n^r)$ for some $r > 1$. As in our previous analysis we neglect the constant factor and assume that CH takes time at most $n^r$ for an $n$ point problem. For the parameter $\sigma > 0$, defining $\alpha = \alpha(n)$ by (2), we choose $\sigma > 2(r-2)$. Analogously to (1) we obtain

$$\begin{aligned} E[T_{\mathrm{CH}}] &\le \sum_{k=1}^{n} k^r P(|S| = k) + n^r P(\mathrm{conv}(S) \not\supseteq G_\alpha) \\ &= E[|S|^r] + n^r P(\mathrm{conv}(S) \not\supseteq G_\alpha). \end{aligned}$$

As will be shown in the Appendix (Lemma A.1), and in view of the fact that $E[|S|] \to \infty$ as $n \to \infty$, there is a constant $C_r$ (not depending on $n$) such that $E[|S|^r] \le C_r \cdot (E[|S|])^r$. By Lemmas 2.2 and 2.3 we obtain

$$\begin{aligned} E[T_{\mathrm{CH}}] &\le C_r[4\sqrt{e}(2+\sigma)(\ln n)^2]^r \\ &\quad + 8[4(2+\sigma)]^{1+\sigma/2}(\ln n)^{2+\sigma/2} \cdot n^{r-1-\sigma/2}, \end{aligned}$$

which is sublinear because $r - 1 - \sigma/2 < 1$. □

Note that this is a consequence of the very advantageous stochastic model used here. Under different stochastic assumptions such a statement would not hold (cf. §3). But, in contrast to usual methods for determining the convex hull, our method is able to exploit this property and achieves a dramatic reduction of computation time.

## 3. Analysis for the unit disk

Let $A = \{a_1, a_2, \ldots, a_n\}$ be $n$ random points generated independently from the uniform distribution on the unit disk $D(0, 1) = \{x = (x_1, x_2) \in$

$\mathbf{R}^2 | x_1^2 + x_2^2 \leq 1\}$. We shall argue along the same lines as in §2. This time we define $h: D(0, 1) \to [0, 1]$ by

$$h(x) = h(x_1, x_2) = \sqrt{1 - (x_1^2 + x_2^2)}$$

so that we have complete rotational symmetry with respect to (0,0). In addition to the representation of a point $x \in D(0, 1)$ as $x = (x_1, x_2)$ in the Cartesian coordinate system, we also use polar coordinates, $x = (r, \gamma)$, where $r = \sqrt{x_1^2 + x_2^2} \in [0, 1]$ and $\gamma \in [0, 2\pi]$ is the angle between $x$ and the $x_1$-axis.

We define $F_\alpha$, $F_\beta$, and $G_\alpha$ analogously to the previous section, where $0 < \beta < \alpha < 1$ will be given below, and as before, we will derive a sufficient condition for the success of Phase 1 (see Lemma 3.1).

First, we make sure that the origin (0,0) is an interior point of $\operatorname{conv}(S)$, otherwise we declare the failure of Phase 1. Then we check if all line segments of adjacent vertices of $\operatorname{conv}(S)$ computed in Phase 1 are contained in $F_\alpha$. If this is true then $\operatorname{conv}(S) = \operatorname{conv}(A)$ and we are done.

The number $\alpha$ that will ensure sublinear expected time spent in running algorithm CH is defined as follows. We choose $\sigma > 0$ and

$$\alpha = \alpha(n) = \sqrt{1 - \left[1 - \left(\frac{2\pi}{\lfloor n^{1/3} \rfloor}\right)^2\right] [1 - \beta^2]}, \tag{3}$$

where $\beta = \beta(n) = n^{-1/3}\sqrt{(\frac{1}{3} + \sigma)\ln n}$. Throughout this section we will assume that $n$ is large enough to satisfy $(\frac{1}{3}+\sigma)\ln n < n^{2/3}$ and $2\pi < \lfloor n^{1/3} \rfloor$, so that $\alpha$ and $\beta$ are well defined and $0 < \beta < \alpha < 1$. Obviously the set $F_\beta$ is an annulus around the origin (0,0) with inner radius $\sqrt{1 - \beta^2}$ and outer radius 1. We partition $F_\beta$ into $\lfloor n^{1/3} \rfloor$ equally sized cells $T_i$ by defining angles $t_0, t_1, t_2, \ldots, t_{\lfloor n^{1/3} \rfloor}$ via $t_i = i \cdot 2\pi / \lfloor n^{1/3} \rfloor$ and

$$T_i = \left\{(r, \gamma) | \sqrt{1 - \beta^2} \leq r \leq 1,\ t_{i-1} \leq \gamma \leq t_i\right\}$$

for $i = 1, 2, \ldots, \lfloor n^{1/3} \rfloor$. Then every cell has area

$$\Lambda(T_i) = \frac{1}{\lfloor n^{1/3} \rfloor} \cdot \Lambda(F_\beta) = \frac{1}{\lfloor n^{1/3} \rfloor} \pi \beta^2 = \frac{1}{\lfloor n^{1/3} \rfloor} \cdot \pi n^{-2/3} \left(\frac{1}{3} + \sigma\right) \ln n.$$

LEMMA 3.1. *If every* $T_i$, $i = 1, 2, \ldots, \lfloor n^{1/3} \rfloor$, *contains at least one point of* $A$, *then* $\operatorname{conv}(A) \supseteq G_\alpha$.

PROOF. We argue as in the previous section. Thus, we have to show that the line segment between any two points from adjacent cells is contained in $F_\alpha$. Because of rotational symmetry we can restrict to the cells

$T_1$ and $T_{\lfloor n^{1/3}\rfloor}$. Let $u = (u_1, u_2) \in T_1$, $v = (v_1, v_2) \in T_{\lfloor n^{1/3}\rfloor}$, and $x = (x_1, x_2) \in [u, v]$. Obviously both $u_1$ and $v_1$ are not less than $\sqrt{1-\beta^2}\cdot \cos(2\pi/\lfloor n^{1/3}\rfloor)$, and hence $x_1 \geq \sqrt{1-\beta^2}\cdot\cos(2\pi/\lfloor n^{1/3}\rfloor)$. Observing that

$$\cos\left(\frac{2\pi}{\lfloor n^{1/3}\rfloor}\right) = \sqrt{1-\sin^2\left(\frac{2\pi}{\lfloor n^{1/3}\rfloor}\right)} > \sqrt{1-\left(\frac{2\pi}{\lfloor n^{1/3}\rfloor}\right)^2},$$

we obtain

$$(h(x))^2 = 1 - x_1^2 - x_2^2 \leq 1 - x_1^2 \leq 1 - (1-\beta^2)\left(1-\left(\frac{2\pi}{\lfloor n^{1/3}\rfloor}\right)^2\right) = \alpha^2,$$

and hence $x \in F_\alpha$. □

LEMMA 3.2. $P(\operatorname{conv}(S) \not\supseteq G_\alpha) \leq n^{-\sigma}$.

PROOF. We obtain the following from Lemma 3.1 along the same lines as in the proof of Lemma 2.2:

$$\begin{aligned}
P(\operatorname{conv}(S) \not\supseteq G_\alpha) &\leq P(A \cap T_i = \emptyset \quad \text{for at least one } i \in \{1, 2, \ldots, \lfloor n^{1/3}\rfloor\}) \\
&\leq \sum_{i=1}^{\lfloor n^{1/3}\rfloor} P(A\cap T_i = \emptyset) = \sum_{i=1}^{\lfloor n^{1/3}\rfloor}\left(1-\frac{\Lambda(T_i)}{\pi}\right)^n \\
&= \lfloor n^{1/3}\rfloor \cdot \left(1 - \frac{(1/3+\sigma)\ln n}{\lfloor n^{1/3}\rfloor n^{2/3}}\right)^n \\
&\leq n^{1/3}\left(1-\frac{(1/3+\sigma)\ln n}{n}\right)^n \\
&\leq n^{1/3} e^{-(1/3+\sigma)\ln n} = n^{-\sigma}. \quad \square
\end{aligned}$$

LEMMA 3.3. $E[|S|] \leq n^{1/3}[(\frac{1}{3}+\sigma)\ln n + (2\pi+1)^2]$.

PROOF. Since $|S|$ is binomially distributed with parameters $n$ and $p = \Lambda(F_\alpha)/\pi = \alpha^2$, we obtain by (3)

$$\begin{aligned}
E[|S|] &= n \cdot p \\
&\leq n\left[1 - \left[1-\left(\frac{2\pi}{\lfloor n^{1/3}\rfloor}\right)^2\right][1-\beta^2]\right] \\
&\leq n\left[\beta^2 + \left(\frac{2\pi}{\lfloor n^{1/3}\rfloor}\right)^2\right] \\
&= n^{1/3}\left[\left(\frac{1}{3}+\sigma\right)\ln n + \left(\frac{n^{1/3}}{\lfloor n^{1/3}\rfloor}\right)^2 (2\pi)^2\right] \\
&\leq n^{1/3}\left[\left(\frac{1}{3}+\sigma\right)\ln n + (2\pi+1)^2\right]. \quad \square
\end{aligned}$$

By Lemmas 3.2 and 3.3, and (1) we can estimate the expected running time of algorithm CH, provided that its worst case running time is $O(n \ln n)$.

LEMMA 3.4. $E[T_{CH}] \leq n^{1/3} \ln n[(\frac{1}{3} + \sigma) \ln n + (2\pi + 1)^2] + n^{1-\sigma} \ln n$. □

So we have obtained the following result for the case of the unit disk.

THEOREM 3.5. *The expected running time of the two-phase algorithm is linear if the given points are generated independently from a uniform distribution on the unit disk.* □

Also in the case of the unit disk we can achieve linear expected running time for construction methods of higher complexity. But this time this possibility is restricted to the set of algorithms CH with worst case time complexity $O(n^r)$, where $r < 3$. This restriction is obvious, because the expected number of vertices of the convex hull is $O(n^{1/3})$.

THEOREM 3.6. *For any algorithm CH with worst case time complexity $O(n^r)$, where $r < 3$, there is a version of our two-phase algorithm with linear expected running time.*

PROOF. Suppose that CH is of worst case time complexity $O(n^r)$, where $1 \leq r < 3$. Without loss of generality we assume that CH needs less than $n^r$ time for solving problems with $n$ points.

Depending on $r$ we choose $\sigma > r - 1$. Then we have (as in 2.6)

$$\begin{aligned} E[T_{CH}] &\leq \sum_{k=1}^{n} k^r P(|S| = k) + n^r P(\operatorname{conv}(S) \not\supseteq G_\alpha) \\ &= E[|S|^r] + n^r P(\operatorname{conv}(S) \not\supseteq G_\alpha). \end{aligned}$$

Again we refer to Lemma A.1 of the Appendix, which states that there is a constant $\widetilde{C}_r$ such that $E[|S|^r] \leq \widetilde{C}_r E[|S|]^r$. Exploiting the results of Lemmas 3.2 and 3.3 we obtain

$$E[T_{CH}] \leq \widetilde{C}_r[(\tfrac{1}{3} + \sigma) \ln n + (2\pi + 1)^2]^r n^{r/3} + n^{r-\sigma}.$$

This expression is sublinear in $n$ because $r < 3$, $r - \sigma < 1$, and $(\ln n)^r$ grows slower than $n^{1-r/3}$. □

## 4. Computational experiments

For the theoretical analysis of our method it was sufficient to assume that in case of the failure of Phase 1 the $O(n \ln n)$ worst case time algorithm is applied to the complete set of points. In practice, we can do better. If Phase 1 fails, we compute the minimum $\gamma$ such that the convex hull computed in Phase 1 includes $G_\gamma$. Clearly, it suffices to restrict the computation in Phase 2 to the set $S_\gamma$ of those points contained in $F_\gamma$. In fact, determining the success or failure of Phase 1 reduces to checking if $\gamma > \alpha$. In our computational experiments for the unit square we chose $\alpha = 4 \ln n / n$.

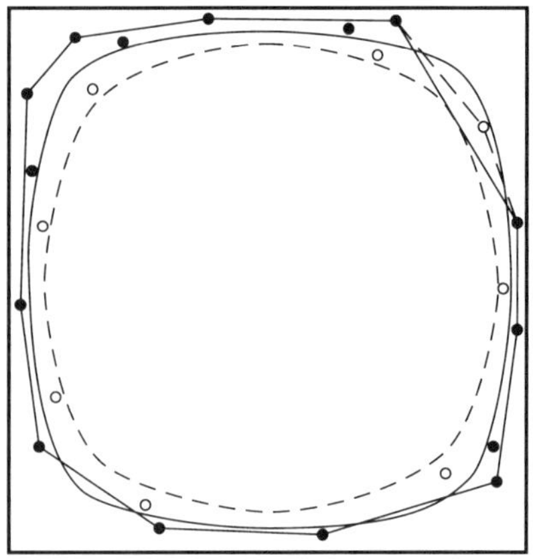

FIGURE 3. Illustration of the algorithm

Summarizing, the implemented version of the algorithm can be stated as follows:

PHASE 1

(1) Set $\alpha = \alpha(n) = 4 \ln n / n$.
(2) Let $S_\alpha = \{a_i \in A | h(a_i) \leq \alpha\}$.
(3) Apply algorithm CH to the subset $S_\alpha$.
(4) Compute the minimum $\gamma$ such that $\operatorname{conv}(S_\alpha) \supseteq \{x \in [0, 1] \times [0, 1] | h(x) > \gamma\}$.
(5) If $\alpha \geq \gamma$ then stop: $\operatorname{conv}(S_\alpha) = \operatorname{conv}(A)$. Otherwise, go to Phase 2.

PHASE 2

(6) Apply algorithm CH to the subset $S_\gamma = \{a_i \in A | h(a_i) \leq \gamma\}$.

Figure 3 gives an illustration of the algorithm. The solid curve bounds the set $F_\alpha$. The set $S_\alpha$ is then given by the solid points (•) and its convex hull by solid lines. The broken curve defines the set $S_\gamma$ and the additional points to be considered in Phase 2 are shown as small circles (○). The extreme points of the correct convex hull resulting from Phase 2 are obtained by adding one point in the north-east corner. Our choice of $\alpha$ ensures that with high probability Phase 2 is not needed.

In this section we compare the practical behavior of various linear expected time algorithms. The computational study consists of five such algorithms which we run on test problems with independently and uniformly distributed points in the unit square, namely

[1] the divide and conquer algorithm of Bentley and Shamos [**3**],
[2] the maximal vectors approach of Devroye [**4**],
[3] the throw-away principle of Devroye and Toussaint [**5**],
[4] the throw-away principle of Akl and Toussaint [**1**], and
[5] the algorithm described above.

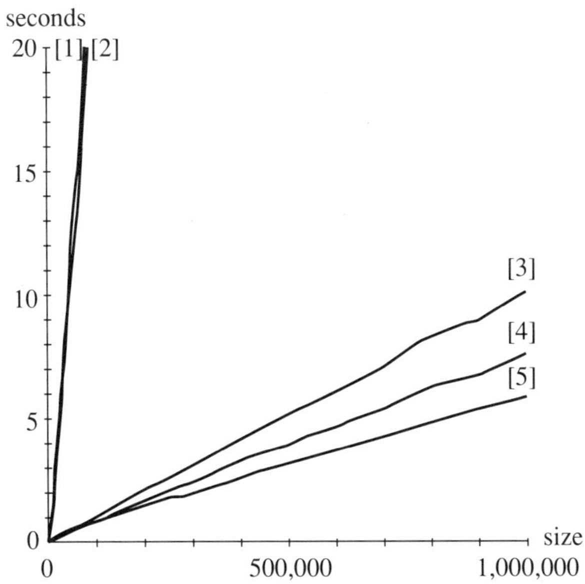


FIGURE 4. Comparison of convex hull algorithms

For algorithms [2], [3], [4], and [5] we apply Graham's scan (Graham **[8]**) to the selected points. All five algorithms have been implemented as Pascal programs on a SUN SPARCstation 1. Of course, the performance is influenced by our programming skills. We have tried to put the same efforts into all five programs to make the comparison as fair as possible. For instance, in all throw-away type algorithms we found that the following trick helped to reduce the computation time. As soon as the elimination area has been determined (a closed polygon in case [3] and [4] and the curve defined by the function $h$ in [5]) we inscribe the biggest possible rectangle with vertical and horizontal sides into this area. Assume that this rectangle (always a square in case [5]) has vertices $(x_1, y_1)$, $(x_2, y_1)$, $(x_2, y_2)$, and $(x_1, y_2)$. The elimination criterion is satisfied for a point with coordinates $(\overline{x}, \overline{y})$ if $x_1 \leq \overline{x} \leq x_2$ and $y_1 \leq \overline{y} \leq y_2$, which takes only four comparisons to check. Only if this criterion fails we have to check the actual elimination criterion which, e.g. in case [5], amounts to checking which quadrant $(\overline{x}, \overline{y})$ lies in, and depending on this, up to two additions, one multiplication, and one comparison.

Figure 4 shows the results. The curves are based on ten sample problems for each problem size $n = 1\,000, 2\,000, \ldots, 10\,000, 20\,000, \ldots, 100\,000, 200\,000, \ldots, 1\,000\,000$. In our opinion, these curves should be interpreted as follows. When doing practical computation the throw-away principle is superior when compared to the divide and conquer algorithms. The four point method suggested in Akl and Toussaint **[1]** is slightly better than the eight point method of Devroye and Toussaint **[5]**.

It would be interesting to check if similar computational results hold in real world applications as well. The fact that the (improbable) failure of Phase 1 of [5] increases the computation time substantially might sometimes not be acceptable.

## 5. Conclusion

We have shown for two special stochastic models how $O(n \ln n)$ worst case time algorithms, and even algorithms of higher worst case time complexity, can be converted into linear average time algorithms. However, it is clear that these methods can also be applied to sets more general than just unit cubes and unit balls.

It follows from the results of Raynaud [**13**] that the expected number of vertices of the convex hull of $n$ uniformly and independently distributed points in the unit disk is $O(n^{1/3})$. For this reason it is clear that no algorithm of more than cubic worst case time complexity can give us the linear expected time result. For the unit square it follows from the results from Rényi and Sulanke [**12**] that $O(\ln n)$ points can be expected to form the vertices of the convex hull. The estimate we have given for the expected number of points to be considered in Phase 1 of our algorithm comes close to these values up to a factor of $\ln n$, both for the unit square and the unit disk. Thus there seems to be some room for improvement; note, however, that such improvement would not affect the overall time complexity.

Finally, we would like to remark that we can generalize our approach to higher dimensions. Note that the computation of the convex hull (as performed by algorithm CH) consists of a description of its facial structure, e.g. in the form of the face lattice. In the two-dimensional case discussed in this paper, the number of facets is equal to the number of vertices, and our analysis shows that both vertices and facets are readily available. In the generalization of our method to higher dimensions, we need the availability of the complete face lattice in order to check the success or failure of Phase 1 algorithmically. The analysis of the expected running time depends not only on the number of points selected in Phase 1 but also on the fact that we are able to check the success or failure of Phase 1 efficiently.

We have carried out the analysis in the three-dimensional case. As in two dimensions, the worst case time complexity for computing the convex hull of $n$ points is $O(n \ln n)$ (cf. Preparata and Hong [**11**]). For the unit cube we have generalized our algorithm to eliminate all but $O((\ln n)^3)$ points. The number of vertices, edges, and facets of the convex hull constructed in Phase 1 have the same order of magnitude. This enables us to obtain linear expected time performance. Similarly, we can handle the unit ball with the difference that in this case $O(n^{2/3} \ln n)$ points are left after the elimination step. Also here the number of vertices, edges, and facets is sufficiently small to obtain linear average case performance again.

For dimensions $d \geq 4$ the best known convex hull algorithms work in worst case time $O(n \ln n + n^{\lfloor (d+1)/2 \rfloor})$ and the worst case number of facets is $\Omega(n^{\lfloor d/2 \rfloor})$ (cf. Edelsbrunner **[6]**). The fact that the worst case running time is polynomial in $n$ and that the expected number of facets of the convex hull of $n$ points in the $d$-dimensional hypercube is $O((\ln n)^{d-1})$ (cf. Rényi and Sulanke **[12]**) makes us confident of obtaining linear expected time behavior by a suitable generalization of our approach. On the other hand, the expected number of facets in the convex hull for $n$ points in the $d$-dimensional ball is $O(n^{(d-1)/(d+1)})$ (cf. Raynaud **[13]**). This makes it impossible to prove expected linearity by our reasoning even for $d = 4$.

ACKNOWLEDGMENT. The authors would like to thank Stefan Thienel for careful implementation of the algorithms and Rudolf Mathar for careful reading of an earlier version of this paper.

## Appendix

For $n \in \mathbf{N}$ and $p \in (0, 1)$ let $X_{n,p}$ be a random variable which is binomially distributed with parameters $n$ and $p$, i.e., $P(X_{n,p} = k) = \binom{n}{k} p^k (1-p)^{n-k}$, $k = 0, 1, \dots, n$. For $r \geq 1$ (not necessarily an integer) we consider the expectation of the $r$th power of $X_{n,p}$, i.e.,

$$E[X_{n,p}^r] = \sum_{k=0}^{n} k^r \binom{n}{k} p^k (1-p)^{n-k}$$

(here we define $0^r = 0$).

LEMMA A.1. *There exists a constant $C_r$ (depending only on $r$) such that for all $n \in \mathbf{N}$ and $p \in (0, 1)$*

$$E[X_{n,p}^r] \leq C_r \cdot [\max\{np, 1\}]^r.$$

PROOF. We consider first the case that $r$ is a positive integer and use induction on $r$. If $r = 1$, then $E[X_{n,p}] = np$ and we take $C_1 = 1$. Assume that for some $r \in \mathbf{N}$ there is a constant $C_r$ satisfying the inequalities from the assertion. Then, if $n \geq 2$,

$$\begin{aligned} E[X_{n,p}^{r+1}] &= \sum_{k=1}^{n} k^{r+1} \binom{n}{k} p^k (1-p)^{n-k} \\ &= np \sum_{k=1}^{n} k^r \binom{n-1}{k-1} p^{k-1} (1-p)^{n-1-(k-1)} \\ &= np \sum_{k=0}^{n-1} (k+1)^r \binom{n-1}{k} p^k (1-p)^{n-1-k} \\ &\leq np \left[ (1-p)^{n-1} + 2^r \sum_{k=1}^{n-1} k^r \binom{n-1}{k} p^k (1-p)^{n-1-k} \right] \\ &\leq np[1 + 2^r E[X_{n-1,p}^r]] \\ &\leq np[1 + 2^r C_r (\max\{(n-1)p, 1\})^r] \\ &\leq (2^r C_r + 1)[\max\{np, 1\}]^{r+1}. \end{aligned}$$

If $n = 1$, then the inequality

$$E[X_{n,p}^{r+1}] \leq (2^r C_r + 1)[\max\{np, 1\}]^{r+1}$$

is trivially true since $E[X_{1,p}^{r+1}] = p$. So we can take $C_{r+1} = 2^r C_r + 1$.

Now let $s - 1 < r < s$ for an integer $s \geq 2$. Then $X_{n,p}^r = Z_{n,p}^{r/s}$, where $Z_{n,p} = X_{n,p}^s$. By the Jensen inequality

$$\begin{aligned} E[X_{n,p}^r] = E[Z_{n,p}^{r/s}] &\leq (E[Z_{n,p}])^{r/s} = (E[X_{n,p}^s])^{r/s} \\ &\leq (C_s \cdot [\max\{np, 1\}]^s)^{r/s} = C_s^{r/s} \cdot [\max\{np, 1\}]^r. \end{aligned}$$

So we can take $C_r = C_s^{r/s}$. □

## REFERENCES

1. G. Akl and G. T. Toussaint, *A fast convex hull algorithm*, Inform. Process. Lett. **7** (1978), 219–222.
2. J. L. Bentley, H. T. Kung, M. Schkolnick, and C. D. Thompson, *On the average number of maxima in a set of vectors and applications*, J. Assoc. Comput. Mach. **25** (1978), 536–543.
3. J. L. Bentley and M. I. Shamos, *Divide and conquer for linear expected time*, Inform. Process. Lett. **7** (1978), 87–91.
4. L. Devroye, *A note on finding convex hulls via maximal vectors*, Inform. Process. Lett. **11** (1980), 53–56.
5. L. Devroye and G. T. Toussaint, *A note on linear expected time algorithms for finding convex hulls*, Computing **26** (1981), 361–366.
6. H. Edelsbrunner, *Algorithms in combinatorial geometry*, Springer-Verlag, Berlin, 1987.
7. M. Golin and R. Sedgewick, *Analysis of a simple yet efficient convex hull algorithm*, Proc. 4th Annual Sympos. on Computational Geometry, Urbana, Illinois, 1988, pp. 153–163.
8. R. L. Graham, *An efficient algorithm for determining the convex hull of a finite planar set*, Inform. Process. Lett. **1** (1972), 132–133.
9. D. G. Kirkpatrick and R. Seidel, *The ultimate planar convex hull algorithm*, SIAM J. Comput. **15** (1986), 287–299.

10. D. T. Lee and F. P. Preparata, *Computational geometry—A survey*, IEEE Trans. Comput. **C-33** (1984), 1072–1101.
11. F. P. Preparata and S. J. Hong, *Convex hulls of finite sets of points in two and three dimensions*, Comm. ACM **20** (1977), 87–93.
12. A. Rényi and R. Sulanke, *Über die konvexe Hülle von $n$ zufällig gewählten Punkten*, Z. Wahrsch. Verw. Gebiete **2** (1963), 75–84.
13. H. Raynaud, *Sur l'enveloppe convexe des nuages de points aléatoires dans $R^n$*, J. Appl. Probab. **7** (1970), 35–48.

Institut für Mathematik, Universität Augsburg, Universitätsstrasse 8, D-8900 Augsburg, West Germany

*E-mail address*: reinelt%augsopt.Uni-Augsburg.De or
reinelt%augsopt6.Uni-Augsburg.DE @relay.cs.net

DIMACS Series in Discrete Mathematics
and Theoretical Computer Science
Volume **4**, 1991

Applied Geometry
and Discrete Mathematics
THE VICTOR KLEE FESTSCHRIFT

# Measures of $F$-Stars in Finitely Starlike Sets

MARILYN BREEN

ABSTRACT. The notion of visibility plays an important role in computational geometry and modeling, especially in problems which involve pattern recognition, hidden line elimination, image processing, and control of robots. (See [2, 3, **18**, **22**].) The purpose of this short paper is to provide an overview of some recent visibility results for finitely starlike sets, both in the plane and in higher dimensions. The results are of Helly type, disclosing information about a set from the behavior of small subsets. In particular, with suitable restrictions on a set $S$, the results make it possible to determine whether $S$ is finitely starlike and, if it is, to determine the Lebesgue measure of certain visibility sets (called $F$-stars) in $S$. Generalizations to unions of relatively starlike sets are considered as well.

## 1. Introduction

We begin with some familiar definitions and results. Let $S$ be a set in $R^d$. For points $x$ and $y$ in $S$, we say $x$ *sees* $y$ via $S$ ($x$ is *visible* from $y$ via $S$) if and only if the corresponding segment $[x, y]$ lies in $S$. Subset $A$ of $S$ is said to be *finitely starlike relative to* $S$ or *relatively starlike in* $S$ if and only if every finite subset $F$ of $A$ sees via $S$ a common point. The intersection of the visibility sets of the points in $F$ is called the $F$-*star* of $S$ or the *visibility set* of $F$ in $S$. Set $S$ is called *finitely starlike* if and only if $S$ itself is finitely starlike relative to $S$. Finally, set $S$ is *starshaped* if and only if there is some point $p$ in $S$ such that $p$ sees via $S$ each point of $S$, and the set of all such points $p$ is called the (convex) *kernel* of $S$.

It is clear that every starshaped set will be finitely starlike. However, there are finitely starlike sets which fail to be starshaped. (Of course, such sets must be noncompact.) The following examples illustrate this.

EXAMPLE 1. Let $S \equiv \{(x, y): y \leq \sin x\}$ in $R^2$. Set $S$ is closed, unbounded, finitely starlike, but not starshaped.

EXAMPLE 2. Let $D$ denote the closed unit disk centered at the origin

1980 *Mathematics Subject Classification* (1985 *Revision*). Primary 52A30, 52A35; Secondary 52A20.

Supported in part by NSF grant DMS-8705336.

in $R^2$, and let $T = D \sim \{(0, 0)\}$. Set $T$ is bounded, nonclosed, finitely starlike, but not starshaped.

A well-known theorem of Krasnosel′skiĭ [**13**] provides the following characterization for compact starshaped sets in $R^d$ : If $S$ is a nonempty compact set in $R^d$, then $S$ is starshaped if and only if every $d+1$ points of $S$ see via $S$ a common point. However, as Examples 1 and 2 above illustrate, Krasnosel′skiĭ's theorem fails for noncompact sets, and the problem of obtaining a Krasnosel′skiĭ-type (Helly-type) theorem to characterize arbitrary starshaped sets is a challenging one. This, in turn, leads to the idea of finitely starlike sets, introduced by Bruce Peterson [**19**]. While Krasnosel′skiĭ's theorem fails for noncompact sets, many of the sets for which it fails satisfy the weaker requirement of being finitely starlike. (Again consider Examples 1 and 2.) Peterson has conjectured that a Krasnosel′skiĭ-type theorem exists to characterize finitely starlike sets. Moreover, since there are Krasnosel′skiĭ-type theorems to predict the dimension of the kernel of a starshaped set, it is natural to look for analogues of these results which give the dimension of the $F$-stars in finitely starlike sets.

This paper will attempt to discuss some recent results concerning finitely starlike sets and the measures of their $F$-stars. §2 will examine the planar case, §3 will look at generalizations in $R^d$, while §4 will present a special argument in $R^3$. Finally, some results on unions of relatively starlike sets will be considered in §5. Since proofs will not be given in detail, appropriate references will be provided when necessary. In addition, the reader is referred to Valentine [**23**] and Lay [**15**] for a discussion of geometric topics, to Danzer, Grünbaum, and Klee [**9**] for a discussion of Helly-type and Krasnosel′skiĭ-type theorems, to Taylor [**21**] for information on the Lebesgue measure, to Nadler [**17**] for information on the Hausdorff metric, and to Alexandroff and Hopf [**1**] and Massey [**16**] for information on homology theory.

Throughout the paper, $\operatorname{conv} S$, $\operatorname{cl} S$, $\operatorname{int} S$, and $\ker S$ will be used to denote the convex hull, closure, interior, and kernel, respectively, of set $S$. For $S$ convex, $\dim S$ will be the dimension of $S$.

## 2. Finitely starlike sets in the plane

We begin with two preliminary lemmas from [**5**]. The first of these concerns some nice properties of the visibility sets in $S$ when $S$ is simply connected, while the second is a Helly-type result for these visibility sets.

LEMMA 1. *Let $S$ be a simply connected set in the plane, and let $\mathscr{V}$ be a family of visibility sets in $S$ of the form $S_x = \{y : [x, y] \subseteq S\}$. For $1 \le k$, every $k$ members of $\mathscr{V}$ have an intersection which (if nonempty) is simply connected and 2-polygonally connected.*

Note that the intersections referred to in Lemma 1 are exactly the visibility sets determined by finite subsets of $S$.

LEMMA 2. *Let $\mathscr{V}$ be a finite family of sets in the plane. Assume that every subfamily of $\mathscr{V}$ has an intersection which (if nonempty) is simply connected and 2-polygonally connected. If every three members of $\mathscr{V}$ have a nonempty intersection, then $\bigcap\{V: V \text{ in } \mathscr{V}\} \neq \emptyset$. The number three is best possible.*

The main portion of this section will present some Krasnosel′skiĭ-type theorems from [**8** and **6**] which allow us to determine whether a certain planar set $S$ is finitely starlike. It is interesting to observe that each of these results involves some kind of restriction on the set of points in $(\operatorname{int} \operatorname{cl} S) \sim S$. (Intuitively, such points "should be" interior to $S$ but have been removed.) In fact, it turns out that these troublesome points play an important role and, in a way, allow or prohibit the finite starlikeness of set $S$.

THEOREM 1. *Let $S \subseteq R^2$ and $(\operatorname{int} \operatorname{cl} S) \sim S = \emptyset$. Then $S$ is finitely starlike if and only if every four points of $S$ sees via $S$ a common point. The number four is best possible.*

COMMENTS ON THE PROOF. The basic idea in Theorem 1 is to show that sets $\operatorname{cl} S$ and $S$ are simply connected, then use Lemmas 1 and 2 above to conclude that every finite family of $S_x$ sets has a nonempty intersection. For details, see [**8**].

It is perhaps surprising that the number four in Theorem 1 is best possible. However, this number is needed to guarantee that the set $\operatorname{cl} S$ be simply connected, as the following example from [**8**] reveals.

EXAMPLE 3. Let $A$ be a square polygon in $R^2$, and let $S = R^2 \sim A$. Then every three points of $S$ see via $S$ a common point, yet an appropriate four points of $S$, chosen close to the midpoints of distinct edges of $A$, cannot see a common point.

A second example from [**8**] illustrates that Theorem 1 fails without the requirement that $(\operatorname{int} \operatorname{cl} S) \sim S$ be empty.

EXAMPLE 4. Let $T$ be the compact set bounded by the polygonal curve in Figure 1, with $p \in \operatorname{int} T$. Define set $S = T \sim \{p\}$, so that $p \in (\operatorname{int} \operatorname{cl} S) \sim S$. Then every four points of $S$ see via $S$ a common point, while the points $x_i$, $1 \leq i \leq 5$, do not.

The importance of the set $(\operatorname{int} \operatorname{cl} S) \sim S$ becomes more apparent in the following example [**8**]. Here $(\operatorname{int} \operatorname{cl} S) \sim S$ is a subset of a line, revealing a relationship between the one-dimensional Lebesgue measure of $(\operatorname{int} \operatorname{cl} S) \sim S$ and the finite starlikeness of $S$.

EXAMPLE 5. Let $A$ denote a square polygon in $R^2$ having its center on line $L$, the $x$ axis, and let $L_1$ and $L_2$ denote distinct open half-planes determined by $L$. Define set $S = A \sim \{(q, 0): q \text{ irrational})\}$. (See Figure 2). It is not too hard to show that there are three points in $L_1 \cap S$ which see via $S$ no point of $L_2 \cap S$. Likewise, there are three points in $L_2 \cap S$ which see no point of $L_1 \cap S$. Choosing these six points, together with any two points on $L \cap S$, we obtain eight points which see via $S$ no common point. Hence $S$ is not finitely starlike.

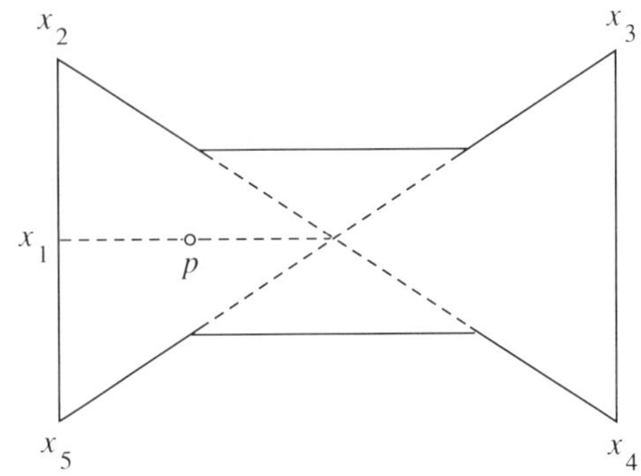


FIGURE 1

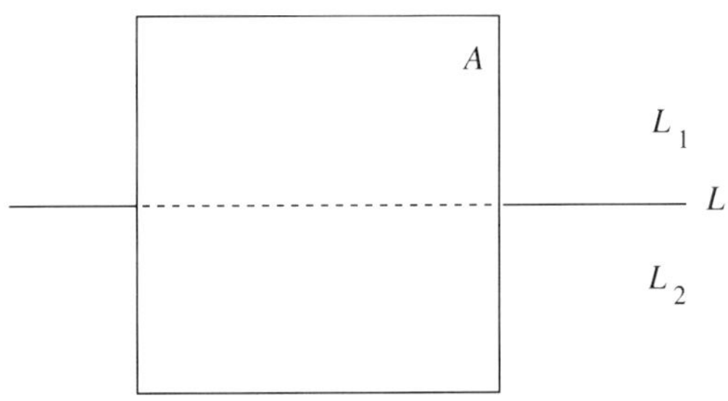


FIGURE 2

However, if we define $S' = A \sim \{(q, 0): q \text{ rational}\}$, then set $S'$ will be finitely starlike. This suggests the following theorem [**6**].

THEOREM 2. *Let $S$ be a set in $R^2$. Assume there is a countable collection of lines $\{L_i: 1 \leq i\}$ with $(\text{int cl}\, S) \sim S \subseteq \bigcup\{L_i: 1 \leq i\}$ and such that $((\text{int cl}\, S) \sim S) \cap L_i$ has one-dimensional Lebesgue measure zero for each $i$. Then the following assertions hold:*

(1) *Every four point subset of $S$ sees via $S$ a set of positive two-dimensional Lebesgue measure if and only if every $F$-star in $S$ contains such a set.*
(2) *Every four point subset of $S$ sees via $S$ a (linear) set of positive one-dimensional Lebesgue measure if and only if every $F$-star in $S$ contains such a set.*
(3) *Every five points of $S$ see via $S$ a common point if and only if $S$ is finitely starlike.*

*In each case, the number four or five is best possible.*

COMMENTS ON THE PROOF. In Theorem 2, the main idea is to "fill in the holes" of set $S$, obtain a proof for the resulting simply connected set $T$, then return to set $S$ without damaging the measures of the required $F$-stars. To accomplish this, define set $T \equiv S \cup ((\text{int cl}\, S) \sim S)$. For $x \in S$, let $T_x$ be the visibility set $\{y: [x, y] \subseteq T\}$ and let $S_x$ be the visibility set $\{y: [x, y] \subseteq S\}$.

By Lemmas 1 and 2, for $F$ any finite subset of $S$, the associated intersection $\bigcap\{T_x : x \in F\}$ must be simply connected, 2-polygonally connected, and nonempty. Let $A$ denote a $k$-dimensional ball in $\bigcap\{T_x : x \in F\}$, where $k$ is maximal, $0 \leq k \leq 2$.

To establish assertion (1), it is necessary to show that the set $A$ is fully two-dimensional. Moreover, for each $x$ in $F$, $A \sim S_x$ will have two-dimensional Lebesgue measure zero, and hence set $A \cap (\bigcap\{S_x : x \text{ in } F\})$ and set $A$ will have the same two-dimensional Lebesgue measure, finishing the proof.

For assertions (2) and (3), it is necessary to examine cases determined by the dimension of set $A$.The details appear in [**6**, **8**].

Notice that the numbers four and five in Theorem 2 are best possible: Example 3 shows that the number four is best in both (1) and (2), while Example 4 indicates that the number five is best in (3). Furthermore, Theorem 2 fails without some restriction on $(\operatorname{int} \operatorname{cl} S) \sim S$, as Example 5 illustrates.

Finally, it is interesting to note that Peterson's original conjecture is still unsolved. That is, no one has either established or disproved the existence of a Krasnosel′skiĭ number to characterize arbitrary finitely starlike sets, even in the plane. An example in [**8**] shows that if such a number exists, it must be at least nine, but no larger bound has been determined.

## 3. Bounded, finitely starlike sets in $R^d$

This section will be concerned with a generalization of Theorem 2 to higher dimensions [**7**]. The family of lines $\{L_i\}$ will be replaced by a family of flats $\{H_i\}$, and corresponding measures will be adapted appropriately. The following quantitative Helly-type lemma will be needed, also.

LEMMA 1. *For each $k$ and $d$, $1 \leq k \leq d$, define $f(d, d) = d + 1$ and $f(d, k) = 2d$ if $1 \leq k \leq d - 1$. Let $\mathscr{M}$ be a uniformly bounded collection of compact convex sets in $R^d$. Then for a fixed $k$ with $1 \leq k \leq d$, $\dim \bigcap\{M : M \text{ in } \mathscr{M}\} \geq k$ if and only if for some $\alpha > 0$, every $f(d, k)$ members of $\mathscr{M}$ contain a common $k$-dimensional set of measure (volume) at least $\alpha$.*

COMMENTS ON THE PROOF. The case for $k = d$ follows from a result by Polikanova and Perel′man [**20**], while the proof for arbitrary $k$, $1 \leq k \leq d$, involves induction on the dimension $d$ of our space. Letting $\mathscr{N}$ denote the closure of $\mathscr{M}$ relative to the Hausdorff metric, it is not hard to show that our hypothesis applies to $\mathscr{N}$ as well as to $\mathscr{M}$. Moreover, by Helly's familiar theorem, $\bigcap\{N : N \text{ in } \mathscr{N}\} \neq \varnothing$. If $\bigcap\{N : N \text{ in } \mathscr{N}\}$ is fully $d$-dimensional, the proof is finished. Otherwise, using results by Falconer [**11**] and Katchalski [**12**], for some $r$ sets $N_1, \ldots, N_r$ in $\mathscr{N}$, $\dim \bigcap\{N_i : 1 \leq i \leq r\} \equiv q \geq k$, where $r \leq 2(d - q)$. The argument is completed by applying the induction hypothesis in the $q$-flat determined by $\bigcap\{N_i : 1 \leq i \leq r\}$. Details are given in [**7**].

Using Lemma 1, we obtain the following generalization of Theorem 2 [**7**].

THEOREM 3. *For each* $k$ *and* $d$, $1 \leq k \leq d$, *define* $f(d, d) = d + 1$ *and* $f(d, k) = 2d$ *if* $1 \leq k \leq d - 1$. *Let* $S$ *be a bounded subset of* $R^d$. *Assume that for some fixed* $k$, $1 \leq k \leq d$, *there exists a countable family of* $(k-1)$-*flats* $\{H_i : i \geq 1\}$ *in* $R^d$ *such that* $\mathrm{cl}\, S \sim S \subseteq \bigcup\{H_i : i \geq 1\}$ *and such that* $(\mathrm{cl}\, S \sim S) \cap H_i$ *has* $(k-1)$-*dimensional measure zero for each* $i$. *If for some fixed* $\alpha > 0$, *every* $f(d, k)$ *points of* $S$ *see via* $S$ *a set of* $k$-*dimensional measure at least* $\alpha$, *then every finite subset of* $S$ *sees via* $S$ *a set of positive* $k$-*dimensional measure.*

COMMENTS ON THE PROOF. Again the idea is to obtain a proof for $\mathrm{cl}\, S$, then restrict it to $S$, preserving the measures of the $F$-stars involved. For convenience, for $x$ in $S$ and $z$ in $\mathrm{cl}\, S$, define visibility sets $S_x \equiv \{y : [x, y] \subseteq S\}$ and $T_z \equiv \{y : [z, y] \subseteq \mathrm{cl}\, S\}$. The first step is to show that every $f(d, k)$ members of $\{\mathrm{conv}\, T_z : z$ in $\mathrm{cl}\, S\}$ contain a common $k$-dimensional set of measure (volume) at least $\alpha$. Then, by the lemma, $\dim \bigcap\{\mathrm{conv}\, T_z : z$ in $\mathrm{cl}\, S\} \geq k$. Moreover, by the proof of Krasnosel′skiĭ's theorem [**9**, pp. 111–112], $\bigcap\{\mathrm{conv}\, T_z : z$ in $\mathrm{cl}\, S\}$ is exactly $\ker \mathrm{cl}\, S$.

Thus for $F$ any finite subset of $S$, we may select a corresponding $k$-dimensional ball $B$ in $\ker \mathrm{cl}\, S$ and disjoint from $F$. For each $x$ in $F$, the subset of $B$ which $x$ fails to see via $S$ has $k$-dimensional measure zero. Hence set $B \cap (\bigcap\{S_x : x$ in $F\})$ and set $B$ have the same $k$-dimensional measure, and the proof is complete. See [**7**] for the details and for a brief discussion of best bounds.

## 4. Open, finitely starlike sets in $R^3$

Observe that in each of the previous theorems, we have employed a Helly-type result to establish the proof. Here we make use of another Helly result, the topological version of Helly's theorem from Debrunner [**10**], to obtain a theorem for certain open, finitely starlike sets in $R^3$. Because this version of Helly's theorem requires neither convex sets nor compact ones, it is a natural choice in finitely starlike problems [**4**].

THEOREM (HELLY, VIA DEBRUNNER). *Let* $\{X_j : j$ *in* $J\}$ *be a finite family of open subsets of* $R^d$ *such that the intersection* $X_{j1} \cap \cdots \cap X_{jr}$ *of each* $r$ *sets of this family is nonempty for* $r \leq d + 1$ *and is even a homology cell for* $r \leq d$. *Then* $\bigcap\{X_j : j$ *in* $J\}$ *is a homology cell (and is therefore nonempty).*

THEOREM 4. *Let* $S$ *be an open subset of* $R^3$ *having the property that for every triangle* $T$ *with edges in* $S$, $\mathrm{conv}\, T \subseteq S$. *Then* $S$ *is finitely starlike if and only if every seven points of* $S$ *see via* $S$ *a common point.*

COMMENTS ON THE PROOF. Again for $x$ in $S$ define $S_x = \{y : [x, y] \subseteq S\}$. Certainly each $S_x$ set must be open. Also, every four of the $S_x$ sets must have a nonempty intersection. The heart of the proof involves showing that for $1 \leq r \leq 3$, every collection of $r$ $S_x$ sets has as its intersection a homology cell. Then by the topological version of Helly's theorem from Debrunner [**10**], every finite collection of the $S_x$ sets has as its intersection a homology cell, also. Using this, it is easy to show that the set $S$ is finitely starlike. The details are given in [**4**].

Observe that since set $S$ is open in Theorem 4, the associated $F$-stars must be fully three-dimensional there.

## 5. Generalizations to unions of finitely starlike sets

In conclusion, some of these Helly-type results, together with a decomposition theorem by Lawrence, Hare, and Kenelly [**14**], may be used to generalize theorems about finitely starlike sets to theorems about unions of relatively starlike sets. This is illustrated by our three final results, one in the plane, one in 3-space, and one in $R^d$.

THEOREM 5. *Let $S$ be a simply connected set in $R^2$, and let $k$ be a fixed natural number. Suppose for every finite subset $F$ of $S$ there is a $k$-partition $F_1, \dots, F_k$ of $F$ such that every three points of $F_i$ see via $S$ a common point, $1 \le i \le k$. Then there is a $k$-partition $S_1, \dots, S_k$ of $S$ such that $S_i$ is finitely starlike relative to $S$, $1 \le i \le k$.*

COMMENTS ON THE PROOF. Using a theorem by Lawrence, Hare, and Kenelly [**14**], obtain a partition $\{S_1, \dots, S_k\}$ of $S$ such that every three points of $S_i$ see via $S$ a common point. For $i$ fixed, consider the collection $\mathscr{V}_i$ of visibility sets $S_x$ in $S$ determined by members $x$ of $S_i$. By Lemmas 1 and 2, $\mathscr{V}_i$ has the finite intersection property, and so $S_i$ is finitely starlike relative to $S$. See [**4**] for the details.

Observe that if in addition set $S$ above is compact, then $S$ will be a union of $k$ starshaped sets.

THEOREM 6. *Let $S$ be an open subset of $R^3$ having the property that for every triangle $T$ with edges in $S$, $\operatorname{conv} T \subseteq S$. Let $k$ be a fixed natural number. Suppose that for every finite subset $F$ of $S$ there exists a $k$-partition $F_1, \dots, F_k$ of $F$ such that every seven points of $F_i$ see via $S$ a common point, $1 \le i \le k$. Then there is a $k$-partition $S_1, \dots, S_k$ of $S$ such that each set $S_i$ is finitely starlike relative to $S$.*

COMMENTS ON THE PROOF. The argument, which is similar to the one in Theorem 5, uses the Helly results for visibility sets from Theorem 4. For details see [**4**].

Finally, we close with a generic result on unions of relatively starlike sets in $R^d$ which holds when the $F$-stars are sufficiently well behaved. It is an easy consequence of [**14, 10**].

THEOREM 7. *Let $k$ be a fixed natural number, and let $S$ be an open set in $R^d$. Assume that for every finite subset $F$ of $S$ there is a $k$-partition $F_1, \dots, F_k$ of $F$ such that the following properties hold:*

(1) *For every* $d$ *or fewer members of* $F_i$, *the corresponding visibility set in* $S$ *is a homology cell.*

(2) *For every* $d+1$ *members of* $F_i$, *the corresponding visibility set in* $S$ *is nonempty.*

*Then there is a* $k$-*partition* $S_1, \ldots, S_k$ *of* $S$ *such that each* $S_i$ *is finitely starlike relative to* $S$.

An interesting problem suggested by Theorem 7 is that of finding suitable conditions to impose on set $S$ which will guarantee that properties (1) and (2) hold, so that a more satisfying result can be obtained.

## REFERENCES

1. Alexandroff and H. Hopf, *Topologie*, Springer, Berlin, 1935.
2. D. Avis and G. T. Toussaint, *An efficient algorithm for decomposing a polygon into star-shaped polygons*, Pattern Recognition **13** (1981), 395–398.
3. ——, *An optimal algorithm for determining the visibility of a polygon from an edge*, IEEE Trans. Comput. C-30 (1981), 910–914.
4. Marilyn Breen, *A Krasnosel′skiĭ theorem for open finitely starlike sets in* $R^3$ (submitted).
5. ——, *Clear visibility, starshaped sets, and finitely starlike sets*, J. Geom. **19** (1982), 183–196.
6. ——, *Finitely starlike sets whose* $F$*-stars have positive measure*, J. Geom. **35** (1989), 19–25.
7. ——, *Krasnosel′skiĭ numbers for bounded finitely starlike sets in* $R^d$, J. Geom. **37** (1990), 48–54.
8. ——, *Some Krasnosel′skiĭ numbers for finitely starlike sets in the plane*, J. Geom. **32** (1988), 1–12.
9. Ludwig Danzer, Branko Grünbaum, and Victor Klee, *Helly's theorem and its relatives*, in Convexity, Proc. Sympos. Pure Math., vol. 7, Amer. Math. Soc., Providence, R.I., 1962, pp. 101–180.
10. H. E. Debrunner, *Helly type theorems derived from basic singular homology*, Amer. Math. Monthly **77** (1970), 375–380.
11. K. J. Falconer, *The dimension of the kernel of a compact starshaped set*, Bull. London Math. Soc. **9** (1977), 313–316.
12. Meir Katchalski, *The dimension of intersections of convex sets*, Israel J. Math. **10** (1971), 465–470.
13. M. A. Krasnosel′skiĭ, *Sur un critère pour qu'un domaine soit étoilé*, Mat. Sb. **(61) 19** (1946), 309–310.
14. J. F. Lawrence, W. R. Hare, Jr., and J. W. Kenelly, *Finite unions of convex sets*, Proc. Amer. Math. Soc. **34** (1972), 225–228.
15. Steven R. Lay, *Convex sets and their applications*, Wiley, New York, 1982.
16. William S. Massy, *Singular homology theory*, Springer-Verlag, New York, 1980.
17. S. Nadler, *Hyperspaces of sets*, Dekker, New York, 1978.
18. Joseph O'Rourke, *Art gallery theorems and algorithms*, Oxford Univ. Press, London, 1987.
19. Bruce Peterson, *Is there a Krasnosel′skiĭ theorem for finitely starlike sets*?, in Convexity and Related Combinatorial Geometry, Dekker, New York, 1982, pp. 81–84.
20. I. V. Polikanova and G. Ya. Perel′man, *A remark on Helly's theorem*, Sibirsk. Mat. Zh. **27** (1986), 191–194, 207.
21. Angus E. Taylor, *General theory of functions and integration*, Blaisdell, New York, 1965.
22. Godfried J. Toussaint, *Computational geometry: recent developments*, in New Advances in Computer Graphics, Springer-Verlag, New York, 1989, pp. 23–51.
23. F. A. Valentine, *Convex sets*, McGraw-Hill, New York, 1964.

DEPARTMENT OF MATHEMATICS, UNIVERSITY OF OKLAHOMA, NORMAN, OKLAHOMA 73069

DIMACS Series in Discrete Mathematics
and Theoretical Computer Science
Volume **4**, 1991

Applied Geometry
and Discrete Mathematics
THE VICTOR KLEE FESTSCHRIFT

# On Sign-Nonsingular Matrices and the Conversion of the Permanent into the Determinant

RICHARD A. BRUALDI AND BRYAN L. SHADER

ABSTRACT. We discuss the relationship between sign-nonsingular matrices and conversions of the permanent into the determinant. We show the equivalence between the known characterizations of convertible $(0, 1)$-matrices and even digraphs. We prove that up to scaling the rows and columns by $\pm 1$, a convertible $(0, 1)$-matrix with total support has a unique conversion. Equivalently, we show that any two sign-nonsingular matrices with total support having the same sign pattern are diagonally equivalent. This property is used to indicate an algorithm which avoids backtracking to determine whether or not a matrix is convertible, and to construct a conversion when possible. We then consider $(0, 1)$-matrices of order $n$ which are maximal with respect to the property of being convertible. Using such matrices, we show how two convertible matrices can be joined to obtain a new convertible matrix. We conclude with some related topics and questions for further research.

## 1. Introduction

Let $A = [a_{ij}]$ $(1 \le i, j \le n)$ be a $(0, 1, -1)$-matrix of order $n$. Then $A$ is called *sign-nonsingular* **[5]** provided that each matrix with the same sign pattern as $A$ is nonsingular. Let $\sigma = (j_1, j_2, \dots, j_n)$ be a permutation of $\{1, 2, \dots, n\}$ such that $a_{1j_1} a_{2j_2} \cdots a_{nj_n} \neq 0$. Let $B = [b_{ij}]$ $(1 \le i, j \le n)$ be a nonnegative matrix with the same zero pattern as $A$ such that $b_{ij_i}$ equals 1 $(i = 1, 2, \dots, n)$ and $b_{rs}$ equals a positive number $\epsilon$ for all other positions $(r, s)$ for which $a_{rs} \neq 0$. Let $A * B$ denote the Hadamard product (entrywise product) of $A$ and $B$. The matrix $A * B$ has the same sign pattern as $A$ and for $\epsilon$ sufficiently small,

$$\operatorname{sign}(\det A * B) = (-1)^{\operatorname{sgn}(\sigma)} a_{1j_1} a_{2j_2} \cdots a_{nj_n}.$$

It now follows that $A$ is sign-nonsingular if and only if each nonzero term

1980 *Mathematics Subject Classification* (1985 *Revision*). Primary 15A15, 05C50.

Research partially supported by National Science Foundation Grants No. DMS-8521521 and DMS-8901445 and National Security Agency Grant No. MDA904-89-H-2060.

$(-1)^{\text{sgn}(\sigma)}a_{1j_1}a_{2j_2}\cdots a_{nj_n}$ in the determinant expansion of $A$ has the same sign. Let $|A|$ denote the $(0, 1)$-matrix obtained from $A$ by replacing each entry with its absolute value. We thus have the following characterization of sign-nonsingular matrices in terms of the permanent and determinant:

(1) The $(0, 1, -1)$-matrix $A$ of order $n$ is sign-nonsingular if and only if $\text{per}\,|A| = |\det A|$.

Let $E$ be a $(0, 1)$-matrix of order $n$ with per $E \neq 0$, and let $M_n(E)$ be the set of all real matrices having 0's in those positions in which $E$ has 0's (and possibly elsewhere). Let $\widehat{E}$ be a matrix obtained from $E$ by affixing minus signs to some of its 1's. Then we say $\widehat{E}$ *converts* the permanent of the matrices in $M_n(E)$ into their determinant provided

$$\text{per}\,X = |\det(\widehat{E} * X)| \quad \text{for all } X \in M_n(E). \tag{2}$$

Taking $X = E$ in (2) we obtain

$$\text{per}\,E = |\det \widehat{E}|. \tag{3}$$

Conversely, (3) implies (2).

The matrix $E$ is called *convertible* provided there exists an $\widehat{E}$ for which (3) holds. The matrix $\widehat{E}$ is then called a *conversion* of $E$. It follows that

(4) $\widehat{E}$ converts the permanent of the matrices in $M_n(|\widehat{E}|)$ into their determinant if and only if $\widehat{E}$ is a sign-nonsingular matrix.

This connection between converting the permanent into the determinant and sign-nonsingular matrices seems to have been little noticed in the literature [**2**].

Let $A = [a_{ij}]$ be a $(0, 1)$-matrix of order $n$. Let $i$ and $j$ be integers such that $a_{ij} = 1$, and suppose that row $i$ and column $j$ each contain exactly two 1's. There exists a unique integer $r \neq i$ such that $a_{rj} = 1$ and a unique integer $s \neq j$ such that $a_{is} = 1$. Suppose that $a_{rs} = 0$. Let $B$ be the matrix of order $n - 1$ obtained from $A$ by replacing the 0 in position $(r, s)$ with a 1 and deleting row $i$ and column $j$. Then $B$ is the matrix obtained from $A$ by *contracting* the 1 in position $(i, j)$.

An identity matrix is denoted by $I$, the identity matrix of order $n$ by $I_n$, and all the 1's matrix of order $n$ by $J_n$. If $E$ and $F$ are matrices of the same size, we write $F \leq E$ provided that each element of $F$ is less than or equal to the corresponding element of $E$.

The following characterization of convertible $(0, 1)$-matrices is a consequence of Corollary 1 of [**7**].

THEOREM 1.1. *The $(0, 1)$-matrix $E$ of order $n$ is convertible if and only if there does not exist a $(0, 1)$-matrix $F \leq E$ whose rows and columns can be permuted to obtain a matrix of the form $I \oplus Y$, where $J_3$ can be obtained from $Y$ by a sequence of contractions.*

For example, let

(5)
$$E = \begin{bmatrix} \mathbf{1} & 0 & 1 & 0 & 0 \\ 1 & 1 & 0 & 1 & 0 \\ 0 & 0 & 1 & 1 & 1 \\ 0 & 1 & 1 & 1 & 0 \\ 0 & 1 & 0 & 0 & \mathbf{1} \end{bmatrix}.$$

Contracting the boldfaced 1's we obtain $J_3$ and hence by Theorem 1.1 ($I$ is an empty matrix in this case) $E$ is not convertible.

Let $A = [a_{ij}]$ be a matrix of order $n$. With $A$ we associate a *digraph* $D(A)$ of order $n$ as follows. The set of vertices of $D(A)$ is $\{1, 2, \ldots, n\}$. For $i \neq j$ there is an arc $(i, j)$ from $i$ to $j$ if and only if $a_{ij} \neq 0$. Notice that the digraph $D(A)$ does not depend on the main diagonal of $A$ ($D(A)$ has no loops). If we assign the weights $a_{ij}$ to the arcs $(i, j)$ of $D(A)$, then we refer to the *weighted digraph* of $A$. The *weight* of a directed cycle and a path is defined to be the product of the weights of their arcs. (Directed cycles and paths have by definition no repeated vertex.)

Suppose $A$ is a sign-nonsingular matrix of order $n$. A matrix obtained from $A$ by row and column permutations and by multiplication of certain rows and columns by $-1'$s is also sign-nonsingular. It follows that in discussing sign-nonsingular matrices one may assume without loss of generality that all elements on the main diagonal equal $-1$. As observed by Bassett, Maybee and Quirk [**1**], a $(0, 1-1)$-matrix $A$ of order $n$ all of whose main diagonal elements equal $-1$ is sign-nonsingular if and only if the weight of each directed cycle of $D(A)$ equals $-1$. It follows that

(6) if $B$ is a $(0, 1)$-matrix of order $n$ all of whose main diagonal's entries equal 1, then $B$ is convertible if and only if it is possible to assign a weight of 1 or $-1$ to each arc of $D(B)$ so that each directed cycle has weight $-1$.

Let $(x, y)$ be an arc of digraph $D$. By *splitting the arc* $(x, y)$ we mean adjoining a new vertex $z$ to $D$ and replacing the arc $(x, y)$ by the two arcs $(x, z)$ and $(z, y)$. A *subdivision* of $D$ is a digraph obtained from $D$ by successively splitting arcs. The digraph $D$ is considered a subdivision of itself. A digraph $D$ is *even* (Seymour and Thomassen [**11**]) provided every subdivision contains a directed cycle of even length. It follows that a digraph $D$ is even if and only if for every weighting of the arcs of $D$ with weights 1 and $-1$ there exists a directed cycle whose weight equals 1.

Let $x$ be a vertex of a digraph $D$. By *splitting the vertex* $x$ we mean adjoining a new vertex $z$ and a new arc $(x, z)$ and then replacing each arc of the form $(x, y)$ with the arc $(z, y)$. A *splitting* of the digraph $D$ is a subdivision of a digraph obtained from $D$ by splitting some of its vertices. The digraph $D$ is considered a splitting of itself.

Let $k \geq 3$ be an integer. The digraph $C_k^*$ is the digraph obtained from an undirected cycle $C_k$ of length $k$ by replacing each edge with two oppositely directed arcs. The following characterization of even digraphs is due to Seymour and Thomassen [**11**].

THEOREM 1.2. *A digraph* $D$ *is even if and only if it contains a splitting of* $C_k^*$ *for some odd integer* $k$.

A splitting of $C_k^*$ is called a *weak* $k$*-double cycle* by Seymour and Thomassen. From Theorem 1.2 we obtain the following characterization of convertible $(0, 1)$-matrices.

COROLLARY 1.3. *Let* $A$ *be a* $(0, 1)$*-matrix of order* $n$ *all of whose main diagonal elements equal* $1$. *Then* $A$ *is convertible if and only if* $D(A)$ *does not contain a splitting of* $C_k^*$ *for any odd integer* $k \geq 3$.

From Theorem 1.1 and Corollary 1.3 we obtain the following corollary which relates the notation of contraction of a $(0, 1)$-matrix with that of splitting of a digraph.

COROLLARY 1.4. *Let* $A$ *be a* $(0, 1)$*-matrix of order* $n$ *all of whose main diagonal elements equal* $1$. *Then* $D(A)$ *contains a splitting of* $C_k^*$ *for some odd integer* $k \geq 3$ *if and only if there exists a* $(0, 1)$*-matrix* $F \leq A$ *whose rows and columns can be permuted to obtain a matrix of the form* $I \oplus Y$, *where* $J_3$ *can be obtained from* $Y$ *by a sequence of contractions.*

It is possible to give a direct proof of the equivalence of the two properties in Corollary 1.4, although the details are somewhat tedious. We indicate how the argument goes.

Let $B$ be a $(0, 1)$-matrix of order $n$ all of whose main diagonal elements equal 1. Let $D'$ be the digraph obtained from $D(B)$ by splitting the arc $(i, j)$. Let the new vertex be labelled $n + 1$. Let $A'$ be the $(0, 1)$-matrix of order $n + 1$ with 1's on its main diagonal whose associated digraph is $D'$. By contracting on the 1 in position $(n + 1, n + 1)$ of $A'$ we obtain the matrix $B$. Thus, splitting an arc corresponds to contracting a 1 on the main diagonal. Now let $D''$ be the digraph obtained from $D(B)$ by splitting the vertex $k$. Let the new vertex be labelled $z$. Let $A''$ be the $(0, 1)$-matrix of order $n + 1$ with 1's on the main diagonal whose associated digraph is $D''$, where the vertices of $D''$ are taken in the order $1, \dots, k, z, k+1, \dots, n$. By contracting on the 1 of $A''$ corresponding to the arc $(k, z)$ we obtain the matrix $B$. Thus, splitting a vertex corresponds to contracting a 1 not on the main diagonal.

Suppose $A$ is a $(0, 1)$-matrix of order $n$ all of whose main diagonal elements equal 1 such that $D(A)$ contains a splitting $D_k^*$ of $C_k^*$ for some odd integer $k$. Let $B$ be the principal submatrix of $A$ corresponding to the vertices of $D_k^*$. Let $C$ be the matrix obtained from $B$ by replacing with 0's those 1's corresponding to arcs of $D(A)$ which are not arcs of $D_k^*$. By contracting $C$ we obtain a matrix $X$ with 1's on its main diagonal such that $D(X)$ is a $C_k^*$. Let $X'$ be the matrix obtained from $X$ by replacing with

0's some set of $k-3$ 1's on its main diagonal corresponding to consecutive vertices of $C_k^*$. Then $X'$ is contractible to $J_3$ and it follows that there exists a $(0, 1)$-matrix $B' \leq B$ which is contractible to $J_3$.

For the other direction, it suffices to consider digraphs $D$ whose adjacency matrix $A$ has the property that there exists a $(0, 1)$-matrix $F \leq I+A$ which can be contracted to $J_3$ and no digraph obtained from $D$ by removing an arc has this property. It can be shown that such digraphs $D$ are splittings of $C_k^*$ for some odd integer $k$.

Let $E = [e_{ij}]$ be a $(0, 1)$-matrix of order $n$. The following concepts are well known. The matrix $E$ has *total support* provided each 1 of $E$ enters into a nonzero term of the determinant expansion (equivalently, the permanent expansion) of $E$. Let $E_{ij}$ be the $(0, 1)$-matrix of order $n$ whose only 1 is a 1 in position $(i, j)$. Then $E$ has total support if and only if $\operatorname{per} E > \operatorname{per}(E - E_{ij})$ for all $(i, j)$ such that $e_{ij} = 1$. In investigating convertible matrices it clearly suffices to consider only nonzero matrices with total support. The nonzero matrix $E$ is *fully indecomposable* provided $E$ has total support and there do not exist permutation matrices $P$ and $Q$ of order $n$ such that

$$PEQ = E' \oplus E'',$$

where $E'$ and $E''$ are square (nonvacuous) matrices. A nonzero matrix of total support is, up to row and column permutations, a direct sum of fully indecomposable matrices. Usually it suffices to consider only fully indecomposable matrices because a matrix of total support is sign-nonsingular if and only if each of its fully indecomposable components is sign-nonsingular. The above considerations apply to more general matrices upon replacing each nonzero element with a 1.

We now summarize the contents of the remaining sections. Let $E$ be a $(0, 1)$-matrix with total support. In §2, we show that up to diagonal equivalence there is at most one matrix which converts the permanent of the matrices in $M_n(E)$ into their determinant. Equivalently, we show that a sign-nonsingular $(0, 1, -1)$-matrix with total support is uniquely determined up to diagonal equivalence by its zero pattern. We then show that there is an algorithm which avoids backtracking that determines whether or not $E$ is convertible and, if $E$ is convertible, constructs a conversion $\widehat{E}$ of $E$. In §3, we consider maximal convertible matrices and show how to combine two convertible matrices to obtain a new convertible matrix. We conclude by discussing some related ideas and questions for further research.

## 2. Convertible matrices

Let $E$ be a $(0, 1)$-matrix with total support. In this section we consider the question of determining when there is a sign-nonsingular matrix $\widehat{E}$ with $|\widehat{E}| = E$. As observed in the introduction this is equivalent to determining when $E$ is a convertible matrix. We first show that if $\widehat{E}$ exists, then $\widehat{E}$ is unique up to diagonal equivalence.

THEOREM 2.1. *Let $E$ be a $(0, 1)$-matrix with total support and let $\widehat{E}$ and $\widetilde{E}$ be sign-nonsingular matrices with $|\widehat{E}| = |\widetilde{E}| = E$. Then there exist diagonal matrices $D_1$ and $D_2$ whose diagonal elements equal $\pm 1$ such that $\widetilde{E} = D_1\widehat{E}D_2$. If $E$ is fully indecomposable, then the matrices $D_1$ and $D_2$ are unique up to a scalar factor of $-1$.*

PROOF. We first consider the case in which $E$ is fully indecomposable. It follows from Theorem 1 of Sinkhorn and Knopp **[12]** (see also Theorems 4.3 and 4.4 of Ryser **[9]** and Theorem 5.7 of Engel and Schneider **[3]**) that there exist nonsingular diagonal matrices $D_1$ and $D_2$ such that $\widetilde{E} = D_1\widehat{E}D_2$. The matrices $D_1$ and $D_2$ are unique up to multiplication of $D_1$ by a nonzero scalar $\theta$ and multiplication of $D_2$ by $\theta^{-1}$. Let $D_1 = \operatorname{diag}(d_1, d_2, \dots, d_n)$ and $D_2 = \operatorname{diag}(d'_1, d'_2, \dots, d'_n)$. Let $R = \{r : |d_r| = |d_1|, 1 \le r \le n\}$ and $S = \{s : |d'_s| = |d_1|^{-1}, 1 \le s \le n\}$. Let $\overline{R}$ and $\overline{S}$ be the complements in $\{1, 2, \dots, n\}$ of $R$ and $S$, respectively. Since $\widehat{E}$ and $\widetilde{E}$ are $(0, 1, -1)$-matrices, it follows that the submatrices $E[R, \overline{S}]$ and $E[\overline{R}, S]$ of $E$ in rows $R$ and columns $\overline{S}$ and rows $\overline{R}$ and columns $S$, respectively, are zero matrices. Because $E$ is fully indecomposable we have $R = S = \{1, \dots, n\}$. Let $D'_1 = d_1^{-1}D_1$ and $D'_2 = d_1D_2$. Then $D'_1$ and $D'_2$ are diagonal matrices whose diagonal elements equal $\pm 1$ satisfying $D'_1\widehat{E}D'_2 = \widetilde{E}$. The conclusion of the theorem in the case $E$ has total support now follows. □

THEOREM 2.2. *Let $E = [e_{ij}]$ be a fully indecomposable $(0, 1)$-matrix of order $n$, and let $\widehat{E} = [\hat{e}_{ij}]$ be a sign-nonsingular matrix with $|\widehat{E}| = E$. Let $F$ be a $(0, 1)$-matrix of order $n$ which is obtained from $E$ by replacing a $0$ in position $(u, v)$ with a $1$. Then the following are equivalent*:

(i) *There exists a sign-nonsingular matrix $\widetilde{F} = [\tilde{f}_{ij}]$ with $|\widetilde{F}| = F$.*

(ii) *There exists a sign-nonsingular matrix $\widehat{F} = [\hat{f}_{ij}]$ with $|\widehat{F}| = F$ such that $\hat{e}_{ij} = \hat{f}_{ij}$ for all $(i, j) \neq (u, v)$.*

(iii) *The matrix obtained from $\widehat{E}$ by deleting row $u$ and column $v$ is a sign-nonsingular matrix.*

PROOF. Clearly (iii) is equivalent to (ii) and (ii) implies (i). Suppose that (i) holds. Let $\widetilde{E}$ be the matrix obtained from $\widetilde{F}$ by replacing $\tilde{f}_{uv}$ with 0. Then $|\widetilde{E}| = E$ and $\widetilde{E}$ is a sign-nonsingular matrix. By Theorem 2.1 there exist diagonal matrices $D_1$ and $D_2$ whose diagonal elements equal $\pm 1$ such that $D_1\widetilde{E}D_2 = \widehat{E}$. But then the matrix $\widehat{F} = D_1\widetilde{F}D_2$ satisfies (ii). □

We now extend Theorem 2.2 to include matrices with total support.

THEOREM 2.3. *Let $E = [e_{ij}]$ be a $(0, 1)$-matrix with total support and assume that $E = E_1 \oplus \cdots \oplus E_k \oplus E_{k+1}$, where $E_1, \dots, E_k$ are fully indecomposable matrices. Let $\widehat{E} = [\hat{e}_{ij}] = \widehat{E}_1 \oplus \cdots \oplus \widehat{E}_k \oplus \widehat{E}_{k+1}$ be a sign-nonsingular* *matrix with* $|\widehat{E}| = E$. *Let*

$$F = \begin{bmatrix} E_1 & O & O & \cdots & O & F_k \\ F_1 & E_2 & O & \cdots & O & O \\ O & F_2 & E_3 & \cdots & O & O \\ \vdots & \vdots & \vdots & \ddots & \vdots & \vdots \\ O & O & O & \cdots & E_{k-1} & O \\ O & O & O & \cdots & F_{k-1} & E_k \end{bmatrix} \oplus E_{k+1},$$

*where the matrix* $F_i$ *is a* $(0, 1)$*-matrix with exactly one* 1 *and this* 1 *is in the position* $(u_i, v_i)$ *of* $F_i$ $(1 \le i \le k)$. *Then the following are equivalent*:

(i) *There exists a sign-nonsingular matrix* $\widetilde{F}$ *with* $|\widetilde{F}| = F$.

(ii) *There exists a sign-nonsingular matrix* $\widehat{F} = [\hat{f}_{ij}]$ *with* $|\widehat{F}| = F$ *such that* $\hat{e}_{ij} = \hat{f}_{ij}$ *for all* $(i, j)$ *for which* $e_{ij} \neq 0$.

(iii) *For* $i = 1, \ldots, k$ *the matrix* $\widehat{E}_i'$ *obtained from* $\widehat{E}_i$ *by deleting row* $u_{i-1}$ *and columns* $v_i$ *of* $\widehat{E}_i$ *is a sign-nonsingular matrix* (*where* $u_0$ *is taken to be* $u_k$).

(Note that $(u_1, v_i)$ refers to the relative position of the 1 in $F_i$ and not the position of that 1 in $F$.)

PROOF. Clearly (ii) implies (i) and (ii) implies (iii). Now suppose that (iii) holds. Then the matrix

$$\widehat{F} = \begin{bmatrix} \widehat{E}_1 & O & O & \cdots & O & \widehat{F}_k \\ F_1 & \widehat{E}_2 & O & \cdots & O & O \\ O & F_2 & \widehat{E}_3 & \cdots & O & O \\ \vdots & \vdots & \vdots & \ddots & \vdots & \vdots \\ O & O & O & \cdots & \widehat{E}_{k-1} & O \\ O & O & O & \cdots & F_{k-1} & \widehat{E}_k \end{bmatrix} \oplus \widehat{E}_{k+1}$$

is a sign-nonsingular matrix for one of $\widehat{F}_k = F_k$ or $\widehat{F}_k = -F_k$. The implication (i) implies (ii) follows as in the proof of Theorem 2.2 by replacing the nonzero elements in the off-diagonal blocks of $\widetilde{F}$ with 0's. □

Let $E$ be a $(0, 1)$-matrix of order $n$ with total support and assume without loss of generality that $I_n \le E$. Starting with the identity matrix $I_n$, we can obtain $E$ by repeatedly applying the constructions of the matrices $F$ in Theorems 2.2 and 2.3 (in general after row and column permutations). Let $A_0 = I_n, A_1, \ldots, A_l = E$ be one such way. Theorems 2.2 and 2.3 imply that given any conversion $\widehat{A}_i$ of $A_i$ there is a conversion of $A_{i+1}$ if and only if there is a conversion $\widehat{A}_{i+1}$ of $A_{i+1}$ which extends the conversion $\widehat{A}_i$. Starting with the conversion $\widehat{A}_0 = I_n$ of $A_0$, we attempt to extend a conversion of $A_i$ to a conversion of $A_{i+1}$. If at some point we are unable to do so, Theorems 2.2 and 2.3 imply that no conversion of $E$ exists. This provides us with an algorithm which avoids backtracking to determine whether or not the exists a sign-nonsingular matrix $\widehat{E}$ with $|\widehat{E}| = E$ and to determine such an $\widehat{E}$ when it exists. We illustrate the algorithm with an example. Let

$$E = \begin{bmatrix} 1 & 0 & 1 & 1 & 0 & 0 \\ 1 & 1 & 1 & 0 & 0 & 0 \\ 0 & 1 & 1 & 0 & 0 & 1 \\ 0 & 1 & 0 & 1 & 0 & 0 \\ 1 & 0 & 0 & 0 & 1 & 0 \\ 0 & 0 & 0 & 0 & 1 & 1 \end{bmatrix}.$$

Starting with $I_6$, we can obtain $E$ as follows:

$$A_0 = I_6,$$

$$A_1 = \begin{bmatrix} 1 & 0 & 1 & 0 & 0 & 0 \\ 1 & 1 & 0 & 0 & 0 & 0 \\ 0 & 1 & 1 & 0 & 0 & 0 \\ 0 & 0 & 0 & 1 & 0 & 0 \\ 0 & 0 & 0 & 0 & 1 & 0 \\ 0 & 0 & 0 & 0 & 0 & 1 \end{bmatrix}, \quad A_2 = \begin{bmatrix} 1 & 0 & 1 & 0 & 0 & 0 \\ 1 & 1 & 1 & 0 & 0 & 0 \\ 0 & 1 & 1 & 0 & 0 & 0 \\ 0 & 0 & 0 & 1 & 0 & 0 \\ 0 & 0 & 0 & 0 & 1 & 0 \\ 0 & 0 & 0 & 0 & 0 & 1 \end{bmatrix},$$

$$A_3 = \begin{bmatrix} 1 & 0 & 1 & 1 & 0 & 0 \\ 1 & 1 & 1 & 0 & 0 & 0 \\ 0 & 1 & 1 & 0 & 0 & 0 \\ 0 & 1 & 0 & 1 & 0 & 0 \\ 0 & 0 & 0 & 0 & 1 & 0 \\ 0 & 0 & 0 & 0 & 0 & 1 \end{bmatrix}, \quad A_4 = E.$$

Successively extending conversions we obtain

$$\widehat{A}_0 = I_6, \quad \widehat{A}_1 = A_1,$$

$$\widehat{A}_2 = \begin{bmatrix} 1 & 0 & 1 & 0 & 0 & 0 \\ 1 & 1 & -1 & 0 & 0 & 0 \\ 0 & 1 & 1 & 0 & 0 & 0 \\ 0 & 0 & 0 & 1 & 0 & 0 \\ 0 & 0 & 0 & 0 & 1 & 0 \\ 0 & 0 & 0 & 0 & 0 & 1 \end{bmatrix}, \quad \widehat{A}_3 = \begin{bmatrix} 1 & 0 & 1 & 1 & 0 & 0 \\ 1 & 1 & -1 & 0 & 0 & 0 \\ 0 & 1 & 1 & 0 & 0 & 0 \\ 0 & 1 & 0 & 1 & 0 & 0 \\ 0 & 0 & 0 & 0 & 1 & 0 \\ 0 & 0 & 0 & 0 & 0 & 1 \end{bmatrix}.$$

Since the submatrix obtained by deleting row 3 and column 1 from $\widehat{A}_3$ is not a sign-nonsingular matrix, $\widehat{A}_3$ cannot be extended to a conversion of $A_4$. Hence the algorithm halts and we conclude that there does not exist a sign-nonsingular matrix $\widehat{E}$ with $|\widehat{E}| = E$.

It is implicit in the proof of Theorem 1.1 as given in Little [7] and the proof of Theorem 1.2 as given in Seymour and Thomassen [**11**] that an algorithm like that described above is valid.

## 3. Maximal convertible matrices

Let $E$ be a convertible $(0, 1)$-matrix of order $n$ with total support. If $E$ is not fully indecomposable, then it follows from Theorem 2.3 that there exists a fully indecomposable $(0, 1)$-matrix $E'$ which is convertible and satisfies $E \leq E'$. We thus define a *maximal convertible matrix* to be a fully indecomposable convertible $(0, 1)$-matrix $E$ which satisfies the property that no matrix obtained from $E$ by replacing a 0 with a 1 is convertible. The matrices

$$[1], \qquad \begin{bmatrix} 1 & 1 \\ 1 & 1 \end{bmatrix}, \quad \text{and} \quad \begin{bmatrix} 1 & 1 & 0 \\ 1 & 1 & 1 \\ 1 & 1 & 1 \end{bmatrix}$$

are easy examples of maximal convertible matrices. The incidence matrix of the projective plane of order 2 given by $E = I + P + P^3$, where $P$ is the permutation matrix corresponding to the cycle $1 \to 2 \to \cdots \to 7 \to 1$, satisfies $\operatorname{per} E = \det E = 24$ (see pp. 124–125 of [**10**]). Thus $\widehat{E} = E$ is a conversion of $E$ and hence $E$ is a sign-nonsingular matrix. It follows easily from Theorem 2.2 that $E$ is a maximal convertible matrix. We note that a conversion of a maximal convertible matrix is a maximal $L$-matrix (sign-nonsingular matrix) in the sense used in Lundgren and Maybee [**8**]. In this section we obtain some properties of maximal convertible matrices and show how such matrices can be used to construct new convertible matrices.

PROPOSITION 3.1. *Let*

$$E = \begin{bmatrix} \overbrace{1 \cdots 1}^{k} & 0 \cdots 0 \\ & \\ & X \\ & \end{bmatrix}$$

*be a maximal convertible* $(0, 1)$*-matrix of order* $n$. *Then the matrix*

$$G = \begin{bmatrix} 1 & \overbrace{1 \cdots 1}^{k} \quad 0 \cdots 0 \\ 1 & \\ 0 & \\ \vdots & E \\ 0 & \end{bmatrix}$$

*is a maximal convertible matrix of order* $n + 1$.

PROOF. The matrix $G$ is fully indecomposable. There exists a conversion $\widehat{E}$ of $E$ such that the first row of $\widehat{E}$ is equal to the first row of $E$. Let

$$\widehat{G} = \begin{bmatrix} 1 & \overbrace{1 \cdots 1}^{k} \quad 0 \cdots 0 \\ -1 & \\ 0 & \\ \vdots & \widehat{E} \\ 0 & \end{bmatrix}.$$

Then $\widehat{G}$ is a conversion of $G$. The maximality of $E$ implies that any matrix obtained from $G$ by replacing a 0 in one of columns $2, \ldots, n+1$ by a 1 is not convertible. Let $G'$ be a matrix obtained from $G$ by replacing a 0 in row $i$ $(3 \leq i \leq n+1)$ and column 1 with a 1. By Theorem 2.2, $G'$ is convertible if and only if the submatrix of $\widehat{G}$ obtained by deleting row $i$ and column 1 is a sign-nonsingular matrix. But this submatrix has determinant equal to 0 and it follows that $G$ is a maximal convertible matrix. □

Conversely, we now show that any maximal convertible matrix having a column with exactly two 1's arises as in the construction given in Proposition 3.1.

PROPOSITION 3.2. *Let $G$ be a maximal convertible $(0, 1)$-matrix of order $n$ whose first column is $[1\ 1\ 0\ \cdots\ 0]^T$. Then the first two rows of $G$ are identical and the matrix obtained from $G$ by deleting row 1 and column 1 is a maximal convertible matrix.*

PROOF. Let $\widehat{G}$ be a conversion of $G$. Let $j$ be an integer with $2 \leq j \leq n$. Then the matrix obtained from $\widehat{G}$ by deleting row 1 and column $j$ is sign-nonsingular if and only if the matrix obtained from $\widehat{G}$ by deleting row 2 and column $j$ is sign-nonsingular. Applying Theorem 2.2, we conclude that the first two rows of $G$ are identical. It now follows from Proposition 3.1 that the matrix obtained from $G$ by deleting row 1 and column 1 is a maximal convertible matrix. □

COROLLARY 3.3. *A maximal convertible $(0, 1)$-matrix of order $n \geq 4$ has at least $3n$ 1's; if $n \geq 5$, then the equality holds only if there are exactly three 1's in every row and column.*

PROOF. It is not difficult to show that up to row and column permutations the maximal convertible matrices of order 4 are

$$\begin{bmatrix} 1 & 1 & 0 & 0 \\ 1 & 1 & 1 & 0 \\ 1 & 1 & 1 & 1 \\ 1 & 1 & 1 & 1 \end{bmatrix}, \quad \begin{bmatrix} 0 & 1 & 1 & 1 \\ 1 & 0 & 1 & 1 \\ 1 & 1 & 0 & 1 \\ 1 & 1 & 1 & 0 \end{bmatrix},$$
$$\begin{bmatrix} 0 & 0 & 1 & 1 \\ 1 & 1 & 0 & 0 \\ 1 & 1 & 1 & 1 \\ 1 & 1 & 1 & 1 \end{bmatrix}, \quad \begin{bmatrix} 0 & 1 & 1 & 1 \\ 0 & 1 & 1 & 1 \\ 1 & 0 & 1 & 1 \\ 1 & 0 & 1 & 1 \end{bmatrix}.$$

Each of these matrices has at least 12 1's. We now let $A$ be a maximal convertible matrix of order $n \geq 5$ and proceed by induction on $n$. If every row and column of $A$ has at least three 1's, the inductive step holds. Now suppose that $A$ has a column with exactly two 1's. Without loss of generality we assume that the first column of $A$ is $[1\ 1\ 0\ \cdots\ 0]^T$. By Proposition 3.2 the first two rows of $A$ are identical and the matrix $A'$ obtained from $A$ by deleting row 1 and column 1 is maximal convertible. The full indecomposability of $A$ now implies that row 1 of $A$ has at least three 1's, and applying the inductive hypothesis to $A'$ we conclude that $A$ has at least $4 + 3(n-1) = 3n+1$ 1's. □

Let

$$(7) \qquad A = \begin{bmatrix} A_1 & A_2 \\ A_3 & C \end{bmatrix} \quad \text{and} \quad B = \begin{bmatrix} C & B_1 \\ B_2 & B_3 \end{bmatrix}$$

be partitioned matrices with common block $C$ as show. The matrix

$$A \star B = \begin{bmatrix} A_1 & A_2 & O \\ A_3 & C & B_1 \\ O & B_2 & B_3 \end{bmatrix},$$

where the matrices $O$ are zero matrices of an appropriate size, is called the *join* of $A$ and $B$ at $C$. The next two theorems show how the convertibility of $A$ and $B$ relates to the convertibility of $A \star B$.

THEOREM 3.4. *Let $A$ and $B$ be fully indecomposable* $(0, 1)$*-matrices of the form* (7). *Assume that all main diagonal entries of $A$ and $B$ equal* 1 *and that $C$ is fully indecomposable. Let*

$$\widehat{A} = \begin{bmatrix} \widehat{A}_1 & \widehat{A}_2 \\ \widehat{A}_3 & \widehat{C} \end{bmatrix} \quad \textit{and} \quad \widehat{B} = \begin{bmatrix} \widehat{C} & \widehat{B}_1 \\ \widehat{B}_2 & \widehat{B}_3 \end{bmatrix}$$

*be conversions of $A$ and $B$, respectively, which agree on $C$. Then $\widehat{A} \star \widehat{B}$ is a conversion of $A \star B$ if and only if $A \star B$ is convertible.*

PROOF. Assume that $A \star B$ is convertible. Let

$$(\widetilde{A \star B}) = \begin{bmatrix} \widetilde{A}_1 & \widetilde{A}_2 & O \\ \widetilde{A}_3 & \widetilde{C} & \widetilde{B}_1 \\ O & \widetilde{B}_2 & \widetilde{B}_3 \end{bmatrix}$$

be a conversion of $A \star B$. Let

$$\widetilde{A} = \begin{bmatrix} \widetilde{A}_1 & \widetilde{A}_2 \\ \widetilde{A}_3 & \widetilde{C} \end{bmatrix} \quad \text{and} \quad \widetilde{B} = \begin{bmatrix} \widetilde{C} & \widetilde{B}_1 \\ \widetilde{B}_2 & \widetilde{B}_3 \end{bmatrix}.$$

Since each diagonal entry of $A \star B$ is 1, $\widetilde{A}$ and $\widetilde{B}$ are conversions of $A$ and $B$, respectively. By Theorem 2.1, there exist diagonal matrices $D_1$ and $D_2$, and $D_3$ and $D_4$ whose main diagonal entries equal $\pm 1$, satisfying $D_1 \widetilde{A} D_2 = \widehat{A}$ and $D_3 \widetilde{B} D_4 = \widehat{B}$. For $i = 1, 2, 3, 4$ let $D_i = D_i' \oplus D_i''$ be the decomposition determined by the partitioned form of $A$ or $B$ as appropriate. Then $D_1'' \widetilde{C} D_2'' = \widehat{C} = D_3' \widetilde{C} D_4'$. It follows from Theorem 2.1 that $D_3' D_1'' = D_4' D_2'' = \pm I$. Hence, either $D_3' = D_1''$ and $D_4' = D_2''$ or $D_3' = -D_1''$ and $D_4' = -D_2''$. By multiplying $D_1$ and $D_2$ by the scalar $-1$ if necessary, we may assume that $D_3' = D_1''$ and $D_4' = D_2''$. But now

$$(D_1' \oplus D_1'' \oplus D_3'')(\widetilde{A \star B})(D_2' \oplus D_2'' \oplus D_4'') = \widehat{A} \star \widehat{B}$$

and we conclude that $\widehat{A} \star \widehat{B}$ is a conversion of $A \star B$. The theorem now follows. □

The following example shows that the full indecomposability of $C$ in Theorem 3.4 is necessary. Let

$$A = \begin{bmatrix} 1 & 1 & 1 \\ 1 & 1 & 0 \\ 1 & 0 & 1 \end{bmatrix}, \quad B = \begin{bmatrix} 1 & 0 & 1 \\ 0 & 1 & 1 \\ 1 & 1 & 1 \end{bmatrix}, \quad \text{and} \quad C = \begin{bmatrix} 1 & 0 \\ 0 & 1 \end{bmatrix}.$$

The matrices

$$\widehat{A} = \begin{bmatrix} -1 & 1 & 1 \\ -1 & -1 & 0 \\ -1 & 0 & -1 \end{bmatrix} \quad \text{and} \quad \widehat{B} = \begin{bmatrix} -1 & 0 & -1 \\ 0 & -1 & -1 \\ 1 & 1 & -1 \end{bmatrix}$$

are conversions of $A$ and $B$, respectively, which agree on $C$. The matrix

$$\begin{bmatrix} 1 & -1 & -1 & 0 \\ 1 & 1 & 0 & 1 \\ 1 & 0 & 1 & -1 \\ 0 & -1 & 1 & 1 \end{bmatrix}$$

is a conversion of $A \star B$. But,

$$\widehat{A} \star \widehat{B} = \begin{bmatrix} -1 & 1 & 1 & 0 \\ -1 & -1 & 0 & -1 \\ -1 & 0 & -1 & -1 \\ 0 & 1 & 1 & -1 \end{bmatrix}$$

is not a conversion of $A \star B$ because the terms in the expansion of the determinant of $\widehat{A} \star \widehat{B}$ corresponding to the permutations $(3, 2, 1, 4)$ and $(3, 1, 4, 2)$ of $(1, 2, 3, 4)$ have opposite signs. Hence, $A \star B$ is convertible, but $\widehat{A} \star \widehat{B}$ is not a conversion of $A \star B$.

THEOREM 3.5. *Let $A$ and $B$ be fully indecomposable* $(0, 1)$*-matrices of the form* (7). *Assume that all main diagonal entries of $A$ and $B$ equal* 1 *and that $C$ is a maximal convertible matrix. Then $A \star B$ is convertible if and only if both $A$ and $B$ are convertible.*

PROOF. The convertibility of $A \star B$ immediately implies the convertibility of $A$ and $B$. Now assume that both $A$ and $B$ are convertible. Let

$$\widehat{A} = \begin{bmatrix} \widehat{A}_1 & \widehat{A}_2 \\ \widehat{A}_3 & \widehat{C}' \end{bmatrix} \quad \text{and} \quad \widehat{B} = \begin{bmatrix} \widehat{C} & \widehat{B}_1 \\ \widehat{B}_2 & \widehat{B}_3 \end{bmatrix}$$

be conversions of $A$ and $B$, respectively, in which all the diagonal elements equal $-1$. Since all main diagonal entries of $A$ and $B$ equal 1, $\widehat{C}'$ and $\widehat{C}$ are conversions of $C$. Because $C$ has total support, it follows from Theorem 2.1 that we may assume $\widehat{C}' = \widehat{C}$. We weight the digraphs $D(A)$, $D(B)$, and $D(A \star B)$ by the entries of $\widehat{A}$, $\widehat{B}$, and $\widehat{A} \star \widehat{B}$, respectively. Since $\widehat{A}$ and $\widehat{B}$ are conversions of $A$ and $B$, respectively, it follows from (6) that each directed cycle of $D(A)$ and $D(B)$ has weight $-1$. It will also follow from (6) that $\widehat{A} \star \widehat{B}$ is a conversion of $A \star B$ if we show that each directed cycle of $D(A \star B)$ has weight $-1$. Both $D(A)$ and $D(B)$ may be viewed as subdigraphs of $D(A\star B)$. Let the vertex set of $D(A\star B)$ be $X\cup Z\cup Y$, where $X\cup Z$ is the vertex set of $D(A)$, $Z\cup Y$ is the vertex set of $D(B)$, and $Z$ corresponds to $C$.

Let $\gamma$ be a directed cycle of $D(A\star B)$. We show that its weight is $-1$ by induction on the length of $\gamma$. If $\gamma$ is a directed cycle in $D(A)$ or in $D(B)$, then it has weight $-1$; in particular, if $\gamma$ has length two, its weight is $-1$. Now assume that $\gamma$ has length greater than two and is not in $D(A)$ or in $D(B)$. There exists a subpath $\gamma_1$ of $\gamma$ with length at least two which, except for its initial vertex $u\in Z$ and terminal vertex $v\in Z$, lies entirely within $Y$. Let $\gamma_2$ be the subpath of $\gamma$ from $v$ to $u$. Since $C$ is fully indecomposable, there exists a path in $Z$ from $v$ to $u$ and each such path can be combined with $\gamma_1$ to obtain a directed cycle in $Z\cup Y$ which therefore has weight $-1$. It follows that all the paths in $Z$ from $v$ to $u$ have the same weight. The maximality property of $C$ now implies that $(u, v)$ is an arc in $D(A\star B)$. Let $\gamma'$ be a path from $v$ to $u$ in $Z$. We may combine the arc $(u, v)$ and the path $\gamma'$ to obtain a directed cycle $(\gamma', u, v)$ which is in $Z$ and therefore has weight $-1$. We may also combine $\gamma_1$ and $\gamma'$ to obtain a directed cycle $(\gamma', \gamma_1)$ which is in $X\cup Y$ and therefore has weight $-1$. The directed cycle $(\gamma_2, u, v)$ has length strictly less than that of $\gamma$ and hence by the inductive hypothesis has weight $-1$. We have

$$\text{wt}(\gamma)=\frac{\text{wt}(\gamma', \gamma_1)\,\text{wt}(\gamma_2, u, v)}{\text{wt}(\gamma', u, v)}=-1, \tag{8}$$

completing the induction. □

The following example shows that the maximality of $C$ is needed in Theorem 3.5. Let

$$A=\begin{bmatrix}1&0&1&0\\1&1&0&1\\0&0&1&1\\0&1&1&1\end{bmatrix},\quad B=\begin{bmatrix}1&0&1&0\\0&1&1&1\\1&1&1&0\\1&0&0&1\end{bmatrix},\quad\text{and}\quad C=\begin{bmatrix}1&0&1\\0&1&1\\1&1&1\end{bmatrix}.$$

The join $A\star B$ is the nonconvertible matrix $E$ in (5). Both $A$ and $B$ are convertible, but their join $A\star B$ is not convertible.

COROLLARY 3.6. *Let $B$ and $A=\left[\begin{smallmatrix}B&D\\C&B\end{smallmatrix}\right]$ be maximal convertible matrices. If $A\star A$ is a maximal convertible matrix, then*

$$A^{(k)}=\overbrace{A\star\cdots\star A}^{k\text{ factors}}=\begin{bmatrix}B&D&O&\cdots&O&O&O\\C&B&D&\cdots&O&O&O\\O&C&B&\cdots&O&O&O\\\vdots&\vdots&\vdots&\ddots&\vdots&\vdots&\vdots\\O&O&O&\cdots&B&D&O\\O&O&O&\cdots&C&B&D\\O&O&O&\cdots&O&C&B\end{bmatrix}$$

*is a maximal convertible matrix for all $k\geq 2$.*

PROOF. By repeated applications of Theorem 3.5, it follows that $A^{(k)}$ is convertible for $k \geq 2$. Let the vertex set of $D(A)$ be partitioned into $X_1 \cup X_2 \cup \cdots \cup X_{k+1}$, where $X_i$ is the set of vertices corresponding to the $i$th diagonal block of $A^{(k)}$. Let $u$ and $v$ be vertices in $X_1$ and $X_{k+1}$ respectively. By the full indecomposability of $B$ there is a vertex $w$ in $X_k$ and a path $\gamma$ from $v$ to $w$ which, except for $w$, lies in $X_{k+1}$. We may identify $D(A^{k-1})$ as the subgraph of $D(A^{(k)})$ with vertex set $X_1 \cup \cdots \cup X_k$. If the digraph obtained from $D(A^{(k)})$ by adjoining the arc $(u, v)$ is not even, then the digraph obtained from $D(A^{(k-1)})$ by adjoining the arc $(u, w)$ is also not even. A similar statement holds when $u$ and $v$ are vertices of $X_{k+1}$ and $X_1$, respectively. An inductive argument now shows that adjoining any new arc to $D(A^{(k)})$ results in an even digraph. It follows that $A^{(k)}$ is a maximal convertible matrix. □

As an example of Corollary 3.6 let

$$B = \begin{bmatrix} 1 & 1 \\ 1 & 1 \end{bmatrix} \quad \text{and} \quad C = D = \begin{bmatrix} 1 & 0 \\ 0 & 1 \end{bmatrix}.$$

Then $B$, $A$, and $A \star A$ are maximal convertible. Hence,

$$A^{(k)} = \begin{bmatrix} J_2 & I_2 & O & \cdots & O & O & O \\ I_2 & J_2 & I_2 & \cdots & O & O & O \\ O & I_2 & J_2 & \cdots & O & O & O \\ \vdots & \vdots & \vdots & \ddots & \vdots & \vdots & \vdots \\ O & O & O & \cdots & J_2 & I_2 & O \\ O & O & O & \cdots & I_2 & J_2 & I_2 \\ O & O & O & \cdots & O & I_2 & J_2 \end{bmatrix}$$

is a maximal convertible matrix.

## 4. Coda

In §3 we considered maximal convertible matrices. It is natural to ask for the largest number, $\tau_n$, of 1's in a convertible $(0, 1)$-matrix of order $n$, equivalently the smallest number, $\sigma_n = n^2 - \tau_n$, of 0's in a sign-nonsingular matrix. In fact, Gibson [**4, Theorem 1**] proved that $\tau_n = (n^2 + 3n - 2)/2$, while Thomassen [**13, Proposition 2.4**] independently proved the equivalent statement that $\sigma_n = \binom{n-1}{2}$. In addition, Gibson showed that every convertible matrix of order $n$ with $\tau_n$ 1's can be obtained from

$$T_n = \begin{bmatrix} 1 & 1 & 0 & \cdots & 0 & 0 & 0 \\ 1 & 1 & 1 & \cdots & 0 & 0 & 0 \\ 1 & 1 & 1 & \cdots & 0 & 0 & 0 \\ \vdots & \vdots & \vdots & \ddots & \vdots & \vdots & \vdots \\ 1 & 1 & 1 & \cdots & 1 & 1 & 0 \\ 1 & 1 & 1 & \cdots & 1 & 1 & 1 \\ 1 & 1 & 1 & \cdots & 1 & 1 & 1 \end{bmatrix}$$

by row and column permutations. Assuming (without loss of generality) there are no 0's on the main diagonal, Thomassen characterized the (unweighted) digraphs of sign-nonsingular matrices with $\sigma_n$ 0's as extended caterpillars.

One particular conversion of $T_n$ is the matrix

$$\widehat{T}_n = \begin{bmatrix} 1 & -1 & 0 & \cdots & 0 & 0 & 0 \\ 1 & 1 & -1 & \cdots & 0 & 0 & 0 \\ 1 & 1 & 1 & \cdots & 0 & 0 & 0 \\ \vdots & \vdots & \vdots & \ddots & \vdots & \vdots & \vdots \\ 1 & 1 & 1 & \cdots & 1 & -1 & 0 \\ 1 & 1 & 1 & \cdots & 1 & 1 & -1 \\ 1 & 1 & 1 & \cdots & 1 & 1 & 1 \end{bmatrix}.$$

Kräuter and Seifter [**6**] showed that $n-1$ is the smallest number of $-1$'s in any conversion of $T_n$ and that if $n \geq 4$, the only conversions with exactly $n-1$ elements equal to $-1$ are the four matrices obtained by multiplying row 1 and column 1 by $\pm 1$. It follows that $\binom{n+1}{2}$ is the largest number of $-1$'s in any conversion of $T_n$. Kräuter and Seifter also showed that every number between $n-1$ and $\binom{n+1}{2}$ is the number of $-1$'s of some conversion of $T_n$. We outline an easier proof of these results.

Let $\widetilde{T}_n$ be any conversion of $T_n$. Deleting rows 1 and row $i$ $(i \geq 2)$ and columns 1 and 2 of $T_n$ results in a matrix with a nonzero permanent. It follows that either the two nonzero elements in row 1 of $\widetilde{T}_n$ have opposite signs or there are at least $n-1$ entries in columns 1 and 2 equal to $-1$. In the latter case, arguing on rows $n-1$ and $n$ we conclude that $\widetilde{T}_n$ has at least $n$ entries equal to $-1$. In the first case, we may assume that the first row of $\widetilde{T}_n$ is $[1\ -1\ 0\ \cdots\ 0]$. Deleting row 1 and column 1 of $T_n$ results in $T_{n-1}$ and of $\widetilde{T}_n$ results in a conversion $\widetilde{T}_{n-1}$ of $T_{n-1}$. By induction, the number of $-1$'s of $\widetilde{T}_{n-1}$ is at least $n-2$. Hence, the number of $-1$'s of $\widetilde{T}_n$ is at least $n-1$ and if equality holds the first column of $\widetilde{T}_n$ is all 1's and, by induction, $\widetilde{T}_{n-1}$ can be obtained from $\widehat{T}_{n-1}$ by multiplying the last column by $\pm 1$. To show that all intermediate values are attained we note that multiplying row $i$ of $\widehat{T}_n$ by $-1$ results in an increase of the number of $-1$'s by $i-1$ if $i = 1, \ldots, n-1$ and an increase by $n$ if $i = n$. Since every integer between 0 and $\binom{n+1}{2} - (n-1) = (n^2 - n + 2)/2$ can be written as a sum of distinct integers in the set $\{1, 2, \ldots, n-2, n\}$, all intermediate values are attained.

In §3 we showed that a maximal convertible matrix of order $n \geq 5$ has at least $3n$ 1's and exactly $3n$ 1's only if every row and column contains three 1's. In that section we noted that the incidence matrix of the projective plane of order 2 is a maximal convertible matrix of order 7. This matrix has three 1's in every row and column.

Conjecture 4.1. The only maximal convertible matrix of order $n \geq 5$ with exactly $3n$ 1's is the incidence matrix of a projective plane of order 2.

In connection with the above conjecture we prove the following.

PROPOSITION 4.2. *If* $A$ *is a maximal convertible matrix of order* $n \geq 5$ *with exactly* $3n$ *1's, then* $A$ *has no submatrix equal to* $J_2$.

PROOF. Assume $J_2$ is a submatrix of $A$. Let $\widehat{J}_2$ be the corresponding submatrix in a conversion $\widehat{A}$ of $A$. First suppose that $\widehat{J}_2$ has an even number of $-1$'s. Then without loss of generality we may assume that $\widehat{J}_2 = J_2$. It follows that the submatrix $B$ of $A$ of order $n-2$ complementary to $J_2$ has permanent equal to zero. By König's theorem (see, e.g., Ryser [**10**]) $B$ has an $r$ by $s$ zero submatrix $O_{rs}$ for some $r$ and $s$ with $r+s = n-1$. Hence, after row and column permutations $A$ has the form

$$\begin{bmatrix} J_2 & * & V \\ * & * & O_{rs} \\ * & * & C \end{bmatrix},$$

where $C$ is an $s-1$ by $s$ matrix. Since $A$ is fully indecomposable, $V$ has a 1 in each of its rows and it follows that the matrix $C$ has exactly $3s-2$ 1's. But $C$ has $s-1$ rows and thus at most $3s-1$ 1's. Hence $\widehat{J}_2$ contains an odd number of $-1$'s. The matrix $A$ cannot have a 2 by 3 or a 3 by 2 submatrix of all 1's, for if it did, $\widehat{A}$ would have a submatrix $\widehat{J}_2$ with an even number of 1's.

It remains to show that $\widehat{A}$ does not have a submatrix $\widehat{J}_2$ with an odd number of $-1'$s. Without loss of generality the principal submatrix of $\widehat{A}$ determined by the first four rows and columns has the form

$$\begin{bmatrix} 1 & -1 & 1 & 0 \\ 1 & 1 & 0 & 1 \\ 1 & 0 & a & b \\ 0 & 1 & c & d \end{bmatrix}.$$

Suppose $a = 0$. We claim that the matrix $\widetilde{A}$ obtained from $\widehat{A}$ by replacing $a$ with $\tilde{a} = -1$ is sign-nonsingular and thereby contradicts the maximality property of $A$. Each nonzero term in the determinant expansion of $\widetilde{A}$ that uses $\tilde{a}$ either uses the 1 in position $(1, 1)$, or the $-1$ in position (1, 2) and the 1 in position (2, 1). In the first case, by considering the 1's in positions $(1, 3)$ and $(3, 1)$, these terms have the same sign as the nonzero terms in the determinant expansion of $\widehat{A}$. In the second case, by considering the 1's in positions $(1, 3)$, (2, 2), and $(3, 1)$, the same conclusion holds. It follows that $\widetilde{A}$ is sign-nonsingular. We therefore conclude that $a = 1$ and in a similar way that $b = c = d = 1$. But then $n = 4$, contrary to our assumption. □

As previous remarked, the incidence matrix $E = I+P+P^3$ of a projective plane of order 2 is a conversion of itself, that is $|\det E| = \operatorname{per} E$. Let $B$ be a $(0, 1)$-circulant of order $n$ with $k$ 1's in each row and column. Tinsley [**14**] proved that if $k \geq 3$, then $\operatorname{per} B > |\det B|$, and if $k = 3$, then $\operatorname{per} B = |\det B|$ if and only if $B$ is the direct sum of the matrix $E$ taken $n/7$ times.

An infinite class of $(0, 1)$-matrices $F$ satisfying $\operatorname{per} F = |\det F|$ can be constructed as follows. Let $n$ be an integer with $n \geq 3$. Let $a$ and $b$ be positive integers with $a+b=n-1$ and let $\vec{K}_{X,Y}$ be an acyclic orientation of the complete bipartite graph with bipartition $X$, $Y$, where $|X| = a$ and $|Y| = b$. Let $D_n(a, b)$ denote the digraph obtained from $\vec{K}_{X,Y}$ by adjoining a new vertex $u$, the arcs from each vertex in $X$ to $u$, and the arcs from $u$ to each vertex in $Y$. Each directed cycle of $D_n(a, b)$ has odd length. It is not difficult to show that $D_n(a, b)$ is strongly connected if and only if the indegrees of the vertices in $X$ and the outdegrees of the vertices in $Y$ are at least $1$. Assume $D_n(a, b)$ is strongly connected, and let $A$ be the adjacency matrix of $D$. Then $I_n + A$ is fully indecomposable and $\operatorname{per}(I_n + A) = \det(I_n + A)$. The number of 1's of $I_n + A$ equals $n-1+ab$. The maximum number of 1's is

$$f(n) = \begin{cases} (n^2 + 6n - 3)/4, & n \text{ odd}, \\ (n^2 + 6n - 4)/4, & n \text{ even}, \end{cases}$$

and occurs when $\{a, b\} = \{\lfloor (n-1)/2 \rfloor, \lceil (n-1)/2 \rceil\}$. The minimum number of 1's is $3n - 3$ and occurs when $\{a, b\} = \{1, n-2\}$.

CONJECTURE 4.3. The maximum number of 1's in a fully indecomposable $(0, 1)$-matrix $F$ of order $n$ satisfying $\operatorname{per} F = |\det F|$ equals $f(n)$ with equality if and only if there exist permutation matrices $P$ and $Q$ of order $n$ such that $PFQ = I_n + A$, where $A$ is the adjacency matrix of a strongly connected $D_n(\lfloor (n-1)/2 \rfloor, \lceil (n-1)/2 \rceil)$ or $D_n(\lfloor (n-1)/2 \rfloor, \lceil (n-1)/2 \rceil)$.

Note that the conjecture implies that the maximum number of arcs in a strongly connected digraph $D$ of order $n$ with no directed cycle of even length equals $f(n) - n$. It does not imply that the only strongly connected digraphs with $f(n) - n$ arcs and no directed cycles of even length are the digraphs of the classes $D_n(\lfloor (n-1)/2 \rfloor, \lceil (n-1)/2 \rceil)$ and $D_n(\lfloor (n-1)/2 \rfloor, \lceil (n-1)/2 \rceil)$. This is because there are in general permutation matrices $R$ and $S$ of order $n$ such that $I_n \leq R(I_n + A)S$, where $R(I_n + A)S - I_n$ is not the adjacency matrix of a member of these classes. This phenomenon seems to us to be the difficulty in a proof by induction on $n$.

Finally, we make the following conjecture.

CONJECTURE 4.4. Let $F$ be a fully indecomposable $(0, 1)$-matrix of order $n$ which is maximal with respect to the property that $\operatorname{per} F = |\det F|$. Then $F$ has at least $3n - 3$ 1's with equality if and only if there are permutation matrices $P$ and $Q$ such that

$$PFQ = \begin{bmatrix} 0 & 1 & \cdots & 1 \\ 1 & & & \\ \vdots & & I_{n-1} & \\ 1 & & & \end{bmatrix}.$$

## REFERENCES


1. L. Bassett, J. Maybee, and J. Quirk, *Qualitative economics and the scope of the corresponding principle*, Econometrica **36** (1968), 544–563.
2. R. A. Brualdi, *Counting permutations with restricted positions: Permanents of* (0, 1)-*matrices. A tale in four parts* (1987 Utah State University Department of Mathematics Conference Report by L. Beasley and E. E. Underwood), Linear Algebra Appl. **104** (1988), 173–183.
3. G. Engel and H. Schneider, *Cyclic and diagonal products on a matrix*, Linear Algebra Appl. **7** (1973), 301–335.
4. P. Gibson, *Conversion of the permanent into the determinant*, Proc. Amer. Math. Soc. **27** (1971), 471–476.
5. V. Klee, R. Ladner, and R. Manber, *Signsolvability revisited*, Linear Algebra Appl. **59** (1984), 131–157.
6. A. Kräuter and N. Seifter, *Some properties of the permanent of* (1, −1)-*matrices*, Linear and Multilinear Algebra **15** (1984), 207–223.
7. C. H. C. Little, *A characterization of convertible* (0, 1)-*matrices*, J. Combin. Theory Ser. B **18** (1975), 187–208.
8. J. Lundgren and J. Maybee, *A class of maximal L-matrices*, Congr. Numer. **44** (1984), 239–249.
9. H. J. Ryser, *Indeterminates and incidence matrices*, Linear and Multilinear Algebra **1** (1973), 149–157.
10. ———, *Combinatorial mathematics*, Math. Assoc. Amer., 1963.
11. P. Seymour and C. Thomassen, *Characterization of even directed graphs*, J. Combin. Theory Ser. B **42** (1987), 36–45.
12. R. Sinkhorn and P. Knopp, *Problems involving diagonal products in nonnegative matrices*, Trans. Amer. Math. Soc. **136** (1969), 67–75.
13. C. Thomassen, *Sign-nonsingular matrices and even cycles in directed graphs*, Linear Algebra Appl. **75** (1986), 27–41.
14. M. F. Tinsley, *Permanents of cyclic matrices*, Pacific J. Math. **10** (1960), 1067–1082.


DEPARTMENT OF MATHEMATICS, UNIVERSITY OF WISCONSIN, MADISON, WISCONSIN 53706
*E-mail address*: brualdi@math.wisc.edu
*E-mail address*: bshader@corral.uwyo.edu

DIMACS Series in Discrete Mathematics
and Theoretical Computer Science
Volume **4**, 1991

Applied Geometry
and Discrete Mathematics
THE VICTOR KLEE FESTSCHRIFT

# Recognizing Properties of Periodic Graphs

EDITH COHEN AND NIMROD MEGIDDO

ABSTRACT. A periodic (dynamic) graph is an infinite graph with a repetitive structure and a compact representation. A periodic graph is represented by a finite directed graph, called the dependence or the static graph, with $d$-dimensional integer vector weights associated with its edges. For every vertex in the dependence graph there corresponds a $d$-dimensional lattice in the periodic graph. For every edge $(u, v)$ in the dependence graph with vector weight $\mathbf{a}$, there are infinitely many edges in the periodic graph, namely, from every point on the lattice corresponding to $u$ to the point shifted by $\mathbf{a}$ on the lattice corresponding to $v$. Periodic graphs are used, for example, to model VLSI circuits and systems of uniform recurrence relations. In this paper we give algorithms to compute weakly connected components, to test bipartiteness, and to compute a minimum average cost spanning tree for $d$-dimensional periodic graphs.

## 1. Introduction

Periodic graphs are infinite graphs with a repetitive structure. They have a finite description (the "period") given by a directed graph with integer vector weights associated with the edges. A more formal definition follows.

DEFINITION 1.1. Given $G = (V, E, \mathbf{f})$, where $\mathbf{f}: E \to Z^d$ is a weight function on the edges of $G$, the *periodic* (*dynamic*) graph $G^* = (V^*, E^*)$ defined by $G$ has:

$$V^* = \{(\mathbf{z}, v) | \mathbf{z} \in Z^d, \ v \in V\},$$
$$E^* = \{((\mathbf{z}, u); (\mathbf{z} + \mathbf{f}(u, v), v)) | (u, v) \in E\}.$$

The *dimension* of $G^*$ is $d$. The graph $G$ is called the *dependence* or the *static* graph. For an edge $(u, v) \in E$, the edges *generated* by $(u, v)$ are $((\mathbf{z}, u); (\mathbf{z} + \mathbf{f}(u, v), v))$ $(\mathbf{z} \in Z^d)$.

See Figures 1 and 2 for examples of one- and two-dimensional periodic graphs, respectively. Note that for our purposes the periodic graph as defined

1980 *Mathematics Subject Classification* (1985 *Revision*). Primary 05CXX; Secondary 68R10, 68Q35.

Research partially supported by NSF PYI Grant CCR-8858097.


1052-1798/91 $1.00 + $.25 per page

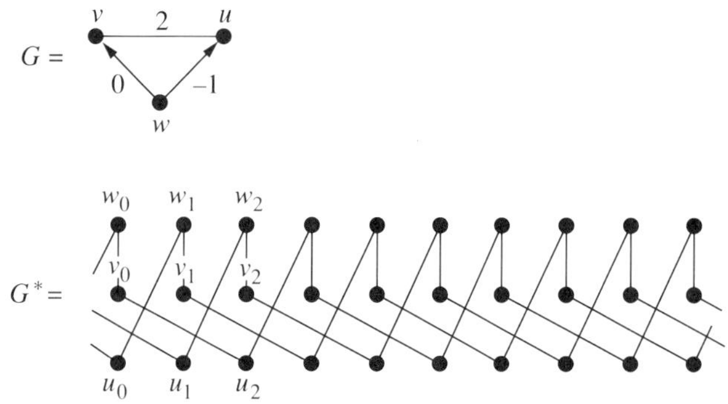


FIGURE 1. Example of a one-dimensional periodic graph.

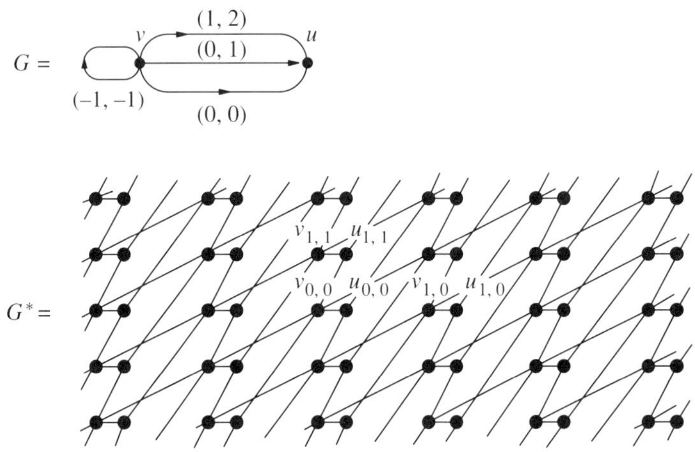


FIGURE 2. Example of a two-dimensional periodic graph.

above may be viewed as undirected. Therefore, a directed edge $(u, v) \in E$ in the dependence graph with $\mathbf{f}(u, v) = \mathbf{z}$ is equivalent to a directed edge $(v, u)$ with weight $\mathbf{f}(v, u) = -\mathbf{z}$. The undirected infinite periodic graph defined by $G$ does not change if $(u, v)$ is replaced by $(v, u)$. Periodic graphs can be used to model a system of uniform recurrence equations [**12**] and VLSI circuits [**8, 9**]. We are interested in testing properties of a periodic graph by working on the defining finite dependence graph. Previous work on periodic graphs was concerned with scheduling [**2, 12, 15**], planarity testing [**10**], cycle detection [**1, 2, 9, 12, 13**], and finding strongly connected components [**2**] of directed periodic graphs. Orlin [**14**] defined the concept of a one-dimensional dynamic/periodic graph and discussed the complexity of recognizing some properties of it. In particular, he considered finding the connected and strongly connected components, recognizing bipartiteness, and computing a minimum average cost spanning tree. Iwano [**7**] studied two-dimensional periodic graphs and gave algorithms that test bipartiteness, and compute connected components.

The current paper generalizes some of the results of Orlin and of Iwano to higher dimensions. We consider here the problems of recognizing connectivity, bipartiteness, and computing a minimum average cost spanning tree. The periodic graph of Figure 1 is bipartite and has three connected components. The periodic graph of Figure 2 is connected but is not bipartite.

§2 gives an algorithm for the connectivity problem. This algorithm is used in the succeeding sections. It is a generalization of the algorithms of [**14**, 7]. The connectivity algorithm computes dependence graphs whose corresponding periodic graphs are isomorphic to the connected components of the original periodic graph. §3 gives an algorithm that checks whether the periodic graph is bipartite. §4 gives an algorithm for finding a tree that spans a given periodic graph and has an asymptotically minimal average cost. The problems of recognizing bipartiteness and of computing a minimum average cost spanning tree in one-dimensional periodic graphs were studied by Orlin [**14**]. The problems of finding connected components and testing bipartiteness in two-dimensional periodic graphs were studied by Iwano [7]. In this paper we generalize these results to higher dimensions.

## 2. Connectivity

PROBLEM 2.1. Given $G = (V, E, \mathbf{f})$, recognize whether or not the periodic graph $G^*$ is connected. If $G^*$ is not connected, find $G_i = (V_i, E_i, \mathbf{f}_i)$, $i = 1, \ldots, r$, such that the corresponding periodic graphs $G_i^*$ are isomorphic to the connected components of $G^*$.

An algorithm that solves Problem 2.1 for one-dimensional periodic graphs was given by Orlin [**14**]. An algorithm for two-dimensional periodic graphs was given by Iwano [7]. Orlin also proved that if the dependence graph $G$ is connected, then all the connected components of $G^*$ are isomorphic, and therefore only one dependence graph suffices for describing the connected components of $G^*$. We show that the same is true in higher dimensions.

The following proposition is a restatement of Lemma 4 of [**14**]. The proof given there works for any dimension.

DEFINITION 2.2. Let $\mathbf{f}\colon E \to R^d$ be a weight function on the edges of a graph. A function $\mathbf{d}\colon V \to R^d$ is called a *potential*. Denote by $\mathbf{f}^{(\mathbf{d})}\colon E \to R^d$ the function $\mathbf{f}^{(\mathbf{d})}(i, j) = \mathbf{f}(i, j) + \mathbf{d}(i) - \mathbf{d}(j)$.

PROPOSITION 2.3. *For any potential function* $\mathbf{d}\colon V \to Z^d$, *if* $G = (V, E, \mathbf{f})$ *and* $G' = (V, E, \mathbf{f}^{(\mathbf{d})})$, *then* $G^*$ *is isomorphic to* $(G')^*$ *and the isomorphism* $\mathscr{H}\colon V^* \to (V')^*$ *is given by* $\mathscr{H}(\mathbf{z}, i) = (\mathbf{z} + \mathbf{d}(i), i)$.

DEFINITION 2.4. If $G = (V, E, \mathbf{f})$ contains a spanning tree $T \subset E$ such that for all the edges $e \in T$, $\mathbf{f}(e) = \mathbf{0}$, then $G$ is in a *basic form*.

The following corollary of Lemma 4 of [**14**] generalizes to vector weights.

COROLLARY 2.5. *Given* $G = (V, E, \mathbf{f})$ *and a spanning tree* $T \subset E$ *of* $G$, *a potential function* $\mathbf{d}$ *can be computed such that for the dependence graph with new weights* $G' = (V, E, \mathbf{f}^{(\mathbf{d})})$, *the periodic graph* $(G')^*$ *is isomorphic to* $G^*$ *and* $\mathbf{f}^{(\mathbf{d})}(e) = \mathbf{0}$ *for all* $e \in T$.

PROOF. This is done by computing the distances on the tree $T$ from one vertex to all the other vertices, and choosing $\mathbf{d}: V \to Z^d$ to be the distance function. □

We refer to the above procedure as a transformation of $G$ into a basic form.

DEFINITION 2.6. Given an integer matrix $\mathbf{A} \in Z^{d\times m}$, we define an equivalence relation $\overset{\mathbf{A}}{\sim}$ on $Z^d$ as follows. Two vectors $\mathbf{a}, \mathbf{b} \in Z^d$ are in the same equivalence class ($\mathbf{a} \overset{\mathbf{A}}{\sim} \mathbf{b}$) if and only if there exists an integer vector $\mathbf{z} \in Z^m$ such that $\mathbf{a} - \mathbf{b} = \mathbf{A}\mathbf{z}$.

REMARK 2.7. Observe that if $d = m = 1$, then the relation $a \overset{\mathbf{A}}{\sim} b$ simply means $a \equiv b \pmod{A_{11}}$. In general, an equivalence class of the relation $\overset{\mathbf{A}}{\sim}$ is simply a translate (by a vector in $Z^d$) of the lattice spanned by the columns of $\mathbf{A}$. When $m < d$, there are infinitely many equivalence classes. When $m = d$ and $\mathbf{A}$ is nonsingular, there are finitely many equivalence classes. When $m > d$, there exists a $d \times d$ matrix $\mathbf{B}$ that defines the same equivalence relations as $\mathbf{A}$ (see Problem 2.8).

PROBLEM 2.8. Let

$$\mathscr{L} = \left\{ \sum_{j=1}^{m} z_j \mathbf{a}^j | z_j \in Z \right\} \subset Z^d$$

be the lattice spanned by $m$ vectors $\mathbf{a}^1, \ldots, \mathbf{a}^m \in Z^d$. Find vectors $\mathbf{b}^1, \ldots, \mathbf{b}^l \in Z^d$, a permutation $(i_1, \ldots, i_d)$ of $\{1, \ldots, d\}$, and vectors $\mathbf{z}^1, \ldots, \mathbf{z}^m \in Z^l$ with the following properties:

(i) $\mathscr{L} = \{\sum_{j=1}^{l} z_j \mathbf{b}^j | z_j \in Z\}$.

(ii) The vectors $\mathbf{b}^j$ are linearly independent.

(iii) The permutation $(i_1, \ldots, i_d)$ is such that for $j = 1, \ldots, l$, $\mathbf{b}^j_{i_k} = 0$ for $k < j$.

(iv) For all $j \in \{1, \ldots, m\}$, $\mathbf{a}^j = \mathbf{B}\mathbf{z}^j$, where $\mathbf{B}$ is the matrix whose columns are the $\mathbf{b}^j$'s.

PROPOSITION 2.9. *Problem* 2.8 *can be solved in* $O(d^2 m B(d \log(d\|\mathbf{A}\|)))$ *time, where* $\|\mathbf{A}\| = \max_{ij} |\mathbf{a}^j_i|$ *and* $B(t)$ *bounds the time required to perform the extended Euclidean algorithm on two* $t$ *bit integers.*

PROOF. Denote by $\mathbf{A} \in Z^{d\times m}$ the matrix whose $j$th column is $\mathbf{a}^j$ ($j = 1, \ldots, m$). Note that the rows of $\mathbf{A}$ are not necessarily independent. We begin by identifying a maximal set of linearly independent rows of $\mathbf{A}$. This can be done in polynomial time by "careful" Gaussian elimination [**4**]. Without loss of generality, suppose that the first $l$ rows comprise such a set (otherwise, the rows of $\mathbf{A}$ can be permuted accordingly), and denote the submatrix by $\mathbf{A}' \in Z^{l\times m}$. Let $\mathbf{A}'' \in Z^{(d-l)\times m}$ be the submatrix consisting of the last $d-l$ rows of $\mathbf{A}$. Moreover, during the process of identifying the submatrices, we also find a $\mathbf{Y} \in Q^{(d-l)\times l}$ such that $\mathbf{A}'' = \mathbf{YA}'$, and the size of each entry of $\mathbf{Y}$ is bounded by the determinants of the square submatrices of $\mathbf{A}$ (that is, $\|\mathbf{Y}\| = O(\|\mathbf{A}\|^l)$). This computation can be done in $O(ldm)$ arithmetic operations on numbers of size $O(\|\mathbf{A}\|^l)$. Let $\mathscr{L}' \subset Z^l$ denote the lattice spanned by the columns of $\mathbf{A}'$. It is easy to verify that $(\mathbf{b}', \mathbf{b}'')$ $(\mathbf{b}' \in R^l, \mathbf{b}'' \in R^{d-l})$ is in $\mathscr{L}$ if and only if $\mathbf{b}' \in \mathscr{L}'$ and $\mathbf{b}'' = \mathbf{Yb}'$. Moreover, if $\mathbf{u}^1, \ldots, \mathbf{u}^l \in \mathscr{L}'$ span the lattice $\mathscr{L}'$, then the vectors $(\mathbf{u}^1, \mathbf{Yu}^1), \ldots, (\mathbf{u}^l, \mathbf{Yu}^l)$ span the lattice $\mathscr{L}$.

Denote the columns of $\mathbf{A}'$ by $\mathbf{a}'^j$. Observe that the lattice $\mathscr{L}'$ spanned by the $\mathbf{a}'^j$'s does not change if any vector $\mathbf{a}'^j$ is replaced by $-\mathbf{a}'^j$, or by $\mathbf{a}'^j + \mathbf{a}'^k$ or $\mathbf{a}'^j - \mathbf{a}'^k$ for some $k \neq j$. By a classical theorem of Hermite **[6]** (note that $\mathbf{A}'$ has a full row rank), the matrix $\mathbf{A}'$ can be triangulated by elementary (unimodular) column operations, i.e., the only operations used in this transformation are additions of integral multiples of one column to another, and exchanges of columns. Neither of these operations alters the lattice spanned by the columns of $\mathbf{A}'$. The matrix $\mathbf{A}'$ is transformed so that the last $m-l$ $(l \leq d)$ columns become zero, and the first $l$ columns comprise an integer lower triangular matrix with nonzero diagonal entries. If the diagonal entries are maximal in their rows, then the triangular matrix is unique and is called the Hermite normal form of $\mathbf{A}'$. It follows from Hermite's proof that this can be accomplished in at most $lm$ applications of the extended Euclidean algorithm for the g.c.d. (of two numbers) plus $O(l^2m)$ arithmetic operations on integers. Kannan and Bachem **[11]** proposed an improvement of this algorithm, where all the operations are done on integers of polynomial size. Hafner and McCurley **[5]** proposed an asymptotically faster algorithm where all integers encountered do not exceed $l^l\|\mathbf{A}'\|^{2l}$. Their algorithm runs in $O(l^2mB(l\log(l\|\mathbf{A}\|)))$ bit operations. Moreover, for the simpler task of triangulating $\mathbf{A}'$, they used fast matrix multiplication to get even better time bounds. Using $O(l^{\theta-1}m\log(2m/l)B(l\log(l\|\mathbf{A}'\|)))$ bit operations (where $\theta < 2.376$, following Coppersmith and Winograd **[3]**), the algorithm computes a triangular matrix whose entries have absolute values bounded by $l^{l/2}\|\mathbf{A}'\|^l$.

Assume $\mathbf{A}'$ has been transformed into triangular form, replace $\mathbf{A}''$ by $\mathbf{YA}'$. Note that the new $\mathbf{A}''$ has integer entries of absolute values $O(l^{l/2+1}\|\mathbf{A}\|^{2l})$. Thus, the new matrix $\mathbf{A}$ has for every $k$, $k = 1, \ldots, l$, $A_{kk} \neq 0$ and $A_{kj} = 0$ for $j > k$. Also, for $k = l+1, \ldots, d$, $A_{kj} = 0$ for all $j > l$. Finally, let the vector $\mathbf{b}^j$ be equal to the $j$th column of the matrix $\mathbf{A}$ $(j = 1, \ldots, l)$. Note that properties (i) and (ii) of Problem 2.8 imply that $l \leq d$ and that the permutation $(i_1, \ldots, i_d)$ and the vectors $\mathbf{z}^1, \ldots, \mathbf{z}^m$ exist. Given the $\mathbf{b}^j$'s, the vectors $\mathbf{z}^j$ can be computed easily. □

COROLLARY 2.10. (i) *If* $\mathbf{A}$ *has full row rank, then Problem* 2.8 *can be solved in*

$$O(l^{\theta-1} m \log(2m/l) B(l \log(l\|\mathbf{A}'\|)))$$

*bit operations (where* $\theta < 2.376$*).*

(ii) *When* $d$ *is fixed, Problem* 2.8 *can be solved in* $O(m \log(\|\mathbf{A}\|))$ *arithmetic operations.*

(iii) *If an integer multiple* $h$ *of* $\det \mathbf{B}$ *is given, then it follows from Theorem* 1 *of* [5] *that Problem* 2.8 *can be solved in* $O(d^2 m B(h))$ *arithmetic operations.*

Algorithm 2.11 solves Problem 2.1. It first transforms the weights of $G$ to a basic form. The algorithm continues and solves Problem 2.8 with respect to the transformed edge weights. The number of connected components of $G^*$ is $\det \mathbf{B}$.

ALGORITHM 2.11. (Connectivity). (i) *Compute a spanning tree* $T \subset E$ *of* $G$ *and the appropriate potential function* $\mathbf{d}$, *and transform* $\mathbf{f}$ *to a basic form, i.e., replace* $\mathbf{f}$ *by* $\mathbf{f}^{(\mathbf{d})}$ *(see Corollary* 2.5*).*

(ii) *Denote by* $\mathbf{a}^1, \ldots, \mathbf{a}^m$ *the vector weights in the range of the (transformed)* $\mathbf{f}$. *Solve Problem* 2.8 *with respect to the* $\mathbf{a}^j$*'s, i.e., compute the vectors* $\mathbf{b}^j, \mathbf{z}^j$.

(iii) *If* $l = d$, *then the number of connected components of* $G^*$ *is* $\prod_{i=1}^d B_{ii}$. *Otherwise, there are infinitely many connected components, each of which is an* $l$*-dimensional periodic graph. Consider the new edge weights* $\tilde{\mathbf{f}}: E \to Z^l$ *given by* $\tilde{\mathbf{f}}(e) = \mathbf{z}^j$, *where* $\mathbf{f}(e) = \mathbf{a}^j$. *Construct the dependence graph* $\widetilde{G} = (V, E, \tilde{\mathbf{f}})$. *The periodic graph* $(\widetilde{G})^*$ *is isomorphic to each of the connected components of* $G^*$. *The isomorphism* $\mathscr{H}: \widetilde{V}^* \to V^*$ *is given by* $\mathscr{H}(\mathbf{a}, u) = (\mathbf{Ba} + \mathbf{z}, u)$, *so that different connected components have different values of* $\mathbf{z} \in Z^d$ *associated with them.*

The correctness of the algorithm is proved in the following proposition.

PROPOSITION 2.12. *Suppose* $G = (V, E, \mathbf{f})$ *is connected and* $\mathbf{f}$ *is in a basic form. Denote by* $\mathbf{B} \in Z^{d \times l}$ *the matrix sought in Problem* 2.8 *with respect to the vectors* $\mathbf{f}(e)$ $(e \in E)$. *Under these conditions:*

(i) *If* $l = d$, *then the number of connected components of* $G^*$ *is* $\prod_{i=1}^d B_{ii}$.

(ii) *The connected components of* $G^*$ *correspond to the equivalence classes of* $\overset{\mathbf{B}}{\sim}$, *i.e., two vertices* $(\mathbf{a}, u), (\mathbf{b}, v) \in V^*$ *are in the same connected component if and only if* $\mathbf{a} \overset{\mathbf{B}}{\sim} \mathbf{b}$.

(iii) *The graph* $(\widetilde{G})^*$ *is connected and isomorphic to each of the connected components of* $G^*$.

PROOF. For the proof of part (i), note that $\det \mathbf{B}$ is the volume of a "cell" of the lattice, which is the same as the number of distinct lattices that can be obtained by an integral translation of the given lattice. To prove part (ii) observe that $(\mathbf{a}, u)$ and $(\mathbf{b}, v)$ are in the same connected component if and only if $\mathbf{a} - \mathbf{b}$ equals an integer combination of the vectors $\mathbf{f}(E)$. Part (iii) follows from part (ii). The isomorphism $\mathscr{H}: \widetilde{V}^* \to V^*$ is given by $\mathscr{H}(\mathbf{a}, u) = (\mathbf{Ba} + z, u)$. □

REMARK 2.13. The connected components of $G^*$ are $l$-dimensional periodic graphs. Observe that when the weights $\mathbf{f}$ are in the basic form, then $l$ is the dimension of the vector space spanned by $\mathbf{f}(E)$.

Denote $m = |E|$ and $\|\mathbf{f}\| = \max_{e \in E} \|\mathbf{f}(e)\|_\infty$. Denote by $\mathrm{conn}(m, \|\mathbf{f}\|, d)$ the number of operations required by the connectivity algorithm.

PROPOSITION 2.14. *Let $B(t)$ be as in Proposition 2.9. We have*

$$\mathrm{conn}(m, \|\mathbf{f}\|, d) = O(d^2 m B(d \log(d\|\mathbf{f}\|))).$$

PROOF. The computation done by Algorithm 2.11 amounts to transforming $G$ to a basic form, and to a solution of an instance of Problem 2.8. Recall that the transformation to a basic form involves a computation of a spanning tree, and computing single source shortest paths on the tree. Therefore, the transformation is done in $O(m)$ operations. The proof follows from Proposition 2.9. □

COROLLARY 2.15. *The following are consequences of Corollary 2.10:*

(i) *If the connected components of $G^*$ are $l$-dimensional periodic graphs, then Algorithm 2.11 requires $O(l^{\theta-1} m \log(2m/l) B(l \log(l\|\mathbf{f}\|)))$ operations ($\theta < 2.376$).*

(ii) *If the dimension $d$ is fixed, the algorithm requires $O(m \log \|\mathbf{f}\|)$ operations.*

(iii) *Consider the problem of recognizing whether the number of connected components of $G^*$ equals $k$, where $k$ is some fixed integer. It follows from Corollary 2.10(iii) that this task can be done in $O(d^2 m)$ time.*

## 3. Bipartiteness

In this section we consider the following problem.

PROBLEM 3.1. Given a dependence graph $G = (V, E, \mathbf{f})$, decide whether or not $G^*$ is bipartite.

DEFINITION 3.2. Given a dependence graph $G$, define $\widetilde{G} = (V, \widetilde{E}, \tilde{\mathbf{f}})$ as follows. Let $\widetilde{E}$ be the set of pairs $(u, v) \in V \times V$ such that there exists $w \in V$ with $\{(u, w), (w, v)\} \subseteq E$. For such pairs let $\tilde{\mathbf{f}}(u, v) = \mathbf{f}(u, w) + \mathbf{f}(w, v)$.

REMARK 3.3. Note that the graph $\widetilde{G}$ is well defined even when $G$ does not have edge weights. In this case the edges of $\widetilde{G}$ correspond to paths of length 2 in $G$. If $G$ is connected, then it is bipartite if and only if $\widetilde{G}$ has two connected components. If $G$ is a nonbipartite connected graph, then $\widetilde{G}$ is connected.

We assume in this section that $G^*$ is connected. If this is not true, then the dependence graphs that define the connected components of $G^*$ should be considered independently.

ALGORITHM 3.4. (Bipartiteness). (i) *Construct* $\tilde{G}$ *as described in Definition* 3.2.

(ii) *If* $\tilde{G}$ *is not connected, then* $G^*$ *is bipartite. Stop.*

(iii) *Otherwise, the graph* $G^*$ *is bipartite if and only if* $(\tilde{G})^*$ *is not connected.*

**Correctness.** Observe that $(\tilde{G})^* = \tilde{G}^*$, that is, $(u, v)$ is an edge of $(\tilde{G})^*$ if and only if there is a path of length two between $u$ and $v$ in $G^*$. Therefore, it follows from Remark 3.3 that $G^*$ is bipartite if and only if the graph $(\tilde{G})^*$ is not connected. This condition is checked in steps (ii) and (iii) of the algorithm.

**Complexity.** The algorithm amounts to constructing the graph $\tilde{G}$, and then testing if the periodic graphs corresponding to $\tilde{G}$ (or to its connected components) have one or two connected components. The construction of $\tilde{G}$ requires $O(m^2)$ time. Observe that $|\tilde{E}| = O(|E|^2)$. Thus, the time complexity of testing if $\tilde{G}^*$ has one or two connected components is $O(d^2m^2)$ (see Corollary 2.15(iii)). It follows that the time complexity of the algorithm is $O(d^2m^2)$.

## 4. Minimum average cost spanning tree

Orlin **[14]** defined a minimum average cost spanning tree of a one-dimensional periodic graph and gave an algorithm that computes one. We extend his definition to higher dimensions as follows.

DEFINITION 4.1. Let $G = (V, E, \mathbf{f}, c)$ be a dependence graph together with edge *costs* given by $c: E \to R^+$.

(i) Extend $c$ to the edges of $G^*$ in the obvious way: $c((\mathbf{z}, u); (\mathbf{z} + \mathbf{f}(u, v), v)) = c(u, v)$.

(ii) For any positive integer $n$, denote by $G^n = (V^n, E^n)$ the finite subgraph of $G^*$ induced by the set of vertices $(\mathbf{z}, v)$ $(v \in V)$ with $|z_i| \le n$ $(i = 1, \dots, d)$.

(iii) A *minimum average cost spanning tree* (MACST) $T^* \subset E^*$ of $G^*$ is a spanning tree for which the limit

$$\lim_{n\to\infty} (2n)^{-d} \sum_{e \in E^n \cap T^*} c(e)$$

exists and is minimal among all trees for which such a limit exists.

(iv) A spanning tree $T^*$ is said to be *optimal* if each finite subtree $T_1^* \subset T^*$ is a minimum cost spanning tree of the subgraph of $G^*$ induced by the nodes of $T_1^*$.

DEFINITION 4.2. Let $\mathbf{B} \in Z^{d\times d}$ and let $\overset{\mathbf{B}}{\sim}$ be the equivalence relation on $Z^d$ defined by $\mathbf{B}$ in Definition 2.6. Let $\mathscr{L}$ be the lattice spanned by the columns of $\mathbf{B}$. For any $\mathbf{a} \in Z^d$, denote by $\mathscr{L}^{(\mathbf{a})}$ the lattice spanned by the columns of $\mathbf{B}$ together with $\mathbf{a}$. Denote by $\mathbf{B}^{(\mathbf{a})} \in Z^{d\times d}$ a matrix whose columns span the lattice $\mathscr{L}^{(\mathbf{a})}$. Denote by $R(\mathbf{B}) \subset Z^d$ any set of representatives of the equivalence classes of the relation $\overset{\mathbf{B}}{\sim}$.

REMARK 4.3. If $\mathbf{B}$ is full rank and triangular, then a triangular matrix $\mathbf{B}^{(\mathbf{a})}$ can be computed within $O(d^2 B(\log\det\mathbf{B}))$ time, which amounts to $O(d^2\log\det\mathbf{B})$ arithmetic operations (see [**5**]).

Algorithm 4.4 below computes an optimal MACST of $G^*$. The algorithm maintains a set of isomorphic trees and a matrix $\mathbf{B}$, such that the equivalence classes of $\overset{\mathbf{B}}{\sim}$ correspond to these trees. The infinite tree $T^*$ computed by the algorithm is represented by a collection of pairs $(e, \mathbf{B})$, where $e \in E$ and $\mathbf{B} \in Z^{d\times i}$ is an integer matrix. Each such pair represents a set of edges of $G^*$:

(i) When $i < d$, each such pair represents infinitely many edges: for $(u, v) \in E$, $((\mathbf{z}, u); (\mathbf{z}+\mathbf{f}(u, v), v)) \in T^*$ if and only if $\mathbf{z} \in R(\mathbf{B})$. The algorithm generates one such pair for each $i < d$.

(ii) When $i = d$, the matrix $\mathbf{B}$ is nonsingular. The algorithm may produce as many as $m - d$ such pairs. Each pair represents finitely many edges: for $(u, v) \in E$, $((\mathbf{z}, u); (\mathbf{z}+\mathbf{f}(u, v), v)) \in T^*$ if and only if $\mathbf{z} \in R(\mathbf{B})$ and $\mathbf{z} \overset{\mathbf{B}}{\not\sim} \mathbf{y}$ for all $\mathbf{y} \in R(\mathbf{B}^{(\mathbf{f}(e))})$. Observe that the number of edges that correspond to such pairs is $|R(\mathbf{B})| - |R(\mathbf{B}^{(\mathbf{f}(e))})|$.

Each edge $e \in E$ can occur in at most one pair $(e, \mathbf{B})$. Algorithm 4.4 below consists of two phases. During the first phase (steps (iii), (iv)), pairs with $i < d$ are selected. During the second phase (steps (v), (vi)), pairs with $i = d$ are selected. Within each phase, the algorithm is greedy and selects the edges (with a certain desired property) according to increasing cost. Without loss of generality, assume $G^*$ is connected.

ALGORITHM 4.4. (Optimal Minimum Average Cost Spanning Tree).

(i) *Find a minimum cost spanning tree $T$ of $G$ and transform $G$ to the basic form corresponding to $T$ (see Corollary* 2.5).

(ii) *Initialize* $T^* \leftarrow \{(t, \{\mathbf{0}\}) | t \in T\}$.

(iii) *Do step* (iv) *for* $i = 1, \ldots, d$, *and then go to* (iv).

(iv) *Find an edge $e_i \in E$ of minimum cost, such that* $\mathbf{f}(e_i)$ *is linearly independent of the weights of previously selected edges:* $\mathbf{f}(e_j)$ $(j < i)$. *Let* $\mathbf{B} \in Z^{d\times i}$ *be the matrix whose columns are the vectors* $\mathbf{f}(e_1), \ldots, \mathbf{f}(e_i)$. *Add the pair* $(e_i, \mathbf{B})$ *to the forest* $T^*$.

(v) *At this point* $\mathbf{B} \in Z^{d\times d}$ *is a square matrix, and* $T^*$ *consists of* $\det\mathbf{B}$ *isomorphic trees. Transform* $\mathbf{B}$ *into a triangular matrix that defines the same lattice (see Proposition* 2.9). *Repeat step* (vi) *until* $\det\mathbf{B} = 1$.

(vi) *Select an edge* $e$ *of minimal cost such that* $\mathbf{f}(e) \neq \mathbf{Bz}$ *for all* $\mathbf{z} \in Z^d$. *Add the pair* $(e, \mathbf{B})$ *to* $T^*$. *Update* $\mathbf{B}$ *to be* $\mathbf{B}^{(\mathbf{f}(\mathbf{e}))}$ (*see Definition* 4.2 *and Remark* 4.3).

**Correctness.** Denote by $e_1, \dots, e_s$ $(d \leq s \leq m)$ the edges selected by the algorithm. The edges $e_1, \dots, e_d$ are selected during step (iv). First we prove that $T^*$ is indeed a tree.

PROPOSITION 4.5. *Throughout the execution of the algorithm, the connected components of* $T^*$ *constitute a collection of isomorphic trees corresponding to the equivalence classes of the relation* $\overset{\mathbf{B}}{\sim}$.

PROOF. Consider any point during the execution of the algorithm and suppose $e_1, \dots, e_k$ are the edges that have been selected so far. Recall that two vectors are in the same equivalence class of $\overset{\mathbf{B}}{\sim}$ if and only if they differ by an integer linear combination of the vectors $\mathbf{f}(e_1), \dots, \mathbf{f}(e_k)$. First, observe that after step (ii) of the algorithm, $T^*$ consists of a collection of finite trees. The vertex sets of these trees have the form $\{(i, v) | v \in V\}$. The proposition follows by induction on $k$. □

The graph $G^*$ is connected. Therefore, if all edges are selected, we must have $\det \mathbf{B} = 1$. Thus the algorithm is guaranteed to terminate with $\det \mathbf{B} = 1$. It follows from Proposition 4.5 that when the algorithm terminates the forest $T^*$ has one connected component.

PROPOSITION 4.6. *The tree* $T^*$ *computed by the algorithm has a minimum average cost.*

PROOF. Observe that the set $\overline{T}$ of edges of $G^*$ corresponding to the set $T \cup \{e_1\}$ is contained in $T^*$ and does not contain a cycle. Thus, $\overline{T}$ is a forest of $G^*$. We will show that it can be converted into a tree by adding an asymptotically small number of edges from $E^*$, so that the limit of the average cost is the same as in $\overline{T}$.

Consider the intersection of $\overline{T}$ with $G^n$. Imagine running any greedy algorithm for a minimum spanning tree of $G^n$. We show that such an algorithm selects all the edges of $\overline{T} \cap G^n$. Obviously, the edges generated by $T$ do not introduce any cycles, and they are of minimum cost. All the other edges of $G$, whose weights equal $\mathbf{0}$, generate edges of $G^*$ that introduce cycles. However, the algorithm selects the next minimum cost edge $e_1 \in E$ such that $\mathbf{f}(e_1) \neq \mathbf{0}$, and adds to $T^*$ all the edges generated by $e_1$. Thus, no cycle is introduced.

By now, we have a forest of isomorphic trees of size $\Omega(n/\|\mathbf{f}\|)$ that cover $G^n$. The asymptotic average cost of the forest is

$$\lim_{n\to\infty} \frac{(2n)^d c(T \cup \{e_1\})}{(2n)^d} = c(T \cup \{e_1\}).$$

This gives a lower bound on the asymptotic average cost of the spanning tree of $G^*$. Observe that it is an upper bound as well, since the number of additional edges needed to complete this forest to a tree is $O((2n)^{d-1}\|\mathbf{f}\|)$. It is easy to see that, regardless of the choice of these additional edges, the asymptotically minimum average cost of the tree is the same as that of the initial forest, i.e., $c(T \cup \{e_1\})$. □

PROPOSITION 4.7. *The tree* $T^*$ *is optimal* (*see Definition* 4.1(iv)).

PROOF. To prove optimality, consider some subtree $T_1^*$ of $T^*$. Denote by $H$ the subgraph of $G^*$ induced by the nodes of $T_1^*$. We need to show that $T_1^*$ is a minimum cost spanning tree of $H$, i.e., the edges of $T$ are the same as the ones picked by some greedy algorithm. Algorithm 4.4 is greedy within each of its two phases. Edges are selected according to increasing cost. When an edge $e$ is selected, the algorithm adds to $T^*$ a maximal subset of the edges generated by $e$, so that $T^*$ remains acyclic. We need to show that during the first phase of the algorithm, no two trees that are connected by an edge can be connected by an edge of a lower cost. Consider the state before choosing a new edge in the first phase. Observe that all vertices of a connected component always lie on a "flat" which is a translate of the subspace spanned by the columns of $\mathbf{B}$ (see Proposition 4.5). The weight of any edge selected by the algorithm during step (iv) is linearly independent of the columns of $\mathbf{B}$. Thus, the new edges must lie between two such "flats". All edges of lower cost are linearly dependent on the columns of $\mathbf{B}$. Therefore, they lie within such flats. It follows that the new edges added to $T^*$ are minimum cost edges, linking connected components of the current $T^*$. Hence, they occur in some minimum cost spanning tree. □

The algorithm amounts essentially to computing a minimum cost spanning tree in $G$, and then computing the matrix $\mathbf{B}$ and updating it at most $m$ times (see Remark 4.3). Thus, the complexity is $O(n \log n + md^2 B(d \log(d\|\mathbf{f}\|)))$.

## REFERENCES


1. E. Cohen and N. Megiddo, *Strongly polynomial and NC algorithms for detecting cycles in dynamic graphs*, Proc. 21st Annual ACM Sympos. on Theory of Computing, ACM, 1989, pp. 523–534.
2. ——, *Strongly polynomial time and NC algorithms for detecting cycles in periodic graphs*, IBM Research Report RJ 7587 (70764), IBM Almaden Research Center, San Jose, Ca., July 1990.
3. D. Coppersmith and S. Winograd, *Matrix multiplication via arithmetic progressions*, Proc. 19th Annual ACM Sympos. on Theory of Computing, ACM, 1987, pp. 1–6.
4. J. Edmonds, *Systems of distinct representatives and linear algebra*, J. Res. Nat. Bur. Standards **71B** (1967), 241–245.
5. J. L. Hafner and K. S. McCurley, *Asymptotically fast triangularization of matrices over rings*, IBM Research Report RJ 6921 (66027), IBM Almaden Research Center, San Jose, Ca., July 1989. Also in: Proc. 1st Annual ACM-SIAM Sympos. on Discrete Algorithms, ACM-SIAM, 1990, 194–200.
6. C. Hermite, *Sur l'introduction des variables continues dans la théorie des nombres*, J. Reine Angew. Math. **41** (1851), 431–451.
7. K. Iwano, *Some problems on doubly periodic infinite graphs*, Tech. Rep. CS-TR-078-87, Princeton University, 1987.

8. K. Iwano and K. Steiglitz, *Optimization of one-bit full adders embedded in regular structures*, IEEE Trans. Acoust. Speech Signal Process., IEEE, New York, 1986.
9. ——, *Testing for cycles in infinite graphs with periodic structure*, Proc. 19th Annual ACM Sympos. on Theory of Computing, ACM, 1987, pp. 46–53.
10. ——, *Planarity testing of doubly connected periodic infinite graphs*, Networks **18** (1988), 205–222.
11. R. Kannan and A. Bachem, *Polynomial algorithms for computing the Smith and Hermite normal forms of an integer matrix*, SIAM J. Comput. **8** (1979), 499–507.
12. R. M. Karp, R. E. Miller, and S. Winograd, *The organization of computations for uniform recurrence equations*, J. Assoc. Comput. Mach. **14** (1967), 563–590.
13. K. S. Kosaraju and G. F. Sullivan, *Detecting cycles in dynamic graphs in polynomial time*, Proc. 27th Annual IEEE Sympos. on Foundations of Computer Science, ACM, 1988, pp. 398–406.
14. J. B. Orlin, *Some problems in dynamic/periodic graphs*, in Progress in Combinatorial Optimization (W. R. Pullyblank, ed.), Academic Press, Orlando, Florida, 1984, pp. 273–293.
15. V. P. Roychowdhury and T. Kailath, *Study of parallelism in regular iterative algorithms*, Proc. 1990 ACM Sympos. on Parallel Algorithms and Architectures, ACM, 1990, pp. 367–376.

Department of Computer Science, Stanford University, Stanford, California 94305 and IBM Almaden Research Center, San Jose, California 95120-6099
*E-mail address*: edith@cs.stanford.edu

IBM Almaden Research Center, San Jose, California 95120-6099 and School of Mathematical Sciences, Tel Aviv University, Tel Aviv, Israel
*E-mail address*: megiddo@ibm.com

DIMACS Series in Discrete Mathematics
and Theoretical Computer Science
Volume 4, 1991

Applied Geometry
and Discrete Mathematics
THE VICTOR KLEE FESTSCHRIFT

# On Generic Global Rigidity

ROBERT CONNELLY

ABSTRACT. It is shown that the conditions, generic redundant rigidity and 4-connectivity, are not sufficient for a graph to be globally rigid in three-space at a generic configuration. The example is $K_{5,5}$.

## 1. Introduction

Given a collection of points in some Euclidean space and the distances between some pairs of the points, when can we find another collection of points that have the same distances between corresponding points? Of course, we can simply move the whole collection by a rigid congruence, but when is there another essentially different configuration? This is what we call the question of global rigidity, and we regard it as a fundamental geometric problem.

Conditions for global rigidity, that is conditions for unique graph embeddings (see Hendrickson [7]), are related to the existence of molecular conformations (see Crippen and Havel [4] as well as the rigidity of frameworks and the structural stability of frameworks.

Here we investigate the problem of global rigidity when the configuration of points is in "generic" position. Roughly speaking, this means that the points are "randomly" chosen so that no special constraints are satisfied. In [7] it is conjectured that certain other well-known conditions on the graph are equivalent to this generic global rigidity. We show in dimension three that for at least one graph, the complete bipartite graph $K_{5,5}$, Hendrickson's conditions are not sufficient.

The author gratefully thanks Bruce Hendrickson, Maria Terrel, Bob Walker, and all three referees for their many helpful comments and corrections.

## 2. Basic definitions

**2.1. A framework.** Let $G$ be a finite graph (with no loops or multiple edges) with $v$ vertices, and let $(\mathbf{p}_1, \mathbf{p}_2, \dots, \mathbf{p}_v)$ be a corresponding set of

1980 *Mathematics Subject Classification* (1985 *Revision*). Primary 51K99; Secondary 51A20, 73G49.

Partially supported by N.S.F. Grant Number MCS-790251.

$v$ points in $\mathbf{R}^d$. Note that these points are not necessarily distinct. We call this labeled collection of points $(\mathbf{p}_1, \mathbf{p}_2, \dots, \mathbf{p}_v) = \mathbf{p}$ a *configuration* in $\mathbf{R}^d$. For every edge in $G$ we imagine that there is a rigid bar joining the corresponding points of the configuration. The graph $G$ together with the configuration $\mathbf{p}$ is denoted by $G(\mathbf{p})$ and is called a *framework in* $\mathbf{R}^d$. We often denote such a framework as in Figure 2.1.1. Note that the bars of the framework are allowed to overlap without necessarily having a vertex at the intersection.

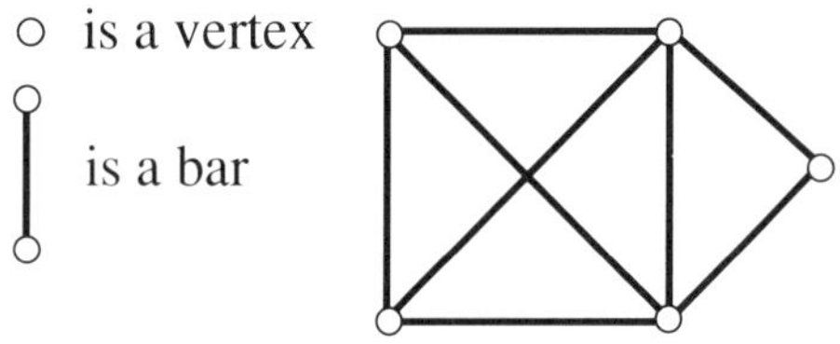


FIGURE 2.1.1

**2.2. Equivalent framework in $\mathbf{R}^d$ and global rigidity.** We say that two frameworks $G(\mathbf{p})$ and $G(\mathbf{q})$ are *equivalent*, written as $G(\mathbf{p}) \cong G(\mathbf{q})$, if whenever $\{i, j\}$ corresponds to an edge of $G$, then

$$|\mathbf{p}_i - \mathbf{p}_j| = |\mathbf{q}_i - \mathbf{q}_j|,$$

where $|\mathbf{p}|$ is the Euclidean norm of the vector $\mathbf{p}$. In other words, corresponding bars of the two frameworks have the same length.

Note, however, that there is a very simple way to obtain equivalent frameworks. Namely we can take the configuration $\mathbf{p}$ and apply any fixed rigid congruence of $\mathbf{R}^d$ to each point $\mathbf{p}_i$ to get $\mathbf{q}_i$ and thus another configuration $\mathbf{q}$. In this case we say that $\mathbf{p}$ is *congruent* to $\mathbf{q}$, written as $\mathbf{p} \approx \mathbf{q}$.

We say that $G(\mathbf{p})$ is *globally rigid in* $\mathbf{R}^d$ if for any configuration $\mathbf{q}$, $G(\mathbf{q}) \cong G(\mathbf{p})$ implies that $\mathbf{p} \approx \mathbf{q}$. In other words, when $G(\mathbf{p})$ is globally rigid, only congruent configurations give equivalent frameworks. For example, in $\mathbf{R}^2$ Figure 2.2.1(a) is equivalent to Figure 2.2.1(b), but they are not congruent. Figure 2.2.1(c) is globally rigid in any $\mathbf{R}^d$ for $d \geq 2$.

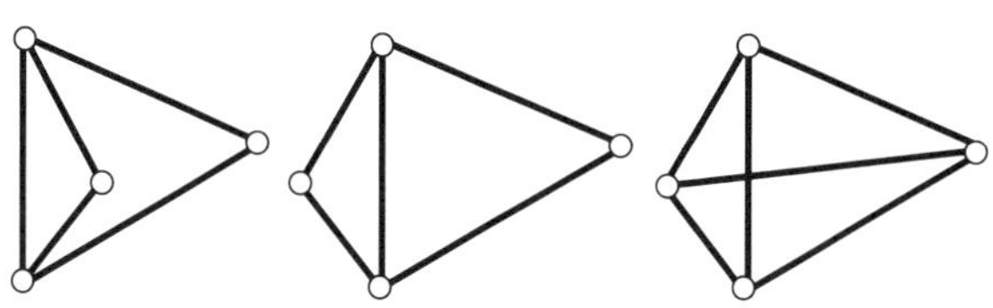

(a) (b) (c)

FIGURE 2.2.1

**2.3. Generic configurations.** Consider any configuration $\mathbf{p}$ in $\mathbf{R}^d$, but regard $\mathbf{p}$ as a point in $\mathbf{R}^{dv}$ by looking at all $dv$ of its coordinates. We say that $\mathbf{p}$ is a *generic configuration* if all of its $dv$ coordinates are algebraically independent over the rationals, or, in other words, if the coordinates of $\mathbf{p}$ do not satisfy any nonzero polynomial equation with integer coefficients.

If a configuration is generic, its points cannot have any undue symmetries and they cannot satisfy any "special" conditions. Whether the configuration $\mathbf{p}$ is generic or not can influence whether $G(\mathbf{p})$ is globally rigid. For example, the framework in Figure 2.3.1(a) is not globally rigid in $\mathbf{R}^2$. However, when the same graph is chosen but with a generic configuration, as in Figure 2.3.1(b), then it is globally rigid.

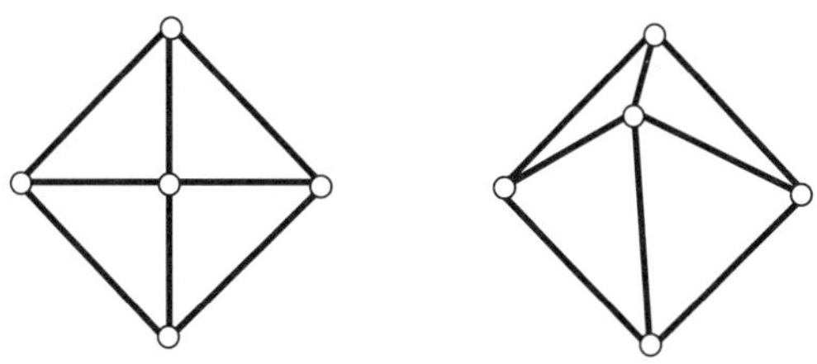

(a) (b)

FIGURE 2.3.1

Our basic problem then is to determine, for a given graph $G$, whether $G(\mathbf{p})$ is globally rigid for all generic configurations $\mathbf{p}$. This also brings up the following question, first raised by Maria Terrell: Is global rigidity a generic property? In other words, if $G(\mathbf{p})$ is globally rigid for one generic configuration $\mathbf{p}$, is $G(\mathbf{q})$ globally rigid for all other generic configurations $\mathbf{q}$? The answer seems to be yes, but we do not have a proof. See §3.2, where a related concept "generic rigidity" is discussed.

## 3. Related basic results and definitions

**3.1. Infinitesimal rigidity.** An important form of rigidity is a linearization known as infinitesimal rigidity. Let $G(\mathbf{p})$ be a framework in $\mathbf{R}^d$. Let $\mathbf{p}' = (\mathbf{p}'_1, \mathbf{p}'_2, \dots, \mathbf{p}'_v)$ be another configuration in $\mathbf{R}^d$ but regarded as a vector $\mathbf{p}'_i$, a "velocity", associated to each point $\mathbf{p}_i$. We say that $\mathbf{p}'$ is an *infinitesimal flex* or an *infinitesimal motion* if for all bars $\{i, j\}$ in $G$, the following holds:

$$(\mathbf{p}_i - \mathbf{p}_j)(\mathbf{p}'_i - \mathbf{p}'_j) = 0, \tag{3.1}$$

where the product of vectors is the usual scalar product.

However, for any configuration $\mathbf{p}$ there are certain infinitesimal flexes that always exist. These *trivial infinitesimal flexes* can be described as the derivative at time 0 of a path of congruences of $\mathbf{R}^d$ (starting at the identity), or explicitly $\mathbf{p}_i' = S\mathbf{p}_i + \mathbf{t}$, where $S$ is a $d$ by $d$ skew symmetric matrix and $\mathbf{t}$ is a vector fixed and independent of $i$.

If a framework $G(\mathbf{p})$ has only trivial infinitesimal flexes, then we say $G(\mathbf{p})$ is *infinitesimally rigid in* $\mathbf{R}^d$. For example, the frameworks of Figures 3.1.1(a) and 3.1.1(b) are infinitesimally rigid in $\mathbf{R}^2$ and $\mathbf{R}^3$, respectively, and the frameworks of Figures 3.1.1(c) and 3.1.1(d) are not infinitesimally rigid in $\mathbf{R}^2$ and $\mathbf{R}^3$, respectively. The four top vertices of Figure 3.1.1(d) are coplanar. The infinitesimal motion or velocities are indicated on Figures 3.1.1.(c) and 3.1.1.(d).

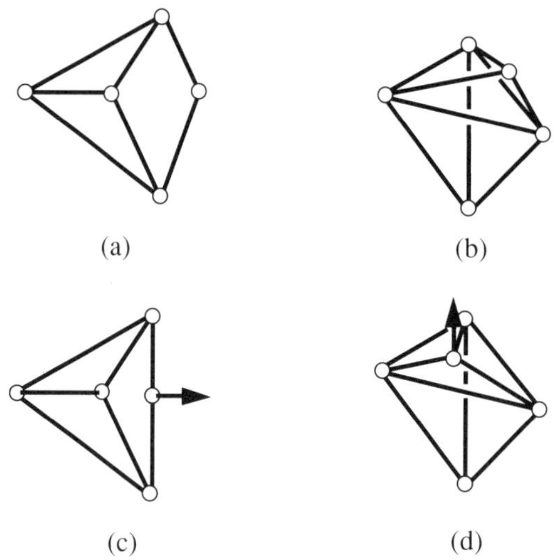


FIGURE 3.1.1

**3.2. Generic rigidity.** If $G(\mathbf{p})$ is infinitesimally rigid for one configuration $\mathbf{p}$, then $G(\mathbf{q})$ may or may not be infinitesimally rigid for another configuration $\mathbf{q}$. Nevertheless, there is a "typical" behavior for a given graph $G$. We say that a graph $G$ is *generically rigid in* $\mathbf{R}^d$ if for some configuration $\mathbf{p}$, $G(\mathbf{p})$ is infinitesimally rigid in $\mathbf{R}^d$. The reason we call this "generic" rigidity is the following basic result from algebraic geometry. (See Gluck [**6**], Asimow and Roth [**2**], or Connelly [**5**] for a proof.)

THEOREM 3.2.1. *If the graph $G$ is generically rigid in $\mathbf{R}^d$ and $\mathbf{p}$ is any generic configuration, then $G(\mathbf{p})$ is infinitesimally rigid in $\mathbf{R}^d$.*

By virtue of generic rigidity being a generic property for configurations, it is only a property of the graph $G$ and not any particular configuration $\mathbf{p}$. See Sugihara [**9**], Lovàsz and Yemini [**8**], or Hendrickson [**7**] for a deterministic polynomial time complexity algorithm that decides when a graph $G$ is generically rigid in $\mathbf{R}^2$. For $\mathbf{R}^3$ no such algorithm is known, and indeed, the search for such an algorithm is a well-known unsolved problem.

**3.3. Graph connectivity.** We say that a graph $G$ is *vertex n-connected* or just *n-connected* if one needs to remove at least $n$ vertices from $G$ to disconnect the remaining vertices of $G$. For example, the graph of Figure 3.3.1(a) is 3-connected but not 4-connected. The graph of Figure 3.3.1(b) is 2-connected but not 3-connected.

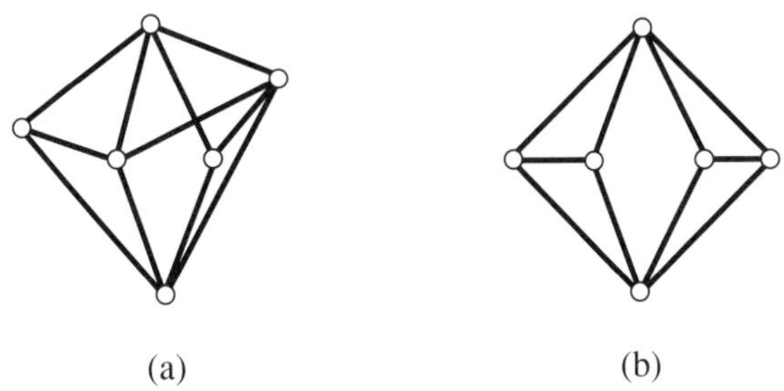


FIGURE 3.3.1

Note that if $G(\mathbf{p})$ is to be globally rigid in $\mathbf{R}^d$, even when $\mathbf{p}$ is generic, $G$ must be at least $(d+1)$-connected. For example, the framework represented by Figure 3.3.1(b) cannot be globally rigid in $\mathbf{R}^2$, since we can reflect part of the vertices of $G(\mathbf{p})$ about the line separating the vertices. This idea works in any dimension.

## 4. Hendrickson's conjecture

How does one determine when a framework $G(\mathbf{p})$ is globally rigid in $\mathbf{R}^d$ at a generic configuration $\mathbf{p}$? We saw in §3.3 that $(d+1)$-connectivity was necessary. The following result of Hendrickson [7] provides more conditions.

We say a graph $G$ is *generically redundantly rigid in* $\mathbf{R}^d$ if for each bar of $G$, $G$ minus that bar is generically rigid in $\mathbf{R}^d$. The *complete graph* on $n$ vertices is the graph with an edge between all pairs of $n$ vertices.

THEOREM 4.1 (Hendrickson). *If a framework $G(\mathbf{p})$ is globally rigid in $\mathbf{R}^d$ for all generic configurations $\mathbf{p}$, then either $G$ is the complete graph with $d+1$ or fewer vertices, or the following conditions hold:*

(i) *the graph $G$ is $(d+1)$-connected, and*
(ii) *the graph $G$ is generically redundantly rigid.*

Hendrickson also conjectured that these two conditions are sufficient.

CONJECTURE (Hendrickson). If a graph $G$ is generically redundantly rigid in $\mathbf{R}^d$ and $G$ is $(d+1)$-connected, then for all generic configurations $\mathbf{p}$, $G(\mathbf{p})$ is globally rigid in $\mathbf{R}^d$.

Hendrickson's Conjecture is easily seen to be true for $d = 1$. It looks very likely to be true for $d = 2$ as well. It will be shown that it is false for $d \geq 3$.

## 5. Review of results for bipartite graphs: Bolker-Roth theory

A *complete bipartite graph* is a finite graph, in which the vertices are partitioned into two sets $A$ and $B$ with edges joining every vertex of $A$ with every vertex of $B$ and no other edges. If $A$ has $n$ vertices and $B$ has $m$ vertices, we call this finite graph $K_{m,n}$.

In Bolker and Roth [**3**] there is a very thorough treatment of conditions which are equivalent to $K_{m,n}(\mathbf{p})$ being infinitesimally rigid. We will implicitly use their results by means of the following interpretation, an alternate point of view presented in Whiteley [**10**].

THEOREM 5.1 (Whiteley [**10**]). *The complete bipartite framework $K_{m,n}(\mathbf{p})$ for $m, n \geq 2$, $d \geq 2$ in $\mathbf{R}^d$ has a nontrivial infinitesimal motion if and only if either*

(i) *the points of $A \cup B$ lie on a quadric surface, or*
(ii) *one side ($A$ or $B$) lies on a hyperplane along with at least one point of the other side, or*
(iii) *one side ($A$ or $B$) lies on a hyperplane $H$ and lies on a quadric surface within $H$.*

We will concentrate on condition (i). Suppose that the quadric surface $S$ is given by

$$S = \{\mathbf{p}_i \in \mathbf{R}^d | (\mathbf{p}_i)^t Q\mathbf{p}_i = 1\},$$

where $Q$ is a symmetric $d$ by $d$ matrix, and $(\ )^t$ is the transpose operation. Then Whiteley gives a very simple description of the nontrivial infinitesimal motion $\mathbf{p}'$. It is given by

$$\mathbf{p}_i' = \begin{cases} Q\mathbf{p}_i & \text{if } i \text{ is in } A, \\ -Q\mathbf{p}_i & \text{if } i \text{ is in } B. \end{cases}$$

See Figure 5.1 for a picture of the infinitesimal motion in case $Q$ is the identity matrix and $S$ is the unit sphere. It is easy to check that it is nontrivial when $m, n \geq 2$, the points are distinct, and $Q$ is nonsingular. The arrows represent the infinitesimal flex. We will call this infinitesimal flex the *innyouty flex.*

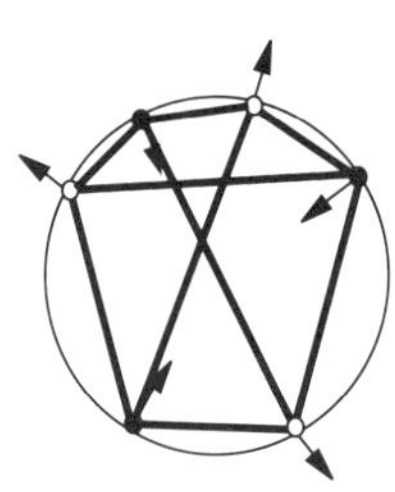

FIGURE 5.1

## 6. The averaging principle

The following is a result relating infinitesimal rigidity and global rigidity. Proofs, which are easy, can be found in Connelly [**5**].

THEOREM 6.1. *Suppose that* $\mathbf{p}'$ *is an infinitesimal flex for a framework* $G(\mathbf{p})$ *in* $\mathbf{R}^d$, *where the points of* $\mathbf{p}$ *do not all lie in a hyperplane. Then*

$$G(\mathbf{p}+\mathbf{p}') \cong G(\mathbf{p}-\mathbf{p}'),$$

*and* $\mathbf{p}+\mathbf{p}'$ *is congruent to* $\mathbf{p}-\mathbf{p}'$ *if and only if* $\mathbf{p}'$ *is a trivial infinitesimal flex.*

For example, in $\mathbf{R}^2$ the framework of Figure 6.1(a) is equivalent to the framework of Figure 6.1(c) using the infinitesimal flex of the framework of Figure 6.1(b). Only one vertex has a nonzero infinitesimal motion.

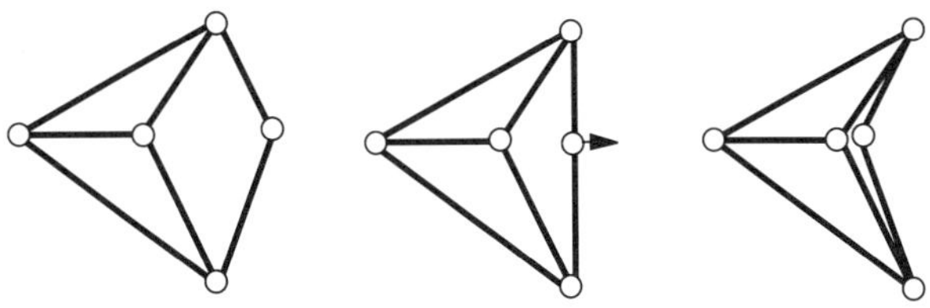

(a) (b) (c)

FIGURE 6.1

## 7. The example

**7.1. The main theorem.** We consider the case $K_{5,5}(\mathbf{p})$ in $\mathbf{R}^3$. A unique quadric surface is determined by nine generic points in $\mathbf{R}^3$. Thus when the ten points of $K_{5,5}(\mathbf{p})$ form a generic configuration, they do not lie on a quadric surface and condition (i) of Whiteley's Theorem 5.1 does not hold. Note that five generic points alone do not lie in a hyperplane, and so conditions (ii) and (iii) do not hold either. Thus $K_{5,5}$ is generically rigid in $\mathbf{R}^3$. Furthermore, $K_{5,5}$ has ten vertices and 25 edges. Thus there are 25 linear equations (3.1) for infinitesimal rigidity, with 30 variables, the coordinates of $\mathbf{p}'$. There is a six-dimensional space of solutions to these equations corresponding to the space of trivial infinitesimal flexes. So one of the equations corresponding to one of the edges is redundant. Thus $K_{5,5}(\mathbf{p})$ is infinitesimally rigid even after the removal of one of its edges. But the symmetries of the abstract graph $K_{5,5}$ are transitive on the edges. Thus, since $\mathbf{p}$ is a generic configuration, $K_{5,5}(\mathbf{p})$ is infinitesimally rigid after the removal of any edge. Thus $K_{5,5}(\mathbf{p})$ is redundantly rigid.

It is also easy to check that $K_{5,5}$ is at least 4-connected. (Indeed, it is 5-connected.) Thus Hendrickson's conjecture says that $K_{5,5}(\mathbf{p})$ should be globally rigid for all generic configurations $\mathbf{p}$. We will show on the contrary the following:

THEOREM 7.1.1. *There is a generic configuration* $\mathbf{p}$ *in* $\mathbf{R}^3$, *where* $K_{5,5}(\mathbf{p})$ *is not globally rigid.*

Before we proceed with the proof, we define the following subset of the space of all configurations:

$$\mathrm{IF}(G) = \{\mathbf{p}|G(\mathbf{p}) \text{ is not infinitesimally rigid in } \mathbf{R}^d\},$$

the space of infinitesimally flexible configurations in $\mathbf{R}^d$ for $G$. Note that the space of all configurations, flexible or not, is $dv$-dimensional. We regard this as $\mathbf{R}^{dv}$.

For the case of $K_{5,5}$ for $\mathbf{R}^3$ we see that $dv = 30$, and conditions (i), (ii), and (iii) define the subset $\mathrm{IF}(K_{5,5})$ of $\mathbf{R}^{30}$. Since the nine points $\mathbf{p}_1, \mathbf{p}_2, \dots, \mathbf{p}_9$ can be chosen arbitrarily, as well as two of the three coordinates of $\mathbf{p}_{10}$, we can choose 29 coordinates freely (in some open set) and have the points lie on a quadric surface. Thus, condition (i) of Theorem 5.1 defines a 29-dimensional subset (an algebraic set) of $\mathbf{R}^{30}$. The other two conditions define a lower-dimensional subset. In any case, $\mathrm{IF}(K_{5,5})$ is 29-dimensional.

The following lemma is critical for showing that our example is a counter-example to Hendrickson's conjecture.

LEMMA 7.1.2. *There exists a configuration of ten points* $(\mathbf{p}_1, \mathbf{p}_2, \dots, \mathbf{p}_{10}) = \mathbf{p}$ *on the unit sphere in* $\mathbf{R}^3$ *and a* $t > 0$ *such that*

$$\mathbf{p} + t\mathbf{p}' \notin \mathrm{IF}(K_{5,5}),$$

*where* $\mathbf{p}'$ *is Whiteley's inny-outy flex for* $K_{5,5}(\mathbf{p})$ *in* $\mathbf{R}^3$.

PROOF. The most direct way to do this problem seems to be to find ten points on the unit sphere and then use equations (3.1) to solve for the infinitesimal motions for $t = 1/2$, say. Of course, the $\mathbf{p}$ used in (3.1) is $\mathbf{p}+t\mathbf{p}'$ here. We omit this calculation, which amounts to solving 25 linear equations in 30 unknowns. (Alternatively, this calculation is equivalent to computing the rank of a 25 by 30 matrix.)

**7.2. The proof of the main result.** Consider the function

$$g\colon \mathrm{IF}(K_{5,5}) \times \mathbf{R} \to \mathbf{R}^{30}$$

defined by $g(\mathbf{p}, t) = \mathbf{p} + t\mathbf{p}'$, where $\mathbf{p}'$ is Whiteley's inny-outy infinitesimal flex for $K_{5,5}(\mathbf{p})$ defined in §5. The domain of $g$ is only the set of those $(\mathbf{p}, t)$ where the configuration $\mathbf{p}$ determines a unique quadric surface. This is an open subset of $\mathrm{IF}(K_{5,5}) \times \mathbf{R}$. We known that $g$ is well defined on this subset by using condition (i) of Theorem 5.1, since that is where $\mathbf{p}'$

is well defined and unique. In any case, the domain and range of $g$ is a 30-dimensional manifold, and $g$ is a smooth algebraic function.

We will show that the image of $g$ contains a generic point. By considering the case when $t = 0$, we see that the image of $g$ contains $\mathrm{IF}(K_{5,5})$. The smooth algebraic image of a smooth manifold near a generic point is a smooth manifold again. By Lemma 7.1.2, we know that the image of $g$ is not in $\mathrm{IF}(K_{5,5})$. Since being infinitesimally rigid is a generic property, it is also clear that the image of $g$ is not in the intersection of $\mathrm{IF}(K_{5,5})$ and any neighborhood of any point in $\mathrm{IF}(K_{5,5})$. Thus the image of $g$ must be greater than 29-dimensional. The image must be 30-dimensional and hence must contain an open set of $\mathbf{R}^{30}$ and a generic point of $\mathbf{R}^{30}$.

Suppose that $\mathbf{p} + t\mathbf{p}'$ is this generic configuration. By the averaging principle, Theorem 6.1, $K_{5,5}(\mathbf{p} + t\mathbf{p}')$ is equivalent to $K_{5,5}(\mathbf{p} - t\mathbf{p}')$. Hence, $K_{5,5}(\mathbf{p} + t\mathbf{p}')$ is not globally rigid at a generic configuration.

**7.3. Remarks and generalizations.** In $\mathbf{R}^3$, $K_{5,5}$ is the only example we know which is a counter-example to Hendrickson's Conjecture. In $\mathbf{R}^d$ for $d \geq 4$, an argument similar to the above (which we omit) shows the following:

THEOREM 7.3.1. *For $m, n \geq d+2$ and $m + n = (d+1)(d+2)/2$, there is a generic configuration $\mathbf{p}$ in $\mathbf{R}^d$ such that $K_{m,n}(\mathbf{p})$ is not globally rigid.*

It is easy to check that $K_{m,n}(\mathbf{p})$ as above is also $(d+1)$-connected and redundantly rigid in $\mathbf{R}^d$. Thus each of the above graphs serve as a counter-example to Hendrickson's Conjecture.

## REFERENCES

1. L. Asimow and B. Roth, *The rigidity of graphs*, Trans. Amer. Math. Soc. **245** (1978), 279–289.
2. ——, *The rigidity of graphs*. II, J. Math. Anal. Appl. **68** (1979), 171–190.
3. E. Bolker and B. Roth, *When is a bipartite graph a rigid framework?*, Pacific J. Math. **90** (1980), 27–44.
4. G. M. Crippen and T. M. Havel, *Distance geometry and molecular conformation*, Wiley, New York, 1988.
5. R. Connelly, *Basic concepts of infinitesimal rigidity*, draft chapter for The Geometry of Rigid Structures (H. Crapo and W. Whiteley, eds.) (to appear) (preprint, Cornell Univ.).
6. H. Gluck, *Almost all simply connected surfaces are rigid*, in Geometric Topology, Lecture Notes in Math., vol. 438, Springer-Verlag, 1975, pp. 225–239.
7. B. Hendrickson, *Conditions for unique graph embeddings*, Technical Report 88-950, Dept. of Computer Science, Cornell Univ., 1988.
8. L. Lovàsz and Y. Yemini, *On generic rigidity in the plane*, SIAM J. Algebraic Discrete Methods **3** (1982), 91–98.
9. K. Sugihara, *On redundant bracing in plane skeletal structures*, Bull. Electrotech. Lab. **44** (1980), 76–88.
10. W. Whiteley, *Infinitesimal motions of bipartite frameworks*, Pacific J. Math. **110** (1984), 233–255.

DEPARTMENT OF MATHEMATICS, CORNELL UNIVERSITY, ITHACA, NEW YORK 14853
*E-mail address*: connelly@mssun7.msi.cornell.edu

DIMACS Series in Discrete Mathematics
and Theoretical Computer Science
Volume 4, 1991

Applied Geometry
and Discrete Mathematics
THE VICTOR KLEE FESTSCHRIFT

# Some Regular Maps and Their Polyhedral Realizations

H. S. M. COXETER AND G. C. SHEPHARD

ABSTRACT. The idea of realizing certain regular maps as regular (or almost regular) polyhedra [**6, pp. 75–105**] was used first for maps with "holes" [**10, p. 109**] and then for maps with "second holes". The earlier treatment of the latter was incomplete. Here we have paid particular attention to three maps having 168 edges. Coordinates for their vertices are fed into a computer, along with instructions to join certain pairs so as to yield significant drawings of projections (in various directions) of the appropriate complex polyhedra. More precisely, our last seven figures were drawn on a CALCOMP 1051 graph-plotter using GINO-F software. We are grateful for the facilities provided by the director and staff of the Computing Centre at the University of East Anglia.

## 1. Regular skew polyhedra

Figure 1 shows the case $p = 3$ of the Cartesian product of two regular $p$-gons, conveniently denoted by $\{p\}\times\{p\}$. This is a four-dimensional polytope having $p^2$ vertices, $2p^2$ edges, $p^2$ square faces, $2p$ $p$-gonal faces, and $2p$ facets (or "cells"): $p$-gonal prisms arranged in two interlocked rings of $p$ (see [**8, p. 37**]). The $p^2$ squares fit together so that every edge is shared by two of them; they are the *faces* of a regular *skew polyhedron*

$$\{4, 4|p\}$$

with $p$-gonal *holes*. This is a geometric realization of a regular *map* $\{4, 4\}_{p,0}$ [**10, p. 109**], derived from the square lattice $\{4, 4\}$ by identifying pairs of points separated by $p$ steps, horizontally or vertically, as in Figure 2. In other words, the $p^2$ vertices are the points whose Cartesian coordinates are the residue-classes modulo $p$.

Such a "hole" is the $p$-gon formed by a path along successive edges leaving, at each vertex, *two* faces on one side, say the right (the same at every stage).

1980 *Mathematics Subject Classification* (1985 *Revision*). Primary 05C25, 20D06, 20F05, 20F32, 52A25, 51D20, 51F15.


1052-1798/91 $1.00 + $.25 per page

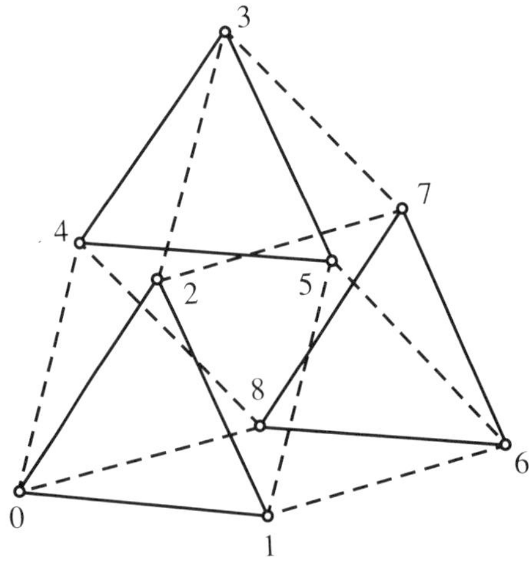


FIGURE 1.
The double prism $\{3\} \times \{3\}$

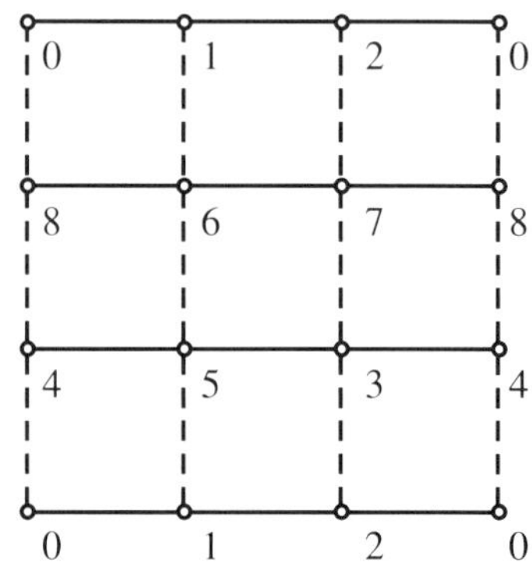


FIGURE 2.
The regular map $\{4, 4|3\}$

Since in this simple case there are four faces at each vertex, the "path" is "straight" and the distinction between right and left is unnecessary; but this distinction becomes important when we generalize to a regular map $\{l, m|p\}$ of $l$-gons, $m$ at each vertex.

The above extraction of $\{4, 4|p\}$ from the double prism

$$\{p\} \times \{p\} = \mathrm{e}\{p, 2, p\} = \mathrm{t}_{1,2}\{p, 2, p\}$$

[**7, pp. 574–575; 8, pp. 16–18**] generalizes to a construction of $\{4, 2q|p\}$ by using the squares

in the "expanded" polytope

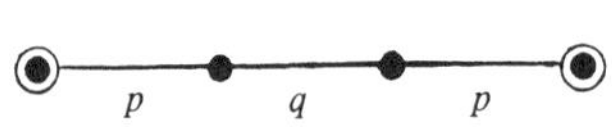 $= \mathrm{e}\{p, q, p\} = \mathrm{t}_{0,3}\{p, q, p\}$

[**2, p. 366; 6, pp. 45, 88**]. This polytope is derived from the regular $\{p, q, p\}$ by Mrs. Boole Stott's process of *expansion*, e or $\mathrm{e}_3$ [**1, p. 139**], whereby the facets $\{p, q\}$ are separated from one another and translated away from the center of the regular polytope through a sufficient distance to allow the two new positions of each face $\{p\}$ to be joined by a uniform $p$-gonal prism

 $= \{p\} \times \{\ \}$

with square side-faces. Again every edge belongs to two squares, and again the holes are the omitted $p$-gons; but now, since the vertex figure of the expansion is a $q$-gonal antiprism, the number of squares surrounding each vertex is $2q$, justifying the symbol

$$\{4, 2q|p\}.$$

In this manner, the regular simplex $\{3, 3, 3\}$ and the 24-cell $\{3, 4, 3\}$ yield regular skew polyhedra $\{4, 6|3\}$ and $\{4, 8|3\}$. The simple $\{4, 4|p\}$ can be included by regarding the double prism $\{p\}\times\{p\}$ as an expanded $\{p, 2, p\}$.

More interestingly, when $\{p, q, p\}$ is the infinite cubic honeycomb $\{4, 3, 4\}$, which is its own expansion **[4, p. 402]**, we obtain $\{4, 6|4\}$: the regular skew apeirohedron discovered by John Flinders Petrie during a memorable country walk in 1926 **[6, p. 76]**. Petrie dualized this to obtain $\{6, 4|4\}$: the infinite member of another family $\{2q, 4|p\}$ whose faces are the $2q$-gons 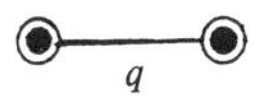 in the "sesqui-truncated" regular polytope

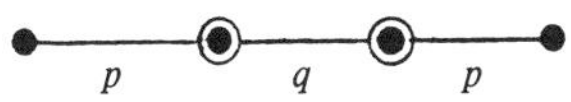 $= \mathrm{t}_{1,2}\{p, q, p\}$

**[5, p. 70]**. This has one facet

 $= \mathrm{t}_{1,2}\{p, q\} = \mathrm{t}_{0,1}\{q, p\} = \mathrm{t}\{q, p\}$

**[9, p. 409]** for each facet $\{p, q\}$ of $\{p, q, p\}$, and again one facet

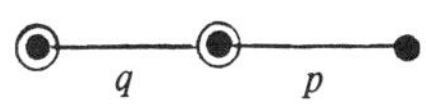 $= \mathrm{t}\{q, p\}$

for each vertex of $\{p, q, p\}$. Thus all the facets are alike and the faces consist of $2q$-gons and $p$-gons. The $2q$-gons fit together to form the faces of the skew polyhedron $\{2q, 4|p\}$ whose holes are the omitted $p$-gons. In this manner, the self-dual regular polytopes

$$\{3, 3, 3\}, \quad \{3, 4, 3\}, \quad \{p, 2, p\}, \quad \text{and} \quad \{4, 3, 4\},$$

yield the skew polyhedra

$$\{6, 4|3\}, \quad \{8, 4|3\}, \quad \{4, 4|p\}, \quad \text{and} \quad \{6, 4|4\}.$$

The self-dual skew apeirohedron $\{6, 6|3\}$, using the hexagons in the "sporadic" honeycomb

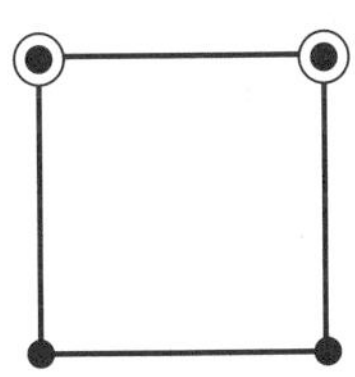 $= \mathrm{q}\{4, 3, 4\} = \mathrm{q}\delta_4$

**[7, p. 9]**, is unique, not belonging to any "family" unless we agree to combine $\{4, 2q|p\}$ and $\{2q, 4|p\}$ into one large family $\{2q', 2q|p\}$ by regarding $\mathrm{t}_{0,3}\{p, q, p\}$ and $\mathrm{t}_{1,2}\{p, q', p\}$ as the cases $q' = 2$ and $q = 2$ of

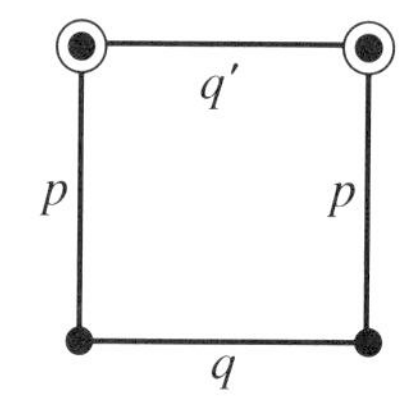

**[11, p. 1182; 18, pp. 242, 250; 12]**.

The vertices of $\{6, 6|3\}$, being the vertices of $q\delta_4$, are the points whose Cartesian coordinates $(x, y, z)$ have residues modulo 4 which either are some permutation of $(1, 2, 3)$ or else one is 0 while the other two are equal; thus $(0, 0, 0)$ is one vertex and the nearest others form the skew hexagon

$$(0, 1, 1)\ (-1, 0, -1)\ (1, 1, 0)\ (0, -1, -1)\ (1, 0, 1)\ (-1, -1, 0).$$

## 2. Regular maps with holes

All these regular skew polyhedra and apeirohedra are geometric realizations of regular maps $\{l, m|p\}$ with $p$-gonal holes **[6, p. 93; 10, p. 109]**. In the instances so far considered, $l$ and $m$ are even; but from the topological standpoint this is not necessarily the case. In particular, the Platonic solids $\{p, 3\}$ and $\{3, p\}$ can be expressed as $\{p, 3|p\}$ and $\{3, p|p\}$ **[6, pp. 82, 96]**, and Poinsot's *great dodecahedron* $\{5, \frac{5}{2}\}$ is a geometric realization of the map $\{5, 5|3\}$ whose triangular holes are the faces of the icosahedron that has the same vertices and edges. Fifteen other regular maps, determined by their holes, are known **[6, Table 1 on p. 104]**.

These finite maps $\{l, m|p\}$ are derived from hyperbolic tessellations $\{l, m\}$ (with $l^{-1} + m^{-1} < \frac{1}{2}$) by making suitable identifications **[10, pp. 53, 102, 109]**. For this purpose we may conveniently use the "rotation group" $[l, m]^+$ with the presentation

$$S^m = T^2 = (ST)^l = 1$$

**[10,** $(4 \cdot 43)$ **on p. 38]**. $S$ may be thought of as the rotation through $2\pi/m$ about the vertex 0 in Figure 2 or Figure 3, while $T$ is the half-turn that reverses the edge 04. Then the product $ST$, taking 1 (via 4) to 0, and 0 to 4, rotates the face $\ldots 1\,0\,4 \ldots$ (negatively). In Figure 3, $S^2T$, taking 9 (via 4) to 0, 0 to 4, and 4 (via 6) to 10, similarly rotates the "hole" $\ldots 9\,0\,4 \ldots$ **[6, Figure xii, p. 94]**. Thus the period of $S^2T$ is $p$, justifying the symbol $\{l, m|p\}$ for the map of type $\{l, m\}$ whose rotation group has the presentation

$$S^m = T^2 = (ST)^l = (S^2T)^p = 1.$$

The substitution $\overline{S} = ST$, implying $S = \overline{S}T$ and $S^2T = \overline{S}T\overline{S}$, allows the same group to be presented as

$$\overline{S}^l = T^2 = (\overline{S}T)^m = (\overline{S}^2T)^p = 1.$$

Thus every map $\{l, m|p\}$ has a *dual* $\{m, l|p\}$.

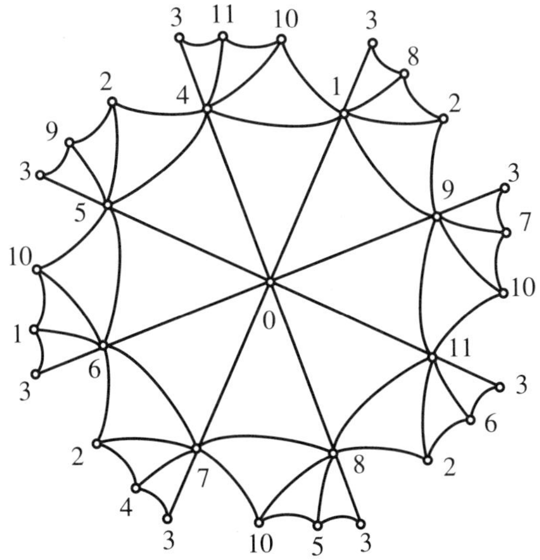


FIGURE 3. The regular map $\{3, 8\}_6 = \{3, 8|, 3\}$, of genus 4

It must not be forgotten that, although $S$ and $ST$ appear in the *map* $\{l, m|p\}$ as *rotations*, when there is a corresponding *polyhedron* or *apeirohedron* (in Euclidean or non-Euclidean 3-space) these operations are no longer simple rotations but *rotatory reflections*. This observation explains why, in these cases, $l$ and $m$ are even.

### 3. Regular maps with second holes

Looking again at Figure 3, we see that the operation $S^3T$, taking 11 (via 4) to 0, 0 to 4, and 4 (via 7) to 11, rotates the *second hole* 11 0 4 11, which is another polygon, say a $q$-gon (in the present case a triangle) formed by a path of successive edges leaving, at each vertex, *three* faces on one side (say the right). Thus the period of $S^3T$ is $q$ (in this case 3) and we may conveniently use the symbol $\{l, m|, q\}$ for the map of type $\{l, m\}$ determined by having $q$-gonal second holes, so that its rotation group has the presentation

$$S^m = T^2 = (ST)^l = (S^3T)^q = 1 \tag{3.1}$$

**[6, (6.2), p. 102]**. In particular, the map $\{3, 8|, 3\}$ is precisely represented in Figure 3 with the peripheral vertices identified according to their numbering (with second holes 11 04 11, 2412, 9059, etc.). However, such maps do not occur in dual pairs.

*Each map $\{l, m|, q\}$, with $m = 8$ or $10$, has a companion $\{q, m|, l\}$ with the same group and the same "skeleton" while the faces of each are the second holes of the other.*

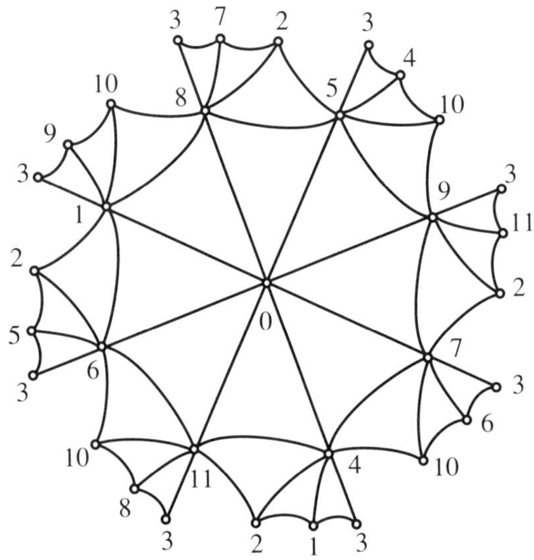


FIGURE 4. The same map with faces and second holes interchanged

(By "*skeleton*" we mean the graph of vertices and edges.) To prove this theorem we make the substitution

$$S' = S^3 . \tag{3.2}$$

Since $m = 3^2 \mp 1$, this implies $S'^3 = S^{\pm 1}$ and allows the group (3.1) to be presented as

$$S'^m = T^2 = (S'T)^q = (S'^3 T)^l = 1 .$$

The substitution (3.2) is equivalent to the facetting operation $\phi_3$ of McMullen and Schulte [**13, p. 7** (§**4.2**)].

The relation between $\{l, m|, q\}$ and $\{q, m|, l\}$ can already be seen in the simple case of $\{3, 8|, 3\}$, even though now $l = q$. Figures 3 and 4 show the same map differently numbered, so that each face in either figure agrees with a second hole of the other. The numbering has been artfully chosen so that either figure is transformed into the other when the numbers, regarded as residues modulo 12, are multiplied by 5. Presumably the same kind of game could be played with the map $\{3, 10|, 3\}$, which as 15 vertices.

When $l = 3$, the group (3.1) can be recognized as $(2, 3, m; q)$ [**10, p. 96**]. In fact, $TST$, being the inverse of $STS$, transforms the commutator $S^{-1}TST$ into $STS \cdot S^{-1}TST \cdot TST = S^3T$.

The three maps $\{l, 6|, 2\}$ $(l = 3, 4, 5)$ can be derived from the trivalent Platonic solids $\{l, 3\}$ by allowing their faces and edges to be repeated so as to form two-sheeted Riemann surfaces with simple branch points at the vertices. In fact, by defining

$$\overline{S} = S^2, \qquad Z = S^3, \qquad \overline{T} = ZT = TZ,$$

we see that the group

$$S^6 = T^2 = (ST)^l = (S^3T)^2 = 1$$

(which is (3.1) with $m = 6$ and $q = 2$) is the direct product

$$[l, 3]^+ \times \mathfrak{C}_2$$

of the polyhedral group $\overline{S}^3 = \overline{T}^2 = (\overline{S}\,\overline{T})^l = 1$ (generated by a rotation $\overline{S}$ and a half-turn $\overline{T}$) and the group of order 2 generated by $Z$. If we represent $Z$ by the central inversion, so that $S$ is a rotatory reflection and $T$ is simply a reflection, this product is:

| | |
|---|---|
| when $l = 3$, | the pyritohedral group $[3^+, 4]$, |
| when $l = 4$ or $5$, | the reflection group $[3, l]$ |

**[7, p. 561]**.

The map $\{3, 6|, 2\}$ is the simplest instance of

$$\{3, 6|, q\} = \{3, 6\}_{q,0} = \{3, 6\}_{2q},$$

whose complete group is $G^{3,6,2q}$ so that its rotation group is $(2, 3, 6; q)$ **[10, pp. 108, 113, 139]**. Since $G^{3,6,2q} \cong G^{3,2q,6}$ the rotation groups $(2, 3, 6; q)$ and $(2, 3, 2q; 3)$ of the maps $\{3, 6|, q\}$ and $\{3, 2q|, 3\}$ have the same order; however, they are not isomorphic (unless $q = 3$). For instance, when $q = 2$, the former is the pyritohedral group $[3^+, 4] \cong \mathfrak{A}_4 \times \mathfrak{C}_2$, as we have seen, while the latter is the octahedral group $[3, 4]^+ \cong \mathfrak{S}_4$ **[10, p. 134]**. In contrast, the two subgroups $(2, 3, 8; 5)$ and $(2, 3, 10; 4)$ of $G^{3,8,10}$, which are the rotation groups of $\{3, 8|, 5\}$ and $\{3, 10|, 4\}$, *are* isomorphic **[3, p. 92]**.

Clearly, the distinction between first and second holes disappears when $m = 5$. Accordingly (since Table I on p. 104 of **[6]** is correct as it stands), the accompanying new table (which supersedes Table II) assumes $m > 5$. Otherwise we might have included three more regular maps:

| | |
|---|---|
| $\{4, 5|, 4\} = \{4, 5|4\}$ | with 32 vertices, |
| $\{4, 5|, 5\} = \{4, 5|5\}$ | with 72 vertices, |
| $\{5, 5|, 4\} = \{5, 5|4\}$ | with 72 vertices (again). |

One striking feature of the table is the occurrence of the pair $\{3, 7|, q\}$ with $q = 6$ or 7. They have the same numbers of vertices, edges, and faces (and consequently the same genus), and the same group $PSL(2, 13)$; yet they differ in their second holes! Another pair of numerically similar regular maps is

$$\{4, 9|3\} \quad \text{and} \quad \{4, 9|, 3\}.$$

They have the same number of vertices, edges, and faces (and consequently the same genus), but the rotation group of the former is $PSL(2, 17)$ while

TABLE OF REGULAR MAPS $\{l, m|, q\}$ determined by their second holes

| Map | Faces | Edges | Vertices | Genus | Rotation group | Order |
|---|---|---|---|---|---|---|
| $\{3, 6\|, 2\}$ | 8 | 12 | 4 | 1 | $\mathfrak{A}_4 \times \mathfrak{C}_2$ | 24 |
| $\{4, 6\|, 2\}$ | 12 | 24 | 8 | 3 | $\mathfrak{S}_4 \times \mathfrak{C}_2$ | 48 |
| $\{5, 6\|, 2\}$ | 24 | 60 | 20 | 9 | $\mathfrak{A}_5 \times \mathfrak{C}_2$ | 120 |
| $\{3, 6\|, q\}$ | $2q^2$ | $3q^2$ | $q^2$ | 1 | $(2, 3, 6; q)$ | $6q^2$ |
| $\{3, 2q\|, 3\}$ | $2q^2$ | $3q^2$ | $3q$ | $\binom{q-1}{2}$ | $(2, 3, 2q; 3)$ | |
| $\{3, 7\|, 4\}$ | 56 | 84 | 24 | 3 | $PSL(2, 7)$ | 168 |
| $\{4, 6\|, 3\}$ | 84 | 168 | 56 | 15 | | |
| $\{4, 8\|, 3\}$ | 84 | 168 | 42 | 22 | $PGL(2, 7)$ | 336 |
| $\{3, 8\|, 4\}$ | 112 | 168 | 42 | 8 | | |
| $\{3, 7\|, 6\}$ $\{3, 7\|, 7\}$ | 364 | 546 | 156 | 14 | $PSL(2, 13)$ | 1092 |
| $\{3, 8\|, 5\}$ | 720 | 1080 | 270 | 46 | | |
| $\{5, 8\|, 3\}$ | 432 | 1080 | 270 | 190 | $(2, 3, 8; 5)$ | 2160 |
| $\{3, 10\|, 4\}$ | 720 | 1080 | 216 | 73 | | |
| $\{4, 10\|, 3\}$ | 540 | 1080 | 216 | 163 | | |
| $\{4, 9\|, 3\}$ | 612 | 1224 | 272 | 171 | | 2448 |
| $\{3, 11\|, 4\}$ | 2024 | 3036 | 552 | 231 | $PSL(2, 23)$ | 6072 |
| $\{3, 9\|, 5\}$ | 3420 | 5130 | 1140 | 286 | $PSL(2, 19) \times \mathfrak{C}_3$ | 10260 |
| $\{3, 7\|, 8\}$ | 3584 | 5376 | 1536 | 129 | $(2, 3, 7; 8)$ | 10752 |

that of the latter is a different group having the same order 2448. This phenomenon was revealed when Asia Ivić Weiss modified the known presentation

$$S^9 = T^2 = (ST)^4 = (S^2T)^3 = 1$$

for $PSL(2, 17)$ **[10, Table 6, p. 138]** by changing $(S^2T)^3$ to $(S^3T)^3$. She used an electronic computer to enumerate the 204 cosets of the tetrahedral subgroup generated by $S^3$ and $T$, and then proved by *reductio ad absurdum* that $S$ and $T$ cannot be represented by unimodular matrices modulo 17.

McMullen and Schulte **[14]** have described some remarkable maps determined by both first and second holes:

$$\{6, 6|3, 2l\} \quad \text{with } 4l^3 \text{ vertices},$$
$$\{6, 8|3, 6\} \quad \text{with } 1280.$$

The former can be derived from $\{6, 6|3\}$ by making suitable identifications. Analogously we can derive, from Petrie's $\{4, 6|4\}$,

$$\{4, 6|4, 2l\} \quad \text{with } 8l^3 \text{ vertices}.$$

## 4. Three complex polyhedra related to Klein's 168-group

H. F. Blichfeldt **[15, pp. 250–251]** enumerated the groups generated by homologies in the complex projective plane. Shephard and Todd **[17, pp. 278, 285–301]** interpreted the homologies as reflections in planes of unitary 3-space and thus obtained complex kaleidoscopes consisting of triads of planar mirrors. Shephard **[16]** showed how Wythoff's construction can be used to derive complex polyhedra which include geometric realizations of various regular maps. In particular, the "companions" $\{3, 8|, 4\}$ and $\{4, 8|, 3\}$ can be regarded as having, for their 42 vertices, the $8+12+24$ permutations of

$$(4.1) \qquad (\pm 2, 0, 0), \qquad (0, \pm c, \pm c), \qquad (\pm\overline{c}, \pm 1, \pm 1).$$

Here $c = \frac{1}{2}(-1 + i\sqrt{7})$, so that $c$ and $\overline{c}$ are the roots of the equation

$$x^2 + x + 2 = 0.$$

($c = -\alpha = \beta^4 + \beta^2 + \beta$, in the notation of Shephard and Todd **[17, p. 295]**; see also **[8,** (14 · 53) **and Figure** 14 · 5B].) The 168 edges are determined by pairs of vertices, such as $(\overline{c}, 1, \pm 1)$, whose Hermitian inner product is $c\overline{c} = 2$.

A typical face of the "Kleinian polyhedron" $\{3, 8|, 4\}$ is the equilateral triangle

$$(-\overline{c}, 1, 1), \qquad (1, -\overline{c}, 1), \qquad (1, 1, -\overline{c}),$$

whose center is $\frac{1}{3}\overline{c}^3(1, 1, 1)$. A typical face of the companion $\{4, 8|, 3\}$ (which has the same vertices and edges as $\{3, 8|, 4\}$) is the square

$$(c, \pm c, 0)\ (c, 0, \pm c)$$

with center $(c, 0, 0)$, that is, $c(1, 0, 0)$. Another face, $(\overline{c}, \pm 1, \pm 1)$, has center $\overline{c}(1, 0, 0)$. In fact, the centers of the 84 square faces of $\{4, 8|, 3\}$ are the $42 + 42$ vertices of two homothetic replicas of the polyhedron itself (or of its companion), related by a dilatation with the complex ratio $c/\overline{c}$.

McMullen's polyhedron $\{4, 6|, 3\}$ **[8,** (14·63)] has, for its 56 vertices, the centers of alternate faces of $\{3, 8|, 4\}$, namely the $8+24+24$ permutations of

$$(4.2) \qquad (\pm\overline{c}, \pm\overline{c}, \pm\overline{c}), \qquad (\pm\overline{c}^2, \pm 1, \pm 1), \qquad (\pm c^2, \pm c, 0).$$

Its 168 edges are determined by pairs of vertices, such as $(\overline{c}^2, 1, \pm 1)$, whose Hermitian inner product is 4. Two typical faces are the squares

$$(c^2, \pm c, 0), \quad (c^2, 0, \pm c), \quad \text{and} \quad (\pm\overline{c}, \pm 1, \pm 1),$$

whose centers, $(c^2, 0, 0)$ and $(\overline{c}^2, 0, 0)$, again correspond to the vertex $(2, 0, 0)$ of $\{3, 8|, 4\}$ or $\{4, 8|, 3\}$; but now the ratio of dilatation is $c^2/\overline{c}^2$.

## 5. Orthogonal projections

To obtain significant drawings of such complex polyhedra, it seems most reasonable to choose a complex "projection vector" $(p_1, p_2, p_3)$ and then to say that the real image $(x, y)$ of a complex vertex $(u_1, u_2, u_3)$ is given by

$$x + yi = u_1\overline{p}_1 + u_2\overline{p}_2 + u_3\overline{p}_3.$$

For instance, the projection vector

$$(p_1, p_2, p_3) = (1, -1 - \varepsilon^3 c, -1 - \varepsilon^2 c) \qquad (\varepsilon = e^{\pi i/7})$$

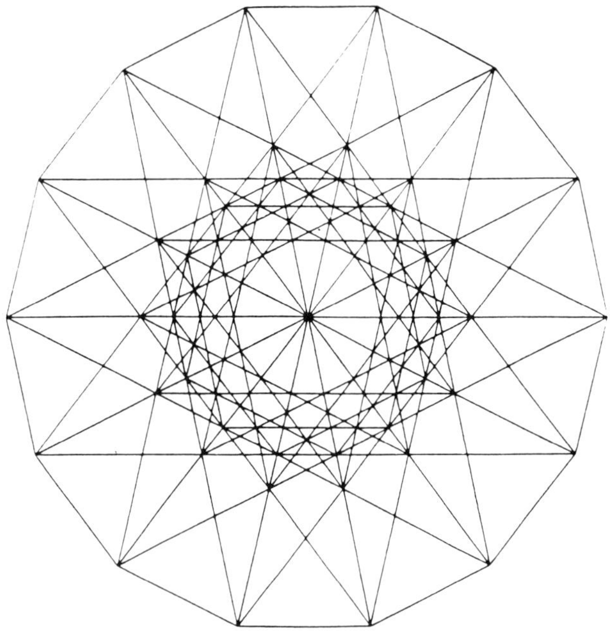

FIGURE 5. The 14-gonal projection of $\{4, 8|, 3\}$

yields the most symmetrical projections of $\{4, 8|, 3\}$ (or $\{3, 8|, 4\}$) and $\{4, 6|, 3\}$ (Figures 5 and 6; compare [**8, Figures** 14 · 5B **and** 14 · 6A]).

We see from §4 that the vectors $(1, 0, 0)$ and $(1, 1, 1)$ may be expected to yield particularly simple projections. Applying $(1, 0, 0)$ to the coordinates (4.1) for $\{4, 8|, 3\}$ or $\{3, 8|, 4\}$, we obtain a "vertex first" projection in which the 42 vertices collapse to 9:

$$x + yi = 0, \ \pm 1, \ \pm 2, \ \pm c, \ \pm \bar{c},$$

as in Figure 7. (Notice the "whiskers" joining 1 to 2, and $-1$ to $-2$.) Although it is a "vertex first" projection, the two opposite vertices $(\pm 2, 0, 0)$ yield points 2 and $-2$, *not* the central point 0. This is also a "face first" projection of $\{4, 8|, 3\}$, whose face-centers $(\pm c, 0, 0)$ and $(\pm \bar{c}, 0, 0)$ all lie on the line $u_2 = u_3 = 0$.

Applying the same projection vector $(1, 0, 0)$ to the coordinates (4.2) for $\{4, 6|, 3\}$, we obtain a "face first" projection, in which the 56 vertices collapse to 11:

$$x + yi = 0, \ \pm 1, \ \pm c, \ \pm \bar{c}, \ \pm c^2, \ \pm \bar{c}^2,$$

as in Figure 8.

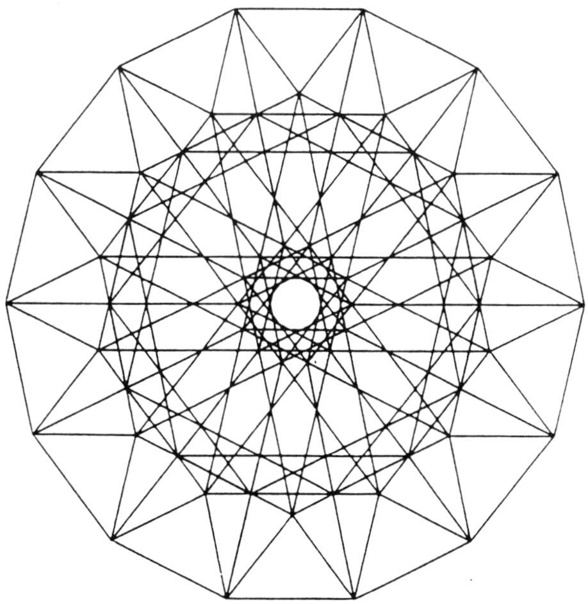

FIGURE 6. The 14-gonal projection of $\{4, 6|, 3\}$

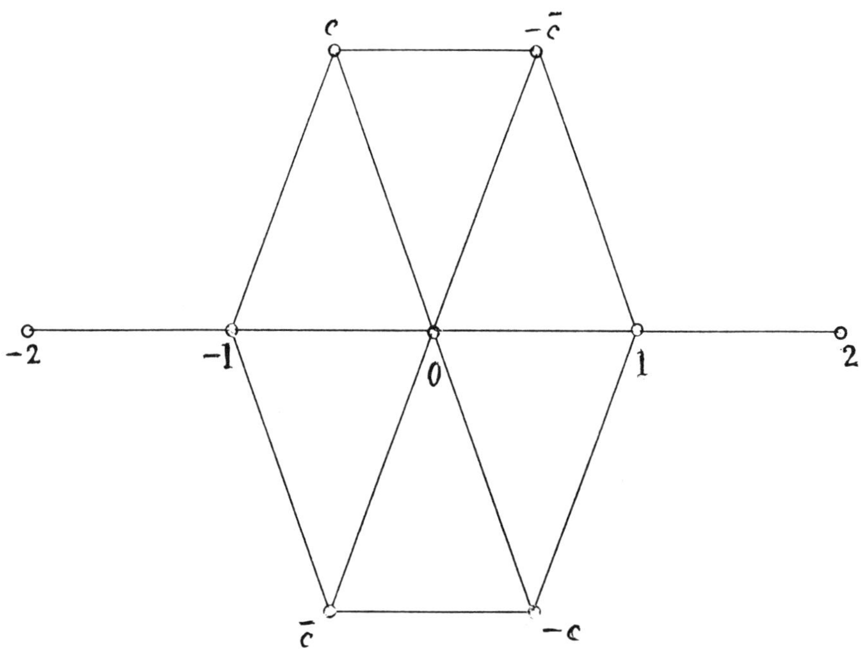


FIGURE 7. Vertex first projection of $\{4, 8|, 3\}$

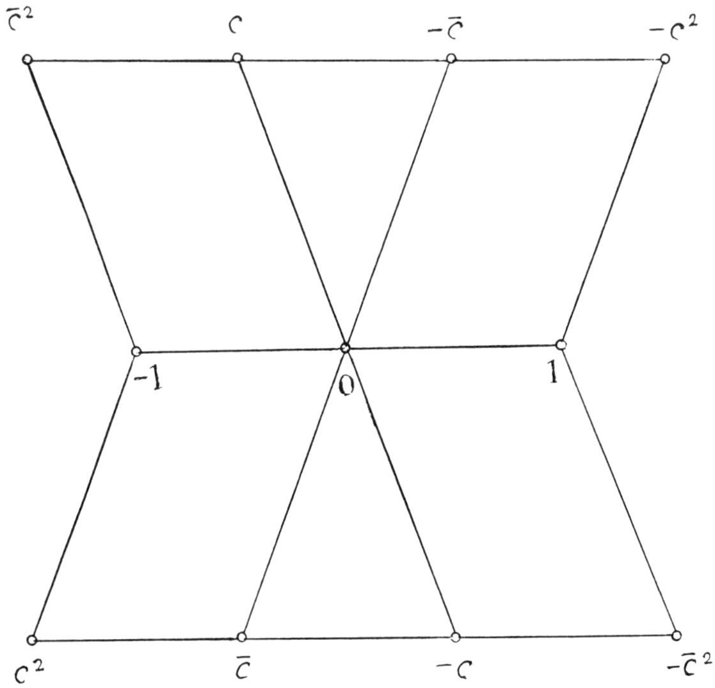


FIGURE 8. Face first projection of $\{4, 6|, 3\}$

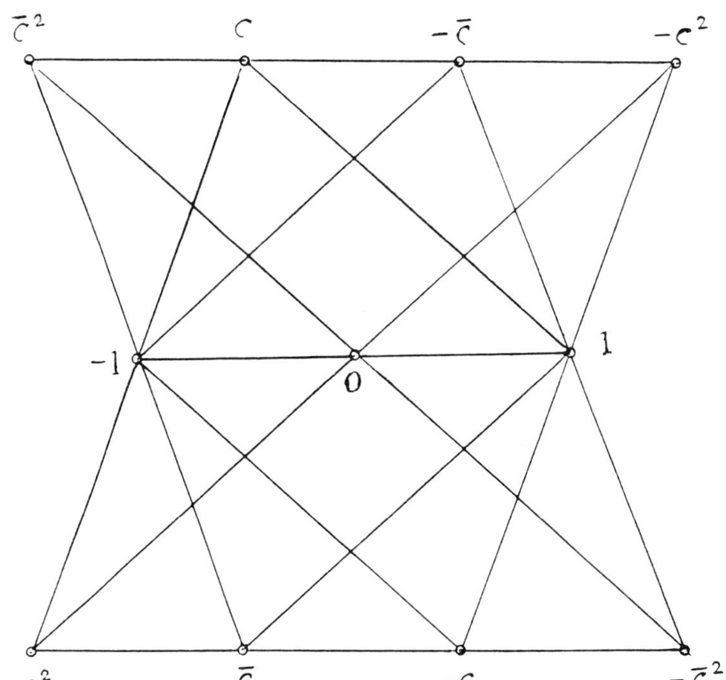


FIGURE 9. Face first projection of $\{3, 8|, 4\}$

The same eleven points, differently joined (as in Figure 9), arise in the "face first" projection of $\{3, 8|, 4\}$ if we apply a complex dilatation (which becomes a rotation of the projected image). In fact, if we multiply the coordinates (4.1) by $c$ so as to obtain

$$(\pm 2c, 0, 0), \qquad (0, \pm c^2, \pm c^2), \qquad (\pm 2, \pm c, \pm c),$$

the projection vector $(\frac{1}{2}, \frac{1}{2}, \frac{1}{2})$ yields those same eleven points. (For instance, $\frac{1}{2}(-2+c+c) = \bar{c}^2$.)

Applying the same dilatation to the coordinates (4.2), we have

$$(\pm 2, \pm 2, \pm 2), \qquad (\pm 2\bar{c}, \pm c, \pm c), \qquad (\pm c^3, \pm c^2, 0).$$

Thus $\{4, 6|, 3\}$ has a "vertex first" projection in which the 56 vertices collapse to 12:

$$x + yi = \pm 1, \pm 2, \pm 3, \pm 3, \pm c, \pm \bar{c}, \pm i\sqrt{7},$$

as in Figure 10. (For instance, $\frac{1}{2}(c^3 - c^2) = 2$.)

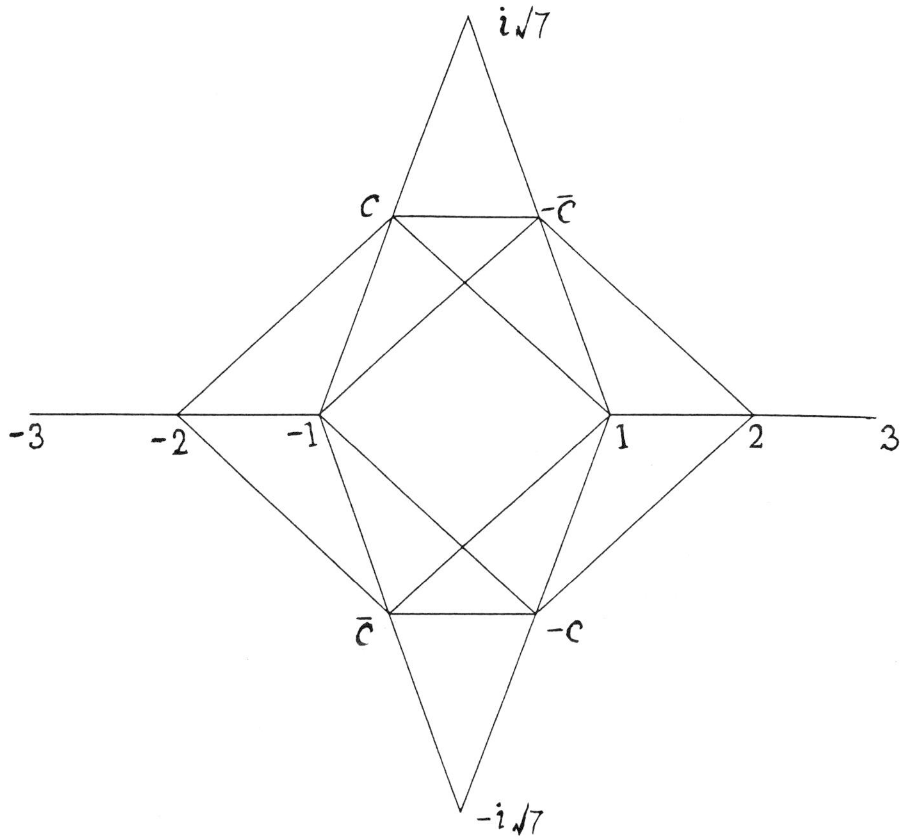


FIGURE 10. Vertex first projection of $\{4, 6|, 3\}$

The technique of computer graphics enables us to see what happens when the projection vector is gradually changed from $v = (1, 1, 1)$ to $w =$

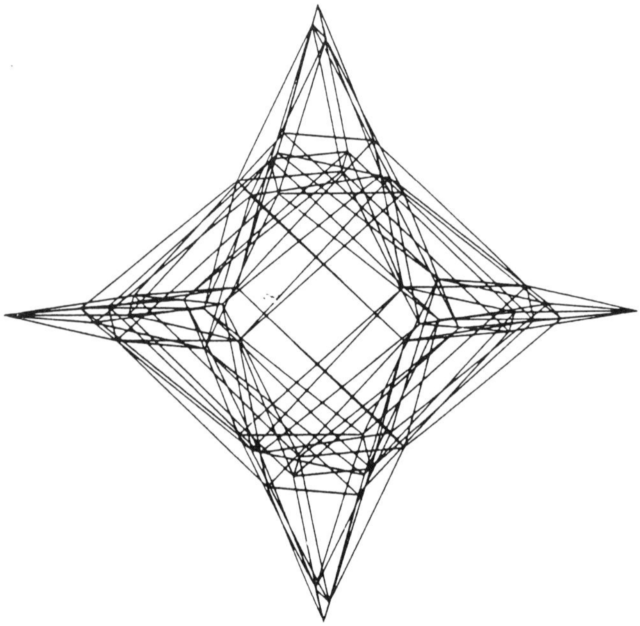

FIGURE 11. $\{4, 6|, 3\}$, projection vector $(5v + w)/6$

$(1, \omega, -\omega^2)$, where $\omega = e^{2\pi i/3}$. Figures 10–16 show $\{4, 6|, 3\}$ as projected by the vector $(1 - \lambda)v + \lambda w$ with $\lambda = 0, \frac{1}{6}, \frac{1}{3}, \frac{1}{2}, \frac{2}{3}, \frac{5}{6}, 1$, respectively.

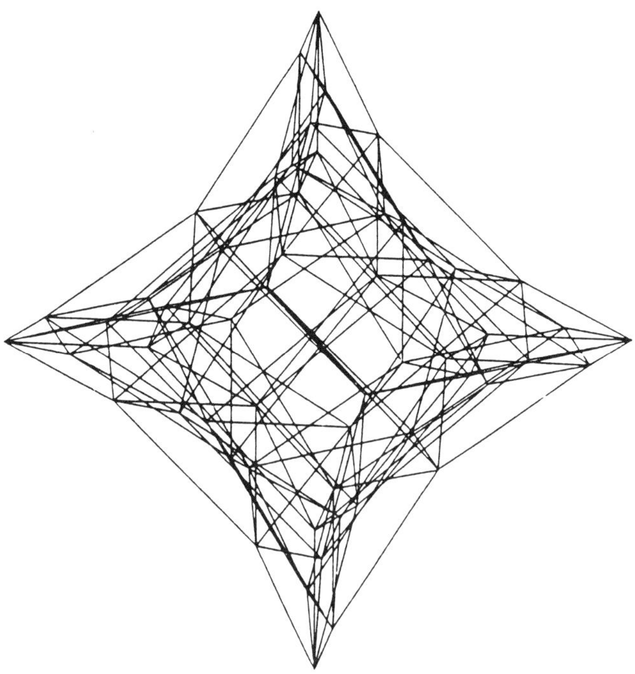

FIGURE 12. $\{4, 6|, 3\}$, projection vector $(2v + w)/3$

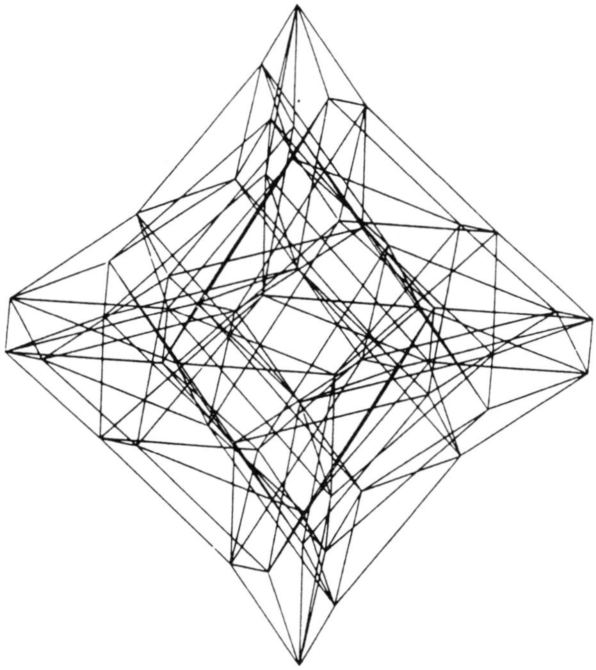

FIGURE 13. $\{4, 6|, 3\}$, projection vector $(v + w)/2$

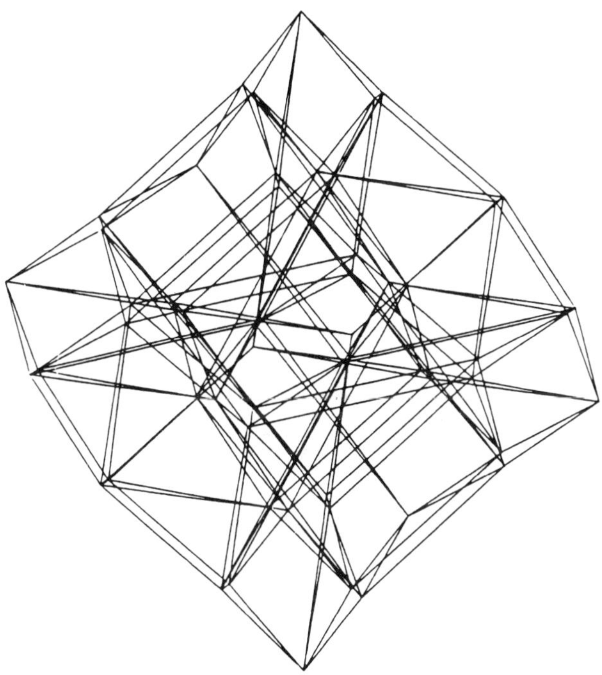

FIGURE 14. $\{4, 6|, 3\}$, projection vector $(v + 2w)/3$

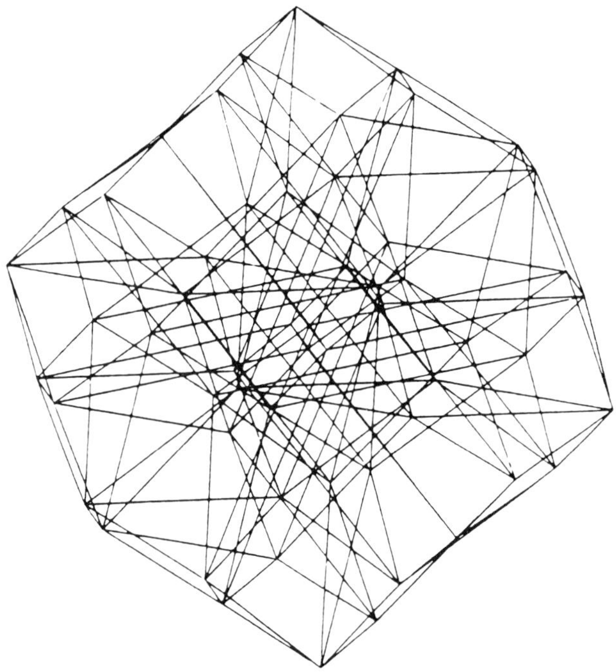

FIGURE 15. $\{4, 6|, 3\}$, projection vector $(v + 5w)/6$

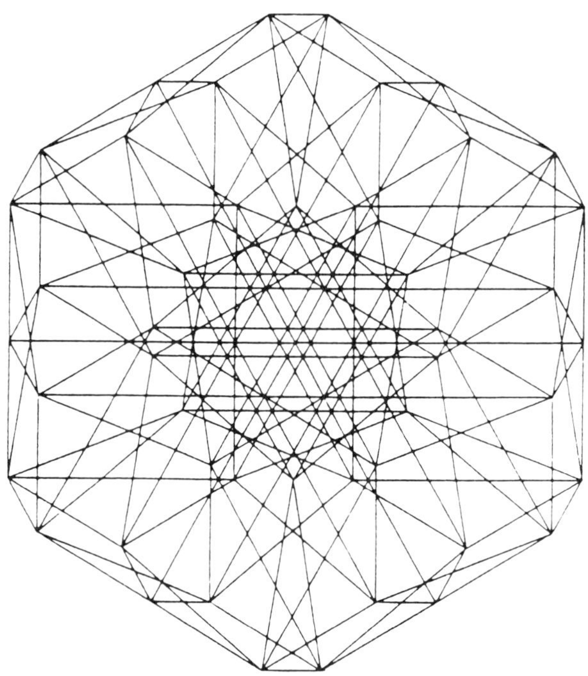

FIGURE 16. $\{4, 6|, 3\}$, projection vector $w = (1, \omega, -\omega^2)$

Finally, in Figure 17, the same vector $w = (1, \omega, -\omega^2)$ is applied to $\{4, 8|, 3\}$ or $\{3, 8|, 4\}$.

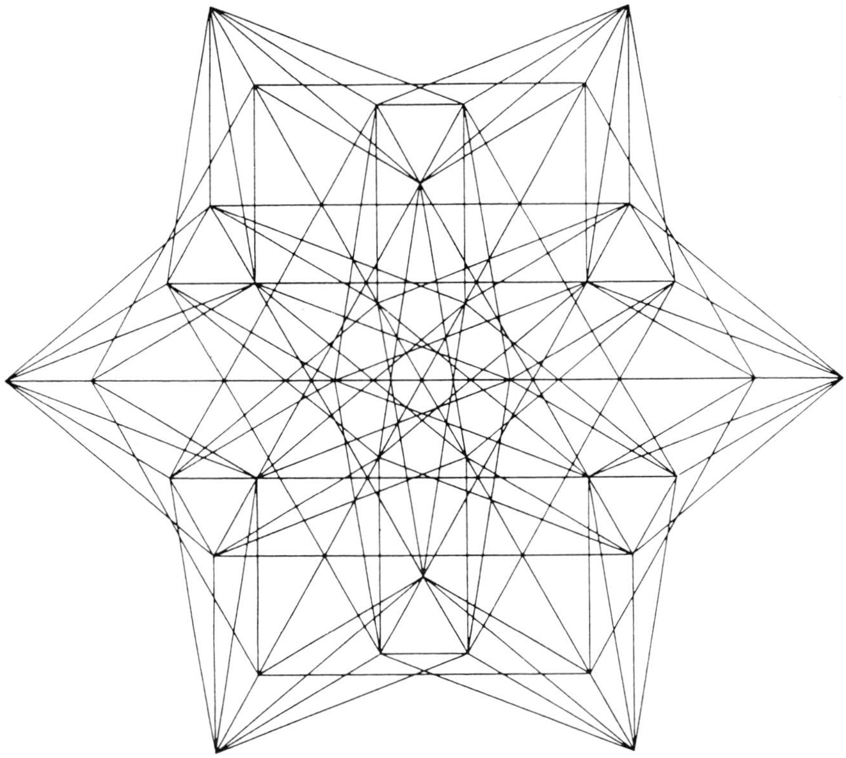

FIGURE 17. $\{4, 8|, 3\}$ or $\{3, 8|, 4\}$, projection vector $w = (1, \omega, -\omega^2)$

## REFERENCES


1. W. W. Rouse Ball and H. S. M. Coxeter, *Mathematical recreations and essays*, 13th ed., Dover, New York, 1987.
2. H. S. M. Coxeter, *The polytopes with regular-prismatic vertex figures*. I, Philos. Trans. Roy. Soc. London Ser. A **229** (1930), 329–425.
3. ——, *The abstract group* $G^{m,n,p}$, Trans. Amer. Math. Soc. **45** (1939), 73–150.
4. ——, *Regular and semi-regular polytopes*. I, Math. Z. **46** (1940), 380–407.
5. ——, *Symmetrical definitions for the binary polyhedral groups*, Proc. Sympos. Pure Math., no. 1, Amer. Math. Soc., Providence, R. I., 1959, pp. 64–87.
6. ——, *Twelve geometric essays*, Southern Illinois Univ. Press, Carbondale, IL, 1968.
7. ——, *Regular and semi-regular polytopes*. II, Math. Z. **188** (1985), 559–591.
8. ——, *Regular complex polytopes*, 2nd ed., Cambridge Univ. Press, 1990.
9. H. S. M. Coxeter, M. S. Longuet-Higgins, and J. C. P. Miller, *Uniform polyhedra*, Philos. Trans. Roy. Soc. London Ser. A **246** (1954), 401–450.
10. H. S. M. Coxeter and W. O. J. Moser, *Generators and relations for discrete groups*, 4th ed., Springer, Berlin, 1980.
11. C. W. L. Garner, *Regular skew polyhedra in hyperbolic three-space*, Canad. J. Math. **19** (1967), 1179–1186.
12. Peter McMullen and Egon Schulte, *Self-dual regular 4-polytopes and their Petrie-Coxeter polyhedra*, Results in Math. **12** (1987), 366–375.

13. ——, *Constructions for regular polytopes*, J. Combin. Theory **53** (1990), 1–28.
14. ——, *Regular polytopes from twisted Coxeter groups and unitary reflection groups*, Adv. in Math. **82** (1990), 35–87.
15. G. A. Miller, H. F. Blichfeldt, and L. E. Dickson, *Theory and applications of finite groups*, 1st ed., Wiley, New York, 1916.
16. G. C. Shephard, *Unitary groups generated by reflections*, Canad. J. Math. **5** (1953), 364–383.
17. G. C. Shephard and J. A. Todd, *Finite unitary reflection groups*, Canad. J. Math. **6** (1954), 274–304.
18. A. I. Weiss and Z. Lučić, *Regular polyhedra in hyperbolic three-space*, Mitt. Math. Sem. Giessen **165** (1984), 237–252.

University of Toronto, Toronto, Canada M5S 1A1

University of East Anglia, Norwich, England NR4 7TJ
*E-mail address*: k101%uk.ac.uea.cpc865@ukacrl

DIMACS Series in Discrete Mathematics
and Theoretical Computer Science
Volume 4, 1991

Applied Geometry
and Discrete Mathematics
THE VICTOR KLEE FESTSCHRIFT

# Volumes of a Random Polytope in a Convex Set

L. DALLA AND D. G. LARMAN

ABSTRACT. The classical result of Blaschke on the expected areas of triangles in a plane convex body is extended to the expected areas of the convex hull of $n$ points in a plane convex body.

Let $K$ be a convex body in Euclidean $d$-space $R^d$. If $n$ points $\mathbf{x}_1, \ldots, \mathbf{x}_n$ are randomly and independently selected from $K$, the convex hull $K(\mathbf{x}_1, \ldots, \mathbf{x}_n)$ of these points can be interpreted as a random polytope with at most $n$ vertices. The expected value of the volume of this polytope is defined by

$$m(n, K) = (\operatorname{Vol} K)^{-n} \int_{\mathbf{x}_1 \in K} \cdots \int_{\mathbf{x} \in K} \operatorname{Vol}(K(\mathbf{x}_1, \ldots, \mathbf{x}_n))\, d\mathbf{x}_1, \ldots, d\mathbf{x}_n .$$

In **[1]**, H. Groemer has shown that for fixed volume $m(n, K)$ attains its minimum value when, and only when, $K$ is an ellipsoid. In **[2]**, I. Barany and D. G. Larman show that for $n$ large

$$(1) \qquad (1 - c_1 n^{-2/d+1}) \operatorname{Vol} K \le m(n, k) \le \left(1 - c_2 \frac{(\log n)^{d-1}}{n}\right) \operatorname{Vol} K,$$

where $c_1$ depends on $d$ and $c_2$ depends on $K$. Further, for $n$ large, polytopes behave like the r.h.s. of (1) and ellipsoids like the l.h.s. of (1).

Consequently it is reasonable to conjecture that for fixed volume the maximum of $m(n, K)$ will be obtained at a polytope and perhaps even at a simplex. We shall prove that for $d = 2$ the maximum is attained at a triangle and, in general, that $m(n, P)$ is a maximum taken over all $d$-polytopes with at most $d + 2$ vertices when $P$ is a $d$-simplex. It is worth remarking that Blaschke **[5]** showed that the maximum of $m(3, K)$ in two dimensions is attained exactly when $K$ is a triangle. A comprehensive survey is contained in a recent article by R. Schneider **[6]**.

---

1980 *Mathematics Subject Classification* (1985 *Revision*). Primary MR52, 60D, 46B.

THEOREM 1. *Let $K$ be a convex body in $R^2$. Then, if $T^2$ denotes a triangle of the same area as $K$,*

$$m(n, K) \leq m(n, T^2)$$

*with strict inequality if $K$ is any polygon other than a triangle.*

REMARK. It seems reasonable to conjecture that equality holds in Theorem 1 only if $K$ is a triangle. Also the methods of Theorem 1 can be applied to show that $m(n, C) \leq m(n, T^3)$ for all convex cylinders $C$ in $R^3$ of the same volume as the simplex $T^3$.

THEOREM 2. *Let $P$ be a $d$-polytope in $R^d$ with at most $d+2$ vertices and let $T^d$ be a $d$-simplex of the same volume as $P$. Then*

$$m(n, P) \leq m(n, T^d)$$

*with equality if and only if $P$ is a $d$-simplex.*

REMARK. The methods used in Theorem 2 can be applied to many other polytopes. The first case in which some modification of the method cannot obviously be used seems to be the dodecahedron in $R^3$.

LEMMA 1. *If $n, p$ are positive integers greater than 2, then there exists a convex polygon $K_0$, of area 1, with at most $p$ vertices and*

$$m(n, K) \leq m(n, K_0)$$

*for any other convex polygon $K$, of area 1, with at most $p$ vertices.*

PROOF. By a result of F. John [**3**], we may suppose that each convex polygon $K$ considered contains a disk of radius $\frac{1}{3}\sqrt{2/\sqrt{3}}$ and is contained in a concentric disk of radius $\frac{2}{3}\sqrt{2/\sqrt{3}}$. So the convex bodies being considered form a compact metric space in the Hausdorff metric. As $m(n, K)$ is a continuous function of $K$ in the Hausdorff metric, Lemma 1 follows.

REMARK. The existence of maximal bodies from a given class in other situations considered in Theorem 2 may be proved in a similar way.

We next need a lemma proved by H. Groemer [**1**]. If $\mathbf{q} = (q_1, \ldots, q_{d-1})$ is a point of $R^{d-1}$ and if $z \in \mathbb{R}$, the point $(q_1, \ldots, q_{d-1}, z)$ of $R^d$ will be denoted by $(\mathbf{q}, z)$.

LEMMA 2. *Let $H$ be the hyperplane which consists of all points $x_d = 0$ and let $\mathbf{c}_1, \ldots, \mathbf{c}_n$ be $n$ given point in $H$, where $n > d$. If $z_1, \ldots, z_n$ are real numbers, put $Z = (z_1, \ldots, z_n)$ and*

$$V(Z) = \mathrm{Vol}(\mathrm{conv}((\mathbf{c}_1, z_1), \ldots, (\mathbf{c}_n, z_n))). \tag{2}$$

*Then, if $Z'$ and $Z''$ are two points of $R^n$,*

$$V(\tfrac{1}{2}Z' + \tfrac{1}{2}Z'') \leq \tfrac{1}{2}V(Z') + \tfrac{1}{2}V(Z'').$$

For completeness we give the proof of Lemma 2.

PROOF. Let

$$K = \operatorname{conv}((\mathbf{c}_1', \tfrac{1}{2}z_1' + \tfrac{1}{2}z_1''), \dots, (\mathbf{c}_n, \tfrac{1}{2}z_n' + \tfrac{1}{2}z_n'')),$$
$$K' = \operatorname{conv}((\mathbf{c}_1, z_1'), \dots, (\mathbf{c}_n, z_n')),$$
$$K'' = \operatorname{conv}((\mathbf{c}_1, z_1''), \dots, (\mathbf{c}_n, z_n'')).$$

The sets $K, K', K''$ have the same orthogonal projection $C = \operatorname{conv}(\mathbf{c}_1, \dots, \mathbf{c}_n)$ on $H$. If $\mathbf{x} \in C$, let $(\mathbf{x}, \overline{x}')$ and $(\mathbf{x}, \underline{x}')$ denote the upper and lower points in $K'$, and similarly let $(\mathbf{x}, \overline{x}'')$ and $(\mathbf{x}, \underline{x}'')$ denote the upper and lower points in $K''$. Let $K^*$ denote the convex body whose upper and lower points are $(\mathbf{x}, \frac{1}{2}(\overline{x}' + \overline{x}''))$ and $(\mathbf{x}, \frac{1}{2}(\underline{x}' + \underline{x}''))$, respectively, $\mathbf{x} \in C$. Then

$$\operatorname{Vol} K^* = \int (\tfrac{1}{2}(\overline{x}' - \underline{x}') + \tfrac{1}{2}(\overline{x}'' - \underline{x}''))\, d\mathbf{x} = \tfrac{1}{2}V(Z') + \tfrac{1}{2}V(Z'').$$

Also, as $(\mathbf{c}_i, \frac{1}{2}z_i' + \frac{1}{2}z_i'') \in K$, $i = 1, \dots, n$,

$$K \subset K^*,$$

and hence $\operatorname{Vol} K = V(\frac{1}{2}Z' + \frac{1}{2}Z'') \le \frac{1}{2}V(Z') + \frac{1}{2}V(Z'')$, as required.

PROOF OF THEOREM 1. We shall assume that all bodies considered have area 1. By Lemma 1, let $Q$ be a polygon with at most $p$ vertices with $m(n, K) \le m(n, Q)$ for any other polygon $K$, of area 1, with at most $p$ vertices.

We shall show that $Q$ is a triangle. Suppose not. Then $Q$ has vertices $P_1P_2P_3 \cdots P_r$, $r \ge 4$, with $P_1P_2, P_2P_3, \dots, P_rP_1$ consecutive edges. We may suppose that there is a Cartesian coordinate system $(x_1, x_2)$ with $P_2P_r$ in the positive direction of the $x_2$-axis and $P_1 \equiv (0, 0)$. The line $P_3P_2$ extended will cut the $x_2$-axis at $(0, -\beta)$ and the line $P_{r-1}P_r$ extended will cut the $x_2$-axis at $(0, \alpha)$. We suppose, without loss of generality, that $0 < \alpha \le \beta$.

Let $P_1^+ = (0, \alpha)$ and $P_1^- = (0, -\alpha)$. Consider now the two polygons

$$Q^+ = \operatorname{conv}(P_1^+, P_2, \dots, P_{r-1}), \qquad Q^- = \operatorname{conv}(P_1^-, P_2, \dots, P_r).$$

Notice that $Q^+$ has only $r-1$ vertices and that $Q^-$ has $r$ (or possibly $r-1$ if $\alpha = \beta$) vertices. By the definition of $Q$,

$$m(n, Q^+) \le m(n, Q), \qquad m(n, Q^-) \le m(n, Q). \tag{3}$$

Now $Q$, $Q^+$, and $Q^-$ have the same orthogonal projection $(0, \gamma)$ onto the $x_1$-axis. Let $(P_2, P_r)$ extended meet the $x_1$-axis in $(0, \delta)$, where $0 < \delta < \gamma$. Let $c_1, \dots, c_n$ be any choice of $n$ numbers in the interval $[0, \gamma]$. Let the vertical line through $(c_i, 0)$ meet $Q, Q^-, Q^+$ in the intervals $[(c_i, \alpha_i - l_i), (c_i, \alpha_1 + l_i)]$, $[(c_i, \alpha_i^- - l_i), (c_i, \alpha_i^- + l_i)]$, $[(c_i, \alpha_i^+ - l_i), (c_i, \alpha_i^+ + l_i)]$, respectively, $i = 1, \dots, n$. Of course, if $\delta \le c_i \le \gamma$, then all three intervals are equal. In any case, all three intervals will have the same length. Let

$t_i$, $i = 1, \ldots, n$, satisfy $|t_i| \le l_i$ and consider the following points, where $\mathbf{t} = (t_i, \ldots, t_n)$:

$$Z(\mathbf{t}) = ((\alpha_1 + t_1), \ldots, (\alpha_n + t_n)),$$
$$Z^+(\mathbf{t}) = ((\alpha_1^+ + t_1), \ldots, (\alpha_n^+ + t_n)),$$
$$Z^-(\mathbf{t}) = ((\alpha_1^- + t_1), \ldots, (\alpha_n^- + t_n)),$$

Then $Z(\mathbf{t}) = \frac{1}{2}Z^+(\mathbf{t}) + \frac{1}{2}Z^-(\mathbf{t})$ and so, by Lemma 2,

$$V(Z(\mathbf{t})) \le \tfrac{1}{2}V(Z^+(\mathbf{t})) + \tfrac{1}{2}V(Z^-(\mathbf{t})). \tag{4}$$

Consequently, integrating (4) over $t_1, \ldots, t_n$ and then over $c_1, \ldots, c_n$, we conclude from (4) that

$$m(n, Q) \le \tfrac{1}{2}m(n, Q^+) + \tfrac{1}{2}m(n, Q^-). \tag{5}$$

However, suppose $P_s$ is a vertex of $Q$ whose orthogonal projection onto the $x_1$-axis is $(0, \gamma)$ and we choose points $(c_1, t_1), \ldots, (c_n, t_n)$ on $(P_1, P_s)$ with $c_1 < \delta$, $c_n > \delta$. Then, if $\mathbf{t}^* = (t_1^*, \ldots, t_n^*)$, $V(Z(\mathbf{t})) = 0$, $V(Z^+(\mathbf{t}^*)) > 0$, and $V(Z^-(\mathbf{t}^*)) > 0$. So, by continuity, we have an improvement on (5) to

$$m(n, Q) < \tfrac{1}{2}m(n, Q^+) + \tfrac{1}{2}m(n, Q^-). \tag{6}$$

Thus (6) contradicts the maximality of $m(n, Q)$. Consequently $Q$ is a triangle. Finally, since any convex body in $R^2$ can be approximated arbitrarily closely from within and without by polygons, we conclude that Theorem 1 holds.

PROOF OF THEOREM 2. We first note that every $d$-polytope $P$ with $d + 2$ vertices is either (i) pyramidal about one of its vertices or (ii) (affinely equivalent to) the convex hull of an $r$-simplex $T^r$ and a $(d - r)$-simplex $T^{d-r}$, $1 \le r \le d - 1$, in orthogonal $r$- and $(d - r)$-dimensional subspaces, respectively, such that $\mathbf{0} \in \operatorname{relint} T^r \cap \operatorname{relint} T^{d-r}$ (see Grünbaum **[4, p. 97]**).

So let $P$ be a $d$-polytope of unit volume with at most $d + 2$ vertices at which $\max m(n, P')$ is achieved, where the maximum is taken over all $d$-polytopes $P'$ of unit volume with at most $d+2$ vertices. We suppose that $P$ has form (ii) and hence show that there exists another polytope $P^*$ of form (i) at which the maximum is also achieved.

(7)

We suppose that $T^r$ is embedded in the $r$-space determined by the first $r$ coordinates and that $T^{d-r}$ is embedded in the $(d - r)$-subspace determined by the last $d - r$ coordinates. Let $\mathbf{e}_d$ denote the $d$th unit vector, which in our case lies in $\operatorname{aff} T^{d-r}$. Consider $T^r + t\mathbf{e}_d$, $t$ real. There exists $\alpha > 0$ such that $T^r + t\mathbf{e}_d$ meets the relative interior of $T^{d-r}$ for $|t| < \alpha$ but (say) $T^r + \alpha\mathbf{e}_d$ meets the relative boundary of $T^{d-r}$.

We shall show that

$$P^+ = \operatorname{conv}(T^r + \alpha \mathbf{e}_d, T^{d-r})$$

satisfies $m(n, P^+) = m(n, P)$ which will establish (5). Let

$$P^- = \operatorname{conv}(T^r - \alpha \mathbf{e}_d, T^{d-r})$$

and let $Q$ denote the orthogonal projection of $P$, $P^+$, and $P^-$ onto the $(d-1)$-subspace $x_d = 0$. Let $\mathbf{q} = (q_1, \dots, q_{d-1}, 0) \in Q$. Then the line $\ell_q = \{(x_1, \dots, x_d): x_1 = q_1, \dots, x_{d-1} = q_{d-1}\}$ meets $P$ in a line segment determined by $q_d^- \le x_d \le q_d^+$, say, and $q^- = (q_1, \dots, q_{d-1}, q_d^-)$ and $q^+ = (q_1, \dots, q_{d-1}, q_d^+)$ are boundary points of $P$. Now every boundary point $\mathbf{z}$ of $P$ has the unique representation $\mathbf{z} = \lambda\mathbf{x} + (1-\lambda)\mathbf{y}$, where $\mathbf{x} \in \operatorname{relbd} T^r$ and $y \in \operatorname{relbd} T^{d-r}$. Consequently $\mathbf{q}^- = \lambda\mathbf{x} + (1-\lambda)\mathbf{y}$, for unique $\mathbf{x} \in \operatorname{relbd} T^r$ and $\mathbf{y} \in \operatorname{relbd} T^{d-r}$, and $\mathbf{q}^+ = \lambda\mathbf{x} + (1-\lambda)\mathbf{y}'$, for unique $\mathbf{y}' \in \operatorname{relbd} T^{d-r}$, $\mathbf{y}' - \mathbf{y}$ in the direction of $\mathbf{e}_d$.

Consider $P^+$ and the same point $\mathbf{q} \in Q$. The line $\ell_q$ meets $P^+$ in a line segment determined by $q_d^{-+} \le x_d \le q_d^{++}$, and $\mathbf{q}^{-+} = (q_1, \dots, q_{d-1}, q_d^{-+})$, and $\mathbf{q}^{++} = (q_1, \dots, q_{d-1}, q^{++})$ are boundary points of $P^+$. Then

$$\mathbf{q}^{-+} = \lambda\mathbf{x} + (1-\lambda)\mathbf{y} + \lambda\alpha\mathbf{e}_d, \qquad \mathbf{q}^{++} = \lambda\mathbf{x} + (1-\lambda)\mathbf{y}' + \lambda\alpha\mathbf{e}_d$$

and hence

$$(8) \qquad q_d^{-+} = q_d^- + \lambda\alpha, \qquad q_d^{++} = q_d^+ + \lambda\alpha.$$

Similarly, for the corresponding $q^{--}$, $q^{+-}$ of $P^-$,

$$(9) \qquad q_d^{--} = q_d^- - \lambda\alpha, \qquad q_d^{+-} = q_d^+ - \lambda\alpha.$$

So, using (8) and (9), the line $\ell_q$ cuts $P^-$, $P^+$, and $P$ in line segments of the same length, with $P^-$ and $P^+$ also having unit volume. Thus if $\mathbf{q}_1, \dots, \mathbf{q}_n$ is any choice of points in $Q$, if $Z = (z_1, \dots, z_n)$, where $z_i = q_{id}^- + t_i$, $0 \le t_i \le q_{id}^+ - q_{id}^-$, $i = 1, \dots, n$, if $Z^+ = (z_1^+, \dots, z_n^+)$, where $z_i^+ = q_{id}^- + t_i + \lambda\alpha$, $i = 1, \dots, n$, and if $Z^- = (z_1^-, \dots, z_n^-)$, where $z_i^- = q_{id}^- + t_i - \lambda\alpha$, $i = 1, \dots, n$, then $Z = \frac{1}{2}Z' + \frac{1}{2}Z''$ and so, by Lemma 2,

$$V(Z) \le \tfrac{1}{2}V(Z') + \tfrac{1}{2}V(Z'').$$

Hence, as in Theorem 1,

$$m(n, P) \le \tfrac{1}{2}m(n, P^+) + \tfrac{1}{2}m(n, P^-)$$

and hence $m(n, P^+) = m(n, P^-) = m(n, P)$, which establishes (7).

We repeat this process within the base of the polytope $P$ which is pyramidal about one of its vertices and for which $m(n, P)$ is maximal until the base becomes a (two-dimensional) square. A further application reduces the square to a triangle and hence $P$ to a $d$-simplex as required. We may produce strict inequality by arguing as in (6) of the proof of Theorem 1.

## REFERENCES

1. H. Groemer, *On the mean value of the volume of a random polytope in a convex set*, Arch. Math. **25** (1974), 86–90.
2. I. Barany and D. G. Larman, *Convex bodies, economic cap coverings, random, polytopes*, Mathematika **35** (1988), 274–291.
3. F. John, *Extremum problems with inequalities as subsidiary conditions*, Courant Anniv. Volume, 1948, pp. 187–204.
4. B. Grünbaum, *Convex polytopes*, Interscience, 1967.
5. W. Blaschke, Leipzig. Ber **69** (1917), 436–453.
6. R. Schneider, *Random approximation of convex sets*, J. Microscopy **151** (1988), 211–227.

DEPARTMENT OF MATHEMATICS, UNIVERSITY OF ATHENS, ATHENS, GREECE

DEPARTMENT OF MATHEMATICS, UNIVERSITY COLLEGE LONDON, GOWER STREET, LONDON, WC1E 6BT ENGLAND

DIMACS Series in Discrete Mathematics
and Theoretical Computer Science
Volume 4, 1991

Applied Geometry
and Discrete Mathematics
THE VICTOR KLEE FESTSCHRIFT

# Bodies of Constant Width in Riemannian Manifolds and Spaces of Constant Curvature

B. V. DEKSTER

ABSTRACT. A set $C$ in a Riemannian manifold $M^n$ is called *definitely convex* if (i) $C$ is compact and each two points of $C$ can be connected in $C$ by a rectifiable curve; (ii) any shortest path in $C$ between two points of $C$ is a geodesic; (iii) any geodesic in $C$ is the unique shortest path in $C$ between its ends; (iv) each geodesic segment in $C$ contains no pair of conjugate points along this segment.

Let $K$ be a definitely convex set in $M^n$ homeomorphic to a ball. It is called a *body of constant width* $W > 0$, if for any $p \in \partial K$ and any exterior unit normal $n$ of $\partial K$ at $p$, there exists the geodesic $pq$ of length $W$ having direction $-n$, such that $K \supset pq$ but $K$ contains no longer geodesic $pq' \supset pq$.

We show that $\operatorname{diam} K = W$ and prove the following

THEOREM. *Let $K$ be a body of constant width $W$ in an $n$-dimensional sphere, Euclidean or hyperbolic space of curvature $k$ satisfying $W \leq \pi/2\sqrt{k}$ when $k > 0$. Let $a$, $b \in \partial K$, $ab = \Delta \geq 0$ and $n_a$, $n_b$ be exterior normals at $a$ and $b$ respectively. Then the angle between $n_a$ and $n_b$ after their parallel translation along the chord $ab$ to the same point does not exceed certain specified quantity depending on $\Delta$, $W$ and $k$.*

Bodies of constant width in spherical and hyperbolic spaces were considered in [**2, 7–10**]. A short review of these results can be found in the second half of §11 in [**4**]. Here we extend the notion of bodies of constant width to Riemannian manifolds $M^n$. Many facts about these bodies in $R^n$ can easily be reestablished in $M^n$ (see Remarks 3 and 4). We study here mostly curvature of the boundary of such a body in $M^n$. The normal curvatures at a point on the boundary of a body of constant width $W$ in $R^n$ are greater than or equal to $\frac{1}{W}$ since the body is supported by a sphere of radius $W$ at each of its boundary points. This argument is extended to $M^n$ in Remark 4. An upper bound for the boundary curvature obviously does not exist; however, certain integral restrictions of this type can still be observed. Consider

1980 *Mathematics Subject Classification* (1985 *Revision*). Primary 53C20, 53A35, 52A20.
The author was supported by a Canadian NSERC Grant.

for instance a plane figure of constant width. The normal to its boundary can jump through an angle of $60^\circ$ at a point $\alpha$. Further along the boundary, at a point $\beta$, it can jump this much again (recall the Reuleaux triangle). One might feel however that the points $\alpha$ and $\beta$ cannot be too close to each other because then the diameter of the figure would be too large. In our main theorem, we quantify this notion for spaces of constant curvature. This result is probably new in $R^n$, too. For more results in $M^n$, see **[6]** and the Bibliography there.

The author thanks Cathy Baker for numerous useful suggestions.

We will assume that $M^n$ is regular but not necessarily complete. A set $C \subset M^n$ will be called *definitely convex* if

(i) $C$ is compact and each two points of $C$ can be connected in $C$ by a rectifiable curve.
(ii) any shortest path within $C$ between two points of $C$ is a geodesic (the path exists due to (i));
(iii) any geodesic segment in $C$ is the unique shortest path within $C$ between its ends;
(iv) each geodesic segment in $C$ contains no pair of conjugate points along this segment.

For instance, let $M^2$ be a unit sphere in $R^3$ and $C$ be a geodesic of length $\frac{3}{2}\pi$; then $C$ is not definitely convex because (iv) fails even though (i), (ii), and (iii) hold. Let $M^2$ be a cylinder of unit radius in $R^3$ and $C \subset M^2$ be a circular arc of length $\frac{3}{2}\pi$; then $C$ is definitely convex. A circumference on $M^2$ is not definitely convex since (iii) fails while (i), (ii), and (iv) hold. The domain $y \leq x^2$ of the $xy$-plane is not definitely convex since (i) and (ii) fail while the other two conditions hold.

By a shortest path, we will mean a shortest path within the definitely convex set $C$ under consideration; by the diameter of $C$, we mean its diameter with respect to the metric $\rho(x, y)$ being the length of a shortest path in $C$ between $x$ and $y \in C$. Thus, the diameter of the arc of length $\frac{3}{2}\pi$ in the first two examples above is $\frac{3}{2}\pi$.

The ambient space $M^n \backslash C$ does not play a major role in the above definition of a definitely convex set $C$: if $C$ is definitely convex in $M^n$, then $C$ is definitely convex in any $\varepsilon$-neighbourhood of $C$. However, if $C$ is not full-dimensional, then $C$ can be definitely convex in $M^n$ but not definitely convex when imbedded in another manifold. The first two examples together illustrate this. We will need an ambient space in order to let our geometric constructions stretch a little beyond the set $C$ (see, for example, Remark 3 and the proof of Remark 4). Conditions (iii) and (iv) of the definition will hold, for instance, if $M^n$ is complete and the diameter of $C$ is less than a lower bound for the injectivity radius over $M^n$. (This was noted by Walter Vannini.)

Let $K$ be a definitely convex set in $M^n$ which is homeomorphic to a ball.

Its boundary, diameter, and interior will be denoted by $\partial K$, $\operatorname{diam} K$, and $\operatorname{int} K$ respectively. We will call $K$ a *body of constant width* $W > 0$ if, for any $p \in \partial K$ and any exterior unit normal $n$ of $\partial K$ at $p$, there exists a geodesic $pq$ of length $W$ having direction $-n$, such that $K \supset pq$ but $K$ contains no longer geodesic $pq' \supset pq$ (then obviously $q \in \partial K$ and, by (iii), $pq$ is a shortest path in $K$); thus $D \overset{\text{def}}{=} \operatorname{diam} K \geq W$. (For the definition of exterior normal see [**3**, §**4.5**].) We use $K$, $W$, and $D$ as defined above throughout the rest of the paper.

REMARK 1. The path $pq$ does not necessarily cross $\operatorname{int} K$. Consider, for example, a triangle $pqr$ on a unit sphere with $pq = qr = pr = \frac{\pi}{2}$. (The notation $pq$ refers to both the path and its length.)

REMARK 2. The assumption that $K$ is homeomorphic to a ball can actually be omitted from the definition of a body of constant width and proved as a property. The point is that a convex set is a totally geodesic submanifold of $M^n$ (see [**5**]). The condition $W > 0$ readily implies that the set is $n$-dimensional. Let $a \in \operatorname{int} K$ and $r$ be a geodesic starting at $a$. By the compactness of $K$ and (iii), $r$ crosses $\partial K$. Now the mapping $\exp_a$ yields an appropriate homeomorphism.

Let $C$ be a closed definitely convex set in $M^n$ that consists of more than a single point and let $p \in \partial C$. Let $S$ be a regular hypersurface passing through $p$. Suppose $p$ has a neighbourhood $U$ divided by $S$ into two parts $U'$ and $U''$ which satisfy $U' \cup U'' = U$, $U' \cap U'' = S \cap U$, and $U \cap C = U' \cap C$. Then we say that $S$ *supports* $C$ at $p$. It is known that the normal $n$ of $S$ at $p$ directed towards $\operatorname{int} U''$ is one of the exterior normals of $\partial C$ at $p$. Let $\kappa$ be the minimum normal curvature of $S$ at $p$ on the side of $-n$. Then we say that $C$ is *$\kappa$-supported at $p$*. If $\partial C$ is regular at $p$, this means that the normal curvature of $\partial C$ on the side of $-n$ in each direction at the point $p$ is $\geq \kappa$. Obviously, $\kappa$-supported implies $\kappa'$-supported for any $\kappa' < \kappa$. If $C$ is $\kappa$-supported at each $p \in \partial C$, we will say that the set $C$ is *$\kappa$-supported.*

REMARK 3. Now we show that $D = W$. Let $x, y \in K$, $xy = D$. Obviously, $x, y \in \partial K$, because otherwise the geodesic $xy$ could be extended within $K$ while remaining a shortest path. Due to (iii) and (iv), the metric sphere of radius $D$ centered at $y$ is a surface which is regular at $x$ and supports $K$ at $x$. The direction $n$ of its radius $yx$ at $x$ is a normal to the sphere and $n$ is obviously an exterior normal of $\partial K$ at $x$. By the definition of constant width, $xy \leq W$. Thus $xy = D = W$.

It is clear now that the geodesic $pq$ in the definition of a body $K$ of constant width $W$ has the following property: its direction at $q$ is an exterior normal of $\partial K$ at $q$. This follows from the fact that the sphere of radius $W = D$ centered at $p$ supports $K$ at $q$.

REMARK 4. Let $k$ be the maximum sectional curvature in $K$. Let $W$ satisfy $W < \pi/\sqrt{k}$ if $k > 0$. Denote by $M_k^2$ a plane, hyperbolic plane, or sphere of curvature $k$. We show that $K$ is $\lambda$-supported for any $\lambda < \kappa$, where $\kappa = \kappa(k, W)$ is the curvature of a circumference of radius $W$ in $M_k^2$, i.e.,

$$(1) \qquad \kappa(k, W) = \begin{cases} \sqrt{k}\cot\sqrt{k}W & \text{for } k > 0,\ 0 < W < \pi/\sqrt{k}; \\ 1/W & \text{for } k = 0; \\ \sqrt{-k}\coth\sqrt{-k}W & \text{for } k < 0. \end{cases}$$

When $k > 0$ and $\pi/2\sqrt{k} < W < \pi/\sqrt{k}$, the curvature $\kappa < 0$ and thus Remark 4 provides no information. This happens, for instance, for $K$ a circle of radius $W/2$ in $M_k^2$. On the other hand, the estimate in Remark 4 is exact for Reuleaux triangles on a unit sphere which are smaller than the triangle $pqr$ in Remark 1.

PROOF OF REMARK 4. We will consider only the case $k > 0$, leaving the rest to the reader. Denote by $H = H(\varepsilon)$ the supremum of sectional curvature in an $\varepsilon$-neighbourhood of $K$. Let $pq$ be the chord of length $W$ mentioned in the definition of $K$. Since $\operatorname{diam} K = W$ (Remark 3), the metric sphere $S$ of radius $W$ centered at $q$ supports $K$ at $p$. The sphere is regular at $p$ by (iii) and (iv). Let $V_0$ be a unit vector tangent to $S$ at $p$. Consider a rotation of the radius $qp$ about $q$ under which $p$ slides along $S$ with the velocity $V_0$. Denote by $V(u)$ the related Jacobi field so that $V(0) = 0$, $V(W) = V_0$. According to **[1, p. 248, Problem 28(b)]**, the normal curvature $\alpha$ of $S$ in the direction $V_0$ satisfies

$$(2) \qquad \alpha = \langle V'(W), V(W)\rangle.$$

Since $W < \pi/\sqrt{k}$, one has $W < \pi/\sqrt{H}$ when $\varepsilon$ is sufficiently small. Now due to the last formula in the proof of Theorem 13 in **[1, p. 249]**,

$$(3) \qquad \alpha \geq \sqrt{H}\sin(\sqrt{H}W)\cos(\sqrt{H}W)\langle E(W), E(W)\rangle,$$

where $E(u)$, $u \in [0, W]$, is a parallel vector field along a normal geodesic in a sphere of curvature $H$ satisfying

$$(4) \qquad |\sin(\sqrt{H}W)E(W)| = |V(W)| = 1.$$

(In **[1]**, the authors denote the radius of $S$ by $r$ instead of the $W$ used here. Condition (4) is stated in **[1, p. 249]** erroneously as $E(r) = V(r)$.) Now (3), (4), and (1) imply that $\alpha \geq \sqrt{H}\cot\sqrt{H}W = \kappa(H, W)$. Then $\alpha \geq \kappa(k, W) - |\kappa(k, W) - \kappa(H, W)| > \lambda$ if $\varepsilon$ is sufficiently small. Thus, $K$ is $\lambda$-supported.

REMARK 5. The bodies of constant width in $M^n$ still differ considerably from Euclidean ones. A somewhat unexpected example is as follows. Let $M^n$ be the unit sphere $S^2$. Consider a regular triangle $abc$ with side length $l$ satisfying $\frac{\pi}{2} < l < \frac{2\pi}{3}$. This is a definitely convex body whose diameter $D$ (the altitude) satisfies $l < D < \pi$. Its circumcircle is a body of constant width $W > D$, $W < \pi$. However, in contrast to the Euclidean case **[4, §4(A), (B)]**, there is no body $K$ of constant width $D$ in $S^2$ which contains the triangle $abc$. Indeed, suppose $K$ exists. By Remark 3, $\operatorname{diam} K = D$.

The six ends of the altitudes $aa'$, $bb'$, $cc'$ of the triangle $abc$ should lie on $\partial K$ because their lengths are $D$. Since the ends of the segment $ab$ and its midpoint $c'$ lie on $\partial K$ and $K$ is convex, the whole segment $ab \subset \partial K$. Similarly, both $bc$, $ac \subset \partial K$. Then $K = abc$, which is impossible since $abc$ is clearly not of constant width.

A natural question arises: Given some $D < \pi$, what is the least number $W = W(D)$ such that each convex body of diameter $D$ in $S^n$ can be covered by a suitable body of constant width $W$?

We will denote by $M_k^n$ a spherical, hyperbolic, or Euclidean $n$-dimensional space of curvature $k$. Consider in $M_k^2$ a quadrangle $abcd$ such that

(i) $a$ is symmetric to $b$ and $c$ is symmetric to $d$ with respect to a geodesic in $M_k^2$;
(ii) $a$ and $d$ are on the same side of the geodesic (see Figure 1);
(iii) $0 < cd = ac = bd = W$, where $W < \pi/\sqrt{k}$ if $k > 0$;
(iv) $0 \leq ab = \Delta \leq W$. In addition,

$$\Delta \geq 3W - 2\pi/\sqrt{k} \quad \text{when } k > 0. \tag{5}$$

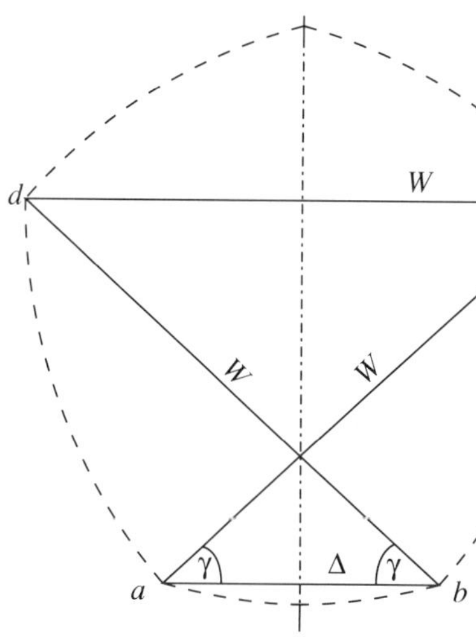


FIGURE 1

Denote by $\gamma = \gamma(\Delta, W, k)$ the angle $cab = dba$ or its limiting value if $\Delta = 0$. One can easily check that the quadrangle exists, and that $\gamma < \frac{\pi}{2}$. (Condition (5) guarantees the existence for $W > 2\pi/3\sqrt{k}$.) Trigonometry shows that

(6) $$\cos\gamma = \tan\frac{\Delta\sqrt{k}}{2}\cot W\sqrt{k} + \frac{1}{2}\sec\frac{\Delta\sqrt{k}}{2}\sec\frac{W\sqrt{k}}{2} \quad \text{if } k > 0;$$

(7) $$\cos\gamma = \frac{W+\Delta}{2W} \quad \text{if } k = 0;$$

$$\cos\gamma = \tanh\frac{\Delta\sqrt{-k}}{2}\coth W\sqrt{-k} + \frac{1}{2}\operatorname{sech}\frac{\Delta\sqrt{-k}}{2}\operatorname{sech}\frac{W\sqrt{-k}}{2}$$

(8) $\quad$ if $k < 0$.

THEOREM. *Let $K$ be a body of constant width $W$ in $M_k^n$, $n \geq 2$, satisfying*

(9) $$W \leq \pi/2\sqrt{k} \quad \textit{when } k > 0.$$

*Let $a, b \in \partial K$, $ab = \Delta \geq 0$, and $n_a$, $n_b$ be exterior normals at $a$ and $b$ respectively. Then the angle $\xi$ between $n_a$ and $n_b$ after their parallel translation along the chord $ab$ to the same point satisfies*

(10) $$\xi \leq \pi - 2\gamma(\Delta, W, k).$$

The equality in (10) can hold for any triple $(\Delta, W, k)$ (see the area in Figure 1 bounded by the five dotted arcs of radius $W$).

PROOF. Let the chords $ap$ and $bb'$ of length $W$ have directions $-n_a$ and $-n_b$. If $\Delta = 0$, then $a = b$ and relation (10) follows from the isosceles triangle $ab'p$ with $b'p \leq W$. If $\Delta = W$, then $\gamma = 0$ and (10) becomes trivial. Let $0 < \Delta < W$ and denote by $M_k^2$ the 2-plane determined by the triangle $abb'$. Let $bb_0 \subset M_k^2$ be a segment with interior point $a$ and let $ab'' \subset M_k^2$ be another segment that lies with $bb'$ on the same side of $bb_0$ and forms the angle

(11) $$\angle b''ab_0 = \angle b'bb_0$$

(see Figure 2). ($\angle$ denotes angle.) Obviously, $\angle b''ap = \xi$. By the triangle inequality,

(12) $$\xi \leq \angle b''ab' + \angle b'ap.$$

Now let $a' \in M_k^2$ be such that $a'b' = pb' = \Delta' \geq 0$, $a'a = W$, and $a'$ and $b$ lie on the same side of $ab'$. We show that $aa'$ and $bb'$ intersect. Since $\Delta > 0$ and $\angle abb' \leq \frac{\pi}{2}$ as it is an angle between an interior normal and a chord, the triangle $abb'$ degenerates if and only if $\angle abb' = 0$. In this degenerate case, one obviously has $aa' \cap bb' \ni a$. Now suppose that the triangle $abb'$ is nondegenerate. Since $\Delta < W$, one has $a \neq b'$. Denote by $g$ the geodesic orthogonal to $ab'$ passing through its midpoint (see Figure 3(a)).

First, we deal with the case $\Delta' \leq \Delta$. Note that $b$ and $b'$ are separated by $g$ since $\Delta < W$. Denote by $a_*$ the point in $M_k^2$ symmetric to $b$ with

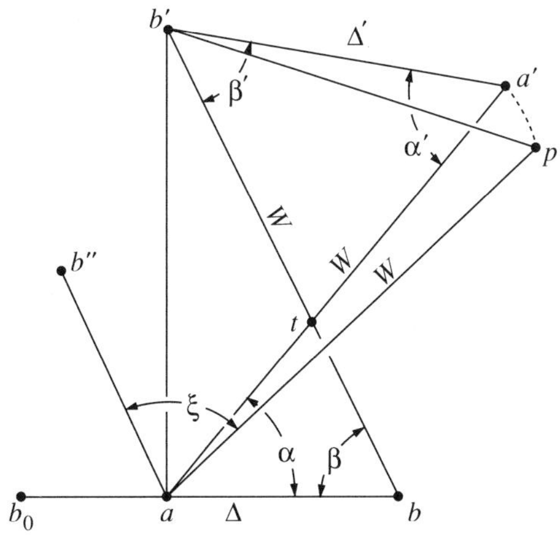


FIGURE 2

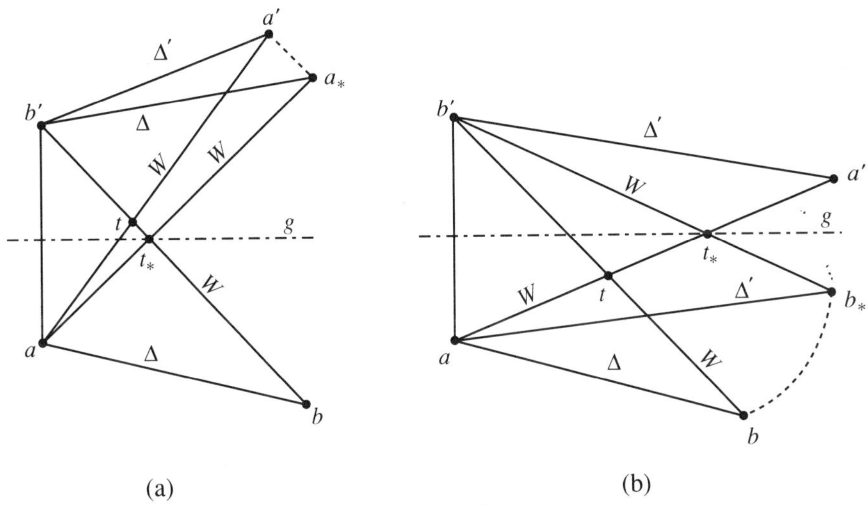


FIGURE 3

respect to $g$. The side $aa_*$ of the triangle $aa_*b'$ crosses $bb'$ at a point $t_* \in g$. Since $\Delta' \leq \Delta$, one has

$$\angle b'aa' \leq \angle b'aa_*. \tag{13}$$

The segments $ax$, $x \in bb'$, lying in $K$ are not longer than $W$; therefore the segment $aa'$ of length $W$ crosses $bb'$ at a point $t \in t_*b'$.

The case $\Delta' > \Delta$ is a little different. Since $\Delta' \leq W$, the points $a$ and $a'$ are separated by $g$ (although $a' \in g$ is now possible). Denote by $b_*$ the point symmetric to $a'$ with respect to $g$. The side $b'b_*$ of the triangle $b'b_*a$ crosses $aa'$ at a point $t_* \in g$ (see Figure 3(b)). Since $\Delta < \Delta'$, one has

$$\angle ab'b < \angle ab'b_*. \tag{14}$$

Take an $x \in aa'$ and let $y \in ap$ satisfy $ay = ax$. Since $b'y \subset K$, one has $b'y \le W$. The equality of the triangles $ab'a'$ and $ab'p$ implies

$$b'x = b'y \le W.$$

Together with (14), it follows that the segment $b'b$ of length $W$ crosses $aa'$ at a point $t \in t_*a$.

We will need a few other relations. Put

$$\alpha = \angle a'ab, \quad \beta = \angle b'ba, \quad \alpha' = \angle aa'b', \quad \beta' = \angle bb'a' \tag{15}$$

(see Figure 2). Obviously, $\beta \le \frac{\pi}{2}$ as it is the angle between an interior normal and a chord. Similarly, $\angle bap$, $\angle b'pa$, and $\angle bb'p$ are $\le \frac{\pi}{2}$. Since $aa'$ crosses $bb'$, by the triangle inequality, we obtain

$$\alpha = \angle bab' - \angle a'ab' = \angle bab' - \angle pab' \le \angle bap \le \frac{\pi}{2};$$
$$\beta' = \angle ab'a' - \angle ab'b = \angle ab'p - \angle ab'b \le \angle bb'p \le \angle\frac{\pi}{2}.$$

Moreover, $\alpha' = \angle b'pa \le \frac{\pi}{2}$. Thus,

$$\alpha, \beta, \alpha', \beta' \in \left[0, \frac{\pi}{2}\right]. \tag{16}$$

Now recall estimate (12). For $k \le 0$, it is obvious that

$$\angle b_0ab' > \beta = \angle b_0ab''. \tag{17}$$

By (9), this is also true for $k > 0$, since $ab' \le W$. Indeed,

$$\frac{\sin \angle b_0ab'}{\sin \beta} = \frac{\sin W\sqrt{k}}{\sin(ab' \cdot \sqrt{k})} \ge 1.$$

This implies (17), as $\beta \le \frac{\pi}{2}$.

Due to (17),

$$\angle b''ab' = \angle b_0ab' - \angle b_0ab'' = \angle b_0ab' - \beta. \tag{18}$$

Since $aa'$ crosses $bb'$,

$$\angle b'aa' = \angle b'ab - \alpha = \angle b'ap. \tag{19}$$

Plugging (18) and (19) into (12), one has

$$\xi \le \angle b_0ab' - \beta + \angle b'ab - \alpha = \pi - (\alpha + \beta). \tag{20}$$

Suppose first that $\alpha = \beta$. Note that (9) implies (5) (with negative right-hand side) and therefore $\gamma = \gamma(\Delta, W, k)$ is defined for any $k$. Since $\Delta' \le W$, one has $\alpha = \beta \ge \gamma$. Now (10) follows from (20).

In the rest of the paper, we prove that $\alpha + \beta$ takes on its minimum $2\alpha$ when $\alpha = \beta$. Then $\xi \le \pi - 2\alpha \le \pi - 2\gamma$, which is (10).

Let $\alpha \neq \beta$. One may assume

(21) $$\alpha < \beta.$$

Then $at > bt$, $a't = W - at < W - bt = b't$, and thus

(22) $$\beta' < \alpha'.$$

We now show that

(23) $$\beta' > 0.$$

Suppose to the contrary that $\beta' = 0$. Then $a' = t$. At the same time $t \neq b$, since otherwise $\Delta = W$ which was ruled out at the beginning of the proof. In the triangle $aa'b$,

(24) $$\angle aa'b = \pi - \alpha' \geq \pi - \frac{\pi}{2} = \frac{\pi}{2},$$

(25) $$\angle aba' = \beta \leq \frac{\pi}{2}.$$

The relations (24), (25) imply that

(26) $$\angle aba' \leq \angle aa'b.$$

This is impossible because

$$aa' = W > \Delta = ab.$$

Thus, by (21), (22), and (23),

(27) $$\alpha \geq 0, \quad \beta > 0, \quad \alpha' > 0, \quad \beta' > 0.$$

The closed polygonal line $aa'b'ba$ in $M_k^2$, with sides of fixed lengths and points $a$, $a'$ separated by $bb'$, is determined (to within isometries) by the value of $\alpha$ (although this value cannot be prescribed arbitrarily: consider the case of short $\Delta'$). Let us vary $\alpha$ and denote by $\alpha_1$ its original value. One can easily check that, under the original conditions (16) and (27), the line $aa'b'ba$ with $aa'$ crossing $bb'$ exists for $\alpha \in [\alpha_1, \alpha_1 + \varepsilon]$ with some $\varepsilon > 0$. Moreover, the angles $\beta$, $\alpha'$, and $\beta'$ are regular functions $\beta(\alpha)$, $\alpha'(\alpha)$, and $\beta'(\alpha)$ of $\alpha \in [\alpha_1, \alpha_1 + \varepsilon]$.

We now show that

(28) $\beta(\alpha)$ and $\alpha'(\alpha)$ are strictly decreasing functions

and

(29) $\beta'(\alpha)$ is a strictly increasing function

of $\alpha \in [\alpha_1, \alpha_1 + \varepsilon]$. It is enough to prove (28), (29) for $\alpha \in (\alpha_1, \alpha_1 + \varepsilon)$. For sufficiently small $\varepsilon$, the relations (21), (22), (23), and (27) still hold for $\alpha \in [\alpha_1, \alpha_1 + \varepsilon]$.

As $\alpha$ increases with unit angular speed, the line $aa'b'ba$ moves in $M_k^2$. In that process, any one of its four sides can obviously be considered immovable.

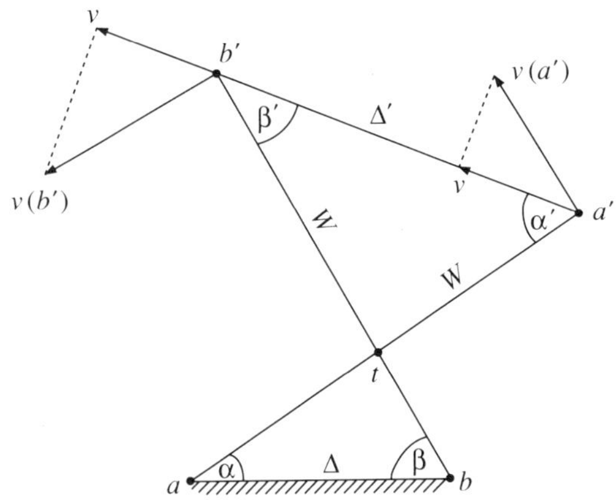


FIGURE 4

Let us first fix $ab$ and consider some $\alpha \in (\alpha_1, \alpha_1 + \varepsilon)$. Denote by $v(a')$ and $v(b')$ the velocities of the points $a'$ and $b'$ in this motion. They are orthogonal to $aa'$ and $bb'$ respectively (see Figure 4). Since $aa' = W < \pi/\sqrt{k}$ for $k > 0$, one has $v(a') \neq 0$. By (27), $\alpha'(\alpha) > 0$; therefore the projection $v$ of the vector $v(a')$ onto the direction of the segment $a'b'$ satisfies

$$v = |v(a')| \sin \alpha'(\alpha) > 0. \tag{30}$$

Projection of $v(b')$ onto the direction of the segment $a'b'$ at the end $b'$ should also be $v$; otherwise, the length $a'b'$ will change. Since $\beta'(\alpha) \neq 0$ (27), a finite vector $v(b')$ with the projection $v > 0$ exists. Since $aa'$ crosses $bb'$, the vector $v(b')$ points into the half-plane of $M_k^2$ where $a$ is located (see Figure 4), and

$$v = |v(b')| \sin \beta'(\alpha). \tag{31}$$

The relations (30) and (31) imply

$$|v(b')| = |v(a')| \frac{\sin \alpha'(\alpha)}{\sin \beta'(\alpha)} > 0. \tag{32}$$

Therefore,

$$\frac{d}{d\alpha} \beta(\alpha) < 0, \qquad \alpha \in (\alpha_1, \alpha_1 + \varepsilon). \tag{33}$$

Similarly, fixing the side $aa'$, we can see that

$$\frac{d}{d\alpha} \alpha'(\alpha) < 0, \qquad \alpha \in (\alpha_1, \alpha_1 + \varepsilon), \tag{34}$$

and fixing the side $bb'$, we get

$$\frac{d}{d\alpha} \beta'(\alpha) > 0, \qquad \alpha \in (\alpha_1, \alpha_1 + \varepsilon). \tag{35}$$

In the last case, note that $\frac{d\beta'}{d\alpha} = \frac{d\beta'}{d\beta} \cdot \frac{d\beta}{d\alpha}$. Since $\frac{d\beta}{d\alpha} < 0$ (33), it is enough to show that $\frac{d\beta'}{d\beta} < 0$. This is quite similar to (34). The condition $\alpha'(\alpha) \neq 0$ (27) will guarantee the existence of a finite velocity $v(a')$. This proves statements (28) and (29).

Since $\beta(\alpha)$ decreases, by (16) and (21) one has

$$0 \leq \alpha_1 < \alpha < \beta(\alpha) < \beta(\alpha_1) \leq \frac{\pi}{2} \tag{36}$$

for $\alpha \in (\alpha_1, \alpha_1 + \varepsilon)$ with sufficiently small $\varepsilon > 0$. This implies of course that the polygonal line $abb'a'a$ with $aa'$ crossing $bb'$ exists for any $\alpha$ in the interval above; otherwise, $\beta(\alpha)$ would not be defined. The angles $\alpha'(\alpha)$ and $\beta'(\alpha)$ are thus also well defined. Since (22) follows from (21), it is clear that $\beta'(\alpha) < \alpha'(\alpha)$ in the same interval. Since $\alpha'(\alpha)$ decreases while $\beta'(\alpha)$ increases, by (23) we have

$$0 < \beta'(\alpha_1) < \beta'(\alpha) < \alpha'(\alpha) < \alpha'(\alpha_1) \leq \frac{\pi}{2} \tag{37}$$

in the same interval $(\alpha_1, \alpha_1 + \varepsilon)$.

Now let $\alpha_2$ be the greatest number such that (36) holds for $\alpha \in (\alpha_1, \alpha_2)$; then (37) holds there, too. The functions $\beta$, $\alpha'$, and $\beta'$ are obviously defined at $\alpha = \alpha_2$. We show that

$$\beta(\alpha_2) = \alpha_2. \tag{38}$$

Due to (36), $\alpha_2 \leq \beta(\alpha_2)$. If $\alpha_2 < \beta(\alpha_2)$, then one can treat $\alpha_2$ as $\alpha_1$ above and construct an interval $(\alpha_2, \alpha_2 + \varepsilon) \ni \alpha$ in which

$$0 \leq \alpha_2 < \alpha < \beta(\alpha) < \beta(\alpha_2) \leq \frac{\pi}{2}. \tag{39}$$

Thus (36) holds in the interval $\alpha \in (\alpha_1, \alpha_2 + \varepsilon)$ which contradicts the definition of $\alpha_2$.

Combining (38) and (36) one has

$$0 \leq \alpha_1 < \alpha < \alpha_2 = \beta(\alpha_2) < \beta(\alpha) < \beta(\alpha_1) \leq \frac{\pi}{2}. \tag{40}$$

As was noted, (37) holds in $(\alpha_1, \alpha_2)$ as well. The line $abb'a'a$ with $\alpha = \alpha_2$ is obviously symmetric about the geodesic orthogonal to $ab$ which passes through its midpoint. By symmetry, (38) implies

$$\beta'(\alpha_2) = \alpha'(\alpha_2). \tag{41}$$

Combining this with (37) one gets

$$\begin{aligned} 0 &< \beta'(\alpha_1) < \beta'(\alpha) < \beta'(\alpha_2) \\ &= \alpha'(\alpha_2) < \alpha'(\alpha) < \alpha'(\alpha_1) \leq \frac{\pi}{2}, \qquad \alpha \in (\alpha_1, \alpha_2). \end{aligned} \tag{42}$$

Getting closer to estimate (20), we prove that $f(\alpha) = \alpha + \beta(\alpha)$ decreases in $(\alpha_1, \alpha_2)$. By (42), $\beta'(\alpha) < \alpha'(\alpha)$. Then (32) yields

$$|v(b')| > |v(a')| > 0. \tag{43}$$

Obviously,

$$|v(a')| = c, \qquad |v(b')| = -\frac{d\beta}{d\alpha}c, \tag{44}$$

where $c = c(k, W) > 0$ is a quantity determined by $k$ and $W$. By (44) and (43),

$$\frac{df}{d\alpha} = 1 + \frac{d\beta}{d\alpha} = 1 - \frac{|v(b')|}{|v(a')|} < 0,$$

which was required.

Obviously, $f(\alpha_2) = 2\alpha_2 \geq 2\gamma(\Delta, W, k)$.

Now, by (20),

$$\xi \leq \pi - f(\alpha_1) < \pi - f(\alpha_2) \leq \pi - 2\gamma(\Delta, W, k),$$

which is (10). □

## BIBLIOGRAPHY


1. R. L. Bishop and R. J. Crittenden, *Geometry of manifolds*, Academic Press, New York-London, 1964.
2. W. Blaschke, *Einige Bemerkungen über Kurven und Flächen von konstanter Briete*, Ber. Verh. Sächs. Akad. Leipzig **67** (1915), 290–297.
3. Yu. D. Burago and V. A. Zalgaller, *Convex sets in Riemannian spaces of non-negative curvature*, Russian Math. Surveys **32** (1977), no. 3, 1–57.
4. G. D. Chakerian and H. Groemer, *Convex bodies of constant width*, in Convexity and its Application, Birkhäuser, Basel-Boston, 1983, pp. 49–96.
5. J. Cheeger and D. Gromoll, *On the structure of complete manifolds of nonnegative curvature*, Ann. of Math. **96** (1972), 413–443.
6. B. V. Dekster, *In- and circumcenters of manifolds of constant width*, Geometriae Dedicata, to appear.
7. J. P. Fillmore, *Barbier's theorem in the Lobachevski plane*, Proc. Amer. Math. Soc. **24** (1970), 705–709.
8. L. A. Santaló, *Propriedades de las figuras convexas sobre la esfera*, Math. Notae **4** (1944), 11–40.
9. ——, *Note on convex curves on the hyperbolic plane*, Bull. Amer. Math. Soc. **51** (1945), 405–412.
10. ——, *On parallel hypersurfaces in the elliptic and hyperbolic n-dimensional space*, Proc. Amer. Math. Soc. **1** (1950), 325–330.


DEPARTMENT OF MATHEMATICS, MOUNT ALLISON UNIVERSITY, SACKVILLE, NEW BRUNSWICK, E0A 3C0 CANADA

DIMACS Series in Discrete Mathematics
and Theoretical Computer Science
Volume **4**, 1991

Applied Geometry
and Discrete Mathematics
THE VICTOR KLEE FESTSCHRIFT

# Uniquely Remotal Hulls

DUANE DETEMPLE, JACK ROBERTSON, AND GRAHAM WOOD

ABSTRACT. A convex compact set $K$ in Euclidean space is uniquely remotal with respect to its complement (has property $U$) if for each $y \notin K$, there is a unique $x \in K$ so that $d(x, y) = \sup_{z \in K} d(z, y)$. If $K$ is a convex compact set then $S$ is a uniquely remotal hull of $K$ ($U$-hull of $K$) if $S$ has property $U$, contains $K$, and if $K \subseteq S' \subseteq S$, where $S'$ also has property $U$, then $S' = S$.

The (nonunique) existence of $U$-hulls for convex compact sets in the plane is established. Hulls for polygons are studied. Property $U$ is preserved by Minkowski sums but the sum of $U$-hulls for two sets need not be a $U$-hull of the sum of the sets.

## 1. Introduction

Assume $K$ is a nonempty subset of a metric space $(X, d)$. If for each $p \in X$ there is some point $k \in K$ farthest from $p$, i.e., $d(p, k) \geq d(p, x)$ for all $x \in K$, then $K$ is said to be *remotal*. If, in addition, for each $p$ the farthest point $k$ is unique then $K$ is said to be *uniquely remotal*. The following result of Motzkin, Straus, and Valentine can be found in standard references (for example, [**3**]).

THEOREM 1.1. *Let $K$ be closed in Euclidean space $E^n$. Then $K$ is uniquely remotal if and only if $K$ is a single point.*

The main interest in this theorem is not in the characterization of a single point in $E^n$, but rather that in infinite-dimensional spaces the status of the corresponding statement is unresolved (see [**4**]).

In Euclidean space a more interesting concept is the following.

DEFINITION 1.2. A convex compact set $K \subset E^n$ is *uniquely remotal with respect to its complement* if for each $p \in E^n - K$ there is a unique $k \in K$ such that for all $x \in K$, $d(p, k) \geq d(p, x)$.

Throughout the paper, the property of being uniquely remotal with respect to the complement will be shortened to *property* $U$. A characterization of

1980 *Mathematics Subject Classification* (1985 *Revision*). Primary 52A10, 52A20.


1052-1798/91 $1.00 + $.25 per page

sets in $E^n$ having property $U$ has been given by Jessen [2] which requires the following definition.

DEFINITION 1.3. Let $K$ be a nonempty convex compact set in $E^n$ which is not a singleton. Let $q$ be a point on the boundary $\partial K$ of $K$, $\alpha$ a support plane to $K$ at $q$, and $H$ the ray from $q$ normal to $\alpha$ and directed toward $K$. Each $q' \in \partial K$, $q' \neq q$, determines a unique center $c'$ on $H$ of the sphere through $q$ and $q'$ and tangent to $\alpha$. (If $q'$ also lies in $\alpha$ then $c'$ is a point at infinity.) If $c = \lim_{q' \to q} c'$ then $c$ is an *osculating center* for the set $K$ with respect to the pair $(q, \alpha)$.

When $\partial K$ is smooth the set of osculating centers is just the evolute of $\partial K$. The osculating centers of a Reuleaux triangle are its three vertices.

THEOREM 1.4 (Jessen). *A convex compact set in $E^n$ has property $U$ if and only if it is a single point or it has interior points and contains all its osculating centers.*

DEFINITION 1.5. Let $K$ be a convex compact set in $E^n$. A compact set $S$ is called a *uniquely remotal hull* (or $U$-hull) of $K$ provided:

(i) $K \subset S$,
(ii) $S$ has property $U$, and
(iii) If $S'$ has property $U$ and $K \subset S' \subset S$ then $S' = S$.

The main result of this paper establishes the (nonunique) existence of $U$-hulls for nonempty convex compact subsets of the Euclidean plane. Hulls of polygons are considered. It is also shown that property $U$ is preserved under Minkowski sums.

## 2. Existence of uniquely remotal hulls

If $K \subset E^n$ has property $U$ then there are two associated functions which are well defined. Suppose for $x \in E^n - K$ that $y_x$ is the unique farthest point in $K$ from $x$. Define $f: E^n - K \to K$ and $F: E^n - K \to R$ by $f(x) = y_x$ and $F(x) = d(x, y_x)$, where $d$ is the Euclidean distance. The following theorem of Motzkin, Straus, and Valentine (generalized by Ahiya, Narang, and Trehan [4]) is used later.

THEOREM 2.1. *Both $f$ and $F$ defined above are continuous.*

Now denote by $\mathscr{K}$ the class of compact sets in $E^n$, and for $K_1, K_2 \in \mathscr{K}$ define $\delta(K_1, K_2) = \sup\{d(x, K_2) : x \in K_1\}$ and $\Delta(K_1, K_2) = \max(\delta(K_1, K_2), \delta(K_2, K_1))$. Then $\Delta$ is the Hausdorff metric on $\mathscr{K}$ [1]. Set $\mathscr{C} = \{C \in \mathscr{K} : C$ is convex$\}$ and $\mathscr{C}_U = \{C \in \mathscr{C} : C$ has property $U\}$. Finally let $\mathscr{C}_1 = \mathscr{C} - \mathscr{C}_U$.

THEOREM 2.2. *Let* $\Lambda$ *be a directed set and for* $\alpha \in \Lambda$ *assume* $C_\alpha \in \mathscr{C}$. *Also assume either*

(i) *for* $\alpha > \beta$, $C_\alpha \subset C_\beta$ *and* $C = \bigcap_{\alpha\in\Lambda} C_\alpha$, *or*

(ii) *for* $\alpha < \beta$, $C_\alpha \subset C_\beta$ *and* $C = \overline{\bigcup_{\alpha\in\Lambda} C_\alpha}$ *is compact.*

*Then* $\{C_\alpha\} \to C$ *in the Hausdorff metric.*

PROOF. Assume $\varepsilon > 0$ is given; we need an $\alpha_0$ so that for $\alpha \geq \alpha_0$, $\Delta(C, C_\alpha) \leq \varepsilon$. It suffices to show that, for $\alpha > \alpha_0$, (a) if $x \in \partial C$ then $d(x, \partial C_\alpha) \leq \varepsilon$, and (b) if $y_\alpha \in \partial C_\alpha$ then $d(y_\alpha, \partial C) \leq \varepsilon$.

For (i), if (a) fails then for any $\alpha_0$ there is an $\alpha > \alpha_0$ and $x \in \partial C$ with $B_\varepsilon(x) \subset (C_\alpha)^\circ$, the interior of $C_\alpha$, which requires $B_\varepsilon(x) \subset C$ so that $x \notin \partial C$. If (b) fails there is a net $\{y_\alpha\}$, $y_\alpha \in \partial C_\alpha$, with $d(y_\alpha, \partial C) \geq \varepsilon$ and we can assume $\{y_\alpha\} \to y_0$. Then $y_0 \in C$ and $d(y_0, \partial C) \geq \varepsilon$, which would mean $y_0 \in C^\circ$. But $y_0$ must belong to $\partial C$ since $y_\alpha \in \partial C_\alpha$.

For (ii), if (a) fails then for any $\alpha_0$ there is an $\alpha > \alpha_0$ and $x_\alpha \in \partial C$ with $d(x_\alpha, \partial C_\alpha) > \varepsilon$. Such $x_\alpha$, $\alpha \in \Lambda'$, form a net with $\Lambda'$ coterminal in $\Lambda$. We may assume $\{x_\alpha\} \to x \in \partial C$. Thus there is an $\alpha_0$ so that if $\alpha > \alpha_0$, $d(x_\alpha, x) < \varepsilon/2$. By the triangle inequality, $d(x, \partial C_\alpha) \geq \varepsilon/2$ for $\alpha > \alpha_0$. This means $x \notin C$, a contradiction.

If (b) fails for (ii), then for any $\alpha_0$ there is an $\alpha > \alpha_0$ and $x_\alpha \in \partial C_\alpha$ with $d(x_\alpha, \partial C) > \varepsilon$. As before, we may assume $\{x_\alpha\}_{\alpha\in\Lambda'} \to x \in C$. In fact $x \in C^\circ$ since $d(x_\alpha, \partial C) \geq \varepsilon$ for $\alpha \in \Lambda'$ and $d(x, x_\alpha) \to 0$. We observe that $x \notin (C_\alpha)^\circ$ for any $\alpha \in \Lambda'$ since it is a limit of $\{x_\alpha\}$ with $x_\alpha \in \partial C_\alpha$.

Suppose $B_\delta(x) \subset C$. Take $n+1$ points (where $n$ is the dimension of $E^n$) $y_1, y_2, \ldots, y_{n+1}$ and $\gamma > 0$ so that $B_\gamma(y_j) \subset B_\delta(x)$ for all $j$, and $x \in (\operatorname{conv}\{z_1, z_2, \ldots, z_{n+1}\})^\circ$ for any choice of $z_i \in B_\gamma(y_i)$. There must be some $\alpha_j \in \Lambda'$ with $C_{\alpha_j} \cap B_\gamma(y_j) \neq \varnothing$. Choose $z_j$ in that intersection. Then $\{z_1, \ldots, z_{n+1}\} \subset C_\beta$ for some $\beta \in \Lambda'$ so that $x \in (C_\beta)^\circ$ contradicting the observation above. $\square$

For the remainder of the paper we work in the plane $E^2$.

THEOREM 2.3. *The space* $\mathscr{C}_U$ *is closed in* $\mathscr{C}$ *under the Hausdorff metric.*

PROOF. Let $C \in \mathscr{C} - \mathscr{C}_U$. Then for some point $o \notin C$ and for some $p_1$, $p_2 \in C$, $p_1 \neq p_2$, $d(o, p_1) = d(o, p_2) \geq d_{p\in C}(o, p)$ (see Figure 1). Then $C$ lies in the disk $D$ of radius $d(o, p_1)$ centered at $o$ and we may assume the closed clockwise arc $p_1p_2$ on $\partial D$ is less than a half circle and the open clockwise arc $\widehat{p_2p_1}$ does not intersect $C$. Let $p_3$ be on $\partial D$ and the ray from $o$ which is the perpendicular bisector of the segment $p_1p_2$. Let $H_1$ and $H_2$ be the two closed half planes given by the extension of the ray with $p_1 \in H_1$ and $p_2 \in H_2$. Also let $H_3$ be the closed half plane not containing $p_1$ or $p_2$ which is determined by the line through $o$ perpendicular to $op_3$. Set $C' = H_3 \cap C$.

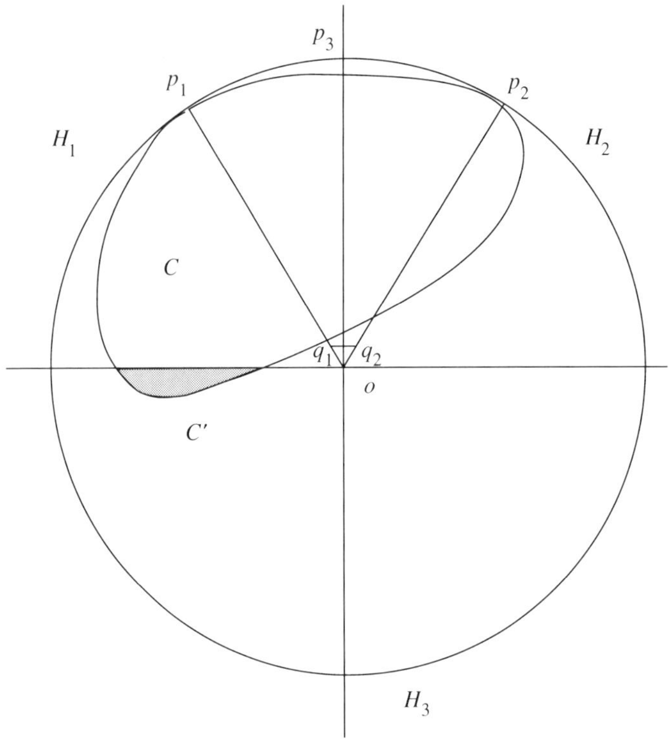


FIGURE 1

Let $D'$ be a sufficiently small disk centered at $o$ so that $D' \cap C = \varnothing$ and for all $x \in D'$ and some $\eta_0 > 0$, $\min(d(x, p_1), d(x, p_2)) - \sup_{y \in C'} d(x, y) \geq \eta_0 > 0$. (If $C' = \varnothing$, the last condition on $D'$ is unnecessary.) For $i = 1, 2$ let $q_i \in D'$, where $q_i$ is on the segment $op_i$, $q_i \neq o$, and the segment $q_1 q_2$ is perpendicular to $op_3$. By a routine geometric argument, there is an $\eta_1 > 0$ so that

$$d(q_1, p_2) - \max_{x \in (H_1 \cap C)} d(q_1, x) \geq \eta_1 > 0,$$

and

$$d(q_2, p_1) - \max_{x \in (H_2 \cap C)} d(q_2, x) \geq \eta_1 > 0.$$

Another geometric argument shows that there is an $\eta_2 > 0$ so that for $x$ on the segment $q_1 q_2$

$$\max(d(x, p_1), d(x, p_2)) - d(x, p_3) \geq \eta_2 > 0.$$

It follows from the above inequalities that if $C_1 \in \mathscr{C}_U$ is sufficiently close to $C$ in the Hausdorff metric, the segment $q_1 q_2$ lies outside $C_1$, and if $f$ is the function of Theorem 2.1 associated with $C_1$ then $f(q_1) \in (H_2)^\circ$, $f(q_2) \in (H_1)^\circ$, and $f(x)$ is not on the line containing $op_3$ for any $x$ on $q_1 q_2$. This contradicts the continuity of $f$ on the segment $q_1 q_2$. □

COROLLARY 2.4. *Every compact set $K$ in the plane possesses a $U$-hull. The $U$-hull is unique only when $K$ has property $U$, and then $K$ is its own $U$-hull.*

PROOF. Partially order sets having property $U$ and containing $K$ by inclusion. By Theorems 2.2(i) and 2.3 every descending chain has a lower bound. Zorn's Lemma applies to give the $U$-hull. If $K$ does not have property $U$ and $x \notin K$, some disk $D$ contains $K$ but not $x$. Applying Zorn's Lemma to a chain with $D$ as a member generates a $U$-hull $K'$ of $K$ not containing $x$. But since $K$ fails to have property $U$ there is a $y \in K' - K$. Applying the same argument at $y$ that we just used for $x$ generates a second $U$-hull of $K$ which fails to contain $y$. □

COMMENT 2.5. Theorem 2.3 is very likely true in higher dimensions. The argument of Theorem 2.3 uses special properties of $E^2$. Jessen's characterization of sets with property $U$ in $E^n$ gives an approach in higher dimensions which we have not been able to exploit without restrictive smoothness conditions (which may not be preserved in a limit). The difficulties seem to be in the rather involved process of finding osculating centers.

## 3. Uniquely remotal hulls of polygons

In this section we look at examples of $U$-hulls of polygons and give a method for constructing them. The following example illustrates the arguments used.

EXAMPLE 3.1. The lens in Figure 2(a) whose boundary is two circular arcs with centers at $c$ and $d$ and the Reuleaux triangle in Figure 2(b) are both $U$-hulls of the segment $ab$. It is easy to see they both have property $U$. If the lens has any point of its boundary, therefore a disk, removed as shown in Figure 2(c) then for a segment $eg$ as shown sufficiently close to $d$, $f(e) = b$, $f(g) = a$, and $f(x)$ is never on the line $l$ for any $x$ on $eg$ (see Figure 2(c)). The reduced set fails to have property $U$ by the continuity of $f$.

For the Reuleaux triangle, any point $x$ in the open segment $cd$ must be in any $U$-set contained in the Reuleaux triangle which contains segment $ab$, otherwise $x$ would be in the complement and have two farthest points $a$ and $b$. Since a $U$-hull is convex, the triangle $abc$ must be in any $U$-hull of segment $ab$ which is a subset of the Reuleaux triangle. Any missing part of a circular arc, say $\widehat{ab}$, gives a segment near $c$ which leads to the same contradiction as we saw for the lens.

These examples suggest looking at sets having property $U$ whose boundaries are made up of a finite number of circular arcs. We have the following.

THEOREM 3.2. *If a convex compact set $K$ has a boundary made up of a finite number of $C^3$ curves and all osculating centers of $K$ are on $\partial K$, then each smooth portion of the boundary is a circular arc.*

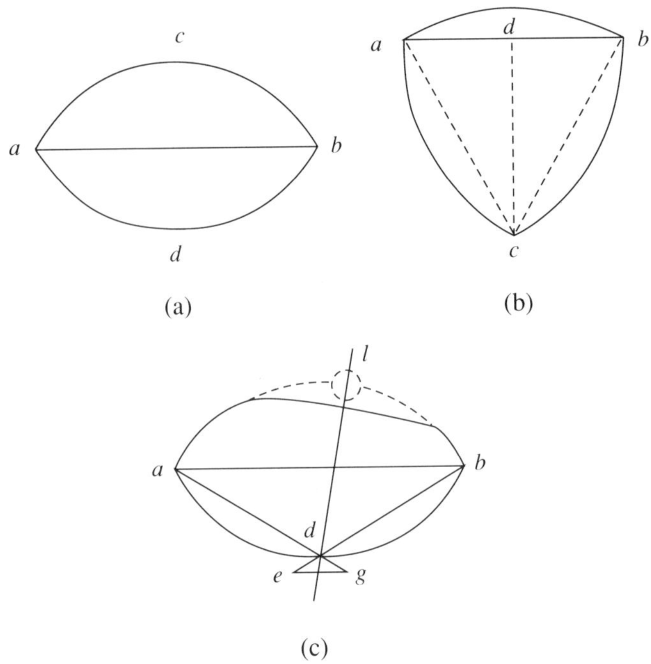


FIGURE 2

PROOF. Let $\vec{r} = \vec{r}(s)$ be a piecewise smooth representation of $\partial K$ parametrized with respect to arclength $s$. Then the locus of centers of curvature can be given in the form $\vec{R} = \vec{r} + \rho\vec{n}$, where $\rho$ is the reciprocal of curvature $\kappa$ and $\vec{n}$ is normal to $\partial K$. Using the Frenet formula we have $\vec{n}' = -\kappa\vec{t}$, where $\vec{t} = \vec{r}'$ is the unit tangent to $\partial K$. Thus

$$\vec{R}' = \vec{t} + \rho\vec{n}' + \rho'\vec{n} = \vec{t} + \rho(-\kappa\vec{t}) + \rho'\vec{n} = \rho'\vec{n}.$$

If $\rho' \neq 0$ then $\vec{R}'$ must be both tangent and normal to $\partial K$. Thus on intervals of smoothness $\rho' = 0$, $\rho$ is constant, and that portion of $\partial K$ is a circular arc. □

Given a convex polygon we next show how to construct $U$-hulls whose boundaries are made up of a finite number of circular arcs.

LEMMA 3.3. *Let $P$ be a convex polygon in $E^2$ and $F$ its set of vertices. Suppose $X$ is a $U$-hull of $P$ whose boundary is made up of a finite number of circular arcs. Then*

(i) *no center of an arc lies in $X^0$, and*
(i) *all arc intersections on $\partial X$ are either centers of other arcs lying on $\partial X$ or are members of $F$.*

PROOF. By Theorem 1.4 the finite number of centers lie in $X$. Assume center $c_i \in X^0$ corresponds to arc $A_i$ on $\partial X$. Some subarc $\widehat{ab}$ of $A_i$ contains no members of $F$ nor any center. But now the portion of the boundary $\widehat{ab}$ with center $c_i$ can be replaced by another circular arc through $a$ and $b$ with larger radius and center $c_i'$ lying sufficiently near $c_i$ and on the perpendicular bisector of segment $ab$. This produces a smaller set $X'$ having property $U$, with $P \subset X' \subset X$, which means $X$ could not be a $U$-hull for $P$.

If an arc intersection $x$ is neither a center of another arc nor a member of $F$ then the same is true for all points in some neighborhood of $x$. But by proper choice of a new center $c \in X$, a new circular arc with center $c$ can be scribed which cuts off the corner at $x$ removing from $X$ only $x$ and points of the neighborhood. Again this reduced set prevents $X$ from being a $U$-hull for $P$. □

DEFINITION 3.4. Let $F$ and $P$ be as in Lemma 3.3, and let $X$ be a set with property $U$ containing $P$ whose boundary is made up of a finite number of circular arcs. A point $x \in \partial X$ is a *critical point* if $x \in F$ or $x$ is a center of an arc lying on $\partial X$. Set $F_0 = F$ and recursively let $F_{i+1} = F_i \cup F_i'$, where $F_i' = \{x : x$ is an arc center with at least two members of $F_i$ on the corresponding arc$\}$. A critical point is essential if it belongs to the (finite) union $\bigcup F_i$.

THEOREM 3.5. *Let $F$ and $P$ be as in Lemma 3.3, and let $X$ be a set with property $U$ containing $P$ whose boundary is made up of a finite number of circular arcs. Then $X$ is a $U$-hull of $P$ if each arc center and each arc intersection is an essential critical point.*

PROOF. We first show that if $P \subset X' \subset X$, where $X'$ has property $U$, then $X'$ must contain all essential critical points of $X$. If $x \in F_0 = F$ (as defined in Definition 3.4) then clearly $x \in X'$. If $x \in F_0'$ then the corresponding arc $A_x$ has at least two members of $F_0$. Let $x_1$ and $x_2$ be the least and most clockwise points of $F$ lying on $A_x$, respectively (Figure 3(a)), and let line $l$ be the perpendicular bisector of segment $x_1x_2$. If $x$ lies outside of $X'$ so does a disk $D$ centered at $x$. Then, reasoning as in the proof of Theorem 2.3, the unique farthest point function $f$ fails to be continuous on segment $q_1q_2$ shown in Figure 3(a) since it takes no values on $l$. Thus $F_1 = F_0 \cup F_0' \subset X'$. The argument showing $F_i \subset X'$ for $i > 1$ is exactly the same as for $F_1$.

We can now show $X' = X$ as required by showing $X'$ contains all of $\partial X$. Suppose that $x$ is an essential center for arc $A_x$ and that $y \in (X - X') \cap A_x$. If $q_1q_2$ is a segment sufficiently close to $x$, as shown in Figure 3(b), then the unique farthest point function $f$ is not continuous on segment $q_1q_2$ because $f$ takes no values on the line containing $xy$. □

With this result many $U$-hulls of polygons are easily constructed. Figure 4 shows four types of $U$-hulls for the segment $ab$. The lens of (a) is the only $U$-hull constructed using Theorem 3.5 with two circular arcs. In this case $F_0' = \{x, y\}$. In (b) the $U$-hull is made up of three arcs. Here $F_0' = \{x\}$

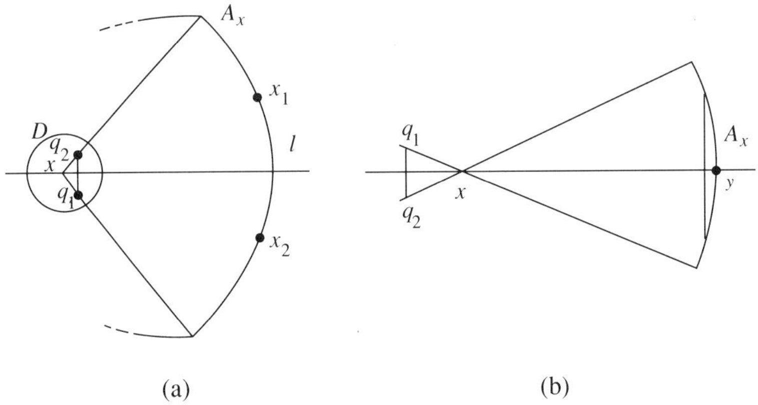


FIGURE 3

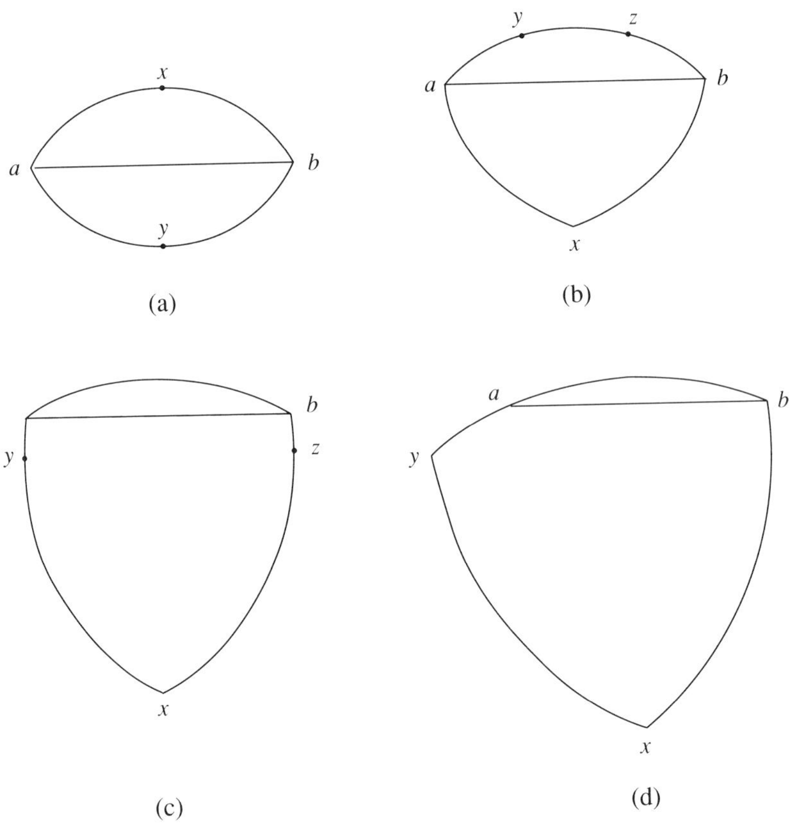


FIGURE 4

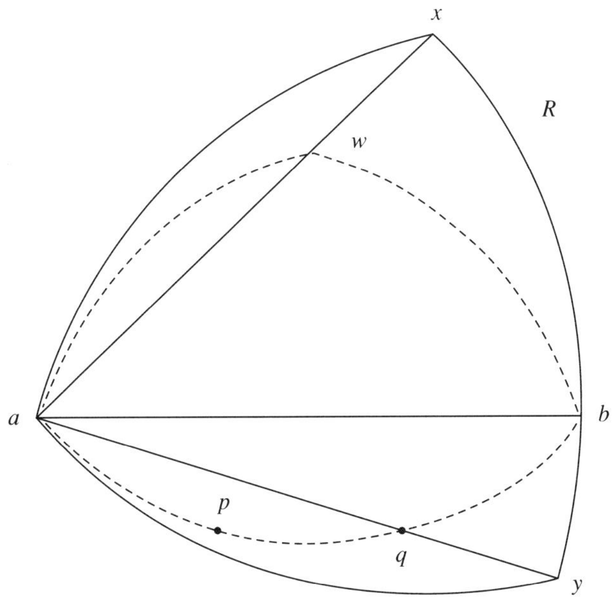


FIGURE 5

and $F_1' = \{y, z\}$, where $y$ and $z$ lie on the circular arc $\widehat{ab}$. The $U$-hulls of the type shown in (c) have three arcs, $F_0' = \{x\}$, $F_1' = \{y, z\}$, and $xyz$ is an equilateral triangle. All $U$-hulls of the type shown in (d) are Reuleaux triangles. This family of hulls of types shown in Figure 4 for a segment $ab$ in fixed position forms a subset of the space $\mathscr{C}$ under the Hausdorff metric which is contractible to a point. In particular, a continuous family of deformations is easily visualized which retracts the entire family of hulls to the Reuleaux triangle with one of its arcs the arc $\widehat{ab}$. This latter Reuleaux triangle lies between hulls of type (b) and (c).

Figure 5 illustrates that if the segment $ab$ is inscribed in a Reuleaux triangle $R$ with vertices $a$, $x$, and $y$ with $b$ on the interior of the arc $xy$ then $R$ is *not* a $U$-hull for the segment $ab$. The included figure (from class (b) in Figure 4) with vertices at $a$, $b$, and $w$ and arc centers at $w$, $p$, and $q$ is a $U$-hull of the segment contained properly in $R$. (If $b$ is at the midpoint of $\widehat{xy}$ the included $U$-hull becomes the lens.)

We shall find the evolute useful in our next theorem. For a smooth boundary arc, the evolute is the locus of the centers of curvature.

THEOREM 3.6. *Suppose $X$ is a $U$-hull of a polygon $P$ whose boundary $\partial X$ is made up of a finite number of smooth arcs each of which has a continuous evolute. Then the smooth arcs on $\partial X$ are circular arcs.*

PROOF. By Theorem 3.2 it suffices to show that the evolute of each arc lies in $\partial X$. Assume to the contrary that $x \in X^\circ$ is on the evolute corresponding to the smooth arc $A_j$ on $\partial X$. Let $A'_j$ be a nondegenerate subarc of $A_j$ such that the centers of curvature for all points on $A'_j$ lie in $X^\circ$. If some point $z \in A'_j$ fails to be in the evolute of $\partial X$, some nondegenerate subarc $A''_j$ of $A'_j$ contains no members of the evolute of $\partial X$ and also no points of $P$. This insures that there exists a point $q$ in the interior of arc $A''_j$ with corresponding center of curvature $p \in X^\circ$ and a neighborhood $N$ of $q$ containing no points of the evolute or of $P$.

Alter $X$ locally at $q$ as follows. Pick a new center $p' \in X$ past $p$ on the direct ray $\overrightarrow{qp}$ and scribe a circular arc of radius at least $|pq|$ and center $p'$ which cuts off $q$ but intersects $\partial X$ at points $r$ and $s$ inside $N$. Let $X'$ have the same boundary as $X$ except replace the arc on $\partial X$ between $r$ and $s$ by the new circular arc. By Theorem 1.4 $X'$ has property $U$, it contains $P$, and is a proper subset of $X$, contradicting the fact that $X$ is a $U$-hull.

It follows that every point of the nondegenerate arc $A'_j$ is in the evolute of $\partial X$, thus some nondegenerate subarc $A''_j$ of $A'_j$ contains centers of curvature corresponding to points on a single smooth arc $A_i$ on $\partial X$. There is no loss of generality to assume no subarc of $A_i$ is circular. The evolute of $A_i$ can turn through an angle no greater than $2\pi$ as $A_i$ is traced since the tangent to the evolute at a point is always the same line as the normal to the original curve at the associated point. Thus the trace of the evolute can leave $A''_j$ at most finitely many times, so a nondegenerate subarc of $A''_j$ is traced in the evolute monotonely as $A_i$ is traced. But by Theorem 3.2 this is impossible unless $A_i$ is a circular arc, contradicting the nondegeneracy of the monotonely traced subarc of $A''_j$. The result follows. □

Some tantalizing questions remain. Note that centers $x$ and $y$ in Figure 5 are not essential critical points for the segment $ab$, supporting our conjecture that the conditions of Theorem 3.5 are also necessary for a set with a boundary composed of a finite number of circular arcs to be a $U$-hull of $P$. If the conjecture holds, then by routine calculations we find that the lens having the segment for its diameter is the piecewise smooth $U$-hull of the segment of minimum area and perimeter. On the other hand, hulls of the type shown in Figure 4(c) and (d) can have arbitrarily large areas and perimeters.

COMMENT 3.7. Any minimal area $U$-hull of a segment $S$ must have $S$ as a diameter. If $X$ has property $U$ and $\operatorname{diam} X > |S|$ then $X$ can be shrunk by a similarity transformation to a set $X'$ having property $U$ which has area less than that of $X$, yet $X'$ still covers a segment of length $|S|$. Now $X'$ contains a $U$-hull of this segment with area smaller than the area of $X$.

## 4. Property $U$ for sums of convex sets

The Minkowski sum of two convex sets $C_1$ and $C_2$ is the set $C_1 + C_2 = \{x + y : x \in C_1, y \in C_2\}$ **[5]**. Our purpose in this section is to show that property $U$ is preserved (in $E^n$) by such sums. We also look at an example to see that the sum of $U$-hulls of two convex sets need not be a $U$-hull of their sum.

THEOREM 4.1. *If* $C_1$ *and* $C_2$ *have property* $U$ *then so does* $C_1 + C_2$.

PROOF. Using Jessen's characterization in Theorem 1.4 we show that if $C_1$ and $C_2$ contain all their osculating centers, $C_1 + C_2$ will contain its centers as well. Thus assume $\alpha$ is a support plane to $C_1 + C_2$ at $q$ and $c$ is an osculating center lying on the ray $H$ normal to $\alpha$ and directed from $q$ toward $C_1 + C_2$. For $i = 1, 2$, there exist unique points $q_i \in C_i$ lying on support planes $\alpha_i$ with normal rays $H_j'$ both in the same direction as $H$. It follows that $\alpha_1 + \alpha_2$ is parallel to $\alpha$ and supports $C_1 + C_2$ since $\alpha_1$ supports $C_1$ and $\alpha_2$ supports $C_2$. Since two distinct points belonging to a convex set with property $U$ cannot lie in any support plane, it also follows that $\alpha_1 + \alpha_2 = \alpha$ and $q_1 + q_2 = q$. Also if $q = x + y$, where $x \in C_1$ and $y \in C_2$, then $x = q_1$ and $y = q_2$. If $x \neq q_1$ then $x$ belongs to the *open* half plane determined by $\alpha_1$ and the sum $x + y$ is in the open half plane determined by $\alpha$, hence $x + y \neq q$.

Assume that center $c = \lim c_i$, where $c_i \in H$ is the center of the circle passing through $q$ and $x_i \in \partial(c_1 + c_2)$ with $\lim x_i = q$. From above, this uniquely defines corresponding sequences $\{y_i\}$ in $\partial C_1$ and $\{z_i\}$ in $\partial C_2$ with $x_i = y_i + z_i$. Without loss of generality we may assume $y_i \to y$ and $z_i \to z$. Thus $q = \lim x_i = \lim(y_i + z_i) = y + z$ and this requires $y = q_1$ and $z = q_2$ since the representation of $q$ as a sum in $C_1 + C_2$ is unique.

Let $c_i' \in H_1$ be the center of the circle passing through $q_1$ and $y_i$, and $c_i'' \in H_2$ be the center of the circle passing through $q_2$ and $z_i$. Since $C_1$ has property $U$, from Theorem 1.4, we can see that some subsequence $c_{i_j}' \to c' \in H_1 \cap C_1$. Similarly a subsequence of $c_{i_j}'' \to c'' \in H_2 \cap C_2$. It now follows that $c_{i_j} = c_{i_j}' + c_{i_j}''$ so that $c = \lim c_{i_j} = \lim(c_{i_j}' + c_{i_j}'') = c' + c'' \in C_1 + C_2$ as required. □

The following is immediate from Theorems 4.1 and 2.3.

COROLLARY 4.2. *Suppose* $C_i$, $i = 1, 2, 3, \dots$, *all have property* $U$ *and* $D_n = \sum_{j=1}^n C_i$. *If* $\overline{\bigcup_{n=1}^\infty D_n}$ *is compact then it has property* $U$.

EXAMPLE 4.3. Suppose $C_1 = \{(x, y) : x = 0, \ -1 \leq y \leq 1\}$ and $C_2 = \{(x, y) : -1 \leq x \leq 1, \ y = 0\}$. Let $L_1$ and $L_2$ be the corresponding lens hulls. Then $L_1 + L_2$ is a set with property $U$ containing $C_1 + C_2$, but it is not a $U$-hull of $C_1 + C_2$. The circle with center at the origin passing through $(1, 1)$ contains $C_1 + C_2$ and is properly contained in $L_1 + L_2$.

The authors, two of whom had the privilege of spending sabbatical time with Professor Victor Klee, would like to thank him for his counsel during the course of this work.

## BIBLIOGRAPHY

1. J. Dugundji, *Topology*, Allyn and Bacon, Boston, 1966.
2. B. Jessen, *Two theorems on convex point sets*, Math. Tiddsskr. **3** (1940), 66–70. (Danish)
3. S. Lay, *Convex sets and their applications*, Wiley, New York, 1982.
4. T. D. Narang, *A study of farthest points*, Nieuw Arch. Wisk. (3) **25** (1977), 54–79.
5. I. M. Yaglom and V. G. Boltyanskiĭ, *Convex figures*, Holt, Rinehart, and Winston, New York, 1961.

(D. DeTemple and J. Robertson) DEPARTMENT OF MATHEMATICS, WASHINGTON STATE UNIVERSITY, PULLMAN, WASHINGTON 99164-3113

(G. Wood) DEPARTMENT OF MATHEMATICS, UNIVERSITY OF CANTERBURY, CHRISTCHURCH, NEW ZEALAND

*E-mail address*: DETEMPLE @ WSUMATH. BITNET

DIMACS Series in Discrete Mathematics
and Theoretical Computer Science
Volume **4**, 1991

Applied Geometry
and Discrete Mathematics
THE VICTOR KLEE FESTSCHRIFT

# The Symmetries of the Cut Polytope and of Some Relatives

M. DEZA, V. P. GRISHUKHIN, AND M. LAURENT

ABSTRACT. We show that the symmetry group of the cut polytope $P_n$, $n \neq 4$, is the central quotient of the symmetry group of the $n$-dimensional cube and that, for $n = 4$, it is the automorphism group of the complete bipartite graph $K_{4,4}$. The symmetry group of the following relatives of $P_n$, the equicut polytope, the $k$-cut polytope $(3 \leq k \leq n)$, and the boolean quadric polytope, is the symmetric group on $n$ points. The action of the symmetry group of $P_n$ on its face lattice is also considered and used for the enumeration of the orbits of the facets of $P_n$.

## 0. Introduction

Let $K_n$ denote the complete graph on the $n$ nodes $[1, n] = \{1, 2, \ldots, n\}$, i.e., with edges all pairs $(i, j)$ with $1 \leq i < j \leq n$. Given a subset $S$ of $[1, n]$, the *cut vector* determined by $S$ is the vector $\delta(S)$ of $R^{n(n-1)/2}$ defined by $\delta(S)_{ij} = 1$ if $i \in S$, $j \notin S$ or $i \notin S$, $j \in S$ and $\delta(S)_{ij} = 0$ otherwise, for all $1 \leq i < j \leq n$; so the cut vectors are the incidence vectors of the cuts of $K_n$. The cut vector $\delta(S)$ is called an *equicut* vector if $|S| = \lfloor n/2 \rfloor$ or $\lceil n/2 \rceil$. The *intersection* vector determined by $S$ is the vector $\pi(S)$ of $R^{n(n+1)/2}$ defined by $\pi(S)_{ij} = 1$ if $i, j \in S$ and $\pi(S)_{ij} = 0$ otherwise, for all $1 \leq i \leq j \leq n$. Given a partition of $[1, n]$ into $k$ classes $S_1, \ldots, S_k$, the *k-cut* vector determined by this partition is the vector $\delta(S_1, \ldots, S_k)$ of $R^{n(n-1)/2}$ defined by $\delta(S_1, \ldots, S_k)_{ij} = 1$ if $i \in S_h$, $j \in S_l$ for distinct $h, l$ in $[1, k]$ and $\delta(S_1, \ldots, S_k)_{ij} = 0$ otherwise, for all $1 \leq i < j \leq n$. For $k = 2$, a 2-cut vector $\delta(S_1, S_2)$ can be alternatively denoted by $\delta(S_1)$ or $\delta(S_2)$.

We consider the *cut polytope* $P_n$, the *equicut polytope* $EP_n$, the *k-cut polytope* $MC_k^{\leq}(n)$, and the *boolean quadric polytope* $QP^n$ which are, respectively, the convex hull of all cut vectors, all equicut vectors, all $h$-cut vectors for $2 \leq h \leq k$, and all intersection vectors. The polytopes $P_n$ and

1980 *Mathematics Subject Classification* (1985 *Revision*). Primary 05C25, 52A25.

$MC_k^{\leq}(n)$ are full dimensional polytopes in $R^{n(n-1)/2}$ with $P_n = MC_2^{\leq}(n)$ having $2^{n-1}$ vertices. The polytope $EP_n$ is the face of $P_n$ induced by the inequality $\sum_{1\leq i<j\leq n} x_{ij} \leq \lfloor n/2 \rfloor \lceil n/2 \rceil$ and its dimension is $n(n-1)/2 - 1$ for $n$ odd and $n(n-1)/2 - n$ for $n$ even. Finally, $QP^n$ is a full dimensional polytope in $R^{n(n+1)/2}$ having $2^n$ vertices and, in fact, $QP^n$ is the image of $P_{n+1}$ under a linear bijective transformation (see §III.1 for details). Those polytopes are core objects for the formulation of some combinatorial optimization problems.

Given a polytope $P$ in $R^n$, $v \in R^n$, and a scalar $v_0$, the inequality $v \cdot x \leq v_0$ is valid for $P$ if $v \cdot x \leq v_0$ holds for all points $x \in P$, i.e., for all vertices $x$ of $P$. The set $F = \{x \in P: v \cdot x = v_0\}$ is then the face of $P$ induced by the inequality $v \cdot x \leq v_0$. The set $F$ is a facet of $P$ if its dimension is one less than the dimension of $P$. The vertices of $P$ lying on a face $F$ are called the *roots of* $F$. The symmetry group $\mathrm{Is}(P)$ of the polytope $P$ is the group of all isometries $f$ of $R^n$ preserving $P$, i.e., $f(P) = P$. The group of permutations on $n$ points is denoted by $\mathrm{Sym}(n)$. We denote by $\|\cdot\|$ the euclidian norm in $R^n$, i.e., $\|x\|^2 = \sum_{1\leq i\leq n} x_i^2$ for $x = (x_i)_{1\leq i\leq n}$ in $R^n$.

Two operations on the faces of $P_n$ are known:

–The *permutation operation.* Given $\sigma \in \mathrm{Sym}(n)$ and a face $F$ of $P_n$ induced by the inequality $v \cdot x \leq v_0$, the inequality $\sigma(v) \cdot x \leq v_0$ induces a face of $P_n$ denoted as $\sigma(F)$, where $\sigma(v) = (v_{\sigma(i)\sigma(j)})_{1\leq i<j\leq n}$. Of course, the same assertion holds for the polytopes $EP_n$, $MC_k^{\leq}(n)$, and $QP^n$.

–The *switching by cut operation.* Given a cut vector $\delta(S)$ and a face $F$ of $P_n$ induced by the inequality $v \cdot x \leq v_0$, the inequality $v^S \cdot x \leq v_0 - v \cdot \delta(S)$ induces a face of $P_n$ denoted as $r_{\delta(S)}(F)$, where $v^S$ is the vector defined by $v_{ij}^S = v_{ij}$ if $i, j \in S$ or $i, j \notin S$ and $v_{ij}^S = -v_{ij}$ otherwise, for $1 \leq i < j \leq n$. Of course, the linear bijection between $P_{n+1}$ and $QP^n$ will produce an analogue of this switching operation on the faces of $QP^n$. However, there is no analogue of switching for $MC_k^{\leq}(n)$ and $EP_n$.

The permutation and switching operations preserve the dimension of faces, i.e., act on the face lattice of the polytope. Obviously, the permutation operation is an isometry and hence a symmetry of our four polytopes. The switching by cut $\delta(S)$ operation corresponds also to an isometry, in fact to a reflection denoted as $r_{\delta(S)}$ (see §II for details), giving therefore a symmetry of $P_n$. However, the analogue of switching for $QP^n$ is not an isometry.

After recalling the symmetry groups of the $n$-dimensional cube and half-cube in the first section, we show in the second section that permutations and reflections $r_{\delta(S)}$ are the only symmetries of the cut polytope $P_n$ for $n \neq 4$. More precisely, for $n \neq 4$, $\mathrm{Is}(P_n)$ is the central quotient of the symmetry group of the $n$-dimensional cube $Q_n$ and thus is the Weyl group $W(D_n)$ for $n$ odd, while $\mathrm{Is}(P_4)$ is isomorphic to the automorphism group of the complete bipartite graph $K_{4,4}$. In §III, we show that the symmetry group of the three relatives of $P_n$, $QP^n$, $MC_k^{\leq}(n)$, and $EP_n$, is the symmetric group $\mathrm{Sym}(n)$. In the last section, we study how the symmetry group of $P_n$ acts on the family of facets of $P_n$. We consider double symmetries of facets, i.e., those coming simultaneously from permutations and switchings. As an application, we derive some enumeration formulas for counting of the facets that are switching and/or permutation equivalent to a given facet of $P_n$.

The cut polytope was studied, in different terms, by many authors, see, e.g., [**BM, DL1**] and references mentioned there; for $k$-cut polytopes, see, e.g., [**CR, DGL**] and references mentioned there; for the equicut polytope, see [**CRS, DFL**]; and for the boolean quadric polytope, see [**P**]. The switching operation in the cut polytope was introduced in [**BM**]; the switching by root operation (in the context of the cut cone) was introduced in [**D**]. Actually, switching was given in [**BM**] for the cut polytope $P_c(G)$ of an arbitrary graph $G$ and, in fact, it is, more generally, a property of the cycle polytope $P_c(M)$ of any binary matroid $M$ [**BG**].

In fact, given a graph $G$, every automorphism of $G$ induces a symmetry of the cut polytope $P_c(G)$; so, it would be interesting to determine whether the symmetry group of $P_c(G)$ consists only of those isometries induced by automorphisms of $G$ and switchings. We see here (cf. Theorem 2.6) that this is the case for the complete graph $G = K_n$, $n \neq 4$. In [**DL3**], a similar result is proved for the even cut polytope (resp. odd cut polytope) which is defined as the convex hull of all even (resp. odd) cuts $\delta(S)$, i.e., having both sets $S$ and $[1, n]\text{-}S$ of even (resp. odd) cardinality. Namely, for $n$ even, $n \neq 6$, the only symmetries of the even and odd cut polytopes are those induced by the permutations and the switchings by even cuts. Another interesting question is the characterization of the binary matroids $M$ for which the symmetry group of the cycle polytope $P_c(M)$ consists only of the isometries induced by switchings.

## I. The symmetries of the cube $Q_n$ and the halfcube $hQ_n$

We describe here the well-known symmetries of the cube $Q_n$ and of the halfcube $hQ_n$, since we will need them for the description of the symmetry group of the cut polytope $P_n$.

The cube $Q_n$ is the $n$-dimensional cube, i.e., the polytope whose vertices are the $2^n$ binary sequences of length $n$, i.e., all incidence vectors of subsets $S$ of $[1, n]$. Every permutation $\sigma$ of $\mathrm{Sym}(n)$ induces an isometry of $R^n$, still denoted by $\sigma$, and defined by $\sigma(x) = (x_{\sigma(1)}, \dots, x_{\sigma(n)})$ for $x = (x_1, \dots, x_n) \in R^n$. Note that the isometry induced by transposition $(ij)$ is the reflection around the hyperplane $x_i = x_j$. For $i \in [1, n]$, denote by $r_i$ the reflection around the hyperplane $x_i = 1/2$, so $r_i(x) = (x_1, \dots, x_{i-1}, 1 - x_i, x_{i+1}, \dots, x_n)$. For a subset $S$ of $[1, n]$, set $r_S = \prod_{i \in S} r_i$ (no ambiguity here since $r_i$, $r_j$ commute). It is well known (see, for instance, [**B**]) that the symmetry group of the cube $Q_n$ is the group generated by the permutations $\sigma$ of $\mathrm{Sym}(n)$ and the reflections $r_S$ for $S \subseteq [1, n]$. Hence, $\mathrm{Is}(Q_n)$ is the semi-direct product of $\mathrm{Sym}(n)$ by the group $\{r_S: S \subseteq [1, n]\}$ and its order is $2^n n!$. We indicate the commutation rules in $\mathrm{Is}(Q_n)$.

(1.1) $$r_S r_T = r_{S\triangle T} \quad \text{for } S, T \subseteq [1, n],$$

(1.2) $$r_S\sigma = \sigma r_{\sigma(S)} \quad \text{for } \sigma \in \mathrm{Sym}(n),\ S \subseteq [1, n],$$

where $S\triangle T$ denotes the symmetric difference of the sets $S$, $T$. The group $\mathrm{Is}(Q_n)$ can also be described as the abstract group having $n$ generators $\sigma_1, \dots, \sigma_{n-1}, \rho$ together with the relation

$$(\sigma_i\sigma_{i+1})^3 = 1 \quad \text{for } i = 1, 2, \dots, n-2, \qquad (\sigma_1\rho)^4 = 1,$$

where $\sigma_i$ denotes the transposition $(i, i+1)$ and $\rho = r_1$. This group, also denoted as $[3^{n-2}, 4]$, [**CM**], is the Coxeter group $W(B_n)$ with diagram:

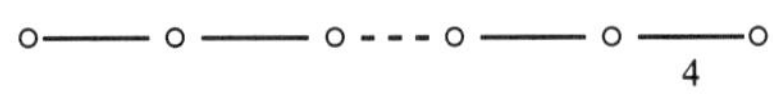


The halfcube $hQ_n$ is obtained from the cube $Q_n$ by selecting exactly one vertex on each edge, i.e., $hQ_n$ is the polytope whose vertices are the incidence vectors of the subsets $S$ of $[1, n]$ of even cardinality. The symmetry group of $hQ_n$ is the semi-direct product of $\mathrm{Sym}(n)$ and the group $\{r_S: S \subseteq [1, n]$ and $|S|$ even$\}$; so, $\mathrm{Is}(hQ_n)$ is a subgroup of $\mathrm{Is}(Q_n)$ of index 2, $|\mathrm{Is}(hQ_n)| = 2^{n-1}n!$. The group $\mathrm{Is}(hQ_n)$ can also be described as the abstract group having $n$ generators $\sigma_1, \dots, \sigma_{n-1}, \rho'$ together with the relations

$$(\sigma_i\sigma_{i+1})^3 = 1 \quad \text{for } i = 1, \dots, n-2, \qquad (\sigma_1\rho')^3 = 1,$$

where $\sigma_i$ denotes the transposition $(i, i+1)$ and $\rho' = \sigma_1 r_{\{1,2\}}$. So, this group, also denoted as $[3^{n-1,1,1}]$, [**CM**], is the Coxeter group $W(D_n)$ with diagram:

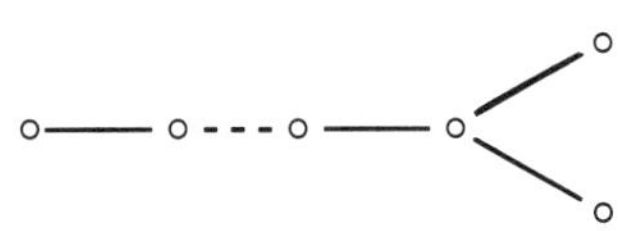

The center of the group $\mathrm{Is}(Q_n)$ is the subgroup $\langle r_N\rangle$ of order 2 generated by the reflection $r_N$, setting $N = [1, n]$, So the central quotient group $\mathrm{Is}(Q_n)/\langle r_N\rangle$ has order $2^{n-1}n!$ and its abstract presentation is obtained from that of $\mathrm{Is}(Q_n)$ by adjoining the relation $(\sigma_{n-1}\sigma_{n-2}\cdots\sigma_1\rho)^n = 1$, since $r_N = (\sigma_{n-1}\sigma_{n-2}\cdots\sigma_1\rho)^n$. For $n$ odd, the reflection $r_N$ does not belong to $\mathrm{Is}(hQ_n)$ and thus:

(1.3) $\mathrm{Is}(hQ_n)$ and $\mathrm{Is}(Q_n)/\langle r_N\rangle$ are isomorphic groups, for $n$ odd.

## II. The symmetry group of the cut polytope $P_n$

In this section, we describe the symmetry group of the cut polytope $P_n$. As we shall see through the proof, the case $n = 4$ has to be treated separately. Recall that $P_n$ is the convex hull of the cut vectors $\delta(S)$ for $S \subseteq [1, n]$.

First, as in the case of the cube $Q_n$, one constructs easily a group $G_n$ of $2^{n-1}n!$ symmetries of $P_n$ corresponding, in fact, to the known permutation and switching operations.

Every permutation $\sigma$ of $\mathrm{Sym}(n)$ induces naturally a permutation on the pairs $(i, j)$ of $[1, n]$ and thus an isometry of $R^{n(n-1)/2}$, still denoted as $\sigma$. Note that $\sigma(\delta(S)) = \delta(\sigma^{-1}(S))$ for any subset $S$ of $[1, n]$ and thus every permutation $\sigma$ induces a symmetry of $P_n$. For any cut $\delta(S)$, set $r_{\delta(S)} = \prod_{(ij)\in\delta(S)} r_{ij}$, where $r_{ij}$ denotes the reflection around the hyperplane $x_{ij} = 1/2$. Note that $r_{\delta(S)}(\delta(T)) = \delta(S\triangle T)$ for all $S$, $T$ subsets of $[1, n]$ and thus every reflection $r_{\delta(S)}$ induces a symmetry of $P_n$. When applied to faces of $P_n$, the reflection $r_{\delta(S)}$ corresponds to the switching by cut $\delta(S)$ operation. Namely, let $F$ be a face of $P_n$ and let $v \cdot x \leq v_0$ be an inequality supporting $F$; then the inequality $v^S \cdot x \leq v_0 - v\cdot\delta(S)$ is valid and supports the face $r_{\delta(S)}(F)$. Analogues of relations (1.1), (1.2) are

$$(2.1)\qquad r_{\delta(S)}r_{\delta(T)} = r_{\delta(S\triangle T)} \quad \text{for } S, T \subseteq [1, n],$$

$$(2.2)\qquad r_{\delta(S)}\sigma = \sigma r_{\delta(\sigma(S))} \quad \text{for } S \subseteq [1, n],\ \sigma \in \mathrm{Sym}(n).$$

Let $G_n$ denote the group generated by all permutations and the reflections $r_{\delta(S)}$ for $S \subseteq [1, n]$. Hence, $G_n$ is a subgroup of order $2^{n-1}n!$ of $\mathrm{Is}(P_n)$. We prove below that these two groups coincide for $n \neq 4$.

For $k \in [1, n-1]$, we define the graph $H_k(n)$ whose vertices are the $2^{n-1}$ cuts $\delta(S)$ for $S \subseteq [1, n]$ and two cuts $\delta(S)$, $\delta(T)$ are adjacent if $\|\delta(S) - \delta(T)\|^2 = k(n-k)$, i.e., $\|\delta(S\triangle T)\|^2 = k(n-k)$ or, equivalently, $|S\triangle T| = k$ or $n-k$. Obviously, every symmetry of $P_n$ induces an automorphism of the graph $H_k(n)$ and, hence, $\mathrm{Is}(P_n)$ is (up to isomorphism) a subgroup of $\mathrm{Aut}(H_k(n))$. So, we have the following inclusions:

$$(2.3)\qquad G_n \subseteq \mathrm{Is}(P_n) \subseteq \mathrm{Aut}(H_k(n)) \quad \text{for } 1 \leq k \leq n-1.$$

**Lemma 2.4.** *$G_n \approx \mathrm{Is}(Q_n)/\langle r_N\rangle$ for $n \geq 3$.*

**Proof.** Consider the group homomorphism $f$ from $\mathrm{Is}(Q_n)$ to $\mathrm{Is}(Q_{n(n-1)/2})$ defined by $f(\sigma) = \sigma$ for $\sigma \in \mathrm{Sym}(n)$ and $f(r_S) = r_{\delta(S)}$ for $S \subseteq [1, n]$. So the range of $f$ is precisely the group $G_n$. To conclude the proof, we need only verify that the kernel of $f$ is $\{\mathrm{id}, r_N\}$. For this, take $g \in \mathrm{Is}(Q_n)$, $g \neq \mathrm{id}$, such that $f(g) = \mathrm{id}$. We have $g = \sigma r_S$ for some $\sigma \in \mathrm{Sym}(n)$, $S \subseteq [1, n]$ and $S \neq \varnothing$. Then $f(g) = \sigma r_{\delta(S)} = \mathrm{id}$ implies that $\sigma = r_{\delta(S)}$ and thus $\sigma(\delta(S)) = 0$, yielding $\delta(S) = 0$, i.e., $S = N$. □

**Lemma 2.5.** *$|\mathrm{Aut}(H_1(n))| = 2^{n-1}n!$ for $n \neq 4$.*

PROOF. Set $K = \mathrm{Aut}(H_1(n))$ and let $K_0$ denote the stabilizer of vertex $\delta(\emptyset) = 0$ in $K$. Since $K$ contains $G_n$, the group $K$ is vertex transitive, implying the equality $|K| = |K_0| 2^{n-1}$. We conclude the proof by showing that $K_0 \approx \mathrm{Sym}(n)$. Obviously, $\mathrm{Sym}(n)$ is contained in $K_0$. Take now $g \in K_0$; then $g$ preserves globally the set $\{\delta(\{i\}) : i \in [1, n]\}$ of neighbors of $\delta(\emptyset)$. So, $g$ induces a permutation, say $\sigma$, on $[1, n]$. Set $g' = \sigma^{-1} g$; so, $g' \in K_0$ and $g'$ fixes every cut $\delta(\{i\})$, $1 \le i \le n$. This implies that $g'$ fixes every cut $\delta(\{i, j\})$, $1 \le i < j \le n$. Indeed, assume that $g(\delta(\{1, 2\})) = \delta(S)$; then $\delta(S)$ is adjacent to $\delta(\{1\})$ and $\delta(\{2\})$ in $H_1(n)$, i.e., each of $|S \triangle \{1\}|$ and $|S \triangle \{2\}|$ is either 1 or $n-1$. For instance, $S = \{1, a\}$ for some $a \in [2, n]$ and, in fact, $a = 2$, else, $|S \triangle \{2\}| = 3 = n - 1$, yielding $n = 4$ which contradicts the assumption. One obtains easily by induction on $|S| \ge 2$ that $g'(\delta(S)) = \delta(S)$ for all cuts $\delta(S)$. So, $g' = \mathrm{id}$, i.e., $g = \sigma \in \mathrm{Sym}(n)$. □

The next result follows directly from relations (1.3), (2.3), and Lemmas 2.4 and 2.5. In other words, the only symmetries of $P_n$ for $n \ne 4$ correspond to permutations and switchings.

THEOREM 2.6. *For* $n \ne 4$, $\mathrm{Is}(P_n) \approx G_n \approx \mathrm{Is}(Q_n)/\langle r_N \rangle$. *Hence, for n odd,* $\mathrm{Is}(P_n) \approx \mathrm{Is}(hQ_n) \approx W(D_n)$.

REMARK 2.7. The graph $H_1(n)$ is, in fact, isomorphic to the *folded n-cube* whose vertices are the partitions $(S, N - S)$ of $N = [1, n]$ with two partitions adjacent if and only if their common refinement is a partition of $N$ into sets of sizes $1, n-1, 1, n-1$. The automorphism group of the folded $n$-cube is $2^{n-1}\mathrm{Sym}(n)$ for $n > 4$, $(\mathrm{Sym}(4) \times \mathrm{Sym}(4))\,\mathrm{Sym}(2)$ for $n = 4$ and $\mathrm{Sym}(4)$ for $n = 3$ (see, e.g., **[BCN, p. 264]**). However, we preferred to give a direct proof of Lemma 2.5 in order, in particular, to point out how the special case $n = 4$ occurs. Note that, for $n = 3$, $P_3$ is a regular 3-dimensional simplex and $G_3 = \mathrm{Is}(P_3)$ indeed coincides with $\mathrm{Sym}(4)$.

We now consider the case $n = 4$. Then $\mathrm{Is}(P_4) \approx \mathrm{Aut}(H_1(4))$ contains $G_4$ as proper subgroup, i.e., $P_4$ admits symmetries which are neither permutations of $\mathrm{Sym}(4)$ nor switchings by cuts. For $n = 4$, $H_1(4)$ is the complete bipartite graph $K_{4,4}$ with node partition $X_1 \cup X_2$, where $X_1 = \{\delta(\{i\}) : 1 \le i \le 4\}$ and $X_2 = \{\delta(\emptyset), \delta(\{1, i\}), 2 \le i \le 4\}$. Set $x_i = \delta(\{i\})$ for $i \in [1, 4]$, $y_1 = \delta(\emptyset)$ and $y_i = \delta(\{1, i\})$ for $2 \le i \le 4$. Recall that the automorphism group of the complete bipartite graph $K_{4,4}$ with node partition $X_1 \cup X_2$ is the semi-direct product of $\mathrm{Sym}(4) \times \mathrm{Sym}(4)$ by $\mathrm{Sym}(2)$; namely, its generators are the following elements:

(i) the permutations $\theta_{(\alpha, \beta)}$ of $X_1 \cup X_2$ mapping $x_i$ onto $x_{\alpha(i)}$ and $y_i$ onto $y_{\beta(i)}$ for $1 \le i \le 4$, for all $\alpha, \beta \in \mathrm{Sym}(4)$;

(ii) the permutation (involution) $\xi$ of $X_1 \cup X_2$ defined by $\xi(x_i) = y_i$, $\xi(y_i) = x_i$ for $1 \le i \le 4$ with the commutation rule $\theta_{(\alpha, \beta)}\xi = \xi\theta_{(\beta, \alpha)}$.

THEOREM 2.8. $\mathrm{Is}(P_4) \approx \mathrm{Aut}(K_{4,4}) \approx (\mathrm{Sym}(4) \times \mathrm{Sym}(4))\,\mathrm{Sym}(2)$.

PROOF. It remains only to show that every element of $\mathrm{Aut}(H_1(4))$ induces a symmetry of $P_4$. We keep the above notation for $X_1$, $X_2$ and their elements. First, the involution $\xi$ of $\mathrm{Aut}(H_1(4))$ identifies with the reflection $r_{\delta(\{1\})}$ of $\mathrm{Is}(P_4)$. We verify now that, for all $\alpha, \beta \in \mathrm{Sym}(4)$, $\theta_{(\alpha,\beta)}$ indeed defines an isometry of $R^6$. This will imply that $\theta_{(\alpha,\beta)}$ is a symmetry of $P_4$, since $\theta_{(\alpha,\beta)}$ preserves the vertices of $P_4$ and any isometry is "linear on convex combinations" (i.e., for an isometry $g$, $g(ax+(1-a)y) = ag(x)+(1-a)g(y)$ for any scalar $a$, $0 \le a \le 1$, and any vectors $x, y$).

We consider the vector spaces $V_1 = \mathrm{Vect}(\delta(\{i\}) - e/2 \colon i = 1, 2, 3, 4)$, where $e = (1, 1, \dots, 1)$ and $V_2 = \mathrm{Vect}(\delta(\{1,2\}), \delta(\{1,3\}), \delta(\{1,4\}))$. One easily checks that $V_1$, $V_2$ both have dimension 3 and are orthogonal spaces (since $\delta(\{i\})\cdot\delta(\{1,j\}) = e\cdot\delta(\{1,j\})/2 = 2$ for all $i, j$). Hence, $R^6$ is the orthogonal sum of $V_1$ and $V_2$.

Given a permutation $\alpha \in \mathrm{Sym}(4)$, we show below how $\theta_{(\alpha,\mathrm{id})}$ defines an isometry of $V_1$. Take $x \in V_1$, so $x = \sum_{1\le i\le 4} a_i(\delta(\{i\}) - e/2)$ for some scalars $a_i$; we set $\theta_{(\alpha,\mathrm{id})}(x) = \sum_{1\le i\le 4} a_i(\delta(\{\alpha(i)\}) - e/2)$. We verify first that this mapping is well defined. One checks easily that $x = \sum_{1\le i\le 4} a'_i(\delta(\{i\}) - e/2)$ is another decomposition of $x$ if and only if $a'_i = a_i + a$ for some scalar $a$. Therefore, $\sum_{1\le i\le 4} a_i(\delta(\{\alpha(i)\}) - e/2) = \sum_{1\le i\le 4} a'_i(\delta(\{\alpha(i)\}) - e/2)$, because $\sum_{1\le i\le 4}(\delta(\{\alpha(i)\}) - e/2) = 0$. Next, we verify that $\theta_{(\alpha,\mathrm{id})}$ induces an isometry of $V_1$, i.e., $\|x-y\| = \|\theta_{(\alpha,\mathrm{id})}(x) - \theta_{(\alpha,\mathrm{id})}(y)\|$ for all $x, y$ in $V_1$. Take $x = \sum_{1\le i\le 4} a_i(\delta(\{i\}) - e/2)$, $y = \sum_{1\le i\le 4} b_i(\delta(\{i\}) - e/2)$, and set $c = \sum_{1\le i\le 4}(b_i - a_i)/2$, $c_i = a_i - b_i$ for $1 \le i \le 4$. Then $\|x-y\| = \|ce + \sum_{1\le i\le 4} c_i\delta(\{i\})\|$ and $\|\theta_{(\alpha,\mathrm{id})}(x) - \theta_{(\alpha,\mathrm{id})}(y)\| = \|ce + \sum_{1\le i\le 4} c_i\delta(\{\alpha(i)\})\|$. The above two quantities coincide, since

$$\left\| ce + \sum_{1\le i\le 4} c_i\delta(\{i\}) \right\| = \|(c_1+c_2+c,\, c_1+c_3+c,\, c_1+c_4+c,\, c_2+c_3+c,\, c_2+c_4+c,\, c_3+c_4+c)\|$$

is a symmetric function in $c_1, c_2, c_3, c_4$.

Given a permutation $\beta \in \mathrm{Sym}(4)$, we show now how $\theta_{(\mathrm{id},\beta)}$ defines an isometry of $V_2$. Take $x \in V_2$, so $x$ can be uniquely written as $x = \sum_{2\le i\le 4} a_i\delta(\{1,i\})$. We set

$$\theta_{(\mathrm{id},\beta)}(x) = \left(1 - \sum_{2\le i\le 4} a_i\right)\beta(\delta(\varnothing)) + \sum_{2\le i\le 4} a_i\beta(\delta(\{1,i\}))$$

(recall that $\beta$ is viewed as permutation of the set $X_2$). We verify that $\theta_{(\mathrm{id},\beta)}$ defines an isometry of $V_2$. For this, take $x = \sum_{2\le i\le 4} a_i\delta(\{1,i\})$, $y = \sum_{2\le i\le 4} b_i\delta(\{1,i\})$, and set $c_i = a_i - b_i$ for $2 \le i \le 4$ and $c_1 =$

$\sum_{2\le i\le 4}(b_i - a_i)$, so $\sum_{1\le i\le 4} c_i = 0$. Then $\|x-y\| = \|\sum_{2\le i\le 4} c_i\delta(\{1, i\})\|$ and

$$\|\theta_{(\mathrm{id}, \beta)}(x) - \theta_{(\mathrm{id}, \beta)}(y)\| = \|c_1\beta(\delta(\emptyset)) + \sum_{2\le i\le 4} c_i\beta(\delta(\{1, i\}))\|.$$

So, $\|x-y\|^2 = 2((c_2+c_3)^2 + (c_2+c_4)^2 + (c_3+c_4)^2)$ while $\|\theta_{(\mathrm{id}, \beta)}(x) - \theta_{(\mathrm{id}, \beta)}(y)\|^2 = 2((c_i+c_j)^2 + (c_i+c_k)^2 + (c_j+c_k)^2)$, where $i, j, k$ are the indices of $[1, 4]$ such that $\{i, j, k\} = \{2, 3, 4\}$ if $\beta(\delta(\emptyset)) = \delta(\emptyset)$, and $\{1, 2, 3, 4\} - \{i, j, k\} = \{h\}$ with $h$ being the index such that $\beta(\delta(\{1, h\})) = \delta(\emptyset)$ otherwise. Using the relation $\sum_{1\le i\le 4} c_i = 0$, one observes easily that $(c_2+c_3)^2+(c_2+c_4)^2+(c_3+c_4)^2 = (c_i+c_j)^2+(c_i+c_k)^2+(c_j+c_k)^2$ for $1\le i<j<k\le 4$. So the relation $\|x-y\| = \|\theta_{(\mathrm{id}, \beta)}(x) - \theta_{(\mathrm{id}, \beta)}(y)\|$ holds.

Finally, given permutations $\alpha$, $\beta$ of Sym(4), we define $\theta_{(\alpha, \beta)}$ as an isometry of $R^6$ as follows. Given $x$ in $R^6$, $x$ can be uniquely decomposed as $x = y + z$ with $y \in V_1$, $z \in V_2$. We set $\theta_{(\alpha, \beta)}(x) = \theta_{(\alpha, \mathrm{id})}(y) + \theta_{(\mathrm{id}, \beta)}(z)$. Since $V_1$, $V_2$ are orthogonal subspaces, $\theta_{(\alpha, \beta)}$ is indeed an isometry. □

## III. The symmetry groups of some relatives

The cut polytope $P_n$ presents a high degree of symmetry which is lost for its relatives considered here, namely,

–the boolean quadric polytope $QP^n$ which is a linear bijective image of $P_{n+1}$ and, hence, admits an analogue of the switching operation which however is not an isometry,

–the $k$-cut polytope $MC_k^{\le}(n)$ $(2\le k\le n)$ which is a generalization of the cut polytope $P_n$ (i.e., case $k=2$) and has no analogue of switching at all if $k\ge 3$,

–the equicut polytope $EP_n$ which is a face (facet for $n$ odd) of $P_n$ and has no switching either.

Obviously, permutations of Sym($n$) induce symmetries of all three polytopes $QP^n$, $MC_k^{\le}(n)$, and $EP_n$ and we show below that they are indeed their only symmetries.

**III.1. The symmetries of the boolean quadric polytope.** Recall that the boolean quadric polytope $QP^n$ is the convex hull of the intersection vectors $\pi(S)$ for all subsets $S$ of $[1, n]$, so $QP^n \subseteq R^{n(n+1)/2}$. Let $P_{n+1}$ denote the cut polytope of the complete graph $K_{n+1}$ on nodeset $[0, n] = \{0, 1, 2, \dots, n\}$, so $P_{n+1}$ is the convex hull of the cut vectors $\delta(S)$ for all subsets $S$ of $[1, n]$ and $P_{n+1} \subseteq R^{n(n+1)/2}$ (note that $\delta(S)$ denotes here the cut vector of $S$ considered as a subset of $[0, n]$). Given vectors $y = (y_{ij})_{1\le i\le j\le n}$ and $z = (z_{ij})_{0\le i<j\le n}$ in $R^{n(n+1)/2}$, consider the function

$y = \theta(z)$ defined by

$$y_{ii} = z_{0i} \quad \text{for } i = 1, \dots, n,$$
$$y_{ij} = (z_{0i} + z_{0j} - z_{ij})/2 \quad \text{for } 1 \le i < j \le n.$$

Then $\theta$ is a linear bijective mapping on $R^{n(n+1)/2}$ and one checks easily that, for all subsets $S$ of $[1, n]$, $\theta(\delta(S)) = \pi(S)$ holds. Therefore, $QP^n = \theta(P_{n+1})$ holds. This simple but interesting connection was discovered, in different terms, independently by several authors; for references, see, e.g., **[DL1]**. Hence, the face lattices of both polytopes $QP^n$ and $P_{n+1}$ are in bijective correspondence and knowledge of the linear description of $P_{n+1}$ yields explicitly that of $QP^n$ and vice versa. In particular, the switching by cut operation on facets of the cut polytope $P_{n+1}$ yields a corresponding operation on facets of $QP^n$. Switching facets by cut $\delta(S)$ in $P_{n+1}$ comes from the pointwise isometry $r_{\delta(S)}$, while the corresponding operation on facets of $QP^n$ comes from the pointwise mapping $\theta r_{\delta(S)} \theta^{-1}$. Note, however, that $\theta_S = \theta r_{\delta(S)} \theta^{-1}$ is not an isometry. Indeed, for instance, for $S = \{1\}$, $\theta_S(\pi(\{1\})) = 0$, and $\theta_S(\pi(\{2\})) = \pi(\{1, 2\})$, and so $\|\pi(\{1\}) - \pi(\{2\})\|^2 = 2$ while $\|\theta_S(\pi(\{1\})) - \theta_S(\pi(\{2\}))\|^2 = 3$. So, from the geometric point of view, the cut polytope is much more symmetric than the boolean quadric polytope. For instance, both $P_3$, and $QP^2$ are 3-dimensional simplices, but $P_3$ is regular (with edges of length $\sqrt{3}$) while $QP^2$ is not regular (it has two edges of length 1, three of length $\sqrt{2}$, and one of length $\sqrt{3}$). We prove below that the only symmetries of the boolean quadric polytope are the permutations.

THEOREM 3.1. $\text{Is}(QP^n) \approx \text{Sym}(n)$ *for* $n \ge 2$.

LEMMA 3.2. *Given subsets* $S, T$ *of* $[1, n]$, $\|\pi(S) - \pi(T)\| = 1$ *if and only if* $S = \varnothing$, $|T| = 1$ *or vice versa.*

PROOF. Observe that $\|\pi(S) - \pi(T)\|^2$ is equal to the number of pairs $(i, j)$, $1 \le i \le j \le n$, such that either $(i, j \in S$ and $i \notin T$ or $j \notin T)$ or $(i, j \in T$ and $i \notin S$ or $j \notin S)$ and, thus,

$$(*) \qquad \|\pi(S) - \pi(T)\|^2 = s(s+1)/2 + t(t+1)/2 - a(a+1),$$

setting $s = |S|$, $t = |T|$, and $a = |S \cap T|$. Then, using relation $(*)$, one obtains easily that Lemma 3.2 holds. □

PROOF OF THEOREM 3.1. Take an isometry $g$ of $\text{Is}(QP^n)$. We prove that $g$ is induced by some permutation $\sigma$, i.e., $g(x) = (x_{\sigma(i)\sigma(j)})_{1 \le i \le j \le n}$ for all $x \in R^{n(n+1)/2}$. First, $g$ fixes the origin. Indeed, assume that $g(0) = \pi(A)$ with $|A| \ge 1$. Both sets $\{\pi(S) \colon \|\pi(S)\| = 1\}$ and $\{\pi(S) \colon \|\pi(S) - \pi(A)\| = 1\}$ have the same cardinality since $g$ is a symmetry of $QP^n$. But, from Lemma 3.2, the first has cardinality $n$ while the second has cardinality at most 1, yielding a contradiction. So, $g(0) = 0$ and, thus, $g$ is linear.

Take $i \in [1, n]$ and set $g(\pi(\{i\})) = \pi(S)$ for some $S \subseteq [1, n]$. Then $\|\pi(\{i\})\| = \|\pi(S)\| = 1$, implying, from Lemma 3.2, that $|S| = 1$. Therefore, there exists a permutation $\sigma$ such that $g(\pi(\{i\})) = \pi(\{\sigma(i)\})$ for all $i \in [1, n]$. Take distinct $i, j$ in $[1, n]$, set $g(\pi(\{i, j\})) = \pi(S)$ for some $S \subseteq [1, n]$. Then $\|\pi(\{i, j\})\| = \|\pi(S)\| = \sqrt{3}$, yielding easily that $|S| = 2$, e.g., $S = \{a, b\}$. Also, $\|\pi(\{i\}) - \pi(\{i, j\})\|^2 = \|\pi(\sigma(i)) - \pi(\{a, b\})\|^2$ which, using relation $(*)$ in the proof of Lemma 3.2, easily implies that $\sigma(i) \in \{a, b\}$. Similarly, $\sigma(j) \in \{a, b\}$ and, thus, $g(\pi(\{i, j\})) = \pi(\{\sigma(i), \sigma(j)\})$.

Finally, if $e_{ij}$ denotes the unit vector in $R^{n(n+1)/2}$ for $1 \le i \le j \le n$, one has that $e_{ii} = \pi(\{i\})$ for $i \in [1, n]$ and $e_{ij} = \pi(\{i, j\}) - \pi(\{i\}) - \pi(\{j\})$ for $1 \le i < j \le n$. Then, from the above observations and the fact that $g$ is linear, we have that $g(e_{ii}) = e_{\sigma(i)\sigma(i)}$ for $i \in [1, n]$ and $g(e_{ij}) = e_{\sigma(i)\sigma(j)}$ for $1 \le i < j \le n$. □

**III.2. The symmetries of the $k$-cut polytope.** Recall that the $k$-cut polytope $MC_k^{\le}(n)$ is the convex hull of the $h$-cuts $\delta(S_1, \dots, S_h)$ for the partitions of $[1, n]$ into $h \le k$ parts $S_1, \dots, S_h$; so $MC_2^{\le}(n) = P_n$. We prove below that the permutations of $\mathrm{Sym}(n)$ are the only symmetries of $MC_k^{\le}(n)$ if $k \ge 3$. So, the nice switching property which, in particular, permits us to obtain all facets of $P_n$ from those containing the origin is lost for the $k$-cut polytope with $k \ge 3$.

THEOREM 3.3. *For* $3 \le k \le n$, $\mathrm{Is}(MC_k^{\le}(n)) \approx \mathrm{Sym}(n)$.

PROOF. We show that every symmetry $g$ of $\mathrm{Is}(MC_k^{\le}(n))$ corresponds to some permutation $\sigma$ of $\mathrm{Sym}(n)$, i.e., $g(x) = (x_{\sigma(i)\sigma(j)})_{1 \le i < j \le n}$ for all $x = (x_{ij})_{1 \le i < j \le n}$ in $R^{n(n-1)/2}$.

LEMMA 3.4. $g(0) = 0$.

PROOF. Given a multicut $x = \delta(S_1, \dots, S_h)$ of $MC_k^{\le}(n)$, we denote by $B_{n-2}(x)$ the set of multicuts $y$ of $MC_k^{\le}(n)$ such that $\|x - y\|^2 \in [1, n-2]$ (since $x, y$ are zero-one vectors, $\|x - y\|^2$ is equal to the Hamming distance between $x, y$). We verify below that $B_{n-2}(0) = \varnothing$ while $B_{n-2}(x) \ne \varnothing$ for every nonzero multicut $x$ of $MC_k^{\le}(n)$, which therefore implies that $g(0) = 0$. First note that $\|\delta(S)\|^2 \ge n - 1$ for every nonzero cut (with equality if and only if $|S| = 1$) and $\|\delta(S_1, \dots, S_h)\|^2 > \|\delta(S_1)\|^2 \ge n - 1$ for every $h$-cut with $h \ge 3$. Hence, $B_{n-2}(0)$ is indeed empty. Now take a nonzero cut $\delta(S)$; we can suppose that $|S| \ge 2$. Given $i \in S$,

$$\|\delta(S - \{i\}, \{i\}, [1, n] - S) - \delta(S)\|^2 = |S| - 1 \le n - 2,$$

implying that $B_{n-2}(\delta(S))$ is not empty. Finally, given an $h$-cut $\delta(S_1, \dots, S_h)$ with $3 \le h \le k$ and $i \in S_1$,

$$\|\delta(S_1 - \{i\}, S_2 \cup \{i\}, S_3, \dots, S_h) - \delta(S_1, \dots, S_h)\|^2 = |S_1| + |S_2| - 1 \le n - 2,$$

implying that $B_{n-2}(\delta(S_1, \dots, S_h))$ is not empty. □

LEMMA 3.5. $g(\delta(\{i\})) = \delta(\{\sigma(i)\})$ *for* $1 \le i \le n$ *and* $g(\delta(\{i, j\})) = \delta(\{\sigma(i), \sigma(j)\})$ *for* $1 \le i < j \le n$, *for some* $\sigma$ *in* $\mathrm{Sym}(n)$.

PROOF. Since $g(0) = 0$, we have that $\|\delta(\{i\})\|^2 = \|g(\delta(\{i\}))\|^2 = n - 1$ which, as we saw above, implies that $g(\delta(\{i\}))$ is a cut $\delta(S)$ with $|S| = 1$. Hence, $g$ induces a permutation $\sigma$ on $[1, n]$ so that $g(\delta(\{i\})) = \delta(\{\sigma(i)\})$ for all $1 \le i \le n$. We now compute, for instance, $g(\delta(\{1, 2\}))$. Suppose first that $g(\delta(\{1, 2\})) = \delta(S_1, \dots, S_h)$ with $3 \le h \le k$. Then $\|\delta(S_1, \dots, S_h)\|^2 = \|\delta(\{1, 2\})\|^2 = 2(n-2) > \|\delta(S_i)\|^2$ for all $1 \le i \le h$, implying that $|S_i| = 1$ for all $1 \le i \le h$, i.e., $h = n$ and $\|\delta(S_1, \dots, S_h)\|^2 = n(n-1)/2$, yielding a contradiction. Therefore, $g(\delta(\{1, 2\}))$ is a cut $\delta(S)$ for some $S \subseteq [1, n]$; set $s = |S|$. We have that $\|\delta(S)\|^2 = s(n-s) = 2(n-2)$, implying that $s = 2$ or $n-2$, i.e., $g(\delta(\{1, 2\})) = \delta(\{a, b\})$ for some $1 \le a < b \le n$. Finally, using the fact that $\|\delta(\{1\}) - \delta(\{1, 2\})\|^2 = \|\delta(\{\sigma(1)\}) - \delta(\{a, b\})\|^2 = n - 1$ and $\|\delta(2) - \delta(\{1, 2\})\|^2 = \|\delta(\{\sigma(2)\}) - \delta(\{a, b\})\|^2 = n - 1$, we deduce that $g(\delta(\{1, 2\}) = \delta(\{a, b\}) = \delta(\{\sigma(1), \sigma(2)\})$. □

Noting that, for all $1 \le i < j \le n$, the unit vector $e_{ij}$ can be written as

$$e_{ij} = (\delta(\{i\}) + \delta(\{j\}) - \delta(\{i, j\}))/2,$$

we deduce from Lemmas 3.4 and 3.5 that $g$ is linear and $g(e_{ij}) = e_{\sigma(i)\sigma(j)}$ for all $1 \le i < j \le n$, concluding the proof of Theorem 3.3. □

**III.3. The symmetries of the equicut polytope.** In §II, we computed the symmetries of $P_n$, i.e., those isometries of $R^{n(n-1)/2}$ preserving globally $P_n$. More specifically, one can look for the symmetries preserving a given face $F$ of $P_n$. In general, finding $\mathrm{Is}(F)$ might be a difficult problem, since the faces of $P_n$ can be rather complicated, also $\mathrm{Is}(F)$ may have no relation with $\mathrm{Is}(P_n)$. However, some faces have an easily computable symmetry group; this is the case for the equicut face $EP_n$ of $P_n$. The equicut face $EP_n$ is the face of $P_n$ supported by the inequality $\sum_{1\le i<j\le n} x_{ij} \le \lfloor n/2\rfloor \lceil n/2\rceil$, i.e., $EP_n$ is the convex hull of all equicuts, i.e., cuts $\delta(S)$ with $|S| = \lfloor n/2\rfloor$ or $\lceil n/2\rceil$. For $n$ odd, $EP_n$ is a facet of $P_n$ while, for $n$ even, $EP_n$ is a face of $P_n$ of dimension $n(n-1)/2 - n$ **[CRS]**. Obviously, every permutation induces a symmetry of $EP_n$ and they are, in fact, the only symmetries of $EP_n$.

THEOREM 3.6. $\mathrm{Is}(EP_n) \approx \mathrm{Sym}(n)$.

PROOF. Consider first the even case $n = 2m$. We define the graph $H_2'(2m)$ whose nodes are the equicuts $\delta(S)$ for $S \subseteq [1, 2m]$ with $|S| = m$, and two equicuts $\delta(S), \delta(T)$ are adjacent if $\|\delta(S) - \delta(T)\|^2 = 2(2m-2)$, i.e., $\|\delta(S \triangle T)\|^2 = 2(2m-2)$. So, $H_2'(2m)$ is the subgraph of the graph $H_2(2m)$ (considered in §II) induced by the equicuts. The graph $H_2'(2m)$ is, in fact, known as the *folded Johnson graph* whose automorphism group is $\mathrm{Sym}(2m)$ (see, e.g., **[BCN, p. 259]**). Since every symmetry of $EP_{2m}$ induces an automorphism of the graph $H_2'(2m)$, we deduce that $\mathrm{Is}(EP_{2m}) \approx \mathrm{Sym}(n)$.

Consider now the odd case $n = 2m + 1$. Then we consider the graph $H_1'(2m+1)$ whose nodes are the equicuts $\delta(S)$ with, for instance, $|S| = m$ and two equicuts $\delta(S)$, $\delta(T)$ are adjacent if $\|\delta(S) - \delta(T)\|^2 = n - 1 = 2m$, i.e., $\|\delta(S \triangle T)\|^2 = 2m$. So, $H_1'(2m+1)$ is the subgraph of the graph $H_1(2m+1)$ induced by the equicuts. The graph $H_1'(2m+1)$ is known as the *odd graph* whose automorphism group is $\mathrm{Sym}(2m+1)$ (see, e.g., **[BCN, p. 259]**). Since $\mathrm{Is}(EP_{2m+1}) \subseteq \mathrm{Aut}(H_1'(2m+1))$, we deduce that $\mathrm{Is}(EP_{2m+1}) \approx \mathrm{Sym}(2m+1)$. □

## IV. Double symmetries of facets

This section is a development of some ideas and results from **[DL2, §3]**. For a more detailed description of the facets $Par_7$ and $Gr_7$ considered here, see **[DL1]**, and for the description of their switchings, see **[DDL]**.

The symmetry group of $P_n$ acts on the facets of $P_n$. For any facet $F$ of $P_n$, denote by $O(F)$ its orbit under the action of $\mathrm{Is}(P_n)$, so $O(F) = \{g(F) \colon g \in \mathrm{Is}(P_n)\} = \{r_{\delta(S)}\sigma(F) \colon \sigma \in \mathrm{Sym}(n),\ S \subseteq [1, n]\}$. We are interested in giving an explicit description of $O(F)$.

We denote by $O_p(F)$ the suborbit of $O(F)$ obtained by considering only permutations, i.e., $O_p(F) = \{\sigma(F) \colon \sigma \in \mathrm{Sym}(n)\}$, and by $O_s(F)$ the suborbit of $O(F)$ obtained by considering only switchings, i.e., $O_s(F) = \{r_{\delta(S)}(F) \colon S \subseteq [1, n]\}$. We define the *automorphism group* $\mathrm{Aut}(F)$ of $F$ as the subgroup of $\mathrm{Sym}(n)$ preserving $F$, i.e., $\mathrm{Aut}(F) = \mathrm{Is}(F) \cap \mathrm{Sym}(n) = \{\sigma \in \mathrm{Sym}(n) \colon \sigma(F) = F\}$; note that $\sigma \in \mathrm{Aut}(F)$ if and only if $\sigma(v) = v$, where $v \cdot x \leq v_0$ is an inequality supporting $F$. An example of a facet admitting the largest possible automorphism group, namely $\mathrm{Sym}(n)$, is the equicut facet (Theorem 3.6). One easily deduces the cardinality of $O_p(F)$:

(4.1) $$|O_p(F)| = n!/|\mathrm{Aut}(F)| \quad \text{for any facet } F \text{ of } P_n.$$

Any two switchings of $F$ are distinct and so $|O_s(F)| = 2^{n-1}$. However, among these $2^{n-1}$ switchings, some of them are permutation equivalent. Let us denote by $\nu(F)$ the number of switchings of $F$ which are pairwise nonpermutation equivalent, and by $F_0 = F, F_1, \ldots, F_{\nu(F)-1}$ these nonpermutation equivalent switchings. In other words, $\nu(F)$ is the number of orbits of the set $O(F)$ under the action of $\mathrm{Sym}(n)$.

Proposition 4.2. *$O(F)$ is partitioned into*

$$O_p(F_0) \cup O_p(F_1) \cup \cdots \cup O_p(F_{\nu(F)-1}),$$

*and, hence,*

$$|O(F)| = \sum_{0 \leq i \leq \nu(F)-1} n!/|\mathrm{Aut}(F_i)|.$$

Proof. Take $F' = \sigma r_{\delta(S)}(F)$ in $O(F)$. The facet $r_{\delta(S)}(F)$ is permutation equivalent to $F_i$ for some $i \in [0, \nu(F) - 1]$ and so $r_{\delta(S)}(F) = \sigma'(F_i)$

for some $\sigma' \in \mathrm{Sym}(n)$. Therefore, $F' = \sigma\sigma'(F_i)$ belongs to $O_p(F_i)$. The enumeration formula then follows using (4.1). □

We give an upper bound for the number $\nu(F)$ of nonpermutation equivalent switchings of $F$ and also for the number $\nu_0(F)$ of nonpermutation equivalent switchings by roots of $F$, where the *roots* of $F$ are the cuts $\delta(S)$ lying in $F$; their set is denoted by $R(F)$. If $\delta(S)$ is a root of $F$ and $\delta(T)$ is not a root of $F$, then $r_{\delta(S)}(F)$ and $r_{\delta(T)}(F)$ are not permutation equivalent, since the first facet contains the origin while the second does not. Hence, among the $\nu(F)$ classes $O_p(F_i)$, $\nu_0(F)$ ones correspond to switchings by roots of $F$. Denote by $w(F)$ (resp. $w_0(F)$) the number of orbits of the set $X$ (resp. $R(F)$) of all cuts (resp. of all roots of $F$) under the action of $\mathrm{Aut}(F)$.

PROPOSITION 4.3. (i) $\nu(F) \le w(F)$ *and* $\nu_0(F) \le w_0(F)$.
(ii) *If* $\mathrm{Is}(F) \cap \mathrm{Is}(P_n) \subseteq \mathrm{Sym}(n)$, *then equality holds:* $\nu(F) = w(F)$ *and* $\nu_0(F) = w_0(F)$.

PROOF. We first prove (i). Take two cuts $\delta(S)$ and $\delta(T)$ which are in the same orbit under $\mathrm{Aut}(F)$, i.e., $\delta(T) = \delta(\sigma(S))$ for some $\sigma \in \mathrm{Aut}(F)$. Then, using (2.2), $r_{\delta(T)}(F) = r_{\delta(\sigma(S))}(F) = \sigma^{-1}\sigma r_{\delta(\sigma(S))}(F) = \sigma^{-1} r_{\delta(S)}(F)$ and, thus, the facets $r_{\delta(S)}(F)$ and $r_{\delta(T)}(F)$ are permutation equivalent. Therefore, $\nu(F) \le w(F)$ and $\nu_0(F) \le w_0(F)$. We now prove (ii). Given two cuts $\delta(S)$ and $\delta(T)$, observe that the facets $r_{\delta(S)}(F)$ and $r_{\delta(T)}(F)$ are permutation equivalent if and only if there exists $\sigma \in \mathrm{Sym}(n)$ such that $r_{\delta(T)}(F) = \sigma r_{\delta(S)}(F)$, i.e., using (2.1), (2.2) and setting $\tau = \sigma^{-1}$, $F = r_{\delta(T\triangle\tau(S))}(F)$, or equivalently, $r_{\delta(T\triangle\tau(S))}\sigma \in \mathrm{Is}(F)$. If $\mathrm{Is}(F) \cap \mathrm{Is}(P_n) \subseteq \mathrm{Sym}(n)$, then this last condition is equivalent to $T = \sigma^{-1}(S)$ and $\sigma \in \mathrm{Aut}(F)$, i.e., $\delta(S)$ and $\delta(T)$ are in the same orbit of the set of cuts under action of $\mathrm{Aut}(F)$ and thus $\nu(F) = w(F)$. The same reasoning gives $\nu_0(F) = w_0(F)$. □

Among the facets which are permutation equivalent to $F$, of special interest are those which are also switching equivalent to $F$; this leads us to define the following set:

(4.4) $$PS(F) = \{\sigma \in \mathrm{Sym}(n) \colon \sigma(F) = r_{\delta(S)}(F) \text{ for some } S \subseteq [1, n]\}.$$

So, the set of switchings of $F$ which are permutation equivalent to $F$ is equal to $O_{ps}(F) = \{\sigma(F) \colon \sigma \in PS(F)\}$.

PROPOSITION 4.5. (i) $PS(F)$ *is a subgroup of* $\mathrm{Sym}(n)$ *containing* $\mathrm{Aut}(F)$.
(ii) *If* $F'$ *is switching equivalent to* $F$, *then* $PS(F') = PS(F)$ *and*

$$|PS(F)| = |\mathrm{Aut}(F')|\,|O_{ps}(F')|.$$

PROOF. We first prove (i). If $\sigma \in \mathrm{Aut}(F)$, then $\sigma(F) = F = r_{\delta(\varnothing)}(F)$ yielding that $\sigma \in PS(F)$ and so $\mathrm{Aut}(F) \subseteq PS(F)$. Take $\sigma, \sigma'$ in $PS(F)$; so $\sigma(F) = r_{\delta(S)}(F)$, $\sigma'(F) = r_{\sigma(S')}(F)$ for some $S, S' \subseteq [1, n]$. Then, using

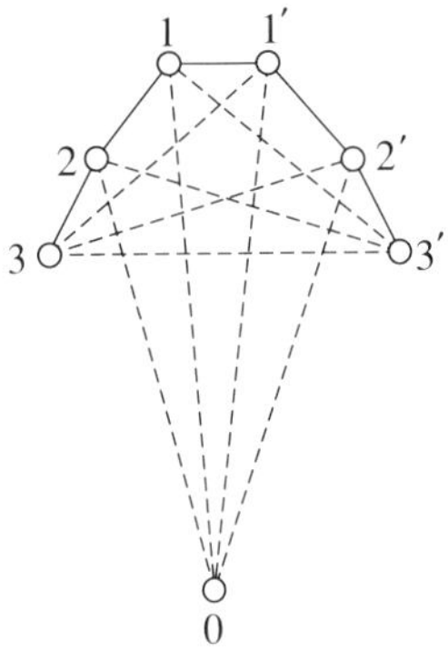


FIGURE 1

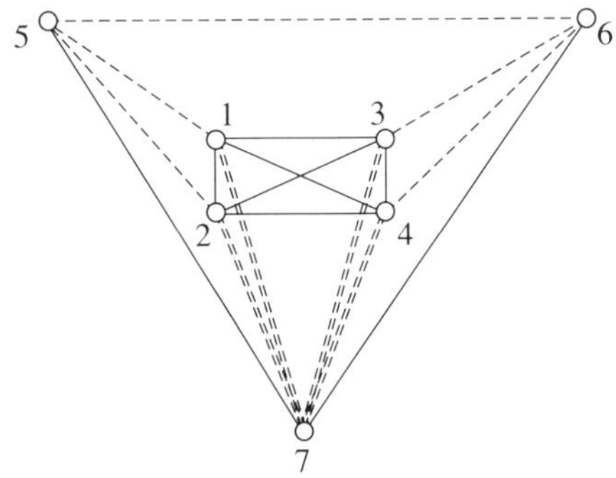


FIGURE 2

(2.1), (2.2) and setting $\tau = \sigma^{-1}$, $\sigma\sigma'(F) = \sigma r_{\delta(S')}(F) = r_{\delta(\tau(S'))}\sigma(F) = r_{\delta(\tau(S')\triangle S)}(F)$ and thus $\sigma\sigma' \in PS(F)$. Also, if $\sigma(F) = r_{\delta(S)}(F)$, then $F = \sigma^{-1}r_{\delta(S)}(F) = r_{\delta(\sigma(S))}\sigma^{-1}(F)$, implying that $\sigma^{-1}(F) = r_{\delta(\sigma(S))}(F)$ and thus $\sigma^{-1} \in PS(F)$, yielding that $PS(F)$ is a group. We now prove (ii). Take $F' = r_{\delta(A)}(F)$, $\sigma \in PS(F)$, so $\sigma(F) = r_{\delta(S)}(F)$ for some $S \subseteq [1, n]$, and set $\tau = \sigma^{-1}$; then $\sigma(F') = \sigma r_{\delta(A)}(F) = r_{\delta(\tau(A))}\sigma(F) = r_{\delta(\tau(A))}r_{\delta(S)}r_{\delta(A)}(F') = r_{\delta(\tau(A)\triangle S\triangle A)}(F')$ and so $\sigma \in PS(F')$. Hence, $PS(F) \subseteq PS(F')$ and the reverse inclusion holds similarly. The relation in (ii) then follows easily. □

The group $PS(F)$ is called the group of *double symmetries* of the facet $F$. So we can refine the description of $O(F)$, namely

$$O(F) = \bigcup_{0\le i\le\nu(F)-1} O_p(F_i),$$

with each $O_p(F_i)$ being partitioned into the set $O_{ps}(F_i)$ of cardinality $|PS(F)|/|\operatorname{Aut}(F_i)|$ and its complement $O_p(F_i) - O_{ps}(F_i)$ of cardinality $(n! - |PS(F)|)/|\operatorname{Aut}(F_i)|$.

The following is an observation which might help us find the group of double symmetries $PS(F)$ of a facet $F$. Let $v \cdot x \le v_0$ be an inequality supporting $F$. Then $\sigma \in PS(F)$ if and only if, for some $S \subseteq [1, n]$, the inequalities $\sigma(v)\cdot x \le v_0$ and $v^S \cdot x \le v_0 - v\cdot\delta(S)$ support the same facet, i.e., $v \cdot \delta(S) = 0$ and $v^S = \sigma(v)$, implying $v_{\sigma(i)\sigma(j)} = v_{ij}$ or $-v_{ij}$ for all $1 \le i < j \le n$. Let $G(v)$ denote the supporting graph of $v$, i.e., $G(v)$ is the weighted graph on nodes $[1, n]$ with edges the pairs $(i, j)$ such that $v_{ij} \ne 0$ and edge $(i, j)$ is assigned weight $v_{ij}$. Similarly, $G(|v|)$ is the weighted graph having edge weights $|v_{ij}|$ and we set $I(v, \sigma) = \{(i, j)\colon v_{\sigma(i)\sigma(j)} = -v_{ij}\}$. Then we have

(4.6) $\sigma \in PS(F)$ if and only if $(\sigma \in \operatorname{Aut}(G(|v|))$ and $I(v, \sigma)$ is a cut such that $v \cdot I(v, \sigma) = 0)$.

As an application, we describe the group of double symmetries of two examples of facets, $F_1 = Par_7$ and $F_2 = Gr_7$, of $P_7$. They are supported by inequalities of the form $v_i \cdot x \leq 0$, where the supporting graphs of $v_i$, $i = 1, 2$, are shown in Figures 1 and 2, respectively. (In the figures a plain line indicates weight 1 on the edge, a dashed line indicates weight $-1$, and a double dashed line indicates weight $-2$.) Note that $G(|v_1|)$ is $K_7 - C_7$, namely the complement of the cycle $(0, 3, 1, 2', 2, 1', 3', 0)$. Therefore, from (4.6), $PS(F_1) \subseteq \mathrm{Aut}(K_7 - C_7) = D_7$, the dihedral group of order 7. The group $D_7$ has two generators. The first generator is the reflection $\sigma = (11')(22')(33')$, which, in fact, leaves $v_1$ invariant and thus $\sigma \in \mathrm{Aut}(F_1) \subseteq PS(F_1)$, and the second generator is the permutation $\rho = (0 \to 3 \to 1 \to 2' \to 2 \to 1' \to 3' \to 0)$ for which $I(v_1, \rho) = \delta(\{2'\})$ and so, from (4.6), $\rho \in PS(F_1)$. Therefore, for facet $F_1 = Par_7$, $PS(F_1) = D_7$. Using (4.6), one checks easily that $PS(F_2)$ is the group generated by the three involutions $(13)(24)(56)$, $(14)(23)(56)$, and $(12)(34)$ which all belong to $\mathrm{Aut}(F_2)$, i.e., $PS(F_2) = \mathrm{Aut}(F_2) \approx \mathrm{Sym}(2) \times \mathrm{Sym}(2) \times \mathrm{Sym}(2)$ for facet $F_2 = Gr_7$.

## References


[BG] F. Barahona, and M. Grötschel, *On the cycle polytope of a binary matroid*, J. Combin. Theory Ser. B **40** (1986), 40–62.

[BM] F. Barahona and A. R. Mahjoub, *On the cut polytope*, Math. Programming **36** (1986), 157–173.

[B] M. Berger, *Geometrie* 3: *convexes et polytopes, polyèdres réguliers, aires et volumes*, Fernand Nathan, Cedic, 1978.

[BCN] A. E. Brouwer, A. M. Cohen, and A. Neumaier, *Distance regular graphs*, Springer-Verlag, 1989.

[CR] S. Chopra and M. R. Rao, *The partition problem*, I, II, 1989 (submitted).

[CRS] M. Conforti, M. R. Rao, and A. Sassano, *The equipartition polytope*. I, II, Report no. 194–195, IASI-CNR, Rome 1987; Math. Programming **49**, no. 1 (1991) (to appear).

[CM] H. S. M. Coxeter and W. O. J. Moser, *Generators and relations for discrete groups*, Springer-Verlag, 1972.

[DDL] C. De Simone, M. Deza, and M. Laurent, *Collapsing and lifting for the cut cone*, Report no. 265, IASI-CNR, Rome, 1989.

[D] M. Deza, *Matrices de formes quadratiques non négatives pour des arguments binaires*, C. R. Acad. Sci. Paris **277** (1973), 873–875.

[DFL] M. Deza, K. Fukuda, and M. Laurent, *The inequicut cone*, Report no. 89–04, University of Tsukuba, GSSM, Tokyo, 1989.

[DGL] M. Deza, M. Grötschel, and M. Laurent, *Clique-web facets for multicut polytopes*, Report no. 186, Schwerpunktprogramm der Deutschen Forschungsgemeinschaft, Universität Augsburg, 1989.

[DL1] M. Deza and M. Laurent, *Facets for the cut cone*. I, research memorandum RMI 88–13, University of Tokyo, 1988, to appear in Mathematical Programming.

[DL2] ——, *New results on facets of the cut cone*, Report no. B-227, Tokyo Institute of Technology, Tokyo, 1989; Proc. Conf. Integer Programming and Combinatorial Optimization, University of Waterloo, Canada, 1990.

**[DL3]** ——, *The even and odd cut polytopes*, Report no. B-231, Tokyo Institute of Technology, Tokyo, 1990.

**[P]** M. Padberg, *The boolean quadric polytope: some characteristics, facets and relatives*, Math. Programming **45** (1989), 139–172.

CNRS, Université Paris VII, 17, Passage de l'Industrie, 75010 Paris, France
CEMI, Academy of Sciences of USSR, Krasikova 32, Moscow 117418, USSR

CNRS, Lamsade, Université Paris Dauphine, Place Du Maréchal de Lattre de Tassigny, 75775 Paris, Cedex 16, France
*E-mail address*: or907 at dbnuor1.earn

DIMACS Series in Discrete Mathematics
and Theoretical Computer Science
Volume 4, 1991

Applied Geometry
and Discrete Mathematics
THE VICTOR KLEE FESTSCHRIFT

# Complete Descriptions of Small Multicut Polytopes

M. DEZA, M. GRÖTSCHEL, AND M. LAURENT

ABSTRACT. We introduce various new classes of polytopes that are associated with certain cut, equicut, multicut, equimulticut, and balanced multicut problems on graphs. We have computed complete and nonredundant descriptions of these polytopes numerically for the complete graphs of order 4 and 5, and we list and classify in this paper the facet-defining inequalities that we found. Quite a number of new classes of such inequalities arose. We generalize some of the inequalities and prove that they define facets for certain of these polytopes, but many are still waiting for a proper understanding.

## 1. Introduction

The area of polyhedral investigations of certain cut (or equivalently partitioning) problems has received considerable attention recently. The reason is that many real world problems can be formulated as cut (or partitioning) problems and that cutting plane methods that are based on polyhedral results have shown to be highly efficient algorithms for the solution of these problems.

Without aiming at completeness we mention a few references that corroborate our statements. Long lists of practical problems that can be phrased as certain cut problems can be found, for instance, in **[W, GW1, BJR, BGJR]**. Polyhedral results on various cut, equicut, or multicut polytopes are contained, for example, in **[BMa, DL, CRS, DFL, W, GW2, CR1, CR2]**. Reports about the computational experience with cutting plane algorithms that are based on such polyhedral investigations are, for instance, **[BGJR, BJR, BM, GW1]**.

There are many more interesting practical problems that can be modelled as certain cut problems, in particular, a number of questions that are addressed in clustering and qualitative data analysis. The aim of this paper is to introduce some of the cut problems that arise in these applications, to

1980 *Mathematics Subject Classification* (1985 *Revision*). Primary 52A25, 90C27.


1052-1798/91 $1.00 + $.25 per page

define the associated polyhedra, and to present complete and nonredundant descriptions of these polyhedra for small dimensions. These linear descriptions have been calculated by means of a computer program. We classify the facet-defining inequalities that we found. Some of these belong to well-known classes of inequalities, but quite a number are new and are waiting for proper understanding and generalization.

That is, the main purpose of this paper is to provide instructive material (in particular many interesting classes of facets) for those polyhedral combinatorialists who are working in the area of cut polytopes. We believe that starting with these numerically computed classes of inequalities large new classes of facet-defining inequalities may be found and we hope that these inequalities will help to solve practical problems by means of cutting plane methods.

The paper is organized as follows. The polytopes we consider are introduced in §2. In §3 we discuss these polytopes for the complete graph on five nodes and give a complete and nonredundant description of each of these. A list and partial classification of all the classes of facets that we determined can be found in §4 together with a large table summarizing the results.

In §5 we report about some observations we made by analyzing the material of §4 and give a few theoretical explanations. In particular, we generalize some of the inequalities we found and prove that they define facets for certain multicut polytopes. Complete and nonredundant descriptions of the multicut polytopes for the complete graph on four nodes are provided in §6.

## 2. Multicut and balanced multicut polytopes

Let $G = (V, E)$ be an (undirected) graph with node set $V$ and edge set $E$. The number $n$ will always denote the cardinality of $V$. If $S \subseteq V$ we denote the set of edges of $E$ with both endnodes in $S$ by $E(S)$, i.e., $E(S) = \{uv \in E | u, v \in S\}$. It is customary to denote the cut induced by $S$, i.e., the set $\{uv \in E | u \in S, v \in V \setminus S\}$, by $\delta(S)$. In order to have smooth notation available for more general types of cuts we do not use this symbol and introduce the following variation. Let $S_1, \dots, S_k$ be a partition of $V$, i.e., $S_i \cap S_j = \emptyset$ for $i \neq j$, $S_i \neq \emptyset$ for $i = 1, \dots, k$, and $\bigcup_{i=1}^k S_i = V$. Then

$$\delta(S_1, \dots, S_k) := \{uv \in E | \exists i, j \text{ with } i \neq j \text{ such that } u \in S_i \text{ and } v \in S_j\}.$$

We call the edge set $\delta(S_1, \dots, S_k)$ a *multicut* and the sets $S_1, \dots, S_k$ the *shores* of the multicut. If we want to stress that the partition of $V$ consists of exactly $k$ sets (shores) we will call $\delta(S_1, \dots, S_k)$ a $k$*-multicut* or just a $k$*-cut*. There is only one 1-cut, namely $\delta(V) = \emptyset$, and one $n$-cut, namely $\delta(\{v_1\}, \dots, \{v_n\}) = E$. The (usual) cuts are our $k$-cuts with $k \in \{1, 2\}$.

The most intensively studied optimization problem in connection with these problems certainly is the max-cut problem (i.e., given weights $c_e$ for all $e \in E$ find a $k$-cut with $k \in \{1, 2\}$ of maximum weight; see, e.g.,

[**BMa**]). Also, the problem of finding a multicut of maximum weight is of particular practical interest. This problem is equivalent to the clique partitioning problem since a clique partitioning (in the complete graph $K_n$) is nothing but the complement $E \setminus \delta(S_1, \dots, S_k)$ of a multicut (see [**GW1, GW2**]).

There are further types of multicuts of practical interest.

DEFINITION 1. Let $G = (V, E)$ be a graph and $s$ be an integer with $0 \le s \le |V| - 1$. A multicut $\delta(S_1, \dots, S_k)$, where $k \ge 2$, is called *s-balanced* if

$$||S_i| - |S_j|| \le s \quad \text{for } 1 \le i < j \le k.$$

For two sets $S_i, S_j$ the number $||S_i| - |S_j||$ is called their *discrepancy.*

To turn optimization problems for multicuts into linear programs and questions about multicuts into questions about polyhedra we introduce polyhedra associated with certain sets of multicuts.

Let $G = (V, E)$ be a graph and $\mathbf{R}^E$ the vector space of vectors $x = (x_e)_{e \in E}$, where the components of $x$ are indexed by the edges in $E$. For every subset $F$ of $E$, we define its *incidence vector* $\chi^F \in \mathbf{R}^E$ by $\chi^F_e = 1$ if $e \in F$ and $\chi^F_e = 0$ if $e \notin F$. We are particularly interested in incidence vectors $\chi^{\delta(S_1, \dots, S_k)}$ of multicuts $\delta(S_1, \dots, S_k)$. To be able to vary names a bit we will also call $\chi^{\delta(S_1, \dots, S_k)}$ a multicut vector or $k$-cut vector. We will now introduce the multicut polyhedra considered in this paper.

DEFINITION 2. Let $G = (V, E)$ be a graph with $n$ nodes. Let $k$ and $s$ be integers with $1 \le k \le n$ and $0 \le s \le n - 1$. Set

$$\begin{aligned}
\mathrm{MC}(G) &:= \mathrm{conv}\{\chi^{\delta(S_1, \dots, S_h)} \in \mathbf{R}^E | \delta(S_1, \dots, S_h) \text{ a multicut of } G\},\\
\mathrm{MC}^{\le}_k(G) &:= \mathrm{conv}\{\chi^{\delta(S_1, \dots, S_h)} \in \mathbf{R}^E | \delta(S_1, \dots, S_h) \text{ an } h\text{-cut of } G \\
&\qquad\qquad \text{with } h \le k\},\\
\mathrm{MC}^{\ge}_k(G) &:= \mathrm{conv}\{\chi^{\delta(S_1, \dots, S_h)} \in \mathbf{R}^E | \delta(S_1, \dots, S_h) \text{ an } h\text{-cut of } G \\
&\qquad\qquad \text{with } h \ge k\},\\
\mathrm{MC}^{=}_k(G) &:= \mathrm{conv}\{\chi^{\delta(S_1, \dots, S_h)} \in \mathbf{R}^E | \delta(S_1, \dots, S_k) \text{ a } k\text{-cut of } G\},\\
s\text{-}\mathrm{BMC}(G) &:= \mathrm{conv}\{\chi^{\delta(S_1, \dots, S_h)} \in \mathbf{R}^E | \delta(S_1, \dots, S_h) \text{ an } s\text{-balanced} \\
&\qquad\qquad \text{multicut of } G\},\\
s\text{-}\mathrm{BMC}^{\le}_k(G) &:= \mathrm{conv}\{\chi^{\delta(S_1, \dots, S_h)} \in \mathbf{R}^E | \delta(S_1, \dots, S_h) \text{ an } s\text{-balanced} \\
&\qquad\qquad h\text{-cut of } G \text{ with } h \le k\},\\
s\text{-}\mathrm{BMC}^{\ge}_k(G) &:= \mathrm{conv}\{\chi^{\delta(S_1, \dots, S_h)} \in \mathbf{R}^E | \delta(S_1, \dots, S_h) \text{ an } s\text{-balanced} \\
&\qquad\qquad h\text{-cut of } G \text{ with } h \ge k\},\\
s\text{-}\mathrm{BMC}^{=}_k(G) &:= \mathrm{conv}\{\chi^{\delta(S_1, \dots, S_h)} \in \mathbf{R}^E | \delta(S_1, \dots, S_h) \text{ an } s\text{-balanced} \\
&\qquad\qquad k\text{-cut of } G\}.
\end{aligned}$$

In case $G$ is the complete graph $K_n$ we will write $\mathrm{MC}(n)$, etc. instead of $\mathrm{MC}(K_n)$, etc. We call $\mathrm{MC}(G)$ *the multicut polytope* and $s$-$\mathrm{BMC}(G)$ the *$s$-balanced multicut polytope* of $G$. $\mathrm{MC}_k^{\leq}(G)$ is called the *$^{\leq}k$-cut polytope*, $\mathrm{MC}_k^{\geq}(G)$ the *$^{\geq}k$-cut polytope*, $\mathrm{MC}_k^{=}(G)$ the *$k$-cut polytope*, $s$-$\mathrm{BMC}_k^{\leq}(G)$ the *$s$-balanced $^{\leq}k$-cut polytope*, $s$-$\mathrm{BMC}_k^{\geq}(G)$ the *$s$-balanced $^{\geq}k$-cut polytope*, and $s$-$\mathrm{BMC}_k^{=}(G)$ the *$s$-balanced $k$-cut polytope* of $G$. In case it is not necessary to be precise we will simply speak of a multicut polytope $\mathrm{MC}_k^{\leq}(G)$, etc.

Note that the standard *cut polytope* (i.e., the convex hull of all incidence vectors of (usual) cuts) of a graph is our $^{\leq}2$-cut polytope $\mathrm{MC}_2^{\leq}(G)$, and the *equicut polytope* (i.e., the convex hull of all incidence vectors of cuts, where both shores differ in cardinality by at most one) of a graph is our 1-balanced $^{\leq}2$-cut polytope 1-$\mathrm{BMC}_2^{\leq}(G)$.

The polytopes most intensively studied in the literature are

- the cut polytope $\mathrm{MC}_2^{\leq}(G)$ (see, e.g., **[BMa, DL]** and the references therein),
- the multicut polytope $\mathrm{MC}(G)$ (see **[CR1, DGL, GW2, W]**),
- the equicut polytope 1-$\mathrm{BMC}_2^{\leq}(G)$ (see **[CRS, DFL]**),
- the $^{\geq}k$-cut and the $^{\leq}k$-cut polytopes $\mathrm{MC}_k^{\leq}(G)$ and $\mathrm{MC}_k^{\geq}(G)$ (see **[CR1, CR2, DGL]**),

while not much work has been done yet on all the other polytopes. To provide examples and ideas for generalizations we give complete descriptions of all polytopes mentioned above for the complete graphs on 4 and 5 nodes. Let us mention that similar work has been done for small equicut and inequicut cones in **[DFL]**.

## 3. The multicut polytopes for $n = 5$

We will now discuss the multicut polytopes introduced in Definition 2 for the complete graph $K_5 = (V, E)$ and provide complete and nonredundant descriptions of all polytopes of interest. For the complete graph $K_4$, similar lists can be found in §6.

By making all possible combinations of the parameters one can see that there are almost 100 cases to be considered for $n = 5$. However, some can be easily derived from others for theoretical reasons and, due to the small dimension, quite a number are trivial or very simple. We will now rule out all trivial cases, discuss the simple cases, and list the interesting ones that will be investigated later.

There is no obvious order of the polytopes since there are many containment relations. We follow here and later the following lexicographic order of the polytopes. We always treat $\mathrm{MC}(n)$ first. Then the multicut polytopes $\mathrm{MC}_k^*(n)$ are studied. Here and further we use the symbol "$*$" to denote one of the symbols "$\leq$", "$\geq$", "$=$". The parameter $k$ specifying a bound on shores is considered in increasing order, "$\leq$" is treated before "$\geq$" and this, in turn, before "=". For the $s$-balanced multicut polytopes we treat the polytopes with largest $s$ first and, within this class, follow the lexicographic rules for $\mathrm{MC}_k^*(n)$. Then we decrease $s$ by 1 and continue this way.

To be able to give short definitions of all multicut polytopes we now list and number all 52 multicut vectors for $n = 5$. We simply specify the partitions $S_1, \dots, S_k$ from which the incidence vectors $\chi^{\delta(S_1, \dots, S_k)} \in \mathbf{R}^E$ can be derived and number them as follows.

| | | |
|---|---|---|
| 1 : 12345 | 2 : 1, 2345 | 3 : 2, 1345 |
| 4 : 3, 1245 | 5 : 4, 1235 | 6 : 5, 1234 |
| 7 : 12, 345 | 8 : 13, 245 | 9 : 14, 235 |
| 10 : 15, 234 | 11 : 23, 145 | 12 : 24, 135 |
| 13 : 25, 134 | 14 : 34, 125 | 15 : 35, 124 |
| 16 : 45, 123 | 17 : 1, 2, 345 | 18 : 1, 3, 245 |
| 19 : 1, 4, 235 | 20 : 1, 5, 234 | 21 : 2, 3, 145 |
| 22 : 2, 4, 135 | 23 : 2, 5, 134 | 24 : 3, 4, 125 |
| 25 : 3, 5, 124 | 26 : 4, 5, 123 | 27 : 1, 23, 45 |
| 28 : 1, 24, 35 | 29 : 1, 25, 34 | 30 : 2, 13, 45 |
| 31 : 2, 14, 35 | 32 : 2, 15, 34 | 33 : 3, 12, 45 |
| 34 : 3, 14, 25 | 35 : 3, 15, 24 | 36 : 4, 12, 35 |
| 37 : 4, 13, 25 | 38 : 4, 15, 23 | 39 : 5, 12, 34 |
| 40 : 5, 13, 24 | 41 : 5, 14, 23 | 42 : 1, 2, 3, 45 |
| 43 : 1, 2, 4, 35 | 44 : 1, 2, 5, 34 | 45 : 1, 3, 4, 25 |
| 46 : 1, 3, 5, 24 | 47 : 1, 4, 5, 23 | 48 : 2, 3, 4, 15 |
| 49 : 2, 3, 5, 14 | 50 : 2, 4, 5, 13 | 51 : 3, 4, 5, 12 |
| 52 : 1, 2, 3, 4, 5 | | |

To explain the notation, consider number 29. The partition is $S_1 = \{1\}$, $S_2 = \{2, 5\}$, $S_3 = \{3, 4\}$ which gives the 3-cut vector $\chi^{\delta(S_1, S_2, S_3)} = (\chi_{12}, \chi_{13}, \dots, \chi_{45}) = (1, 1, 1, 1, 1, 1, 0, 0, 1, 1)$, i.e., this is our vector number 29. To shorten notation, we will now write equations like $\mathrm{MC}_3^{\geq}(5) = \mathrm{conv}\{17, 18, \dots, 52\}$ which means that in order to get the polytope $\mathrm{MC}_3^{\geq}(5)$ we have to take all convex combinations of the multicut vectors with numbers $17, 18, \dots, 52$.

After all these preparations we are now going to discuss the multicut polytopes introduced in Definition 2. For each polytope, we give the number $v$ of vertices and the number $f$ of facets and list the classes of facets that are needed to describe the polytope completely and nonredundantly. The classes of facets are numbered (1), (2), ..., (47). These are defined and discussed in §4. Most of these polytopes have dimension $|E| = 10$. We mention whenever a polytope has dimension less than 10 or is particular in some way.

$$\mathrm{MC}(5) = \mathrm{conv}\{1, 2, \dots, 52\}, \quad v = 52, \quad f = 242,$$

facets : (1), (8), (9), (10), (11), (13), (20), (44), (45), (46).

This polytope, of course, contains all other multicut polytopes. Since the containment relations between the polytopes are quite obvious we do not make any further remarks on this issue. All inequalities valid for $\mathrm{MC}(n)$ are also valid for all other multicut polytopes.

Polytopes of the type $\mathrm{MC}_1^*(n)$ are irrelevant. $\mathrm{MC}_1^{\geq}(n) = \mathrm{MC}(n)$ and $\mathrm{MC}_1^{\leq}(n)$, $\mathrm{MC}_1^{=}(n)$ consist of the single point 0.

$$\mathrm{MC}_2^{\leq}(5) = \mathrm{conv}\{1, 2, \dots, 16\}, \quad v = 16, \quad f = 56,$$
$$\text{facets}: (4), (8), (9), (12), (15).$$

This is the cut polytope of $K_5$. In fact, a complete description of the cut polytope is known for every complete graph up to seven nodes (for details see [**DL**]).

$$\mathrm{MC}_2^{\geq}(5) = \mathrm{conv}\{2, 3, \dots, 52\}, \quad v = 51, \quad f = 264,$$
$$\text{facets}: (1), (8), (9), (10), (11), (13), (17), (20), (27), (44), (45), (46).$$

This polytope is obtained from $\mathrm{MC}(5)$ by deleting point 1 (the zero vector). The neighbors of zero are, however, not on a common hyperplane and so $\mathrm{MC}_2^{\geq}(5)$ cannot be derived from $\mathrm{MC}(5)$ by adding one "cutting plane". Compared with $\mathrm{MC}(5)$ no facet-defining inequality of $\mathrm{MC}(5)$ disappears, but the two new classes (17) and (27) appear.

$$\mathrm{MC}_2^{=}(5) = \mathrm{conv}\{2, 3, \dots, 16\}, \quad v = 15, \quad f = 68,$$
$$\text{facets}: (4), (8), (12), (15), (17), (27).$$

$\mathrm{MC}_2^{=}(5)$ arises from the cut polytope $\mathrm{MC}_2^{\leq}(5)$ by removing one vertex, the zero vector. The class (9) of certain hypermetric facets disappears and two new classes of inequalities, (17) and (27), come up. Observe the difference to the transition from $\mathrm{MC}(5)$ to $\mathrm{MC}_2^{\geq}(5)$. Here also the zero vector is removed, facets (17) and (27) appear, but (9) remains a class of facets of $\mathrm{MC}_2^{\geq}(5)$.

$$\mathrm{MC}_3^{\leq}(5) = \mathrm{conv}\{1, 2, \dots, 41\}, \quad v = 41, \quad f = 333,$$
$$\text{facets}: (1), (3), (8), (9), (10), (11), (14), (20), (38), (41), (43), (44), (45), (46),$$
$$\mathrm{MC}_3^{\geq}(5) = \mathrm{conv}\{17, 18, \dots, 52\}, \quad v = 36, \quad f = 218,$$
$$\text{facets}: (1), (6), (8), (11), (13), (16), (23), (24), (25), (30), (31), (40),$$
$$\mathrm{MC}_3^{=}(5) = \mathrm{conv}\{17, 18, \dots, 41\}, \quad v = 25, \quad f = 99,$$
$$\text{facets}: (1), (3), (6), (14), (16), (30), (31), (38), (40),$$
$$\mathrm{MC}_4^{\leq}(5) = \mathrm{conv}\{1, 2, \dots, 51\}, \quad v = 51, \quad f = 243,$$
$$\text{facets}: (1), (2), (8), (9), (10), (11), (13), (20), (44), (45), (46).$$

$\mathrm{MC}_4^{\leq}(5)$ arises from MC(5) by deleting vertex 52, i.e., the all-ones vector $\chi^E$. Using the adjacency characterization on the clique partitioning polytope of [W], which, by complementation, gives an adjacency characterization of the vertices of the multicut polytope $\mathrm{MC}(n)$, one can show that the neighbors of $\chi^E$ are the 4-cut vectors $42, 43, \ldots, 51$. These points lie on the hyperplane $\{x|x(E) = 9\}$. So it follows that $\mathrm{MC}_4^{\leq}(5) = \mathrm{MC}(5) \cap \{x|x(E) \leq 9\}$. (In fact, this argument holds for general $n$, see §5.) It may be, though, that some inequalities defining a facet of MC(5) are redundant with respect to $\mathrm{MC}_4^{\leq}(5)$. But it is easy to see that $\chi^E$ is only contained in the facets defined by the upper bound constraints (1) $x_e \leq 1$ which also define facets of $\mathrm{MC}_4^{\leq}(5)$ (see §5). So $\mathrm{MC}_4^{\leq}(5)$ has one vertex less and one facet more than MC(5). (For that reason we will not discuss $\mathrm{MC}_4^{\leq}(5)$ further in §4.)

$$\mathrm{MC}_4^{\geq}(5) = \mathrm{conv}\{42, 43, \ldots, 52\}, \quad v = 11, \quad f = 11.$$

$\mathrm{MC}_4^{\geq}(5)$ is a full-dimensional simplex in $\mathbf{R}^E$. The facets are the upper bounds (1) $x_e \leq 1$ and the cardinality constraint $x(E) \geq 9$. $\mathrm{MC}_{n-1}^{\geq}(n)$ is a simplex for any $n$ (see §5). We do not discuss this polytope $\mathrm{MC}_4^{\geq}(5)$ further.

$$\mathrm{MC}_4^{=}(5) = \mathrm{conv}\{42, 43, \ldots, 51\}, \quad v = 10, \quad f = 10.$$

$\mathrm{MC}_4^{=}(5)$ is a 9-dimensional simplex in $\mathbf{R}^E$ arising from $\mathrm{MC}_4^{\geq}(5)$ by turning $x(E) \geq 9$ into an equation. We will not consider $\mathrm{MC}_4^{=}(5)$ in the sequel any more.

Based on this analysis, the multicut polytopes of interest are MC(5), $\mathrm{MC}_2^{\leq}(5)$, $\mathrm{MC}_2^{\geq}(5)$, $\mathrm{MC}_2^{=}(5)$, $\mathrm{MC}_3^{\leq}(5)$, $\mathrm{MC}_3^{\geq}(5)$, $\mathrm{MC}_3^{=}(5)$.

We now turn our attention to the balanced multicut polytopes $s$-$\mathrm{BMC}_k^*(5)$.

Note that for any $k$-cut $\delta(S_1, \ldots, S_k)$ in $K_5$, $k \geq 2$, the discrepancy $||S_i| - |S_j||$ of any two sets $S_i \neq S_j$ is never larger than 3. So, any polytope 4-$\mathrm{BMC}_k^*(5)$ is the same as the polytope 3-$\mathrm{BMC}_k^*(5)$ for $k = 2, \ldots, 5$. Moreover, there is no $k$-cut $\delta(S_1, \ldots, S_k)$, $k \in \{2, \ldots, 4\}$, in $K_5$ with discrepancy $||S_i| - |S_j|| = 0$ for all $i \neq j$. Thus the polytopes 0-$\mathrm{BMC}_k^*(5)$, $k = 2, 3, 4$, are empty. The polytope 0-$\mathrm{BMC}_5^*(5)$ consists of the single point $\chi^E$ and is also irrelevant. Furthermore, any polytope $s$-$\mathrm{BMC}(n)$, $s \in \{0, 1, \ldots, n-1\}$, is equal to the polytope $s$-$\mathrm{BMC}_{n-1}^{\leq}(n)$, so we do not consider the former polytopes in the sequel.

Thus the remaining range of parameters of interest is $s = 3, 2, 1$ and $k = 2, 3, 4, 5$. We will analyze these cases. It will turn out that some of the polytopes are equal to polytopes already considered. In these cases we state this fact without discussing the polytopes further.

$$\begin{aligned}
&\text{3-BMC}_2^{\leq}(5) = \text{MC}_2^{\leq}(5),\\
&\text{3-BMC}_2^{\geq}(5) = \text{MC}_2^{\geq}(5),\\
&\text{3-BMC}_2^{=}(5) = \text{MC}_2^{=}(5),\\
&\text{3-BMC}_3^{\leq}(5) = \text{conv}\{2, 3, \ldots, 41\}, \quad v = 40, \quad f = 355,\\
&\qquad = \text{facets}: (1), (3), (8), (9), (10), (11), (14), (17),\\
&\qquad (20), (27), (38), (41), (43), (44), (45), (46).
\end{aligned}$$

This polytope has the second largest number of facets among all polytopes investigated in this paper.

$$\begin{aligned}
&\text{3-BMC}_3^{\geq}(5) = \text{MC}_3^{\geq}(5),\\
&\text{3-BMC}_3^{=}(5) = \text{MC}_3^{=}(5),\\
&\text{3-BMC}_4^{\leq}(5) = \text{conv}\{2, 3, \ldots, 51\}, \quad v = 50, \quad f = 265,\\
&\qquad \text{facets}: (1), (2), (8), (9), (10), (11), (13),\\
&\qquad (17), (20), (27), (44), (45), (46).
\end{aligned}$$

$\text{3-BMC}_4^{\leq}(5)$ arises from $\text{MC}(5)$ by deleting the "top vertex" $\chi^E = \mathbf{1}$ and the "bottom vertex" $\chi^{\varnothing} = 0$. Analogously, $\text{3-BMC}_4^{\leq}(5)$ can be derived from $\text{MC}_4^{\leq}(5)$ by deleting the zero vector, or from $\text{MC}_2^{\geq}(5)$ by deleting $\chi^E$. As the computation shows (and one can verify theoretically) $\text{3-BMC}_4^{\leq}(5)$ inherits all facets of $\text{MC}_2^{\geq}(5)$, $\text{MC}_4^{\leq}(5)$, and of $\text{MC}(5)$.

$$\begin{aligned}
\text{3-BMC}_4^{\geq}(5) &= \text{MC}_4^{\geq}(5),\\
\text{3-BMC}_4^{=}(5) &= \text{MC}_4^{=}(5),\\
\text{2-BMC}_2^{\leq}(5) &= \text{1-BMC}_2^{\leq}(5).
\end{aligned}$$

Note that the 2-cuts $\delta(S, V \setminus S)$ in $K_5$ have discrepancy $||S| - |V \setminus S||$ equal either to one or to three. For this reason the last equation above holds and thus we discuss this polytope later.

$$\begin{aligned}
\text{2-BMC}_2^{\geq}(5) &= \text{conv}\{7, 8, \ldots, 52\}, \quad v = 46, \quad f = 308,\\
&\quad \text{facets}: (1), (5), (8), (11), (13), (18), (20), (21),\\
&\quad (23), (29), (32), (35), (36), (39), (42),\\
\text{2-BMC}_2^{=}(5) &= \text{1-BMC}_2^{=}(5) \quad (\text{see remark after 2-BMC}_2^{\leq}(5)),\\
\text{2-BMC}_3^{\leq}(5) &= \text{conv}\{7, 8, \ldots, 41\}, \quad v = 35, \quad f = 379,\\
&\quad \text{facets}: (1), (3), (5), (8), (11), (14), (18), (20), (21),\\
&\quad (23), (29), (32), (35), (36), (38), (39), (42), (43).
\end{aligned}$$

This is the polytope with the largest number of facets considered here. There are 18 different classes of facets.

$$\begin{aligned}
&2\text{-}\mathrm{BMC}_3^{\geq}(5) = \mathrm{MC}_3^{\geq}(5)\,,\\
&2\text{-}\mathrm{BMC}_3^{=}(5) = \mathrm{MC}_3^{=}(5)\,,\\
&2\text{-}\mathrm{BMC}_4^{\leq}(5) = \mathrm{conv}\{7\,,\,8\,,\,\dots\,,\,51\}\,, \quad v = 45\,, \quad f = 309\,,\\
&\qquad \text{facets}: (1)\,,\,(2)\,,\,(5)\,,\,(8)\,,\,(11)\,,\,(13)\,,\,(18)\,,\,(20)\,,\\
&\qquad (21)\,,\,(23)\,,\,(29)\,,\,(32)\,,\,(35)\,,\,(36)\,,\,(39)\,,\,(42)\,.
\end{aligned}$$

$2\text{-}\mathrm{BMC}_4^{\leq}(5)$ arises from $2\text{-}\mathrm{BMC}_2^{\geq}(5)$ by deleting the vertex $\chi^E$. All facets of $2\text{-}\mathrm{BMC}_2^{\geq}(5)$ are inherited, and the cardinality constraint (2) $x(E) \leq 9$ appears. The behavior can be explained with the same arguments given in the discussion of the polytope $\mathrm{MC}_4^{\leq}(5)$.

$$\begin{aligned}
&2\text{-}\mathrm{BMC}_4^{\geq}(5) = \mathrm{MC}_4^{\geq}(5)\,,\\
&2\text{-}\mathrm{BMC}_4^{=}(5) = 1\text{-}\mathrm{BMC}_4^{=}(5)\,.
\end{aligned}$$

There is no 4-cut in $K_5$ where two shores have discrepancy 2. So all 2-balanced 4-cuts are in fact 1-balanced, and thus we discuss this polytope later.

$$\begin{aligned}
&1\text{-}\mathrm{BMC}_2^{\leq}(5) = \mathrm{conv}\{7\,,\,8\,,\,\dots\,,\,16\}\,, \quad v = 10\,, f = 10\,,\\
&\qquad \text{facets}: (12) \text{ or equivalently } (29), \text{ dimension } = 9\,.
\end{aligned}$$

This polytope is a 9-dimensional simplex in $\mathbf{R}^E$. It lies on the hyperplane $\{x|x(E) = 6\}$. All inequalities (12) and (29) are facet-defining but the two classes are equivalent in this case. $1\text{-}\mathrm{BMC}_2^{\leq}(5)$ is known in the literature as the *equicut polytope*.

$$\begin{aligned}
&1\text{-}\mathrm{BMC}_2^{\geq}(5) = \mathrm{conv}\{7\,,\,8\,,\,\dots\,,\,16\,,\,27\,,\,28\,,\,\dots\,,\,52\}\,, \quad v = 36\,, \quad f = 113\,,\\
&\qquad \text{facets}: (1)\,,\,(5)\,,\,(8)\,,\,(20)\,,\,(26)\,,\,(33)\,,\,(34)\,,\,(37)\,,\\
&1\text{-}\mathrm{BMC}_2^{=}(5) = 1\text{-}\mathrm{BMC}_2^{\leq}(5)\,,\\
&1\text{-}\mathrm{BMC}_3^{\leq}(5) = \mathrm{conv}\{7\,,\,8\,,\,\dots\,,\,16\,,\,27\,,\,28\,,\,\dots\,,\,41\}\,, \quad v = 24\,, \quad f = 169\,,\\
&\qquad \text{facets}: (1)\,,\,(3)\,,\,(5)\,,\,(8)\,,\,(15)\,,\,(20)\,,\,(26)\,,\,(28)\,,\,(34)\,,\,(37)\,,\\
&\qquad (47)\,,\\
&1\text{-}\mathrm{BMC}_3^{\geq}(5) = \mathrm{conv}\{27\,,\,28\,,\,\dots\,,\,52\}\,, \quad v = 26\,, \quad f = 26\,,\\
&\qquad \text{facets}: (1)\,,\,(7)\,,\,(19)\,,\,(22)\,,\\
&1\text{-}\mathrm{BMC}_3^{=}(5) = \mathrm{conv}\{27\,,\,28\,,\,\dots\,,\,41\}\,, \quad v = 15\,, \quad f = 25\,,\\
&\qquad \text{facets}: (1)\,,\,(14) \text{ or equivalently } (22), (19) \text{ or}\\
&\qquad \text{equivalently } (28), \text{ dimension } = 9\,.
\end{aligned}$$

This polytope lies in the hyperplane $x(E) = 9$, the inequalities (19) are in this case equivalent to the inequalities (28), and the class (14) is equivalent to (22).

$$1\text{-}\mathrm{BMC}_4^{\leq}(5) = \mathrm{conv}\{7, 8, \ldots, 16, 27, 28, \ldots, 51\}, \quad v = 35, \quad f = 114,$$
$$\text{facets}: (1), (2), (5), (8), (20), (26), (33), (34), (37).$$

$1\text{-}\mathrm{BMC}_4^{\leq}(5)$ arises from $1\text{-}\mathrm{BMC}_2^{\geq}(5)$ by removing vertex $\chi^E = \mathbf{1}$. As in two previous cases (cf. the discussion behind $\mathrm{MC}_4^{\leq}(5)$) $1\text{-}\mathrm{BMC}_4^{\leq}(5)$ keeps all facets of $1\text{-}\mathrm{BMC}_2^{\geq}(5)$ and only one new facet, (2) $x(E) \leq 9$, appears.

$$1\text{-}\mathrm{BMC}_4^{\geq}(5) = \mathrm{MC}_4^{\geq}(5),$$
$$1\text{-}\mathrm{BMC}_4^{=}(5) = \mathrm{MC}_4^{=}(5).$$

This finishes our discussion of the multicut polytopes for $n = 5$. For convenient reference we list in Table 1 which of the polytopes are equal and which are of further interest.

TABLE 1

| | |
|---|---|
| 1. | $\mathrm{MC}(5)$ |
| 2. | $\mathrm{MC}_2^{\leq}(5) = 3\text{-}\mathrm{BMC}_2^{\leq}(5)$ |
| 3. | $\mathrm{MC}_2^{\geq}(5) = 3\text{-}\mathrm{BMC}_2^{\geq}(5)$ |
| 4. | $\mathrm{MC}_2^{=}(5) = 3\text{-}\mathrm{BMC}_2^{=}(5)$ |
| 5. | $\mathrm{MC}_3^{\leq}(5)$ |
| 6. | $\mathrm{MC}_3^{\geq}(5) = 3\text{-}\mathrm{BMC}_3^{\geq}(5) = 2\text{-}\mathrm{BMC}_3^{\geq}(5)$ |
| 7. | $\mathrm{MC}_3^{=}(5) = 3\text{-}\mathrm{BMC}_3^{=}(5) = 2\text{-}\mathrm{BMC}_3^{=}(5)$ |
| 8. | $3\text{-}\mathrm{BMC}_3^{\leq}(5)$ |
| 9. | $3\text{-}\mathrm{BMC}_4^{\leq}(5)$ |
| 10. | $2\text{-}\mathrm{BMC}_2^{\geq}(5)$ |
| 11. | $2\text{-}\mathrm{BMC}_3^{\leq}(5)$ |
| 12. | $2\text{-}\mathrm{BMC}_4^{\leq}(5)$ |
| 13. | $1\text{-}\mathrm{BMC}_2^{\leq}(5) = 1\text{-}\mathrm{BMC}_2^{=}(5) = 2\text{-}\mathrm{BMC}_2^{\leq}(5) = 2\text{-}\mathrm{BMC}_2^{=}(5)$ |
| 14. | $1\text{-}\mathrm{BMC}_2^{\geq}(5)$ |
| 15. | $1\text{-}\mathrm{BMC}_3^{\leq}(5)$ |
| 16. | $1\text{-}\mathrm{BMC}_3^{\geq}(5)$ |
| 17. | $1\text{-}\mathrm{BMC}_3^{=}(5)$ |
| 18. | $1\text{-}\mathrm{BMC}_4^{\leq}(5)$ |

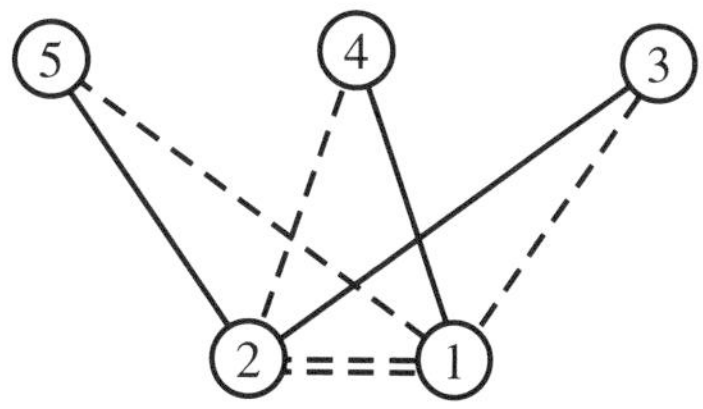


FIGURE 1

The following polytopes will not be considered further.

$$\mathrm{MC}_4^{\leq}(5)$$
$$\mathrm{MC}_4^{\geq}(5) = 3\text{-}\mathrm{BMC}_4^{\geq}(5) = 2\text{-}\mathrm{BMC}_4^{\geq}(5) = 1\text{-}\mathrm{BMC}_4^{\geq}(5),$$
$$\mathrm{MC}_4^{=}(5) = 3\text{-}\mathrm{BMC}_4^{=}(5) = 2\text{-}\mathrm{BMC}_4^{=}(5) = 1\text{-}\mathrm{BMC}_4^{=}(5).$$

## 4. A dictionary of the facets of the multicut polytopes for $n = 5$

We will now list all classes of inequalities that came up in our numerical calculations and that define facets of at least one of our multicut polytopes listed in Table 1. As before, $K_5 = (V, E)$ denotes the complete graph on five nodes and $V = \{1, 2, \ldots, 5\}$. For a convenient description of the inequalities it seems more appropriate to use numbers as indices rather than letters. Instead of saying that the inequality $x_{ij} \leq 1$ defines a facet for every edge $ij \in E$, we will simply state that $x_{12} \leq 1$ defines a facet. This is justified since the property of an inequality to define a facet is preserved under permutation of indices. In other words, writing that $x_{12}+x_{23}+x_{13} \leq 2$ defines a facet means that the inequalities $x_{ij}+x_{jk}+x_{ik} \leq 2$ define facets for all $i, j, k \in V$, $i \neq k \neq j \neq i$. Whenever we list a representative of a class of inequalities we list in brackets the number of permutations of $\{1, \ldots, 5\}$ that lead to different facet-defining inequalities.

We will also present some drawings of the (weighted) support graphs of the inequalities. The convention is that a solid line represents a $+1$-coefficient, two parallel solid lines a $+2$-coefficient, etc., while a broken line represents a $-1$-coefficient and two parallel broken lines represent a $-2$-coefficient, etc. For example, the inequality

$$-2x_{12} - x_{13} + x_{14} - x_{15} + x_{23} - x_{24} + x_{25} \leq 0$$

is depicted by the graph shown in Figure 1.

With each class of inequalities we provide a list of those polytopes for which this class defines facets. We restrict reference to the 18 (interesting) polytopes of Table 1.

In the sequel we will drop the "(5)" in the names of the multicut polytopes to avoid unnecessary symbols, i.e., instead of "$\mathrm{MC}_2^{\geq}(5)$" we will simply write "$\mathrm{MC}_2^{\geq}$" etc.

**(4.1) Upper bounds.** Since all multicut polytopes are 0/1-polytopes, the upper bound inequalities

(1) $$x_{12} \leq 1 \qquad \text{(10 permutations)}$$

are valid for these polytopes. In fact, they define facets for all the 18 polytopes except $\mathrm{MC}_2^{\leq}$, $\mathrm{MC}_2^{=}$, and $1\text{-}\mathrm{BMC}_2^{\leq}$. It is interesting to note that no lower bound inequality $x_{12} \geq 0$ ever defines a facet.

**(4.2) Cardinality constraints.** We call constraints of the type $x(E) \leq u$, where $u$ is some positive integer, an upper cardinality constraint, and similarly we call $x(E) \geq l$ a lower cardinality constraint. Such constraints appear as facet-defining inequalities with different right-hand sides as follows:

(2) $$x(E) \leq 9 \qquad \text{(1 permutation)}$$

is facet-defining for $3\text{-}\mathrm{BMC}_4^{\leq}$, $2\text{-}\mathrm{BMC}_4^{\leq}$, $1\text{-}\mathrm{BMC}_4^{\leq}$;

(3) $$x(E) \leq 8 \qquad \text{(1 permutation)}$$

defines a facet of $\mathrm{MC}_3^{\leq}$, $\mathrm{MC}_3^{=}$, $3\text{-}\mathrm{BMC}_3^{\leq}$, $2\text{-}\mathrm{BMC}_3^{\leq}$, $1\text{-}\mathrm{BMC}_3^{\leq}$;

(4) $$x(E) \leq 6 \qquad \text{(1 permutation)}$$

defines a facet of $\mathrm{MC}_2^{\leq}$, $\mathrm{MC}_2^{=}$. The lower bound

(5) $$x(E) \geq 6 \qquad \text{(1 permutation)}$$

defines a facet of $2\text{-}\mathrm{BMC}_2^{\geq}$, $2\text{-}\mathrm{BMC}_3^{\leq}$, $2\text{-}\mathrm{BMC}_4^{\leq}$, $1\text{-}\mathrm{BMC}_2^{\geq}$, $1\text{-}\mathrm{BMC}_3^{\leq}$, $1\text{-}\mathrm{BMC}_4^{\leq}$;

(6) $$x(E) \geq 7 \qquad \text{(1 permutation)}$$

is facet-defining for $\mathrm{MC}_3^{\geq}$ and $\mathrm{MC}_3^{=}$; and

(7) $$x(E) \geq 8 \qquad \text{(1 permutation)}$$

defines a facet of $1\text{-}\mathrm{BMC}_3^{\geq}$.

**(4.3) Hypermetric inequalities.** We first define some symbols to make notation easier. If $b = (b_1, \dots, b_5)$ is a vector of integers and $x \in \mathbf{R}^E$ then we set

$$Q(b) \cdot x := \sum_{ij \in E} b_i b_j x_{ij}.$$

Moreover, if $\sigma := \sum_{i=1}^{5} b_i$ then the inequality

$$Q(b) \cdot x \leq \tfrac{1}{2}\sigma(\sigma - 1)$$

is called the *hypermetric inequality* associated with $b$. For any integral vector $b$, the hypermetric inequality associated with $b$ is valid for MC and thus for all multicut polytopes considered here (see **[DGL]**). We now list the hypermetric inequalities that come up for $n = 5$:

(8) $$Q(1, 1, -1, 0, 0) \cdot x \leq 0 \qquad \text{(30 permutations)}$$

(8) can be written (in a less sophisticated way) as $x_{12} - x_{13} - x_{23} \leq 0$; this inequality is also known under the name *triangle inequality*. It is facet-defining for all multicut polytopes except $\mathrm{MC}_3^=$, 1-$\mathrm{BMC}_2^{\leq}$, 1-$\mathrm{BMC}_3^{\geq}$, 1-$\mathrm{BMC}_3^=$.

$$(9) \qquad Q(1, 1, 1, -1, -1) \cdot x \leq 0 \qquad \text{(10 permutations)}$$

is facet-defining for MC, $\mathrm{MC}_2^{\leq}$, $\mathrm{MC}_2^{\geq}$, $\mathrm{MC}_3^{\leq}$, 3-$\mathrm{BMC}_3^{\leq}$, 3-$\mathrm{BMC}_4^{\leq}$;

$$(10) \qquad Q(1, 1, 1, 1, -2) \cdot x \leq 1 \qquad \text{(5 permutations)}$$

is facet-defining for MC, $\mathrm{MC}_2^{\geq}$, $\mathrm{MC}_3^{\leq}$, 3-$\mathrm{BMC}_3^{\leq}$, 3-$\mathrm{BMC}_4^{\leq}$;

$$(11) \qquad Q(1, 1, 1, -1, 0) \cdot x \leq 1 \qquad \text{(20 permutations)}$$

is facet-defining for MC, $\mathrm{MC}_2^{\geq}$, $\mathrm{MC}_3^{\leq}$, $\mathrm{MC}_3^{\geq}$, 3-$\mathrm{BMC}_3^{\leq}$, 3-$\mathrm{BMC}_4^{\leq}$, 2-$\mathrm{BMC}_2^{\geq}$, 2-$\mathrm{BMC}_3^{\leq}$, 2-$\mathrm{BMC}_4^{\leq}$;

$$(12) \qquad Q(1, 1, 1, 0, 0) \cdot x \leq 2 \qquad \text{(10 permutations)}$$

is facet-defining for $\mathrm{MC}_2^{\leq}$, $\mathrm{MC}_2^=$, 1-$\mathrm{BMC}_2^{\leq}$;

$$(13) \qquad Q(1, 1, 1, 1, -1) \cdot x \leq 3 \qquad \text{(5 permutations)}$$

is facet-defining for MC, $\mathrm{MC}_2^{\geq}$, $\mathrm{MC}_3^{\geq}$, 3-$\mathrm{BMC}_4^{\leq}$, 2-$\mathrm{BMC}_2^{\geq}$, 2-$\mathrm{BMC}_4^{\leq}$.

**(4.4) $Q$-inequalities.** It turns out that the left-hand sides of some inequalities we found can be written in the form $Q(b) \cdot x$ but that the right-hand sides differ from the value $\frac{1}{2}\sigma(\sigma - 1)$ of the hypermetric inequalities. We will call these inequalities *$Q$-inequalities*.

$$(14) \qquad Q(1, 1, 1, 1, 0) \cdot x \leq 5 \qquad \text{(5 permutations)}$$

can also be viewed as an upper bound on any complete subgraph $K_4$ of $K_5$. It is facet-defining for $\mathrm{MC}_3^{\leq}$, $\mathrm{MC}_3^=$, 3-$\mathrm{BMC}_3^{\leq}$, 2-$\mathrm{BMC}_3^{\leq}$, 1-$\mathrm{BMC}_3^=$.

$$(15) \qquad Q(1, 1, 1, 1, -1) \cdot x \leq 2 \qquad \text{(5 permutations)}$$

is facet-defining for $\mathrm{MC}_2^{\leq}$, $\mathrm{MC}_2^=$, 1-$\mathrm{BMC}_3^{\leq}$.

**(4.5) Lower bounds on cycles.** The next inequalities are of the form $x(C) \geq l$, where $C$ is a 3-, 4-, or 5-cycle in $K_5$ and $l$ is some positive integer.

$$(16) \qquad x_{12} + x_{23} + x_{34} + x_{45} + x_{15} \geq 3 \qquad \text{(12 permutations)}$$

specifies a lower bound of 3 on any pentagon in $K_5$. This inequality defines a facet of $\mathrm{MC}_3^{\geq}$, $\mathrm{MC}_3^=$. Similarly,

$$(17) \qquad x_{12} + x_{23} + x_{34} + x_{45} + x_{15} \geq 2 \qquad \text{(12 permutations)}$$

specifies a lower bound of 2 on any 5-cycle in $K_5$. It is facet-defining for $\mathrm{MC}_2^{\geq}$, $\mathrm{MC}_2^=$, 3-$\mathrm{BMC}_3^{\leq}$, 3-$\mathrm{BMC}_4^{\leq}$.

$$(18) \qquad x_{12} + x_{23} + x_{34} + x_{14} \geq 2 \qquad \text{(15 permutations)}$$

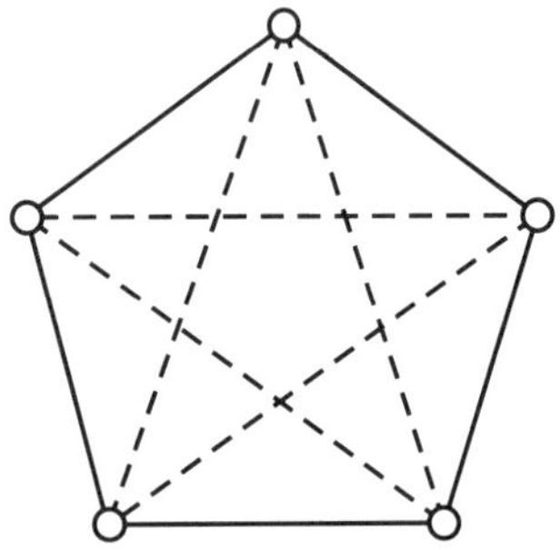

FIGURE 2

requires that each feasible multicut contains at least two edges of every 4-cycle. (There is only one such 4-cycle inequality.) (18) is facet-defining for 2-BMC$_2^{\geqq}$, 2-BMC$_3^{\leqq}$, 2-BMC$_4^{\leqq}$.

(19) $$x_{12} + x_{23} + x_{13} \geq 2 \qquad \text{(10 permutations)}$$

specifies a lower bound of 2 on any triangle. (19) defines a facet of 1-BMC$_3^{\geqq}$, 1-BMC$_3^{=}$.

**(4.6) 2-chorded odd cycle inequalities.** The next inequality can be viewed as a combination of two 5-cycles. It comes from the general class of 2-chorded odd cycle inequalities introduced in [**W, GW2**]. Its form, in our special case, is as follows. Let $C$ be any 5-cycle in $K_5$ and $\overline{C}$ its complement (which is also a 5-cycle). Then

(20) $$x(C) - x(\overline{C}) \leq 2 \qquad \text{(12 permutations)}$$

is valid for MC and thus for all multicut polytopes. (20) defines a facet of MC, MC$_2^{\geqq}$, MC$_3^{\leqq}$, 3-BMC$_3^{\leqq}$, 3-BMC$_4^{\leqq}$, 2-BMC$_2^{\geqq}$, 2-BMC$_3^{\leqq}$, 2-BMC$_4^{\leqq}$, 1-BMC$_2^{\geqq}$, 1-BMC$_3^{\leqq}$, 1-BMC$_4^{\leqq}$. A picture of a 2-chorded 5-cycle inequality is shown in Figure 2.

**(4.7) Degree conditions.** Some polytopes have degree constraints requiring that among all edges incident to any node at least a certain number have to be in a feasible multicut. The first lower bound on a star is of the form

(21) $$x_{12} + x_{13} + x_{14} + x_{15} \geq 2 \qquad \text{(5 permutations)}$$

and is facet-defining for 2-BMC$_2^{\geqq}$, 2-BMC$_3^{\leqq}$, and 2-BMC$_4^{\leqq}$. The second degree bound reads

(22) $$x_{12} + x_{13} + x_{14} + x_{15} \geq 3 \qquad \text{(5 permutations)}$$

and defines a facet of 1-BMC$_3^{\geqq}$ and 1-BMC$_3^{=}$.

**(4.8) Crowns.** A *crown* in $K_5$ is an edge set $C = E \setminus T$, where $T$ is a triangle. (Figures 3 and 4 make it clear why such edge sets are called crowns.) Crowns are supports of seven different classes of facet-defining inequalities

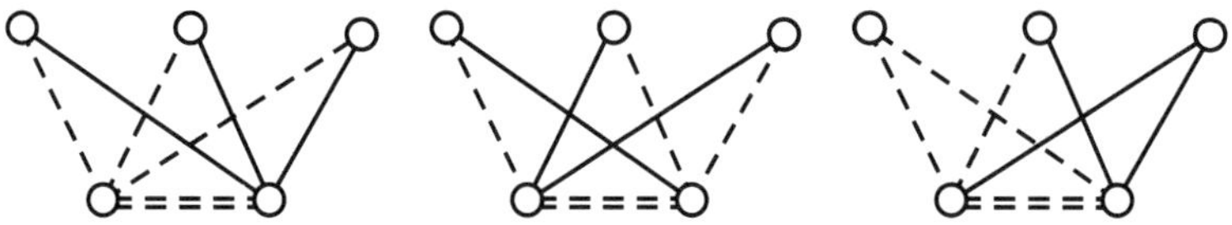

(a) (b) (c)

FIGURE 3

for multicut polytopes. The first three classes are characterized by the fact that the two nodes of the "basis" of the crown are linked by an edge with weight $-2$ and that exactly three of the edges linking these two nodes to the other three (nonbasis) nodes have weight $+1$ and $-1$, respectively.

The first class, shown in Figure 3(a), is of the form

$$(23)\quad -2x_{12} - x_{13} - x_{14} - x_{15} + x_{23} + x_{24} + x_{25} \leq 0 \qquad \text{(20 permutations)},$$

i.e., all edges linking one node of the basis to the three nonbasis nodes have weight $-1$ while the edges leaving the other basis node have weight $+1$. These inequalities define facets of $\mathrm{MC}_3^{\geq}$, 2-$\mathrm{BMC}_2^{\geq}$, 2-$\mathrm{BMC}_3^{\leq}$, and 2-$\mathrm{BMC}_4^{\leq}$.

In the second class of crown inequalities, one of the basis nodes is linked to exactly one nonbasis node by an edge with weight $-1$ while the other basis node is linked to the other two nonbasis nodes by edges with weight $-1$ (see Figure 3(b)), i.e., these inequalities are of the form

$$(24)\quad -2x_{12} - x_{13} + x_{14} + x_{15} + x_{23} - x_{24} - x_{25} \leq 0 \qquad \text{(30 permutations)}.$$

These inequalities induce facets of $\mathrm{MC}_3^{\geq}$.

In the third class of inequalities (see Figure 3(c)) one basis node is linked by two edges of weight $-1$ to two nonbasis nodes, the other basis node is linked by one edge of weight $-1$ to one of the nonbasis nodes that is already incident to a negative edge, so one nonbasis node is always incident to two positive edges. Thus, these inequalities are of the form

$$(25)\quad -2x_{12} - x_{13} - x_{14} + x_{15} - x_{23} + x_{24} + x_{25} \leq 0 \qquad \text{(30 permutations)}.$$

They define facets of $\mathrm{MC}_3^{\geq}$.

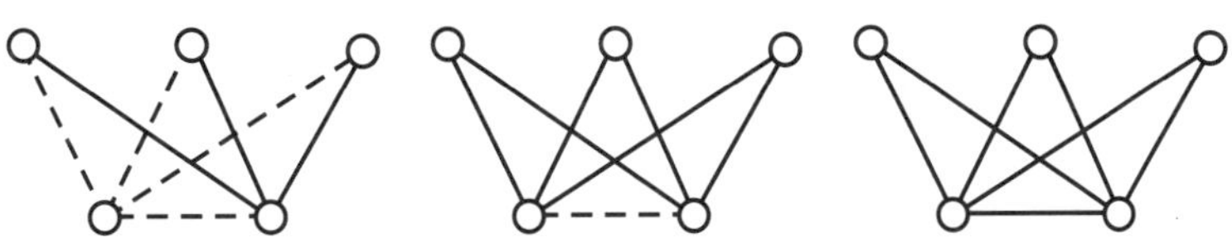

(a) (b) (c)

FIGURE 4

The other four classes of crown inequalities have nonzero coefficients $+1$ and $-1$. In the first of these, the two basis nodes are linked by a $-1$-edge, all

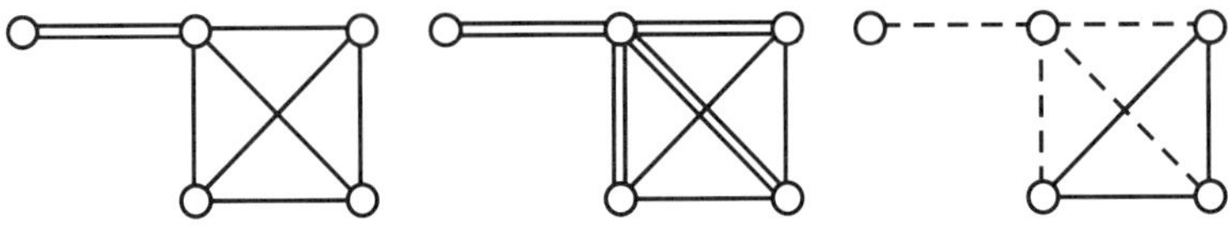

(a) (b) (c)

FIGURE 5

other edges leaving one of the basis nodes have weight $-1$, the other edges leaving the other basis node have weight $+1$ (see Figure 4(a)). Thus, they can be written as

$$(26)\quad -x_{12}-x_{13}-x_{14}-x_{15}+x_{23}+x_{24}+x_{25}\leq 0 \qquad \text{(20 permutations)}.$$

They define facets of $1\text{-BMC}_2^{\geq}$, $1\text{-BMC}_3^{\leq}$, $1\text{-BMC}_4^{\leq}$.

In the second version of this type of crown inequalities, the two basis nodes are linked by a $-1$-edge, all other crown edges have weight $+1$ (see Figure 4(b)), i.e., we have

$$(27)\quad -x_{12}+x_{13}+x_{14}+x_{15}+x_{23}+x_{24}+x_{25}\geq 2 \qquad \text{(10 permutations)}.$$

Inequalities (27) define facets of $\text{MC}_2^{\geq}$, $\text{MC}_2^{=}$, $3\text{-BMC}_3^{\leq}$, $3\text{-BMC}_4^{\leq}$.

Finally, there are two types of inequalities that specify an upper bound and a lower bound, respectively, on the edges in a crown $C$ shown in Figure 4(c). The first cardinality constraint reads

$$(28)\qquad x(C)\leq 6 \qquad \text{(10 permutations)}$$

and defines a facet of $1\text{-BMC}_3^{\leq}$ and $1\text{-BMC}_3^{=}$; the lower bound constraint requires

$$(29)\qquad x(C)\geq 4 \qquad \text{(10 permutations)}$$

and is facet-defining for $2\text{-BMC}_2^{\geq}$, $2\text{-BMC}_3^{\leq}$, $2\text{-BMC}_4^{\leq}$, and $1\text{-BMC}_2^{\leq}$.

**(4.9) Bipartite inequalities.** A further class of inequalities specifies a lower bound on the edge set of any complete bipartite graph $K_{2,3}$ contained in $K_5$. It is of the form

$$(30)\quad x_{13}+x_{14}+x_{15}+x_{23}+x_{24}+x_{25}\geq 4 \qquad \text{(10 permutations)}$$

and defines a facet of $\text{MC}_3^{\geq}$ and $\text{MC}_3^{=}$. It is a special case of the class of bipartite inequalities introduced in **[CR1]** for $\text{MC}_k^{\geq}(n)$.

**(4.10) Casserole inequalities.** We call the next type of inequalities casserole inequalities since their supports resemble a casserole (see Figure 5).

They come in three classes. The first one, shown in Figure 5(a), specifies a lower bound on the edges of a casserole, but where the edges of the "handle" are counted twice, i.e.,

$$(31)\quad 2x_{12}+x_{23}+x_{24}+x_{25}+x_{34}+x_{35}+x_{45}\geq 5 \qquad \text{(20 permutations)}.$$

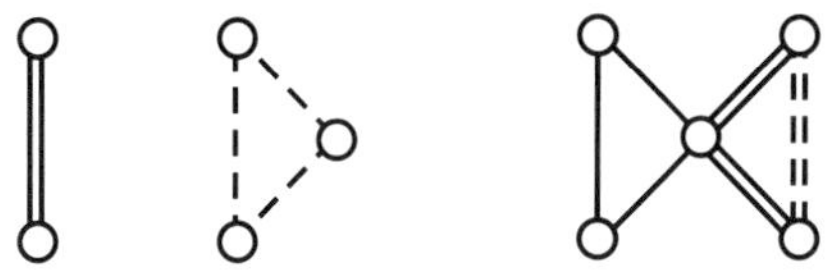

(a) (b)

FIGURE 6

These inequalities define facets of $\mathrm{MC}_3^{\geq}$ and $\mathrm{MC}_3^{=}$. The second one (see Figure 5(b)) is of the form

$$(32)\quad 2x_{12}+2x_{23}+2x_{24}+2x_{25}+x_{34}+x_{35}+x_{45}\geq 6 \qquad \text{(20 permutations)}$$

and is facet-defining for $2\text{-}\mathrm{BMC}_2^{\geq}$, $2\text{-}\mathrm{BMC}_3^{\leq}$, and $2\text{-}\mathrm{BMC}_4^{\leq}$.

The third class of casserole inequalities—the support is shown in Figure 5(c)—is of the form

$$(33)\quad -x_{12}-x_{23}-x_{24}-x_{25}+x_{34}+x_{35}+x_{45}\leq 0 \qquad \text{(20 permutations)}$$

and is facet-defining for $1\text{-}\mathrm{BMC}_2^{\geq}$ and $1\text{-}\mathrm{BMC}_4^{\leq}$.

Observe that the left-hand side of (31) can be written as $2x_{12}+x(E(\{2,3,4,5\}))$, while the left-hand sides of (32) and (33) can be viewed as $2x(\delta(2))+x(T)$ and $x(T)-x(\delta(2))$, respectively, where $T$ is the triangle on the nodes $3, 4, 5$.

**(4.11) Inequalities with low-connectivity support.** One usually expects the support graph of a facet-defining inequality to be highly (at least two) connected. The casserole inequalities already provide a support graph that is not 2-connected. There are two more classes of inequalities with such low-connectivity support graph. In the first case (see Figure 6(a)), the support graph is disconnected and can be viewed as the sum of twice an edge minus a triangle, i.e.,

$$(34)\qquad 2x_{12}-x_{34}-x_{35}-x_{45}\leq 0 \qquad \text{(10 permutations)}.$$

These inequalities define facets of $1\text{-}\mathrm{BMC}_2^{\geq}$, $1\text{-}\mathrm{BMC}_3^{\leq}$, and $1\text{-}\mathrm{BMC}_4^{\leq}$.

The second class has a 1- but not 2-connected support graph (see Figure 6(b)), and reads

$$(35)\qquad x_{12}+x_{13}+x_{23}+2x_{34}+2x_{35}-2x_{45}\geq 2 \qquad \text{(30 permutations)}.$$

These inequalities are facet-defining for $2\text{-}\mathrm{BMC}_2^{\geq}$, $2\text{-}\mathrm{BMC}_3^{\leq}$, and $2\text{-}\mathrm{BMC}_4^{\leq}$.

**(4.12) Further inequalities.** From now on we will make no attempt towards classifying the remaining inequalities further. Their common characteristic is that their support graph is the complete graph with at most two edges missing. With two exceptions all inequalities have coefficients that are different from $0, \pm 1$. In order to be able to draw some of the support graphs a little more nicely and to display structure we will sometimes group the nodes into "supernodes" (large ellipses) and link ellipses by edges. If two ellipses are

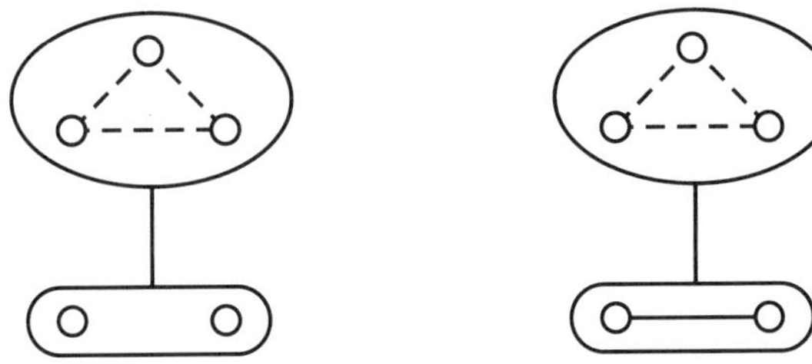

(a) (b)

FIGURE 7

linked by an edge, then all nodes of one ellipse are linked to all nodes of the other ellipse by an edge (of the same weight). Nodes inside an ellipse are not necessarily connected by edges.

We begin with the two classes of pure inequalities (all coefficients are $0, \pm 1$). The first one, displayed in Figure 7(a), reads

$$\begin{aligned} &-x_{12} - x_{13} - x_{23} + x_{14} + x_{15} + x_{24} \\ &\quad + x_{25} + x_{34} + x_{35} \geq 1 \qquad \text{(10 permutations)} \end{aligned} \tag{36}$$

and is facet-defining for $2\text{-}\mathrm{BMC}_2^{\geq}$, $2\text{-}\mathrm{BMC}_3^{\leq}$, and $2\text{-}\mathrm{BMC}_4^{\leq}$.

The second pure class is almost identical. Just one more positive coefficient, the last missing nonzero, appears. It is of the form (see Figure 7(b))

$$\begin{aligned} &-x_{12} - x_{13} - x_{23} + x_{14} + x_{15} + x_{24} + x_{25} \\ &\quad + x_{34} + x_{35} + x_{45} \geq 2 \qquad \text{(10 permutations)} \end{aligned} \tag{37}$$

and is facet-defining for $1\text{-}\mathrm{BMC}_2^{\geq}$, $1\text{-}\mathrm{BMC}_3^{\leq}$, and $1\text{-}\mathrm{BMC}_4^{\leq}$. If we denote the triangle on nodes $1, 2, 3$ by $T$, the edges of $K_{2,3}$ (with $1, 2, 3$ on one side) by $F$, and the edge between 4 and 5 by $e$, we can write (36) and (37), respectively, as

$$-x(T) + x(F) \geq 1,$$
$$-x(T) + x(F) + x_e \geq 2.$$

Another way to view (36) and (37) in our $Q$-notation is

$$-Q(1, 1, 1, -1, -1) \cdot x + x_e \geq 1,$$
$$-Q(1, 1, 1, -1, -1) \cdot x + 2x_e \geq 2.$$

All further inequalities have some coefficients whose absolute value is at least 2. The largest coefficient appearing has absolute value 4.

The support of the next class of inequalities is the complete graph $K_5$. All coefficients have value $+2$, except for one triangle, where all edges have value 1. In $Q$-notation, it reads as follows:

$$2Q(1, 1, 1, 1, 1) \cdot x - Q(1, 1, 1, 0, 0) \cdot x \leq 14 \qquad \text{(10 permutations)} \tag{38}$$

and is facet-defining of $\mathrm{MC}_3^{\leq}$, $\mathrm{MC}_3^{=}$, $3\text{-}\mathrm{BMC}_3^{\leq}$, and $2\text{-}\mathrm{BMC}_3^{\leq}$.

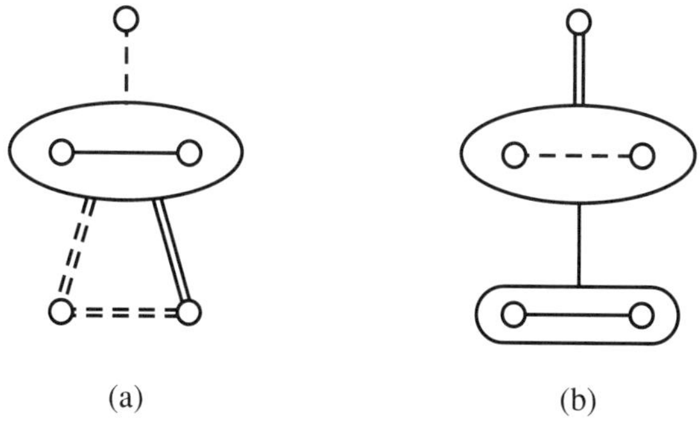

FIGURE 8

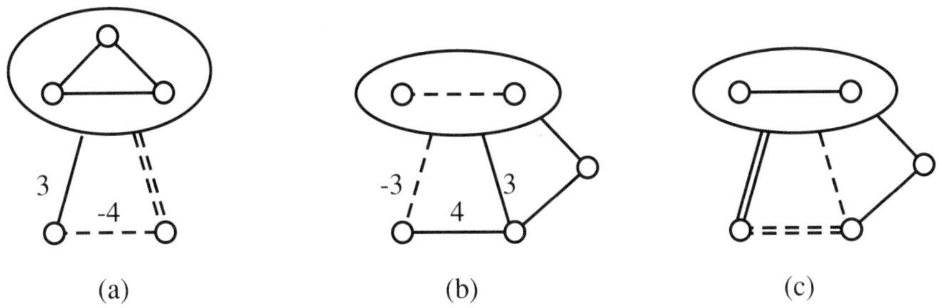


FIGURE 9

The support graph of inequality

$$-x_{12} - x_{13} + x_{23} + 2x_{24} - 2x_{25} + 2x_{34} - 2x_{35} - 2x_{45} \le 0 \qquad \text{(60 permutations)} \tag{39}$$

is shown in Figure 8(a). Inequalities of type (39) define facets of 2-$\mathrm{BMC}_2^{\geq}$, 2-$\mathrm{BMC}_3^{\leq}$, and 2-$\mathrm{BMC}_4^{\leq}$.

The inequality depicted in Figure 8(b) reads

$$2x_{12} + 2x_{13} - x_{23} + x_{24} + x_{25} + x_{34} + x_{35} + x_{45} \ge 5 \qquad \text{(30 permutations)} \tag{40}$$

and defines a facet of $\mathrm{MC}_3^{\geq}$ and $\mathrm{MC}_3^{=}$.

The support graphs of the next three classes of inequalities are shown in Figure 9.

The inequality depicted in Figure 9(a) reads

$$-4x_{12} + 3x_{13} + 3x_{14} + 3x_{15} - 2x_{23} - 2x_{24} - 2x_{25} + x_{34} + x_{35} + x_{45} \le 5 \qquad \text{(20 permutations)} \tag{41}$$

and defines a facet of $\mathrm{MC}_3^{\leq}$ and 3-$\mathrm{BMC}_3^{\leq}$. In $Q$-notation we can write it as

$$Q(3, -2, 1, 1, 1) \cdot x + 2x_{12} \le 5.$$

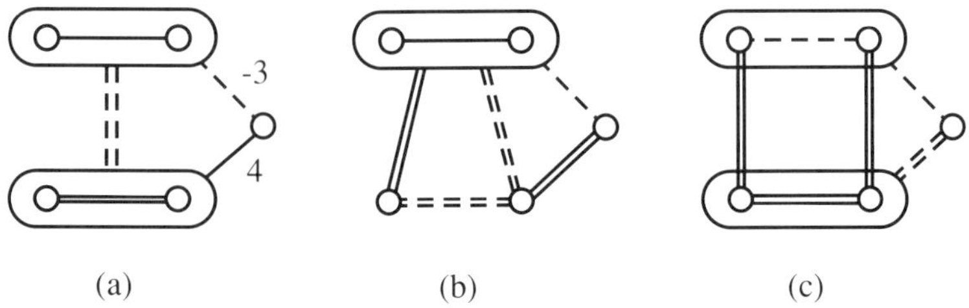


FIGURE 10

The inequality shown in Figure 9(b) is

$$\begin{aligned}(42)\qquad & 4x_{12} - 3x_{13} - 3x_{14} + 3x_{23} + 3x_{24} + x_{25} \\ & - x_{34} + x_{35} + x_{45} \geq 1 \qquad \text{(60 permutations)}\end{aligned}$$

and is facet-defining for $2\text{-BMC}_2^{\geq}$, $2\text{-BMC}_3^{\leq}$, and $2\text{-BMC}_4^{\leq}$.

The inequality belonging to Figure 9(c) can be written as

$$\begin{aligned}(43)\qquad & -2x_{12} + 2x_{13} + 2x_{14} - x_{23} - x_{24} + x_{25} \\ & + x_{34} + x_{35} + x_{45} \leq 5 \qquad \text{(60 permutations)}.\end{aligned}$$

In $Q$-notation, we can view it as

$$Q(2, -1, 1, 1, 1) \cdot x - 2x_{15} + 2x_{25} \leq 5.$$

It is facet-defining for $\text{MC}_3^{\leq}$, $3\text{-BMC}_3^{\leq}$, and $2\text{-BMC}_3^{\leq}$.

Figure 10 shows three further weighted support graphs. The inequality belonging to Figure 10(a) reads

$$\begin{aligned}(44)\qquad & 4x_{12} + 4x_{13} - 3x_{14} - 3x_{15} + 2x_{23} - 2x_{24} - 2x_{25} \\ & - 2x_{34} - 2x_{35} + x_{45} \leq 2 \qquad \text{(30 permutations)}\end{aligned}$$

and can be written in $Q$-notation as

$$Q(3, 2, 2, -1, -1) \cdot x - 2Q(1, 1, 1, 0, 0) \cdot x \leq 2.$$

It is facet-defining for MC, $\text{MC}_2^{\geq}$, $\text{MC}_3^{\leq}$, $3\text{-BMC}_3^{\leq}$, and $3\text{-BMC}_4^{\leq}$ and is a special case of the generalized cycle inequalities introduced in [**CR2**] for $\text{MC}_k^{\leq}(n)$.

The next inequality

$$\begin{aligned}(45)\qquad & 2x_{12} - x_{14} - x_{15} - 2x_{23} - 2x_{24} - 2x_{25} \\ & + 2x_{34} + 2x_{35} + x_{45} \leq 2 \qquad \text{(60 permutations)}\end{aligned}$$

is depicted in Figure 10(b) and can be written in $Q$-notation as

$$Q(-1, -2, 2, 1, 1) \cdot x + 2(x_{13} + x_{23}) \leq 2.$$

It is facet-defining for MC, $\text{MC}_2^{\geq}$, $\text{MC}_3^{\leq}$, $3\text{-BMC}_3^{\leq}$, and $3\text{-BMC}_4^{\leq}$. The inequality shown in Figure 10(c) has the form

$$\begin{aligned}(46)\qquad & -2x_{12} - 2x_{13} - x_{14} - x_{15} + 2x_{23} + 2x_{25} \\ & + 2x_{34} - x_{45} \leq 2 \qquad \text{(60 permutations)}\end{aligned}$$

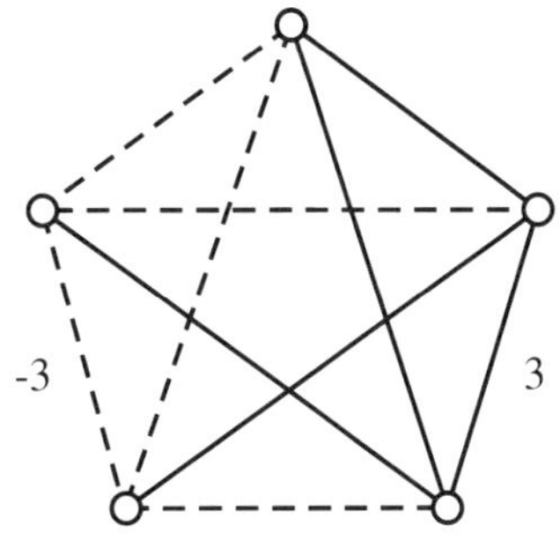


FIGURE 11

and is facet-defining for MC, $\mathrm{MC}_2^{\geq}$, $\mathrm{MC}_3^{\leq}$, 3-$\mathrm{BMC}_3^{\leq}$, and 3-$\mathrm{BMC}_4^{\leq}$. So the classes (44), (45), and (46) are facet-defining for the same multicut polytopes.

Finally, the class of inequalities represented by

$$\begin{aligned} & x_{12} + x_{13} - x_{14} - x_{15} + 3x_{23} + x_{24} - x_{25} \\ & \quad - x_{34} + x_{35} - 3x_{45} \leq 2 \qquad \text{(60 permutations)} \end{aligned} \tag{47}$$

and shown in Figure 11 is facet-defining for 1-$\mathrm{BMC}_3^{\leq}$.

This finishes our list of inequalities. In Table 2 we give a comprehensive overview over all multicut polytopes of $K_5$ and their facets. The polytopes of Table 1 index the first 18 rows of the table. The column labeled $v$ contains the number of vertices, the column labeled $f$ the number of facets, and the column labeled $d$ the dimension of each polytope. The columns labeled (1), (2), (3), ..., (47) correspond to the 47 classes of facets listed above. An entry "x" in the table denotes the fact that all inequalities of the class of facets of its column define a facet of the polytope of its row. The last three rows specify whether inequality is of "$\leq$" or "$\geq$" type, the value of the right-hand side of each inequality in the class, and the number of different inequalities in the class, respectively.

Without going into too much detail we mention a few artifacts that can be read from Table 2. The range of right-hand sides of the inequalities (assuming that they are in "$\leq$" form with integral and coprime coefficients) is $-8$ to 14. Only few facets appear frequently, namely the upper bounds (1), the triangle inequalities (8), and the 2-chorded 5-cycle inequalities (20). Four classes are facet-defining for just one polytope: (7), (24), (25), and (47). And there are clusters of facets that always appear together. For instance,

classes (18), (21), (29), (32), (35), (36), (39), (42)

define facets of 2-$\mathrm{BMC}_2^{\geq}$, 2-$\mathrm{BMC}_3^{\leq}$, and 2-$\mathrm{BMC}_4^{\leq}$ but not of any other polytope (disregarding the lower-dimensional polytope 1-$\mathrm{BMC}_2^{\leq}$). These polytopes, in fact, have 303 common facets in total. Moreover,

classes (6), (16), (30), (31), (40)

TABLE 2

| polytopes | v | f | d | 1 | 2 | 3 | 4 | 5 | 6 | 7 | 8 | 9 | 10 | 11 | 12 | 13 | 14 | 15 | 16 | 17 | 18 | 19 | 20 | 21 | 22 | 23 | 24 | 25 | 26 | 27 | 28 | 29 | 30 | 31 | 32 | 33 | 34 | 35 | 36 | 37 | 38 | 39 | 40 | 41 | 42 | 43 | 44 | 45 | 46 | 47 |
|---|---|---|---|---|---|---|---|---|---|---|---|---|---|---|---|---|---|---|---|---|---|---|---|---|---|---|---|---|---|---|---|---|---|---|---|---|---|---|---|---|---|---|---|---|---|---|---|---|---|---|
| MC | 52 | 242 | 10 | x | | | | | | | x | x | x | x | | x | | | | | | | x | | | | | | | | | | | | | | | | | | | | | | | | x | x | x | |
| $MC_2^{\leq}$ | 16 | 56 | 10 | | | | x | | | | x | x | | | x | | | x | | | | | | | | | | | | | | | | | | | | | | | | | | | | | | | | |
| $MC_2^{\geq}$ | 51 | 264 | 10 | x | | | | | | | x | x | x | x | | x | | | | x | | | x | | | | | | | x | | | | | | | | | | | | | | | | | x | x | x | |
| $MC_2^{=}$ | 15 | 68 | 10 | | | | x | | | | x | | | | x | | | x | | x | | | | | | | | | | x | | | | | | | | | | | | | | | | | | | | |
| $MC_3^{\leq}$ | 41 | 333 | 10 | x | | x | | | | | x | x | x | x | | | x | | | | | | x | | | | | | | | | | | | | | | | | | x | | | x | | x | x | x | x | |
| $MC_3^{\geq}$ | 36 | 218 | 10 | x | | | | | x | | x | | | x | | x | | | x | | | | | | | x | x | x | | | | | x | x | | | | | | | | | x | | | | | | | |
| $MC_3^{=}$ | 25 | 99 | 10 | x | | x | | | x | | | | | | | | x | | x | | | | | | | | | | | | | | x | x | | | | | | | x | | x | | | | | | | |
| 3-$BMC_3^{\leq}$ | 40 | 355 | 10 | x | | x | | | | | x | x | x | x | | | x | | | x | | | x | | | | | | | x | | | | | | | | | | | x | | | x | | x | x | x | x | |
| 3-$BMC_4^{\leq}$ | 50 | 265 | 10 | x | x | | | | | | x | x | x | x | | x | | | | x | | | x | | | | | | | x | | | | | | | | | | | | | | | | | x | x | x | |
| 2-$BMC_2^{\geq}$ | 46 | 308 | 10 | x | | | | x | | | x | | | x | | x | | | | | x | | x | x | | x | | | | | | x | | | x | | | x | x | | | x | | | x | | | | | |
| 2-$BMC_3^{\leq}$ | 35 | 379 | 10 | x | | x | | x | | | x | | | x | | | x | | | | x | | x | x | | x | | | | | | x | | | x | | | x | x | | x | x | | | x | x | | | | |
| 2-$BMC_4^{\leq}$ | 45 | 309 | 10 | x | x | | | x | | | x | | | x | | x | | | | | x | | x | x | | x | | | | | | x | | | x | | | x | x | | | x | | | x | | | | | |
| 1-$BMC_2^{\leq}$ | 10 | 10 | 9 | | | | | | | | | | | | x | | | | | | | | | | | | | | | | | x | | | | | | | | | | | | | | | | | | |
| 1-$BMC_2^{\geq}$ | 36 | 113 | 10 | x | | | | x | | | x | | | | | | | | | | | | x | | | | | | x | | | | | | | x | x | | | x | | | | | | | | | | |
| 1-$BMC_3^{\leq}$ | 24 | 169 | 10 | x | | x | | x | | | x | | | | | | | x | | | | | x | | | | | | x | | x | | | | | | x | | | x | | | | | | | | | | x |
| 1-$BMC_3^{\geq}$ | 26 | 26 | 10 | x | | | | | | x | | | | | | | | | | | | x | | | x | | | | | | | | | | | | | | | | | | | | | | | | | |
| 1-$BMC_3^{=}$ | 15 | 25 | 9 | x | | | | | | | | | | | | | x | | | | | x | | | x | | | | | | x | | | | | | | | | | | | | | | | | | | |
| 1-$BMC_4^{\leq}$ | 35 | 114 | 10 | x | x | | | x | | | x | | | | | | | | | | | | x | | | | | | x | | | | | | | x | x | | | x | | | | | | | | | | |
| inequality type | | | | ≤ | ≤ | ≤ | ≤ | ≥ | ≥ | ≥ | ≤ | ≤ | ≤ | ≤ | ≤ | ≤ | ≤ | ≤ | ≥ | ≥ | ≥ | ≥ | ≤ | ≥ | ≥ | ≤ | ≤ | ≤ | ≤ | ≥ | ≤ | ≥ | ≥ | ≥ | ≥ | ≤ | ≤ | ≥ | ≥ | ≥ | ≤ | ≤ | ≥ | ≤ | ≥ | ≤ | ≤ | ≤ | ≤ | ≤ |
| right-hand side | | | | 1 | 9 | 8 | 6 | 6 | 7 | 8 | 0 | 0 | 1 | 1 | 2 | 3 | 5 | 2 | 3 | 2 | 2 | 2 | 2 | 2 | 3 | 0 | 0 | 0 | 0 | 2 | 6 | 4 | 4 | 5 | 6 | 0 | 0 | 2 | 1 | 2 | 14 | 0 | 5 | 5 | 1 | 5 | 2 | 2 | 2 | 2 |
| no. of inequalities | | | | 10 | 1 | 1 | 1 | 1 | 1 | 1 | 30 | 10 | 5 | 20 | 10 | 5 | 5 | 5 | 12 | 12 | 15 | 10 | 12 | 5 | 5 | 20 | 30 | 30 | 20 | 10 | 10 | 10 | 10 | 20 | 20 | 20 | 10 | 30 | 10 | 10 | 10 | 60 | 30 | 20 | 60 | 60 | 30 | 60 | 60 | 60 |

are only facet-defining for $\mathrm{MC}_3^{\geq}$ and $\mathrm{MC}_3^{=}$,

classes (10), (44), (45), (46)

are only facet-defining for MC, $\mathrm{MC}_2^{\geq}$, $\mathrm{MC}_3^{\leq}$, 3-$\mathrm{BMC}_3^{\leq}$, and 3-$\mathrm{BMC}_4^{\leq}$ and do not appear anywhere else. Similarly,

classes (7), (19), (22)

are only facets of 1-$\mathrm{BMC}_3^{\geq}$ and do not appear anywhere else (disregarding the lower-dimensional polytope 1-$\mathrm{BMC}_3^{=}$),

classes (14), (38)

are only facet-defining for $\mathrm{MC}_3^{\geq}$, $\mathrm{MC}_3^{=}$, 3-$\mathrm{BMC}_3^{\leq}$, and 2-$\mathrm{BMC}_3^{\leq}$, and finally

classes (24), (25)

only define facets of the polytope $\mathrm{MC}_3^{\geq}$ and of no other multicut polytope.

## 5. Some observations

**(5.1) On $(n-1)$-cut polytopes.** We observed that the multicut polytopes $\mathrm{MC}_4^{=}(5)$ and $\mathrm{MC}_4^{\geq}(5)$ are both simplices in $\mathbf{R}^{10}$ of dimension 9 and 10, respectively; also, that the polytope $\mathrm{MC}_4^{\leq}(5)$ can be deduced from the polytope MC(5) simply by adding the following cardinality constraint: $\sum_{ij\in E} x_{ij} \leq 9$. In fact, these observations remain valid more generally for the multicut polytopes $\mathrm{MC}(n)$, $\mathrm{MC}_{n-1}^{=}(n)$, $\mathrm{MC}_{n-1}^{\geq}(n)$, and $\mathrm{MC}_{n-1}^{\leq}(n)$ defined, respectively, as the convex hull of all incidence vectors of multicuts, all $(n-1)$-cuts, all $k$-cuts with $k \geq n-1$, and all $k$-cuts with $k \leq n-1$, of the complete graph $K_n = (V, E)$ on $n$ nodes.

The polytope $\mathrm{MC}_{n-1}^{\geq}(n)$ has $\binom{n}{2}+1$ vertices, all linearly independent, so $\mathrm{MC}_{n-1}^{\geq}(n)$ is the $\binom{n}{2}$-simplex in $\mathbf{R}^E$ defined completely and nonredundantly by

$$(48)\quad \mathrm{MC}_{n-1}^{\geq}(n) = \left\{ x \in \mathbf{R}^E : x_{ij} \leq 1 \text{ for all } ij \in E,\ \sum_{ij\in E} x_{ij} \geq \binom{n}{2} - 1 \right\}.$$

The polytope $\mathrm{MC}_{n-1}^{=}(n)$ has precisely one vertex less than $\mathrm{MC}_{n-1}^{\geq}(n)$, namely the incidence vector of the $n$-cut, i.e., the vector $\chi^E$ whose coordinates are all equal to 1, is a vertex of $\mathrm{MC}_{n-1}^{\geq}(n)$ but not of $\mathrm{MC}_{n-1}^{=}(n)$. So, $\mathrm{MC}_{n-1}^{=}(n)$ is the $(\binom{n}{2}-1)$-simplex defined completely and nonredundantly by

$$(49)\quad \mathrm{MC}_{n-1}^{=}(n) = \left\{ x \in \mathbf{R}^E : x_{ij} \leq 1 \text{ for all } ij \in E,\ \sum_{ij\in E} x_{ij} = \binom{n}{2} - 1 \right\}.$$

Moreover, the polytope $\mathrm{MC}_{n-1}^{\leq}(n)$ has precisely one less vertex than $\mathrm{MC}(n)$; namely, the only multicut vector that is not a vertex of $\mathrm{MC}_{n-1}^{\leq}(n)$

is the $n$-cut vector $\chi^E$. The cardinality condition $\sum_{e\in E} x_e \le \binom{n}{2} - 1$ clearly defines a facet of $\mathrm{MC}_{n-1}^{\le}(n)$, so the following inclusion holds: $\mathrm{MC}_{n-1}^{\le}(n) \subseteq \mathrm{MC}(n) \cap \{x \in \mathbf{R}^E \mid \sum_{ij\in E} x_{ij} \le \binom{n}{2} - 1\}$. Using the fact [W] that the vertices of $\mathrm{MC}(n)$ which are adjacent to $\chi^E$ on $\mathrm{MC}(n)$ are precisely the $(n-1)$-cut vectors, i.e., the multicut vectors lying on the hyperplane defined by the equation $\sum_{ij\in E} x_{ij} = \binom{n}{2} - 1$, we deduce that $\mathrm{MC}_{n-1}^{\le}(n)$ indeed coincides with $\mathrm{MC}(n) \cap \{x \in \mathbf{R}^E \mid \sum_{ij\in E} x_{ij} \le \binom{n}{2} - 1\}$. Furthermore, every inequality defining a facet of $\mathrm{MC}(n)$ also defines a facet of $\mathrm{MC}_{n-1}^{\le}(n)$. This is clear for facets that do not contain $\chi^E$. Note that the only facets of $\mathrm{MC}(n)$ containing $\chi^E$ are those defined by the upper bound conditions (1) $x_{ij} \le 1$. To see this, take a facet $F$ of $\mathrm{MC}(n)$ containing $\chi^E$. Then the vertices of $F$ adjacent to $\chi^E$ together with $\chi^E$ form a set of rank $\binom{n}{2} - 1$. Since, as mentioned above, the multicut vectors adjacent to $\chi^E$ are the $(n-1)$-cut vectors, we deduce that $F$ contains a set of rank $\binom{n}{2} - 1$ consisting of $\chi^E$ and $(n-1)$-cut vectors and hence $F$ is indeed supported by some upper bound condition (1). Summarizing these observations we obtain

PROPOSITION 1. *For $n \ge 3$, the facets of $\mathrm{MC}_{n-1}^{\le}(n)$ are those of $\mathrm{MC}(n)$ together with the cardinality condition $\sum_{ij\in E} x_{ij} \le \binom{n}{2} - 1$. Moreover, the following equality holds*:

$$\mathrm{MC}_{n-1}^{=}(n) = \mathrm{MC}_{n-1}^{\ge}(n) \cap \mathrm{MC}_{n-1}^{\le}(n).$$

One can observe, by looking at the table of facets (Table 2), that the following equality holds:

$$s\text{-}\mathrm{BMC}_4^{\le}(5) = s\text{-}\mathrm{BMC}_2^{\ge}(5) \cap \left\{ x \in \mathbf{R}^E \,\middle|\, \sum_{ij\in E} x_{ij} \le 9 \right\} \quad \text{for } s = 1, 2, 3.$$

Generally, the polytope $s\text{-}\mathrm{BMC}_{n-1}^{\le}(n)$ has the same vertices as the polytope $s\text{-}\mathrm{BMC}_2^{\ge}(n)$ except the $n$-cut vector $\chi^E$ which is cut off by the cardinality condition $\sum_{ij\in E} x_{ij} \le \binom{n}{2} - 1$. It seems that the equality

$$s\text{-}\mathrm{BMC}_{n-1}^{\le}(n) = s\text{-}\mathrm{BMC}_2^{\ge}(n) \cap \left\{ x \in \mathbf{R}^E \,\middle|\, \sum_{ij\in E} x_{ij} \le \binom{n}{2} - 1 \right\}$$

holds for any $n \ge 5$ and that the facets of $s\text{-}\mathrm{BMC}_{n-1}^{\le}(n)$ are those of $s\text{-}\mathrm{BMC}_2^{\ge}(n)$ together with the cardinality condition $\sum_{ij\in E} x_{ij} \le \binom{n}{2} - 1$.

**(5.2) Connectivity of the facet support graphs.** One can observe that, among the inequalities listed in §4, those defining a facet of some multicut polytope $\mathrm{MC}_k^{\le}(5)$ for some $k \in \{2, 3, 4, 5\}$ have a support graph that is 2-connected. More generally, one has the following result.

PROPOSITION 2. *The support graph of an inequality defining a facet of the polytope* $\mathrm{MC}_k^{\leq}(n)$, *for some* $2 \leq k \leq n$, *is* 2*-connected, i.e., the smallest number of nodes whose removal disconnects the graph is greater than or equal to* 2.

PROOF. The proof is a direct extension of the proof given in **[D]** for the case $k = 2$, i.e., for the cut polytope case. Take an inequality $v^T x \leq \alpha$ which defines a facet of $\mathrm{MC}_k^{\leq}(n)$ and let us assume that its support graph $G_v = (V, E(v))$ is not 2-connected. Suppose first that $G_v$ admits an articulation node $i_0$. So $V = V_1 \cup V_2$, $E(v) = E_1 \cup E_2$ with $V_1 \cap V_2 = \{i_0\}$, and $E_i$ is the edge set of $G_v$ induced by $V_i$ for $i = 1, 2$. Denote by $v_i$ the vector of $\mathbf{R}^E$ whose projection on $\mathbf{R}^{E_i}$ coincides with the projection of $v$ and whose projection on $\mathbf{R}^{E \setminus E_i}$ is the zero vector, for $i = 1, 2$. Thus, $v = v_1 + v_2$ holds.

Set $\alpha_i := \max\{v_i^T x | x \in \mathrm{MC}_k^{\leq}(n)\}$. Then the inequality $v_i^T x \leq \alpha_i$ is valid for $\mathrm{MC}_k^{\leq}(n)$ for $i = 1, 2$. We prove that $\alpha = \alpha_1 + \alpha_2$ holds. First, inequality $\alpha \leq \alpha_1 + \alpha_2$ holds trivially. To prove the converse inequality, take a multicut vector $x_i$ of $\mathrm{MC}_k^{\leq}(n)$ satisfying $v_i^T x_i = \alpha_i$, for $i = 1, 2$. So $x_1$ is the incidence vector of an $h$-cut of the form $\delta(S_1, S_2, \dots, S_h)$ with $h \leq k$, and $x_2$ is the incidence vector of an $l$-cut $\delta(T_1, \dots, T_l)$ with $l \leq k$. We may assume that $h \leq l$ and that node $i_0$ belongs to $S_1 \cap T_1$. Consider the multicut vector $x$ whose shores are the sets $(S_j \cap V_1) \cup (T_j \cap V_2)$ for $1 \leq j \leq h$, and $T_j \cap V_2$ for $h+1 \leq j \leq l$. So $x$ is the incidence vector of a multicut having at most $l \leq k$ shores and satisfying $v_i^T x = v_i^T x_i = \alpha_i$ and, thus, $v^T x = \alpha_1 + \alpha_2 \leq \alpha$. Therefore, the inequality $v^T x \leq \alpha$ can be written as the sum of two valid inequalities, contradicting the fact that it is facet inducing. The proof is analogous in the case when the support graph is disconnected. □

This result does not extend to the case of the other multicut polytopes $\mathrm{MC}_k^{\geq}(n)$, $\mathrm{MC}_k^{=}(n)$ or the $s$-balanced multicut polytopes. For instance, the casserole inequalities (31), (32), (33) are connected, but not 2-connected. (See also the example of the facets from Proposition 4.) Moreover, the support graph of inequality (34) is disconnected.

**(5.3) Cycle inequalities.** We introduce a class of inequalities containing inequalities (17), (18), (19) as special cases. We call them *cycle inequalities* since their supporting graph is a cycle $C$ and they are of the form:

$$\sum_{ij \in C} x_{ij} \geq 2. \tag{50}$$

Note that a multicut vector violates inequality (50) if and only if one of its shores contains all nodes of the cycle $C$. This implies easily that, for $2 \leq p \leq n-1$, if $C$ is a cycle of length $p+1$, then the cycle inequality (50) is valid for the polytopes $\mathrm{MC}_{n-p+1}^{\geq}(n)$ and $(p-1)$-$\mathrm{BMC}_{n-p}^{\geq}$. Denote by $1, 2, \ldots, p+1$ the nodes of $C$ and by $p+2, \ldots, n$ the remaining nodes. One checks easily that the roots of inequality (50) in $\mathrm{MC}^{\geq}_{n-p}(n)$, i.e., the multicut vectors $x$ of $\mathrm{MC}^{\geq}_{n-p}(n)$ that satisfy inequality (50) at equality, are, setting $I_u := \{1, 2, \ldots, u-1, u\}$ for $1 \leq u \leq p$, as follows.

(50a) $x$ is a multicut vector with shores $I_u$, $\{1, \ldots, p+1\} \backslash I_u$, and the singletons $\{i\}$ for $p+2 \leq i \leq n$. So $x$ is the incidence vector of an $(n-p+1)$-cut that is $(p-1)$-balanced.

(50b) $x$ is a multicut vector with shores $I_u \cup \{j\}$, $\{1, \ldots, p+1\} \backslash I_u$, and the singletons $\{i\}$ for $p+2 \leq i \leq n$, $i \neq j$ with $p+2 \leq j \leq n$. So $x$ is the incidence vector of an $(n-p)$-cut and is $(p-1)$-balanced if $u \neq p$.

Hence, if $p < n-1$, inequality (50) does not define a facet of $\mathrm{MC}^{\geq}_{n-p+1}(n)$, since the only multicut vectors in $\mathrm{MC}^{\geq}_{n-p+1}(n)$ satisfying inequality (50) with equality are those described in (50a). But these also satisfy the equation $x_{1n} = 1$. The next result indicates that (50) is indeed facet-defining in the remaining cases.

PROPOSITION 3. 1. *Let $C$ be a cycle of length $n$. Then the cycle inequality* (50) *defines a facet of the polytopes* $\mathrm{MC}^{=}_{2}(n)$ *and* $\mathrm{MC}^{\geq}_{2}(n)$.

2. *Let $C$ be a cycle of length $p+1$, $2 \leq p \leq n-1$. Then the cycle inequality* (50) *defines a facet of the polytope* $(p-1)$-$\mathrm{BMC}^{\geq}_{n-p}(n)$.

PROOF. We shall use the description given in (50a), (50b) of the roots of inequality (50).

1. Take an inequality $v^T x \geq \alpha$ that is valid for $\mathrm{MC}^{=}_{2}(n)$ and is satisfied with equality by all the roots of inequality (50) in $\mathrm{MC}^{=}_{2}(n)$. Set $S := \{2, \ldots, u\}$ for $2 \leq u \leq p-1$; then we have

$$v^T \chi^{\delta(S)} = v^T \chi^{\delta(S \cup \{1\})} = v^T \chi^{\delta(S \cup \{u+1\})} = v^T \chi^{\delta(S \cup \{1, u+1\})} = \alpha,$$

from which one easily deduces that $v_{1, u+1} = 0$. Hence, by symmetry, $v_{ij} = 0$ for every edge $ij$ that is not an edge of $C$. Using the fact that $v^T \chi^{\delta(\{1\})} = v^T \chi^{\delta(\{2\})} = v^T \chi^{\delta(\{1,2\})} = \alpha$, one deduces the $v_{12} = v_{23} = \alpha/2$. Therefore, we obtain that inequality $v^T x \geq \alpha$ is a multiple of inequality (50), implying that inequality (50) indeed defines a facet of $\mathrm{MC}^{=}_{2}(n)$ and thus of $\mathrm{MC}^{\geq}_{2}(n)$ too.

2. Take again an inequality $v^T x \geq \alpha$ that is valid for $(p-1)$-$\mathrm{BMC}^{\geq}_{n-p}(n)$ and admits the same roots in $(p-1)$-$\mathrm{BMC}^{\geq}_{n-p}(n)$ as inequality (50). Set $S := \{2, \ldots, u\}$ for $2 \leq u \leq p-1$ and $S' := \{u+2, \ldots, p+1\}$. Let $\delta_1$ (respectively, $\delta_2, \delta_3, \delta_4$) denote the multicut whose shores are the sets $S$ and $S' \cup \{1, u+1\}$ (respectively, $S \cup \{1\}$ and $S' \cup \{u+1\}$, $S \cup \{1, u+1\}$ and $S'$, $S \cup \{u+1\}$ and $S' \cup \{1\}$) together with the singletons $\{i\}$ for

$p+2 \le i \le n$. So $\delta_1, \delta_2, \delta_3, \delta_4$ are of type (50a). Using the fact that $v^T\chi^{\delta_1} - v^T\chi^{\delta_2} = v^T\chi^{\delta_3} - v^T\chi^{\delta_4} = 0$, one deduces easily that $v_{1u+1} = 0$. Take $j \in \{p+2, \dots, n\}$. Then both multicuts with shores $\{1\}, \{2, \dots, p+1\}$ and singletons $\{i\}$ for $p+2 \le i \le n$ and with shores $\{1, j\}, \{2, \dots, p+1\}$ and singletons $\{i\}$ for $p+2 \le i \le n$, $i \ne j$, are roots (of type (50b)) yielding that $v_{1j} = 0$. One now deduces easily that $v^Tx \ge \alpha$ is a multiple of inequality (50), which indeed shows that (50) defines a facet of $(p-1)\text{-BMC}^{\ge}_{n-p}(n)$. □

Finally note that, if $C$ is a cycle of length $n$, then the inequality $\sum_{ij \in C} x_{ij} \ge k$ is trivially valid on $\mathrm{MC}^{\ge}_k(n)$.

**(5.4) A class of "casserole" facets.** We saw that the casserole inequality (31) defines a facet of the polytope $\mathrm{MC}^{\ge}_3(5)$. The following inequality can be seen as an extension of inequality (31):

$$\sum_{1 \le i < j \le n-1} x_{ij} + 2x_{1n} \ge \binom{n-1}{2} - 1 \tag{51}$$

Proposition 4. *Inequality* (51) *defines a facet of the polytopes* $\mathrm{MC}^{\ge}_{n-2}(n)$ *and* $\mathrm{MC}^{=}_{n-2}(n)$ *for* $n \ge 5$.

Proof. Validity of inequality (51) for $\mathrm{MC}^{\ge}_{n-2}(n)$ is easy to check. Its roots are the following multicut vectors $x$:

(1) $x$ is the incidence vector of an $(n-2)$-cut whose shores are singletons except two shores that are pairs $\{1, n\}, \{i, j\}$ for $2 \le i < j \le n-1$.
(2) $x$ is the incidence vector of an $(n-2)$-cut whose shores are singletons except one shore that is a triple which is either $\{1, i, j\}$ with $2 \le i < j \le n-1$, or $\{1, n, i\}$ with $2 \le i \le n-1$, or $\{i, j, k\}$ with $2 \le i < j < k \le n-1$.

Using the above description of the roots, it is not difficult to show that any inequality $v^Tx \ge \alpha$ which is valid for $\mathrm{MC}^{=}_{n-2}(n)$ and is satisfied at equality by the multicut vectors defined in (1), (2), (3) above is indeed a multiple of inequality (51). □

**(5.5) Some classes of "crown" facets.** We introduce two classes of inequalities (52), (53), which contain as special cases the "crown" inequalities (23), (24), and (27), respectively.

Proposition 5. *Consider a partition of* $\{1, 2, \dots, n\}$ *into subsets* $A, A'$, *and* $\{1, n\}$. *Then the inequality*

$$-(n-3)x_{1n} + \sum_{i \in A}(x_{1i} - x_{in}) + \sum_{i \in A'}(x_{in} - x_{1i}) \le 0 \tag{52}$$

*defines a facet of* $\mathrm{MC}^{\ge}_3(n)$ *for* $n \ge 5$.

PROOF. One can easily check that inequality (52) is valid for $\mathrm{MC}_3^{\geq}(n)$ and that its roots in $\mathrm{MC}_3^{\geq}(n)$ are the incidence vectors $x$ of the multicuts for which either

- both nodes $1, n$ belong to the same shore of the multicut, or
- the multicut is of the form $\delta(A' \cup \{1\}, A \setminus \{i\} \cup \{n\}, \{i\})$ for some $i \in A$, or
- the multicut is of the form $\delta(A' \setminus \{i\} \cup \{1\}, A \cup \{n\}, \{i\})$ for some $i \in A'$.

Then one can verify that the set of roots of (52) has rank $\binom{n}{2} - 1$ implying that inequality (52) defines a facet for $\mathrm{MC}_3^{\geq}(n)$. □

Note that inequality (52) is not valid for $\mathrm{MC}_2^{=}(n)$. It is, in fact, violated by the incidence vector of the 2-cut $\delta(A \cup \{n\})$ but by no other 2-cut vector. Note also that, for $n = 5$, inequality (52) coincides with (23) if $|A| = 3$ and with (24) if $|A| = 1$ or 2.

PROPOSITION 6. *The inequality*

$$(n-4)x_{1n} - \sum_{2 \leq i \leq n-2} (x_{1i} + x_{in}) \leq -2 \tag{53}$$

*defines a facet of the polytopes* $\mathrm{MC}_2^{=}(n)$, $\mathrm{MC}_2^{\geq}(n)$.

PROOF. It is easy. We simply mention that the roots of inequality (53) in $\mathrm{MC}_2^{=}(n)$ are the incidence vectors of the 2-cuts $\delta(\{i\})$ for $2 \leq i \leq n-1$, and $\delta(S \cup \{n\})$ for any subset $S$ of $\{2, 3, \ldots, n-1\}$. □

Note that inequality (53) for $n = 5$ coincides with inequality (27).

**(5.6) On the polytopes** $1\text{-}\mathrm{BMC}_{n-2}^{\geq}(n)$, $\mathrm{MC}_{n-2}^{=}(n)$**.** The balanced multicut polytope $1\text{-}\mathrm{BMC}_3^{\geq}(5)$ has 26 vertices and 26 facets which are of one of the following four types:

- upper bound condition (1) $x_{12} \leq 1$,
- cardinality condition (7) $\sum_{ij \in E} x_{ij} \geq 8$,
- lower bound on 3-cycle condition (19) $x_{12} + x_{13} + x_{23} \geq 2$,
- degree condition (22) $\sum_{i=2}^{5} x_{1i} \geq 3$.

All of them extend to the general balanced multicut polytope $1\text{-}\mathrm{BMC}_{n-2}^{\geq}(n)$. The polytope $1\text{-}\mathrm{BMC}_{n-2}^{\geq}(n)$ has $\frac{1}{2}\binom{n}{2}\binom{n-1}{2} + \binom{n}{2} + 1$ vertices that are incidence vectors of the $n$-cut, the $(n-1)$-cuts, and the $(n-2)$-cuts whose shores are singletons except two shores which are pairs. First, the upper bound condition (1) clearly defines a facet of $1\text{-}\mathrm{BMC}_{n-2}^{\geq}(n)$. The cardinality condition for general $n$ is as follows:

$$\sum_{1 \leq i < j \leq n} x_{ij} \geq \binom{n}{2} - 2. \tag{54}$$

One can easily check that inequality (54) indeed defines a facet of $1\text{-}\mathrm{BMC}^{\geq}_{n-2}(n)$. We already proved in Proposition 3 that the lower bound on a 3-cycle defines a facet of $1\text{-}\mathrm{BMC}^{\geq}_{n-2}(n)$. Finally, the degree condition for general $n$ is as follows:

$$\sum_{i=2}^{n} x_{1i} \geq n-2, \tag{55}$$

and it is easily verified that inequality (55) defines a facet of $1\text{-}\mathrm{BMC}^{\geq}_{n-2}(n)$.

We finally mention another multicut polytope for which all its facets admit an easy generalization. This is the case for the polytope $\mathrm{MC}^{=}_{3}(5)$ that has nine types of facets, namely those defined by inequalities (1), (3), (6), (14), (16), (30), (31), (38), and (40). All of them extend to the polytope $\mathrm{MC}^{=}_{n-2}(n)$ for any $n \geq 5$. The polytope $\mathrm{MC}^{=}_{n-2}(n)$ has as vertices the incidence vectors of the $(n-2)$-cuts, i.e., the multicuts whose shores are all singletons except two pairs, or all singletons except one triple. For each of the facet-defining inequalities of $\mathrm{MC}^{=}_{3}(5)$, we can state a general inequality that is valid for $\mathrm{MC}^{=}_{n-2}(n)$ as follows:

- as an extension of the cardinality constraints (3), (6), the inequalities

$$\binom{n}{2} - 3 \leq \sum_{1 \leq i < j \leq n} x_{ij} \leq \binom{n}{2} - 2$$

  are valid for $\mathrm{MC}^{=}_{n-2}(n)$,
- as an extension of (14), inequality

$$\sum_{1 \leq i < j \leq n-1} x_{ij} \leq \binom{n-1}{2} - 1$$

  is valid for $\mathrm{MC}^{\leq}_{n-2}(n)$,
- as an extension of (16), inequality

$$\sum_{ij \in C} x_{ij} \geq n-2,$$

  where $C$ is a cycle of length $n$, is valid for $\mathrm{MC}^{\geq}_{n-2}(n)$,
- as an extension of (30), inequality

$$\sum_{1 \leq i < j \leq n} x_{ij} - (x_{12} + x_{13} + x_{23} + x_{45}) \geq \binom{n}{2} - 6$$

  is valid for $\mathrm{MC}^{=}_{n-2}(n)$,
- as an extension of (31), we have the casserole inequality (51) which is facet-defining for $\mathrm{MC}^{=}_{n-2}(n)$ (see Proposition 4),

- as an extension of (38), inequality

$$2 \sum_{1\le i<j\le n} x_{ij} - (x_{12} + x_{13} + x_{23}) \le 2\binom{n}{2} - 6$$

is valid for $\mathrm{MC}^{\le}_{n-2}(n)$,

- as an extension of (40), inequality

$$\sum_{1\le i<j\le n-1} x_{ij} + 2(x_{1n} + x_{2n} - x_{12}) \ge \binom{n-1}{2} - 1$$

is valid for $\mathrm{MC}^{\ge}_{n-2}(n)$.

## 6. Complete descriptions of the multicut polytopes for $n = 4$

For completeness we now list also all multicut polytopes introduced in Definition 2 for the complete graph $K_4 = (V, E)$. Our numerical computations did not produce any new and interesting class of inequalities. All classes of inequalities that appear as facet-defining inequalities for at least one of the polytopes belong to some general type.

As in §3 we number the 15 different multicut vectors to be able to give short definitions of the polytopes of interest. We do this as follows:

| | | |
|---|---|---|
| 1 : 1234 | 2 : 1, 234 | 3 : 2, 134 |
| 4 : 3, 124 | 5 : 4, 123 | 6 : 12, 34 |
| 7 : 13, 24 | 8 : 14, 23 | 9 : 1, 2, 34 |
| 10 : 1, 3, 24 | 11 : 1, 4, 23 | 12 : 2, 3, 14 |
| 13 : 2, 4, 13 | 14 : 3, 4, 12 | 15 : 1, 2, 3, 4 |

So if we sort the edges $ij$ of $E$ as 12, 13, 14, 23, 24, 34, then the 4th multicut vector, for instance, is the vector $\chi^{(\delta\{3\},\{1,2,4\})} = (0, 1, 0, 1, 0, 1)$.

We now list all polytopes that arise for $n = 4$, $k = 2, 3$ and $s = 0, 1, 2, 3$ leaving out trivial cases of polytopes that are empty, single points, or otherwise trivial or those that are equal to some other polytope listed. The facets mentioned will be introduced in the sequel. The ordering follows the same lexicographic scheme as explained for $n = 5$.

$$\mathrm{MC}(4) = \mathrm{conv}\{1, 2, \dots, 15\}, \quad v = 15, \quad f = 22,$$
facets: (56), (64), (65),

$$\mathrm{MC}_2^{\leq}(4) = \mathrm{conv}\{1, 2, \dots, 8\}, \quad v = 8, \quad f = 16,$$
facets: (59), (64),

$$\mathrm{MC}_2^{\geq}(4) = \mathrm{conv}\{2, 3, \dots, 15\}, \quad v = 14, \quad f = 25,$$
facets: (57), (61), (64), (65),

$$\mathrm{MC}_2^{=}(4) = \mathrm{conv}\{2, 3, \dots, 8\}, \quad v = 7, \quad f = 7,$$
facets: (59), (61),

$$\mathrm{MC}_3^{\leq}(4) = \mathrm{conv}\{1, 2, \dots, 14\}, \quad v = 14, \quad f = 23,$$
facets: (57), (58), (64), (65),

$$\mathrm{MC}_3^{\geq}(4) = \mathrm{conv}\{9, 10, \dots, 15\}, \quad v = 7, \quad f = 7,$$
facets: (57), (60),

$$\mathrm{MC}_3^{=}(4) = \mathrm{conv}\{9, 10, \dots, 14\}, \quad v = 6, \quad f = 6, \quad \text{dimension} = 5,$$
facets: (57), equation: (56),

$$2\text{-}\mathrm{BMC}_3^{\leq}(4) = \mathrm{conv}\{2, 3, \dots, 14\}, \quad v = 13, \quad f = 26,$$
facets: (57), (58), (61), (64), (65),

$$1\text{-}\mathrm{BMC}_2^{\geq}(4) = \mathrm{conv}\{6, 7, \dots, 15\}, \quad v = 10, \quad f = 14,$$
facets: (57), (62), (63),

$$1\text{-}\mathrm{BMC}_3^{\leq}(4) = \mathrm{conv}\{6, 7, \dots, 14\}, \quad v = 9, \quad f = 14,$$
facets: (57), (58), (62), (63).

The polytopes $\mathrm{MC}_2^{=}(4)$, $\mathrm{MC}_3^{\geq}(4)$, and $\mathrm{MC}_3^{=}(4)$ are simplices.

The following equation defines an affine space that is spanned by one of the polytopes:

(56) $$x(E) = 5.$$

**Upper bounds on edge sets**

(57) $x_{12} \leq 1$ (6 permutations),

(58) $x(E) \leq 1$ (1 permutation),

(59) $x(T) \leq 2$ ($T$ a triangle) (4 permutations).

**Lower bounds on edge sets**

(60) $x(E) \geq 5$ (1 permutation),

(61) $x(C) \geq 2$ ($C$ a 4-cycle) (3 permutations),

(62) $x(T) \geq 2$ ($T$ a triangle) (4 permutations),

(63) $x(\delta(i)) \geq 2$ (4 permutations).

### Hypermetric inequalities

(64) $Q(1, 1, -1, 0) \cdot x \leq 0$ (triangle inequalities) (12 permutations),

(65) $Q(1, 1, 1, -1) \cdot x \leq 1$ (4 permutations).

This finishes our list of facet-defining inequalities for multicut polytopes of $K_4$.

## References

**[BJR]** F. Barahona, M. Jünger, and G. Reinelt, *Experiments in quadratic 0-1 programming*, Math. Programming **44** (1989), 127–137.

**[BGJR]** F. Barahona, M. Grötschel, M. Jünger, and G. Reinelt, *An application of combinatorial optimization to statistical physics and circuit layout design*, Oper. Res. **36** (1988), 493–513.

**[BM]** F. Barahona and E. Maccioni, *On the exact ground states of three-dimensional Ising spin glasses*, J. Phys. A **15** (1982), L611–L615.

**[BMa]** F. Barahona and A. R. Mahjoub, *On the cut polytope*, Math. Programming **36** (1986), 157–173.

**[CR1]** S. Chopra and M. R. Rao, *The partition problem* (submitted).

**[CR2]** ——, *Facets of the k-partition polytope* (submitted).

**[CRS]** M. Conforti, M. R. Rao, and A. Sassano, *The equipartition polytope*. I, II, Math. Programming **49** (1990/91), 49–70 and 71–90.

**[D]** C. De Simone, *Lifting facets of the cut polytope*, Operations Research Letters **9** (1990), 341–344.

**[DFL]** M. Deza, K. Fukuda and M. Laurent, *The inequicut cone*, Research report No. 89-04, The University of Tsukuba, Tokyo. Japan, 1989.

**[DGL]** M. Deza, M. Grötschel and M. Laurent, *Clique-Web facets for multicut polytopes*, Report No. 186, Schwerpunktprogramm der Deutschen Forschungsgemeinschaft, Universität Augsburg, Germany, 1989.

**[DL]** M. Deza and M. Laurent, *Facets for the cut cone*. I, Research Memorandum RMI 88-13, University of Tokyo (1988); Math. Programming (to appear).

**[GW1]** M. Grötschel and Y. Wakabayashi, *A cutting plane algorithm for a clustering problem*, Math. Programming B **45** (1989), 59–96.

**[GW2]** ——, *Facets of the clique partitioning polytope*, Math. Programming **47** (1990), 367–387.

**[W]** Y. Wakabayashi, *Aggregation of binary relations: algorithmic and polyhedral investigations*, Ph.D. Thesis, Universität Augsburg, West Germany, 1986.

CNRS, Université de Paris VII, Paris, France

Institut für Mathematik, Universität Augsburg, Universitätsstrasse 8, D-8900 Augsburg, Germany
*E-mail address*: groetschel@augsopt.uni-augsburdg.de

CNRS, Université de Paris IX, LAMSADE, Place du Maréchal de Lattre de Tassigny, 75016, Paris, France
*E-mail address*: or907@dbnuor1

DIMACS Series in Discrete Mathematics
and Theoretical Computer Science
Volume **4**, 1991

Applied Geometry
and Discrete Mathematics
THE VICTOR KLEE FESTSCHRIFT

# A Hyperplane Incidence Problem with Applications to Counting Distances

HERBERT EDELSBRUNNER AND MICHA SHARIR

ABSTRACT. This paper proves an $O(m^{2/3}n^{2/3} + m + n)$ upper bound on the number of incidences between $m$ points and $n$ hyperplanes in four dimensions, assuming all points lie on one side of each hyperplane and the points and hyperplanes satisfy certain natural general position conditions. This result has applications to various three-dimensional combinatorial distance problems. For example, it implies the same upper bound for the number of bichromatic minimum distance pairs in a set of $m$ blue and $n$ red points in three-dimensional space. This improves the best previous bound for this problem.

## 1. Introduction

Combinatorial distance problems for finite point sets in Euclidean spaces are classical topics in discrete geometry. Most popular is probably the question of how often the unit distance can occur in a set of $m$ points in two or three dimensions; this question was originally studied by Erdös [**10, 11**]. The current best upper bound in the plane is $O(m^{4/3})$ (see [**15**] and also [**4**]), and in three dimensions Clarkson et al. [**4**] proved $O(m^{3/2}\beta(m))$, where $\beta(m)$ is an extremely slowly growing function related to the inverse of Ackermann's function. No matching lower bounds are known.

This paper considers three distance problems in three dimensions and improves the best previous upper bound in each case. The results are as follows. The number of bichromatic minimum distance pairs in a set of $m$ blue and

1980 *Mathematics Subject Classification* (1985 *Revision*). Primary 68U05.

*Key words and phrases.* Discrete geometry, counting incidences and distances, bichromatic minimum distance, nearest and furthest neighbors, Delaunay triangulations, convex polytopes, three and four dimensions, geometric transforms.

Research of the first author was supported by the National Science Foundation under grants CCR-8714565 and CCR-8921421. Work of the second author was supported by Office of Naval Research Grants DCR-83-20085 and CCR-89-01484, and by grants from the U.S.–Israeli Binational Science Foundation, the NCRD—the Israeli National Council for Research and Development, and the Fund for Basic Research in Electronics, Computers and Communication administered by the Israeli Academy of Sciences.

$n$ red points is $O(m^{2/3}n^{2/3} + m + n)$. The same bound holds for the number of bichromatic nearest neighbor pairs if no three points are collinear. For a (monochromatic) set of $m$ points, the number of furthest neighbor pairs is $O(m^{4/3})$ if no three points are collinear. All three bounds are corollaries of the following more general result on the number of incidences between $m$ points and $n$ hyperplanes in four dimensions. If no three points are collinear, no three hyperplanes intersect in a common 2-flat, and each hyperplane bounds a closed half-space that contains all points, then the number of incidences is $O(m^{2/3}n^{2/3} + m + n)$. Besides the application of this result to three-dimensional distance problems we also consider an application to three-dimensional Delaunay triangulations.

This paper is organized as follows. §2 proves the upper bound on the hyperplane incidence problem, §3 explains how this problem relates to counting incidences between points and spheres in three dimensions, §4 discusses the combinatorial distance problems, and §5 concludes the paper.

## 2. The hyperplane incidence problem

This section proves the main result of this paper, an upper bound for the following combinatorial geometry problem.

Problem specification. Let $P$ be a set of $m$ points and $H$ a set of $n$ hyperplanes in four-dimensional Euclidean space satisfying the following conditions.

(H.i) No three points of $P$ are collinear.
(H.ii) No three hyperplanes of $H$ intersect in a common 2-flat.
(H.iii) Each hyperplane in $H$ bounds a closed half-space that contains $P$.

What is the maximum number of incidences between the points and the hyperplanes, in terms of $m$ and $n$, where the maximum is taken over all sets $P$ of size $m$ and $H$ of size $n$ satisfying the three conditions? We define $I(m, n)$ to be equal to this number.

We prove an upper bound on $I(m, n)$ by adapting the methods of Clarkson et al. [**4**]. This is done in two steps. First, bounds are obtained which are tight for the cases when $m$ is much smaller than $n$ $(m < \sqrt{n})$ and when $n$ is much smaller than $m$ $(n^2 < m)$; using the terminology of [**4**] we call these bounds Canham thresholds. Second, these bounds together with the methods of [**4**] are used to establish an upper bound for the remaining case $(\sqrt{n} \le m \le n^2)$. To prove the Canham thresholds (Lemma 2.1) we make use of the fact that the problem is self-dual, that is, if we apply a dual transform we arrive at the same problem, only with the roles of $m$ and $n$ interchanged. Indeed, choose the origin in the interior of the intersection of half-spaces that contains all points of $P$. For a point $p$ different from the origin let

$$p^* = \{x \mid \langle x, p\rangle = 1\}$$

be the *polar hyperplane*, and for a hyperplane $h$ avoiding the origin define the *polar point* $h^*$ so that $h = (h^*)^*$. Using straightforward algebraic manipulations it can easily be verified that $p \in h$ iff $h^* \in p^*$. Furthermore, three points are collinear iff their polar hyperplanes intersect in a common 2-flat. In other words, if we map all points to their polar hyperplanes and all hyperplanes to their polar points we get a point/hyperplane incidence problem satisfying conditions (H.i) and (H.ii). In addition to the incidence preserving property the polar transform preserves sidedness relative to the origin, which implies that also condition (H.iii) is satisfied after polarization. Hence, $I(m, n) = I(n, m)$.

**The Canham thresholds.** The intersection of half-spaces defined by the hyperplanes is a closed convex polyhedron $\mathscr{C}$ with at most $n$ facets. The upper bound theorem (see **[3]** or **[8]**) implies that $\mathscr{C}$ has at most $O(n^2)$ vertices, edges, and ridges.[1] Let the *degree* of a face (vertex, edge, ridge, or facet) be the number of hyperplanes in $H$ that contain it. Clearly, the degree of any facet is $1$, by assumption (H.ii) the degree of any ridge is $2$, and the degree of any edge or vertex can be arbitrary. Still, the sum of degrees, over all faces, is at most $O(n^2)$ by the following argument. Take each hyperplane that contains a vertex or edge of $\mathscr{C}$ and push it inwards (towards the origin) by a tiny amount so that each such hyperplane now supports a facet. Call the new polytope $\mathscr{C}'$. Next we slightly perturb the hyperplanes, on a much smaller scale than before, in order to transform $\mathscr{C}'$ into a simple polyhedron $\mathscr{C}''$, again without decreasing the sum of face degrees. Now every edge has degree 3 and every vertex has degree 4 which implies the claim by applying the upper bound theorem to $\mathscr{C}''$. After these introductory remarks we are ready to prove the Canham thresholds.

LEMMA 2.1. *$I(m, n) = O(m\sqrt{n} + n)$ and $I(m, n) = O(n\sqrt{m} + m)$.*

PROOF. Because of the self-dual property of the incidence problem we can restrict ourselves to proving $I(m, n) = O(n\sqrt{m}+m)$. Let $P$ be the set of $m$ points, $H$ the set of $n$ hyperplanes, and $\mathscr{C}$ the polyhedron defined above which contains $P$. If a point $p \in P$ lies in the interior of $\mathscr{C}$ it contributes $0$ to the number of incidences, if it lies in the relative interior of a facet it contributes $1$, and if it lies in the relative interior of a ridge it contributes $2$. The total number of incidences involving such point is therefore $O(m)$. Each edge of $\mathscr{C}$ contains at most two points of $P$, by condition (H.i), and each vertex contains at most one point of $P$. By the claim established earlier, the total number of incidences involving points on edges and vertices of $\mathscr{C}$ is therefore $O(n^2)$ which gives a combined bound of $O(m + n^2)$. To get $I(m, n) = O(n\sqrt{m}+m)$ we simply partition $H$ into $\lceil n/\sqrt{m} \rceil$ subsets of size at most $\sqrt{m}$ each. For each subset we get only $O(m)$ incidences with points of $P$ and therefore we get at most $O(n\sqrt{m} + m)$ incidences altogether. □

[1] In general, a *ridge* is a $(d - 2)$-face of a $d$-polyhedron. Because $\mathscr{C}$ is a 4-polyhedron a ridge of $\mathscr{C}$ is a 2-face.

REMARKS. (1) Conditions (H.i) and (H.ii) can be relaxed to allow up to some constant number of points that are collinear and hyperplanes that meet in a common 2-flat without sacrificing the asymptotic bounds in Lemma 2.1. However, if we drop (H.i) or (H.ii) we can find examples with $\Omega(mn)$ incidences. The problem that results when we drop (H.iii) but not (H.i) and (H.ii) is more difficult. The best lower bound known to the authors is $\Omega(m^{4/3} \log\log m)$ for the case $m = n$ and is based on an example of Erdös [**11**] (see also [**4**]).

(2) Conditions (H.i) and (H.ii) imply that no three hyperplanes can all be incident to three common points. Using a standard extremal graph lemma (see [**2**] or [**4**]) this can be used to prove $O(mn^{2/3} + n)$ and $O(nm^{2/3} + m)$ as upper bounds on $I(m, n)$. These bounds are significantly weaker than the bounds of Lemma 2.1, which illustrates the importance of condition (H.iii) and the use of the upper bound theorem.

**The main result.** We introduce some notation. For a hyperplane $h \in H$ let $h^+$ be the closed half-space bounded by $h$ that contains the origin. Recall that we assume that the origin lies in the interior of $\mathscr{C}$ which implies that $\mathscr{C} = \bigcap_{h \in H} h^+$. We now present a sequence of arguments that add up to a proof of the main result of this section.

First, choose a random sample $R \subseteq H$ of size $r$. Form the convex polyhedron $\mathscr{R} = \bigcap_{h \in R} h^+$ and note that $\mathscr{R}$ contains all points of $P$ and also the origin. It will be convenient to assume that $\mathscr{R}$ is actually a polytope (that is, it is bounded) which can be achieved by intersecting it with a sufficiently large tetrahedron.

Second, we triangulate $\mathscr{R}$ as follows. Choose a directed line so that no hyperplane normal to this line contains two vertices of $\mathscr{R}$ and call the direction defined by this line *vertical*. It thus makes sense to talk about a point being *higher* than another point. A ridge (that is, 2-face) of $\mathscr{R}$ is triangulated by connecting its highest vertex to all other vertices. Each triangle is thus bounded by two edges incident to the highest vertex of the ridge and a third edge which is an original edge of the ridge. Similarly, a facet (that is, 3-face) of $\mathscr{R}$ is triangulated by connecting its highest vertex to all vertices, edges, and triangles in the (triangulated) boundary of the facet. Thus, each tetrahedron is incident to the highest vertex and is bounded by a triangle not incident to this vertex. Finally, the interior of $\mathscr{R}$ is triangulated by connecting the origin to all vertices, edges, triangles, and tetrahedra in the (triangulated) boundary of $\mathscr{R}$. The number of 4-simplices generated is equal to the number of tetrahedra in the boundary of $\mathscr{R}$. In turn, the number of tetrahedra is at most twice the number of triangles because any ridge is incident to only two facets. Finally, the number of triangles used to triangulate the ridges is bounded from above by the number of edge/ridge incidences. The number of ridges incident to a single edge is at most the degree of the edge, and we noted earlier that the sum of edge degrees is at most quadratic in the number of hyperplanes. Thus, if $k$ is the number of 4-simplices in the triangulation then $k = O(r^2)$.

For the application of a probabilistic counting result due to Clarkson and Shor [5] in the fifth step below, it is important that each 4-simplex is uniquely determined by some constant number of hyperplanes. Indeed, a triangle lies in a plane (the intersection of two hyperplanes) and each vertex is defined by two lines (the intersection of this plane with two other hyperplanes). At least two of the three vertices share a line because at least one of the edges of the triangle is chosen to lie on a line. Thus, we need at most seven hyperplanes to define a triangle. A tetrahedron connects a triangle to a vertex (the intersection of four hyperplanes, one of which contains the triangle) and is thus defined by at most ten hyperplanes. A 4-simplex just connects a tetrahedron with the origin and is therefore determined by at most ten hyperplanes and a point that is fixed independent of $\mathscr{R}$.

Another necessary property for the application of the probabilistic counting result is that a 4-simplex defined by the origin and at most ten hyperplanes in the way described above is in the triangulation *if and only if* it does not intersect any other hyperplane. Unfortunately, this is not strictly true because the hyperplanes are not in general position which allows for the possibility that different sets of at most ten hyperplanes define the same 4-simplex. However, we can simulate an arbitrarily small perturbation of the hyperplanes to get them into general position and define the triangulation with respect to this perturbation. The perturbation is used merely for the purpose of assigning proper sets of hyperplanes to the 4-simplices of the triangulation—it is ignored as far as point/hyperplane incidences are concerned. Before we go on with the proof we remark that the origin is an arbitrary point in the interior of $\mathscr{C}$. We may therefore assume that each point of $P$ that does not lie on the boundary of $\mathscr{R}$ lies in the interior of a 4-simplex of the triangulation.

The third step of the proof bounds the number of incidences involving points in the interior of $\mathscr{R}$. For $1 \le i \le k$, let $\sigma_i$ be the interior of the $i$th 4-simplex, set $m_i = |P \cap \sigma_i|$, and let $n_i$ be the number of hyperplanes in $H$ that have nonempty intersection with $\sigma_i$. Using the first bound in Lemma 2.1 we thus get $\sum_{i=1}^{k} O(m_i\sqrt{n_i} + n_i)$ as an upper bound for the number of incidences that happen in the interior of $\mathscr{R}$.

Fourth, we bound the number of incidences involving points on the boundary of $\mathscr{R}$. For a hyperplane $h \in R$ define $P_h = P \cap h$ and $m_h = |P_h|$. The convex hull of $P_h$ is a convex polytope of dimension 3 or less. Because of its low dimensionality, it has at most $O(m_h)$ edges and 2-faces. Since no three points in $P_h$ are collinear (condition (H.i)) and no three hyperplanes in $H$ meet in a common 2-flat (condition (H.ii)), the total number of incidences involving points of $P_h$ is $O(m_h + n)$. If we sum this bound over all $r$ hyperplanes $h \in R$ we get $O(I(m, r) + nr)$ since a point $p$ belongs to as many sets $P_h$ as there are incident hyperplanes $h$ in $R$. We have $I(m, r) = O(r\sqrt{m} + m)$ using the second bound in Lemma 2.1.

In the fifth and final step we make use of the probabilistic counting result of Clarkson and Shor **[5]**. It implies that the expected value of $\sum_{i=1}^{k} O(m_i\sqrt{n_i}+n_i)$ is $O(m\sqrt{n/r}+kn/r)$. Note that $kn/r = O(nr)$ and that the above sum is an upper bound on the number of incidences that happen in the interior of $\mathscr{R}$. Thus, there exists a subset $R \subseteq H$ of size $r$ so that the number of incidences in the interior of $\mathscr{R}$ is at most $O(m\sqrt{n/r} + nr)$. Adding to this bound the incidences on the boundary of $\mathscr{R}$ we get

$$I(m, n) = O(m\sqrt{n/r} + nr + r\sqrt{m} + m)$$

as an upper bound on the total number of incidences. If we choose $r$ equal to $\lceil m^{2/3}/n^{1/3}\rceil$ this bound becomes $O(m^{2/3}n^{2/3} + (m^{7/6}/n^{1/3}) + m)$. This choice of $r$ is meaningful if $\sqrt{n} \leq m \leq n^2$ in which case $m^{7/6}/n^{1/3} \leq m$. For the remaining cases, $m < \sqrt{n}$ and $n^2 < m$, Lemma 2.1 shows that the number of incidences is at most $O(m + n)$. This concludes the proof of the main result of this section which we now state.

THEOREM 2.2. *The maximum number of incidences between $m$ points and $n$ hyperplanes in four dimensions satisfying conditions* (H.i)–(H.iii) *is* $I(m, n) = O(m^{2/3}n^{2/3} + m + n)$.

REMARKS. (1) Remark (1) after Lemma 2.1 also applies to the bound in Theorem 2.2.

(2) Note that the only way condition (H.ii) is exploited in the above proofs is that more than some constant number of points cannot be incident to more than some constant number of common hyperplanes. Another condition achieving the same goal is

(H.ii′) No four points of $P$ lie in a common 2-flat.

In other words, Theorem 2.2 can also be shown if we replace (H.ii) by (H.ii′).

(3) We state as an open problem to prove any superlinear lower bound for $I(m, n)$.

## 3. The sphere incidence problem

Using Theorem 2.2 and a fairly standard geometric lifting transform (see, for example, **[8]**), we can derive a good upper bound for an incidence problem involving points in three dimensions. To simplify the notation we use the word *sphere* to either mean a sphere in the common Euclidean sense, or a plane. We define a (*generalized*) *ball* as a closed ball, a closed half-space, or the complement of an open ball in three dimensions. Thus, each sphere bounds two balls and for each ball there is a sphere that bounds it. This slightly nonstandard use of terms will be restricted to the scope of this section.

PROBLEM SPECIFICATION. Let $P$ be a set of $m$ points and $S$ a set of $n$ spheres in three-dimensional Euclidean space satisfying these conditions.

(S.i) No three spheres in $S$ intersect in a common circle or line.
(S.ii) Each sphere $s \in S$ bounds a ball $s^+$ so that $P \subseteq s^+$.

What is the maximum number of incidences between the points and spheres, where the maximum is taken over all sets $P$ of size $m$ and $S$ of size $n$ satisfying (S.i) and (S.ii)?

We derive an upper bound on the number of incidences by mapping $P$ to a set of $m$ points and $S$ to a set of $n$ hyperplanes in four dimensions so that conditions (H.i)–(H.iii) are satisfied. More specifically, a point $p = (\pi_1, \pi_2, \pi_3)$ is mapped to the point $\bar{p} = (\pi_1, \pi_2, \pi_3, \pi_1^2+\pi_2^2+\pi_3^2)$. Note that $\bar{p}$ is the vertical projection of $p$, a point in the $x_1x_2x_3$-space, onto the paraboloid of revolution specified by the equation $x_4 = x_1^2 + x_3^2 + x_3^2$. Similarly, a sphere $s$ in the $x_1x_2x_3$-space is mapped to a four-dimensional hyperplane $\bar{s}$ by vertically projecting each of its points onto the same paraboloid and defining $\bar{s}$ as the (unique) hyperplane that contains all these points. The crucial property of this transform is that a point $p$ lies outside, on, or inside a sphere $s$ iff $\bar{p}$ lies vertically above, on, or below $\bar{s}$. (If $s$ is a plane then $\bar{s}$ is a vertical hyperplane (parallel to the $x_4$-axis) and sidedness is maintained as in the general case.) It thus follows that the sets $\overline{P} = \{\bar{p} | p \in P\}$ and $\overline{S} = \{\bar{s} | s \in S\}$ satisfy condition (H.iii). Condition (H.i) holds because no three points of the paraboloid are collinear, and (H.ii) follows from (S.i). What we said above includes as a special case that $p \in s$ iff $\bar{p} \in \bar{s}$, which implies the following upper bound for the number of point/sphere incidences.

THEOREM 3.1. *The number of incidences between $m$ points and $n$ spheres in three dimensions satisfying conditions* (S.i) *and* (S.ii) *is* $O(m^{2/3}n^{2/3} + m + n)$.

REMARKS. (1) In agreement with remark (1) after Lemma 2.1, condition (S.i) can be relaxed to allow up to some constant number of spheres intersecting a common circle or line.

(2) As noted in remark (2) after Theorem 2.2, it is possible to replace condition (H.ii) by (H.ii′) without sacrificing the $O(m^{2/3}n^{2/3} + m + n)$ upper bound. By the same reason it is possible to replace (S.i) by

(S.i′) No four points of $P$ lie on a common circle or line

without sacrificing the upper bound given in Theorem 3.1.

(3) No superlinear lower bound for the sphere incidence problem of this section is currently known to the authors; see also remark (3) after Theorem 2.2.

## 4. Combinatorial distance problems

In this section we apply Theorem 3.1 to get upper bounds on problems about repeated distances between points in three-dimensional Euclidean space. Many such problems were originally posed by Paul Erdös and were considered before in the literature. We refer to the problem collection of Moser and Pach **[14]** as a general source of relevant information.

**4.1. Bichromatic minimum distance.** Given a set $P$ of $m$ blue points and a set $Q$ of $n$ red points in three-dimensional Euclidean space, what is the maximum number of pairs $(p, q) \in P \times Q$ that realize the minimum distance between points of different color? The best previous bound, derived in **[4]**, is $O(m^{3/4}n^{3/4}\beta(m, n) + m + n)$, with $\beta(m, n) = 2^{\Theta(\alpha(m^3/n)^2)}$ and $\alpha$ the inverse of Ackermann's function. The reduction to Theorem 3.1 should be obvious: around each blue point draw a sphere with radius equal to the minimum bichromatic distance. Condition (S.i) is satisfied because all spheres are equally large, and condition (S.ii) holds because all red points lie on or outside all spheres.

THEOREM 4.1. *The number of bichromatic minimum distance pairs in a set of* $m$ *blue and* $n$ *red points in three-dimensional Euclidean space is* $O(m^{2/3}n^{2/3} + m + n)$.

REMARK. In the case of a monochromatic set of $m$ points in three dimensions the maximum number of minimum distance pairs is $\Theta(m)$. This is because if 13 or more points have the same distance to a point $p$ then at least two of them are closer to each other than to $p$. It follows that each point $p$ belongs to at most 12 minimum distance pairs. This packing argument fails in the bichromatic case because points of the same color can be arbitrarily close to each other.

**4.2. Bichromatic nearest neighbors.** Given sets $P$ and $Q$ as before, call $(p, q) \in P \times Q$ a *(blue/red) nearest neighbor pair* and $q$ a *nearest neighbor* of $p$ if $q \in Q$ minimizes the Euclidean distance from $p$ to $Q$. Let $N(p)$ be the number of nearest neighbors of $p$ and consider $\sum_{p \in P} N(p)$. We will assume that either no three points of $P$ are collinear or that no four points of $Q$ are cocircular. The best previous upper bound for this sum is $O(m^{3/4}n^{3/4}\beta(m, n) + m + n)$ (see **[4]**). By drawing a sphere around each blue point $p$, with radius equal to the distance between $p$ and its nearest neighbors, we again reduce the distance counting problem to the problem of §3. If no three points of $P$ are collinear we get condition (S.i) and if no four points of $Q$ are cocircular we get condition (S.i$'$) (see remark (2) after Theorem 3.1). In both cases we obtain the following upper bound.

THEOREM 4.2. *The number of blue/red nearest neighbor pairs in a set of* $m$ *blue and* $n$ *red points in three-dimensional Euclidean space is* $O(m^{2/3}n^{2/3} + m + n)$ *if no three blue points are collinear or no four red points are cocircular.*

REMARKS. (1) As in the case of repeated minimum distance pairs, the maximum number of nearest neighbor pairs is $\Theta(m)$ for a monochromatic set of $m$ points in three dimensions. Again, the packing argument used to prove the upper bound fails in the bichromatic case.

(2) The case where all points of $Q$ lie on a circle in three-dimensional space and all points of $P$ lie on the axial line of this circle shows that the bound of Theorem 4.2 does not hold without restrictions on the locations of the points.

**4.3. Furthest neighbors.** Let $P$ be a (monochromatic) set of $m$ points in three dimensions, and call $(p, q)$ a *furthest neighbor pair* and $q$ a *furthest neighbor* of $p$ if $q \in P$ maximizes the Euclidean distance from $p$. For each $p \in P$ let $F(p)$ be the number of furthest neighbors and consider $\sum_{p\in P} F(p)$. Assuming no three points are collinear, the best previous bound on this sum can be found in [**4**] and is $O(m^{3/2}\beta(m))$, with $\beta(m) = 2^{\Theta(\alpha(m)^2)}$, improving earlier bounds given in [**6, 9**]. If we draw around each point $p$ the sphere with radius equal to the distance between $p$ and its furthest neighbors we get an instance of the problem in §3. Indeed, condition (S.ii) is satisfied because no point lies outside any of these spheres, and we get condition (S.i) if no three points are collinear and (S.i$'$) if no four points are cocircular. This implies the following result.

THEOREM 4.3. *The number of furthest neighbor pairs in a set of $m$ points in three-dimensional Euclidean space is $O(m^{4/3})$ if no three points are collinear or no four points are cocircular.*

REMARKS. (1) An example similar to the one in remark (2) after Theorem 4.2 shows that the maximum number of furthest neighbor pairs is quadratic in $m$ if no condition on the location of the points is imposed.

(2) It is worthwhile to note that the maximum number of maximum distance pairs is $\Theta(m)$, namely $2m-2$, as shown independently in [**12, 13, 16**].

**4.4. Delaunay triangulations.** There is a relation between three-dimensional Delaunay triangulations and the incidence problems considered in §§2 and 3. Let $P$ be a set of $m$ points in three dimensions. We call $P' \subseteq P$ a *proper Delaunay subset* if there is a unique sphere so that the points of $P'$ lie on the sphere and all other points of $P$ lie outside the sphere, and we call the sphere a *Delaunay sphere*. The *Delaunay triangulation* of $P$, denoted by $\mathscr{D}(P)$, is the cell complex whose bounded cells are the convex hulls of the proper Delaunay subsets of $P$ and whose unbounded cell is the complement of the convex hull of $P$ (see [7] or [**8**]). We note that $\mathscr{D}(P)$ is not necessarily a triangulation because there may be bounded cells that are not tetrahedra. In fact, the problem we study below is interesting only if many of the cells are not tetrahedral.

Let the *degree* of a cell of $\mathscr{D}(P)$ be its number of vertices; the number of edges of the cell is at most three times the degree minus 6 and the number of 2-faces is at most twice the degree minus 4. We consider the problem of bounding the sum of degrees of subsets of the collection of Delaunay cells.

Let $S$ be the set of all Delaunay spheres. By definition, all points of $P$ lie on or outside any Delaunay sphere. In other words, $P$ and $S$ satisfy condition (S.ii) of §3. We now argue that $S$ also satisfies (S.i), that is, no three Delaunay spheres intersect in a common circle. Suppose to the contrary that there are three Delaunay spheres, $s_1$, $s_2$, and $s_3$, that meet in a common circle. It follows that the centers of the three spheres lie on a common line. Assume that the center of $s_2$ lies between the other two centers. But then all points of $s_2$ that do not belong to the common circle lie either inside $s_1$ or inside $s_3$. Because no points of $P$ can lie inside $s_1$ or $s_3$ it follows that all points of $P \cap s_2$ lie on a circle which contradicts the assumption that $s_2$ is a Delaunay sphere. Using Theorem 3.1 we thus get the following bound on the sum of degrees. Note that since the Delaunay triangulation of $P$ has at most $O(m^2)$ cells, the term $n$ in the bound of Theorem 3.1 is always subsumed by the leading term.

THEOREM 4.4. *The sum of degrees of $n$ cells in the Delaunay triangulation of any $m$ points in three-dimensional Euclidean space is* $O(m^{2/3}n^{2/3} + m)$.

REMARK. Because of the dual correspondence between Delaunay triangulations and Voronoi diagrams (see [**8**]), Theorem 4.4 implies that the total number of edges incident to $n$ vertices of the Voronoi diagram of any $m$ points in three dimensions is $O(m^{2/3}n^{2/3} + m)$.

## 5. Conclusions

This paper studies a number of combinatorial distance problems in three-dimensional Euclidean space and derives improved upper bounds for all problems considered. The bounds are direct applications of an $O(m^{2/3}n^{2/3} + m + n)$ bound on the number of incidences between $m$ points and $n$ hyperplanes in four dimensions which holds if no three points are collinear, no three hyperplanes intersect in a common 2-flat, and each hyperplane bounds a closed half-space that contains all points. It is possible to extend the techniques and results of this paper to $d \geq 5$ dimensions. However, because of the many different general position assumptions possible we face a multitude of different problems with different answers. It seems worthwhile to come up with a classification of the different cases to guide future efforts in this direction.

Coincidentally, the above upper bound is the same as for the number of incidences between $m$ points and $n$ lines in the plane, without restriction on the points and lines. However, unlike in the planar problem where the upper bound is known to be tight, there is no superlinear lower bound known for the hyperplane incidence problem. To close the gap between the current upper and lower bounds is the most important open problem suggested by the results of this paper.

We would like to point out that although the results of this paper are mainly combinatorial, the techniques also have algorithmic applications. For example, [**1**] gives a randomized algorithm inspired by our constructive proof that finds a bichromatic minimum distance pair for a set of $m$ blue and red points in three dimensions in expected time $O(m^{4/3} \log^{4/3} m)$. This algorithm is used to construct a minimum spanning tree of $m$ points in three dimensions in the same amount of time.

## REFERENCES


1. P. Agarwal, H. Edelsbrunner, O. Schwarzkopf, and E. Welzl, *Euclidean minimum spanning trees and bichromatic closest pairs*, Proc. Sixth Ann. Sympos. Comput. Geom. 1990, pp. 203–210.
2. B. Bollobás, *Extremal graph theory*, London Math. Soc. Monographs, no. 11, Academic Press, London, 1978.
3. A. Brønsted, *An introduction to convex polytopes*, Graduate Texts in Math., Springer-Verlag, New York, 1983.
4. K. L. Clarkson, H. Edelsbrunner, L. J. Guibas, M. Sharir, and E. Welzl, *Combinatorial complexity bounds for arrangements of curves and spheres*, Discrete Comput. Geom. **5** (1990), 99–160.
5. K. L. Clarkson and P. W. Shor, *Applications of random sampling in computational geometry*. II, Discrete Comput. Geom. **4** (1989), 387–421.
6. F. R. K. Chung, *Sphere-and-point incidence relations in high dimensions with applications to unit distances and furthest-neighbor pairs*, Discrete Comput. Geom. **4** (1989), 183–190.
7. B. Delaunay, *Sur la sphère vide*, Izv. Akad. Nauk SSSR Otdel. Mat. Estestvennyka Nauk **7** (1934), 793–800.
8. H. Edelsbrunner, *Algorithms in combinatorial geometry*, Springer-Verlag, Heidelberg, 1987.
9. H. Edelsbrunner and S. S. Skiena, *On the number of furthest neighbor pairs in a point set*, Amer. Math. Monthly **96** (1989), 614–618.
10. P. Erdös, *On sets of distances of n points*, Amer. Math. Monthly **53** (1946), 248–250.
11. ——, *On sets of distances of n points in Euclidean space*, Magyar Tud. Akad. Mat. Kutaló Int. Kozl. **5** (1960), 165–169.
12. B. Grünbaum, *A proof of Vázsonyi's conjecture*, Bull. Res. Council Israel Sect. A **6** (1956), 77–78.
13. A. Heppes, *Beweis einer Vermutung von A. Vázsonyi*, Acta Math. Acad. Sci. Hungar. **7** (1956), 463–466.
14. W. O. J. Moser and J. Pach, *Research problems in discrete geometry*, Manuscript, Dept. Math., McGill Univ., Montreal, Quebec, 1986.
15. J. Spencer, E. Szemerédi, and W. T. Trotter, Jr., *Unit distances in the Euclidean plane*, in Graph Theory and Combinatorics, Academic Press, London, 1984, pp. 293–303.
16. S. Straszewicz, *Sur un problème geometrique de P. Erdös*, Bull. Acad. Polon. Sci. Cl. III **5** (1957), 39–40.
17. E. Szemerédi and W. T. Trotter, Jr., *Extremal problems in discrete geometry*, Combinatorica **3** (1983), 381–392.


(H. Edelsbrunner) DEPARTMENT OF COMPUTER SCIENCE, UNIVERSITY OF ILLINOIS AT URBANA-CHAMPAIGN, URBANA, ILLINOIS 61801
*E-mail address*: edels@cs.uiuc.edu

(M. Sharir) COURANT INSTITUTE OF MATHEMATICAL SCIENCES, NEW YORK UNIVERSITY, NEW YORK, NEW YORK 10012 AND SCHOOL OF MATHEMATICAL SCIENCES, TEL AVIV UNIVERSITY, TEL AVIV 69978, ISRAEL
*E-mail address*: sharir@taurus.bitnet

DIMACS Series in Discrete Mathematics
and Theoretical Computer Science
Volume 4, 1991

Applied Geometry
and Discrete Mathematics
THE VICTOR KLEE FESTSCHRIFT

# Gaps in Difference Sets, and the Graph of Nearly Equal Distances

PAUL ERDŐS, ENDRE MAKAI, JÁNOS PACH AND JOEL SPENCER

ABSTRACT. Given $n$ real numbers so that any two of them differ by at least 1, let $1 \leq d_1 \leq d_2 \leq \cdots \leq d_{\binom{n}{2}}$ denote the increasing sequence of all differences determined by them. We prove that $\sum_{1\leq j<\binom{n}{2}} (d_{j+1} - d_j)^2 > c \cdot \log n$ for a suitable constant $c > 0$, and this bound is asymptotically tight. In fact, we give sharp lower bounds for $\sum_{1\leq j<\binom{n}{2}} \phi(d_{j+1} - d_j)$, where $\phi: \mathbf{R} \to \mathbf{R}$ belongs to a large class of monotone increasing convex functions with $\phi(0) = 0$. We also prove that, for any system of $n$ points in the plane with minimal distance at least 1 and for any real $t$, the number of those pairs of points whose distance is between $t$ and $t+1$ cannot exceed $[n^2/4]$, provided that $n$ is sufficiently large. Furthermore, this bound can be attained for every $n$. In the case when the diameter of our point set is at most $d$, and $1 \leq i \leq d/10$, we can show that the maximum number of point pairs whose distance is between $d-i$ and $d-i+1$ is at most $10^3(id)^{3/2}$. This bound is also tight, apart from the value of the constant.

## 1. Introduction

Given $n$ real numbers so that any two of them differ by the least 1, let us order the $\binom{n}{2}$ positive differences determined by them: $1 \leq d_1 \leq d_2 \leq \cdots \leq d_{\binom{n}{2}}$. We are interested in the distribution of the gaps in this sequence. In particular, we wish to estimate the functions

$$f_c(n) = \min \sum_{j=1}^{\binom{n}{2}-1} (d_{j+1} - d_j)^c,$$

---

1980 *Mathematics Subject Classification* (1985 *Revision*). Primary 52A37; Secondary 05B10.

Work by the second and third authors has been supported by Hungarian Science Foundation Grant OTKA-1814. Work by the third author has also been supported by National Science Foundation Grant CCR-89-01484. The first and the third authors also wish to thank the DIMACS Center at Rutgers for additional support under National Science Foundation Grant STC-88-09648.

where $c > 1$ is fixed, $n$ tends to infinity, and the minimum is taken over all sets of $n$ numbers with minimal difference $d_1 \geq 1$.

This choice of function to be minimized may at first appear to be capricious. Suppose, however, that we fix $d_1 = 1$ (which we can obviously assume by a scaling argument) and the *diameter* $d_{\binom{n}{2}} = d$. Then, if the $\binom{n}{2}$ differences could be chosen arbitrarily, one would minimize $\sum_j (d_{j+1} - d_j)^c$ by having the $d_j$ equally spaced. The bounds on $f_c(n)$ reflect our ability to achieve this even spacing.

We shall use the notation $f(n) = \Theta(g(n))$ when there are appropriate positive constants $c_1$, $c_2$ such that $c_1 g(n) \leq f(n) \leq c_2 g(n)$ holds for every sufficiently large $n$.

THEOREM 1. (i) $f_c(n) = \Theta(n^{2-c})$ *if* $1 < c < 2$;
(ii) $f_2(n) = \Theta(\log n)$;
(iii) $f_c(n) = \Theta(1)$ *if* $c > 2$.

This result is an immediate consequence of the following more general bounds, which will be obtained in the next section.

THEOREM 2. *Given a set* $P$ *of* $n$ *real numbers whose differences are* $1 \leq d_1 \leq d_2 \leq \cdots \leq d_{\binom{n}{2}}$, *and an increasing convex function* $\phi\colon \mathbf{R} \to \mathbf{R}$ *with* $\phi(0) = 0$, *let*

$$f_\phi(P) = \sum_{j=1}^{\binom{n}{2}-1} \phi(d_{j+1} - d_j).$$

*Let* $f_\phi(n) = \min_{|P|=n} f_\phi(P)$, *where the minimum is taken over all* $P$ *with* $d_1 \geq 1$. *Then*

$$\sum_{i=1}^{n-2} i\phi\left(\frac{1}{i}\right) \leq f_\phi(n) \leq 10 \sum_{i=0}^{\lceil \log_2 n \rceil} 2^{2i} \phi\left(\frac{1}{2^{i-1}}\right).$$

A crucial step in the proof of this result is that we can give an easy upper bound on the number of $d_j$'s in a given interval of length 1 (Lemma 2.1). This suggests several possible generalizations to higher dimensions. For the sake of simplicity, we will mainly be concerned with planar problems. Instead of differences between numbers, we shall consider distances between points.

THEOREM 3. *Assume* $t > 0$ *and a set of* $n$ *points* $p_1, p_2, \ldots, p_n$ *in the plane with minimal distance at least* 1 *are given. Then the number of pairs* $p_i p_j$, $i < j$, *whose distance is between* $t$ *and* $t + 1$ *cannot exceed* $[n^2/4]$, *provided* $n \geq n_0$. *Moreover, this bound can be attained for every* $n$ *and every sufficiently large* $t > t_0(n)$.

An old and probably quite deep conjecture of Erdős **[E1, E3]** states that the maximum number of times the same distance can occur among $n$ points in the plane is $O(n^{1+c/\log\log n})$. The best currently known bound for this problem is $O(n^{4/3})$ which was obtained by Spencer, Szemerédi, and Trotter

[SST] and Clarkson et al. [CEGSW]. Theorem 3 shows what happens if, instead of equal distances, we count the number of those pairs that determine nearly the same distance. To avoid the trivial case, when all the $\binom{n}{2}$ distances are nearly 0, we assume that our points are separated.

We do not know the smallest integer $n_0$ for which Theorem 3 is valid. The vertices of a regular hexagon together with its center illustrate that one can choose seven points in the plane so that all distances determined by them are between 1 and 2. Bateman and Erdős [BE] proved that there are no eight points in the plane with this property.

In 1934, Pannwitz and Hopf [PH] showed that the maximum number of times the *maximal* distance (the *diameter*) can occur among $n$ points in the plane is $n$. Related problems on the number of occurrences of the $k$th largest distance $(k \geq 2)$ were considered by Vesztergombi [V] and Erdős, Lovász, and Vesztergombi [ELV]. For further results on the distribution of distances among $n$ points in a metric space see [E2], [EMST], [T].

THEOREM 4. *Let $p_1, p_2, \dots, p_n$ be a set of points in the plane with minimal distance at least 1 with diameter $d$, and let $1 \leq i \leq d/10$. Then the number of pairs $p_jp_k$, $j < k$, for which $d - i \leq d(p_j, p_k) \leq d - i + 1$ is at most $10^3(id)^{3/2}$. This bound is tight, apart from the value of the constant.*

In particular, for $i = 1$ Theorem 4 states that the number of near-diameters is $O(d^{3/2})$. This bound is asymptotically sharp, as is shown by the following example: Put as many points as possible on a circle of diameter $d$ so that no two of them are closer than 1, and all but at most one distances between consecutive pairs are equal to 1. For $1 \leq i \leq d/10$, we repeat this construction on a set of concentric circles of diameter $d, d-2, d-4, \dots$. This shows that Theorem 4 is tight.

## 2. Gaps in difference sets

In this section we are going to establish Theorem 2.

First we prove the lower bound. Given $n$ real numbers so that any two of them differ by at least 1, let $1 \leq d_1 \leq d_2 \leq \cdots \leq d_{\binom{n}{2}} = d$ be the increasing sequence of all pairwise differences determined by them. We call $d$ the *diameter* of our set of numbers. We need the following simple observation.

LEMMA 2.1. *For any $0 < i < [d]$, the number of $d_j$'s in the interval $(d-i, d-i+1]$ is at most $i$.*

PROOF. Assume without loss of generality that all our numbers $p_s$ are between 0 and $d$. If $p_t - p_s \in (d-i, d-i+1]$, then $p_s \in [0, i)$. Thus the number of all $p_s$ for which there is a $p_t$ such that $p_t - p_s$ lies in the given interval is at most $i$. However, for every $p_s$ there is only at most one $p_t$ having this property. □

Hence, in order to establish the lower bound, it is sufficient to show the following.

LEMMA 2.2. *Let* $\phi: [0, d] \to \mathbf{R}$ *be an increasing convex function,* $\phi(0) = 0$. *Assume further that* $1 = d_1 \leq d_2 \leq \cdots \leq d_N = d$ *are real numbers with the property that any interval* $(d-i, d-i+1]$ *contains at most* $i$ *elements* $d_j$. *Then*

$$\sum_{j=1}^{N-1} \phi(d_{j+1} - d_j) \geq \sum_{i=1}^{[d]-1} i\phi(1/i).$$

*This bound can be attained if* $d$ *is an integer and the* $d_j$ *in each interval* $[d-i, d-i+1]$ *form an arithmetic progression with difference* $1/i$, *including the endpoints.*

PROOF. Without violating the conditions and without increasing $\sum \phi(d_{j+1} - d_j)$, we shall gradually replace the members of our sequence by new elements, if necessary, so that in each interval $(d-i, d-i+1]$ the elements of the final sequence form an arithmetic progression of length $i$, whose last element is $d-i+1$ and whose difference is $1/i$.

Clearly, this condition is originally satisfied for $i = 1$. Assume now that $i > 1$, and we have already reached the desired situation for the intervals $(d-i+1, d-i+2]$, $(d-i+2, d-i+3]$, $\ldots$, $(d-1, d]$. We can suppose without loss of generality that our sequence has exactly $i$ elements in the interval $(d-i, d-i+1]$, $1 \leq i < d$ because the insertion of a new element can only decrease the sum $\sum_{j=1}^{N-1} \phi(d_{j+1} - d_j)$. Let $d_h, d_{h+1}, \ldots, d_{h+i-1}$ belong to $(d-i, d-i+1]$. We can also assume that these elements form an arithmetic progression, otherwise, keeping $d_h$ and $d_{h+i-1}$ fixed, we could rearrange the intermediate elements so that the sum $\sum_{j=h}^{h+i-2} \phi(d_{j+1} - d_j)$ does not increase. This follows immediately from Jensen's inequality.

We distinguish two cases. If $d_{h+i-1} < d-i+1$, then

$$\begin{aligned} d_{h+i-1} - d_{h+i-2} &= \frac{d_{h+i-1} - d_h}{i-1} < \frac{1}{i-1} \\ &= d_{h+i} - (d-i+1) < d_{h+i} - d_{h+i-1}. \end{aligned}$$

Setting $d'_j = d_j + (j-h)((d-i+1) - d_{h+i-1})/(i-1)$ for all $h \leq j < h+i$ and $d'_j = d_j$ otherwise, we obtain

$$\begin{aligned} \sum_{j=h}^{h+i-1} \phi(d'_{j+1} - d'_j) &= (i-1)\phi\left(\frac{d-i+1-d_h}{i-1}\right) + \phi(d_{h+i} - (d-i+1)) \\ &\leq (i-1)\phi\left(\frac{d_{h+i-1} - d_h}{i-1}\right) + \phi(d_{h+i} - d_{h+i-1}) \\ &= \sum_{j=h}^{h+i-1} \phi(d_{j+1} - d_j). \end{aligned}$$

(We used the fact that, if $\phi$ is a convex function, $x_1, \ldots, x_i$ are reals, and $\overline{x} = \frac{1}{i}(x_1 + \cdots + x_i)$, then $\sum_{j=1}^{i} \phi((1-\lambda)x_j + \lambda\overline{x})$ is a decreasing function of $\lambda$, $0 \leq \lambda \leq 1$.) Furthermore, in this new sequence $d'_{h+i-1} = d-i+1$.

Hence, it is enough to discuss the second case, i.e., $d_{h+i-1} = d - i + 1$. If $(d_{h+i-1} - d_h)/(i-1) = 1/i$, then we are done, and we can pass to the next $i$. If $(d_{h+i-1} - d_h)/(i-1) < 1/i$, then replacing $d_j$ by $d'_j = d - i + (j-h+1)/i$ for every $h \le j < h+i$, $\sum_{j=h-1}^{h+i-1} \phi(d_{j+1} - d_j)$ cannot increase. The case $d_{h-1} = d - i$ can be settled similarly.

Thus, we can assume that $(d_{h+i-1} - d_h)/(i-1) > 1/i$ and $d_{h-1} < d - i$. The elements of our sequence in $(d-i-1, d-i]$ are $d_{h-i-1}, d_{h-i}, \dots, d_{h-1}$. The average length of the $i$ intervals $[d_j, d_{j+1}]$, $h-i-1 \le j < h-1$, is $(d_{h-1} - d_{h-i-1})/i < 1/i$. Keeping $d_{h-i-1}$, $d_{h+i-1}$, and $d_h - d_{h-1}$ fixed, we start increasing the length of each interval $[d_j, d_{j+1}]$, $h-i-1 \le j < h-1$, by the same amount. In the meantime, we uniformly shrink the intervals $[d_j, d_{j+1}]$, $h \le j < h+i-1$. We stop if either of the following events occurs.

(a) $d_{h-1}$ becomes equal to $d - i$, or
(b) $(d_{h+i-1} - d_h)/(i-1)$ becomes equal to $1/i$.

It is obvious that during the whole process $\sum \phi(d_{j+1} - d_j)$ cannot increase. Since our procedure must terminate, we end up with a case we have settled already.

Finally, we obtain a new sequence $(d_j)_{j=1}^N$ having the properties described in the first sentence of this proof. In particular, for every $1 \le i < [d]$, there are exactly $i$ elements $d_j$ in the interval $(d-i, d-i+1]$, and they form an arithmetic progression of difference $1/i$, whose last element is $d - i + 1$. The result follows. □

Next we establish the upper bound in Theorem 2. Assume for the sake of simplicity that $n = 2^k$ for some integer $k$. Put $p_1 = 0$. If $p_1 < p_2 < \cdots < p_{i-1}$ have already been defined for some $2^t < i \le 2^{t+1}$, then let

$$p_i = p_{i-1} + 1 + \begin{cases} 2^{-t} & \text{if } t < k-1, \\ 0 & \text{if } t = k-1. \end{cases}$$

The diameter of this set of numbers is $d = 2^k + k - 2$.

Let $\Delta_i = p_n - p_i$, $1 \le i \le n$. Obviously, $\Delta_1 = d$. Fix now an $i$, $2^{l-1} < i \le 2^l$ for some $2 \le l \le k-1$, and consider the sequence

$$\Delta_t^{(i)} = p_{n-t} - p_{i-t}, \qquad 0 \le t \le i - 2^{l-2}.$$

Then $\Delta_0^{(i)} = \Delta_i$, $\Delta_{i-2^{l-2}}^{(i)} \ge \Delta_{i-1}$, and it is easy to see that

$$0 \le \Delta_{t+1}^{(i)} - \Delta_t^{(i)} \le \frac{1}{2^{l-2}} \quad \text{for all } 0 \le t < i - 2^{l-2}.$$

Similarly, for a fixed $i > 2^{k-1}$, choose an integer $l$ so that $2^{l-1} < n - i \le 2^l$, and consider the sequence

$$\Delta_t^{(i)} = p_{n-t} - p_{i-t}, \qquad 0 \le t \le i - 2^{l-1}.$$

Now $\Delta_0^{(i)} = \Delta_i$, $\Delta_{i-2^{l-1}}^{(i)} \geq \Delta_{i-1}$, and

$$0 \leq \Delta_{t+1}^{(i)} - \Delta_t^{(i)} \leq \frac{1}{2^{l-1}} \quad \text{for all } 0 \leq t < i - 2^{l-1}.$$

Evidently, all numbers $\Delta_t^{(i)}$ defined above appear in the sequence $(d_j)$. Hence,

$$\begin{aligned}\sum_{j=1}^{\binom{n}{2}-1} \phi(d_{j+1} - d_j) &\leq \sum_{l=2}^{k-1} (\Delta_{2^{l-1}} - \Delta_{2^l}) 2^{l-2} \phi\left(\frac{1}{2^{l-2}}\right) + 2\phi(1) \\ &\quad + \sum_{l=1}^{k-1} (2^l - 2^{l-1}) 2^{l-1} \phi\left(\frac{1}{2^{l-1}}\right) + 2\phi(1) \\ &\leq 8 \sum_{l=0}^{k} 2^{2l} \phi\left(\frac{1}{2^l}\right),\end{aligned}$$

as desired.

If $n$ is not a power of 2, say $2^k < n < 2^{k+1}$, then we take a similar copy of the above construction with $2^k$ points so that the minimum distance between them is 2. Let us put the remaining $n - 2^k$ points in some of the midpoints of the intervals determined by consecutive elements of our system. The new sequence of distances $(d_j)$ obviously satisfies

$$\begin{aligned}\sum_{j=1}^{\binom{n}{2}-1} \phi(d_{j+1} - d_j) &= \sum_{d_j < 2} \phi(d_{j+1} - d_j) + \sum_{d_j \geq 2} \phi(d_{j+1} - d_j) \\ &\leq 10 \sum_{l=0}^{k} 2^{2l} \phi\left(\frac{1}{2^{l-1}}\right).\end{aligned}$$

This completes the proof of Theorem 2. □

REMARK. The upper and lower bounds in Theorem 2 differ only in a multiplicative constant for all functions $\phi$ meeting the requirements and satisfying

$$\max_{0 \leq t \leq 1} \frac{\phi(2t)}{\phi(t)} < C$$

for some $C > 0$.

## 3. The distribution of distances determined by $n$ points in the plane

For the sake of simplicity, in this section we shall discuss only planar problems, but our methods readily generalize to higher dimensions.

PROOF OF THEOREM 3. Let $p_1, p_2, \ldots, p_n$ be any system of $n$ points in the plane with minimum distance $\min_{i \neq j} d(p_i, p_j) \geq 1$, and let $t$ be any positive number. We define a graph $G$ on the vertex set $\{p_1, p_2, \ldots, p_n\}$ by joining $p_i$ and $p_j$ with an edge if and only if $t \leq d(p_i, p_j) \leq t + 1$. Assume, in order to obtain a contradiction, that $G$ has more than $[n^2/4]$

edges. Then we can apply the following old (but not too difficult) result of Erdős, which generalizes Turán's theorem (see e.g. [**B**]).

LEMMA 3.1. *There exists a positive constant* $c$ *such that any graph* $G$ *with* $n$ *vertices and at least* $[n^2/4]+1$ *edges has at least* $cn$ *triangles sharing a common edge.*

Suppose without loss of generality that $p_1p_2$ is an edge contained in at least $cn$ triangles of $G$, whose third vertices are $p_3, \dots, p_{\lceil cn\rceil+2}$. Then $t \le d(p_1, p_2) \le t+1$, and there are exactly two points $q_1$ and $q_2$ in the plane that are at distance $d(p_1, p_2)$ from both $p_1$ and $p_2$. By the definition of our graph, all points $p_i$, $3 \le i \le \lceil cn\rceil + 2$, must lie in the intersection of two ring-like regions $R_1$ and $R_2$, where $R_j$ is centered at $p_j$ and its inner and outer radii are $d(p_1, p_2)-1$ and $d(p_1, p_2)+1$, respectively $(j = 1, 2)$. However, it is easy to see that every point of $R_1 \cap R_2$ is closer than 3 to $q_1$ or to $q_2$, i.e., all $p_i$, $3 \le i \le \lceil cn\rceil+2$, can be covered by two disks of radius 3. The maximum number of points that can be taken within a circle of radius 3 so that no two of them are closer than 1 is at most $(3+\frac{1}{2})^2\pi/(\frac{1}{2})^2\pi = 49$. Thus, we obtain that $cn \le 98$, a contradiction if $n$ is sufficiently large. □

Bollobás and Erdős conjectured and Edwards proved (unpublished) that the largest constant $c$ for which Lemma 3.1 remains valid is $1/6$.

A straightforward generalization of the above argument gives the following.

THEOREM 5. *Given* $t \ge 1$ *and a system of* $n$ *points* $P = \{p_1, p_2, \dots, p_n\}$ *in* $k$*-dimensional space with minimum distance at least* 1, *let* $h_k(P, t)$ *denote the maximum number of pairs* $\{p_i, p_j\}$, $i<j$, *whose distance is between* $t$ *and* $t+1$. *Let* $h_k(n) = \max_{P,t} h_k(P, t)$. *Then there exists* $n_0(k)$ *such that*

$$h_k(n) = t_k(n) \quad \text{for every } n \ge n_0(k),$$

*where* $t_k(n)$ *is the maximum number of edges a graph with* $n$ *vertices can have without containing a complete subgraph of* $k+1$ *vertices.*

It is well known that $t_k(n) = n^2(1-(1/k)+o(1))/2$, and that the only $K_{k+1}$-free graphs with $n$ vertices and $t_k(n)$ edges are the complete $k$-partite graphs whose classes are as equal as possible. The inequality $h_k(n) \ge t_k(n)$ can be shown by the following construction. Let $x^i = (x^i_1, x^i_2, \dots, x^i_{k-1}, 0)$, $1 \le i \le k$, be the vertices of a $(k-1)$-dimensional regular simplex of sidelength $n^2$, and let $x^i(j) = (x^i_1, x^i_2, \dots, x^i_{k-1}, j)$. Let $n = n_1 + n_2 + \cdots + n_k$ be a balanced partition of $n$, i.e., any two parts $n_i$ and $n_{i'}$ differ by at most 1. Then the point system $P = \{x^i(j)\colon 1 \le i \le k,\ 1 \le j \le n_i\}$ meets the requirements, and $h_k(P, n^2) = t_k(n)$.

PROOF OF THEOREM 4. Let $P = \{p_1, p_2, \dots, p_n\}$ be a set of $n$ points in the plane with minimum distance at least 1 and maximum distance (diameter) at most $d$, and let $1 \le i \le d/10$. Let $\gamma$ denote the boundary of the convex hull of $P$. It is well known that diameter$(P) \le d$ implies that the length of $\gamma$ is at most $\pi d$. We construct a graph $G$ on the vertex set $P$ by joining $p_j$ and $p_k$ with an edge if and only if $d - i \le d(p_j, p_k) \le d - i + 1$. Let $D_j$ denote the open disk of radius $1/2$ centered at $p_j$. Since no two points of $P$ are closer than 1 to each other, the disks $D_j$ $(1 \le j \le n)$ are pairwise disjoint. Furthermore, if $p_j$ is a nonisolated vertex in $G$, then $D_j$ is entirely contained in the region $R$ between the inner and outer parallel curves of $\gamma$ at distance $i + \frac{1}{2}$ and $\frac{1}{2}$ from $\gamma$, respectively. It follows from the fact that the length of $\gamma$ is at most $\pi d$ that the area of $R$ is at most $(i+1)\pi d + (\frac{1}{2})^2 \pi$. This, in turn, implies the number of nonisolated vertices of $G$:

$$m \le \frac{(i+1)\pi d + (\frac{1}{2})^2 \pi}{(\frac{1}{2})^2 \pi} < 10id.$$

Our theorem can now be deduced by repeated application of the following assertion.

Lemma 3.2. *If $G$ is not empty, then it has a nonisolated vertex whose degree is at most $\Delta = 100(id)^{1/2}$.*

Assume, for contradiction, that the degree of every nonisolated vertex of $G$ is larger than $\Delta$. Let $p_n$ be such a vertex, and suppose that $p_1, p_2, \dots, p_{[\Delta]+1}$ are neighbors of $p_n$, listed in clockwise order. We can assume, by symmetry, that $p_{[\Delta/2]+1}$ has at least $[\Delta/2]$ neighbors on the same side of the line $p_n p_{[\Delta/2]+1}$ as $p_{[\Delta]+1}$ is. Let these be denoted by $q_1, q_2, \dots, q_{[\Delta/2]}$, in clockwise order.

Observe that, by the definition of $G$, the disks $D_1, D_2, \dots, D_{[\Delta/2]}$ (centered at $p_1, p_2, \dots, p_{[\Delta/2]}$, respectively) all lie in the annulus $A$ around $p_n$ whose inner and outer radii are $d - i - \frac{1}{2}$ and $d - i + \frac{3}{2}$, respectively. The area of $A$ is

$$(d - i + \tfrac{3}{2})^2 \pi - (d - i - \tfrac{1}{2})^2 \pi < 4d\pi.$$

Since the total area of $D_1, D_2, \dots, D_{[\Delta/2]}$ is $[\Delta/2](\frac{1}{2})^2\pi$, we conclude that the smallest section of $A$ containing all of them cannot be too small. In particular,

$$\sphericalangle p_1 p_n p_{[\Delta/2]+1} \ge \frac{[\Delta/2](\frac{1}{2})^2 \pi}{8d\pi} > \frac{\Delta}{70d} = \frac{10}{7}\left(\frac{i}{d}\right)^{1/2}.$$

Similarly, we obtain that

$$\sphericalangle p_n p_{[\Delta/2]+1} q_1 \ge \frac{10}{7}\left(\frac{i}{d}\right)^{1/2}.$$

Thus, it follows by easy calculation that

$$d(p_1, q_1) \ge 2\sqrt{\left((d-i)\cos\frac{10}{7}\left(\frac{i}{d}\right)^{1/2} - \frac{d-i+1}{2}\right)^2 + \left((d-i)\sin\frac{10}{7}\left(\frac{i}{d}\right)^{1/2}\right)^2} > d,$$

contradicting the assumption that the diameter of our set $P$ is at most $d$. This completes the proof of Lemma 3.2, and hence the theorem. □

## REFERENCES


[BE] P. Bateman and P. Erdős, *Geometrical extrema suggested by a lemma of Besicovitch*, Amer. Math. Monthly **58** (1951), 306–314.

[B] B. Bollobás, *Extremal graph theory*, Academic Press, London, 1978.

[CEGSW] K. L. Clarkson, H. E. Edelsbrunner, L. J. Guibas, M. Sharir, and E. Welzl, *Combinatorial complexity bounds for arrangements of curves and surfaces*, Discrete Comput. Geom. **5** (1990), 99–160.

[E1] P. Erdős, *On sets of distances of $n$ points*, Amer. Math. Monthly **53** (1946), 248–250.

[E2] P. Erdős, *Aufgabe 250* (Solution by H. Debrunner), Elem. D. Math. **11** (1956), 137.

[E3] ——, *On sets of distances of $n$ points in Euclidean space*, Publ. Math. Inst. Hungar. Acad. Sci. **5** (1960), 165–169.

[ELV] P. Erdős, L. Lovász, and K. Vesztergombi, *On the graph of large distances*, Discrete Comput. Geom. **4** (1989), 541–549.

[EMST] P. Erdős, A. Meir, V. T. Sós, and P. Turán, *On some applications of graph theory*, I. Discrete Math. **2** (1972), 207–228 (Corrigendum, Discrete Math. **4** (1973), 90; II. Studia in Pure Math. (presented to R. Rado), Academic Press, London, 1971, 89–99; III. Canad. Math. Bull. **15** (1972), 27–32.

[PH] E. Pannwitz and H. Hopf, Aufgabe Nr. 167, Jahresber. Deutsch. Math.-Verein. **43** (1934), 114.

[SST] J. Spencer, E. Szemerédi, and W. T. Trotter, *Unit distances in Euclidean plane*, in Graph Theory and Combinatorics, Academic Press, London, 1984, pp. 293–303.

[T] P. Turán, *Applications of graph theory to geometry and potential theory*, in: Combinatorial Structures and their Applications (Proc. Calgary Internat. Conf. 1969), Gordon and Breach, New York, 1970, 423–434.

[V] K. Vesztergombi, *On the distribution of distances in finite sets in the plane*, Discrete Math. **57** (1985), 129–146.



(P. Erdős, E. Makai, and J. Pach) MATHEMATICAL INSTITUTE OF THE HUNGARIAN ACADEMY OF SCIENCES, BUDAPEST, 1364 P.O. BOX 127, HUNGARY

(J. Pach and J. Spencer) COURANT INSTITUTE OF MATHEMATICAL SCIENCES, NEW YORK UNIVERSITY, NEW YORK, NEW YORK 10011

DIMACS Series in Discrete Mathematics
and Theoretical Computer Science
Volume 4, 1991

Applied Geometry
and Discrete Mathematics
THE VICTOR KLEE FESTSCHRIFT

# Remarks on 5-Neighbor Packings and Coverings with Circles

G. FEJES TÓTH AND L. FEJES TÓTH

ABSTRACT. Two members of a packing touching each other, or two members of a covering whose interiors intersect are called *neighbors*. We investigate *five-neighbor packings* and *coverings* with circles, that is packings and coverings each member of which has at least five neighbors. We find the thinnest covering with congruent circles. Further, we show that a five-neighbor covering in which the ratio of the radii of any two circles is greater than $h_0 = \frac{1}{8}\{(23+\sqrt{7424/27})^{1/3} + (23-\sqrt{7424/27})^{1/3} - 1\} = 0.53329\ldots$ has positive density. This result is sharp, as there exists a five-neighbor packing with density zero consisting of unit circles and of circles of radius $h_0$.

## 1. Introduction

Lately several investigations have been aimed at $k$-neighbor packings with convex bodies, that is, at packings at which each body is touched by a prescribed number $k$ of other bodies [**1, 2, 3**]. We recall a simple fact according to which the density of a 5-neighbor packing of equal circles is at least $\pi\sqrt{3}/7$ [**2**]. We shall prove an analogous result for coverings.

By a *k-neighbor covering* of the Euclidean plane with convex discs we mean a covering in which the interior of each disc is intersected by exactly $k$ other discs. Two discs of the covering whose interiors intersect are called *neighbors*. We are looking for the thinnest $k$-neighbor coverings with equal circles. It is easy to see that a 3-neighbor covering with equal circles does not exist. The cases when $n = 4$ and $n = 6$ can be settled easily: the thinnest 4- and 6-neighbor coverings with equal circles are given by the circumcircles of the tiles of the regular tilings $\{4, 4\}$ and $\{6, 3\}$, respectively.

Let $P = p_1p_2p_3p_4p_5$ be a pentagon such that $p_1p_2p_3p_5$ is a rectangle with side lengths $p_1p_2 = \sqrt{(35+\sqrt{73})/18}$ and $p_2p_3 = \sqrt{(37-\sqrt{73})/18}$, and

1980 *Mathematics Subject Classification* (1985 *Revision*). Primary 52A45.
Research supported by the Hungarian National Foundation for Scientific Research under grant no. 1817.

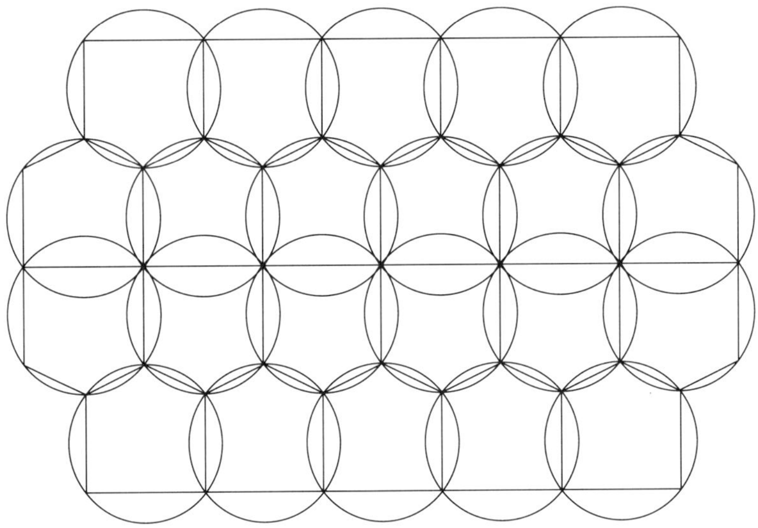

FIGURE 1

$p_3p_4p_5$ is an isosceles triangle with side lengths $p_3p_4 = p_4p_5 = \sqrt{(144-\sqrt{35+\sqrt{73}})/72}$. $P$ is inscribed in a unit circle and an easy calculation shows that its area is $\sqrt{(827+73^{3/2})288}$. Let congruent copies of $P$ form an edge-to-edge tiling. We claim that the circumcircles of these pentagons constitute a thinnest 5-neighbor covering with equal circles (Figure 1). This is expressed by the following

THEOREM 1. *If in a covering with equal circles each circle has at most five neighbors, then the* (*lower*) *density of the covering is at least*

$$\frac{12\pi\sqrt{2}}{\sqrt{827+73^{3/2}}}.$$

The *homogeneity* of a set of circles is defined as the infimum of the quotients of the radii of any two circles from the set. It is obvious that the density of a 5-neighbor packing of circles with sufficiently small homogeneity can be zero. On the other hand, every 5-neighbor packing of circles whose homogeneity is close to 1 has positive density. In fact, it is easy to see that homogeneity greater than $3-2\sqrt{2}+(2-\sqrt{2})^{3/2}=0.619\ldots$ guarantees a positive density. Thus the question arises: What is the maximum real number $h_0$ for which there exists a 5-neighbor packing of circles with homogeneity $h_0$ and density 0? The answer to this question is given by the following

THEOREM 2. *If the homogeneity of a packing of circles exceeds*

$$h_0=\tfrac{1}{8}\{(23+\sqrt{7424/27})^{1/3}+(23-\sqrt{7424/27})^{1/3}-1\}=0.53329\ldots$$

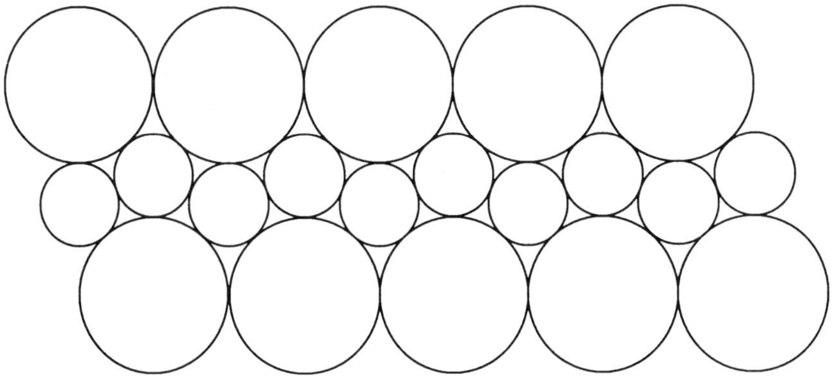

FIGURE 2

*and each circle has at least five neighbors, then the (lower) density of the packing is positive.*

Observe that $h_0$ is the unique real root of the equation $8h_0^3 + 3h_0^2 - 2h_0 - 1 = 0$. For us, the following geometric interpretation of $h_0$ will be more important: Let $C_0$, $C_1$, and $C_5$ be unit circles and let $C_2$, $C_3$, and $C_4$ be circles of radius $h_0$ such that $C_0, \ldots, C_5$ have mutually disjoint interiors, $C_i$ touches $C_0$ for $i = 1, \ldots, 5$ and $C_i$ touches $C_{i+1}$ for $i = 1, \ldots, 4$. Then the centers of $C_0$, $C_1$, and $C_5$ are collinear. This gives rise to a 5-neighbor packing of circles of homogeneity $h_0$ and density 0, as shown in Figure 2.

We conclude this section by mentioning some open problems. By enlarging slightly the circles in the 5-neighbor covering considered above we obtain a 7-neighbor covering. Similarly, enlarging slightly the circumcircles of the tiles in $\{4, 4\}$ we get an 8-neighbor covering. These remarks suggest the question: Is the infimum of the density of a $k$-neighbor covering with equal circles for $k = 7$ and $k = 8$ equal to $\pi/a(P)$ and $\pi/2$, respectively?

As we mentioned above, the density of the thinnest 5-neighbor packing of congruent circles is $\pi\sqrt{3}/7$. We do not know of an example of a 5-neighbor packing of circles with homogeneity $h > h_0$ and density less than $\pi\sqrt{3}/7$. Is it true that the thinnest 5-neighbor packing of circles with homogeneity $h > h_0$ consists of congruent circles?

A packing $\mathscr{P}$ is called *connected* if the closure of the union of the members of $\mathscr{P}$ is connected. The following simple argument by W. Kuperberg shows that a 5-neighbor packing of congruent circles is connected. Suppose that $\mathscr{P}$ is a 5-neighbor packing of congruent circles which is not connected. Consider an arbitrary circle $C$ of $\mathscr{P}$ and enlarge it concentricly until it touches a circle $C'$ which belongs to a component of $\mathscr{P}$ not containing $C$. Then $C'$ cannot have five neighbors, which is a contradiction. We point out that even a 5-neighbor packing of circles with homogeneity exceeding $h_1 = 0.7231$

is connected. To see this, we observe that in a disconnected packing either there are pairs of circles belonging to different components of the packing arbitrarily close to one another or we can move all the circles of one component until one of its circles touches a circle belonging to another component. Thus, from the existence of a disconnected 5-neighbor packing of circles with homogeneity $h$, follows the existence of a circle packing with the same homogeneity in which two touching circles have together ten neighbors. It is easy to see that this is impossible for $h > h_1$. Taking two disjoint copies of the circle packing depicted in Figure 2 yields a disconnected 5-neighbor packing with homogeneity $h_0$. Is it true that any 5-neighbor packing of circles whose homogeneity is greater than $h_0$ is connected?

## 2. Proof of Theorem 1

Let the unit circles $C_1, C_2, \dots$ with centers $c_1, c_2, \dots$ form a covering such that no circle has more than five neighbors. Let $D_i$ be the *Dirichlet cell of* $C_i$, that is, the set of those points of the plane which are nearer to $c_i$ than to any other center $c_j$ $(j \neq i)$. The sets $D_i$ are convex polygons and they form an edge-to-edge tiling $\mathcal{T}$. Obviously, $D_i \subset C_i$.

Let $v$ be a 4-valent vertex of $\mathcal{T}$ generated by the circles $C_i, C_j, C_k$, and $C_l$ in a position such that $C_i$ and $C_j$ as well as $C_k$ and $C_l$ touch each other at $v$. Then we say that $v$ is a *regular* 4-*valent vertex*. Note that the adjacent edges emanating from $v$ are orthogonal.

Let $U = U(r)$ be the closed hull of the union of those tiles of $\mathcal{T}$ which belong to circles contained in the circle $C(r)$ of radius $r$ with a fixed center 0. $U$ is a simple polygon of, say, $n$ sides. We shall give an upper bound for the area $a(U)$. The main tool to do this is given by Lemma 1, which, roughly speaking, states that, in the average, the Dirichlet cells contain at least two regular 4-valent vertices.

LEMMA 1. *Let $k_i$ be the number of regular* 4-*valent vertices of $\mathcal{T}$ lying on the boundary of $D_i$. Let $t = t(r)$ be the number of tiles of $\mathcal{T}$ contained in $U(r)$. Then*

$$\limsup_{r\to\infty} \frac{1}{t(r)} \sum_{D_i \subset U} k_i \geq 2.$$

We observe that if the Dirichlet cells $D_i$ and $D_j$ share an edge of $\mathcal{T}$, then the circles $C_i$ and $C_j$ are neighbors. If $D_i$ and $D_j$ meet at a vertex, then $C_i$ and $C_j$ are neighbors except in the case when $C_i$ and $C_j$ touch each other. It immediately follows that any Dirichlet cell has at most five sides. Furthermore, if seven circles have a point in common then at least one of them has more than five neighbors, a contradiction. A closer investigation shows that $\mathcal{T}$ cannot contain a 6-valent vertex either. (It is likely that even a 5-valent vertex cannot occur, but in the proof we can avoid making use of this.)

Let $v_i$ be the number of $i$-valent vertices lying in $U$, disregarding the edges which are outside of $U$ when counting the valencies, and let $v$ be the total number of vertices of $\mathcal{T}$ belonging to $U$. Let $t_i$ be the number of $i$-gonal tiles, let $t$, as defined in the lemma, be the total number of tiles, and let $e$ be the total number of edges of $\mathcal{T}$ contained in $U$. Of course, all these numbers are functions of $r$, as $U$ itself is, but we shall not indicate this dependence in our notation. We have

(1) $$v = v_2 + v_3 + v_4 + v_5,$$

(2) $$t = t_3 + t_4 + t_5,$$

(3) $$2v_2 + 3v_3 + 4v_4 + 5v_5 = 2e$$

and

(4) $$3t_3 + 4t_4 + 5t_5 = 2e - n.$$

Relations (1) and (3) yield

$$2e = 3v - v_2 + v_4 + 2v_5.$$

Combining this with (2) and with Euler's theorem $t+v = e+1$ for polyhedral maps, we obtain

$$e = 3t_3 + 3t_4 + 3t_5 + v_2 - v_4 - 2v_5 - 3.$$

Hence, by (4) and (2)

(5) $$2v_4 = t - n - 6 + 2t_3 + t_4 + 2v_2 - 4v_5.$$

We consider a Dirichlet cell $D_i$ and a vertex $v$ of it with the following property: $v$ is a 4-valent vertex of the tiling $\mathcal{T}$ which is not regular, and moreover, if the Dirichlet cells meeting at $v$ other than $D_i$ are $D_j$, $D_k$, and $D_l$, then none of the circles $C_j$, $C_k$, and $C_l$ touches $C_i$. Let $k_i^*$ be the number of vertices of $D_i$ with these properties. Further, let $m_i$ be the number of vertices of $D_i$ which are 5-valent vertices of $\mathcal{T}$ and let $s_i$ be the number of sides of $D_i$. Since the number of neighbors of $C_i$ is at least $k_i^* + m_i + s_i$ and at most 5, we have

$$\sum_{D_i \subset U} (k_i^* + m_i + s_i) \leq \sum_{D_i \subset U} 5 = 5(t_3 + t_4 + t_5).$$

Observing that $\sum_{D_i \subset U} m_i = 5v_5$ and $\sum_{D_i \subset U} s_i = 3t_3 + 4t_4 + 5t_5$ we obtain

(6) $$\sum_{D_i \subset U} k_i^* \leq 2t_3 + t_4 - 5v_5 \leq 2t_3 + t_4 - 4v_5.$$

We have

$$\sum_{D_i \subset U} (k_i + 2k_i^*) \geq 4v_4.$$

Combining this with relations (5) and (6) we get

$$\sum_{D_i \subset U} k_i \geq 2t - 2n - 12. \tag{7}$$

To each side of $U$ there is a Dirichlet cell $D_i$ not contained in $U$ such that the respective side is common to $U$ and $D_i$. The circle $C_i$ corresponding to $D_i$ intersects the boundary of the circle $C(r)$. This means that $C_i$ is contained in the ring $C(r+2)\setminus C(r-2)$. There are at most $56r$ circles of the covering lying in this ring. For, if there were more, then there would be a point of the ring belonging to the interior of seven circles, which is impossible. This implies that $n = O(r)$. On the other hand, we have, obviously, $t = O(r^2)$. Using these bounds, the Lemma immediately follows by (7).

Now the proof of Theorem 1 is relatively easy. Let $A(k)$ be the maximum possible area of a tile containing $k$ regular vertices. For $k = 0$ the maximum is attained by a regular pentagon inscribed in a unit circle:

$$A(0) = \frac{5}{2} \sin \frac{2\pi}{5} = 2.3776\ldots.$$

For $k = 1$ and $k = 2$ we have to find a pentagon of maximum area inscribed in a unit circle with one and two right angles, respectively. For $k = 1$ the pentagon is the union of a quadrilateral of side lengths $2, 1, 1, 1$ and an isosceles right triangle:

$$A(1) = \frac{\sqrt{27}}{4} + 1 = 2.2990\ldots.$$

A maximum area pentagon with two right angles inscribed in a unit circle is congruent to the pentagon $P$ described in the introduction:

$$A(2) = a(P) = \frac{\sqrt{827 + 73^{3/2}}}{12\sqrt{2}} = 2.2443\ldots.$$

For $k = 3$ and $k = 4$ the pentagon degenerates to a rectangle; thus, we have

$$A(3) = A(4) = 2.$$

Let $\overline{A}(k)$ be the least function among concave functions defined for $0 \leq k \leq 4$ such that for $k = 0, 2$, and $4$ we have $\overline{A}(k) = A(k)$. Since

$$A(1) < [A(0) + A(2)]/2 = \overline{A}(1)$$

and

$$A(3) < [A(2) + A(3)]/2 = \overline{A}(3),$$

we have

$$a(U) \leq \sum_{D_i \subset U} A(k_i) \leq \sum_{D_i \subset U} \overline{A}(k_i).$$

Now Jensen's inequality implies

$$\frac{1}{t} a(U) \leq \overline{A}\left(\frac{1}{t} \sum_{D_i \subset U} k_i\right).$$

Since $\overline{A}$ is a decreasing function, we have, by Lemma 1,

$$\limsup_{r\to\infty} \frac{1}{t} a(U) \le \overline{A}(2) = A(2) = a(P) = \frac{\sqrt{827 + 73^{3/2}}}{12\sqrt{2}},$$

which, in accordance with Theorem 1, yields the inequality $d \ge \pi/A(2) = 12\pi\sqrt{2}/\sqrt{827+73^{3/2}}$ for the lower density $d = \limsup_{r\to\infty} \pi t/a(U)$ of the covering.

## 3. Main idea of the proof of Theorem 2

Let $\mathscr{P} = \{C_i\}$ be a packing of circles of homogeneity $h > h_0$ such that each circle of $\mathscr{P}$ has at least five neighbors. Let $c_i$ be the center and let $r_i$ be the radius of the circle $C_i \in \mathscr{P}$. Without loss of generality we assume that $\sup r_i = 1$. With each circle $C_i \in \mathscr{P}$ we associate the set $D_i$ consisting of those points $p$ of the plane whose power $pc_i^2 - r_i^2$ with respect to $C_i$ is less than or equal to their power with respect to any other circle of $\mathscr{P}$. $D_i$ is called the *generalized Dirichlet cell* of $C_i$, in short, the *Dirichlet cell* of $C_i$. Let $C_i(r)$ be the circle of radius $r$ concentric with $C_i$. We shall show that there is a number $r(h)$, depending only on $h$, such that $D_i \subset C_i(r(h))$ for all $C_i \in \mathscr{P}$. This will immediately yield the lower bound $[h/r(h)]^2$ for the lower density of $\mathscr{P}$.

Two circles $C$ and $C'$ of $\mathscr{P}$ are called *neighbors of order* $k$, or $k$th *neighbors* if there are $k-1$ circles $C_1, \dots, C_{k-1}$ of $\mathscr{P}$, but not less than $k-1$, such that in the sequence $C = C_0, \dots, C_{k-1}, C_k = C'$ the circle $C_{i-1}$ is a neighbor of $C_i$ for $i = 1, \dots, k$. The idea for finding the number $r(h)$ is the following: We shall prove that there exists a real number $\vartheta(h) < \pi/2$ and a natural number $k(h)$, both depending only on $h$, such that for all circles $C_i$ of $\mathscr{P}$ any angular region of measure $2\vartheta(h)$ with vertex $c_i$ contains the center of a circle from $\mathscr{P}$ which is at most a $k(h)$th order neighbor of $C_i$. This will imply that the number $r(h) = hk(h)\csc\vartheta(h)$ will have the property that $D_i \subset C_i(r(h))$ for all $C_i \in \mathscr{P}$. To see this, we choose an arbitrary point $p$ whose distance from $c_i$ is greater than $hk(h)\csc\vartheta(h)$, and show that it cannot belong to $D_i$. Indeed, according to the statement above, the angular region of measure $2\vartheta(h)$ with vertex $c_i$ whose bisector is the ray $\overrightarrow{c_ip}$ contains the center $c_j$ of a circle $C_j$ from $\mathscr{P}$. The radical axis of the circles $C_i$ and $C_j$ separates the cells $D_i$ and $D_j$. The distance of the radical axis of $C_i$ and $C_j$ from $c_i$ is at most $hk(h)$ while the angle of this radical axis with the ray $\overrightarrow{c_ip}$ is at most $\pi/2 - \vartheta(h)$. Thus the distance of $c_i$ from the point of intersection of the ray $\overrightarrow{c_ip}$ and the radical axis of the circles $C_i$ and $C_j$ is at most $hk(h)\csc\vartheta(h)$. Points of $\overrightarrow{c_ip}$ which lie further from $c_i$ than $hk(h)\csc\vartheta(h)$ cannot belong to $D_i$.

In order to give the exact values of the quantities $r(h)$, $k(h)$, and $\vartheta(h)$, we introduce some auxiliary functions. Let $\Delta_1$ and $\Delta_2$ be isosceles triangles with side lengths $1+h$, $1+h$, and $2h$ and $1+h$, $1+h$, and $2$, respectively.

Let

$$\alpha(h) = 2\arcsin\frac{h}{1+h}$$

be the angle of $\Delta_1$ opposite the side of length $2h$ and let

$$\beta(h) = \arccos\frac{1}{1+h}$$

be the base angle of $\Delta_2$. The functions $\alpha(h)$ and $\beta(h)$ are obviously increasing for $h > 0$. We also observe that

$$\alpha(h_0) = \beta(h_0) = \pi/2;$$

thus, defining $\vartheta(h)$ as

$$\vartheta(h) = \pi - \alpha(h) - \beta(h_0),$$

we have $\vartheta(h) < \pi/2$.

For a positive number $r$, let $\Delta_3$ be a triangle with side lengths $1+h_0$, $1+r$, and $r+h_0$. Let

$$\gamma(r) = \arccos\frac{1+r+h_0-h_0r}{1+r+h_0+h_0r}$$

be the angle of $\Delta_3$ opposite to the side of length $r+h_0$. The function $\gamma(r)$ is increasing in $r$, and we have $\gamma(1) = \beta(h_0)$. Let

$$r^*(h) = \gamma^{-1}(\pi - \beta(h_0) - 2\alpha(h)).$$

Since

$$\gamma(1) = \beta(h_0) = \pi - \beta(h_0) - 2\alpha(h_0) > \pi - \beta(h_0) - 2\alpha(h),$$

we have $r^*(h) < 1$ for $h > h_0$. We can now define $k(h)$ as

$$k(h) = [\ln h/\ln r^*(h)] + 2.$$

We recapitulate what remains to be shown in order to finish the proof of Theorem 2 as

LEMMA 2. *If $\mathcal{P}$ is a packing of circles of homogeneity $h > h_0$ such that every circle from $\mathcal{P}$ has at least five neighbors, then each angular region of measure $2\vartheta(h)$ whose vertex is the center $c$ of a circle $C \in \mathcal{P}$ contains the center of a circle $C' \in \mathcal{P}$ such that the order of the neighborhood of $C$ and $C'$ is at most $k(h)$.*

## 4. Proof of Lemma 2

Let $A$ be an angular region of measure $2\vartheta(h)$ with vertex at the center $c_0$ of the circle $C_0 \in \mathcal{P}$. Let $\overline{A}$ be the complement of $A$. We assume, indirectly, that $\overline{A}$ contains the centers of all circles from $\mathcal{P}$ which are at most $k$th order neighbors of $C_0$. Let $R$ and $R^-$ be the two rays bounding $A$. We assign in the plane a positive sense of orientation so that $R$ is carried into $R^-$ by a rotation through the angle $2\vartheta(h)$. Let $C_0^1$, $C_0^2$, $C_0^3$, $C_0^4$, and

$C_0^5$ be five neighbors of $C_0$ enumerated so that a ray which is rotated around $c_0$ from $R$ to $R^-$ in the positive direction meets the centers of these circles in the order $c_0^1, c_0^2, c_0^3, c_0^4, c_0^5$. We observe that

$$\angle c_0^2 c_0 c_0^4 = \angle c_0^2 c_0 c_0^3 + \angle c_0^3 c_0 c_0^4 \geq 2\alpha(h).$$

Let $x$ and $x^-$ be points on $R$ and $R^-$, respectively, at distance greater than $4k(h)$. Recalling the supposition that the points $c_0^1, \dots, c_0^5$ lie in $\overline{A}$ and the fact that the angle $\angle x^- c_0 x$ is $2\pi - 2\vartheta(h) = 2\alpha(h) + 2\beta(h_0)$, we conclude that either $\angle x^- c_0 c_0^2$ or $\angle c_0^4 c_0 x$ is at most $\beta(h_0)$. Without loss of generality we may assume that

$$\angle c_0^4 c_0 x \leq \beta(h_0).$$

We now choose circles $C_i$, $C_i^1$, $C_i^2$, $C_i^3$, $C_i^4$, and $C_i^5$ for $i = 1, \dots, k(h)$, inductively so that they satisfy the following conditions (Figure 3(a)):

(i) $C_i = C_{i-1}^5$, $C_i^1$, $C_i^2$, $C_i^3$, $C_i^4$, *and* $C_i^5$ *are neighbors of* $C_i$ *enumerated in the positive direction of rotation, and, furthermore,* $C_i^1 = C_{i-1}$.

In what follows we shall consider only the packing $\mathcal{P}^*$ consisting of the circles $C_i$ and $C_i^j$, $0 \leq i \leq k(h) - 1$, $1 \leq j \leq 5$. In accordance with our previous notation we write $c_i$, $r_i$, $c_i^j$, and $r_i^j$ for the center and radius of $C_i$ and $C_i^j$, respectively. We observe that the order of the neighborhood of $C_0$ and any circle of $\mathcal{P}^*$ is at most $k(h)$. Thus, it follows by our indirect supposition that

(ii) *the center of each circle from* $\mathcal{P}^*$ *lies in* $\overline{A}$.

Our objective is to obtain a contradiction by showing that conditions (i) and (ii) imply that the homogeneity of $\mathcal{P}^*$ is smaller than $h$. It is difficult to compare the radii of the circles from $\mathcal{P}^*$ directly. We observe however, that in the case when all the centers $c_i$, $i = 0, 1, \dots, k(h)$, lie on $R$, the comparison of the radii is easy. Therefore we shall replace $\mathcal{P}^*$, without decreasing the homogeneity and without violating conditions (i) and (ii), by another set of circles such that the corresponding centers $c_i$ lie on $R$.

We shall give the transformation mentioned above in $k(h)$ steps. It will not lead to confusion, but will simplify the notation, if we preserve the notation for the circles, their centers, and their radii throughout the procedure. In the $i$th step the circles with subscripts less than $i$ and the circle $C_i^2$ remain unchanged. The rest of the circles are transformed as follows: The circle $C_i$ is replaced by the circle which touches the (unchanged) circles $C_i^1$ and $C_i^2$, and the center of which lies on $R$. $C_i^3$, $C_i^4$, and $C_i^5$ are replaced by circles of the same size touching the new circle $C_i$ such that the distances $c_i^2 c_i^3$, $c_i^3 c_i^4$, and $c_i^4 c_i^5$ remain unchanged. The effect of this transformation to the circle $C_i^5$ is a certain translation $t$. We now apply this translation $t$

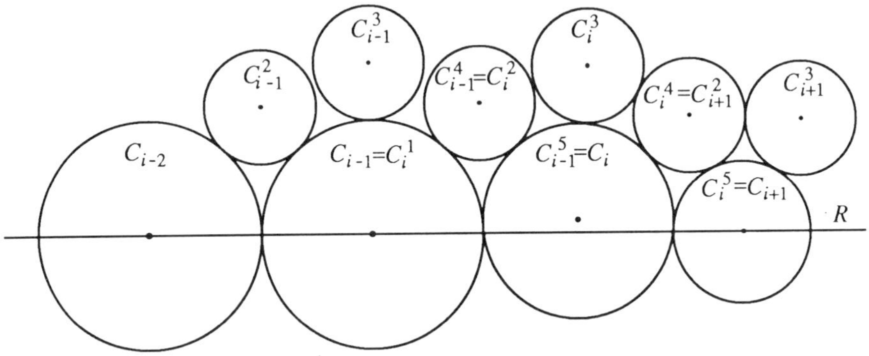


(a)

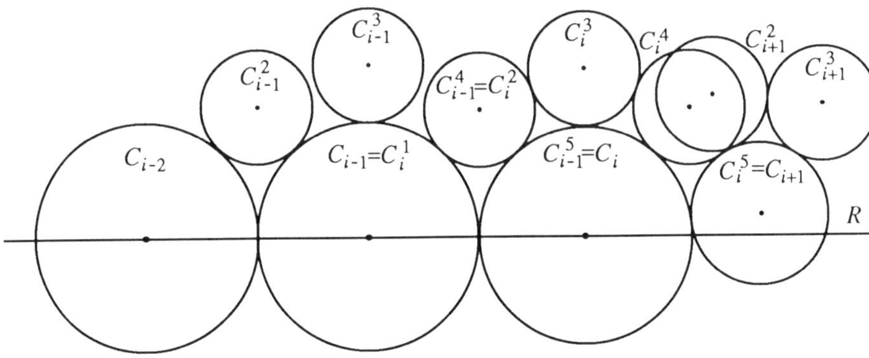


(b)

FIGURE 3

to the circles with subscripts greater than $i$, with the exception of the circle $C_i = C_{i+1}^1$, in order to obtain the new circles corresponding to them in the $i$th step of the transformation (Figure 3(b)).

We have to prove that this transformation does not decrease the homogeneity of $\mathscr{P}^*$, and that the new circles still satisfy conditions (i) and (ii). We note that the new circles need not form a packing. For instance, it may occur that for some $i$ the circles $C_i^4$ and $C_{i+1}^2$ originally coincide, while the new circles corresponding to them are distinct overlapping circles. However, the important fact is that the packing property remains true for the neighbors of $C_i$:

(iii) *The interiors of the circles* $C_i^j$, $j = 1, \ldots, 5$, *are mutually disjoint for all* $i = 0, 1, \ldots, k(h)$.

We observe that the only circles which change their size through the transformation are the circles $C_i$ with $i = 1, \ldots, k(h)$. Thus, in order to see that the transformation does not decrease the homogeneity of $\mathscr{P}^*$, it suffices to show the following:

(iv) *In the* $i$*th step of the transformation the radius* $r_i$ *of* $C_i$ *does not decrease, and the new radii satisfy the inequalities*

$$r_1 \le r_0, \qquad r_i \le r^*(h) r_{i-1} \quad \textit{for } i = 2, \dots, k(h).$$

Obviously, in step 0, that is, originally, the packing $\mathcal{P}^*$ satisfies properties (i)–(iii), while the statement of (iv) is void in this case. Suppose that we finished the $(i-1)$st step $(i \ge 1)$ of the transformation and start the $i$th step with an arrangement of circles satisfying conditions (i)–(iv). In particular, we have $c_{i-1} \in R$ and $r_{i-1} \le r_{i-2}$. This implies that, for $i \ge 2$, we have

$$\angle c_{i-2} c_{i-1} c_i = \angle c_{i-1}^1 c_{i-1} c_{i-1}^5 \ge \beta(h_0) + 2\alpha(h_0) + \gamma(r_i).$$

Using the relation $\pi/2 = \alpha(h_0) + \beta(h_0)$, we get

$$\angle c_i c_{i-1} x = \pi - \angle c_{i-1}^1 c_{i-1} c_{i-1}^5 \le \beta(h_0) - \gamma(r_i).$$

We observe that in view of

$$\angle c_1 c_0 x \le \angle c_0^4 c_0 x - \gamma(r_1) \le \beta(h_0) - \gamma(r_1),$$

the inequality

$$\angle c_i c_{i-1} x \le \beta(h_0) - \gamma(r_i)$$

holds for all $i \ge 1$. Recalling the assumption that each circle from $\mathcal{P}^*$ has radius at most 1, we conclude that the distance of $c_i$ from $R$ is at most

$$(1 + r_i) \sin \angle c_i c_{i-1} x = (1 + r_i) \sin(\beta(h_0) - \gamma(r_i)).$$

A rough estimate shows that the right side of the last inequality is at most

$$2 \sin(\beta(h_0) - \gamma(h_0)) < 0.3 < h_0 < h.$$

This means that $C_i$ intersects $R$. Let $p_i$ be the point of intersection of $C_i$ and $R$ further from $C_0$. It is easy to see that the circles of radii $r_{i-1}\sqrt{2}$ and $r_i\sqrt{2}$ concentric with $C_{i-1}$ and $C_i$, respectively, cover the triangle $c_{i-1}c_i p_i$. Since $h > h_0 > \sqrt{2} - 1$, it follows that the circles with radii $r_{i-1} + h$ and $r_i + h$ concentric with $C_{i-1}$ and $C_i$ also cover this triangle. This yields that

*The center of no circle from* $\mathcal{P}^*$ *is contained in the triangle* $c_{i-1}c_i p_i$.

We now consider the circle $C_i$ and its neighbors $C_i^1 = C_{i-1}$, $C_i^2$, $C_i^3$, $C_i^4$, and $C_i^5$. We take $c_{i-1}$ as the origin of a coordinate system, and denote the vector pointing from $c_{i-1}$ to a point $q$ by $\mathbf{q}$. Let $\mathbf{v}$ be the unit vector perpendicular to $R$ pointing in the region $\overline{A}$. We imagine our transformation as a continuous change enlarging $C_i$ until its center reaches $R$, and consider its effect on the vectors $\mathbf{c}_i^3$, $\mathbf{c}_i^4$, and $\mathbf{c}_i^5$. We observe that the condition that $c_i^j \in \overline{A}$ is equivalent to $\mathbf{v}\mathbf{c}_i^j > 0$ $(3 \le j \le 5)$. Moreover, the distance of $c_i^j$ from $R$ is given by $\mathbf{v}\mathbf{c}_i^j$. Therefore we shall prove that $\mathbf{v}\mathbf{c}_i^j$ $(3 \le j \le 5)$ increases while we enlarge the circle $C_i$.

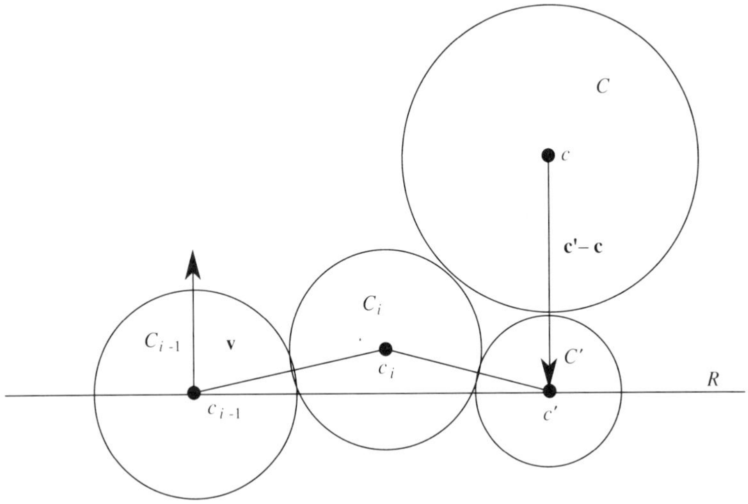


FIGURE 4

We have $\mathbf{v}\mathbf{c}_i^j = \mathbf{v}(\mathbf{c}_i^j - \mathbf{c}_i^{j-1}) + \mathbf{v}\mathbf{c}_i^{j-1}$; thus, in our view of the fact that $C_i^2$ remains unchanged, it suffices to show that $\mathbf{v}(\mathbf{c}_i^j - \mathbf{c}_i^{j-1})$ $(3 \leq j \leq 5)$ increases. It is easily seen that enlarging $C_i$ while maintaining the distances $c_i^{j-1}c_i^j$ causes all the angles $\angle c_i^{j-1}c_ic_i^j$ and $\angle c_i^jc_ic_i^{j-1}$ to increase. This means that the angle enclosed by $\mathbf{v}$ and any of the vectors $\mathbf{c}_i^j - \mathbf{c}_i^{j-1}$ $(3 \leq j \leq 5)$ decreases. Since $|\mathbf{c}_i^j - \mathbf{c}_i^{j-1}|$ remains unchanged, the monotonicity of $\mathbf{v}(\mathbf{c}_i^j - \mathbf{c}_i^{j-1})$ $(3 \leq j \leq 5)$ will be proved by showing that all these angles are positive, but smaller than $\pi$.

It is easy to see that the angle enclosed by $\mathbf{v}$ and $\mathbf{c}_i^2 - \mathbf{c}_i^1$ is positive. Since the angle between $\mathbf{v}$ and $\mathbf{c}_i^j - \mathbf{c}_i^{j-1}$ increases with $j$ $(2 \leq j \leq 5)$, it remains to show that these angles do not exceed $\pi$. For a given position of the circle $C_i$ we vary two circles $C$ and $C'$ so that they touch $C_i$, they are disjoint from one another and from $C_{i-1}$, their centers $c$ and $c'$ lie in $\overline{A}$ but not in the triangle $c_{i-1}c_ip_i$, the (directed) angle $\angle cc_ic'$ is positive, and, finally, each of them has radius at least $h_0$, but at most 1. It is easy to see that under these conditions the angle of $\mathbf{v}$ and $\mathbf{c}' - \mathbf{c}$ attains its maximum if $C$ and $C'$ touch each other, $C$ has radius 1, $C'$ has radius $h_0$, and the center of $C'$ is on $R$ (Figure 4). In this case the angle of $\mathbf{v}$ and $\mathbf{c}' - \mathbf{c}$ is equal to

$$\pi/2 + \angle c_{i-1}c'c = \pi/2 + \angle c_{i-1}c'c_i + \angle c_ic'c,$$

where

$$\angle c_ic'c = \arccos\frac{h_0^2 + h_0r_i + h_0 - r_i}{h_0^2 + h_0r_i + h_0 + r_i}.$$

Recalling that the distance from $c_i$ to $R$ is at most $(1+r_i)\sin(\beta(h_0)-\gamma(r_i))$, we obtain the inequality

$$\angle c_{i-1}c'c_i \le \arcsin\left[\frac{1+r_i}{r_i+h_0}\sin(\beta(h_0)-\gamma(r_i))\right]$$
$$= \arcsin\left[\frac{1+r_i}{r_i+h_0}\sin\left(\arccos\frac{1}{1+h_0} - \arccos\frac{1+r_i+h_0-h_0r_i}{1+r_i+h_0+h_0r_i}\right)\right].$$

One can check that

$$\arccos\frac{h_0^2+h_0r+h_0-r}{h_0^2+h_0r+h_0+r}$$
$$+\arcsin\left[\frac{1+r}{r+h_0}\sin\left(\arccos\frac{1}{1+h_0}-\arccos\frac{1+r+h_0-h_0r}{1+r+h_0+h_0r}\right)\right] < \frac{\pi}{2}$$

for all $r \in [h_0, 1]$. Thus the angle of $\mathbf{v}$ and $\mathbf{c}-\mathbf{c}'$ and similarly the angle between $\mathbf{v}$ and any of the vectors $(\mathbf{c}_i^j - \mathbf{c}_i^{j-1})$, $3 \le j \le 5$, is at most $\pi$. Therefore, each of the circles $C_i^j$, $3 \le j \le 5$, is transformed in the $i$th step so that the distance of its center from $R$ increases. The same is true then for all circles with subscripts greater than $i$. This shows that condition (ii) is preserved during the transformation.

It is obvious by the construction that condition (i) is true for the new set of circles. In order to see that condition (iii) holds for the new circles, it suffices to show that the interiors of the new circles $C_{i+1}^1 = C_i$ and $C_{i+1}^2$ are disjoint. We have to distinguish two cases. If originally we had $C_1^4 = C_{i+1}^2$, then we recall that the angle between $\mathbf{v}$ and the vector $\mathbf{c}_i^4 - \mathbf{c}_i^5 = \mathbf{c}_i^4 - \mathbf{c}_{i+1}$ decreases in the transformation, while the angle between $\mathbf{v}$ and $\mathbf{c}_{i+1}^2 - \mathbf{c}_{i+1}$ remains unchanged. Thus, in the new position of the circles, we have $\angle c_ic_{i+1}c_i^4 \le \angle c_ic_{i+1}c_{i+1}^2$. Since $C_i$ and $C_i^4$ have disjoint interiors, it follows that so do $C_i$ and $C_{i+1}^2$. In a similar way one can see that (iii) remains true when $C_i^4 = C_{i+1}^2$.

It remains to show that (iv) holds. It is obvious from the construction that $r_i$ does not decrease. We have for $i > 2$

$$\angle c_{i-2}c_{i-1}c_{i-1}^4 \ge \beta(h_0) + 2\alpha(h).$$

Hence, at the end of the $i$th step, we have

$$\angle c_{i-1}^4c_{i-1}c_{i-1}^5 = \angle c_{i-1}^4c_{i-1}x \le \pi - \angle c_{i-2}c_{i-1}c_{i-1}^4 \le \pi - \beta(h_0) + 2\alpha(h).$$

In view of the definitions of the functions $\gamma(r)$ and $r^*(h)$, this implies the inequality $r_i \le r^*(h)r_{i-1}$ for $i = 2, \dots, k(h)$. For $i = 1$ we recall that $\angle c_0^4c_0x \le \beta(h_0)$, which, in a similar way, yields the inequality $r_1 \le r_0$.

Now the inequality $r_i \le r^*(h)r_{i-1}$ together with the definition of $k(h)$ implies that $r_{k(h)} < h$, which is a contradiction. This completes the proof of Lemma 2.

## References

1. I. Bárány, Z. Füredi, and J. Pach, *Discrete convex functions and proof of the six circle conjecture of L. Fejes Tóth*, Canad. J. Math. **36** (1984), 569–576.
2. L. Fejes Tóth, *Five-neighbour packing of convex discs*, Periodica Math. Hungar. **4** (1973), 221–229.
3. E. Makai, Jr., *Five-neighbour packing of convex plates*, in Intuitive Geometry (Siófok, 1985), Colloq. Math. Soc. János Bolyai, no. 48, North-Holland, Amsterdam and New York, 1987, pp. 373–381.

Mathematical Institute of the Hungarian Academy of Sciences, P.O.B. 127, H-1364 Budapest, Hungary

DIMACS Series in Discrete Mathematics
and Theoretical Computer Science
Volume **4**, 1991

Applied Geometry
and Discrete Mathematics
THE VICTOR KLEE FESTSCHRIFT

# Symmetric Solutions to Isoperimetric Problems for Polytopes

P. FILLIMAN

ABSTRACT. An algebraic method is described to solve the problem of maximizing the volume of a symmetric polytope $P$ which has a fixed combinatorial type, and is the projection of a regular $(n-1)$-dimensional simplex, $n = |\operatorname{vert} P|$. The square, pentagon, and cube are shown to be maximal projections. Surprisingly, the regular hexagon is a degenerate critical point, and the regular $n$-gon, $n \geq 7$, is a saddle point.

## Introduction

L. Fejes Tóth proved that each Platonic solid has larger volume than any polytope of the same combinatorial type and the same circumradius. A natural question to ask is whether the same result holds for Archimedean solids. In this paper, we shall develop a method for maximizing volume which utilizes the symmetry group of the figures being considered. This method can therefore be applied to both regular and Archimedean solids.

Let $P$ be a symmetric polytope and let $Q \approx P$ mean that $Q$ has the same combinatorial type as $P$. The method we shall use to show that $P$ has greater volume than $Q$ is based on exterior algebra (see **[12]**), and will be outlined in §§1 and 2. This approach naturally applies to three classes of polytopes.

$\Pi = \{Q \approx P : Q$ is a projection of a regular simplex with edge length $\sqrt{2}\}$,

$\Delta = \{Q \approx P : \sum_{v_i \in \operatorname{vert} Q} |v_i|^2 = k\}$, and

$\Omega = \{Q \approx P : |v_i| \leq \sqrt{k/n} \;\; \forall v_i \in \operatorname{vert} Q\}$.

Theorem 6 in **[12]** implies that a symmetric polytope which maximizes volume in $\Pi$ also maximizes volume in $\Delta$ and $\Omega$. Therefore, in this paper we

1980 *Mathematics Subject Classification* (1985 *Revision*). Primary 52A25, 52A40; Secondary 51M20.

Partly supported by the U. S. Army Research Office through the Mathematical Sciences Institute of Cornell University.

will deal only with polytopes which are projections of the regular simplex. In particular, all regular and Archimedean solids are projections of the regular simplex (see **[5, 13.91]**).

The following results are established in §§3 and 5.

THEOREM 1. *If $P$ is a regular n-gon, $n \leq 5$, or a regular cube, then the volume of $P$ satisfies $V(P) > V(Q)$ for any $Q$ in $\Pi$, $\Delta$, or $\Omega$ different from $P$.*

§4 covers the cubeoctahedron, which turns out to be a saddle point in $\Pi$. Some related results can be found in **[13, 16]**.

## 1. Exterior algebra and projections

The main result in **[12]** is that the volume of the projection of a polytope is a piecewise linear function on the Grassmann manifold. The precise description of this fact is given below after some preliminary notation.

Let $\{e_1, \dots, e_n\}$ be the standard basis of $\mathbf{R}^n$, and let $T^{n-1} = \operatorname{conv}\{e_1, \dots, e_n\}$ be the regular $(n-1)$-dimensional simplex. An oriented $k$-plane $L \subset \mathbf{R}^n$ will be identified with the simple, unit $k$-vector $\xi = x_1 \wedge \cdots \wedge x_k$, where $\{x_1, \dots, x_k\}$ is an oriented orthonormal basis of $L$. This map from $L$ to $\xi$ is the classical Plücker embedding of the Grassmann manifold $G(k, n)$ into the exterior algebra $\bigwedge_k \mathbf{R}^n$.

To each plane, there corresponds a polytope $P = \Pi(T^{n-1} : L)$, the orthogonal projection of $T^{n-1}$ into $L$. We may also write $P$ as $\operatorname{conv}\{v_1, \dots, v_n\}$, $v_i = \Pi(e_i : L)$. If we fix a triangulation $\Gamma$ of the boundary of $P$, we can create a "*projection form*" $\phi \in \bigwedge_k \mathbf{R}^n$ which is a sum of terms $e_\lambda = e_{\lambda_1} \wedge \cdots \wedge e_{\lambda_k}$, one for each facet $\operatorname{conv}\{v_1, \dots, v_k\}$ in $\Gamma$. Each term is multiplied by $\operatorname{sign}_\Gamma(\lambda) \in \{\pm 1\}$ which is the orientation of the facet $\lambda$ of $\Gamma$. Altogether,

$$\phi = \sum_{\lambda \in \Gamma} \operatorname{sign}_\Gamma(\lambda) e_\lambda. \tag{1.1}$$

Note that the orientation $\operatorname{sign}_\Gamma(\lambda) = \operatorname{sign}\langle \xi, e_\lambda \rangle$.

The piecewise linearity of the volume of $\Pi(T^{n-1} : L)$ follows from

THEOREM 2. *The volume of $P = \Pi(T^{n-1} : L)$ is*

$$V(P) = \frac{1}{k!} \langle \phi, \xi \rangle, \tag{1.2}$$

*where $\phi$ is given by* (1.1). *Formula* (1.2) *holds throughout a semi-algebraic subset of $G(k, n)$ in which the projections have the same combinatorial type as $P$.*

We use this theorem to find the biggest projection by maximizing $\langle \xi, \phi \rangle$ for all unit $\xi \in G(k, n)$. The maximum of $\langle \cdot, \phi \rangle$ is known in geometric measure theory as "*comass*", and is denoted by

$$\|\phi\| = \sup\{\langle \xi, \phi \rangle | \xi \in G(k, n) \text{ and } |\xi| = 1\} \tag{1.3}$$

(see [**10, p. 38**]). If we are lucky, the projection in the maximal plane will have the combinatorial type corresponding to $\phi$ and our problem will be solved.

## 2. Invariant subspaces of $\mathbf{R}^n$

In the preceding section, we transformed the problem of finding the biggest polytope to that of determining the comass of the projection form $\phi$ in (1.1). In most cases, we have not gained much since it is extremely difficult to find the comass of arbitrary forms in $\bigwedge_k \mathbf{R}^n$. However, for symmetric polytopes the comass is often easier to find. The idea is to choose the "right" basis of $\mathbf{R}^n$, i.e., the one in which $\phi$ has its simplest expression. This basis will consist of vectors in the invariant spaces, or $G$-spaces, of the symmetry group $G$ of $P$. In this section, we shall briefly cover that part of the representation theory of finite groups needed to find explicit coordinates for these $G$-spaces.

Let $G$ be a finite group, and let $T: G \to \mathrm{GL}(n)$ be a representation of $G$ in the group of linear isomorphisms of $\mathbf{R}^n$. It suffices for our purposes to require $T$ to be a transitive permutation representation on the standard basis of $\mathbf{R}^n$. Define

$$H = \{g \in G | T(g)(e_1) = e_1\}, \tag{2.1}$$

and choose a set of representatives $\{g_1 = 1, g_2, \dots, g_n\}$ for the left cosets $G/H$ so that $g_j(e_1) = e_j$.

Let $L_1, \dots, L_s$ be a full set of irreducible $\mathbf{R}G$-modules, with characters $\chi_1, \dots, \chi_s$. Then the character $\tau$ of $T$ can be written as

$$\tau = a_1\chi_1 + \cdots + a_s\chi_s, \tag{2.2}$$

where each $a_i$ is a nonnegative integer. According to [**8, (30.14)**], (2.1) determines $T$ uniquely up to equivalence. In fact, $\mathbf{R}^n$ is isomorphism as an $\mathbf{R}G$-module to

$$\underbrace{L_1 \oplus \cdots \oplus L_1}_{a_1} \oplus \cdots \oplus \underbrace{L_s \oplus \cdots \oplus L_s}_{a_s}. \tag{2.3}$$

Thus $\mathbf{R}^n$ has a decomposition into irreducible $G$-spaces, $a_1$ isomorphic to $L_1$, $a_2$ isomorphic to $L_2$, etc. It remains to find coordinates for these $G$-spaces.

Suppose we are given an irreducible representation $U: G \to \mathrm{O}(p)$, which maps $G$ into the orthogonal group in $\mathbf{R}^p$. Our problem is to produce a $p \times n$ matrix $\mathbf{X}$ whose rows are an orthonormal basis for a $G$-space $L$ corresponding to $U$. Since $T(g)$ restricted to $L$ equals $U(g)$, we have

$$\mathbf{XT}(g) = \mathbf{U}(g)\mathbf{X} \tag{2.4}$$

for all $g \in G$. But, by hypothesis, $T(g)$ simply permutes the columns of $\mathbf{X}$. Hence the $j$th column of $\mathbf{X}$ equals $\mathbf{U}(g_j)u$ for some $u \in \mathbf{R}^p$. A shorthand notation for such an $\mathbf{X}$ will be

$$\mathbf{X} = [\mathbf{U}(g_j)u], \qquad j = 1, \dots, n. \tag{2.5}$$

This formula shows how to obtain the coordinates of a basis of a $G$-space from the representation $U$. Simply choose a vector $u \in \mathbf{R}^p$ which is fixed by $U(H)$, and form the matrix $\mathbf{X}$.

Let $L'$ be another $G$-space, spanned by the rows of

$$\mathbf{Y} = [\mathbf{V}(g_j)v], \qquad j = 1, \dots, n, \quad v \in \mathbf{R}^q, \tag{2.6}$$

with $V: G \to \mathrm{O}(q)$ a representation of $G$. The angles between the bases of $L$ and $L'$ can be tabulated neatly in the matrix $\mathbf{M} = \mathbf{X}\mathbf{Y}^t$. Everything we will need to know about $\mathbf{M}$ is contained in a paper by Coxeter and Brauer **[6]**.

THEOREM 3. *Suppose* $\mathbf{X}$ *and* $\mathbf{Y}$ *are given by* (2.5) *and* (2.6), *respectively. If* $U$ *and* $V$ *are inequivalent, irreducible representations of* $G$ *then*

$$\mathbf{X}\mathbf{Y}^t = (0). \tag{2.7}$$

*If* $U$ *and* $V$ *are equivalent, absolutely irreducible representations of* $G$ *then*

$$\mathbf{X}\mathbf{Y}^t = \frac{n}{p}\langle u, v\rangle \mathbf{I}. \tag{2.8}$$

## 3. Regular polygons

It is well known that the regular $n$-gon $\{n\}$ has greater area than any other convex $n$-gon in a disk. Remarkably, $\{n\}$ is a maximal projection of $T^{n-1}$ only if $n \leq 5$. We shall see below that this result is a simple consequence of the theory developed in §§1 and 2. The notation for regular polytopes and their symmetry groups is taken from **[5]**.

Let $P$ be a convex $n$-gon with the origin in its interior. Label the vertices $v_1, \dots, v_n$ of $P$ in counter-clockwise order about the origin. We assume $P = \Pi(T^{n-1} : L)$, and $v_i = \Pi(e_i : L)$, for some 2-plane $L$ in $\mathbf{R}^n$. By (1.1), the projection form of $P$ is

$$\phi = e_{12} + e_{23} + \cdots + e_{n-1,n} + e_{n,1}. \tag{3.1}$$

Theorem 2 states that the area of $P$ is $\frac{1}{2}\langle \xi, \phi\rangle$, where $\xi$ is the simple, unit 2-vector corresponding to $L$. Following §2, we express $\phi$ in a new basis of $\mathbf{R}^n$ derived from certain invariant spaces described below.

Suppose $T^+$ is the regular representation of the rotation group $[n]^+$ of $\{n\}$ on the basis $\{e_1, \dots, e_n\}$. From **[4, Theorem (15.2)]**, every irreducible representation of a group appears as a constituent of the regular representation. The irreducible representations of $[n]^+$ in $\mathbf{R}^k$ are:

$k = 1$. *The identity representation, and if* $N$ *is even, the map* $z^j \to (-1)^j$, *where* $z$ *is any generator of* $[n]^+$.

$k = 2$. *For* $1 \leq d < n/2$, *a representation of* $[n]^+$ *as the rotations of the regular star* $n$*-gon* $\{n/d\}$ *of density* $d$ *(see* **[11, p.** 107]). *These representations are not absolutely irreducible. They each split into two one-dimensional representations* $z \to e^{di\theta}$ *and* $z \to e^{-di\theta}$, $\theta = 2\pi/n$, *over* $\mathbf{C}$.

The representations given above for $[n]^+$ are all inequivalent, so by Theorem 3 they appear in completely orthogonal spaces. Therefore, there exists an orthonormal basis $\{f_1, \ldots, f_n\}$ of $\mathbf{R}^n$, where

(3.2) $$\Pi(T^{n-1} : \operatorname{span}\{f_{2d-1}, f_{2d}\}) \text{ is the regular star polygon } \{n/d\}$$

(see also [**5, 13.3**]). The expression for $\phi$ in the $f$-basis is

$$\phi = \sum_{i<j} \langle \phi, f_{ij} \rangle f_{ij}.$$

Either by direct calculation, or by [**13, Theorem 1**], we find that $\langle \phi, f_{ij} \rangle = 0$ unless $f_{ij}$ is one of the planes in (3.2). Hence

(3.3) $$\phi = \sum_{1 \le d < n/2} \langle \phi, f_{2d-1,2d} \rangle f_{2d-1,2d},$$

According to Theorem 2, $\langle \phi, f_{2d-1,2d} \rangle$ is twice the "area" of $\{n/d\}$ (see [**11, p. 107**]). A simple calculation of this area gives

(3.4) $$\phi = \sum_{1 \le d < n/2} 2 \sin d\theta\, f_{2d-1,2d}, \qquad \theta = \frac{2\pi}{n}.$$

This expression is easier to use to find $\|\phi\|$ than (3.1) because the terms $f_{2d-1,2d}$ correspond to completely orthogonal subspaces of $\mathbf{R}^n$. In fact, the critical points of $\langle \cdot, \phi \rangle$ on $G(2, n)$ are just the points $f_{2d-1,2d}$, $1 \le d < n/2$ (see [**10, p. 29**]). If $n$ is odd, these points are isolated critical points. However, if $n$ is even, each pair $f_{2d-1,2d}$ and $f_{n-2d-1,n-2d}$ belong to an infinite level set isomorphic to the 2-sphere (see [**18, Corollary 3.4**]). It follows that the comass of $\phi$ is

(3.5) $$\|\phi\| = \max\left\{ \sin \frac{2\pi d}{n} \,\middle|\, 1 \le d < \frac{n}{2} \right\}.$$

If $n \le 5$, the maximum occurs precisely when $d = 1$, and so $\{n\}$ maximizes area in $\Pi$. The regular hexagon is included in an infinite set of maximal polygons. Finally, if $n \ge 7$, $\{n\}$ is a saddle point for area.

## 4. Cubeoctahedron

In this section we prove that the cubeoctahedron is not a maximal projection of $T^{11}$. Although we obtain a negative result for the cubeoctahedron, it is a nice example of the application of the methods in §§1 and 2 to Archimedean solids.

The symmetry group of the cubeoctahedron is the same as the group of the cube, denoted by $[3, 4]$. This group is not the best for our purposes since its order 48 does not match the number of vertices of the cubeoctahedron. Fortunately, an appropriate subgroup exists. The vertices of the cube can be partitioned into two sets whose convex hulls are oppositely oriented tetrahedra (see [**5, Figure 3.6B**]). Thus, the symmetry group $[3, 3]$ of the tetrahedron is a subgroup of $[3, 4]$. Its rotation group $[3, 3]^+$ has order 12 and is just the group we need.

The group $G = [3, 3]^+$ is also isomorphic to $A_4$, the alternating group of degree 4 (see **[5, p. 48]**). As above, we produce a new basis of $\mathbf{R}^{12}$ from the $G$-spaces of the regular representation $T$ of $G$. First we list the irreducible representations of $G$ in $\mathbf{R}^k$.

$k = 1$. *The identity representation.*

$k = 2$. *A representation $V$ as the rotations of the regular triangle. This representation is not absolutely irreducible.*

$k = 3$. *The representation $U$ as* $[3, 3]^+$.

The representation $U$ is absolutely irreducible, and thus appears three times in the decomposition of $\mathbf{R}^{12}$ in (2.2) (see **[8, (27.18)]**). A count of dimensions in (2.2) gives $1 + a_2 + 3 \cdot 3 = 12$ and $a_2 = 1$, so the degree 2 representation must occur only once.

The next step is to find coordinates for these $G$-spaces. A presentation of $A_4$ as

(4.1) $$\{\langle a, b\rangle | a^3 = b^2 = (ab)^2 = 1\}$$

is given in **[4, (1.57)]**. The representation $U$ may be defined by letting $U(a)$ be a rotation of $2\pi/3$ about $(1, 1, 1)$, and $U(b)$ a rotation of $\pi$ about $(1, 0, 0)$. Comparing with Figure 1, we see that $T$ is generated by

(4.2) $$T(a) = (123)(456)(789)(10\ 11\ 12),$$ $$T(b) = (14)(27)(3\ 10)(68)(9\ 11)(5\ 12).$$

We shall obtain coordinates of $G$-spaces of $T$ by using (2.5). This requires that we label the elements of $G$ as $\{g_1, \dots, g_{12}\}$ so that $U(g_j)(v_1) = v_j$, $v_j$ being the $j$th vertex of Figure 1. This list is

(4.3) $$(1, a, a^2, b, ab, a^2b, ba, aba, a^2ba, ba^2, aba^2, a^2ba^2).$$

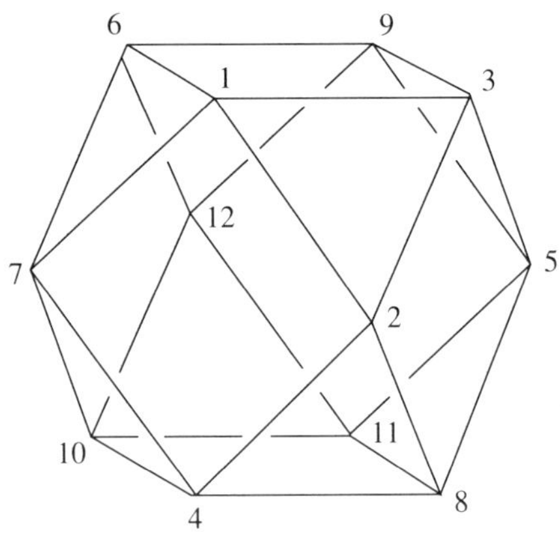


FIGURE 1

Letting $u = (x, y, z)$ in (2.5), we obtain the matrix

(4.4)

$$\mathbf{X}(x, y, z) = \begin{bmatrix} x & z & y & x & -z & -y & z & -y & -x & y & -x & -z \\ y & x & z & -y & x & -z & -x & z & -y & -z & y & -x \\ z & y & x & -z & -y & x & -y & -x & z & -x & -z & y \end{bmatrix}.$$

The totality of all such matrices gives rise to $G$-spaces lying in a nine-dimensional subspace of $\mathbf{R}^{12}$. The best possible spanning set would be a collection of three such spaces which are all critical points of $\langle\cdot, \phi\rangle$, where $\phi$ is a projection form of the cubeoctahedron. We proceed to calculate these invariant critical points.

The form $\phi$ is not unique, but depends on how we triangulate the square faces of the cubeoctahedron. The natural choice is to draw in both diagonals on each square face so that the triangulation is invariant under $G$. The corresponding form, which is also invariant under $G$, is

$$(4.5)\qquad \phi = \sum_G \frac{1}{3}T(g)(e_{123}) + \frac{1}{3}T(g)(e_{167}) - \frac{1}{2}T(g)(e_{124}) - \frac{1}{2}T(g)(e_{127}).$$

Note that

$$T(g)(e_{ijk}) = T(g)(e_i) \wedge T(g)(e_j) \wedge T(g)(e_k)$$

is the natural map of $T(g)$ on $\bigwedge_3 \mathbf{R}^{12}$.

The critical equations for $\langle\cdot, \phi\rangle$ on $G(k, n)$ are given in [**12, Proposition 9**]. Assuming the critical point is obtained from an invariant plane spanned by the rows of $\mathbf{X}(x, y, z)$, the critical equations reduce to three polynomials in four variables:

$$(4.6)\qquad \begin{aligned} \alpha x &= 2x^2 + 2xz - y^2 + z^2, \\ \alpha y &= -2xy - 2xz, \\ \alpha z &= 2z^2 + 2xz - y^2 + x^2. \end{aligned}$$

It is quite easy to solve this system. One obtains the following as invariant critical points of $\langle\cdot, \phi\rangle$:

$$\begin{aligned} \mathbf{X}(1, 0, 1) &\sim \text{cubeoctahedron}, \\ \mathbf{X}(-1, 1, 1) &\sim \text{tetrahedron with doubled vertices}, \\ \mathbf{X}(1, 3, 1) &\sim \text{truncated tetrahedron}. \end{aligned}$$

Since $(1, 0, 1) \perp (-1, 1, 1)$, Theorem 3 implies the first two figures occur in completely orthogonal 3-spaces in $\mathbf{R}^{12}$. However the truncated tetrahedron is skew to the others. We shall have to reposition this space to get an orthonormal basis of $\mathbf{R}^{12}$. At this stage, we know the cubeoctahedron is a critical point. We prove below that it is a saddle point by calculating the eigenvalues of the Hessian matrix.

The degree 2 representation $V$ of $G$ is defined by letting $V(a)$ be a rotation of $\mathbf{R}^2$ by $2\pi/3$, and letting $V(b)$ be the identity. Coordinates for this $G$-space are given by the $2 \times 12$ matrix $\mathbf{Y}$ whose columns equal the image of $(1/\sqrt{6}, 0)$ under the mappings obtained by applying $V$ to (4.3). Altogether, we have an orthonormal basis $\{f_1, \dots, f_{12}\}$ of $\mathbf{R}^{12}$, where:

$$f_1, f_2, \text{ and } f_3 \text{ are the rows of } \mathbf{X}\left(\frac{1}{2\sqrt{2}}(1, 0, 1)\right),$$
$$f_4, f_5, \text{ and } f_6 \text{ are the rows of } \mathbf{X}\left(\frac{1}{2\sqrt{3}}(-1, 1, 1)\right),$$
$$f_7, f_8, \text{ and } f_9 \text{ are the rows of } \mathbf{X}\left(\frac{1}{2\sqrt{6}}(-1, -2, 1)\right),$$
$$f_{10} \text{ and } f_{11} \text{ are the rows of } \mathbf{Y}, \text{ and}$$
$$f_{12} = \frac{1}{\sqrt{12}}(1, \dots, 1).$$

The remaining results in this section were calculated with the aid of the Maple computer algebra system. The expression of $\phi$ in the $f$-basis is:

(4.8)
$$\phi = \frac{5\sqrt{2}}{4} f_{123} + \frac{1}{2}\sum_{S_3} \operatorname{sgn}\sigma f_{\sigma(1,2,3)+(0,3,6)} - \frac{\sqrt{2}}{4}\sum_{A_3} \operatorname{sgn}\sigma f_{\sigma(1,2,3)+(0,6,6)}$$
$$+ \frac{1}{4}\left(\sum_{i=1}^{3} f_i \wedge f_{i+3} \wedge V(a^i)(3, -\sqrt{3}) + \sum_{i=1}^{3} f_i \wedge f_{i+6} \wedge V(a^i)(-3\sqrt{2}, \sqrt{6})\right.$$
$$\left. + \sum_{i=1}^{3} f_{i+3} \wedge f_{i+6} \wedge V(a^i)(-\sqrt{3}, -3)\right).$$

Unfortunately, it was not possible to calculate $\|\phi\|$ from (4.8). However, we can get the information we need about the point $f_{123}$. The tangent space to $G(3, 12)$ at $\xi = f_{123}$ is

$$T_\xi G(3, 12) = \operatorname{span}\{f_{ijk} : |\{i, j, k\} \cap \{1, 2, 3\}| = 2\}. \tag{4.9}$$

Since $\phi$ is orthogonal to $T_\xi G(3, 12)$, $f_{123}$ is a critical point of $\langle \cdot, \phi\rangle$. The Hessian of $\langle \cdot, \phi\rangle$ at $\xi$ can be represented by a $27 \times 27$ matrix $\mathbf{H}$ which acts on $T_\xi G(3, 12)$. This matrix decomposes into four $6 \times 6$ blocks and three $1 \times 1$ blocks in the $f$-basis. It turns out $\mathbf{H}$ has 24 negative eigenvalues and three positive eigenvalues, and so $\xi$ is a saddle point for $\langle \cdot, \phi\rangle$.

One final calculation needs to be made to show $\xi$ is also a saddle point for projections in $\Pi$. Because the cubeoctahedron has square faces, the figures of the same combinatorial type do not correspond to an open set of $G(3, 12)$. In a neighborhood of $\xi$, the appropriate 3-planes belong to

$$\Pi' = \{\eta \in G(3, 12) | \langle \eta, T(g)(\psi)\rangle = 0 \ \forall g \in G\}, \tag{4.10}$$

where $\psi = e_{174} - e_{742} + e_{421} - e_{217}$. Two of the positive eigenvalues of $\mathbf{H}$ have eigenvectors in $\Pi'$, so the cubeoctahedron is indeed a saddle point for $\Pi$.

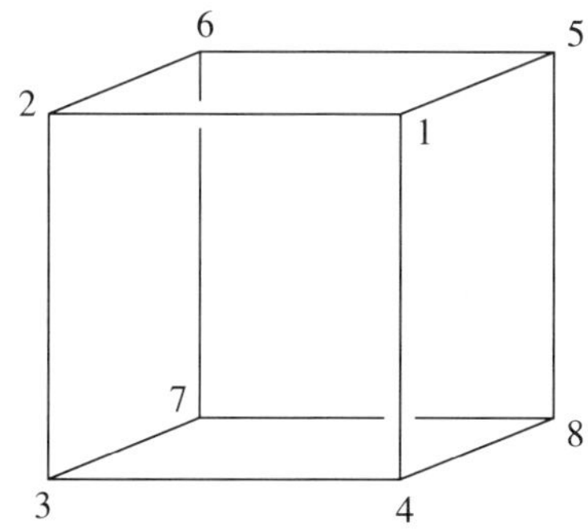


FIGURE 2

## 5. Cube

The final polytope we shall investigate is the cube. Its rotation group $G = [3, 4]^+$ is isomorphic to $S_4$ (see **[5, p. 49]**). As a presentation of $S_4$, we choose

$$\{\langle a, b\rangle | a^3 = b^3 = (ab)^2 = 1\} \tag{5.1}$$

(see **[7, p. 9]**).

In the natural representation $U = [3, 4]^+$, $U(a)$ is a rotation of $2\pi/3$ about a vertex, and $U(b)$ is a rotation of $\pi/2$ about an adjacent face. In Figure 2, these correspond to the permutations

$$T(a) = (245)(386) \quad \text{and} \quad T(b) = (1234)(5678). \tag{5.2}$$

This defines a representation $T$ of $G$ as permutations of $\{e_1, \dots, e_8\}$. Note that $T$ is not the regular representation of $G$.

TABLE 1. Character table of $S_4$

| | $C_1$ | $C_2$ | $C_3$ | $C_4$ | $C_5$ |
|---|---|---|---|---|---|
| $\chi_1$ | 1 | 1 | 1 | 1 | 1 |
| $\chi_2$ | 1 | 1 | 1 | $-1$ | $-1$ |
| $\chi_3$ | 2 | 2 | $-1$ | 0 | 0 |
| $\chi_4$ | 3 | $-1$ | 0 | 1 | $-1$ |
| $\chi_5$ | 3 | $-1$ | 0 | $-1$ | 1 |

In order to find the $G$-spaces of $T$ in $\mathbf{R}^8$, we must first find the character $\tau$ of $T$. The conjugacy classes of $G$ appearing in Table 1 are as follows: $C_1 = \{1\}$, $C_2 = \{\text{rotations about a face by } \pi\}$, $C_3 = \{\text{rotations about a vertex}\}$, $C_4 = \{\text{rotations about a face by } \pi/2\}$, and $C_5 = \{\text{rotations about}$

an edge}. The integer $\tau(C_i)$ is simply the number of vertices of the cube fixed by any element of $C_i$. Hence $\tau(C_1) = 8$, $\tau(C_3) = 2$, and $\tau$ equals 0 on the other classes. From Table 1, the expression of $\tau$ as a sum of irreducible characters of $G$ is

$$\tau = \chi_1 + \chi_2 + \chi_4 + \chi_5. \tag{5.3}$$

The corresponding irreducible representations are (see **[8, pp. 332–333]**):

$U_1$, the identity.

$U_2$, the representation $\sigma \to \operatorname{sgn}\sigma$, $\sigma \in S_4$. In terms of the generators in (5.1), $U_2(b)$ becomes central inversion of the real line.

$U_4$, the representation $[3, 4]^+$.

$U_5$, the representation $[3, 3]$, which is the full symmetry group of $T^3$.

In terms of our generators, $U_5(b) = U_4(b) \circ c$, where $c$ is central inversion about the origin in $U^3$. This follows from the fact $U_5 = U_2 \otimes U_4$ (see **[8, p. 333]**).

The coordinates for the $G$-spaces of $T$ in $\mathbf{R}^8$ are, as usual, given by (2.5). However, this requires a list of representatives of $G/H$, where $H = \{1, a, a^2\}$. The list which follows the orbit of $v_1$ in Figure 2 is

$$Q = (1, b, b^2, b^3, ab^3, bab^3, b^2ab^3, b^3ab^3). \tag{5.4}$$

The degree 3 $G$-spaces are then given by the matrices

$$\mathbf{X} = [\mathbf{U}_4(Q_j)(1, 1, 1)], \tag{5.5}$$

$$\mathbf{Y} = [\mathbf{U}_5(Q_j)(1, 1, 1)], \qquad j = \{1, \dots, 8\}. \tag{5.6}$$

Altogether, our basis $\{f_1, \dots, f_8\}$ of $\mathbf{R}^8$ consists of the rows of

$$\mathbf{F} = \frac{1}{2\sqrt{2}} \begin{bmatrix} 1 & 1 & 1 & 1 & -1 & -1 & -1 & -1 \\ 1 & -1 & -1 & 1 & 1 & -1 & -1 & 1 \\ 1 & 1 & -1 & -1 & 1 & 1 & -1 & -1 \\ 1 & -1 & 1 & -1 & 1 & -1 & 1 & -1 \\ 1 & 1 & -1 & -1 & -1 & -1 & 1 & 1 \\ 1 & -1 & -1 & 1 & -1 & 1 & 1 & -1 \\ 1 & -1 & 1 & -1 & -1 & 1 & -1 & 1 \\ 1 & 1 & 1 & 1 & 1 & 1 & 1 & 1 \end{bmatrix}. \tag{5.7}$$

Rows 1–3 of $\mathbf{F}$ equal $\mathbf{X}$, rows 4–6 are $\mathbf{Y}$, row 7 corresponds to $U_2$, and row 8 to $U_1$.

The final step is to express a projection form $\phi$ in this basis. Adding in both diagonals of each square face of the cube gives a triangulation invariant under $G$. The corresponding form is

$$\phi = \sum_G \frac{1}{2} T(g)(e_{123}). \tag{5.8}$$

Calculating the products $\langle \phi, f_{ijk} \rangle$ gives

$$\phi = \frac{3}{\sqrt{2}} f_{123} - \frac{1}{\sqrt{2}} \sum_{A_3} f_{\sigma(1,2,3)+(0,3,3)}. \tag{5.9}$$

The comass of $\phi$ can be easily determined using a result of Dadock and Harvey **[9]**. One finds that $\|\phi\| = 3/\sqrt{2}$, and it is attained only by the 3-plane $f_{123}$. This proves the cube is the maximum projection of $T^7$ among figures of the same type.

## REFERENCES


1 A. Aitken, *Determinants and matrices*, Oliver and Boyd, London, 1956.
2. J. D. Berman and K. Hanes, *Volumes of polyhedra inscribed in the unit sphere in* $E^3$, Math. Ann. **188** (1970), 78–84.
3. B. Buchberger, *Gröbner bases: an algorithmic method in polynomial ideal theory*, in Multidimensional Systems Theory (N. K. Bose, ed.), Reidel, 1985, pp. 184–232.
4. M. Burrow, *Representation theory of finite groups*, Academic Press, New York, 1965.
5. H. S. M. Coxeter, *Regular polytopes*, Dover, New York, 1973.
6. H. S. M. Coxeter and R. Brauer, *A generalization of theorems of Schönhardt and Mehmke on polytopes*, Trans. Royal Soc. Canada (3) **34** (1940), 29–34.
7. H. S. M. Coxeter and W. O. J. Moser, *Generators and relations for discrete groups*, Springer-Verlag, New York, 1980.
8. C. Curtis and I. Reiner, *Representation theory of finite groups and associative algebras*, Wiley, New York, 1962.
9. J. Dadok and R. Harvey, *Calibrations on* $\mathbf{R}^6$, Duke Math. J. **50** (1983), 1231–1243.
10. H. Federer, *Geometric measure theory*, Springer-Verlag, Berlin, 1969.
11. L. Fejes Tóth, *Regular figures*, MacMillan, New York, 1964.
12. P. Filliman, *Exterior algebra and projections of polytopes*, Discrete Comput. Geom. **5** (1990), 305–322.
13. ——, *A metrical characterization of symmetric polytopes*, Geom. Dedicata **30** (1989), 81–91.
14. R. Gardner, *Some applications of the retraction theorem in exterior algebra*, J. Differential Geom. **2** (1968), 25–31.
15. J. Goodman and R. Pollack, *Upper bounds for configurations and polytopes in* $\mathbf{R}^d$, Discrete Comp. Geom. **1** (1986), 219–227.
16. J. Linhart, *Volume and circumradius of simplicial polytopes*, Discrete Geom. Convexity (1985), 97–105.
17. M. Marcus, *Finite dimensional multilinear algebra*, Part I, Dekker, New York, 1973.
18. F. Morgan, *The exterior algebra* $\bigwedge_k \mathbf{R}^n$ *and area minimization*, Linear Algebra Appl. **66** (1985), 1–28.
19. H. Weyl, *Classical groups*, Princeton Univ. Press, Princeton, N. J., 1946.



DEPARTMENT OF MATHEMATICS, WHITE HALL, CORNELL UNIVERSITY, ITHACA, NEW YORK 14853

*E-mail address*: filliman@mssun7.msi.cornell.edu

DIMACS Series in Discrete Mathematics
and Theoretical Computer Science
Volume 4, 1991

Applied Geometry
and Discrete Mathematics
THE VICTOR KLEE FESTSCHRIFT

# A Global Newton Method

A. A. GOLDSTEIN

ABSTRACT. In the setting defined below, a global Newton method for finding a zero of a vector-valued function $F$ is formulated. The method always converges, and automatically becomes an ordinary Newton method on the tail end of the sequence. An iteration count is given for the number of steps required to reduce the norm of $F$ to less than any prescribed tolerance $\varepsilon$. The iteration count is given in terms of a parameter $\alpha$ which is generally unknown; nevertheless, the result is of interest for the perspective it furnishes. Moreover, to use the algorithm no knowledge of $\alpha$ is necessary.

A special setting for a global Newton method is that of minimizing a strongly convex function $f$ defined on a real Hilbert space $H$. By strongly convex we mean that $f$ is twice continuously differentiable with a Hessian that is bounded from above, and bounded from below by a positive number. The ratio of the upper to lower bound will be called the condition number.

Such Hessians are invertible, so that Newton's method is well defined for every point in $H$. Moreover, a strongly convex function achieves a unique minimum, where $\nabla f(x) = 0$. However, Newton's method may not converge to the root of $\nabla f(x) = 0$ from arbitrary points in $H$. This is a *raison d'être* for Global Newton methods. Some such methods, whose ingredients contain Newton steps, generate sequences that converge for every strongly convex function and any starting point in $H$. An early history of this subject may be found in Polak [**10**]. Algorithms for this problem have been considered by Goldstein [**4**], Dunn [**1, 2, 3**], and others.

These methods give an estimated asymptotic superlinear rate of convergence, which, unfortunately, has no practical significance. By this we mean that given a set of strongly convex functions, say $\mathscr{F}$, whose condition number is bounded above by $Q$, given a positive integer $N$, and given a starting point $x_0$, there exists a function $f \in \mathscr{F}$, having a minimum value $f(z)$, such that any sequence generated from $f$ by one of the above global Newton

1980 *Mathematics Subject Classification* (1985 *Revision*). Primary 58C15, 68C25.
The author was partially supported by grant NIH RR01243.


1052-1798/91 \$1.00 + \$.25 per page

methods satisfies for a certain fixed positive number $\rho$ the inequality

$$(e^{-1/\rho})^k \leq (f(x_k) - f(z))/(f(x_0) - f(z)) \quad \text{for } 1 \leq k \leq N.$$

When the dimension of $H$ is $n$, $\rho$ depends on $Q$ and $n$, and is independent of $f$, $x_0$, and $N$. Roughly speaking, the class $\mathscr{F}$ contains functions that "resist" superlinear convergence (see Nemerovsky and Yuden **[8]**, and Goldstein **[5]**). On the other hand, for the global Newton algorithm below, those functions in a family such that $\mu^2/4L \geq \alpha$ (see below for definitions) commence their superlinear convergence in no more than a fixed number of steps, that number depending on the starting point $x_0$ and the parameter $\alpha$.

"Backtracking" introduced in **[4]** and used also in **[1, 2, 3]** will also be used in the new algorithm. Backtracking is much simpler to implement than the classical global Newton method of continuation. A history and discussion of continuation, which goes back to at least 1934, may be found in Ortega and Rheinboldt **[9]**. Current references may be found in Smale **[11]**.

In general, we regard a global Newton method as any algorithm incorporating Newton steps that generates a finite sequence terminating in an "approximate root". In Smale's terminology, this is a point from which the ordinary Newton's method will converge at a specified quadratic rate of convergence.

I am indebted to Brad Bell, Lucien Polak, and the referees for their generosity and helpful criticisms.

NOTATION, SETTING, AND PRELIMINARIES. Let $F$ be a map from a Banach space $D$ to a Banach space $E$. When it exists, let $F'(x)$ denote the Fréchet derivative of $F$ at $x \in D$. The differential of $F$ at $x$ with increment $h$ will be written as $F'(x)h$. Let $x_0$ be an arbitrary point in $D$, and let $S = \{x \in D\colon \|F(x)\| \leq \|F(x_0)\|\}$. If $F'(x)$ exists, we assume that it is invertible and satisfies for some $\mu > 0$ and all $h \in D$ the inequalities

$$\mu\|h\| \leq \|F'(x)h\| \leq \lambda\|h\|.$$

Let

$$\phi(x) = F'_{-1}(x)F(x).$$

The notation $F'_{-1}(x)$ is used for the inverse of $F'(x)$. If $F'$ is Lipschitz continuous on $S$ with constant $L$ then the Kantorovich condition [7] for $x_0$ is

$$(*) \qquad \|F'_{-1}(x_0)F(x_0)\|\|F'_{-1}(x_0)\|L = h_0 < 1/2.$$

A point $x_0$ is called an *approximate root* if the inequality $(*)$ above holds with $h_0 \leq 1/4$. Thus $x_0$ is an approximate root if

$$\|F(x_0)\| \leq \mu^2/4L.$$

If $x_0$ is an approximate root then a zero $z$ of $F$ lies in the ball $B = \{x \in D\colon \|x - x_0\| \leq 0.293\mu/L\}$, the Newtonian iterates defined by $x_{n+1} = x_n - F'_{-1}(x_n)F(x_n)$ lie in $B$, and $\|x_n - z\| \leq 2^{-n+1}(2h_0)^{2^n-1}\|x_1 - x_0\|$ **[7, 9]**. Thus, $\|x_n - z\| \leq 2(1/2)^{2^n+1}\|x_1 - x_0\| \leq (1/2)2^{2^n+1}(\mu/L)$.

In the first part of the lemma below we consider the case when $F'$ is Lipschitz continuous, while in the second part $F'$ is allowed some jump discontinuities.

LEMMA. *Assume that* $x \in S$ *and* $S$ *is bounded. Let* $I(x) = \{x - t\phi(x): \|F(x - t\phi(x))\| \leq \|F(x)\|$ *and* $t \geq 0$ *is minimal*$\}$. *That is,* $t$ *is to be taken sufficiently small to ensure that* $I(x)$ *is connected. Assume that* $F'$ *satisfies a Lipschitz condition with constant* $2L$ *on* $S$. *Let* $2l(x)$ *denote the Lipschitz constant for* $F'$ *on the segment* $I(x)$. *If* $x - \gamma\phi(x) \in I(x)$ *then*

(a)
$$\|(1-\gamma)F(x) - F(x - \gamma\phi(x))\| \leq \gamma^2 l(x)\|\phi(x)\|^2$$

*and*

(b)
$$\sigma(\gamma) - \frac{\gamma l(x)\|\phi(x)\|^2}{\|F(x)\|} \leq \frac{\|F(x)\| - \|F(x - \gamma\phi(x))\|}{\gamma\|F(x)\|} \leq \sigma(\gamma) + \frac{\gamma l(x)\|\phi(x)\|^2}{\|F(x)\|}.$$

*In this formula,* $\sigma(\gamma) = 1$ *if* $0 < \gamma \leq 1$, *and* $\sigma(\gamma) = (2-\gamma)/\gamma$ *if* $\gamma \geq 1$.

For the second part of the lemma assume that, for each $x$ in $S$, the restriction of $F$ to $I(x)$ is piecewise, twice-continuously differentiable. Whenever $F''$ is defined on $S$ assume that $\|F''\| \leq M$. If $F'$ is discontinuous at $y \in I(x)$, define $F'(y_+)$ as the right-hand limit of $F'(y - t\phi(x))$ as $t$ tends to $0_+$. The following sup is finite:

$$\sup\left\{\int_0^1 \frac{\|F'(x - \gamma s\phi(x)) - F'(x)\|}{\gamma\|\phi(x)\|}\, ds: x - \gamma\phi(x) \in I(x)\right\} = \Lambda(x).$$

The formulas of parts (a) and (b) above hold with $l(x)$ replaced with $\Lambda(x)$, and $\Lambda(x)$ is bounded on $S$.

PROOF. It follows from the differentiability of $F$ and the triangle inequality that the segment $I(x)$ is nonempty, and $I(x)$ is bounded because $S$ is bounded. From the differentiability we have:

$$\frac{\|F(x - t\phi(x)) - F(x) - F'(x)(-t\phi(x))\|}{t\|\phi(x)\|} < \varepsilon.$$

Take $\varepsilon \leqq \|F(x)\|/2\|\phi(x)\|$. Then $\|F(x)\| - \|F(x - t\phi(x))\| \geq t(\|F(x)\| - \varepsilon\|\phi(x)\|) > 0$. By Theorems 1, 2 and 4 of Graves **[6]**, one has

$$F(x + th) - F(x) - F'(x)(th) = \int_0^1 [F'(x + sth)(th) - F'(x)(th)]\, ds,$$

a generalized Riemann integral. Thus

$$\|F(x + th) - F(x) - F'(x)(th)\| \leq \int_0^1 \|F'(x) - F'(x + sth)\|t\|h\|\, ds = \theta.$$

For (a) and (b) we get, using the Lipschitz condition, that

$$\theta \leq t^2\|h\|^2 l(x).$$

Replace $th$ by $-\gamma\phi(x)$, and replace $F'(x)(\gamma\phi(x))$ by $\gamma F(x)$ to get (a). By the triangle inequality:

$$\text{(A)} \qquad -\theta \leq |1-\gamma|\,\|F(x)\| - \|F(x-\gamma\phi(x))\| \leq \theta.$$

If $\gamma \geq 1$, $|1-\gamma| = \gamma - 1$ and (A) becomes, after dividing by $\gamma\|F(x)\|$,

$$\frac{2-\gamma}{\gamma} - \frac{\theta}{\gamma\|F(x)\|} \leq \frac{\|F(x)\| - \|F(x-\gamma\phi(x))\|}{\gamma\|F(x)\|} \leq \frac{2-\gamma}{\gamma} + \frac{\theta}{\gamma\|F(x)\|}.$$

If $\gamma \leq 1$, the values $(2-\gamma)/\gamma$ in the above formula are replaced by unity. This proves the first part. To prove the second part notice that

$$\frac{\theta}{\gamma\|F(x)\|} = \gamma\frac{\|\phi(x)\|^2}{\|F(x)\|}\int_0^1 \frac{\|F'(x) - F'(x-\gamma s\phi(x))\|}{\gamma\|\phi(x)\|}\,ds.$$

Bound the numerator on each piece where $F'$ is continuous by using the mean value theorem for vector-valued functions. Integrating piecewise we find that the above integral is bounded for each $x \in S$ by $M/2$. Comparing with the first part we see that $M/2$ plays the same role as $2L$.

When a formula below is followed by the word "steps", we mean that the formula is to be rounded up to the nearest integer.

ALGORITHM 1. *Case* 1. *Assume the hypotheses of the first part of the lemma. Let*

$$g(x,\gamma) = \frac{\|F(x)\| - \|F(x-\gamma\phi(x))\|}{\gamma\|F(x))\|}.$$

*Choose* $x_0 \in D$ *and* $\delta \in (0, 1/\sqrt{8}]$. *Let* $\alpha = \mu^2/4L$. *If* $g(x_k, 1) \geq \delta$ *set* $\gamma_k = 1$. *Otherwise take* $\gamma_k$ *to satisfy* $\delta \leq g(x_k, \gamma_k) \leq 1-\delta$. *Set* $x_{k+1} = x_k - \gamma_k\phi(x_k)$.

CLAIM 1. *The algorithm obtains an approximate root in no more than* $K = (\|F(x_0)\| - \alpha)/4\delta^2\alpha$ *steps. When an approximate root has been reached, the method automatically switches to the Newton method and then reduces* $\|F\|$ *to less than* $\varepsilon$ *in at most* $\log_2\log_2(\mu\lambda/L\varepsilon)$ *steps.*

*If* $2^{-\mathscr{L}} = \varepsilon$ *then the number of steps to achieve maximum possible accuracy grows as* $\log_2\mathscr{L}$. *Here* $\mathscr{L}$ *represents the binary word length in the computer.*

*Case* 2. *Assume the hypothesis of the second part of the lemma. By* (c) *of the lemma*

$$\sup\{\Lambda(x) : x \in S\} = M/2 \quad \textit{and} \quad \alpha = \mu^2/M.$$

*We assume that* $F'$ *exists at* $x_0$. *If a subsequent iterate is a point* $x_k$ *where* $F'(x_k)$ *is not defined use instead the operator* $F'(x_{k+})$. *Here* $\alpha$ *is defined analogously to Case* 1. *We have no Kantorovich type theorem; however,*

CLAIM 2. *When* $\|F(x)\| \leq \alpha$ *the rate of convergence is geometric, with ratio* $1/2$. *The estimate* $K$ *of the number of steps to reduce* $\|F(x)\|$ *to* $\leq \alpha$

*is the same as Case* 1. *Thereafter,* $\|F(x)\|$ *will be reduced to less than* $\varepsilon$ *in*

$$\log_2\left(\frac{\|F(x)\|}{\varepsilon}\right) = \log_2(\|F(x_0)\|) + \mathscr{L} \quad \textit{steps.}$$

PROOF. By the lemma we see that

$$\lim_{\gamma\to 0+} g(x, \gamma) = 1.$$

Since $S$ is closed and bounded there exists a minimal number $\hat{\gamma} > 0$ such that $\|F(x)\| = \|F(x - \hat{\gamma}\phi(x))\|$. Then $g(x, \hat{\gamma}) = 0$. So $g$ takes on all values between 0 and 1. We shall now bound $\gamma_k$ from below. In what follows we assume $\gamma = \gamma_k$.

Using the lemma we have, when $\gamma = 1$ is rejected by the algorithm, that

$$1 - \frac{\gamma\|\phi\|^2 l(x)}{\|F(x)\|} \le 1 - \delta.$$

Whence

$$\gamma \ge \frac{\delta\|F(x)\|}{\|\phi(x)\|^2 l(x)} \ge \frac{\delta\mu^2}{l(x)\|F(x)\|}.$$

Let us now assume that $\|F(x)\| \ge 2\alpha$. Then if $\gamma < 1$, we get $\gamma \ge \delta\mu^2/(l(x)\|F(x)\|)$, while $g(x, \gamma) \ge \delta$. Thus $\gamma\|F(x)\|g(x, \gamma) \ge \delta^2\mu^2/l(x) \ge \delta^2\mu^2/L$. If $\gamma \ge 1$ and $\|F(x)\| \ge 2\alpha$, then

$$\gamma\|F(x)\|g(x, \gamma) \ge 2\alpha\delta \ge \frac{\mu^2\delta}{2L} \ge \frac{\mu^2\delta^2}{L}.$$

If $\|F(x)\| < 2\alpha$, then by the lemma

$$g(x, \gamma) \ge \sigma(\gamma) - \frac{\gamma\|\phi(x)\|^2 L}{\|F(x)\|} \ge \sigma(\gamma) - \frac{\gamma\|F(x)\|L}{\mu^2} > \sigma(\gamma) - \frac{\gamma}{2}.$$

So $g(x, 1) \ge 1/2 \ge \delta$. The algorithm will choose $\gamma = 1$ for this case. Also, $\gamma\|F(x)\|g(x, \gamma) \ge \|F(x)\|/2 \ge \mu^2/(8L) \ge \delta^2\mu^2/L$.

Then, if

$$\|F(x_0)\| - (k + 1)\delta^2\mu^2/L \le \alpha,$$

we have an approximate root. Finally we recall that if $x_0$ is an approximate root then the Newton sequence started at $x_0$ converges to a root $z$ at a rate faster than $\|x_j - z\| \le \frac{\mu}{L}(\frac{1}{2})^{2^j}$. Also $\|F(x_j) - F(z)\| \le \lambda\|x_j - z\|$. Now choose $k$ so that

$$\frac{\mu\lambda}{L}\left(\frac{1}{2}\right)^{2^k} \le \varepsilon.$$

For Case 2, observe that whenever $\|F(x)\| < 2\alpha$ then $\|F(x)\| \le 2^{-k}\|F(x_0)\|$. Now choose $k$ so that $\|F(x)\| \le \varepsilon$. □

Algorithm 2. Claim 3. *For the algorithm below, the number of steps required to compute* $\gamma_k$ *does not exceed the smallest integer that is greater than or equal to*

$$\Gamma \geq \log_2\left(\frac{4\lambda\|F(x_0)\|}{\mu\alpha\delta(1-2\delta)}\right).$$

Algorithm. *In Algorithm* 1 *if* $g(x_k, 1) \geq \delta$, *we set* $\gamma_k = 1$. *Thus we may assume that* $g(x_k, 1) < \delta$, *and* $\gamma_k$ *lies in the interval* $(0, 1)$. *We shall refine our search for it by the method of bisection. Hence, a sequence of subintervals that have a lower bound* $l$ *and an upper bound* $u$ *will be generated. Initially,* $l = 0$ *and* $u = 1$. *Our trial points* $t$ *will be the midpoints of* $l$ *and* $u$. *Thus our first trial is* $t = 1/2$.

(I) *If* $g(x_k, t) > 1 - \delta$ *set* $l = t$.
(II) *If* $g(x_k, t) < \delta$ *set* $u = t$.
(III) *If* $\delta \leq g(x_k, \gamma_k, t) \leq 1 - d$, *stop. Set* $\gamma_k = t$.
(IV) *Else, set* $t = (u + l)/2$ *and go to* (I).

Proof. Let $\Delta t \subset (0, 1)$ be a closed interval such that if $t \in \Delta t$, then (III) is satisfied. Let $m(\Delta t)$ denote the length of $\Delta t$. At each iteration, $l \leq s \leq u$ for all $s \in \Delta t$. Moreover, $u - 1 \leq (0.5)^k$. Thus, when $k$ is any integer greater than or equal to $-\log_2 m(\Delta t)$, either $l$ or $u$ will belong to $\Delta t$. To estimate $\Delta t$, let $g(x, \gamma_1) = 1 - \delta$ and $g(x, \gamma_2) = \delta$. Then $1 - 2\delta \leq \max\{g'(x, \gamma): \gamma \geq \gamma_1 \text{ and } x \in S\} m(\Delta t) = Mm(\Delta t)$. We compute that $m(\Delta t) \geq (1 - 2\delta)/M$,

$$g'(x, \gamma) = \frac{f'(x - \gamma\phi(x))(\phi(x))}{\gamma f(x)} - \frac{g(x, \gamma)}{\gamma},$$

$$\gamma_1 \geq \frac{\delta\mu^2}{\|F(x_0)\|L} \quad \text{and} \quad M = \frac{4\|F(x_0)\|\lambda}{\delta\alpha\mu}.$$

This yields our estimate of $\Gamma$. □

## Bibliography


1. J. C. Dunn, *Newton's method and the Goldstein step length rule for constrained minimization problems*, SIAM J. Control Optim. **18** (1980), 659–674.
2. ——, *Projected Newton methods for nonlinearly-constrained minimization problems* (to appear).
3. J. C. Dunn and G. C. Hughes, *Newton-Goldstein convergence rates for convex constrained minimization problems with singular solutions*, Appl. Math. Optim. **12** (1984), 203–230.
4. A. A. Goldstein, *On Newton's method*, Numer. Math. **7** (1965), 391–393.
5. ——, *How good are global Newton methods*, manuscript, 1989.
6. Lawrence M. Graves, *Riemann integration and Taylor's theorem in general analysis*, Trans. Amer. Math. Soc. **29** (1927), 163–177.
7. L. V. Kantorovich, *Functional analysis and applied mathematics*, Uspeki Mat. Nauk **3** (1948), 89–185; English transl., Rep. 1509, National Bureau of Standards, Washington, D.C., 1952. (Also: L. V. Kantorovich and G. P. Akilov, *Functional analysis in normed spaces*, Fizmatgiz, Moscow, 1959, English transl., Pergamon Press, Oxford, 1964.)
8. A. S. Nemerovsky and D. B. Yudin, *Problem complexity and method efficiency in optimization*, Wiley, 1983.

9. J. M. Ortega and W. C. Rheinboldt, *Iterative solution of nonlinear equations in several variables*, Academic Press, New York, 1970.
10. E. Polak, *A historical survey of computational methods in optimal control* (Twentieth Anniversary of SIAM, Special Lectures), SIAM Rev. **15** (1973), 553–584.
11. Steve Smale, *Algorithms for solving equations*, Proc. Internat. Congr. Math., Vol. 1, Amer. Math. Soc., Providence, R.I., 1986, pp. 172–195.
12. ——, *On the efficiency of algorithms of analysis*, Bull. Amer. Math. Soc. **134** (1985), 1–36.

DEPARTMENT OF MATHEMATICS, UNIVERSITY OF WASHINGTON, SEATTLE, WASHINGTON 98195

*E-mail address*: gold@math.washington.edu

DIMACS Series in Discrete Mathematics
and Theoretical Computer Science
Volume **4**, 1991

Applied Geometry
and Discrete Mathematics
THE VICTOR KLEE FESTSCHRIFT

# Volume Approximation of Convex Bodies by Circumscribed Polytopes

PETER M. GRUBER

ABSTRACT. An asymptotic formula is given for the difference of the volumes of a convex body in $\mathrm{E}^3$ of class $C^2$ and its circumscribed polytopes of minimum volume, as the number of facets of the latter tends to infinity. The proof makes use of a moment lemma of Fejes Tóth dealing with planar Dirichlet-Voronoi tilings.

## 1. Introduction and statement of result

A *convex body* (in Euclidean space $\mathrm{E}^3$) is a compact convex subset of $\mathrm{E}^3$ with nonempty interior. For convex bodies $C, D$ denote by $\delta^s(C, D)$ the volume of their symmetric difference. This defines the *symmetric difference metric* $\delta^s$ on the space of all convex bodies. For a convex body $C$ and for $n = 4, 5, \ldots$, let $\mathscr{P}^c_{(n)} = \mathscr{P}^c_{(n)}(C)$ denote the space of all convex polytopes circumscribed to $C$ and having at most $n$ facets. Thus any facet of a polytope in $\mathscr{P}^c_{(n)}$ touches $C$. Define

$$\delta^s(C, \mathscr{P}^c_{(n)}) = \inf\{\delta^s(C, P) : P \in \mathscr{P}^c_{(n)}\}.$$

The infimum is actually a minimum. The polytopes attaining it are called *best approximating*. Finally, let $\sigma$ be the ordinary surface area measure in $\mathrm{E}^3$.

In this paper the following asymptotic estimate for $\delta^s(C, \mathscr{P}^c_{(n)})$ will be derived:

THEOREM. *Let $C$ be a convex body in $\mathrm{E}^3$ with boundary* bd $C$ *of differentiability class $\mathscr{C}^2$ and positive Gaussian curvature $\kappa$. Then*

$$(1) \qquad \delta^s(C, \mathscr{P}^c_{(n)}) \sim \frac{5}{36\sqrt{3}} \left(\int_{\mathrm{bd}\, C} \kappa^{1/4} d\sigma\right)^2 \frac{1}{n} \quad \textit{as } n \to \infty.$$

1980 *Mathematics Subject Classification* (1985 *Revision*). Primary 52A15, 52A40, 41A60.

The corresponding two-dimensional result is due to McClure and Vitale [**8**]. Their result and our Theorem were first conjectured and made plausible by Fejes Tóth [**3, pp. 43, 152**]. The Theorem supplements a result of the author [**5**] on approximation of convex bodies of class $\mathscr{C}^3$ with positive Gaussian curvature by inscribed polytopes having at most $n$ vertices. Whereas in the proof of the latter result the main difficulty consisted of the solution of an elementary extremum problem and of a sort of double approximation, the proof of the above Theorem heavily relies on a moment lemma of Fejes Tóth [**3**] and uses only simple approximation. For a survey of approximation of convex bodies up to 1983 we refer to the article [**4**].

The integral in (1) is the *affine surface area* of $C$. Hence Blaschke's [**1**] affine isoperimetric inequality (see also [**2**]) implies the following: Among all convex bodies in $E^3$ of given volume with sufficiently differentiable boundary and positive Gaussian curvature precisely the ellipsoids are asymptotically worst approximated by circumscribed polytopes.

It remains an open question whether there is an $\alpha < 2$ such that for all convex bodies $C$ of class $\mathscr{C}^\alpha$ the quantity $\delta^s(C, \mathscr{P}^c_{(n)})$ has a simple asymptotic behaviour as $n \to \infty$. We believe not. In case $\alpha = 1$, for most convex bodies $C$ the asymptotic behaviour of $\delta^s(C, \mathscr{P}^c_{(n)})$ as $n \to \infty$ is very irregular. (It is easy to prove a result for $\delta^s(C, \mathscr{P}^c_{(n)})$ analogous to Theorem 2 in [7].)

Denote by $\mathscr{P}^c_n = \mathscr{P}^c_n(C)$ and $\mathscr{P}^i_{(n)} = \mathscr{P}^i_{(n)}(C)$ the spaces of all convex polytopes circumscribed to a convex body $C$ (respectively inscribed to $C$) and having at most $n$ vertices (respectively facets). The quantities $\delta^s(C, \mathscr{P}^c_n)$ an $\delta^s(C, \mathscr{P}^i_{(n)})$ are defined analogously to $\delta^s(C, \mathscr{P}^c_{(n)})$.

CONJECTURE. *Let $C$ be as in the Theorem. Then, as $n \to \infty$,*

$$\delta^s(C, \mathscr{P}^c_n) \sim \frac{1}{12\sqrt{3}} \left( \int_{\operatorname{bd} C} \kappa^{1/4} \, d\sigma \right)^2 \frac{1}{n}$$

*and*

$$\delta^s(C, \mathscr{P}^i_{(n)}) \sim \frac{7}{36\sqrt{3}} \left( \int_{\operatorname{bd} C} \kappa^{1/4} \, d\sigma \right)^2 \frac{1}{n}.$$

Some remarks on these asymptotic formulae will be made in §3.

In contrast to asymptotic best approximation with respect to the Hausdorff metric, the Banach-Mazur distance, and other notions of distance (see [**6, 9, 10, 11**]) it appears to be difficult to extend our Theorem and the corresponding result in [**5**] to higher dimensions (but it is plausible to conjecture that such extensions exist). One reason for this might be: It is most likely that nontrivial covering tools like Delone dissections or the moment lemma are indispensable for the proofs in the case of the metric $\delta^s$. Unfortunately these are not well understood even in dimension 3.

Denote by $|\ |$ and $\|\ \|$ the area and Euclidean norm in $\mathrm{E}^2$, by $o$ the origin of $\mathrm{E}^2$ or $\mathrm{E}^3$, and by bd, int, relint, det, and diam the boundary, interior, interior relative to the affine hull, discriminant of a quadratic form, and diameter, respectively.

## 2. Proof of the Theorem

**I.** First, some results are stated which are needed later, the main one being the following moment lemma of Fejes Tóth **[3, p. 81]**:

*Let $\varphi: [0, +\infty) \to [0, +\infty)$ be nondecreasing, let $S$ be a convex hexagon in $\mathrm{E}^2$, and let $r_1, \dots, r_n \in \mathrm{E}^2$. Then*

(2) $$\int_S \varphi\left(\min_k\{\|s - r_k\|\}\right) ds \geq n \int_T \varphi(\|s\|)\, ds,$$

*where $T$ is a regular hexagon in $\mathrm{E}^2$ with center $o$ and $|T| = |S|/n$.*

($s = (s^1, s^2)$, $ds = ds^1 ds^2$.) An elementary calculation gives the next proposition.

(3) *Let $T$ be a convex hexagon in $\mathrm{E}^2$ with center $o$ which is regular with respect to the norm $q^{1/2}$, where $q$ is a positive quadratic form on $\mathrm{E}^2$. Then*

$$\int_T q(s)\, ds = \frac{5}{18\sqrt{3}} |T|^2 (\det q)^{1/2}.$$

The inequality between the arithmetic and the quadratic mean will be used in the following form:

(4) Let $\alpha_k \geq 0$, $k = 1, \dots, n$. Then

$$\sum_k \alpha_k \geq \frac{1}{n}\left(\sum_k \alpha_k^{1/2}\right)^2.$$

For the following choose $\varepsilon > 0$.

**II.** Before beginning with the essential steps of the proof, some preparatory remarks are in order: After a translation, if necessary, we may suppose that $o \in \operatorname{int} C$. Then choose a dissection of $\mathrm{E}^3$ into closed convex cones $C_i$, $i = 1, \dots, l$, with common apex $o$. For each $i$ let $H_i$ be a plane with $H_i \cap C = \varnothing$ which intersects and is orthogonal to some ray in $C_i$. Choose a Cartesian coordinate system in $H_i$. Together with the unit normal vector of $H_i$ pointing toward $C$ it forms a Cartesian coordinate system in $\mathrm{E}^3$. For each $i$ represent the "lower" part of $\operatorname{bd} C$ in the form $(u, f_i(u))$ for $u \in C'$, where the $'$ denotes the (orthogonal) projection of $\mathrm{E}^3$ onto $H_i$ and $f_i$ is a convex function on $C'$ such that the restriction $f_i | \operatorname{relint} C'$ is of class

$\mathscr{C}^2$. Since $\kappa > 0$ for any $u \in \operatorname{relint} C'$ the quadratic form

$$q_{iu}(s) = f_{ixx}(u)(s^1)^2 + 2f_{ixy}(u)s^1s^2 + f_{iyy}(u)(s^2)^2 \quad \text{for } s \in \mathrm{E}^2$$

is positive. ($f_{ix}$, $f_{iy}$, $f_{ixx}$, ... denote partial derivatives of $f_i$.) Finally choose $p_i \in D_i = (C_i \cap \operatorname{bd} C)'$ and let $q_i = q_{ip_i}$.

From now on we suppose that the dissection $C_i$, $i = 1, \dots, l$, of $\mathrm{E}^3$ is chosen so fine that the following properties are fulfilled:

$$D_i \subset \operatorname{relint} C', \tag{5}$$

$$\frac{1}{1+\varepsilon} < \frac{q_{iu}(s)}{q_i(s)} < 1+\varepsilon \quad \text{for } u \in D_i,\ s \in \mathrm{E}^2 \backslash \{o\}, \tag{6}$$

$$\frac{1}{1+\varepsilon} \le \frac{\int_{\operatorname{bd} C} \kappa^{1/4}\, d\sigma}{\sum_i |D_i|(1 + f_{ix}(p_i)^2 + f_{iy}(p_i)^2)^{1/2}\kappa(p_i)^{1/4}} \le 1 + \varepsilon. \tag{7}$$

(Observe the definition of surface integrals.) Let

$$\beta_i = \frac{|D_i|(1 + f_{ix}(p_i)^2 + f_{iy}(p_i)^2)^{1/2}\kappa(p_i)^{1/4}}{\sum_i |D_i|(1 + f_{ix}(p_i)^2 + f_{iy}(p_i)^2)^{1/2}\kappa(p_i)^{1/4}}. \tag{8}$$

Then clearly

$$\beta_1 + \cdots + \beta_l = 1. \tag{9}$$

**III.** By carefully estimating volumes from below the following proposition will result:

(10) For $n = 4, 5, \dots$, let $P_{(n)} \in \mathscr{P}^c_{(n)}$ be a best approximating polytope of $C$. Then for all sufficiently large $n$,

$$\delta^s(C, P_{(n)}) \ge \frac{5}{(1+\varepsilon)^5 36\sqrt{3}} \left( \int_{\operatorname{bd} C} \kappa^{1/4}\, d\sigma \right)^2 \frac{1}{n}.$$

We will apply (2). For each $i$ choose in the interior of $D_i$ disjoint closed convex hexagons $S_{ij}$, $j = 1, \dots, j_i$, such that

$$\sum_j |S_{ij}| \ge \frac{|D_i|}{1+\varepsilon}. \tag{11}$$

$\kappa > 0$ implies the strict convexity of $C$. Together with $P_{(n)} \to C$ this yields

$$\max\{\operatorname{diam} F : F \text{ facet of } P_{(n)}\} \to 0 \quad \text{as } n \to \infty.$$

Thus for all sufficiently large $n$ the following holds: For each $i$ and each facet $F$ of $P_{(n)}$ on the lower side of $P_{(n)}$ (with respect to $H_i$) for which $F'$ intersects one of the hexagons $S_{ij}$, the projection $F'$ is contained in $D_i$ and meets no other hexagon. Assume $n$ to be so large that this is fulfilled and consider for each hexagon $S_{ij}$ the facets $F_k = F_{ijk}$, $k = 1, \dots, k_{ij}$, on the lower side of $P_{(n)}$ for which $F_k'$ intersects $S_{ij}$. Clearly,

$$k_{11} + \cdots + k_{1j_1} + k_{21} + \cdots + k_{lj_l} \leq n. \tag{12}$$

By our choice of $C_i$, $H_i$, $D_i$, $S_{ij}$, and $F_{ijk}$,

$$\delta^s(C, P_{(n)}) \geq \sum_i \sum_j \sum_k \{\text{volume of the set in } P_{(n)} \text{ below } C \text{ and above } F_{ijk} \cap S_{ij}\}. \tag{13}$$

Let $r_k = r_{ijk}$ be the projection into $H_i$ of the intersection point of $F_k$ and $C$. Note that $F_k \subset D_i \subset \operatorname{relint} C'$ (see (5)) and $f_i | \operatorname{relint} C'$ is of class $\mathscr{C}^2$. Hence we may represent the innermost sum in (13) in the form

$$\sum_k \{\ \} = \sum_k \int_{F_k' \cap S_{ij}} (f_i(s) - f_i(r_k) - f_{ix}(r_k)(s^1 - r_k^1) - f_{iy}(r_k)(s^2 - r_k^2))\, ds. \tag{14}$$

Now, using that $f_i$ is of class $\mathscr{C}^2$, the convexity of $D_i$, and Taylor's formula we see that the integrand in (14) equals

$$\tfrac{1}{2} q_{ir_k + \xi(s - r_k)}(s - r_k)$$

for suitable $\xi \in [0, 1]$ (depending on $s$). Observing (6) then shows that the integrand is at least

$$\frac{1}{2(1+\varepsilon)} q_i(s - r_k).$$

Hence (14) implies

$$\begin{aligned} \sum_k \{\ \} &\geq \frac{1}{2(1+\varepsilon)} \sum_k \int_{F_k' \cap S_{ij}} q_i(s - r_k)\, ds \\ &\geq \frac{1}{2(1+\varepsilon)} \int_{S_{ij}} \min_k \{q_i(s - r_k)\}\, ds. \end{aligned} \tag{15}$$

Here we have used the fact that $F_k' \cap S_{ij}$, $k = 1, \dots, k_{ij}$, is a dissection of $S_{ij}$. Propositions (2) and (3) and the inequality (15) together yield the following:

$$\sum_k \{\ \} \geq \frac{5}{(1+\varepsilon) 36\sqrt{3}} \frac{|S_{ij}|^2 (\det q_i)^{1/2}}{k_{ij}}. \tag{16}$$

Combining (13), (16), (4), (12), (11), (7) and noting that

$$\kappa(p_i) = \frac{\det q_i}{(1 + f_{ix}(p_i)^2 + f_{iy}(p_i)^2)^2}, \tag{17}$$

we arrive at the desired inequality:

$$\begin{aligned}\delta^s(C, P_{(n)}) &\geq \frac{5}{(1+\varepsilon)36\sqrt{3}} \sum_i \sum_j \frac{|S_{ij}|^2 (\det q_i)^{1/2}}{k_{ij}^2} k_{ij} \\ &\geq \frac{5}{(1+\varepsilon)36\sqrt{3}} \frac{1}{k_{11} + \cdots + k_{1j_1} + \cdots + k_{lj_l}} \\ &\quad \times \left( \sum_i \sum_j \frac{|S_{ij}|(\det q_i)^{1/4}}{k_{ij}} k_{ij} \right)^2 \\ &\geq \frac{5}{(1+\varepsilon)^3 36\sqrt{3}} \frac{1}{n} \left( \sum_i |D_i| (\det q_i)^{1/4} \right)^2 \\ &\geq \frac{5}{(1+\varepsilon)^5 36\sqrt{3}} \left( \int_{\operatorname{bd} C} \kappa^{1/4}\, d\sigma \right)^2 \frac{1}{n}.\end{aligned}$$

This concludes the proof of (10).

**IV.** We construct quite explicitly polytopes $Q_{(n)} \in \mathscr{P}^c_{(n)}$ for large $n$ which are close to being best approximating:

(18) For each sufficiently large integer $n$ there is a polytope $Q_{(n)} \in \mathscr{P}^c_{(n)}$ such that

$$\delta^s(C, Q_{(n)}) \leq \frac{(1+\varepsilon)^5 5}{36\sqrt{3}} \left( \int_{\operatorname{bd} C} \kappa^{1/4}\, d\sigma \right)^2 \frac{1}{n}.$$

For $i = 1, \ldots, l$, choose a closed set in the lower part of $\operatorname{bd} C$ which projects onto a closed set $E_i$ in $H_i$ such that the following properties hold:

(19) $$D_i \subset \operatorname{relint} E_i, \qquad E_i \subset \operatorname{relint} C',$$

(20) (6) holds with $D_i$ replaced by $E_i$,

(21) $E_i$ is Jordan measurable and $|E_i| \leq (1+\varepsilon)|D_i|$.

For $n = 3, 4, \ldots$, consider for each $i$ a tiling of $H_i$ with congruent convex hexagons which are regular in the sense of the Euclidean norm $q_i^{1/2}$ and have area $|E_i|/\beta_i n$. Let $S_{ij}$, $j = 1, \ldots, j_i$, be the hexagons contained in $E_i$. Since the $S_{ij}$'s do not overlap and since

(22) $$|S_{ij}| = \frac{1}{\beta_i n}|E_i|,$$

their number $j_i$ satisfies

(23) $$j_i \leq \beta_i n.$$

Thus by (9)

(24) the total number of hexagons $S_{ij}$ is $\leq n$.

Let $p_{ij}$ be the point on the lower side of $\operatorname{bd} C$ (with respect to $H_i$) which projects onto the center $r_{ij}$ of $S_{ij}$ and let $H(p_{ij})$ and $H(p_{ij})^+$ be the supporting plane and halfspace of $C$ at $p_{ij}$, respectively.

For sufficiently large $n$ the corresponding hexagons $S_{ij}$ almost fill up $E_i$ and $D_i$ is contained "deep" in the relative interior of their union (note (19)) such that the following holds: Any ray in $C_i$ starting at $o$ intersects one of the supporting planes, say $H(p_{ij})$, such that the segment on the ray between $\operatorname{bd} C$ and $H(p_{ij})$ projects into the union of the hexagons in $E_i$. For such $n$ let

$$Q_{(n)} = \bigcap_{i,j} H(p_{ij})^+ .$$

Then (24) shows

(25) $$Q_{(n)} \in \mathscr{P}^c_{(n)} .$$

By construction,

(26)
$$\begin{aligned}\delta^s(C, Q_{(n)}) &\leq \sum_i \sum_j \{\text{volume of the set in } Q_{(n)} \text{ below } C \text{ and above } S_{ij}\} \\ &\leq \sum_i \sum_j \int_{S_{ij}} (f_i(s) - f_i(r_{ij}) - f_{ix}(r_{ij})(s^1 - r^1_{ij}) \\ &\qquad\qquad - f_{iy}(r_{ij})(s^2 - r^2_{ij}))\, ds .\end{aligned}$$

Applying Taylor's formula, (6), and (20) we see that the integrand in (26) is at most

$$\frac{1+\varepsilon}{2} q_i(s - r_{ij}) .$$

This together with (26), (3), (22), (23), (21), (8), (17), and (7) gives

$$\begin{aligned}\delta^s(C, Q_{(n)} &\leq \frac{1+\varepsilon}{2} \sum_i \sum_j \int_{S_{ij}} q_i(s - r_{ij})\, ds \\ &\leq \frac{(1+\varepsilon)5}{36\sqrt{3}} \sum_i j_i \frac{1}{\beta_i^2 n^2} |E_i|^2 (\det q_i)^{1/2} \\ &\leq \frac{(1+\varepsilon)^3 5}{36\sqrt{3}} \sum_i \frac{1}{\beta_i} |D_i|^2 (\det q_i)^{1/2} \frac{1}{n}\end{aligned}$$

$$\leq \frac{(1+\varepsilon)^3 5}{36\sqrt{3}} \sum_i \frac{|D_i|^2(\det q_i)^{1/2}}{|D_i|(1+f_{ix}(p_i)^2+f_{iy}(p_i)^2)^{1/2}\kappa(p_i)^{1/4}}$$
$$\times \sum_i |D_i|(1+f_{ix}(p_i)^2+f_{iy}(p_i)^2)^{1/2}\kappa(p_i)^{1/4}\frac{1}{n}$$
$$= \frac{(1+\varepsilon)^3 5}{36\sqrt{3}} \left(\sum_i |D_i|(1+f_{ix}(p_i)^2+f_{iy}(p_i)^2)^{1/2}\kappa(p_i)^{1/4}\right)^2 \frac{1}{n}$$
$$\leq \frac{(1+\varepsilon)^3 5}{36\sqrt{3}} \left(\int_{\operatorname{bd} C} \kappa^{1/4}\, d\sigma\right)^2 \frac{1}{n}.$$

Considering (25) this concludes the proof of (18).

**V.** Since $\varepsilon > 0$ was arbitrary, (10) and (18) combined yield the Theorem.

## 3. Final remarks

It is plausible that the asymptotic formula for $\delta^s(C, \mathscr{P}_n^c)$ can be proved along the lines of the proof of the Theorem in **[5]**, taking ideas from the proof in §2 above. Instead of using proposition (5) in **[5]** one might apply the following elementary result:

> Denote by $Q$ the paraboloid $\{(x, y, z): z \geq x^2 + y^2\}$ in $\mathrm{E}^3$. Let $T$ be a triangle in the $(x, y)$-plane with $|T| = 1$ which contains $o$. Then the volume of the set "below" $Q$ and "above" $T$ is at least $1/3\sqrt{3}$, where equality holds if and only if $T$ is an equilateral triangle with centroid $o$.

It is not clear whether a proof of the asymptotic formula for $\delta^s(C, \mathscr{P}_{(n)}^i)$ can be related to the proof of the Theorem in the present article.

The proofs in **[5]** and in this paper, the remarks just made, and the difficulties which seem to prevent extensions to higher dimensions exhibit an unexpected fact: Approximation problems for the symmetric difference metric which are similar in appearance might require quite different lines of attack.

ACKNOWLEDGMENT. For their help in the preparation of this article I am obliged to Dr. Müller and Professor Schnitzer.

## REFERENCES

1. W. Blaschke, *Vorlesungen über Differentialgeometrie*. II, Springer-Verlag, Berlin, 1923.
2. A. Deicke, *Über die Finsler-Räume mit* $A_i = 0$, Arch. Math. **4** (1953), 45–51.
3. L. Fejes Tóth, *Lagerungen in der Ebene, auf der Kugel und im Raum*, Springer-Verlag, Berlin, Göttingen, and Heidelberg, 1953.
4. P. M. Gruber, *Approximation of convex bodies*, in Convexity and Its Applications, Birkhäuser-Verlag, Basel, Boston, and Stuttgart, 1983, pp. 131–162.
5. ——, *Volume approximation of convex bodies by inscribed polytopes*, Math. Ann. **281** (1988), 229–245.
6. ——, *Asymptotic best and step by step approximation of convex bodies* (in preparation).

7. P. M. Gruber and P. Kenderov, *Approximation of convex bodies by polytopes*, Rend. Circ. Mat. Palermo (2) **31** (1982), 195–225.
8. D. E. McClure and R. A. Vitale, *Polygonal approximation of plane convex bodies*, J. Math. Anal. Appl. **51** (1975), 326–358.
9. R. Schneider, *Zur optimalen Approximation konvexer Polyeder*, Math. Ann. **256** (1981), 289–301.
10. ——, *Affine-invariant approximation by convex polytopes*, Studia Sci. Math. Hungar. **21** (1986), 401–408.
11. ——, *Polyhedral approximation of smooth convex bodies*, J. Math. Anal. Appl. **128** (1987), 470–474.

Abteilung Für Analysis, Technische Universität Wien, Wiedner Hauptstrasse 8–10/1142, A-1040 Wien, Austria

DIMACS Series in Discrete Mathematics
and Theoretical Computer Science
Volume **4**, 1991

Applied Geometry
and Discrete Mathematics
THE VICTOR KLEE FESTSCHRIFT

# Points Sets with Small Integral Distances

HEIKO HARBORTH AND LOTHAR PIEPMEYER

ABSTRACT. For the smallest diameter $k(d, p)$ of $p$ points in $E^d$, not all in $E^{d-1}$, and with pairwise integral distances it is proved $3 \le k(d, p) \le 4$ for $d+2 \le p \le 2d$, and $k(d, 2d) = 4$. Point sets with $k(d, d+2) = 3$ are known for $d = 3, 6$, and 8.

Nearly everyone knows that at most $d+1$ points in the $d$-dimensional Euclidean space $E^d$ can have pairwise the same (unit) distance. What about $d+2$ or more points in $E^d$, $d \ge 2$, if only 1 and 2 are allowed as pairwise distances?

If $k(d, p)$ denotes the smallest diameter of point sets of $p$ points in $E^d$ not all in $E^{d-1}$, with pairwise integral distances, then we prove here $3 \le k(d, d+2) \le 4$. Thus, $d+2$ points only with distances 1 and 2 do not exist. Moreover, we prove that $k(d, p) \le 4$ for $p \le 2d$ with equality for $p = 2d$. A classification of dimensions $d$ into the two classes with $k(d, d+2) = 3$ and $k(d, d+2) = 4$ remains open. So far point sets with $k(d, d+2) = 3$ are known only for $d = 3, 6$, and 8.

The existence of $k(d, p)$ for any $(d, p)$ is proved in **[6]**. In **[4]** some results concerning $d+2$ points in $E^d$ with odd integral distances are discussed.

The following lemma was already proved in **[5]** (see also **[6]**). Here we present an alternative proof.

LEMMA 1. $k(2, 4) = 4$.

PROOF. The example in Figure 1 proves $k(2, 4) \le 4$. If we assume $k(2, 4) \le 3$, then only the integral triangles $(1, 1, 1)$, $(2, 2, 2)$, $(3, 3, 3)$, $(2, 2, 1)$, $(3, 3, 1)$, $(2, 2, 3)$, and $(3, 3, 2)$ are possible. The angles are $\frac{\pi}{3}$, $\alpha_i$, and $\pi - 2\alpha_i$, $1 \le i \le 4$, with

$$\cos\alpha_1 = \tfrac{1}{4}, \quad \cos\alpha_2 = \tfrac{1}{6}, \quad \cos\alpha_3 = \tfrac{3}{4}, \quad \cos\alpha_4 = \tfrac{1}{3}.$$

1980 *Mathematics Subject Classification* (1985 *Revision*). Primary 52A37.

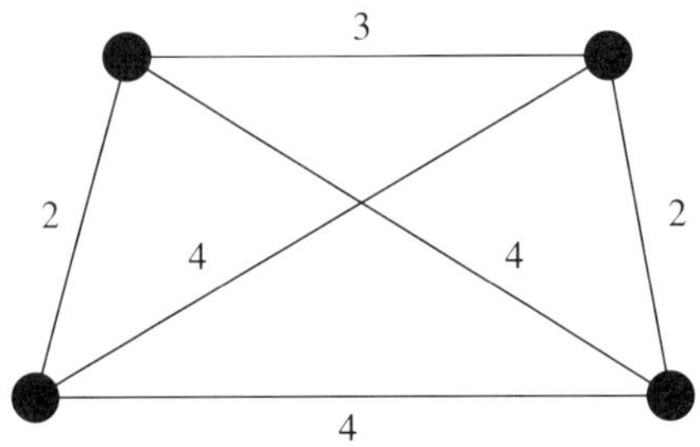


FIGURE 1

At least once an angle of one triangle has to be the sum of two angles of the triangles. This easily can be checked to be impossible using the following inequalities together with $\frac{\pi}{3}$.

$$\begin{aligned} 75.52^\circ &< \alpha_1 < 75.53^\circ, & 28.9^\circ &< 180^\circ - 2\alpha_1 < 29^\circ, \\ 80.4^\circ &< \alpha_2 < 80.5^\circ, & 19.1^\circ &< 180^\circ - 2\alpha_2 < 19.2^\circ, \\ 41.4^\circ &< \alpha_3 < 41.41^\circ, & 97.1^\circ &< 180^\circ - 2\alpha_3 < 97.2^\circ, \\ 70.5^\circ &< \alpha_4 < 70.6^\circ, & 38.94^\circ &< 180^\circ - 2\alpha_4 < 38.95^\circ. \end{aligned}$$

Thus, $k(2, 4) \leq 3$ is impossible, and the proof is complete. □

In the following, the Cayley-Menger determinant,

$$D_p = D(P_1, \dots, P_p) = \begin{vmatrix} 0 & 1 & \dots & 1 \\ 1 & & & \\ \vdots & & d^2(P_i, P_j) & \\ 1 & & & \end{vmatrix},$$

is useful. If $P_1, \dots, P_p$ for $p \geq d+2$ are points in $E^d$, then $D_p = 0$. Moreover, $P_1, \dots, P_{d+2}$ in $E^d$ belong to a $d$-sphere or to the same $E^{d-1}$ if and only if that determinant $D^{1,1}_{d+2}$ obtained from $D_{d+2}$ by deleting the first row and the first column vanishes (see [2]). Furthermore, no $h$ points of $P_1, \dots, P_{d+2}$ are in an $E^{h-2}$, $2 \leq h \leq d+1$, if the Cayley-Menger determinants $D_{d+1}$ of all $(d+1)$-element subsets do not vanish.

THEOREM 1. *Any two distance set in* $E^d$ *with distances* 1 *and* $\sigma > \sqrt{2+\sqrt{3}} = 1.93\dots$ *has at most* $d+1$ *points.*

PROOF. Assume there are $d+2$ or more points with distances 1 and $\sigma$. Any subset $\mathcal{M} = \{P_1, \dots, P_{d+2}\}$ does not contain triangles $(1, 1, \sigma)$ for $\sigma > \sqrt{2+\sqrt{3}}$. If such a triangle $(A, B, C)$ with side $(A, B)$ of length $\sigma$ exists, then, using symmetry, a fourth point $D$ has to be on one of the three hyperspheres determined by the third point of the triangles with $(A, B)$ and additional sides 1 and $\sigma$, $\sigma$ and $\sigma$, or 1 and 1. The corresponding radii are (in this sequence)

$$\rho_1 = \sqrt{1 - \frac{1}{4\sigma^2}}, \quad \rho_2 = \sigma\frac{\sqrt{3}}{2}, \quad \rho_3 = \sqrt{1 - \left(\frac{\sigma}{2}\right)^2}.$$

For the remaining distances $d(C, D)$ we obtain the inequalities

$$1 < \left(\frac{\sigma}{2} - \frac{1}{2\sigma}\right)^2 + (\rho_3 - \rho_1)^2 \le d^2(C, D) \le \left(\frac{\sigma}{2} - \frac{1}{2\sigma}\right)^2 + (\rho_3 + \rho_1)^2 < \sigma^2,$$
$$1 < \rho_2 - \rho_3 \le d(C, D) \le \rho_2 + \rho_3 < \sigma,$$
$$d(C, D) \le 2\rho_3 < 1$$

in the cases $(1, \sigma)$, $(\sigma, \sigma)$, and (1,1), respectively. Since these inequalities are valid at most for $\sigma \le \sqrt{2+\sqrt{3}}$, already a fourth point with distances 1 or $\sigma$ is impossible. For $\sigma = \sqrt{2+\sqrt{3}}$ just two configurations exist, both in the plane (containing an equilateral triangle of side length 1, respectively $\sigma$).

The nonexistence of the triangles $(1, 1, \sigma)$ implies that all unit distances determine $n$ disjoint unit simplices of $j_i \ge 2$ points, $1 \le i \le n$, and possibly $k$ points with distances only $\sigma$ $(k = d + 2 - \sum j_i)$. Then by a suitable notation of the $d+2$ points of $\mathscr{M}$ their distance matrix can be written as follows

$$\begin{pmatrix} A_{j_1} & & & \\ & \ddots & & \sigma \\ \sigma & & A_{j_n} & \\ & & & B_k \end{pmatrix},$$

where

$$A_{j_i} = \begin{pmatrix} 0 & 1 & \dots & 1 \\ 1 & \ddots & \ddots & \vdots \\ \vdots & \ddots & \ddots & 1 \\ 1 & \dots & 1 & 0 \end{pmatrix}, \qquad B_k = \begin{pmatrix} 0 & \sigma & \dots & \sigma \\ \sigma & \ddots & \ddots & \vdots \\ \vdots & \ddots & \ddots & \sigma \\ \sigma & \dots & \sigma & 0 \end{pmatrix}.$$

The computation of the corresponding Cayley-Menger determinant yields

$$D_{d+2} = (-1)^d \sigma^{2k} \left[\frac{k}{\sigma^2} + \sum_{i=1}^{n} \frac{j_i}{1 + j_i(\sigma^2 - 1)}\right] \prod_{\nu=1}^{n} [1 + j_\nu(\sigma^2 - 1)].$$

Since $D_{d+2} \ne 0$, the assumed set $\mathscr{M}$ of $d+2$ points cannot exist for $\sigma > \sqrt{2+\sqrt{3}}$. □

It can be remarked that Theorem 1 for $\sigma > 2$ was proved also in [**3**]. We have to thank L. Danzer for an idea that substantially improved the proof of Theorem 1.

An upper bound for the cardinality of two distance sets is $\binom{d+2}{2}$ (see, for example, [**1**, **7**]). For integral two distance sets we improve this upper bound to $d+2$.

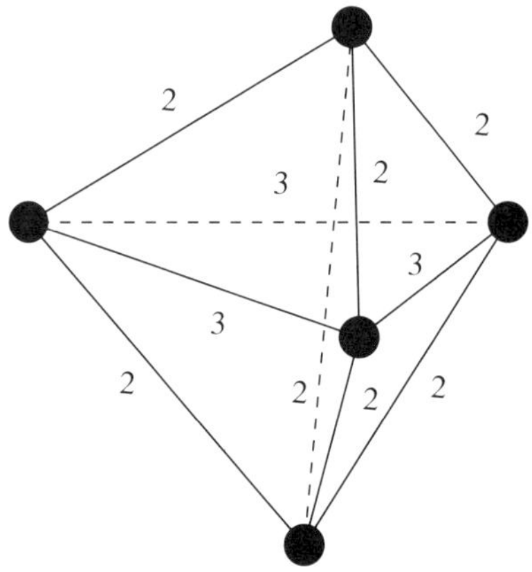


FIGURE 2

THEOREM 2. *An integral two distance set* $\mathscr{M}$ *in* $E^d$ *with distances* $m, n$ *and* $(m, n) = 1$ *has at most* $d+2$ *points. If* $\mathscr{M}$ *has exactly* $d+2$ *points, then there are no* $d+1$ *points in* $E^{d-1}$, *all points of* $\mathscr{M}$ *do not determine a* $d$*-sphere, and* $(m^2 - n^2) | (d+2)$.

PROOF. Assume $d+3$ points in $E^d$ have distances $m$ and $n$, $(m, n) = 1$. Then $D_{d+3} = 0$ and $D_{d+2} = 0$ for any $d+2$ of the $d+3$ points. Because $m^2 \equiv n^2 \pmod{m^2 - n^2}$, we may use $m$ for all distances in $D_{d+3}$ and $D_{d+2}$ to obtain

$$D_{d+3} \equiv (-1)^{d+3} m^{2(d+2)} (d+3) \equiv D_{d+2}$$
$$\equiv (-1)^{d+2} m^{2(d+1)} (d+2) \equiv 0 \pmod{m^2 - n^2}.$$

Since $(m, m^2 - n^2) = 1$, this congruence yields the contradiction $d+3 \equiv d+2 \pmod{m^2 - n^2}$. If $d+2$ points exist, then the right most congruence above implies $d+2 \equiv 0 \pmod{m^2 - n^2}$. At last,

$$D^{1,1}_{d+2} \equiv (-1)^{d+1} m^{2(d+2)} (d+1) \not\equiv 0 \pmod{m^2 - n^2}$$

proves $D^{1,1}_{d+2} \neq 0$, and $D_{d+1} \neq 0$ for all $(d+1)$-element subsets of the $d+2$ points follows from

$$D_{d+1} \equiv (-1)^{d+1} m^{2d} (d+1) \not\equiv 0 \pmod{m^2 - n^2}. \quad \square$$

In $E^3$ only $m = 3$ and $n = 2$ is possible for $(m^2 - n^2) | (d+2)$. The well-known bipyramid (see [**3**, **6**]) in Figure 2 shows the existence of five points with distances 2 and 3. This bipyramid can be generalized for $d$ odd: Take two concentric simplices of side lengths $(d+1)/2$ and $(d+3)/2$ in $E^{(d-1)/2}$ and an orthogonal $E^{(d+1)/2}$, respectively. Then all distances between points of different simplices are $(d+1)/2$ since the center of a unit simplex in $E^h$ has distances $\sqrt{h/2(h+1)}$ to all vertices.

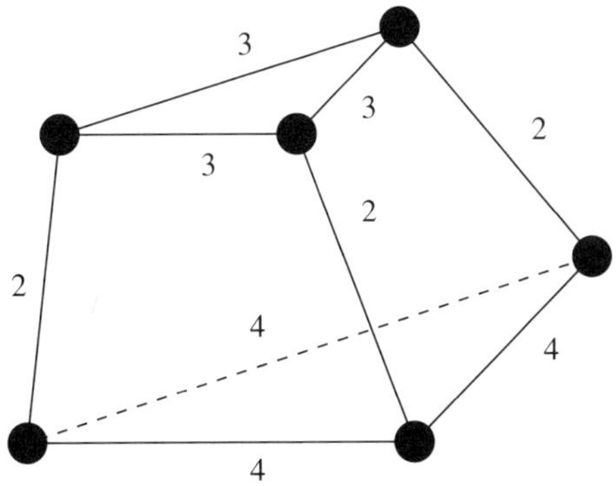


FIGURE 3

In $E^8$ only $m = 3$ and $n = 2$ is possible for $(m^2 - n^2)|(8+2)$. The simplex of length 3 has radius 2 and thus together with its center gives a realization.

Using Theorem 1 up to $d = 11$ no other pairs of integral distances $m$ and $n$ allow integral-two distance sets.

THEOREM 3. *$k(d, 2d) = 4$ for $d \geq 2$.*

PROOF. At first we show $k(d, p) \leq 4$ for all $p \leq 2d$. This is proved by the following $2d$ points in $E^d$:

$$P_i = 3(A + B_i), \quad Q_i = 4(A + B_i), \qquad 1 \leq i \leq d,$$

where

$$A = \frac{\sqrt{7d+1} - 1}{\sqrt{2d}}(1, \dots, 1) \quad \text{and} \quad B_i = \frac{1}{\sqrt{2}}(0, \dots, 0, 1, 0, \dots, 0)$$

with 1 at position $i$. Then $d(P_i, Q_i) = 2$ for $1 \leq i \leq d$, and $d(P_i, P_j) = 3$, $d(P_i, Q_j) = d(Q_i, Q_j) = 4$ for $1 \leq i, j \leq d$, $i \neq j$. For $d = 3$ the resulting truncated pyramid having three quadrangles of Figure 1 as faces is shown in Figure 3.

Now $k(d, 2d) \geq 4$ is proved by induction on $d$. Lemma 1 settles $d = 2$. Assume $k(d, 2d) \geq 4$ is valid. From Theorem 1 we conclude $k(d+1, 2d+2) > 2$ for $d \geq 2$. Assume that in $E^{d+1}$ the $2d+2$ points $A, B, P_1, \dots, P_{2d}$ have diameter 3. Theorem 2 implies that all distances 1, 2, and 3 have to occur. Let $A$ and $B$ have unit distance. Any point $P_i$ with different distances to $A$ and $B$ would be collinear with $A$ and $B$, and then four points with diameter 3 would exist in the plane, contradicting Lemma 1. Thus all $P_i$ are in a subspace $E^d$ orthogonal to $(A, B)$ with distances $\leq 3$, contradicting the induction hypothesis. □

The above constructed two distance sets of $d+2$ points for $d = 3$ and $d = 8$ are examples for $k(d, d+2) = 3$. In $E^6$ a simplex in $E^5$ of length 3 together with a concentric orthogonal line segment of length 1 gives another example for $k(d, d+2) = 3$; however, all distances 1, 2, and 3 occur.—Do further dimensions $d$ exist with $d+2$ points and $k(d, d+2) = 3$ as integral diameter?

## REFERENCES


1. E. Bannai, E. Bannai, and D. Stanton, *An upper bound for the cardinality of an $s$-distance subset in real Euclidean space*. II, Combinatorica **3** (1983), 147–152.
2. M. Berger, *Geometry*. I, Springer-Verlag, Berlin, Heidelberg, New York, 1987.
3. S. J. Einhorn and I. J. Schoenberg, *On euclidean sets having only two distances between points*, Indag. Math. **28** (1966), 479–504.
4. R. L. Graham, B. L. Rothschild, and E. G. Straus, *Are there $n+2$ points in $E^n$ with odd integral distances*?, Amer. Math. Monthly **81** (1974), 21–25.
5. H. Harborth, *Aufgabe* 950, Elem. Math. **42** (1987), 172–173.
6. H. Harborth and A. Kemnitz, *Diameters of integral point sets* in Colloq. Math. Soc. Jànos Bolyai, vol. 48 (Siófok, Hungary, 1985), Intuitive Geometry, North-Holland, 1987, pp. 255–266.
7. D. G. Larman, C. A. Rogers, and J. J. Seidel, *On two-distance sets in Euclidean space*, Bull. London Math. Soc. **9** (1977), 261–267.


TECHNISCHE UNIVERSITÄT BRAUNSCHWEIG, D-3300 BRAUNSCHWEIG, WEST GERMANY
*E-mail address*: i1031801 at DBSTU1.bitnet

DIMACS Series in Discrete Mathematics
and Theoretical Computer Science
Volume 4, 1991

Applied Geometry
and Discrete Mathematics
THE VICTOR KLEE FESTSCHRIFT

# Convex Minimizers of Variational Problems

ERHARD HEIL

ABSTRACT. Without assuming any growth condition, we show that

$$\int_a^b [g(y') + h(x, y)]\, dx,$$

$g$ convex, $h$ nondecreasing with respect to $y$, possesses convex minimizers, possibly discontinuous at $a$ or $b$.

## 1. Introduction

We consider the basic problem in the calculus of variations

$$\int_a^b f(x, y(x), y'(x))\, dx \to \min, \qquad y(a) = y_a, \; y(b) = y_b,$$

$y\colon [a, b] \to \mathbb{R}$ absolutely continuous. We assume $f$ to be convex in its last argument. General existence as well as regularity theorems for absolute minimizers assume a coercivity condition such as

$$f(x, y, u) \geq \alpha |u|^p + \beta, \qquad \alpha > 0, \; \beta \in \mathbb{R}, \; p > 1 \tag{1.1}$$

(see, for instance, Clarke **[5]** and Akhiezer **[1, Chapter 9]**). Without any coercivity condition neither existence nor regularity (smoothness) can be assured in general. The best known example is the problem of rotational minimal surfaces where

$$f(x, y, u) = y\sqrt{1 + u^2}.$$

No continuous solution exists if the bounding circles are too small or too far apart from each other. Here we consider problems with integrands of the form

$$f(x, y, u) = g(u) + h(x, y), \tag{1.2}$$

$g$ convex, $h$ nondecreasing in its second argument, *without any coercivity condition.* The special form (1.2) is motivated by an extremum problem

1980 *Mathematics Subject Classification* (1985 *Revision*). Primary 49A05; Secondary 52A40.

for convex bodies which is described in §4. In general a minimum problem with integrand (1.2) has no continuous minimizer, but we will show, loosely speaking, that nothing happens in the open interval $(a, b)$. If we assume that the graph of the minimizer lies within a region $B$ which forbids jumps at $a$ or $b$, then the minimizer will exist and be regular and therefore will obey the Euler-Lagrange equation in the interior of $B$.

The idea of proof is the following: From (1.2) we conclude that there is a minimal sequence of convex functions $y_n$, a subsequence of which converges to a convex function $y$ which may have jump discontinuities at $a$ or $b$. If this is not allowed by the form of $B$, then $y$ is absolutely continuous in $[a, b]$ and Lipschitz-continuous in every compact subinterval of $(a, b)$. Thus, as is well known, in the interior of $B$ we have $y \in C^2$ and $y$ obeys the Euler-Lagrange equation.

The form (1.2) may seem very special, but our procedure may probably be extended to more general cases. For instance one may ask if the situation is similar if all solutions of the Euler-Lagrange equation are convex. Here a result of Tonelli implies regularity, whereas the question of existence remains open.

## 2. Existence of minimizers

Let $B$ denote a compact convex region in $\mathbb{R}^2$ such that

$$(a, \alpha), (b, \beta) \in B \Leftrightarrow \alpha = y_a, \quad \beta = y_b. \tag{2.1}$$

The integral to be minimized is

$$I[y] := \int_a^b [g(y'(x)) + h(x, y(x))]\, dx, \tag{2.2}$$

where $g: \mathbb{R} \to \mathbb{R}$ is convex, and $h: B \to \mathbb{R}$ is continuous and nondecreasing in its second argument. The set of admissible functions is

$$\mathscr{A} := \{y | y: [a, b] \to \mathbb{R} \text{ absolutely continuous; } y(a) = y_a, \; y(b) = y_b; \; \text{graph } y \subset B, \; I[y] \text{ exists as a Lebesgue integral}\}.$$

$\mathscr{A} \neq \emptyset$ since there is an affine function in $\mathscr{A}$.

THEOREM 1. *Under the above assumptions*

$$\inf_{y \in \mathscr{A}} I[y] > -\infty, \tag{2.3}$$

*and there exists a convex minimizer* $y_0 \in \mathscr{A}$:

$$I[y_0] = \inf_{y \in \mathscr{A}} I[y]. \tag{2.4}$$

PROOF. After showing (2.3) we will exhibit a minimizing sequence of convex functions in $\mathscr{A}$ which has a convergent subsequence by Blaschke's selection theorem. A lower semicontinuity theorem of Tonelli will establish (2.4). We split the proof into several propositions.

PROPOSITION 1. (2.3) *is valid, and without loss of generality we may assume* $g(u) \geq 0$ *for* $u \in \mathbb{R}$.

PROOF. Since $g$ is convex, its graph is bounded from below by some line of support:

$$g(u) \geq k \cdot u + g(0), \qquad k \in \partial g(0).$$

Since $h$ is continuous on the compact region $B$, $h$ is bounded from below by some constant $m$. Thus,

$$I[y] \geq k \cdot (y_b - y_a) + (g(0) + m)(b - a).$$

This gives (2.3). By a common device we set

$$\overline{g}(u) := g(u) - k \cdot u - g(0) \geq 0$$

and have

$$\int_a^b [\overline{g}(y'(x)) + h(x, y(x))]\, dx = I[y] - k(y_b - y_a) - g(0)(b - a).$$

Without loss of generality we therefore may assume $g(u) \geq 0$.

PROPOSITION 2. *A continuous convex function* $f$ *on a compact interval* $[a, b]$ *is absolutely continuous.*

PROOF. This is an immediate consequence of Corollary 24.2.1 in Rockafellar **[9, p. 231]**.

For any $y \in \mathscr{A}$ we form its convexification

$$\tilde{y}(x) := \sup_s s(x), \tag{2.5}$$

where sup is taken over all affine functions $s$ with

$$s(u) \leq y(u) \quad \text{for all } u \in [a, b].$$

PROPOSITION 3. *If* $y \in \mathscr{A}$, *then* $\tilde{y} \in \mathscr{A}$.

PROOF. Any point $(u, \tilde{y}(u))$, $u \in [a, b]$, not belonging to $B$ could be separated strictly from $B$ by a straight line, contradicting the definition of $\tilde{y}$. Thus, graph $\tilde{y} \subset B$ and by (2.1) $\tilde{y}(a) = y_a$ and $\tilde{y}(b) = y_b$. Again by (2.1), the convex function $\tilde{y}$ may have no jump discontinuity at $a$ or $b$. Therefore $\tilde{y}$ is absolutely continuous by Proposition 2. Since $h(\cdot, \tilde{y}(\cdot))$ is continuous, it remains to be shown that $g(\tilde{y}'(\cdot))$ is integrable on $[a, b]$. This will be done in Proposition 6.

PROPOSITION 4. *For* $I_g[y] := \int_\alpha^\beta g(y'(x))\, dx$ *and* $\mathscr{A}^* := \{y | y\colon [\alpha, \beta] \to \mathbb{R}$ *absolutely continuous*; $y(\alpha) = y_\alpha$, $y(\beta) = y_\beta$; $I_g[y]$ *exists*$\}$, *the affine function* $y_0 \in \mathscr{A}^*$ *gives the absolute minimum of* $I_g$:

$$I_g[y_0] \leq I_g[y] \quad \textit{for } y \in \mathscr{A}^*.$$

PROOF. Since $g$ is convex,

$$g(u) \geq g(u_0) + k \cdot (u - u_0) \quad \text{for } k \in \partial g(u_0).$$

For $y, y_0 \in \mathscr{A}^*$, $y_0$ affine,

$$\begin{aligned} I_g[y] &= \int_\alpha^\beta g(y'(x))\,dx \geq \int_\alpha^\beta [g(y_0') + k\cdot(y'(x) - y_0')]\,dx \\ &= \int_\alpha^\beta g(y_0')\,dx + k[(y(\beta) - y(\alpha)) - (y(\beta) - y(\alpha))] = I_g[y_0]. \end{aligned}$$

(Proposition 4 is contained in Troutman [**13, Theorem, p. 57**]. The above proof shows that in this special case we need no differentiability assumption on $g$.)

Again for $y \in \mathscr{A}$ let $\tilde{y}$ denote its convexification (2.5). If at $x_0 \in [a, b]$

$$\tilde{y}(x_0) < y(x_0),$$

then we have a maximal open (relative to $[a, b]$) interval containing $x_0$ in which this is true. There are at most countably many such intervals since the rationals are countable. We denote them by $I_j$, $j = 1, 2, 3, \ldots$ (or $j = 1, \ldots, N$), and define for $x \in [a, b]$

$$\eta_j(x) = \begin{cases} \tilde{y}(x) & \text{for } x \in \bigcup_{i=1}^j I_i, \\ y(x) & \text{otherwise}. \end{cases} \tag{2.6}$$

**Proposition 5.** *$\eta_j' \to \tilde{y}'$ a.e.*

**Proof.** Since $y$ and $\tilde{y}$ are absolutely continuous, $y'$ and $\tilde{y}'$ exist a.e. We consider $x_0 \in [a, b]$ such that $y'(x_0)$ and $\tilde{y}'(x_0)$ exist. Proposition 5 will be proved if we show for some $j_0$

$$\eta_j'(x_0) = \tilde{y}'(x_0) \quad \text{for } j \geq j_0. \tag{2.7}$$

Case 1: $x_0 \in \bigcup_i I_i$. Then there is $j_0$ with $x_0 \in I_{j_0}$ and for $j \geq j_0$

$$\eta_j(x) = \tilde{y}(x) \quad \text{for } x \in I_{j_0}.$$

This implies (2.7) in Case 1.

Case 2: $x_0 \notin \bigcup_i I_i$. Then $\tilde{y}(x_0) = y(x_0)$. Now $\tilde{y}$ is convex and $\tilde{y}'(x_0)$ exists. Therefore the tangent at $(x_0, y(x_0))$ exists and supports graph $\tilde{y}$. Since $y(\tilde{x}) \leq y(x)$ and $y'(x_0)$ exists, we therefore have

$$\tilde{y}'(x_0) = y'(x_0).$$

From $\tilde{y}(x) \leq \eta_j(x) \leq y(x)$ we now conclude that $\eta_j'(x_0)$ exists and equals $\tilde{y}'(x_0)$ for $j = 1, 2, 3 \ldots$. Thus we have (2.7) in Case 2, too.

**Proposition 6.** *For $y \in \mathscr{A}$ the function $g(\tilde{y}'(\cdot))$ is integrable and*

$$I[\tilde{y}] \leq I[y].$$

**Proof.** $\psi_j(\cdot) := g(\eta_j'(\cdot))$ arises from $g(y'(\cdot))$ by replacing its values on the intervals $I_1, \ldots, I_j$ by constants. Since $g(y'(\cdot))$ is integrable for $y \in \mathscr{A}$,

$\psi_j$ is integrable, too. By Proposition 1 we may assume $\psi_j \geq 0$. Since $\eta_j$ is an affine function on $I_1, \ldots, I_j$, we conclude from (2.6) and Proposition 4

$$\int_a^b \psi_j(x)\,dx = \int_a^b g(\eta_j'(x))\,dx \leq \int_a^b g(y'(x))\,dx < \infty. \tag{2.8}$$

Therefore all hypotheses of Fatou's lemma are fulfilled so that $\underline{\lim}\,\psi_j$ is integrable and

$$\int_a^b \underline{\lim}\,\psi_j(x)\,dx \leq \underline{\lim} \int_a^b \psi_j(x)\,dx.$$

Since $g$ is continuous, $\underline{\lim}\,\psi_j = \underline{\lim}\,g(\eta_j') = g(\underline{\lim}\,\eta_j')$, and from Proposition 5 we have

$$\underline{\lim}\,\psi_j(x) = g(\tilde{y}'(x)) \quad \text{a.e.}$$

Thus $g(\tilde{y}'(\cdot))$ is integrable, and from (2.8) we have

$$\int_a^b g(\tilde{y}'(x))\,dx = \int_a^b \underline{\lim}\,\psi_j(x)\,dx \leq \underline{\lim} \int_a^b \psi_j(x)\,dx \leq \int_a^b g(y'(x))\,dx.$$

From $\tilde{y} \leq y$ and the monotonicity of $h$ we have

$$\int_a^b h(x, \tilde{y}(x))\,dx \leq \int_a^b h(x, y(x))\,dx.$$

Together this gives $I[\tilde{y}] \leq I[y]$.

PROPOSITION 7. *There is a minimizing sequence* $(y_i)$ *such that*

$$y_i \in \mathscr{A}, \qquad I[y_i] \to \inf_{y \in \mathscr{A}} I[y],$$

*where the* $y_i$ *are convex functions converging uniformly to a continuous convex function* $y_0$.

PROOF. There is a sequence of functions $v_i \in \mathscr{A}$ such that

$$I[v_i] \to \inf_{y \in \mathscr{A}} I[y].$$

We take the convexified functions $w_i := \tilde{v}_i$ (cf. (2.5)). By Proposition 3 $w_i \in \mathscr{A}$, and by Proposition 6 $(w_i)$ is a minimizing sequence, too. Let

$$W_i := (\text{epigraph } w_i) \cap B.$$

By Blaschke's selection theorem **[3, p. 34]** there is a subsequence $(Y_i)$ of $(W_i)$ converging (with respect to Hausdorff distance) to a convex set $Y_0$, which determines a convex function $y_0$. From $(\text{graph } y_0) \subset B$ we conclude that $y_0$ has no jump discontinuities at $a$ or $b$ and is continuous. Therefore graph $y_0$ is compact and has positive distance from graph $(y_0 + \varepsilon)$ and graph $(y_0 - \varepsilon)$. From this we conclude that $Y_i \to Y_0$ (with respect to Hausdorff distance) implies $y_i \to y_0$ uniformly.

PROPOSITION 8. $y_0 \in \mathscr{A}$ *and* $I[y_0] = \inf_{y\in\mathscr{A}} I[y]$.

PROOF. Since $y_0$ is convex and continuous, it is absolutely continuous (Proposition 2). In order to show $y_0 \in \mathscr{A}$, all that is left to be shown is that $I[y_0]$ exists. We write (2.2) $I = I_1 + I_2$.

$$I_2[y_0] = \int_a^b h(x, y_0(x))\,dx$$

exists since $h$ and $y_0$ are continuous. Moreover $h$ is equicontinuous on the compact set $B$, and therefore $I_2[\cdot]$ is continuous with respect to uniform convergence.

The existence of

$$I_1[y_0] = \int_a^b g(y_0'(x))\,dx$$

is not so clear because $y_0'(a)$ or $y_0'(b)$ may be infinite. Since $y_0'$ is monotone and $g$ as a convex function is piecewise monotone, $g(y_0'(\cdot))$ is measurable. By Proposition 1 we may assume $g(u) \geq 0$ and therefore have to ask if $I_1[y_0] = +\infty$ is possible. Accordingly to Tonelli [**12, §§6 and 7**], we then would have $I_1[y_n] \to \infty$ and, by the continuity of $I_2$, also $I[y_n] \to \infty$. This contradicts the fact that $y_n$ is a minimizing sequence for $I$. We have thus shown that $I[y_0]$ exists. According to Tonelli [**12, §§3 and 4**], $I_1$ is lower semicontinuous with respect to uniform convergence. Since $I_2$ is continuous, $I = I_1 + I_2$ is lower semicontinuous. We therefore have

$$\inf_{y\in\mathscr{A}} I[y] = \lim_{n\to\infty} I[y_n] \geq I[y_0] \geq \inf_{y\in\mathscr{A}} I[y],$$
$$I[y_0] = \inf_{y\in\mathscr{A}} I[y].$$

This proves Proposition 8, which is the last step in the proof of Theorem 1.

REMARK 1. Tonelli assumed in his two theorems just quoted that $g$ is differentiable, but he remarked that convexity is sufficient. The seminormality he supposed means in our case that $g$ is not an affine function. But this is the trivial case where $I_1[y]$ is independent of $y \in \mathscr{A}$.

REMARK 2. We may replace the continuity of $h$ by weaker conditions which guarantee the lower semicontinuity of $I_2$. See for instance Ekeland and Teman [**7, Chapter VIII, Proposition 1.4**].

REMARK 3. If $h$ is nonincreasing in its second argument, the situation is essentially the same. There will be a minimizing sequence of concave functions and the minimizer in Theorem 1 will be concave. Theorem 2 (in §3) remains unchanged.

REMARK 4. Instead of uniform ($L^\infty$) convergence, today usually weak convergence in $H^{1,p}$ is used. Then $\int f(x, y, y')\,dx$ is lower semicontinuous if $f$ is convex in $y'$ (and the seminormality condition can be dropped). In our case $y_n$ is weakly convergent in $H^{1,1}$ but not in $H^{1,p}$, $p > 1$. To prove this one has to show in addition that $y_n'$ is weakly convergent in $L^1$.

## 3. Regularity of minimizers

We first remark that a convex minimizer $y_0$, which exists according to Theorem 1, has a derivative $y_0'$ a.e., and this is bounded in any interval $[c, d] \subset (a, b)$. As is well known, $y_0$ then has further smoothness properties if (2.2) has some natural properties. More precisely, we have

THEOREM 2. *In addition to the assumptions adopted at the beginning of* §2 *and in Theorem* 1 *let the integrand* $f = g + h$ *of* (2.2) *be twice continuously differentiable on* $B \times \mathbb{R}$. *Assume*

$$g''(u) > 0, \qquad u \in \mathbb{R}.$$

*Then any minimizer* $y_0 \in \mathscr{A}$ *of* $I$ (2.2) *is convex. If for* $(\alpha, \beta) \subset [a, b]$ *the graph of* $y_0$ *does not meet the boundary* $\partial B$, *then* $y_0$ *is twice continuously differentiable on* $(\alpha, \beta)$ *and obeys the Euler-Lagrange equation*

$$\frac{d}{dx} f_{y'} = f_y \quad on\ (\alpha, \beta).$$

PROOF. Since $g'' > 0$ the minimizers of $\int g(y')\,dx$ are unique, namely affine functions. Therefore $I[y]$ can strictly be decreased for any nonconvex function $y$. Thus any minimizer $y_0$ of $I$ is convex. If $\alpha = a$ or $\beta = b$ we have to restrict ourselves to a smaller interval in order to have a bounded derivative $y_0'$. But we then can exhaust $(\alpha, \beta)$ by smaller intervals. We therefore may assume that $y_0'$ is bounded on $(\alpha, \beta)$. Also $y_0$ is a minimizer of $I$ relative to the boundary conditions $y(\alpha) = y_0(\alpha)$, $y(\beta) = y_0(\beta)$. By Theorem 2.2.i in Cesari [**4, p. 30**] $y_0$ obeys the Euler-Lagrange equation a.e. in $(\alpha, \beta)$ in the sense made precise there. From Tonelli's theorem (Cesari [**4, p. 58**]) we see that $y_0$ is continuously differentiable on $(\alpha, \beta)$. From Weierstrass' theorem [**4, p. 60**] we have that $y_0$ is twice continuously differentiable and obeys the Euler-Lagrange equation throughout in $(\alpha, \beta)$. This proves Theorem 2.

One may try to generalize Theorems 1 and 2 in several directions, for instance: If $f: B \times \mathbb{R} \to \mathbb{R}$ is $C^2$ and if all solutions of the Euler-Lagrange equation are convex, does there exist, in a region $B$ as in §2, a regular convex minimizer? We do not know anything about its existence. But regularity is easily established by means of the following theorem of Tonelli [**11, §17**].

THEOREM (Tonelli). *Let* $y_0$ *be a minimizer of*

$$I[y] = \int_a^b f(x, y, y')\,dx$$

*in a given class of absolutely continuous functions with fixed boundary conditions. Assume* $f \in C^3$ *and* $f_{y'y'} > 0$. *Then there is a relatively open set* $\Omega \subset [a, b]$ *of full measure such that* $y_0 \in C^2(\Omega)$, $y_0$ *obeys the Euler-Lagrange equation on* $\Omega$, *and the plane curve represented by* $y_0$ *has a continuous tangent everywhere. (The tangent will be vertical at points corresponding to* $[a, b]\backslash\Omega$.)

(There is an earlier version of this theorem in **[10, p. 359]** which does not include the continuity of the extended function $y_0'$. See also Ball and Mizel **[2, p. 333]** and Clarke and Vinter **[6, Corollaries 1 and 2]**. Tonelli **[11, §21]**, gives an example where $\Omega \neq [a, b]$.)

THEOREM 3. *Assume that for the problem in Tonelli's theorem*

$$f_y - f_{y'x} - f_{y'y}y' \geq 0 \quad (\leq 0). \tag{3.1}$$

*Then any minimizer* $y_0$ *is twice continuously differentiable in* $(a, b)$.

PROOF. Let $(\alpha, \beta)$ be a maximal interval in $\Omega$. According to Tonelli's theorem $y_0$ obeys the Euler-Lagrange equation in $(\alpha, \beta)$. If (3.1) holds (with $\geq 0$, e.g.), then $y_0$ is convex in $(\alpha, \beta)$ and $y_0'(\beta) = +\infty$ (if not $\beta = b$). If there were an adjacent interval of $\Omega$, then $y_0'(\beta) = -\infty$. If $\beta$ were an accumulation point of intervals in $\Omega$, then $\beta$ would be an accumulation point of points with $y_0' = -\infty$ and, by the extended continuity of $y_0'$, $y_0'(\beta) = -\infty$ too. Thus $\beta = b$, and in the same way we conclude $\alpha = a$.

REMARK 1. It suffices that (3.1) is valid if $|y'| > M$ for some $M \in \mathbb{R}$.

REMARK 2. We included Theorem 3 in this paper because it is a direct consequence of Tonelli's partial regularity theorem. But there is another theorem of Tonelli closely related to Theorem 3. In **[11]** he first gives a condition (§1, see also §2) which insures that any minimizer is an *estremaloide*, i.e., obeys the integrated Euler-Lagrange equation. Then he shows (§9) that under his conditions (21) or (22) every *estremaloide* has a finite derivative, except possibly at the endpoints. Our convexity condition (3.1) is just a special case of (21), (22).

## 4. Application to a geometric problem

For a plane convex body with area $F$ and circumference $L$ Bonnesen showed that

$$F - xL + x^2\pi \leq 0$$

is true for $x = R$ (circumradius) and $x = r$ (inradius) (see Bonnesen and Fenchel **[3, p. 112]**). The discriminant of the quadratic polynomial is nonnegative, and this is the isoperimetric inequality $L^2 - 4\pi F \geq 0$. Now

$$L = \int_0^{2\pi} h(\varphi)\, d\varphi = \pi \overline{B},$$

where $h$ is the support function, $\varphi$ is the support angle, and $\overline{B}$ is the mean width **[3, p. 65]**. This suggests the following form of Bonnesen's inequality in euclidean space $E^n$, $n \geq 3$. For a convex body $K$ we consider its volume and its mean width, i.e., the quermassintegrals $W_0$ and $W_{n-1}$ **[3, p. 49]**, together with $W_n$, the volume of the unit ball. We claim that for centrally symmetric convex bodies in $E^n$ there is $\beta_n > 1$ so that

$$W_0 - \beta_n R^{n-1} W_{n-1} + (\beta_n - 1) R^n W_n \leq 0. \tag{4.1}$$

We choose $\beta_n$ maximal since then (4.1) is sharpest. In the proof we may assume $R = 1$, and by a symmetrization process we reduce the problem to bodies of revolution. Thus we arrive at a one-dimensional variational problem

$$\beta_n W_{n-1} - W_0 \to \min, \tag{4.2}$$

*where the solution is to be found in the half-circle with radius* $R = 1$. For details see **[8]**. We solve this variational problem for arbitrary $\beta_n$ and, in order to get (4.1), then choose $\beta_n$ such that

$$\min(\beta_n W_{n-1} - W_0) = (\beta_n - 1) W_n. \tag{4.3}$$

*The variational problem* (4.2) *in the upper half-circle is of the type* (2.2) *considered above.* For instance, in $E^3$ it takes the form

$$\int_{-1}^{1} [\gamma y' \arctan y' - y^2]\, dx,$$

where $\gamma$ is a constant related to $\beta_3$. Note that the integrand is growing linearly with respect to $y'$. Theorems 1 and 2 show that minimizers exist and are composed of boundary parts and extremal arcs. Checking all possibilities one solves (4.2) and finds the $\beta_n$ in (4.3) and (4.1). It turns out that equality in (4.1) is attained exactly for balls and straight segments. Details will be published elsewhere.

## References


1. N. I. Akhiezer, *The calculus of variations*, Harwood, London, 1988.
2. J. M. Ball and V. J. Mizel, *One-dimensional variational problems whose minimizers do not satisfy the Euler-Lagrange equation*, Arch. Rational Mech. Anal. **90** (1985), 325–388.
3. T. Bonnesen and W. Fenchel, *Theorie der konvexen Körper*, Springer-Verlag, Berlin, 1934. (Nachdruck 1974.)
4. L. Cesari, *Optimization—Theory and Applications*, Springer-Verlag, New York, 1983.
5. F. H. Clarke, *Tonelli's regularity theory in the calculus of variations: Recent progress*, Lecture Notes in Math., vol. 1190, Springer-Verlag, New York, 1986, pp. 163–179.
6. F. H. Clarke and R. B. Vinter, *Existence and regularity in the small in the calculus of variations*, J. Differential Equations **59** (1985), 336–354.
7. I. Ekeland and R. Temam, *Convex analysis and variational problems*, North-Holland, Amsterdam, 1976.
8. E. Heil, *Extensions of an inequality of Bonnesen to D-dimensional space and curvature conditions for convex bodies*, Aequationes Math. **34** (1987), 35–60.
9. R. T. Rockafellar, *Convex analysis*, Princeton Univ. Press, Princeton, N. J., 1970.
10. L. Tonelli, *Fondamenti di calcolo delle variazioni*, vol. 2. Zanichelli, Bologna, 1923.
11. ——, *Sulle proprietà delle estremanti*, Ann. Scuola Norm. Sup. Pisa **3** (1934), 213–237. *Opere Scelte* III, Cremonese, Roma, 1962, pp. 160–191.
12. ——, *Su gli integrali del calcolo delle variazioni in forma ordinaria*, Ann. Scuola Norm. Sup. Pisa **3** (1934), 401–450. *Opere Scelte* III, pp. 192–254.
13. J. L. Troutman, *Variational calculus with elementary convexity*, Springer-Verlag, New York, 1983.


FACHBEREICH MATHEMATIK, TECHNISCHE HOCHSCHULE DARMSTADT, SCHLOSSGARTENSTRASSE 7, 6100 DARMSTADT, FEDERAL REPUBLIC OF GERMANY

DIMACS Series in Discrete Mathematics
and Theoretical Computer Science
Volume 4, 1991

Applied Geometry
and Discrete Mathematics
THE VICTOR KLEE FESTSCHRIFT

# Flattening a Rooted Tree

PAUL HILFINGER, EUGENE L. LAWLER, AND GÜNTER ROTE

ABSTRACT. The following is the graph theoretic formulation of a problem arising in the compilation of block-structured computer languages. Let $T = (V, E)$ be a rooted tree and let $A$ and $C$ be sets of additional arcs with the following properties:

(1) If $(v, w) \in A$, then $w$ is a proper ancestor of $v$.
(2) If $(v, w) \in C$, then $w$ is the child of an ancestor of $v$ (allowing $v$ to be an ancestor of itself.)

It is desired to find a rooted tree $T^* = (V, E^*)$ in which the depth of each node is as small as possible, subject to conditions (1), (2) and

(3) $w$ is an ancestor of $v$ in $T^*$ only if $w$ is an ancestor of $v$ in $T$.

A simple, almost linear-time, algorithm for constructing an optimal tree $T^*$ is described. The constructed tree $T^*$ has the property that node $w$ is an ancestor of node $v$ in $T^*$ only if this is true in every tree satisfying conditions (1)–(3). Hence the depth of each and every node in $T^*$ is simultaneously minimized. The running time of the algorithm is $O(m\alpha(m, n))$, where $n$ is the number of nodes in $T$, $m \geq n$ is the total number of arcs in $A$ and $C$, and $\alpha$ is a very slowly growing function related to the functional inverse of Ackermann's function.

## 1. Introduction

Compilers for block-structured programming languages with nested, potentially recursive subprograms generally allow subprograms to access variables defined in enclosing subprograms. The classical implementation is to maintain a *display*, which in this context means a vector of pointers to all the stack frames containing variables visible to the currently executing subprogram. Alternatively, the display vector may be implemented as a linked list of stack frames, sometimes known as a *static chain*.

Long static chains can result in long access times for nonlocal variables. Although large displays do not have this problem, they must sometimes be copied and stored, as in the case of languages that permit subprogram-valued

1980 *Mathematics Subject Classification* (1985 *Revision*). Primary 68R10, 05C05.
The second author's research was supported in part by NSF Grant CCR-8704184.

arguments or variables. Hence it is advantageous to keep both displays or static chains short.

In straightforward implementations, the size of the display or static chain needed for a given subprogram is determined by the number of subprograms that textually enclose its definition—its *lexical level*. In general, a subprogram $P$ must be nested inside a subprogram $Q$ in order to access variables of $Q$ or to call subprograms (other than $Q$ itself) that themselves must be nested inside $Q$. The scope rules of block-structured languages guarantee this constraint by imposing a stronger one: $P$ must be nested within $Q$ if either

(i) $P$ accesses the local variables of $Q$, or
(ii) $P$ calls a subprogram that is nested immediately within $Q$.

NOTE. A literal subprogram may be passed as a parameter or stored in a variable, as well as being called. Both of these situations should be treated as calls.

It is permissible to nest subprograms where accessibility of variables does not require it, and it is sometimes expedient or aesthetically desirable to do so. A programmer may group subprograms together for the sake of readability, without concern for efficiency. However, a compiler should be able to reorganize or "flatten" the nesting of subprograms so as to minimize lexical levels, and hence display sizes.

Let us represent a feasible nesting of subprograms by a rooted tree $T = (V, E)$ in which the root $r$ represents the "main" or "outer level" program, each of the other nodes in $V$ represents a subprogram, and the edges in $E$ represent "defined immediately within" relations. Two additional sets of arcs represent the relations "accesses local variables of" $(A)$ and "calls" $(C)$. The following is a graph-theoretic formulation of the minimization problem we should like a compiler to solve.

*Given*: A tree $T = (V, E)$ with root $r \in V$, and sets $A$ and $C$ of additional arcs between nodes in $V$ such that

(1) If $(v, w) \in A$, then $w$ is a proper ancestor of $v$.
(2) If $(v, w) \in C$, then $w$ is the child of an ancestor of $v$ (allowing $v$ to be an ancestor of itself).

*Find*: A rooted tree $T^* = (V, E^*)$ that minimizes the depth of each node, subject to constraints (1) and (2), for the given sets $A$, $C$, and the constraint that

(3) $w$ is an ancestor of $v$ in $T^*$ only if $w$ is an ancestor of $v$ in $T$.

It is not immediately evident that there exists a tree $T^*$ that simultaneously minimizes the depth, i.e., the distance from the root, of each and every node. However, this is indeed the case. In particular, there is a tree $T^*$ with the property that a node $w$ is an ancestor of a node $v$ only if $w$ is an ancestor of $v$ in every tree that satisfies conditions (1)–(3). We will describe a simple, almost linear-time, algorithm for constructing such an optimal tree $T^*$.

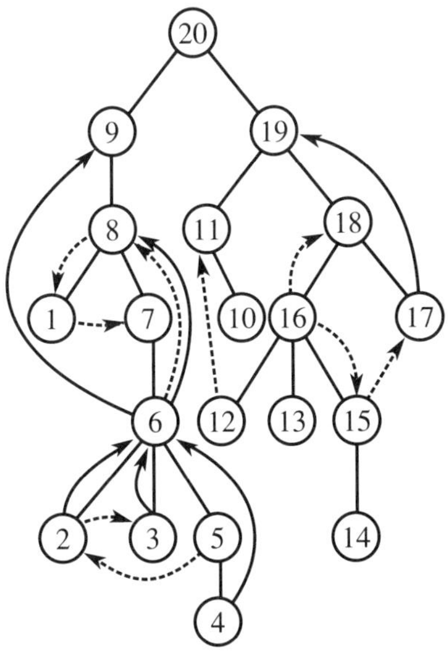


FIGURE 1

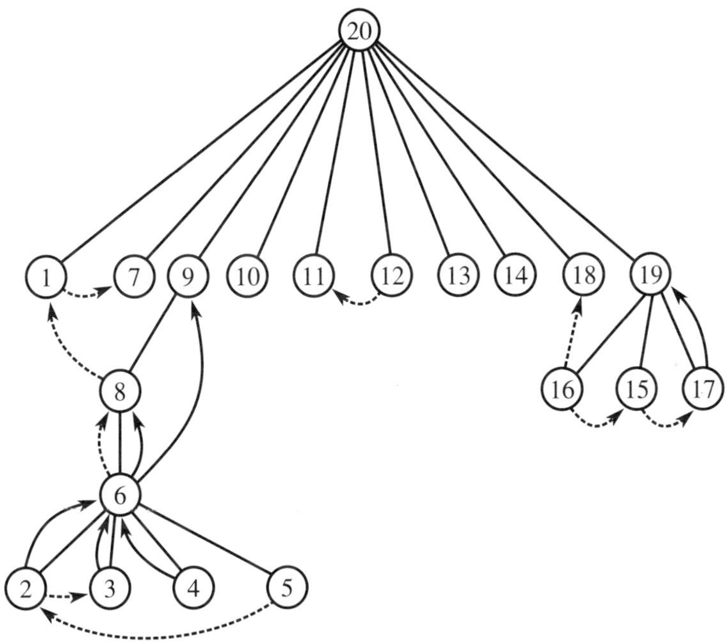


FIGURE 2

An instance of a tree $T = (V, E)$ and additional sets of arcs $A$ and $C$ is shown in Figure 1, with nodes of $T$ numbered in postorder. Arcs in $A$ are (2,6), (3,6), (4,6), (6,8), (6,9), (17,19). Arcs in $C$ are (1,7), (2,3), (5,2), (6,8), (8,1), (12,11), (15,17), (16,15), (16,18); these are indicated by broken lines. Figure 2 indicates an optimal "flattening" of $T$.

## 2. The special case when $C$ is empty

For simplicity, we begin by considering the case in which the set $C$ is empty. In order to ensure that our algorithm produces a rooted tree $T^*$, we assume that the set $A$ contains an arc $(v, r)$ from each node $v$ to $r$, where $r$ is the root of $T$. (If necessary, add arcs to $A$ so that this is true.) We also assume that for each node $w$ of $T$ there is an (arbitrarily ordered) list of the arcs $(v, w)$ of $A$ directed into $w$.

Our algorithmic strategy is to process the nodes of $T$ in *postorder*. This means that at the time a node $w$ is processed, all of its descendants in $T$ will have already been processed. As a result of this prior processing, the descendants of $w$ will have been arranged in subtrees. We shall assume by induction that the ancestor relationships that hold in these subtrees are required to hold in all trees satisfying conditions (1) and (3).

To process node $w$, we do the following. For each arc $(v, w)$ in $A$, we find the root $v'$ of the subtree in which $v$ is contained. If $v' = w$, then $w$ has already been made an ancestor of $v$ in $T^*$. If not, we add the edge $(v', w)$ to $T^*$. A pseudocode description of this procedure is as follows:

```
flatten(T, w):
   for each child w' of w in T,
      flatten(T, w');
   for each arc (v, w) in A,
      find the root v' of the subtree of T* containing v;
      if v' ≠ w then
         add an edge (v', w) to T*;
```

Let us prove that this procedure is valid. We have only to show that the newly inserted edges introduce no ancestor relationships that are not required by conditions (1) and (3). The existence of the arc $(v, w)$ in $A$ implies that $w$ must be an ancestor of $v$. By the inductive assumption, $v'$ must also be an ancestor of $v$. Hence condition (3) forces us to make $w$ an ancestor of $v'$. It follows that the new edge added to $T^*$ introduces no ancestor relations that are not required by conditions (1) and (2).

## 3. The general case

Let us now consider the general case in which the set $C$ is nonempty. Our strategy is to proceed as in the special case, dealing with arcs in $C$ only when we are forced to. This occurs when we make $v'$ the child of $w$ and there is an arc $(x, v')$ in $C$ with $x \neq w$. When this happens, it is necessary to make $w$ an ancestor of $x$ in $T^*$, else condition (2) will be violated. The easiest way to accomplish this is to impute the existence of an arc $(x, w)$ in $A$. A pseudocode description of our procedure now looks as follows:

```
flatten(T, w):
  for each child w' of w in T,
    flatten(T, w');
  while there is an arc (v, w) in A,
    remove (v, w) from A;
    find the root v' of the subtree of T* containing v;
    if v' ≠ w then
      add an edge (v', w) to T*;
      for each arc (x, v') in C with x ≠ w,
        add an arc (x, w) to A;
```

The justification of this procedure lies in the fact that an arc in $A$ is imputed if and only if its existence is required by condition (2).

We have seen that if $C$ is empty then $T^*$ observes the ancestor relations of $T$. To see that this is true in the general case, it is necessary only to observe that whenever the existence of an arc $(x, w)$ in $A$ is imputed, it is necessarily the case that $w$ is an ancestor of $x$ in $T$.

## 4. Implementation with union-find operations

An efficient implementation of our procedure requires a good method for finding the root $v'$ of the subtree of $T^*$ in which a given node $v$ is contained. The easiest way to do this is to implement a union-find data structure for the nodes of $T$. Such a data structure begins with $n$ singleton sets and supports a sequence of union and find operations. Specifically, *find*$(v)$, where $v$ is any element, returns the representative (root) $v'$ of the set (subtree) in which $v$ is contained. The operation union$(v', w)$ joins the two sets whose representatives are $v'$ and $w$. We shall assume that $w$, the second argument, is made the representative of the new set, i.e., $w$ is the root of the new subtree formed by adding the arc $(v', w)$ to $T^*$. This gives us the following pseudocode:

```
flatten(T, w):
  for each child w' of w in T,
    flatten(T, w');
  while there is an arc (v, w) in A,
    remove (v, w) from A;
    v' := find(v);
    if v' ≠ w then
      add an arc (v', w) to T*;
      union(v', w);
      for each arc (x, v') in C with x ≠ w,
        add an arc (x, w) to A;
```

Tarjan [1] has shown that the running time of a simple union-find implementation is bounded by $O(m\alpha(m, n))$, where $n$ is the number of elements and $m \geq n$ is the number of unions and finds performed. Here $\alpha$ is a very slowly growing function related to the functional inverse of Ackermann's function; for all practical purposes, the value of $\alpha(m, n)$ is bounded by a small constant. Exactly $n-1$ unions are performed in the course of the computation. The number of finds is equal to the number of (given and imputed) arcs in $A$. Hence the overall running time of the algorithm is $O(m\alpha(m, n))$, where $n$ is the number of nodes in $T$ and $m \geq n$ is the total number of arcs in $A$ and $C$.

## Reference

1. R. E. Tarjan, *Efficiency of a good but not linear set union algorithm*, J. Assoc. Comput. Mach. **22** (1975), 215–225.

(P. Hilfinger and E. L. Lawler) Computer Science Division, University of California, Berkeley, California 94720

*E-mail addresses:* Hilfinger@Berkeley.EDU
lawler@ernie. Berkeley.EDU

(G. Rote) Institut für Mathematik, Technische Universität Graz, Kopernikusgasse 24, A-8010 Graz, Austria

*E-mail address:* ROTE@KOP.TU-GRAZ. ADA.AT

DIMACS Series in Discrete Mathematics
and Theoretical Computer Science
Volume **4**, 1991

Applied Geometry
and Discrete Mathematics
THE VICTOR KLEE FESTSCHRIFT

# The Geometric Complementarity Problem and Transcending Stationarity in Global Optimization

REINER HORST AND HOANG TUY

ABSTRACT. We address the problem: Given two convex sets $C, D \subset \mathbb{R}^n$, find a point $x \in D \backslash C$ (geometric complementarity problem). It is shown that, for large classes of global optimizations which include concave minimization, reverse convex programming, DC optimization, and Lipschitz programming, the fundamental problem of transcending stationarity (in the sense of finding a better feasible solution than a given one which is usually a stationary point or local minimum) can be reduced to a geometric complementarity problem. An algorithm is proposed to solve this problem, and its convergence properties are demonstrated.

## 1. Introduction

This paper addresses the following problem which will be referred to as the *Geometric Complementarity Problem* (GCP):

(GCP) *Given two convex sets $C, D$ in $\mathbb{R}^n$ ($C$ open or closed, $D$ closed), find a point $x \in D \backslash C$.*

This problem is a generalization of the classical *linear complementarity problem* which has been a subject of intensive research over the past two decades (cf., e.g., Murty [**3**] and references there). Indeed, given an $n \times n$ matrix $Q$ and an $n$-vector $q$, the linear complementarity problem is to find a couple of $n$-vectors $x, y$ such that

$$(1) \qquad y = Qx + q, \qquad x \geq 0, \ y \geq 0, \ x^T y = 0.$$

Setting $C = \{x: \sum_{i=1}^n \min\{x_i, (Qx+q)_i\} > 0\}$, $D = \{x: Qx+q \geq 0, \ x \geq 0\}$, one easily sees that a couple $x, y = Qx+q$ solves the linear complementarity problem if and only if $x \in D \backslash C$.

The generalization of the linear complementarity problem to problem GCP is not of purely theoretical interest. Actually, our main motivation to study

1980 *Mathematics Subject Classification* (1985 *Revision*). Primary 49D37.
*Key words and phrases.* Global optimization, complementarity.

GCP stems from the development of numerical techniques for large classes of multiextremal global optimization problems.

Standard nonlinear programming techniques have not been successful for solving these problems. Their deficiency is due to the intrinsic multiextremality of the formulation and not to the lack of smoothness or continuity. One can observe that local tools such as gradients, subgradients, and second-order constructions such as Hessians, cannot be expected to yield more than local solutions. One finds, for example, that a stationary point (satisfying certain first-order optimality conditions) is often detected for which there is even no guarantee of local minimality. For these reasons global solution methods must be significantly different from standard nonlinear programming techniques, and they can be expected to be and are much more expensive computationally. However, the enormous practical need for solving global optimization problems coupled with a rapidly advancing computer technology has allowed one to consider problems which a few years ago would have been considered computationally intractable. As a consequence, we are seeing the creation of a large and increasing number of diverse algorithms for solving a wide variety of multiextremal global optimization problems. Most of these procedures are designed for special problem types where helpful specific structures can be exploited. Moreover, in many practical global optimizations, the multiextremal feature involves only a small number of variables, and additional structure is amenable to large scale solutions. For a comprehensive treatment of the diverse methods and the underlying general concepts, see Horst and Tuy [**2**].

Other methods which have been proposed for solving very general and difficult global problems that possess little additional structure can handle only small problem sizes with sufficient accuracy. To solve larger general global optimization problems, one could conceive a two phase local-global method where a standard local procedure alternates with a global one which, given a stationary point (or local minimum), would find a better feasible point. Since, moreover, a closer look at existing global optimization methods which do not use local nonlinear programming techniques shows that in many cases such a global phase is involved, we see that the core of global optimization is a strategy for *transcending stationarity* in the above sense. Surprisingly enough, it turns out that for a very wide class of global optimization problems of practical importance, transcending stationarity can be reduced to solving a GCP. Therefore, the study of GCP is not only of relevant interest for the linear complementarity problem but may also serve as the basis for transcending stationarity in global optimization.

The paper is subdivided in six sections. In §§2 and 3 we shall show how GCP arises when transcending stationarity in many important classes of optimization problems. In §§4 and 5 we shall describe a general method for solving GCP and discuss the subdivision strategy which plays an important role in this kind of branch and bound procedure. The convergence of the method will be established in §6.

## 2. Transcending stationarity in DC optimization

In an abundant class of constrained global optimization, convexity is present in a *reverse sense.* In this direction our research focusses on the following main problem classes:

- minimization of concave functions subject to linear and convex constraints (concave minimization),
- minimization over the intersection of convex sets and complements of convex sets (reverse convex programming),
- global optimization of functions that can be expressed as the difference of two convex functions (DC programming).

**2.1. Concave minimization.** One of the most important nonconvex global optimizations with abundant applications, and which also serves as prototype global problem (in the sense that main solution concepts devised to solve it are often amenable to extensions to more general problems) is the *concave minimization* problem

$$\text{minimize } f(x) \text{ s.t. } x \in D, \tag{2}$$

where $D$ is a closed convex set in $\mathbb{R}^n$ and $f(x)$ is a concave function on $\mathbb{R}^n$ (cf. Horst and Tuy [**2**]).

Let $x^0$ be a local minimum. If we set $C := \{x: f(x) \geq f(x^0)\}$, then clearly

> $x^0$ *achieves the global minimum if and only if* $D \subset C$. *Any point* $x \in D \setminus C$ *is a better feasible solution than* $x^0$,

so that transcending local minimality in the concave program (2) actually amounts to solving GCP.

This way of looking at concave minimization has been particularly successful in the linearly constrained case when $D$ is a polyhedron. In this case, a local minimum of $f$ is always attained at a vertex of $D$ (cf., e.g., Rockafellar [**4**]) and can be determined by standard pivoting techniques (cf. [**2**]). Various cutting plane methods for solving the linearly constrained concave minimization problem can be interpreted as successive solutions of GCP that use polyhedral outer approximations $P$ of $D\setminus C$ satisfying $x^0 \notin P$ (cf. [**2**]).

**2.2. Convex programs with an additional reverse convex constraint.** A *reverse convex* inequality is an inequality of the form $g(x) \geq 0$, where $g(x)$ is a convex function. When such an inequality is added to the constraints of a convex program the resulting nonconvex problem is

$$\text{minimize } f(x) \text{ s.t. } x \in \Omega, \quad g(x) \geq 0, \tag{3}$$

where $f(x)$, $g(x)$ are convex functions on $\mathbb{R}^n$ and $\Omega$ is a closed convex set in $\mathbb{R}^n$. Problems of the form (3) arise from various applications (cf. [**2**]); for example, a zero-one restriction $x_i \in \{0, 1\}$ is obviously equivalent to $0 \le x_i \le 1$, $x_i^2 - x_i \ge 0$.

When the constraint $g(x) \ge 0$ is not redundant one can assume that a point $w$ is available such that

$$w \in \Omega, \quad g(w) < 0, \quad f(w) < \min\{f(x): x \in \Omega, \ g(x) \ge 0\}$$

(for example, $w$ is an optimal solution of the convex program $\min\{f(x): x \in \Omega\}$). Moreover, assume the regularity condition

$$\inf\{f(x): x \in \Omega, \ g(x) \ge 0\} = \inf\{f(x): x \in \Omega, \ g(x) > 0\}.$$

Then it can be shown (see, e.g., [**2**]) that transcending stationarity of $x^0$ amounts to solving a GCP, with $C = \{x: g(x) \le 0\}$, $D = \{x \in \Omega: f(x) \le f(x^0)\}$. Specifically:

> $x^0$ *is a global optimal solution of* (3) *if and only if* $D \subset C$. *If* $x \in D \backslash C$, *then the intersection* $\pi(x)$ *of the line segment* $[w, x]$ *with the surface* $g(x) = 0$ *is a better feasible solution than* $x^0$.

Again, transcending stationarity reduces to solving a GCP. In important applications, problem (3) reduces to a linear program with an additional reverse convex constraint ($f(x)$ affine, $\Omega$ polyhedron). In this case, one can show that an optimal solution is attained at the intersection of an edge of $\Omega$ with the surface $g(x) = 0$, and the above observation can be elaborated to quite efficient algorithms (cf. [**2**] and references there).

**2.3. DC optimization problems.** A *DC function* is a function which can be expressed as a difference of two convex functions. A *DC inequality* is an inequality of the form $g(x) \ge 0$, where $g(x)$ is a DC function. A *DC optimization* problem is an optimization problem in which the objective function is a DC function, while all the constraints are DC inequalities. As far as optimization theory is concerned, a nice property of DC functions is that the pointwise maximum or the pointwise minimum of a finite collection of DC functions is still a DC function (cf., e.g., [**2**] and references there). Due to this property, any DC optimization problem, with a finite number of DC constraints, can in principle be converted, at the expense of additional variables, into a convex program with exactly one additional reverse convex constraint (the "canonical DC optimization problem"; see, e.g., [**2**]). But we have seen that, under a mild regularity assumption, transcending stationarity in problem (3) reduces to solving a GCP. For example, the DC optimization problem

$$\text{minimize } f(x) - g(x) \text{ subject to } x \in \Omega$$

($f$, $g$ convex functions, $\Omega$ convex set) can be rewritten as

$$\text{minimize } f(x) - t \text{ subject to } x \in \Omega, \ \ g(x) \ge t.$$

Hence, transcending stationarity of a feasible solution $(x^0, t^0)$ reduces to finding a point $(x, t) \in D\backslash C$, where $D = \{(x, t): x \in \Omega,\ f(x) - t \leq f(x^0) - t^0\}$ and $C = \{(x, t): g(x) - t \leq 0\}$ (a stationary point must satisfy $g(x) - t = 0$; **[2]**).

### 3. Transcending stationarity in continuous optimization

Consider now the general problem

$$\text{minimize } f(x) \text{ subject to } x \in D, \tag{4}$$

where $f(x)$ is a continuous function on $\mathbb{R}^n$ and $S$ is a compact set in $\mathbb{R}^n$. Despite its generality, this problem is amenable to the $DC$ structure in the following way. For each value $\alpha \in (-\infty, +\infty)$ define

$$S_\alpha := \{x \in S: f(x) < \alpha\}, \qquad \overline{S}_\alpha = \{x \in S: f(x) \geq \alpha\}.$$

A lower semicontinuous function $r(\alpha, y): (-\infty, +\infty] \times \mathbb{R}^n \to R_+$ is called a *separator for* $(f, S)$ if:

(i) $r(\alpha, y) \leq d(y, \overline{S}_\alpha)$, the distance from $y$ to $\overline{S}_\alpha$;
(ii) $r(\alpha_1, y) \geq r(\alpha_2, y)$ whenever $\alpha_1 \leq \alpha_2$;
(iii) $r(\alpha, y) > 0$ whenever $y \notin \overline{S}_\alpha$.

A trivial example of separator is the distance function $d(y, \overline{S}_\alpha)$ itself which is, of course, very difficult to compute. However, in many cases of practical importance, a separator is available that is readily computable. For example, if $f(x)$, $g_i(x)$ $(i = 1, \ldots, m)$ satisfy the Lipschitz condition with constants $L_0$, $L_i$, then it can be verified (see **[5, 2]**) that a separator for $(f, S)$, $S = \{x: g_i(x) \leq 0\}$, is

$$r(\alpha, y) = \max\{(f(x) - \alpha)^+/L_0,\ (g_i(x))^+/L_i,\ i = 1, \ldots, m\}$$

(as usual, $A^+$ denotes $\max(0, A)$).

Assume the regularity condition

$$\inf\{(f(x): x \in S\} = \inf\{f(x): x \in \operatorname{int} S\}$$

(which necessarily holds if $S = \operatorname{cl}(\operatorname{int} S)$). If for each $\alpha \in (-\infty, +\infty]$ we define the convex function

$$h(\alpha, x) := \sup\{r^2(\alpha, y) + 2x^T y - \|y\|^2: y \notin S_\alpha\},$$

then the following properties hold (see **[5, 2]**):

(i) $\min\{h(\alpha, x) - \|x\|^2: x \in \mathbb{R}^n\} = 0$ if and only if $\alpha = \min f(S)$, and then any $x^0 \in \operatorname{argmin}\{h(\alpha, x) - \|x\|^2: x \in \mathbb{R}^n\}$ is a global optimal solution of (4).
(ii) If $\min\{h(\alpha, x) - \|x\|^2: x \in \mathbb{R}^n\} < 0$, then any $x$ such that $h(\alpha, x) - \|x\|^2 < 0$ is a feasible solution of (4) with $f(x) < \alpha$.
(iii) If $\min\{h(\alpha, x) - \|x\|^2: x \in \mathbb{R}^n\} > 0$, then $\alpha < \min f(S)$.

(It follows from these conditions that if $x^0 \in S$ and $f(x^0) = \alpha$, then $\min\{h(\alpha, x) - \|x\|^2 : x \in \mathbb{R}^n\} \leq 0$ and $h(\alpha, x^0) - \|x^0\|^2 = 0$.)

Therefore, for any given value $\alpha \in (-\infty, +\infty]$, finding a feasible solution $x$ such that $f(x) < \alpha$ reduces to finding an $x$ satisfying $h(\alpha, x) - \|x\|^2 < 0$, i.e., a couple $(x, t) \in D \setminus C$, where $D = \{(x, t) : h(\alpha, x) \leq t\}$ and $C = \{(x, t) : \|x\|^2 \leq t\}$. This is again a GCP.

To sum up, GCP serves as a basis for a unifying approach to the numerical solution of a wide variety of optimization problems, ranging from concave programming to continuous (including Lipschitz) optimization.

## 4. An algorithm for solving GCP

Special GCPs have been treated as subproblems in several global optimization algorithms (see, e.g., [**2**]). In this section we present a general algorithm for GCP which does not require boundedness of $D$ or $C$ and uses a more satisfactory subdivision strategy than the special algorithms treated in [**2**].

Let us assume that:

(i) $0 \in D \cap (\operatorname{int} C)$;
(ii) $D = \{x : g(x) \leq 0\}$ with $g(x)$ a convex function, and a point $w \in C$ is available such that $g(w) < 0$ (i.e., $w \in C \cap \operatorname{int} D$).

We start with constructing a collection $\mathscr{C}$ of polyhedral cones whose union contains $D$. Each cone of this collection will be generated by the $n$ vertices of certain facets $M$ of an $n$-simplex $T$, $0 \notin M$, $T \subset C$. An $n$-simplex $T$ satisfying $T \subset C$ exists because of assumption (i). For its construction, we distinguish two cases.

Case (a): $D$ is polyhedral and $0$ is a vertex of $D$. This case is of particular interest in applications (cf., e.g., the discussion in §2 with $x^0 = 0$). When 0 is a nondegenerate vertex of $D$ we simply set $\mathscr{C} = \{\operatorname{con}(D)\}$, where $\operatorname{con}(D)$ denotes the conical hull of $D$ (which is the cone generated by the $n$ edges of $D$ emanating from 0). Let $s^i \neq 0 \in C$ denote an arbitrary point in $C$ on the $i$th edge of $D$ emanating from 0 $(i = 1, \dots, n)$. Then, obviously, $T = [0, s^1, \dots, s^n]$ is a suitable $n$-simplex (with vertices $0, s^1, \dots, s^n$), and its facet $M = [s^1, \dots, s^n]$ generates $\operatorname{con}(D)$, i.e., $\operatorname{con}(D) = \operatorname{con}(M)$. When 0 is a degenerate vertex of $D$, several (slightly more involved) constructions are available which are similar to those used in cutting plane techniques for solving linearly constrained concave minimization problems. A detailed discussion can be found in [**2**, Chapters III and V].

Case (b): $D$ is an arbitrary convex set, and $0$ is an arbitrary point in $D$. In this case, a suitable simplex $T$ can be determined in many ways. For example, choose $T = [v^1, \dots, v^n, v]$, where $v^i \neq 0$ is an arbitrary point in $C$ on the $i$th unit vector $e^i$ $(i = 1, \dots, n)$, and $v$ is an arbitrary point in $C$ on the line segment $(0, -e]$, where $e = (1, 1, \dots, 1) \in \mathbb{R}^n$. (Note that the requirement $T \subset C$, which is convenient for a simple formulation of the algorithm below, is not essential. In computation, one can simply choose $T = [e^1, \dots, e^n, -e]$.) Let $\mathscr{M}_0$ denote the collection of the $n+1$ facets. Then $\mathscr{C} = \{\text{con}(M) : M \in \mathscr{M}_0\}$ clearly satisfies $D \subset \bigcup_{M \in \mathscr{M}_0} \text{con}(M) = \mathbb{R}^n$.

The proposed algorithm is suggested by the following property of GCP.

PROPOSITION 1. *If GCP is solvable, then at least one solution is at an extreme point of $D$.*

PROOF. Indeed, if all vertices of $D$ lie in $C$, then, since $D$ is convex, we must have $D \subset C$. □

In view of this property, the search procedure should concentrate on the boundary of $D$. If $p(x)$ is the gauge of $C$, then our problem is to find a point $x$ on the boundary of $D$ with $p(x) > 1$ (if $C$ is closed) or $p(x) \geq 1$ (if $C$ is open). We shall do this by applying a branch and bound procedure for maximizing the (convex, positively homogeneous) function $p(x)$ over the convex set $D$. To this end, let us first define the bounding operation.

At any stage of the algorithm we shall have a polyhedron $D_k$ ($k$ = iteration counter) which is an outer approximation of the convex set $D$. Let $M = [s^1, \dots, s^n]$ be any $(n-1)$-simplex contained in one facet of $T$ (this facet is a member of $\mathscr{M}_0$). For each $i$ let

$$\theta_i = \sup\{\theta : \theta s^i \in C\} \tag{5}$$

($\theta_i \geq 1$, since $T \subset C$). Consider the linear program

$$\text{maximize } \sum_{i=1}^{n} \lambda_i / \theta_i \tag{6}$$

$$\text{LP}(M) \qquad \text{s.t. } \sum_{i=1}^{n} \lambda_i s^i \in D_k, \tag{7}$$

$$\lambda = (\lambda_1, \dots, \lambda_n) \geq 0. \tag{8}$$

(In LP($M$) we agree that $1/\theta_i = 0$ if $\theta_i = +\infty$.) Denote by $\mu(M)$ the optimal value and by $\lambda(M)$ an optimal solution of LP($M$) and let $x(M) = \sum_{i=1}^n \lambda_i(M) s^i$ when $\mu(M) < +\infty$. If $H = H(M)$ is the hyperplane through $\theta_i s^i$, $i = 1, \dots, n$ (more accurately, through $\theta_i s^i$ if $\theta_i < +\infty$ and parallel to $s^i$ if $\theta_i = +\infty$), then clearly, $x(M)$ is the point of $D_k$ that stands the farthest possible from $H(M)$.

Since the objective function of LP($M$) is an affine function of $x = \sum_{i=1}^n \lambda_i s^i$ that agrees with $p(x)$ at the points 0, $\theta_i s^i$ $(i = 1, \dots, n)$, it overestimates $p(x)$ for $x \in \text{con}(M)$. Hence, $\mu(M)$ is an upper bound for $\max\{p(x) : x \in D \cap \text{con}(M)\}$. The following lemma results.

LEMMA 1. *If $C$ is open (resp. closed), then $\mu(M) < 1$ (resp. $\mu(M) \leq 1$) implies that $D \cap \text{con}(M) \subset C \cap \text{con}(M)$.*

DEFINITION 1. A simplex $M = [s^1, \ldots, s^n]$ is said to be of *first category* if all $\theta_i$ $(i = 1, \ldots, n)$ defined by (5) and the optimal value $\mu(M)$ of LP$(M)$ are finite; it is said to be of *second category* otherwise.

Moreover, let

$$\theta(M) = \min\{\theta_i : i = 1, \ldots, n\}. \tag{9}$$

Note that when $\mu(M) = +\infty$ the optimal solution $x(M)$ of LP$(M)$ is a point at infinity, i.e., a direction. Given a point or a direction $x$, we shall denote by $\pi(s)$ the point where the half-line from $w$ through $x$ (parallel to $x$, resp.) intersects the boundary of $D$. The proposed algorithm is stated next.

ALGORITHM. *Initialization: Determine a polyhedron* $D_0 \supset D$ (*cf.* [**2**, *Chapter* II]). *Let* $\mathscr{P}_0 = \mathscr{M}_0$, $I_0 = \varnothing$. *Set* $k = 0$.

*Iteration* $k = 0, 1, \ldots$ :

1) *For each* $M \in \mathscr{P}_k$ *solve* LP$(M)$ *obtaining* $\mu(M)$ *and* $x(M)$. *If for some* $M$, *the half-line from* $w$ *through* $x(M)$ (*or parallel to* $x(M)$, *if* $x(M)$ *is a direction*) *contains a point* $y \in D \backslash C$, *then terminate. Otherwise, go to* 2).

2) *In* $\mathscr{M}_k$ *delete all* $M$ *with* $\mu(M) < 1$ *if* $C$ *is open or* $\mu(M) \le 1$ *if* $C$ *is closed. Let* $\mathscr{R}_k$ *be the collection of remaining simplices of* $\mathscr{M}_k$. *If* $\mathscr{R}_k$ *is empty, then terminate: GCP has no solution. Otherwise, go to* 3).

3) *Let* $\mathscr{R}_{k\mathrm{I}}$ *denote the collection of all members of* $\mathscr{R}_k$ *that are of first category. If* $\mathscr{R}_{k\mathrm{I}}$ *is nonempty, then select*

$$M_k \in \operatorname{argmax}\{\mu(M) : M \in \mathscr{R}_{k\mathrm{I}}\}.$$

*Otherwise, select*

$$M_k \in \operatorname{argmin}\{\theta(M) : M \in \mathscr{R}_k\}.$$

4) *If* $x^k := x(M_k) \in D$ *or* $x^k$ *is a recession direction of* $D$, *then set* $I_{k+1} = I_k$. *Otherwise, set* $I_{k+1} = I_k \cup \{k+1\}$,

$$l_{k+1}(x) = (t^k)^T (x - \pi(x^k)), \tag{10}$$

$$D_{k+1} = D_k \cap \{x : l_{k+1}(x) \le 0\}, \tag{11}$$

*where* $t^k \in \partial g(\pi(x^k))$ (*the subdifferential of* $g$ *at* $\pi(x^k)$).

5) *Subdivide* $M_k$ *into a number of* $(n-1)$*-simplices. Let* $\mathscr{P}_{k+1}$ *be the partition of* $M_k$ *obtained by this subdivision* (*cf.* § 5).

6) *Let* $\mathscr{M}_{k+1} = (\mathscr{R}_k \backslash \{M_k\}) \cup \mathscr{P}_{k+1}$. *Set* $k \leftarrow k+1$ *and go to* 1).

REMARK. Step 1) can be carried out by standard line searches to determine whether the half-line intersects the boundary of $C$.

## 5. Normal subdivision process

To complete the description of the above algorithm we still have to stipulate how to subdivide $M_k$ in Step 5).

A common method for partitioning an $(n-1)$-simplex $M = [s^1, \dots, s^n]$ into subsimplices is the following. Let $u = \sum_{i=1}^n \lambda_i s^i$ $(\lambda_i \geq 0, \sum_{i=1}^n \lambda_i = 1)$ be a chosen point of $M$ and $J = \{j : \lambda_j > 0\}$; then it is easily seen that the collection $\{M(j), j \in J\}$, where $M(j)$ is the $(n-1)$-simplex spanned by $u$ and all $s^i$ with $i \neq j$, is a partition of $M$. We refer to the subdivision as the *radial subdivision* of $M$ with respect to $u$ (see, e.g., [**2**]). In the special case where $u$ is the midpoint of a longest edge of $M$, the partition has just two members and is called a *bisection.*

Clearly, using the radial subdivision method, convergence as well as efficiency of the algorithm depend on the choice of the subdivision point $u^k$ in $M_k$ in Step 5). A natural choice is $u^k = \hat{x}^k$, where $\hat{x}^k$ denotes the intersection of $M_k$ with the half-line from 0 through $x(M_k)$ (or parallel to $x(M_k)$, if $x(M_k)$ is a direction). However, convergence of the algorithm cannot be assured in this case. Therefore, to enforce convergence, from time to time we have to depart from this natural choice and take a $u^k \neq \hat{x}^k$ in a way to make the subdivision process "exhaustive". The latter means that any infinite nested sequence $\{M_k\}$ of simplices generated by the subdivision process will shrink to a point, i.e., $\operatorname{diam} M_k \to 0$ (cf. [**2**]).

Denote by $\sigma(M_k)$ the number of ancestors of $M_k$ which can be computed inductively as follows: $\sigma(M) = 0$ for any $M \in \mathscr{M}_0$; for $k = 0, 1, \dots,$ if $M$ is a partition member of $M_k$, then $\sigma(M) = \sigma(M_k) + 1$.

Let $M_k = [s^{k1}, \dots, s^{kn}]$, $\delta(M_k) = \operatorname{diam} M_k$, and let $\rho \in (0, 1)$ be a given constant. We say that a point $u^k \in M_k$ satisfies the *ρ-eccentricity condition* (for $M_k$) if

$$\max\{\|u^k - s^{ki}\| : i = 1, \dots, n\} \leq \rho\delta(M_k). \tag{12}$$

It can be proved that, when $u^k$ is the midpoint of a longest edge of $M_k$, this condition is satisfied with $\rho \in [\sqrt{3}/2, 1)$ (cf. [**6**], see also [**2**]). In the general case, when $u^k$ is not a vertex, (12) always holds when $\rho$ is sufficiently close to 1. Now fix a natural number $N \geq 2$ and a constant $\rho \in (\sqrt{3}/2, 1)$, and apply the following subdivision rule in Step (5).

($*$) *Subdivide $M_k$ with respect to $\hat{x}^k$ (i.e., let $u^k = \hat{x}^k$) if $\hat{x}^k$ satisfies the ρ-eccentricity condition, while $\sigma(M_k)$ is not a multiple of N. Bisect $M_k$ (i.e., take $u^k$ = midpoint of a longest edge of $M_k$) in all other cases.*

(Note that if $u^k$ is a vertex of $M_k$, then the partition of $M_k$ consists of just one simplex $M = M_k$, but with $\sigma(M) = \sigma(M_k) = \sigma(M_k) + 1$.)

PROPOSITION 2. *For any infinite nested sequence of simplices $\{M_k = [s^{k1}, \dots, s^{kn}]\}$ generated by the above subdivision rule we have* $\operatorname{diam} M_k \to 0$ *as $k \to \infty$.*

PROOF. It is clear that the $\rho$-eccentricity condition holds for each $k$ and the sequence involves infinitely many bisections. The property then follows from a result earlier established in [**6**] (see also [**2**]). □

## 6. Convergence properties

We prove convergence of the above algorithm when the subdivision rule $(*)$ is used.

LEMMA 2. *For any infinite nested sequence $\{M_{k_\nu}\}$ of simplices of second category we have $\theta(M_{k_\nu}) \to \infty$ as $\nu \to \infty$.*

PROOF. To simplify the notation, let us write $\nu$ rather than $k_\nu$ for the index in $\{M_{k_\nu}\}$. By Proposition 2, $\mathrm{con}(M_\nu)$ shrinks to a ray $\Gamma$ as $\nu \to \infty$. We first show that $\Gamma \subset C$. Suppose the contrary. Then $\Gamma \backslash C$ must be a half-line. Let $W$ be a ball centered at some point of $\Gamma\backslash C$ such that $W \subset \mathbb{R}^n \backslash C$. For sufficiently large $\nu$, say $\nu \geq \nu^*$, every ray in $\mathrm{con}(M_\nu)$ will intersect $W$, and hence will intersect the boundary $\partial C$ of $C$ at a finite point. It follows that $\theta(M_\nu) < \infty \ \ \forall \nu \geq \nu^*$, and, since all of the $M_\nu$ are of second category, $\mu(M_\nu) = +\infty \ \ \forall \nu \geq \nu^*$ (cf. Definition 1).

As an intermediate result, we prove that $\mu(M_\nu) = +\infty$ implies $\Gamma \subset D$. Indeed, in the contrary case, because $w \in \mathrm{int}\, D$, the half-line $w + \Gamma$ would intersect the boundary $\partial D$ of $D$ at some point $a \neq w$ (recall that $w$ is the point mentioned in assumption (ii)). But $\mu(M_\nu) = +\infty$ implies that $x^\nu$ is a recession direction of $D_\nu \cap \mathrm{con}(M_\nu) \subset \mathrm{con}(M_\nu)$; and hence $x^\nu$ must tend to the direction of $\Gamma$ as $\nu \to \infty$. Furthermore, if $\Gamma \not\subseteq D$, then obviously $\Gamma$ is not a direction of recession of $D$. Since the recession cone of $D$ is closed (cf. [**4**]), it follows that for all sufficiently large $\nu$, $x^\nu$ is not a direction of recession of $D$. Therefore, the half-line originating from $w$ parallel to $x^\nu$ intersects $\partial D$ at a point $\pi(x^\nu)$. Clearly $\pi(x^\nu) \to a$, and hence, by passing to a subsequence if necessary, $t^\nu \to t \in \partial g(a)$, where $t^\nu$ is the subgradient in (10) (for the corresponding properties of sequences of subgradients, see [**4**]). By definition of a subgradient, and since $g(a) = 0 > g(w)$, we have $t^T(a - w) \geq g(a) - g(w) > 0$, and there is $\alpha > 0$ such that $t^T(\alpha(a-w) - a) = \alpha t^T(a-w) - t^T a > 0$. It follows that for sufficiently large $\nu$ we also have $(t^\nu)^T(\alpha(a-w) - \pi(x^\nu)) > 0$. In other words: the half-space $L_\nu := \{x : t^\nu(x - \pi(x^\nu)) \leq 0\}$ (which contains $D_k$ for $k > \nu$) does not contain the point $\alpha(a-w) \in \Gamma$, i.e., $L_\nu \cap \Gamma$ is contained in the line segment $[0, \alpha(a-w)]$. As $\mathrm{con}(M_\nu)$ shrinks to $\Gamma$, we infer that for sufficiently large $\nu$ the set $L_\nu \cap \mathrm{con}(M_\nu)$ (which contains $D_\nu \cap \mathrm{con}(M_\nu)$) is bounded. This contradicts $\mu(M_\nu) = +\infty$; and hence we must have $\Gamma \subset D$.

Now, in proceeding in the proof of the assertion $\Gamma \subset C$, since $\Gamma\backslash C$, is a half-line while $w \in C \cap \mathrm{int}\, D$, we see that $(w + \Gamma)\backslash C$ is also a half-line, and there is a ball $B$ centered at a point $b \in (w+\Gamma)\backslash C$ such that $B \subset D\backslash C$. But since $x^\nu$ tends to a direction of $\Gamma$, it is clear that for sufficiently large $\nu$ the half-line from $w$ parallel to $x^\nu$ must intersect $B$, and hence the algorithm should have terminated in Step 1). This contradiction shows that $\Gamma \subset C$.

Finally, since $0 \in \operatorname{int} C$, for any positive number $\eta$ we can construct a ball $U$ centered at some point of $\Gamma$ such that $U \subset C \setminus \eta T$. As $\operatorname{con}(M_\nu)$ shrinks to $\Gamma$, for sufficiently large $\nu$ any ray in $\operatorname{con}(M_\nu)$ will intersect $U$. This implies that $\theta(M_\nu) \geq \eta$ for all these $\nu$, and this completes the proof. □

PROPOSITION 3. *If the algorithm generates infinitely many $M_k$ of second category, then GCP has no solution.*

PROOF. First observe that, if $M_k$ is of second category, then GCP has no solution in $T_k := \theta(M_k)T$. Indeed, $\mathscr{R}_{k\mathrm{I}}$ is empty and $\theta(M) \geq \theta(M_k)$ for all $M \in \mathscr{R}_k$. If $y \in T_k$, then either $y$ belongs to $\operatorname{con}(M)$ for some simplex $M$ deleted at an iteration $h \leq k$ (in which case $y$ cannot be a solution), or $y$ belongs to $\operatorname{con}(M)$ for some $M \in \mathscr{R}_k$, and then $y \in (\theta(M)T) \cap \operatorname{con} M \subset C$. This proves our observation.

Now, if the algorithm generates infinitely many simplices of second category, then by a simple standard argument, it is easy to see that there exists an infinite nested sequence $\{M_{k_\nu}\}$ of simplices of second category. Then for any $\nu$ no solution of GCP exists in $\theta(M_{k_\nu})T$, and since $\theta(M_{k_\nu}) \to +\infty$ (by Lemma 2), this means that GCP has no solution. □

LEMMA 3. *For any infinite nested sequence $\{M_{k_\nu}\}$ of simplices of first category we have $\mu(M_{k_\nu}) \downarrow 1$ as $\nu \to +\infty$.*

PROOF. It is easy to see that the sequence $\mu(M_k)$ is nonincreasing as $k \uparrow \infty$. Indeed, if $M_h \subset M_k$, then $h > k$; and hence $D_h \subset D_k$. Furthermore, every point $z_{hi} = \theta_{hi} s^{hi}$ $(i = 1, \ldots, n)$ satisfies $\psi_k(z^{hi}) \geq 1$, where $\psi_k(x) = 1$ is the equation of the hyperplane through $z^{k1}, \ldots, z^{kn}$. Since $\psi_h(z^{hi}) = 1$ $(i = 1, \ldots, n)$ and $\psi_h(0) = \psi_k(0)$, it follows from the linearity of $\psi_h(\cdot)$ that $\psi_h(x) \leq \psi_k(x)$ for all $x$ in $\operatorname{con}(M_h)$. Therefore, $\max\{\psi_h(x) : x \in D_h \cap \operatorname{con}(M_h)\} \leq \max\{\psi_k(x) : x \in D_k \cap \operatorname{con}(M_k)\}$, i.e., $\mu(M_h) \leq \mu(M_k)$, proving the monotonicity of $\mu(M_k)$.

Now, let $\{M_\nu\}$ be an infinite nested sequence of simplices of first category. For each $\nu$, denote by $v^\nu$ the intersection of the simplex $[z^{\nu 1}, \ldots, z^{\nu n}]$ with the half-line from 0 through $x^\nu$. Since $\operatorname{con}(M_\nu)$ shrinks to a ray $\Gamma$, it follows that $v^\nu$ and $x^\nu$ tend to a common limit $x^*$ which is the intersection of $\Gamma$ with $\partial C$. (Recall that $z^{\nu i} = \theta_{\nu i} s^{\nu i}$ is the intersection of the half-line from 0 through $s^{\nu i}$ with $\partial C$.) Therefore, $x^\nu - v^\nu \to 0$, and hence $\mu(M_\nu) = \|x^\nu\| / \|v^\nu\| = 1 + (\|x^\nu\| - \|v^\nu\|)/\|v^\nu\| \to 1$. □

PROPOSITION 4. *If the algorithm is infinite but generates only finitely many simplices $M_k$ of second category, then any accumulation point $x^*$ of the sequence $\{z^k\}$ belongs to $\partial C \cap \partial D$.*

PROOF. Let $x^*$ be any accumulation point of the sequence $\{x^k\}$; for example, $x^* = \lim x^k$ $(k \in \Delta,\ k \to \infty)$. There is in $\mathscr{M}_0$ at least an $M_{k_0}$ such that $\mathrm{con}(M_{k_0})$ contains infinitely many $x^k$ with $k \in \Delta$. Then there is among the immediate successors of $M_{k_0}$ at least an $M_{k_1}$ such that $\mathrm{con}(M_{k_1})$ contains infinitely many $x^k$ with $k \in \Delta$; and so forth. Thus we can find an infinite nested sequence $\{M_{k_\nu},\ \nu = 1, 2, \dots\}$ such that $\lim x^{k_\nu} = x^*$. Hence, as seen in the proof of Lemma 3, we have $x^* \in \partial C \cap \Gamma$. It remains to show that $x^* \in \partial D$. To simplify notation, we write again $x^\nu$ instead of $x^{k_\nu}$. It suffices to consider the case when $x^\nu \notin \partial D$ for all $\nu$. Since the sequence $\{x^\nu\}$, and hence the sequence $\{\pi(x^\nu)\}$, are bounded, it follows that the sequence $\{t^\nu \in \partial g\pi(x^\nu)\}$ is bounded, too. Therefore, we may assume, by passing to a subsequence if necessary, that $t^\nu \to t \in \partial g(\pi(x^*))$ (for the corresponding properties of sequences of subgradients, see [**4**]). Since for $\nu' > \nu$ we have $x^{\nu'} \in D_\nu$, and hence $(t^\nu)^T(x^{\nu'} - \pi(x^\nu)) \le 0$, we see, by letting $\nu' \to \infty$ (for $\nu$ fixed), that $(t^\nu)^T(x^* - \pi(x^\nu)) \le 0$. From this inequality follows (for $\nu \to \infty$) that $t^T(x^* - \pi(x^*)) \le 0$. On the other hand, since $x^\nu - \pi(x^\nu) = \alpha(w - \pi(x^\nu))$ for some $\alpha < 0$, we have

$$(t^\nu)^T(x^\nu - \pi(x^\nu)) = \alpha(t^\nu)^T(w - \pi(x^\nu)) \ge -\alpha[g(w) - g(\pi(x^\nu))] > 0,$$

and hence (letting $\nu \to \infty$) $t^T(x^* - \pi(x^*)) \ge 0$. Therefore, $t^T(x^* - \pi(x^*)) = 0$. This implies $x^* = \pi(x^*)$, because otherwise $x^* - \pi(x^*) = \alpha^*(w - \pi(x^*))$ for some $\alpha^* < 0$, and the definition of a subgradient would imply that $t^T(w - \pi(x^*)) \le g(w) - g(\pi(x^*)) = g(w) < 0$. This proves that $x^* \in \partial D$. □

PROPOSITION 5. *Suppose that $C$ and $D$ are bounded. If the algorithm is finite, then $D \subset C$, and every accumulation point $x^*$ of the sequence $\{x^k\}$ belongs to $\partial C \cap \partial D$.*

PROOF. If both $C$ and $D$ are bounded, then every $M_k$ generated during the procedure is of first category and the second assertion follows from Proposition 4.

Now suppose that there exists a point $x \in D \backslash C$. Then $x$ must be contained in some $\mathrm{con}(M)$ with $M \in \mathscr{M}_0$. Clearly, for this simplex $M$, we must have $\mu(M) > 1$, because otherwise $D_0 \cap \mathrm{con}(M) \subset C$. Since $\mu(M_k) \downarrow 1$ (cf. Lemma 3), there must be some $k_1$ such that $M = M_{k_1}$. Let $\Delta = \{k : x \in \mathrm{con}(M_k)\}$. It follows that $\mu(M_k) > 1$ for all $k \in \Delta$. If $\Delta$ were finite, say $\Delta = \{k_1, \dots, k_h\}$, then we would have $\mu(M_{k_h}) \le \mu(M_k)$ for all $k > k_h$ (cf. the definition of $\mu(M)$). But this is impossible, because we know from Lemma 3 that $\mu(M_k) \downarrow 1$ as $k \to \infty$. Therefore, $\Delta$ must be infinite. Obviously the sequence $\{M_k : k \in \Delta\}$ is nested (monotonically decreasing). Let $z^\nu$ denote the intersection of the half-line from 0 through $x$ with the simplex $[z^{k1}, \dots, z^{kn}]$, $k \in \Delta$. Since $x \in D \cap \mathrm{con}(M_k)$, we must have

$$\mu(M_k) - 1 \geq \|x - z^k\|/\|z^k\|.$$

But as $k \to +\infty$ $(k \in \Delta)$, the $z^{ki}$ tends to a common limit $z^* \in \partial C \cap \Gamma$, which is also the limit of $\{z^k\}$. Then the above inequality yields $0 \geq \|x - z^*\|/\|z^*\| > 0$, a contradiction. Therefore, $D \subset C$, which completes the proof. □

We can summarize the above results as follows.

THEOREM 1. *Let $C$ be open. If GCP is solvable, then either the algorithm finds a solution after finitely many steps or it generates an infinite sequence $\{x^k\}$ every accumulation point of which yields a solution.*

PROOF. This follows from Propositions 3 and 4. □

THEOREM 2. *Let $C$ be closed and let $C'$ be any open convex set containing $C$. If instead of (5) we take*

$$\theta_i = \sup\{\theta : \theta s^i \in C'\},$$

*then either $D \subset C'$ or the algorithm finds a solution of GCP after finitely many steps.*

PROOF. The modification amounts to applying the algorithm to $C'$, $D$ instead of $C$, $D$. By the previous theorem, if $D \backslash C'$ is nonempty, then every accumulation point $x^*$ of the sequence $\{x^k\}$ belongs to $D \backslash C'$, hence for sufficiently large $k$ we must have $x^k \in D \backslash C$. □

In practice, $C'$ can be chosen as a certain approximation of $C$. For instance, if $C = \{x : f(x) \leq \gamma\}$, where $f(x)$ is a convex function, then $C' = \{x : f(x) < \gamma + \varepsilon \ (\varepsilon > 0)\}$ and the situation $D \subset C'$ means that $f(x) < \gamma + \varepsilon \ \forall x \in D$.

THEOREM 3. *Let $C$ be open and bounded. If the algorithm is infinite, then GCP has a solution, and every accumulation point of the sequence $\{x^k\}$ solves GCP.*

PROOF. Indeed, $\theta(M_k) < +\infty$ for all $k$ so that, by Proposition 2, the algorithm can generate only finitely many $M_k$ of second category. The conclusion then follows from Proposition 4. □

THEOREM 4. *Let $C$ and $D$ be compact. If GCP is solvable, then the algorithm finds a solution after finitely many steps.*

PROOF. This follows from Proposition 5. □

## REFERENCES

1. R. Horst and N. V. Thoai, *Modification, implementation and comparison of three algorithms for globally solving linearly constrained concave minimization problems*, Computing **42** (1989), 271–289.
2. R. Horst and H. Tuy, *Global optimization (Deterministic approaches)*, Springer-Verlag, Berlin, 1990.

3. K. G. Murty, *Linear complementarity, linear and nonlinear programming*, Heldermann, Berlin, 1988.
4. R. T. Rockafellar *Convex analysis*, Princeton Univ. Press, Princeton, N.J., 1970.
5. P. T. Thach and H. Tuy, *The relief indicator method for constrained global optimization*, Naval Res. Logist. Quart. **37** (1990), 473–497.
6. H. Tuy, *Normal conical algorithm for concave minimization over polytopes*, Preprint, Institute of Mathematics, Hanoi, 1988 (submitted).

FACHBEREICH IV–MATHEMATIK, UNIVERSITY OF TRIER, P.O. BOX 3825, 5500 TRIER, FEDERAL REPUBLIC OF GERMANY
*E-mail address*: horst @ uni-trier.dbp.de

INSTITUTE OF MATHEMATICS, VIEN TOAN HOC, P.O. BOX 631, BO-HO, 10000 HANOI, VIETNAM

DIMACS Series in Discrete Mathematics
and Theoretical Computer Science
Volume **4**, 1991

Applied Geometry
and Discrete Mathematics
THE VICTOR KLEE FESTSCHRIFT

# Every Tree is Graceful (But Some are More Graceful than Others)

T. C. HU AND A. B. KAHNG

ABSTRACT. The Ringel-Kotzig conjecture that "every tree is graceful" remains unsolved. In this note, we propose a new equivalent conjecture, namely, the chain is the hardest tree of $n$ nodes to be gracefully labeled for all $n \geq 7$. Since all chains can be gracefully labeled, this new conjecture may provide an alternative approach to settling the original conjecture.

The Ringel-Kotzig conjecture that "every tree is graceful" was formulated between 1963 and 1966 [**2, 3**] and remains unresolved. For a survey of known results, see Bloom [**1**]. In this note, we propose a new equivalent conjecture, namely, "for $n \geq 7$, the chain is the hardest order-$n$ tree to gracefully label". Since it is well known that all chains can be gracefully labeled, this will add more weight toward a positive result for the original conjecture.

First, define a tree of order $n$ to be a connected graph with $n$ vertices and $n-1$ edges. A chain is a tree with two vertices of degree 1. We assign the labels (numbers) $1, 2, \ldots, n$ to the vertices $v_1, v_2, \ldots, v_n$ of the tree. This induces an edge labeling where each edge receives a label that is the absolute difference of the labels on its two incident vertices. A labeling of the vertices is called "graceful" if all induced edge labels are distinct. The Ringel-Kotzig conjecture is that every finite tree has a graceful labeling.

For example, both trees of order six in Figure 1 have graceful labelings. In Figure 1(a), the vertex $v_1$ receives the label 2, and $v_2$ gets the label 4, while in Figure 1(b), vertex $v_1$ gets the label 1 and $v_2$ is labeled with 6. If we ignore symmetries, then every tree of order $n$ has $n!$ labelings, of which some are graceful and some are not.

1980 *Mathematics Subject Classification* (1985 *Revision*). Primary 05C05.

This work was supported by NSF Grant MIP-8700767 and by a grant from the State of California MICRO program.


1052-1798/91 \$1.00 + \$.25 per page

(a)

(b)

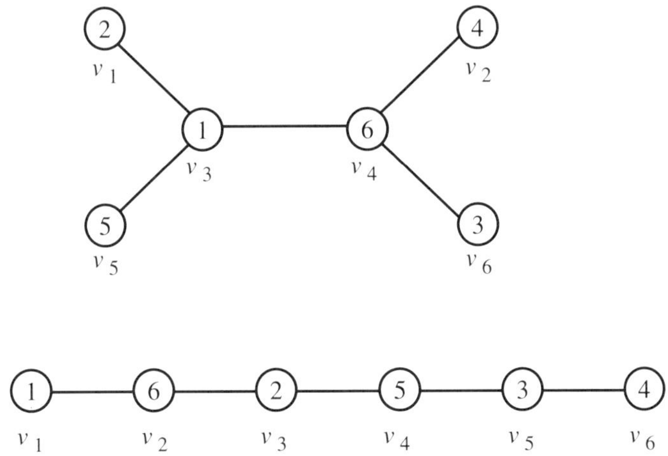


FIGURE 1

For a given tree, let $N_p$ be the number of graceful labelings, and let $N_q$ be the number of nongraceful labelings. Then

$$N_p + N_q = n! \tag{1}$$

and the original conjecture can be stated as

$$\frac{N_p}{n!} \neq 0 \tag{2}$$

for an arbitrary tree of order $n$.

Now let $N_p(C_n)$ be the number of graceful labelings for the chain of $n$ nodes, and let $N_p(T_n)$ be the number of graceful labelings of a tree of $n$ nodes. Our conjecture is that

$$\frac{N_p(T_n)}{n!} \geq \frac{N_p(C_n)}{n!} \quad \text{for } n \geq 7. \tag{3}$$

In other words, the chain is in some sense the most difficult tree to gracefully label. Since we know that

$$\frac{N_p(C_n)}{n!} \neq 0$$

for arbitrary $n$, a proof of (3) will settle the original conjecture.

We have verified (3) empirically for $n$ up to 14 and there are many heuristic justifications for (3) to hold. In particular, the study of $N_p(T_i)$ over sequences of trees

$$\{T_0, T_1, \dots, T_{k-1}, T_k = C_n\},$$

where $T_{i+1}$ is obtained from $T_i$ by pruning and reattaching a single leaf node, seems fruitful.

Finally, we note that a probabilistic result like "Almost all trees are graceful" should be possible using the notion of $N_p(T)$. A summary of sample enumeration results is appended in Table 1.

TABLE 1

Average number of graceful labelings of the $(k)$ $N$-trees which are distance $d$ away from the $N$-chain (distance = minimum number of leaf attachments).

| $N$ | $d=0$ | $d=1$ | $d=2$ | $d=3$ | $d=4$ | $d=5$ | $d=6$ | $d=7$ |
|---|---|---|---|---|---|---|---|---|
| 7 | 32 (1) | 40 (2) | 54.40 (5) | 180 (2) | 1440 (1) | | | |
| 8 | 40 (1) | 92 (3) | 121.142 (7) | 175.5 (8) | 816 (3) | 10080 (1) | | |
| 9 | 120 (1) | 176 (3) | 272.363 (11) | 407.714 (14) | 746.286 (14) | 5536 (3) | 80640 (1) | |
| 10 | 296 (1) | 411.5 (4) | 603.714 (14) | 951.793 (29) | 1536.94 (32) | 3213.33 (21) | 35496 (4) | 725760 (1) |
| 11 | 648 (1) | 1011 (4) | 1523.158 (19) | | | | | |
| 12 | 1328 (1) | 2573.6 (5) | 4103.13 (23) | | | | | |
| 13 | 3200 (1) | 6689.6 (5) | 11149.32 (29) | | | | | |
| 14 | 9912 (1) | 17576.66 (6) | 31003.647 (34) | | | | | |

## REFERENCES

1. G. S. Bloom, *A chronology of the Ringel-Kotzig conjecture and the continuing quest to call all trees graceful*, Ann. New York Acad. Sci. **326** (1979), 32–51.
2. G. Ringel, *Problem* 25: *Theory of graphs and its applications*, Proc. Internat. Sympos. Smolenice (June 1963), Czech. Acad. Sci., Prague, p. 162.
3. A. Rosa, *On certain valuations of the vertices of a graph*, Theory of Graphs, Gordon and Breach, New York, 1967, pp. 349–355.

DEPARTMENT OF COMPUTER SCIENCES AND ENGINEERING, UNIVERSITY OF CALIFORNIA, SAN DIEGO, CALIFORNIA 92093

*E-mail address*: hu@ cs.ucsd.edu.

DEPARTMENT OF COMPUTER SCIENCE, UNIVERSITY OF CALIFORNIA, LOS ANGELES, CALIFORNIA 90024

*E-mail address*: Kahng@ cs.ucla.edu.

DIMACS Series in Discrete Mathematics
and Theoretical Computer Science
Volume **4**, 1991

Applied Geometry
and Discrete Mathematics
THE VICTOR KLEE FESTSCHRIFT

# Qualitative Analysis of Schur Complements

CHARLES R. JOHNSON AND JOHN MAYBEE

ABSTRACT. We present some general techniques for making qualitative statements about particular entries of a Schur complement. These are used to completely determine, in terms of path products, when an entry of a Schur complement is unambiguously sign determined in the purely qualitative case. Uses in the presence of certain quantitative information are also mentioned.

## Introduction

If

$$A = \begin{bmatrix} A_{11} & A_{12} \\ A_{21} & A_{22} \end{bmatrix}$$

is a partitioned square matrix with $A_{22}$ square and nonsingular, then

$$\begin{bmatrix} I & -A_{12}A_{22}^{-1} \\ 0 & I \end{bmatrix} A = \begin{bmatrix} A_{11} - A_{12}A_{22}^{-1}A_{21} & 0 \\ A_{21} & A_{22} \end{bmatrix},$$

so that $\det A = \det A_{22} \det(A_{11} - A_{12}A_{22}^{-1}A_{21})$. Since **[H]** appeared, the matrix

$$S = (A/A_{22}) \equiv A_{11} - A_{12}A_{22}^{-1}A_{21}$$

has been referred to as the *Schur complement* of $A_{22}$ in $A$, as Schur **[S]** explicitly noted the above determinantal identity for $A$. In several guises **[Co]**, the Schur complement has proved an enormously convenient and useful tool in a remarkable variety of applications, including differential and difference equations and systems theory, economic modelling **[BMNV]**, and statistics **[O]**, as well as matrix theory. See the survey **[Ca]** for an algebraic treatment of some aspects of Schur complements, as well as further indications of their ubiquity. The survey **[O]** is very useful.

In general, a Schur complement may be formed with respect to any nonsingular principal submatrix of an $n \times n$ matrix $A$. (A generalized Schur

1980 *Mathematics Subject Classification* (1985 *Revision*). Primary 15A15, 05C50.

The work of the first author was supported in part by National Science Foundation grant number DMS 90-00839 and by Office of Naval Research grant number N00014-90-J-1739.

The work of the second author was supported in part by Office of Naval Research grant number N00014–88-K-0087.

complement with respect to a singular or nonsquare submatrix may also be formed by using a generalized inverse **[Ca, O]**, but we shall not be concerned with that here.) In order to discuss Schur complements with respect to general principal submatrices, we adopt the following notation. Let $N = \{1, 2, \ldots, n\}$. If $\alpha$ and $\beta$ are nonempty subsets of $N$, we denote the submatrix lying in rows $\alpha$ and columns $\beta$ of an $n \times n$ matrix $A = (a_{ij})$ by $A[\alpha, \beta]$; when $\beta = \alpha$, this submatrix is principal, and we abbreviate $A[\alpha, \alpha]$ to $A[\alpha]$. Let $\alpha^c$ denote the complement of $\alpha \subseteq N$ with respect to $N$. If $A[\alpha]$ is nonsingular, then

$$S(\alpha) = (A/A[\alpha]) \equiv A[\alpha^c] - A[\alpha^c, \alpha]A[\alpha]^{-1}A[\alpha, \alpha^c]$$

is the Schur complement of $A[\alpha]$ in $A$. If $i, j \in N$ and $i, j \notin \alpha \subseteq N$, we denote the $i, j$ entry of $S(\alpha)$ as $S(\alpha)_{ij}$, in which we imagine that the rows and columns of $S(\alpha)$ are naturally indexed by $\alpha^c$.

Our purpose here is to describe situations in which certain entries of $S(\alpha)$ have signs unambiguously determined by the signs of the entries of $A \in M_n(\mathbb{R})$. For example, if $n = 2$ and the sign pattern of $A$ is $\left[\begin{smallmatrix} + & - \\ + & + \end{smallmatrix}\right]$, then $S(\{2\}) = (+) - (-[+]+) = +$, unambiguously. Despite several dozen papers about Schur complements (see the references in **[O]** and **[Ca]**, for example) and a number of applications in which such information could be useful, no such qualitative analysis of Schur complements seems to have yet been undertaken. We note in advance that there are many parallels between qualitative analysis of Schur complements and the study of sign nonsingular matrices (or $L$-matrices), as studied by several authors **[BMQ, KLM]**. See the definition to follow.

## Fundamental algebraic formulae

From the definition of the Schur complement, the $i, j$ entry of $S(\alpha)$ is simply

$$(1) \qquad S(\alpha)_{ij} = a_{ij} - A[i, \alpha]A[\alpha]^{-1}A[\alpha, j].$$

Here, as throughout, we benignly identify $i$ with $\{i\}$.

Our strategy is to derive further algebraic formulae for entries of the Schur complement that allow us to deduce sign determinacy from statements about the sign and sparsity pattern of $A$.

We assume that $\alpha \subseteq N$, $i, j \in \alpha^c$, $A$ is an $n \times n$ matrix, and $\det A[\alpha] \neq 0$. A natural starting point is to calculate

$$\begin{aligned} \det \begin{bmatrix} a_{ij} & A[i, \alpha] \\ A[\alpha, j] & A[\alpha] \end{bmatrix} &= a_{ij} \det A[\alpha] - A[i, \alpha] \operatorname{adj} A[\alpha]A[\alpha, j] \\ &= \det A[\alpha](a_{ij} - A[i, \alpha]A[\alpha]^{-1}A[\alpha, j]) \\ &= \det A[\alpha]S(\alpha)_{ij}. \end{aligned}$$

We conclude that

$$\text{(2)} \qquad S(\alpha)_{ij} = \det\begin{bmatrix} a_{ij} & A[i,\alpha] \\ A[\alpha,j] & A[\alpha] \end{bmatrix} \Big/ \det A[\alpha].$$

At this point, it is clear that if $A[\alpha]$ is sign nonsingular (so that $\det A[\alpha]$ is unambiguously signed $+$ or $-$) and if $\det\left[\begin{smallmatrix} a_{ij} & A[i,\alpha] \\ A[\alpha,j] & A[\alpha] \end{smallmatrix}\right]$ is unambiguously signed ($+$, $-$, or 0), then the $i, j$ entry of $S(\alpha)$ is unambiguously signed. It is also clear that $A[\alpha]$ must be sign nonsingular in order that $S(\alpha)$ be defined for all matrices with the same sign pattern as $A$.

In order to assess the role of the sign pattern of $A$ on the sign determinacy of $S(\alpha)_{ij}$, we wish to express

$$\det\begin{bmatrix} a_{ij} & A[i,\alpha] \\ A[\alpha,j] & A[\alpha] \end{bmatrix}$$

in terms of "path products" (from $i$ to $j$ via indices in $\alpha$). First, note that

$$\text{(3)} \qquad \det A[\alpha \cup \{i\}, \alpha \cup \{j\}] = (-1)^{\tau(i)+\tau(j)+1} \det\begin{bmatrix} a_{ij} & A[i,\alpha] \\ A[\alpha,j] & A[\alpha] \end{bmatrix}$$

in which $\tau(k)$ is the position of the positive integer $k$ in the ordered set $\alpha \cup \{i, j\}$. But $\det A[\alpha \cup \{i\}, \alpha \cup \{j\}]$ may also be expressed in terms of path products from $A[\alpha \cup \{i\}, \alpha \cup \{j\}]$, based upon a general formula found in [**MOvW**]. By a path in $A$ from $i$ to $j$ via $\alpha$ we mean a sequence of nonzero entries $a_{it_1}, a_{t_1t_2}, \ldots, a_{t_qj}$ of $A$ in which no index is repeated among $t_1, \ldots, t_q$ and $\{t_1, \ldots, t_q\} \subseteq \alpha$. We also allow the singleton $a_{ij}$. Let $m$ be the number of distinct such paths, listed as $p_1, p_2, \ldots, p_m$ and let $V_\alpha(p_k)$ be the complement, with respect to $\alpha$, of the indices appearing in the $k$th path, $k = 1, \ldots, m$. Also, let $A(p_k)$ be the product of the entries from $A$ associated with the path $p_k$, $k = 1, \ldots, m$. From [**MOvW**] we then have

$$\text{(4)} \qquad \det A[\alpha \cup \{i\}, \alpha \cup \{j\}] = (-1)^{\tau(i)+\tau(j)} \sum_{k=1}^{m} (-1)^{l_k} A(p_k) \det A[V_\alpha(p_k)],$$

in which $l_k$ is the length of the path $p_k$ measured in terms of the number of entries of $A$. Equating the right-hand sides of (3) and (4), we have

$$\text{(5)} \qquad \det\begin{bmatrix} a_{ij} & A[i,\alpha] \\ A[\alpha,j] & A[\alpha] \end{bmatrix} = \sum_{k=1}^{m} (-1)^{l_k+1} A(p_k) \det A[V_\alpha(p_k)],$$

and therefore by (2)

$$\text{(6)} \qquad S(\alpha)_{ij} = \frac{1}{\det A[\alpha]} \sum_{k=1}^{m} (-1)^{l_k+1} A(p_k) \det A[V_\alpha(p_k)].$$

Note that when $j = i$, we have from (2) that $S(\alpha)_{ii} = \det A[\alpha \cup i] / \det A[\alpha]$. In this case the path products $A(p_k)$ are cycle products, and the numerator of the right-hand side of (6) is then an expression for $\det A[\alpha \cup i]$.

Equation (6), and others, can be exploited to obtain information about an entry or entries of a Schur complement in a variety of settings. In the "purely qualitative case" in which nothing about $A$ is assumed except a known sign pattern, the analysis resembles, in part, that of matrices whose determinant is sign determined. However, an important direction for qualitative analysis is the assimilation of some quantitative information. We mention examples of this type of use also.

## The purely qualitative case

By a *sign pattern matrix* we mean a rectangular array each of whose entries is one of the symbols $+$, $-$, or $0$. Naturally associated with a sign pattern matrix is the class of all real matrices, the pattern of the signs of the entries of which is the given sign pattern matrix. If the array is square and if every real matrix in this sign pattern class is nonsingular, we call the sign pattern matrix *sign nonsingular.* By continuity, it is clear that every determinant in the class must have the same sign in this event.

The sign nonsingular matrices have received considerable study [**BMO, KLM**]. It is not hard to show, for example, that for a matrix to be sign nonsingular every term in its determinant expansion must be (weakly) of the same sign. Of course, at least one term must be nonzero. Thus, up to permutation equivalence, all diagonal entries of a sign nonsingular matrix are nonzero, and every cycle product is (weakly) of the same sign as the corresponding diagonal product. It has proved convenient in the study of sign nonsingular matrices to further assume (without loss of generality, via a signature multiplication) that all diagonal entries are negative. We say that such a sign nonsingular matrix is in *standard form* and note that any sign nonsingular matrix may be put in standard form via (right) permutation and signature multiples. Of course, such form is not generally unique and may typically be accomplished in many ways.

Our interest here is in describing, for a given $n$ by $n$ sign pattern matrix $A$, given $\alpha \subseteq N$ and $i, j \in N - \alpha$, when the $i, j$ entry of $S(\alpha)$ is sign determined. This occurrence is not related in any simple way to whether $A$ is sign nonsingular, but concepts from sign nonsingularity are importantly involved. It is clear from (2) or (6) that it is necessary that $A[\alpha]$ be sign nonsingular. For exposition of our primary observation, it is convenient to further assume that $A[\alpha]$ is in standard form. In a sense, there is no loss of generality in this assumption, as any sign nonsingular matrix may be put in this form (e.g., via right multiplication by a product of a permutation and a signature matrix), and the effect of the necessary transformation may then be taken into account. However, it is interesting to further note that right nonsingular multiplication of the $\alpha$ columns of $A$ has no effect upon $S(\alpha)$. For the moment, we make more precise our general Schur complement notation by explicitly designating the object matrix, so that

$$S_A(\alpha) \equiv (A/A[\alpha]).$$

It is then straightforward to check by partitioned calculation that if $X \in M_n$ is nonsingular, with $X[\alpha, \alpha^c] = 0$ and $X[\alpha^c, \alpha] = 0$, then

$$S_{AX}(\alpha) = S_A(\alpha)X[\alpha^c].$$

Thus, if, in addition, $X[\alpha^c] = I$, we have

$$S_{AX}(\alpha) = S_A(\alpha).$$

So, if $X = \left[\begin{smallmatrix} I & 0 \\ 0 & Y \end{smallmatrix}\right]$ with $X[\alpha] = Y$ the product of a permutation and signature matrix that puts $A[\alpha]$ in standard form (for example), then $S(\alpha)$ is unchanged by this normalization.

Our primary result then is

THEOREM 1. *Let $A$ be an $n \times n$ sign pattern matrix, $\alpha \subseteq N$, and $i, j \in \alpha^c$. If $A[\alpha]$ is a sign nonsingular matrix in standard form, then the $i, j$ entry of the Schur complement of $A[\alpha]$ in $A$ is sign determined if and only if every product of entries along a path from $i$ to $j$ via $\alpha$ has the same sign* $(+, -,$ *or* $0)$.

PROOF. Use equation (6). Note first that $S(\alpha)_{ij}$ is unambiguously 0 if and only if $m = 0$. Hence, we may assume $m \geq 1$. Since $A[\alpha]$ is in standard form, the sign of $\det A[V_\alpha(p_k)]$ is $(-1)^{|\alpha|-l_k+1}$, since this exponent is the number of indices in $V_\alpha(p_k)$. Also, the sign of $\det A[\alpha]$ is $(-1)^{|\alpha|}$. According to (6), then

$$\begin{aligned} S(\alpha)_{ij} &= (-1)^{|\alpha|} \sum_{k=1}^{m} (-1)^{l_k+1} (-1)^{|\alpha|-l_k+1} A(p_k) \left| \frac{\det A[V_\alpha(p_k)]}{\det A[\alpha]} \right| \\ &= (-1)^{2(|\alpha|+1)} \sum_{k=1}^{m} (-1)^{l_k - l_k} A(p_k) \left| \frac{\det A[V_\alpha(p_k)]}{\det A[\alpha]} \right| \\ &= \sum_{k=1}^{m} A(p_k) \left| \frac{\det A[V_\alpha(p_k)]}{\det A[\alpha]} \right| . \end{aligned}$$

Thus, if each $A(p_k)$ has a common sign, $S(\alpha)_{ij}$ has that sign unambiguously. Conversely, since $m > 1$, if $S(\alpha)_{ij}$ is unambiguously signed, then, according to (2), $A[\{i\} \cup \alpha, \{j\} \cup \alpha]$ must be sign nonsingular. But, then, since (5) is an expansion of its determinant, every term on the right-hand side of (5) must have the same sign. Since $(-1)^{l_k+1} \det A[V_\alpha(p_k)]$ has the constant sign $(-1)^{|\alpha|}$, this means each $A(p_k)$ must have the same sign, which completes the proof. □

We note that if $a_{ij}$ is signed (+ or −) it constitutes a path from $i$ to $j$ (vacuously) via $\alpha$. In this event, if $A[\alpha]$ is standard form sign nonsingular, then $S(\alpha)_{ij}$ is sign determined exactly when every path product from $i$ to

$j$ via $\alpha$ has the sign of $a_{ij}$. Of course, if $a_{ij} = 0$, several eventualities may occur: no paths, in which case $S(\alpha)_{ij}$ is unambiguously 0; path products of both signs $+$ or $-$, in which case the sign of $S(\alpha)_{ij}$ is ambiguous; or all paths "line up" with products of a common sign ($+$ or $-$), in which case the sign of $S(\alpha)_{ij}$ is unambiguously determined.

## Mixed qualitative quantitative cases

In a variety of situations, some quantitative, information is available. In some such situations, interesting qualitative conclusions may be drawn exploiting, in part, the quantitative information. A familiar such example is the fact that the inverse of an $M$-matrix **[HJ]** is component-wise nonnegative. We wish to note here that our basic observations may also be used for such purposes.

An $n \times n$ real matrix $A$ is called a *P-matrix* if every principal minor of $A$ is positive, i.e., $\det A[\alpha] > 0$ whenever $\emptyset \neq \alpha \subseteq N$. Familiar examples of $P$-matrices include the $M$-matrices, positive definite matrices, and totally positive matrices. Of course, the property of being a $P$-matrix may involve considerable quantitative information, and the Schur complement with respect to any index set is well defined for each $P$-matrix. Application of equations such as (6) is relatively straightforward in the presence of this sort of quantitative information. There is no complication from the sign of the denominator.

We say that the paths in $A$ from $i$ to $j$ via $\alpha$ are *positively aligned* if the sign of every such path product is the same as $(-1)^l$ in which $l$ is the length, or number of entries from $A$, in the path. Similarly, these paths are *negatively aligned* if all the path products have the sign $(-1)^{l+1}$. In either of these cases (or if there are no paths at all), we refer to the paths as *aligned*, and otherwise we call them *nonaligned*. From (6) we see that if $A$ is a $P$-matrix and the paths from $i$ to $j$ via $\alpha$ are positively aligned, then $S(\alpha)_{ij}$ is unambiguously negative. If the paths are negatively aligned, then $S(\alpha)_{ij}$ is positive, and, of course, if there are no paths, then $S(\alpha)_{ij} = 0$. If the paths are nonaligned, then it is possible to produce examples in which positive or negative signs occur. (We note that all of this only uses $\det A[\alpha] > 0$ for the particular $\alpha$ in question.) We summarize these remarks in

THEOREM 2. *Let $B$ be an $n \times n$ sign pattern matrix, $\alpha \subseteq N$, and $i, j \in \alpha^c$. Then, the $i, j$ entry of the Schur complement of $A[\alpha]$ in $A$ is sign determined for every $n \times n$ P-matrix $A$ with the sign pattern of $B$ if and only if the paths in $B$ from $i$ to $j$ via $\alpha$ are aligned.*

In an $M$-matrix (a $P$-matrix with nonpositive off-diagonal entries), the paths from $i$ to $j$ via any $\alpha$ are (clearly) positively aligned. The known fact that Schur complements in $M$-matrices are again $M$-matrices follows immediately.

## REFERENCES

[BMQ] L. Bassett, J. Maybee, and J. Quirk, *Qualitative economics and the scope of the correspondence principle*, Econometrica **26** (1968), 544–563.

[BMNV] F. Brouwer, J. Maybee, P. Nijkamp, and H. Voogd, *Sign-solvability in economic models through plausible restrictions*, Atlanta Economic J. **17** (1989), 21–26.

[Ca] D. Carlson, *What are Schur complements, anyway?*, Linear Algebra Appl. **74** (1986), 257–275.

[Co] R. Cottle, *Manifestations of the Schur complement*, Linear Algebra Appl. **8** (1974), 189–211.

[H] E. Haynsworth, *Determination of the inertia of a partitioned Hermitian matrix*, Linear Algebra Appl. **1** (1968), 73–81.

[HJ] R. Horn and C. R. Johnson, *Topics in matrix analysis*, Cambridge Univ. Press, New York, 1990.

[KL] V. Klee and R. Ladner, *Qualitative matrices: Strong sign solvability and weak satisfiability*, in Computer-Assisted Analysis and Model Simplification (H. Greenberg and J. Maybee, eds.), Academic Press, New York, 1981, pp. 293–320.

[KLM] V. Klee, R. Ladner, and R. Manber, *Sign solvability revisited*, Linear Algebra Appl. **59** (1984), 131–157.

[MOvW] J. Maybee, D. Olesky, P. van den Driessche, and G. Wiener, *Matrices, digraphs and determinants*, SIAM J. Matrix Anal. Appl. **10** (1989), 500–519.

[O] D. Ouellette, *Schur complements and statistics*, Linear Algebra Appl. **36** (1981), 187–295.

[S] J. Schur, *Über Potenzreihen, die im Innern des Einheitskreises beschränkt sind*, J. Reine Angew. Math. **147** (1917), 205–232.

DEPARTMENT OF MATHEMATICS, THE COLLEGE OF WILLIAM AND MARY, WILLIAMSBURG, VIRGINIA 23187-8795

*E-mail address*: CRJOH2@WMMVS.BITNET

DEPARTMENT OF MATHEMATICS, UNIVERSITY OF COLORADO, BOULDER, COLORADO 80309-0426

DIMACS Series in Discrete Mathematics
and Theoretical Computer Science
Volume 4, 1991

Applied Geometry
and Discrete Mathematics
THE VICTOR KLEE FESTSCHRIFT

# Centers and Invariant Points of Convex Bodies

M. J. KAISER, T. L. MORIN, AND T. B. TRAFALIS

ABSTRACT. This survey provides a general introduction to centers and invariant points of convex bodies. In particular, we discuss analytic, Chebyshev, Huard, Santaló, Steiner, and volumetric centers. We concentrate on algorithmic considerations involved in and the computational complexity of computing centers of convex polytopes, which finds its motivation from recent results in linear optimization theory. It is interesting to note that Victor Klee's seminal work **[35]** on the complexity of the simplex algorithm provided impetus for the development of the new nonadjacent extreme point algorithms for linear programming that, in turn, motivated the renewed interest in computing centers of convex polytopes.

## 1.0. Introduction

Interest in centers of convex bodies dates back at least as far as the ancient Greeks who computed centers of triangles and regular convex polygons. For an exhaustive survey of invariant points associated with convex bodies, see **[24]**; for a shorter discussion, see **[9]**; and for a modern applications-oriented viewpoint (specifically for analytic centers) see **[58]**. This paper considers invariants of convex bodies which provide the background to gain additional insight into recent formulations of interior point techniques in mathematical programming. As pointed out by one of the referees, centers can be categorized as being either "geometric" or "analytic," according as they are defined through polytopes or over inequalities, respectively. We will return to this issue again later in this work and make the formulation more exact. Interest in centers of convex sets can also be found in computational geometry **[47]**; control theory **[57]**; design centering **[13, 62]**; economic planning processes **[17]**; mathematical programming **[19, 20, 36, 56, 59, 60]**; and vector optimization **[30, 65]**.

Interior point optimization methods are based on the concept of the analytic center. The most dramatic optimization results have occurred in linear programming. Indeed, recent breakthroughs in linear programming (LP)

---

1980 *Mathematics Subject Classification* (1985 *Revision*). Primary 52A25, 52A30, 90C05.

have literally revolutionized the field. While Klee and Minty [**35**] had demonstrated that the simplex algorithm could require an exponential number of pivots, the computational complexity of LP remained an open question until 1979 when Khachian [**34**] showed that LP could be solved in polynomial time using the ellipsoid algorithm. More recently, Karmarka [**33**] and others [**28**] have demonstrated just how computationally effective interior point methods, which rely on the efficient computation of interior points and analytic centers of polytopes, can be.

This paper then is organized in the following manner. §§1.1 through 1.4 introduce various centers and invariant points that can be associated with a convex body. In §1.5 we briefly discuss Steiner's curvature centroid, and in §1.6 the properties of the Steiner point are defined. Algorithmic considerations provide the main motivation in the second half of the paper. Specifically, §2.0 begins with a general discussion of the center as used in mathematical programming, and in §2.1 we define the Huard center. We continue in §2.2 by investigating the *need* for a point which can be computed in polynomial time with properties similar to that of the center of gravity (the analytic center is such a point). In §§2.3 and 2.4 we describe logarithmic barrier functions and volumetric centers, respectively, and in §2.5 center manifolds are briefly outlined. The paper concludes with a discussion in §3.

### 1.1. The Santaló point

The density for straight lines in the plane invariant under the three-parameter group

$$x' = a_1 x + b_1 y, \qquad y' = a_2 x + b_2 y$$

with $\left|\begin{smallmatrix} a_1 & b_1 \\ a_2 & b_2 \end{smallmatrix}\right| = 1$ is $dG = d\varphi dp/p^3$. If $p = p(\varphi)$ is the support function of a convex curve $K$ containing $O$ in its interior, then the measure of the set of straight lines exterior to $K$ will be

$$\int_{G\cap K=0} dG = \int_0^{2\pi} d\varphi \int_p^\infty \frac{dp}{p^3} = \frac{1}{2}\int_0^{2\pi} \frac{d\varphi}{p^2}.$$

This integral is an invariant of $K$ under the affine transformations defined above. It is called the central affine length of $K$ with respect to the interior point $O$. Furthermore, there exists a unique point $O_M \in K$ that minimizes the value of the integral.

The result can be generalized to convex bodies in $\mathbf{R}^n$. Let $p = p(u_{n-1})$ denote the support function of the convex body $K$ in $\mathbf{R}^n$ with respect to the origin $O \in K$. Then the measure of the set of hyperplanes $L_{n-1}$ exterior to $K$, which is invariant under the homogeneous affine group, is

$$\begin{aligned} I(K, 0) &= m(L_{n-1}; L_{n-1} \cap K = 0) \\ &= \int \frac{dp\, du_{n-1}}{p^{n+1}} = \int_p^\infty \frac{dp}{p^{n+1}} \int_{U_{n-1}} du_{n-1} = \frac{1}{n}\int_{U_{n-1}} \frac{du_{n-1}}{p^n}. \end{aligned}$$

Observe that $I(K, O)$ is the volume of $K^*$, the polar reciprocal body of $K$ with respect to $O$. If $I(K, O)$ is an affine invariant that depends on $K$ and the point $O \in K$, then there exists a unique point $O_M \in K$ such that $I(K, O_M) = \min I(K, O)$ for $O \in K$.

The point $O_M$ will be called the Santaló point of $K$. It can be shown that $O_M$ is an affine invariant point of $K$. That is, a point $p(K) \in \mathbf{R}^n$ defined for all $K$ is affine-invariant if for any nonsingular affine map $g$ of $\mathbf{R}^n$ onto itself

$$g(p(K)) = p(g(K)),$$

where $p(K)$ depends continuously on $K$. See Petty **[46]** for the proof, and **[54]** for more general considerations. Recently, Sonnevend **[57]** has defined the analytic center $c(K)$ of a compact convex set $K$ as that point which becomes the center of gravity of the dual set $K^d$, when it is taken as the center of duality. With this definition he proves that $c(K)$ is uniquely determined.

### 1.2. The centroid point

Let $K$ be an $n$-dimensional convex body, $x$ a point in the interior of $K$, and take $vz$ to be the chord through $x$ (where $v$ and $z$ are on the boundary of $K$). Then

$$x = \lambda v + (1 - \lambda) z,$$

for some $0 < \lambda < 1$. It is clear that $\lambda$ depends on $x$ and on the chord. We call the points $v$ and $z$ on the boundary of $K$ *antipodal* with respect to the interior point $x$. If we turn the chord about $x$, then $\lambda$ varies continuously. Hence, it takes as a minimum $\lambda^*$. This minimum is a continuous positive function of $x$ on the interior of $K$ that approaches zero as $x$ approaches the boundary of $K$. Therefore, $\lambda^*$ achieves a maximum $\mu$ on the interior of $K$, where

$$\mu = \max_{x \in \operatorname{int}(K)} \min_{v \in \partial K} \lambda.$$

The definition of $\mu$ is invariant with respect to affine transformations. Any point $x^*$ which divides a chord in $\mu$ is called a *centroid*.

Newmann **[44]** has proved that for any planar convex body there exists a unique centroid $\mu$ that satisfies $\frac{1}{3} \le \mu \le \frac{1}{2}$. See also **[43]** for further geometric analysis.

Hammer **[25]** has generalized this result for a convex body in $n$-dimensions although the point is no longer unique. If we call a pair of points in the boundary of a convex polytope that are collinear with the analytic center *antipodal*, then we have the following result **[69]**:

THEOREM 1. *If the analytic center of a full-dimensional bounded polyhedron defined by $m$ linear inequalities is expressed as the convex combination of two antipodal points $\nu$ and $z$, then $\lambda \ge 1/m$.*

### 1.3. The Chebyshev point

Let $A$ be a bounded subset of $\mathbf{R}^n$, $x \in \mathbf{R}^n$, and define the minimal radius of a ball centered at $x$ containing $A$ as

$$r(x, A) = \sup_{y \in A} \|y - x\|.$$

For $G \subset \mathbf{R}^n$, the Chebyshev radius of $A$ in $G$ is

$$r_G(A) = \inf_{y \in G} r(y, A),$$

and the Chebyshev center set of $A$ in $G$ is

$$Z_G(A) = \{y \in G : r(y, A) = r_G(A)\}.$$

In [2] there are several characterizations of $r_G(A)$ and the set $Z_G(A)$. In particular, the Chebyshev center for a set $A$ can be regarded as a special case $n = 1$ of the more general best $n$-net for $A$; i.e., the $\mathbf{y} = \{y_1, \ldots, y_n\} \in \mathbf{R}^n$ that minimizes

$$r_{\mathbf{y}}(A) = \sup_{x \in A_i} \min \|x - y_i\|.$$

Chebyshev centers may also be generalized to asymptotic and approximate centers.

### 1.4. The Steiner point

In [**24**] Grünbaum discussed invariant points and sets as related to the symmetry of convex sets. In particular, he mentioned problems associated with Steiner's curvature centroid. Grünbaum's paper subsequently sparked much research on generalizing the curvature centroid to higher dimensions where it is known as the Steiner point. It should not be confused with the definition of Steiner point as used in graph theory. Steiner introduced the curvature centroid as the point with respect to the pedal curve of an oval which had a minimum area. He then showed geometrically that if the oval was considered as loaded with a mass density proportional to its curvature, then its center of gravity would equal its curvature centroid. See [**4**].

For $K$ a closed convex curve, take a point $P \in K$ and define $\theta$ as the angle between the tangent at $P$ and the $x$-axis, and $\kappa(\theta)$ as the curvature of $K$ at $P$. If we consider a point-mass of weight $\kappa(\theta)$ over all $K$, then the center of gravity $\mathbf{s}(K)$ of such a system is given by coordinates

$$x_1 = \frac{1}{2\pi}\int_0^{2\pi} x(\theta)\,d\theta, \qquad y_1 = \frac{1}{2\pi}\int_0^{2\pi} y(\theta)\,d\theta.$$

Then $\mathbf{s}(K) = (x_1, y_1)$ is the Steiner curvature centroid of $K$.

Generalization of this definition to $n$ dimensions is straightforward. For a closed bounded convex set $K \subset \mathbf{R}^n$, the Steiner point of $K$ is defined by

$$\mathbf{s}(K) = \frac{1}{\sigma_n}\int_{S^{n-1}} uH(u)\,d\omega,$$

where $u$ is a variable unit vector, $H(u)$ is the support function of $K$, $d\omega$ is the surface area element of the sphere $S^{n-1}$, and $\sigma_n$ is the volume of the $n$-dimensional unit ball, $U_n$.

### 1.5. Curvature centroids

For a continuous mass distribution with density $\mu$, the centroid of $K$ is defined as $s = \int_K x\mu\, dx / \int_K \mu\, dx$. For a homogeneous or uniform mass distribution, that is, a constant mass per unit volume, $\mu =$ constant and $s = (\int_K x\, dx)/V(K)$. Hence, since $\mu$ does not enter into the problem, instead of referring to a center of mass, we modify the terminology and refer to a center of volume. We make one final change and prefer *centroid* of mass to *center* of mass, since the latter term has a different geometric meaning. The centroid of mass coincides with the center of gravity $g(K)$ and in the special case that $\mu$ is constant, the volume centroid coincides with $g(K)$. We will refer to the volume centroid henceforth as $g(K)$.

The name "center of gravity" arises since the resultant of the pulls of gravity on a body may be regarded as acting through the center of mass. The center of gravity can be thought of as the balance point of a body, but it may not be part of the body. For $K$ convex, however, $g(K) \in \operatorname{int}(K)$. For a triangle $T$, the expected value of the distance from a random point $x \in T$ to a line not passing through $T$ is equal to the distance from the line to the center of gravity of $T$. This also holds for a general convex body $K$. Observe that the mass distribution for the center of gravity is over $K$. Similarly, we can define a uniform mass distribution over the boundary of $K$, $\partial K$, or

$$s = \frac{1}{V(\partial K)} \int_{\partial K} x\, dx,$$

in which case we have a surface centroid.

More generally, for a closed convex hypersurface $K \subset \mathbf{R}^n$, there are $n-1$ principal radii of curvature. These radii of curvature can be combined into $n-1$ elementary symmetric curvature functions. If we let the density $\mu$ be one of these curvature functions, we can cover $\partial K$ with a mass distribution which has a density given by such a function. The centroid in such a case will be called a curvature centroid. Thus, the surface centroid is a special case of the family of curvature centroids, since associated with each point, regardless of the curvature at that point, is a uniform density. In particular, for a plane convex curve we have:

| Centroid | Density |
|---|---|
| (1) (Steiner) curvature centroid | $\mu \propto$ curvature, |
| (2) perimeter centroid | $\mu \sim U(\partial K)$, and |
| (3) area centroid | $\mu \sim U(K)$. |

As just defined, the curvature centroid is the center of mass of the perimeter of the convex curve when each point of the perimeter has a density equal to the curvature at that point. The perimeter centroid is the centroid of the perimeter, where each point of the perimeter has a uniform density. The area centroid is the centroid of area, where each point of the region bounded by the curve has a uniform density.

For $n = 3$ we can define four centroids associated with a closed convex surface:

| Centroid | Density |
|---|---|
| (1) Gaussian curvature centroid | $\mu \propto$ Gaussian curvature, |
| (2) mean curvature centroid | $\mu \propto$ mean curvature, |
| (3) surface centroid | $\mu \sim U(\partial K)$, and |
| (4) volume centroid | $\mu \sim U(K)$. |

The Gaussian curvature centroid is the center of mass of $\partial K$ when each point of $\partial K$ has a density equal to $1/RR'$, where $R$ and $R'$ are the principal radii of curvature. The mean curvature centroid is defined similarly except that each point of $\partial K$ has a density equal to $\frac{1}{2}(\frac{1}{R}+\frac{1}{R'})$. The surface centroid is analogous to the perimeter centroid in the plane: every point of the surface $\partial K$ has a uniform density. Similarly, the volume centroid is analogous to the area centroid in the plane: every point of $K$ has a uniform density.

For the general case, $(1), \dots, (n-1)$ are curvature centroids, where $\mu$ is proportional to the $n-1$ symmetric curvature functions, $(n)$ is the surface centroid where $\mu \sim U(\partial K)$, and $(n+1)$ is the volume centroid where $\mu \sim U(K)$. Call $H_r$ the $r$th elementary symmetric function of the $n-1$ principal curvatures of the boundary $\partial K$. Then the vectorial quermassintegrale of $K$ are defined by

$$q_i(K) = \left(n\binom{n-1}{i-1}\right)^{-1} \int_{\partial K} xH_{i-1}\, d\sigma, \qquad i = 1, \dots, n,$$
$$q_0(K) = \int_K x\, dv,$$

where $dv$ is the volume element of $\mathbf{R}^n$, $d\sigma$ is the area element of $\partial K$, and $x$ denotes the radius vector from the origin to the point. The curvature centroids are then defined as $p_i = q_i/W_i$, $i = 0, \dots, n$, where the $W_i$ are the ordinary quermassintegrale; $i = 0$ corresponds to the volume centroid denoted $g(K)$, $i = 1$ corresponds to the surface area centroid, ..., and $i = n$ corresponds to the Steiner point $\mathbf{s}(K)$ of $K$.

The quermassintegrale, or mean cross-sectional measures, are defined by

$$W_r(K) = \frac{(n-r)O_{n-1}}{nO_{n-r-1}} E(V(K'_{n-r})),$$

where $O_i$ is the surface area of the $i$-dimensional unit sphere, and $E(V(K'_{n-r}))$ represents the mean value of the volume $V(K'_{n-r})$. For a convex set $K$, $K'_{n-r}$ is the orthogonal projection of $K$ into all the $(n-r)$-planes $L_{n-r[0]}$ through 0. See [**51**].

### 1.6. Steiner's curvature centroid

Let us consider the case of a plane and examine in more detail properties of the Steiner curvature centroid. If $K$ is a closed convex curve and $P \in K$ is represented by $(x(\theta), y(\theta))$, $0 \leq \theta < 2\pi$, where $\theta$ is the angle between the tangent to $P$ and the $x$-axis and $\kappa(\theta)$ is the curvature of $K$ at $P$, then the curvature centroid of such a system is calculated from $s = \int_{\partial K} x\mu\, dx / \int_{\partial K} \mu\, dx$, where $\mu = \kappa(\theta)$. Since $\kappa(\theta) = \frac{d\theta}{ds}$,

$$\int_{\partial K} \mu\, ds = \int_{\partial K} \frac{d\theta}{ds} ds = \int_0^{2\pi} d\theta = 2\pi,$$

$$\int_{\partial K} x\mu\, ds = \int_{\partial K} x \frac{d\theta}{ds} ds = \int_0^{2\pi} x(\theta)\, d\theta.$$

The Steiner curvature centroid is thus

$$\mathbf{s} = \frac{1}{2\pi} \int_0^{2\pi} x(\theta)\, d\theta.$$

The form of this integral suggests that for closed convex curves that admit a rational parameterization, $\mathbf{s}$ can be calculated explicitly. For instance, for the unit circle the standard rational parameterization is

$$x(\lambda) = \frac{2\lambda}{1+\lambda^2}, \qquad y(\lambda) = -\frac{1-\lambda^2}{1+\lambda^2}.$$

Hence, if we use approximating bodies that yield nice parameterizations, then the Steiner curvature centroid should be computable in closed form.

The Steiner curvature centroid has some interesting geometric properties which can be found in [**11**]. Returning to the triangle, let $\theta_i$ be the internal angle at vertex $A_i$. Then the Steiner point $\mathbf{s}(P)$ is the centroid of a system of masses, the mass $\pi - \theta_i$ being attached to vertex $A_i$. This definition makes sense in analogy with the continuous case, and holds in general for a convex $p$-gon. Shephard has established an identity similar in nature to the Dehn-Sommerville relations. See [**53**]. Observe that the centroid of a system of equal masses placed at the vertices $A_i$ yields the median point. The center of gravity of the perimeter of a triangle equals the center of the Spieker circle.

### 1.7. Properties of the general Steiner point

The general definition of the Steiner point has thus been motivated. The following properties characterize $\mathbf{s}(K)$:

(i) $\mathbf{s}(K)$ is additive:

$$\mathbf{s}(\lambda_1 K_1 + \lambda_2 K_2) = \lambda_1 \mathbf{s}(K_1) + \lambda_2 \mathbf{s}(K_2).$$

(ii) $\mathbf{s}(K)$ is similarity invariant:

$$\mathbf{s}(TK) = T\mathbf{s}(K),$$

for all similarity transformations $T : \mathbf{R}^n \to \mathbf{R}^n$.

(iii) $\mathbf{s}(K)$ is a continuous function of $K$.

(iv) $\mathbf{s}(K)$ is a relative interior point of $K$ if $\dim K > 0$.

(v) If $K = K_1 \cup \cdots \cup K_n$, where $K_1, \ldots, K_n$ and $K$ are compact convex subsets of $\mathbf{R}^d$, then

$$\mathbf{s}(K) = \sum_{i=1}^{n} \mathbf{s}(K_i) - \sum_{i<j} \mathbf{s}(K_i \cap K_j) + \cdots + (-1)^{n-1}\mathbf{s}(K_1 \cap \cdots \cap K_n).$$

In (i), the addition on the left is vector addition of convex sets, so that if $\mathbf{s}(K_1) = \mathbf{s}(K_2) = z$, then $\mathbf{s}(K_1 + K_2) = z$. A special case of property (ii) was discussed in the last section where it was stated that the Gaussian curvature centroid for parallel convex bodies is a fixed point. For instance, if $\mathbf{s}(K) = z$ is the origin, then $\mathbf{s}(\lambda K) = z$ for all $\lambda$. (iii) follows from the definition and although (iv) is not obvious, the definition of the Steiner point of a closed hypersurface as considered by Flanders **[18]** is. (v) has been shown in **[50]**. Schneider **[52]** has shown that $\mathbf{s}(K)$ is the only vector-valued function defined for all compact, convex sets in $\mathbf{R}^n$ that satisfy (i), (ii), and (iii). See **[6]** for an abstract version of the Steiner point.

In the next section algorithmic considerations play the fundamental role. We now provide the main motivation for developing concepts of analytic centers by considering algorithmic considerations in mathematical programming.

## 2.0. The center in mathematical programming

We describe a general method for solving convex programming problems and follow **[42]**. See [7] for relevant background material and consult **[64]** for recent developments (for example, material from **[38]** will not be discussed). Suppose we have the following convex programming problem

$$\begin{aligned} &\text{minimize } f_0(x) \\ &\text{subject to}: f_i(x) \le 0, \qquad i = 1, \ldots, m, \ x \in G, \end{aligned}$$

where $G$ is a nonempty closed convex set and $f_i$; $i = 0, 1, \ldots, m$, are continuous convex functions defined on $G \subset \mathbf{R}^n$. Let $x_0 \in \operatorname{int}(G)$. Consider $x_0$ as a first approximation to an optimal solution, and take a hyperplane which passes from $x_0$ and cuts $G$ into parts $G^+$ and $G^-$. If $G^+$ contains no approximate solution with the required accuracy, then all the solutions to the problem lie in $G^-$. Hence, we take $G^-$ as the new domain of the functions appearing in the problem, replace $x_0$ by a "suitable" point of $G^-$, and repeat the procedure. After a number of iterations, the localization of the solution becomes progressively smaller and we can then extract a solution with the required accuracy.

The problem that arises is how to select the next point of the iterative procedure. We would like to select the point such that the domain of localization decreases from each step "as rapidly as possible." To measure the domain of localization, we can use the Lebesgue volume. Suppose the current domain of localization is $G$ and we wish to select a point in $G$. A poorly chosen $x$ (e.g., $x \in \partial G$) and a "bad" hyperplane of intersection (e.g., a plane touching $\partial G$ when $x \in \partial G$) will not sufficiently decrease the new domain of localization. Our intuition suggests that the closer $x$ is to the "center" of $G$, the greater will be the decrease in the domain of localization. A natural candidate for a "center" of a convex set $G$ is its center of gravity. Grünbaum **[23]** has shown the following theorem, with a more geometric proof given by Mitiagin **[39]**. See also the collected works of Blaschke; in particular, **[8]**.

THEOREM 2. *Let $G$ be a compact convex set in $\mathbf{R}^n$ with center of gravity $x$, and let $G^+$ and $G^-$ be the sets in which a hyperplane $H$ passing through $x$ divides $G$. Then the volumes $V(G^+)$ and $V(G^-)$ satisfy the inequality*

$$V(G^i) \leq \left[1 - \left(1 - \frac{1}{n+1}\right)^n\right] V(G), \quad \text{where } i = + \text{ or } -.$$

Optimization techniques that follow this general methodology are called centroid methods. See **[36]**, **[45]**, and **[68]**. Centroid methods can be considered as generalizations of the bisection method on the real line. See **[42]** for further extensions, as well as Shor's paper **[54]**, where a reduction rate in volume occurs in geometric progression. A drawback of these methods is that the center of gravity cannot be computed in polynomial time if the dimension of the polytope is greater than 3. However, as shown in **[61]**, by using iterations of the method of inscribed ellipsoids, such a procedure can be completed in polynomial time.

The analytic center for a convex polyhedron defined by linear inequalities was introduced by Sonnevend **[56]** and Bayer and Lagarias **[5]** to remedy the problems of the center of gravity. Renegar **[48]** and Vaidya **[66]** give a polynomial Newton's method-based algorithm to compute the center of a polytope. Since the precursor of the recent interior point methods is the method of centers developed by Huard **[26]**, we next develop the concept of Huard's center.

## 2.1. The Huard center

The concept of a Huard center essentially depends on the choice of a distance function. Given a point $x$ in a convex set $G$, define the distance $d$ of $x$ to the boundary of $G$ by a function $d(x, G)$ which is continuous and satisfies:

(i) $d(x, G) = 0$ for all points on the boundary of $G$,
(ii) $d(x, G) > 0$ for all points in the interior of $G$, and
(iii) if $G'$ is a convex set which is a subset of $G$, then $d(x, G') \leq d(x, G)$ for all $x$ in $G'$.

Given a convex set $G$ and a distance function $d(x, G)$ defined on $G$, we call the point $\hat{x}$ in $\text{int}(G)$ that maximizes $d(x, G)$ the *center* of $G$.

### 2.2. Algorithmic considerations in calculating volume

We observe that it is "hard" to compute the center of gravity of a convex set. We now make this observation more precise. Let $P = \{x \in \mathbf{R}^n : Ax \geq b\}$, where $A \in \mathbf{R}^{m\times n}$ and $b \in \mathbf{R}^m$. Then $\text{vol}(P) = \int_P dy$, where $dy$ is the Lebesgue measure in $\mathbf{R}^n$. The following questions arise:

Q1. *Can* $\text{vol}(P)$ *be computed in polynomial time?*

Lovász conjectured that the answer is negative [**37**]. To compute the volume of *some* convex bodies *in a fixed dimension* is fairly straightforward. Unfortunately, most bodies fall outside this class. It is well known that computing the volume of a convex polytope in polynomial time remains unresolved. Dyer and Frieze [**14**] demonstrate that computing the volume of a bounded polytope, either given as a list of facets or vertices, is as hard as computing the permanent of a matrix, which is itself #**P**-Hard [**20**]. Furthermore, Elekes [**16**] has shown that no polynomial time algorithm can compute the volume of a convex body with less than exponential relative error. Specifically, he shows that the volume $V$ of the convex hull of $m$ points chosen from the $n$-dimensional unit ball $U_n$ is bounded above by $Vm/2^n$.

In [**3**], Bárány and Fürendi show that the Elekes bound is not as bad as it first appears *if* $m \sim 2^{n/2}$. Specifically, they show that for $n > 100$ and $m = 20n^{3/4}2^{n/2}$, Elekes results are quite good. This result was obtained as a special case of determining the probability that the convex hull $H_n$ has exactly $n$ vertices. Bárány and Fürendi have also shown that for $K \subset \mathbf{R}^n$, any polynomial time algorithm that gives an upper and lower bound on the volume of $K$, represented as $\overline{\text{vol}}(K)$ and $\underline{\text{vol}}(K)$, respectively, necessarily has a huge gap between them. They show

$$\frac{\overline{\text{vol}}(K)}{\underline{\text{vol}}(K)} \geq \left(\frac{cn}{\log n}\right)^n,$$

where $c$ is a constant independent of $n$. Stated another way, there does not exist a polynomial time algorithm which would compute $\underline{\text{vol}}(K)$ *and* $\overline{\text{vol}}(K)$ such that

$$\frac{\overline{\text{vol}}(K)}{\underline{\text{vol}}(K)} \leq \left(\frac{cn}{\log n}\right)^n.$$

Hence, *deterministic* algorithms for computing the volume of a convex body $K \subset \mathbf{R}^n$ yield poor worst-case results. However, as shown in [**15**], a *random* polynomial time algorithm can, with high probability, find a good estimate for the volume of $K$. In each of these cases, $K$ is assumed to be given by an "oracle" that describes how $K$ is encoded; i.e., its inherent size. This normally requires knowledge of the dimension of the space in which $K$ is embedded and the radius values of a circumscribing and inscribed ball of

$K$. See **[21]** and **[22]** and the references listed therein for more details. The general idea is rather elementary. $K$ can be represented either as the convex hull of its extreme points or as the intersection of its support hyperplanes. If a point $x$ belongs to $K$, then this is membership. If there exists a point $x$ such that a hyperplane through $x$ contains $K$, then this is validity. Although these are logically equivalent, it remained an open question whether membership and validity were algorithmically equivalent. With the discovery of the ellipsoid method **[34]**, this question was answered in the affirmative.

The answer to the next question is thus obvious.

Q2. *Can the center of gravity $x$ of $P$ be computed in polynomial time?*

Since $x = \int_G y\,dy / \int_G dy$, the computation of $x$ relies on the computation of $\mathrm{vol}(P)$. Hence, a negative answer to Q1 has the same consequence for Q2. Therefore, it seems reasonable to surmise that no polynomial algorithm can compute the center of gravity of a polytope given by linear inequalities.

Now the main idea of recent interior point-techniques (Karmarkar **[33]**, Sonnevend **[55]**) is to find two-sided ellipsoidal approximations of the feasible region or to find an interior point $\overline{x}$ (center) and an ellipsoid $E$ such that

$$\overline{x} + r_i E \subset P \subset \overline{x} + r_0 E,$$

where $r_0 = kr_i$ and $k$ is as small as possible. In particular, we are interested in finding a solution $\overline{x}$, $E$, $r_i$, $r_0$ that depends smoothly on the parameters which define the feasible region $P$. John **[29]** proved that there exists a two-sided ellipsoidal approximation of a compact convex set in $\mathbf{R}^n$. Specifically, he showed

THEOREM 3. *For any convex set $P$ in $\mathbf{R}^n$, we can find an ellipsoid $E$ such that*

$$\overline{x} + r_i E \subset P \subset \overline{x} + r_0 E,$$

*where $r_0 = nr_i$.*

Take $\overline{x} + r_i E$ to be the ellipsoid of largest volume contained in $P$ and set $r_0 = nr_i$. Then $\overline{x}$ is the *inscribed ellipsoid center $i(P)$*.

However, $\overline{x}$ and $E$ in John's construction depend on a subset of "active constraints" which depend discontinuously on the data. $\overline{x} + r_i E$ depends only on the point set $P$ and does not depend smoothly on the parameters describing the set. To construct low complexity optimization algorithms we should therefore avoid John's construction because of the combinatorial difficulties in conventional active set constraints. Low complexity optimization algorithms need continuous and analytic dependence on the data of the optimization problem.[1] Hence, we have briefly motivated the need for a "center"

[1]It is also known that simultaneous consideration of all constraints of a linear programming problem is an essential part of Karmarkar's method.

that can be computed in polynomial time and depends "smoothly" on the data while considering all the problem constraints simultaneously.

### 2.3. A logarithmic barrier function and the analytic center of a polytope

In this section we define the concept of the analytic center of a bounded polyhedron that is defined by linear inequalities. It can be computed in polynomial time [**48**, **66**]. A basic property of the analytic center is that it is invariant with respect to affine transformations. The concept of the analytic center was developed by Sonnevend [**55**] and Bayer and Lagarias [**5**]. This center also induces a nonlinear coordinate system in $\mathrm{int}(P)$. Consider a polytope $P$ that is described by a finite number of inequalities of the form

$$a_i^T x \geq b_i, \qquad i = 1, \dots, m.$$

Let $H_i = \{x : a_i^T x \geq b_i\}$, and consider the Euclidean distance of $x$ from $H_i$. It is easy to see that this distance is invariant under the group of all Euclidean motions in $\mathbf{R}^n$ [**5**]. We have

$$d(x, H_i) = |a_i|^{-1} |a_i^T x - b_i|.$$

Now define the logarithmic barrier function $f_P : \mathrm{int}(P) \to \mathbf{R}$, where $P$ is the polyhedron defined by the given linear inequalities, i.e.,

$$f_P(x) = \sum_{i=1}^{m} \log d(x, H_i).$$

The function $f_P(x)$ can be expressed as

$$f_P(x) = \sum_{i=1}^{m} \log(a_i^T x - b_i) + \alpha,$$

where $\alpha = -\sum_{i=1}^m \log \|a_i\|_2$. Our objective is to maximize this function. The maximizer of this function will be a candidate for the center of a given polytope $P$. The geometric interpretation of the center is that it is an interior point of $P$ that maximizes the log of the product of distances of $x$ to each of the hyperplanes defining the constraint region. Therefore, consider the differential $df_P$ and find the zeros of this differential. The zeros of $df_P$ generally describe a hyperplane which is called a central flat. If $P$ is bounded, then the central flat becomes a point.

The advantage of this definition of a center is that it can be computed analytically and depends continuously on the data. This property is useful to develop polynomial and stable algorithms for convex optimization. In [**5**], Bayer and Lagarias show

THEOREM 4. *If $P$ is a bounded full dimensional polyhedron, then $f_P$ is strictly concave in* $\mathrm{int}(P)$, *and* (*therefore*) *the analytic center of $P$ exists and is unique.*

Now assume that we define the concept of the analytic center as above, but instead of using the Euclidean distance, we use the $l^p$ distance in $\mathbf{R}^n$ with

$1 < p < \infty$. Then, an interesting property of this analytic center is that it is invariant with respect to the particular $l^p$ metric we use. For further details see [**1**].

It can be shown that for a simplex the analytic center coincides with its center of gravity. This is not generally true, however. Recently, Ye [**70**] has established a geometric property for the analytic center similar to Grünbaum's characterization of the center of gravity. The concept of the analytic center can also be generalized by adding a perturbation factor $\gamma$ to the barrier function. See [**69**] for a discussion of $\gamma$-centers.

### 2.4. Volumetric centers

In [**67**], Vaidya developed a new analytic center of a polytope defined by linear inequalities. He named this center the volumetric center. Earlier work by Tarasov et al. [**61**] used a similar concept. This latter material is particularly interesting. Call $E^*$ the unique inscribed ellipsoid of maximum volume in $K$. Denote by $\mu(K)$ the volume of $E^*$ as a function of $K$. Through the center $x^*$ of $E^*$, i.e., the center $x^*$ of $K$, an arbitrary hyperplane is passed. If we denote by $K_g$ one of the two bodies into which the hyperplane divides $K$, then it is shown that

$$\mu(K_g) \leq 0.843\mu(K).$$

The method of proof is by recursion. Using this result, Tarasov et al. [**61**] then describe a convex programming method with polynomial iteration cost optimal with respect to the order of the number of iterations. Recently, Nesterov and Nemirovskii [**40, 41**] have generalized this approach.

If we let $P$ be the bounded full-dimensional polytope $P = \{x \in \mathbf{R}^n : Ax \geq b\}$, where $A \in \mathbf{R}^{m\times n}$ and $b \in \mathbf{R}^m$, then the volumetric center $\omega$ of $P$ is defined as the point which minimizes the potential function

$$F(x) = \tfrac{1}{2}\log(\det H(x)).$$

$H(x)$ is the Hessian of

$$f(x) = -\sum_{i=1}^{m} \log(a_i^T x - b_i)$$

and is positive definite for all $x \in \operatorname{int}(P)$. Let

$$E(x) = \{y \in \mathbf{R}^n : (y-x)^T H(x)(y-x) \leq 1\}.$$

Then $E(x)$ is an ellipsoid such that $E(x) \subseteq P$ and is a local quadratic approximation to $P : E(\omega)$ has the largest volume among all ellipsoids with $x \in \operatorname{int}(P)$. Notice that, as pointed out by one of the referees, the volumetric center is an analytic but not a geometric center. The referee offered the following example. Consider the polytope

$$P = \{x \in \mathbf{R} : x \geq 0, \ldots, x \geq 0,\ x \leq 1\},$$

where $x \geq 0$ is repeated $m-1$ times. The analytic center is $x=(m-1)/m$ and the volumetric center is $x = (m-1)^{1/3}/(1+(m-1)^{1/3})$. Both of them are "bad" if $m$ is large. This example underscores the importance of emphasizing the two types of centers, geometric and analytic.

Finally, we note that the papers by Todd **[63]** and Ye **[70]** discuss the containing and shrinking properties of ellipsoids arising in Karmarkar's algorithm, and illustrate a nice geometric view of the interior point algorithm.

### 2.5. Center manifolds

Higher dimensional versions of the affine trajectories described in **[5]** can be formulated. Consider a $k$-dimensional polyhedral cone $\Lambda$ in $\mathbf{R}^n$ and the set of all its translations. If the system of linear inequalities

$$\Lambda_s^T x \geq 0, \qquad s=1,\dots,p,$$

describes the cone, then any parallel displacement of the cone can be described by the following system of linear inequalities:

$$\Lambda_s^T x \geq \Lambda_s^T \lambda, \qquad s=1,\dots,p.$$

Adjoining these to the original system of inequalities defining the polyhedron $P$ yields the expanded system

$$a_i^T x \geq b_i, \quad i=1,\dots,m, \qquad \Lambda_s^T x \geq \Lambda_s^T \lambda, \quad s=1,\dots,p.$$

Then we have

PROPOSITION 1. *The locus of all analytic centers of such systems, provided they are solvable, forms an algebraic manifold. This manifold is characterized by the system of equations*

$$\sum_{i=1}^{m} \frac{a_i^T}{a_i^T x - b_i} + \sum_{s=1}^{p} \frac{\Lambda_s^T}{\Lambda_s^T x - \Lambda_s^T \lambda} = 0.$$

For a proof, consider the function

$$\phi(x,\lambda) = \sum_{i=1}^{m} \frac{a_i^T}{a_i^T x - b_i} + \sum_{s=1}^{p} \frac{\Lambda_s^T}{\Lambda_s^T x - \Lambda_s^T \lambda}.$$

This function defines a Legendre transform coordinate system in the interior of the polytope $P(\Lambda)$ **[5]**, and is given by a rational function of $x_i$. Since $P(\Lambda)$ is bounded, $\phi$ can be inverted and its inverse $\phi^{-1}$ given by algebraic functions of the Legendre transform space.

### 3.0. Discussion

We saw earlier the need to define an appropriate "center" of $K$. Many such alternative points can be examined. These can be viewed from either a purely mathematical interest or from a computational point of view. The distinction is not entirely artificial. For the special case of a polygon $P$, a "center" is defined as a point that either minimizes

(i) $$\sum_i d(x, H_i)^n, \qquad n = \dots, -2, -1, 1, 2, \dots,$$

or maximizes

(ii) $$\prod_i d(x, H_i)^n, \qquad n = \dots, -2, -1, 1, 2, \dots,$$

where $d(x, H_i) = d$ represents the perpendicular distance from $x$ to line $H_i$. We sum and multiply over the lines that describe $P$. Under the monotonic transformation logarithm, these choices are obviously equivalent (under the correspondence $n = -j$ in (i) and $n = j$ in (ii)). When the logarithm is not applied, however, there is a distinction between solutions to (i) and (ii) for different $n$.

Furthermore, "centers" can be defined depending on *how* the problem is represented and what type of solution is desired. For instance, definitions (i) and (ii) may also be considered for other polynomial systems. The only essential difference is in the representation of the distance function. Consider the following types of polygonal systems in the plane:

- "points" $P_0 = \{A_1, A_2, \dots, A_n\}$,
- "lines" $x \cos\theta_j + y \sin\theta_j - p_j = 0$, $j = 1, \dots, n$, that form $P_1$, or, alternatively, written in the form $P_1 : Ax = b$, and
- "bodies" $P_2 : Ax \geq b$.

For $P_0$ we replace line $H_i$ with point $A_i$, and thus the distance function is a measure to $A_i$. For $P_2$ we understand the body to be "full-dimensional." We can generalize $P_2$ and consider $P_2' = \{x | f_j(x) \leq 0; j = 1, \dots, n\}$, for $f_j(x)$ continuous convex functions. In this case, we replace the distance function with $f_j(x)$. For $j = 1$, under this representation and (ii), we have the general definition of the *analytic* center, which up to a constant is equivalent for arbitrary $j$. For $j = 1$, under the representation $P_2$, we have the standard definition **[66]**. Hence, the use of logarithmic transformation makes the $n$ in (i) and (ii) redundant. So, depending on how a problem is formulated and what type of solution is required will often determine the "proper" definition of a "center" point. For more details in this direction, see **[31]**. In **[32]**, it is shown that under a limit argument the analytic centers corresponding to system $P_0$ yield Sonnevend's recent formulation **[57]**.

Now suppose we want to approximate a triangle by a circle. What is the "best" point of the triangle to take? If by "best" we mean that we want to minimize the symmetric difference of the area between the two bodies, then the point to take is called the Lemoine (or symmedian) point. For more details and the relationship of this point to approximation theory, see **[12]**. Such a point will minimize the sum of the squares of the distances from the point to the sides of the triangle. This can be motivated by geometric or algebraic arguments quite easily. Unfortunately, the symmedian point does not directly correspond to any of the centroids we have discussed, but under an appropriate limit it should. In fact, in an unpublished paper Davis [**11**] has shown the Steiner point to be the best approximating point of the body by a circle. It is obtained as the limit of symmedian points. That is,

$$\lim_{i\to\infty} x_n^i \to p_n = \mathbf{s}(K),$$

where $x_n^i$ is the point of $K$ which minimizes the sum of the distances squared to $\partial K$, i.e., its "symmedian point"; here $i$ represents the number of hyperplanes circumscribing $K$, the direction of the normals being uniformly distributed over $U_n$.

The natural question that arises is what corresponds to the remaining centroids? More precisely, if

$$\lim_{i\to\infty} x_k^i \to p_k, \qquad k = 0, \dots, n,$$

then what is the form of the $(n+1)$-tuple $x = (x_0, \dots, x_n)$? The determination of this vector is an open area of research. Observe that for $k = n$, the form of $x_n$ is known.

Now if our motivation is algorithmic in nature, then the following point of view can be taken. We have defined the analytic, Santaló, and volumetric centers as maximizers or minimizers of certain potential (or barrier) functions. Therefore, the concept of a potential function seems to be a unifying tool to study other possible analytic centers.

An "analytical" center $\hat{x}$ of a convex set $G$ is the unique point in $\operatorname{int}(G)$ that satisfies an equation of the form $\nabla f_G(\hat{x}) = 0$, where $f_G$ is an analytic function of the constraints that define $G$. An interesting area of research would be to identify the class of potential functions $f_G$ such that $\hat{x}$ can be computed in polynomial time. By understanding such functions we could classify centers into "solvable" (computable in polynomial time) and "hard" (no known polynomial time algorithm). An example of a "solvable" center is the centroid of a set of $N$ points in $k$ dimensions. This can be computed in $O(Nk)$ arithmetic operations [**47**]. An example of a "hard" center is the center of gravity of an $n$-dimensional polytope defined by linear inequalities [**42**].

## Acknowledgments

The authors would like to thank the referees for numerous insightful comments and for offering the example in §2.5. This research was supported in part by the Office of Naval Research under Purdue's University Research Initiative in Computational Combinatorics: ONR Contract N00014-86-K-0689.

## References


1. S. Abhyankar, T. Morin, and T. Trafalis, *Efficient faces of polytopes: Interior point algorithms, parameterization of algebraic varieties and multiple objective optimization*, in Mathematical Developments Arising from Linear Programming (J. C. Lagarias and M. J. Todd, eds.), Contemporary Mathematics Series, Amer. Math. Soc., Providence, RI, 1991.
2. D. Amir, *Characterizations of inner product spaces*, Birkhauser Verlag, Basel, 1986.
3. I. Bárány, and Z. Fürendi, *On the shape of the convex hull of random points*, Probab. Theory Related Fields, **77** (1988), 231–240.
4. N. H. Ball, *On ovals*, Bull. Amer. Math. Soc. **36** (1930), 348–353.
5. D. Bayer and J. Lagarias, *The nonlinear geometry of linear programming*. I, II, Trans. Amer. Math. Soc. **314** (1989), 499–526; 527–581.
6. C. Berg, *Abstract Steiner points for convex polytopes*, J. London Math. Soc. **2** (1971), 176–180.
7. R. G. Bland, D. Goldfarb, and M. J. Todd, *The ellipsoid method: A survey*, Oper. Res. **29** (1981), 1039–1091.
8. W. Blaschke and K. Reidemeister, *Vorlesungen über Differentialgeometrie*. II. *Affine Differentialgeometrie*, Springer-Verlag, Berlin, 1923.
9. T. Bonnesen and W. Fenchel, *Theory of convex bodies*, BCS Associates, Moscow, ID, 1987.
10. S. Buchin, *Lectures on differential geometry*, World Scientific, Singapore, 1980.
11. P. J. Davis, *Lemoine approximation and Steiner approximation of convex sets*, manuscript, Brown University, Providence, RI, 1982.
12. P. J. Davis, R. A. Vitale, and E. Ben-Sabar, *On the deterministic and stochastic approximation of regions*, J. Approx. Theory **21** (1977), 60–88.
13. S. W. Director and D. Hachel, *The simplicial approximation approach to design centering*, IEEE Trans. Circuits and Systems **CAS-24** (1977), 363–372.
14. M. E. Dyer and A. M. Frieze, *On the complexity of computing the volume of a polyhedron*, SIAM J. Comput. **17** (1988), 967–974.
15. M. E. Dyer, A. M. Frieze, and R. Kannan, *A random polynomial time algorithm for approximating the volume of convex bodies*, Proceedings of the 21st Annual ACM Symposium on the Theory of Computing (1989), 375–381.
16. G. Elekes, *A geometric inequality and the complexity of computing volume*, Discrete Comput. Geom. **1** (1986), 289–292.
17. I. I. Erlikh, *On the possibilities of use of a certain class of convex programming methods in diagonal planning processes*, Engrg. Cybernetics **19** (1981), 31–36.
18. H. Flanders, *The Steiner point of closed hypersurfaces*, Mathematika **13** (1966), 181–188.
19. K. R. Frisch, *The logarithmic potential method of convex programming*, Memorandum, University Institute of Economics, Oslo, Finland, 1985.
20. M. R. Garey and D. S. Johnson, *Computers and intractability: A guide to the theory of NP-completeness*, Freeman, San Francisco, 1979.
21. M. Grötschel, L. Lovász, and A. Schrijver, *Geometric methods in combinatorial optimization*, in Progress in Combinatorical Optimization (W. R. Pulleyblank, ed.), Proc. Silver Jubilee Conference on Combinatorics (University of Waterloo), Academic Press, New York, 1982, pp. 167–180.
22. ——, *Geometric algorithms and combinatorial optimization*, Springer-Verlag, Berlin, 1988.
23. B. Grünbaum, *Partitions of mass-distributions and of convex bodies by hyperplanes*, Pacific J. Math. **10** (1960), 1257–1261.
24. ——, *Measures of symmetry for convex sets*, in Convexity, Proc. Sympos. Pure Math. Vol. II, Amer. Math. Soc., Providence, RI, 1963, pp. 233–270.
25. P. C. Hammer, *The centroid of a convex body*, Proc. Amer. Math. Soc. **2** (1951), 522–525.
26. P. Huard, *Resolution of mathematical programming with nonlinear constraints by the method of centers*, in Nonlinear Programming (J. Abadie, ed.), North-Holland, Amsterdam, 1967, pp. 209–219.

27. Fl. Jarre, *On the method of analytic centers when applied to convex, quadratic programs*, Preprint, Univ. Würzburg, 1987; Math. Programming (to appear).
28. Fl. Jarre, G. Sonnevend, and J. Stoer, *An implementation of the method of analytic centers*, Lecture Notes in Control and Inform. Sci. (A. Bensoussan and J. L. Lions, eds.), vol. 111, Springer-Verlag, Berlin, 1988, pp. 297–307.
29. F. John, *Extremum problems with inequalities as subsidiary conditions*, in Studies and Essays, Courant Anniversary Volume, Interscience, New York, 1948, pp. 187–204.
30. M. Kaiser, *Applications of integral geometry to vector optimization*, Ph.D. Dissertation, Purdue University, West Lafayette, IN, 1990, (in preparation).
31. M. Kaiser and T. Morin, *Zero-, one-, and two-dimensional analytic centers*, Computational Combinatorics Technical Report CC-90-28, Institute for Interdisciplinary Engineering Studies, Purdue University, West Lafayette, IN, 1990.
32. ——, *On analytic centers*, Computational Combinatorics Technical Report CC-90-29, Institute for Interdisciplinary Engineering Studies, Purdue University, West Lafayette, IN, 1990.
33. N. Karmarkar, *A new polynomial-time algorithm for linear programming*, Combinatorica **4** (1984), 373–395.
34. L. G. Khachian, *A polynomial algorithm for linear programming*, Dokl. Akad. Nauk USSR **244** (1979), 1093–1096.
35. V. Klee and G. L. Minty, *How good is the simplex method?* in Inequalities. III, (O. Shisha, ed.), Academic Press, New York, 1972, pp. 159–175.
36. A. Levin, *On an algorithm for the minimization of convex functions*, Soviet Math. Dokl. **6** (1965), 286–290.
37. L. Lovász, *An algorithmic theory of numbers, graphs, and convexity*, CMBS-NSF Regional Conf. Ser. in Applied Math., vol. 50, SIAM, Philadelphia, PA, 1986.
38. N. Meggido and M. Shub, *Boundary behaviour of interior point algorithms in linear programming*, Math. Oper. Res. **14** (1989), 97–147.
39. B. S. Mitiagin, *Two inequalities for volumes of convex bodies*, Math. Notes **5** (1968), 61–85.
40. Ju. Nesterov and A. Nemirovskii, *A general approach to polynomial-time algorithms design for convex programming*, Central Econ. and Math. Inst., USSR Academy of Sciences, Moscow, 1988.
41. ——, *Self-concordant functions and polynomial-time methods in convex programming*, USSR Academy of Sciences, Moscow, 1989.
42. A. S. Nemirovskii and D. B. Yudin, *Problem complexity and method efficiency in optimization*, Wiley, New York, 1983.
43. B. H. Neumann, *On an invariant of plane regions and mass distributions*, J. London Math. Soc. **20** (1945), 226–237.
44. ——, *On some affine invariants of closed convex regions*, J. London Math. Soc. **14** (1939), 262–272.
45. D. J. Newman, *Location of the maximum on unimodal surfaces*, J. Assoc. Comput. Mach. **12** (1965), 395–398.
46. C. M. Petty, *Geominimal surface area*, Geom. Dedicata **3** (1974), 77–97.
47. F. P. Preparata and M. I. Shamos, *Computational geometry: An introduction*, Springer-Verlag, New York, 1985.
48. J. Renegar, *A polynomial-time algorithm, based on Newton's method for linear programming*, Math. Programming **40** (1988), 59–93.
49. C. Roos and I-Ph. Vial, *A polynomial method of approximate centers for linear programming*, Report 88–68, Technical University Delft, 1988.
50. G. T. Sallee, *A valuation property of Steiner points*, Mathematika **13** (1966), 76–82.
51. L. A. Santaló, *Integral geometry and geometric probability*, Addison-Wesley, Reading, MA, 1976.
52. R. Schneider, *On the Steiner points of convex bodies*, Israel J. Math **9** (1970), 241–249.
53. G. C. Shephard, *The Steiner point of a convex polytope*, Canad. J. Math. **18** (1966), 1294–1300.

54. N. Z. Shor, *Cut-off method with space extension*, in Convex Programming Problems, Cybernetics **13** (1977), 94–96.
55. G. Sonnevend, *An analytical center for polyhedrons and new classes of global algorithms for linear (smooth, convex) programming*, Proc. 12th IFIP Conf. on System Modeling, Budapest, Hungary, 1985.
56. ——, *An analytic center for polyhedrons and new classes of global algorithms for linear (smooth convex) programming*, in Lecture Notes in Control and Inform. Sci., vol. 84, Springer-Verlag, Berlin, 1985, pp. 866–876.
57. ——, *Applications of analytic centers for the numerical solution of semiinfinite convex programs arising in control theory*, Math. Programming Ser. B (Special Issue on Interior Point Algorithms in Linear Programming) (1990) (to appear).
58. ——, *Applications of analytic centers*, in Proc. NATO ASI on Numerical Linear Algebra and Digital Signal Processing, (P. Van Dooren, ed.), Reidel, 1988.
59. ——, *New algorithms in convex programming based on a notion of "centre" (for systems of analytic inequalities) and on rational extrapolation*, in Trends in Mathematical Optimization, (K. H. Hoffmann and J. Zowe, eds.), ISNM, vol. 84, Birkhäuser, Basel, 1988, pp. 311–327.
60. G. Sonnevend and J. Stoer, *Global ellipsoidal approximations and homotopy methods for solving convex analytic programs*, Appl. Math. Optim. **21** (1990), 139–165.
61. S. P. Tarasov, L. G. Khachiyan, and I. I. Erlikh, *The method of inscribed ellipsoids*, Soviet Math. Dokl. **37** (1988), 226–230.
62. D. M. Titterington, *Optimal design: Some geometrical aspects of D-optimality*, Biometrika **62** (1975), 313–320.
63. M. J. Todd, *Improved bounds and containing ellipsoids in Karmarkar's linear programming algorithm*, Math. Oper. Res. **13** (1988), 650–659.
64. ——, *Recent developments and new directions in linear programming*, Technical Report no. 827, College of Engineering, Cornell University, 1988.
65. T. Trafalis, *Efficient faces of a polytope: Interior methods in multiobjective optimization*, Ph.D. Dissertation, Purdue University, West Lafayette, IN, 1989.
66. P. M Vaidya, *A locally well-behaved potential function and a simple Newton-type method for finding the center of a polytope*, Technical Report, AT&T Bell Laboratories, Murray Hill, NJ, 1987.
67. ——, *A new algorithm for minimizing convex functions over convex sets*, Second Workshop on Math. Programming, Monterey, CA, 1990.
68. D. J. Wilde, *Optimization by the method on contour tangents*, Amer. Inst. Chemical Engineers J. **9** (1963), 186–190.
69. Ch. Witzgall, P. T. Boggs, and P. Domich, *On center trajectories and their relatives in linear programming*, Preprint, National Bureau of Standards, Gaithersburg, MD, 1988.
70. Y. Ye, *Karmarkar's algorithm and the ellipsoid method*, Oper. Res. Lett. **4** (1987), 177–182.

(Kaiser, Morin) School of Industrial Engineering and Program in Computational Combinatorics, IIES, Purdue University, West Lafayette, Indiana 47907
*E-mail address*: iies @ ecn.purdue.edu

(Trafalis) School of Industrial Engineering, University of Oklahoma, Norman, Oklahoma 73019

DIMACS Series in Discrete Mathematics
and Theoretical Computer Science
Volume 4, 1991

Applied Geometry
and Discrete Mathematics
THE VICTOR KLEE FESTSCHRIFT

# The Diameter of Graphs of Convex Polytopes and $f$-Vector Theory

GIL KALAI

ABSTRACT. The purpose of this paper is to discuss a relation between the diameter problem for simple polytopes and $f$-vectors of simplicial polytopes and subcomplexes of their boundary complexes. This is done via the recent concept of magnifying properties of graphs. We prove a far-reaching generalization of the upper bound theorem and use this result to prove a magnifying property and a polynomial bound on the diameter for graphs of dual-to-neighborly polytopes. We also prove the generalized lower bound inequalities for a large class of simplicial spheres. This paper contains the following sections:

1. Introduction, 2. Magnifiers and their diameters, 3. $M$-vectors and $K$-vectors, 4. $h$-vectors and $f$-vectors, 5. The face ring, Cohen-Macaulayness, and the Hard Lefschetz theorem, 6. Algebraic shifting, 7. Shifting and the $h$-vector, 8. The generalized lower bound inequalities for a large class of simplicial spheres, 9. The generalized upper bound theorem, 10. Toward a Kruskal-Katona type theorem for embedded complexes, 11. Magnification property and polynomial diameter via the generalized upper bound theorem, and 12. Open problems on the magnification of graphs of polytopes, random walks, and linear programming.

## 1. Introduction

In this paper we hope to establish a relation between two topics which are both among Victor Klee's favorite research areas. One topic is the study of $f$-vectors of polytopes, and especially results related to the upper-bound theorem. The other topic is the diameter problem for graphs of simple polytopes. The connection relies on the recent notion of expansion properties of graphs.

Let $P$ be a $d$-dimensional convex polytope (briefly, a $d$-polytope). Denote by $G(P)$ the *graph* (1-skeleton) of $P$. We mainly consider graphs of simple $d$-polytopes which are $d$-regular $d$-connected graphs. Denote by $f_k(P)$ the number of $k$-dimensional faces of $P$. The vector $(f_0(P), f_1(P), \dots, f_{d-1}(P))$ is called the *$f$-vector* of $P$. For general information on the theory

1980 *Mathematics Subject Classification* (1985 *Revision*). Primary 52A25.
The author was supported in part by a BSF grant, by a GIF grant, and by DIMACS.

of convex polytopes, the reader is referred to the books of Grünbaum [**Gr1**], McMullen and Shephard [**MS**], and Brønsted [**Br**], and to the recent survey [**KK2**].

One of the outstanding open problems in the theory of convex polytopes is to give upper bounds for the diameter of their graphs. This problem was motivated by Dantzig's simplex algorithm for linear programming. For a polytope $P$, $\delta(P)$ denotes the diameter of the graph of $P$. Let $\Delta(d, n)$ be the maximum of $\delta(P)$ over all $d$-polytopes with $n$ facets.

The following conjecture was proposed by W. M. Hirsch in 1957 and reported in Dantzig [**Da**].

THE HIRSCH CONJECTURE. $\Delta(d, n) \leq n - d$.

For further information on the diameter problem we refer the reader to [**Gr1, Chapter 16; Kl1, KW, La**] and to a recent survey by Klee and Kleinschmidt [**KK1**]. Klee and Walkup [**KW**] observed that $\Delta(d, n)$ is attained by simple polytopes.

In view of the failure of several natural extensions of the Hirsch conjecture there are some doubts about the validity of the Hirsch conjecture in high dimensions. The best upper bound for $\Delta(d, n)$, proved by Larman [**La**], is $\Delta(d, n) \leq 2^{d-3}n$. (See note added in proof.) Thus, even the following weak form of the Hirsch conjecture is open and is of great interest.

POLYNOMIAL DIAMETER CONJECTURE. There is a universal constant $c$ so that $\Delta(d, n) \leq n^c$.

The relation between $f$-vectors and the diameter problem depends on an important concept from graph theory. Following Alon and Milman [**AM**] we call a graph $G$ with $n$ vertices a *c-magnifier* if for every subset $X$ of its vertices, $|X| \leq n/2$, the number of vertices which do not belong to $X$ but are adjacent to some vertex of $X$ is at least $c|X|$. By repeatedly applying the magnifying property it follows immediately that the diameter of a $c$-magnifier with $n$ vertices is at most $1 + 2\log_{1+c} n$.

A fundamental theorem of Harper [**Ha**] from 1966 implies that the graph of the $d$-cube is a $(2\pi d)^{-1/2}$-magnifier. This classical result which preceded the general notion of expanders and magnifiers is a cornerstone of extremal combinatorics, and can be regarded as our starting point. See also [**Bo, Chapter 16**].

In this paper we prove a polynomial bound on $\delta(P)$ via a magnifying property of $G(P)$ for a large class of simple polytopes. The description of this class of polytopes and the proof of the magnifying property are both related to the upper bound theorem which was proved by Peter McMullen in 1970, in response to a conjecture dated from 1957 by Theodore Motzkin.

THE UPPER BOUND THEOREM [**Mc1**]. *The maximum number of facets of a d-polytope with n vertices is*

$$g(d, n) = \binom{n - \lfloor (d+1)/2 \rfloor}{n-d} + \binom{n - \lfloor (d+2)/2 \rfloor}{n-d}.$$

This value is attained for $C(d, n)$, the cyclic $d$-polytope with $n$ vertices, and for a large class of simplicial polytopes called *neighborly polytopes* (see **[Gr1, Chapter 7; Sh]**). In a dual form, the upper bound theorem asserts that every $d$-polytope with $n$ facets has at most $g(d, n)$ vertices.

THEOREM 1. *Let $P$ be a $d$-polytope with $n$ facets and $g(d, n)$ vertices. Then, $G(P)$ is a $(d^{-2}(n-d)^{-2})$-magnifier and $\delta(P) \leq d^2(n-d)^2 \log n$.*

In fact, we show that a magnifying property and a polynomial diameter hold whenever the number of vertices of $P$ is close enough to $g(d, n)$.

The proof of Theorem 1 depends on a far-reaching generalization of the upper bound theorem.

THEOREM 2 (The generalized upper bound theorem). *Let $K$ be a $(d-1)$-dimensional subcomplex of the boundary complex of a simplicial $d$-polytope. If $f_{d-1}(K) \geq f_{d-1}(C(d, n))$, then $f_i(K) \geq f_i(C(d, n))$ for every $i \leq d-1$.*

Note that Theorem 2 reduces to the upper bound theorem when $K$ is the entire boundary complex of $P$ and $i = 0$.

How do we deduce Theorem 1 from Theorem 2? Given a set $A$ of $a$ vertices in a simple $d$-polytope, Theorem 2 (more precisely, a finer version of Theorem 2 of Kruskal-Katona type) gives a lower bound $S(a)$ on the number of 2-faces which meet $A$. (We use the case $i = d-3$ of Theorem 2, in a dual form.) The crucial fact is the subadditivity of the function $S(a)$. Let $P$ be a simple polytope and consider a partition of the vertices of $P$ into two parts $A$ and $B$. Put $a = |A|$, $b = |B|$, and let $a \leq b$. From Theorem 2 it follows that there are at least $S(a)$ 2-faces incident to vertices of $A$ and at least $S(b)$ 2-faces incident to vertices of $B$. If $P$ is a dual-to-neighborly polytope, then $P$ has altogether $S(a+b)$ 2-faces and we obtain, by the subadditivity of the function $S(x)$, a lower bound on the number of 2-faces which contain vertices from both $A$ and $B$. This implies the required magnifying property. This type of reasoning is closely related to the work of P. Levy **[Lev]**.

The proof of Theorem 2 depends on recent results and tools from $f$-vector theory. In the last twenty years many results concerning face numbers of polytopes and more general complexes were proved using various techniques. See **[St5, BK2, Bj]** for surveys of these developments. In 1975 Richard Stanley **[St1]** used the Cohen-Macaulay property of a certain commutative ring (the face-ring) associated to simplicial spheres to extend the upper bound theorem to arbitrary simplicial spheres. Around 1980 Stanley **[St3]** and Billera and Lee **[BL1]** proved the $g$-theorem, conjectured by McMullen **[Mc2]**, which gives a complete description of $f$-vectors of simplicial polytopes. In his proof of the necessity part of the $g$-theorem Stanley **[St3]** needed a further crucial property of the face ring of simplicial polytopes which he derived from the hard-Lefschetz theorem for toric varieties.

In order to prove Theorem 2 we need to use the algebraic machinery described above combined with another tool-algebraic shifting, which was introduced by the author in **[Ka1]**, see also **[Ka2, BK1, BK2, Ka4]**. We develop a variant of algebraic shifting based on the face rings.

New applications of the face ring, via algebraic shifting, to the study of $f$-vectors of subcomplexes of boundary complexes of simplicial polytopes is the second purpose of this paper. McMullen and Walkup conjectured in **[MW]** a system of linear inequalities—the "generalized lower bound inequalities" for the $f$-vectors of simplicial polytopes. These inequalities were proved by Stanley **[St3]** as part of his proof for the necessity part of the $g$-theorem. It was repeatedly conjectured that the generalized lower bound inequalities and the entire $g$-theorem apply to arbitrary triangulations of spheres. (See note added in proof.)

THEOREM 3. *The $f$-vector of every simplicial $(d-2)$-sphere which is a subcomplex of the boundary complex of a simplicial $d$-polytope satisfies the generalized lower bound inequalities.*

The structure of the paper is as follows. §§2–5 contain background material on magnifiers, $f$-vectors, $h$-vectors, and face rings. Background material which is not strictly needed for the proofs is given separately at the end of each section. In §6 we develop the notion of algebraic shifting via the face ring. The proofs of our main results are given in §§7, 8, 9, and 11. In §10 we discuss several conjectures on $f$-vectors of embedded complexes, and §12 contains a discussion of open problems related to the magnifying properties of graphs of polytopes, the diameter conjecture, and linear programming.

## 2. Magnifiers and their diameters

Let $G$ be a graph. For a set $X$ of vertices of $G$ denote by $N(X)$ the set of vertices of $G$ which are either in $X$ or are adjacent to some vertex of $X$. Recall that a graph $G$ with $n$ vertices is a $c$-magnifier if $|N(X)| \geq (1+c)|X|$ for every subset $X$ of its vertices, $|X| \leq n/2$.

LEMMA 2.1 **[AM]**. *Let $G$ be a $c$-magnifier with $n$ vertices. Then, $\delta(G) \leq 1 + 2\log_{1+c} n$.*

PROOF. Let $v$ be a vertex of $G$. Denote by $N_k(v)$ the set of all vertices in $G$ whose distance from $v$ is at most $k$. (Thus, $N_0(v) = \{v\}$, and $N_1(v) = N(\{v\})$.) By the magnifying property, if $|N_k(v)| \leq n/2$, then $|N_{k+1}(v)| \geq (1+c)|N_k(v)|$. It follows that for $k = 1+\log_{1+c}(n/2)$, $|N_k(v)| > n/2$. If $w$ is another vertex of $G$, then $N_k(v) \cap N_k(w) \neq \varnothing$, which implies that the distance between $v$ and $w$ is at most $2k-1$. □

The magnifying constant can be replaced by a magnifying function depending on the size of $X$. Let $c(x)$ be a real function on the interval $[0, 1]$. A graph $G$ with $n$ vertices is a *$c(x)$-magnifier* if for every subset $X$ of its vertices $|N(X)| \geq c(|X|/n)|X|$. The proof of Lemma 2.1 applied in such a situation may improve the bound on the diameter. For example,

Harper's theorem [**Ha, Bo**] actually gives the precise magnification function of the $d$-dube. In this case (as in any case where the minimum magnification is achieved by balls in the graph metric), the magnification function implies the correct value of the diameter, while the magnification constant may give a worse bound.

**Further discussion.** *Background of magnifiers.* Starting in 1973 by Pinsker, a rich theory of magnifiers (more precisely, expanders, which are their bipartite analogs) and related objects was developed. Probabilistic methods were used to prove the existence of graphs with superb magnifying properties, and deep algebraic methods were used for explicit constructions of such graphs. Milder magnifying properties were identified in large classes of graphs.

Numerous applications of expanders and magnifiers in various areas of mathematics and computer science were given. The reader is referred for more details and for references to the splendid survey article by Bien [**Bi**].

For a finite graph $G$, let $I(G)$ be the incidence matrix of $G$. The Laplacian $L(G)$ is defined by $L(G) = I(G)I(G)^{tr}$. An important theory on the relation between magnifying properties of a graph $G$ and the second eigenvalue of the Laplacian of the graph, denoted here by $\lambda(G)$, was developed by Tanner, Alon, and Milman. Tanner [**T**] and separately Alon and Milman [**AM**] proved that a graph $G$ with maximal vertex degree $k$ is a $(2\lambda(G)/(k + 2\lambda(G)))$-magnifier. Alon [**A**] proved that for a $c$-magnifier $G$, $\lambda(G) \geq c^2/(4 + 2c^2)$. Alon's theorem implies (as observed by Aldous) that random walks on (even rather mild) magnifiers converge quickly to a limiting distribution.

Direct connections between $\lambda(G)$ and the diameter of $G$ are given by Alon and Milman [**AM**], by Chung [**C**], and (implicitly) by Lubotzky, Phillips, and Sarnak [**LPS**].

## 3. M-vectors and K-vectors

For $n, k \geq 1$ there is a unique expansion

$$n = \binom{a_k}{k} + \binom{a_{k-1}}{k-1} + \cdots + \binom{a_i}{i}, \tag{3.1}$$

such that $a_k > a_{k-1} > \cdots > a_i \geq i \geq 1$. This given, define

$$\partial^{k-1}(n) = \binom{a_k - 1}{k-1} + \binom{a_{k-1} - 1}{k-2} + \cdots + \binom{a_i - 1}{i-1}.$$

Also let $\partial^{k-1}(0) = 0$.

A sequence $f = (f_0, f_1, \ldots, f_{d-1})$ is an M-vector if $f_0 = 1$ and $0 \leq \partial^k(f_k) \leq f_k$.

A collection $M$ of monomials in variables $x_1, x_2, \ldots$ is an *order ideal of monomials* if whenever $m \in M$ and a monomial $u$ divides $m$ then $u \in M$. Let $f_k(M)$ be the number of monomials in $M$ of degree $k$.

MACAULAY THEOREM [**M**]. *Let* $f = (f_0, f_1, \dots)$ *be a sequence of nonnegative integers. Then the following conditions are equivalent*:

(i) $f$ *is an* M-*vector.*
(ii) $f$ *is the* $f$-*vector of some order ideal of monomials.*

For integers $n, k \geq 1$, given the representation (3.1) of $n$, define

$$\partial_{k-1}(n) = \binom{a_k}{k-1} + \binom{a_{k-1}}{k-2} + \cdots + \binom{a_i}{i-1}.$$

Also let $\partial_{k-1}(0) = 0$.

A sequence $f = (f_0, f_1, \dots, f_d, \dots)$ is a K-vector if $0 \leq \partial_k(f_k) \leq f_{k-1}$.

KRUSKAL-KATONA THEOREM. *For an ultimately vanishing integral sequence* $f = (f_0, f_1, \dots, f_d, \dots)$ *the following conditions are equivalent*:

(i) $f$ *is the* $f$-*vector of some simplicial complex,*
(ii) $f$ *is a* K-*vector.*

For references to the original papers, and for short proofs of the Kruskal-Katona theorem the reader is referred to [**Bo, F**]. Lovàsz [**Lo**] observed that if $n = \binom{x}{k}$, then $\partial_k(n) \geq \binom{x}{k-1}$ and similarly (see [**BFS**])

$$\partial^k(n) \geq \binom{x-1}{k-1}. \tag{3.2}$$

This gives slightly weaker but more handy forms of the Macaulay and Kruskal-Katona theorems.

## 4. $f$-vectors and $h$-vectors

We need the notion of $h$-vectors introduced by McMullen and Walkup [**MW**], with seeds going back to Sommerville. $h$-vectors play an important role in McMullen's proof of the upper bound theorem [**Mc1**], in the formulation of the $g$-theorem [**Mc2, BL1, St3**], and in many other developments.

Let $K$ be a $(d-1)$-dimensional simplicial complex and $f = (f_0, f_1, \dots, f_{d-1})$ be the $f$-vector of $K$. The $h$-vector of $K$ is defined by the relation

$$\sum_{k=0}^{d} h_k x^{d-k} = \sum_{k=0}^{d} f_{k-1}(x-1)^{d-k}. \tag{4.1}$$

This is equivalent to

$$h_k = \sum_{j=0}^{k} (-1)^{k-j} \binom{d-j}{d-k} f_{j-1} \quad \text{and} \quad f_{k-1} = \sum_{j=0}^{k} (-1)^{k-j} \binom{d-j}{d-k} h_j.$$

McMullen's proof [**Mc1**] of the upper bound theorem is based on the following sharper inequalities in terms of the $h$-numbers.

McMULLEN'S UPPER BOUND THEOREM. *For every simplicial $d$-polytope $P$ with $n$ vertices*

$$h_k(P) \le \binom{n-d+k-1}{k}, \qquad 0 \le k \le \frac{d}{2}. \tag{4.2}$$

We state now the $g$-theorem (conjectured by McMullen in [**Mc2**]) which gives a complete description of $f$-vectors of simplicial polytopes.

THE $g$-THEOREM (Necessity: Stanley [**St3**]; Sufficiency: Billera and Lee [**BL1**]). *A vector $h = (h_0, h_1, \dots, h_d)$ is the $h$-vector of a simplicial $d$-polytope if and only if the following conditions hold*:

(1) (*Dehn-Sommerville relations*) $h_i = h_{d-i}$.
(2) (*Generalized lower bound inequalities*) $h_0 \le h_1 \le \cdots \le h_{\lfloor d/2 \rfloor}$.
(3) *The vector* $(h_0, h_1 - h_0, \dots, h_{\lfloor d/2 \rfloor} - h_{\lfloor d/2 \rfloor - 1})$ *is an* M-*vector.*

It is conjectured that the $h$-vector of every simplicial sphere satisfies conditions 1–3. It is known that the $h$-vector of every simplicial sphere satisfies the Dehn-Sommerville relations, but conditions 2 and 3 are open.

**Further discussion.** *Geometrical and combinatorial meaning of the $h$-vector.* A *shelling order* of a simplicial $d$-polytope $P$ is an ordering $F_1, F_2, \dots, F_t$ of the facets of $P$ with the property that for every $i$, $1 < i < t$, $G_i = (F_1 \cup F_2 \cup \cdots \cup F_{i-1}) \cap F_i$ is the union of $k$ subfacets of $F_i$ for some $1 \le k \le d$. Define $\deg(F_i) = k$. McMullen proved that for every shelling order $F_1, F_2, \dots, F_t$ of the facets of a simplicial polytope $P$,

$$h_k(P) = |\{i : \deg(F_i) = k\}|.$$

This relation is crucial in McMullen's proof of the upper bound theorem. Bruggesser and Mani [**BrM**] proved that the boundary complex of every polytope is shellable (= admit a shelling order).

Let $P$ be a simplicial $d$-polytope and let $Q$ be its polar. Let $\phi$ be a linear objective function which is not constant on any edge of $Q$. Order the vertices according to their $\phi$ value. (This corresponds to a special type of shelling order on the facets of $P$.) $h_k(P)$ is the number of vertices of $Q$ which have exactly $k$ neighbors with a larger $\phi$ value. (See, e.g., [**Mc1, Br**].)

*Shellability and the Hirsch conjecture.* Provan and Billera [**PB, KK1**] proved the Hirsch conjecture for every simplc polytope whose dual satisfies a very strong form of shellability, called *vertex-decomposability.*

## 5. The face ring, Cohen-Macaulayness, and the hard-Lefschetz theorem

Let $K$ be an abstract $(d-1)$-dimensional simplicial complex on a finite set $V$ of vertices. Assume that $V = [n] = \{1, 2, \dots, n\}$. Consider the polynomial ring over the complex numbers $\mathbf{C}[x_1, x_2, \dots, x_n]$. For a monomial $m \in \mathbf{C}[x_1, \dots, x_n]$, $m = \prod x_i^{k_i}$, define the support of $m$ by $\mathrm{supp}(m) = \{i : k_i > 0\}$. Let $I(K)$ be the ideal defined by

$$I(K) = \{m \in \mathbf{C}[x_1, \dots, x_n] : \mathrm{supp}(m) \notin K\}.$$

The *face ring* of $K$ is defined as the quotient $R(K) = \mathbf{C}[x_1, \dots, x_n]/I(K)$. This ring was first considered by Stanley [**St1**] and Reisner [**R**]. (It is often called the *Stanley-Reisner ring*.) The connection between convex polytopes and commutative algebra was first considered in the fundamental paper by Hochster [**Ho**].

$R(K)$ is a *Cohen-Macaulay ring* if there are $d$ homogeneous linear forms (called a *system of parameters*) $\theta_1, \theta_2, \dots, \theta_d$ and a finite set of monomials $\eta_1, \eta_2, \dots, \eta_t$, so that every element $m$ in $R(K)$ can be written uniquely as

$$m = \sum_{k=1}^{t} p_k(\theta_1, \dots, \theta_d)\eta_k,$$

where $p_k$ is a polynomial in $\theta_1, \dots, \theta_d$. In other words, $\mathscr{H} = R(K)/\langle\theta_1, \dots, \theta_d\rangle$ is a finite-dimensional vector space. We can write $\mathscr{H}$ as a direct sum $\mathscr{H} = \mathscr{H}_0 \oplus \mathscr{H}_1 \oplus \cdots$, where $\mathscr{H}_k$ is the vector space spanned by those $\eta$'s of degree $k$.

Reisner [**R**] found a topological condition on $K$ which is equivalent to the Cohen-Macaulayness of the face ring $R(K)$.

REISNER'S THEOREM [**R**]. *Let $K$ be a simplicial $(d-1)$-dimensional complex. A necessary and sufficient condition for $R(K)$ to be Cohen-Macaulay is*

(CM) $\widetilde{H}_k(\mathrm{lk}(S, K)) = 0$ *for every face $S$ and $k < \dim \mathrm{lk}(S, K)$.*

(*Here* $\mathrm{lk}(S, K)$ *denotes the link of $S$ in $K$ and $\widetilde{H}_k$ are the reduced homology groups with coefficients in* $\mathbf{C}$.)

The Cohen-Macaulayness of face-rings of shellable complexes has a simple inductive proof due to Kind and Kleinschmidt [**KiK**]. Condition (CM) clearly holds for simplicial spheres.

Stanley proved that if $R(K)$ is Cohen-Macaulay, then

$$\dim \mathscr{H}_k(K) = h_k(K).$$

From this interpretation of the $h$-vector combined with Reisner's theorem Stanley proved the upper bound theorem (in fact, inequality (4.2)) for arbitrary simplicial spheres.

Stanley also proved that if $R(K)$ is Cohen-Macaulay, then the $h$-vector of $K$ is an M-vector, and he observed [**St2**] that the necessity of McMullen's conditions follows from the following algebraic hypothesis on face-rings of simplicial spheres.

$\omega$-HYPOTHESIS. Let $K$ be a $(d-1)$-dimensional simplicial sphere. There is a linear form $\omega$ so that for every $k \le \frac{d}{2}$, multiplication by $\omega^{d-2k}$ gives an isomorphism between $\mathscr{H}_k$ and $\mathscr{H}_{d-k}$.

In 1980 Stanley proved the $\omega$-Hypothesis for face-rings of boundary complexes of simplicial polytopes. His proof is based on the Hard-Lefschetz theorem for toric varieties.

We make an essential use of Cohen-Macaulayness and of the $\omega$-Hypothesis for face rings of boundary complexes of simplicial polytopes. We need the following obvious fact.

PROPOSITION 5.1. *Let* $K$ *be a* $(d-1)$*-dimensional simplicial complex. If* $R(K)$ *is Cohen-Macaulay, then the system of parameters* $\theta_1, \dots, \theta_d$ *can be chosen as generic linear forms. If* $R(K)$ *satisfies the* $\omega$*-Hypothesis, then* $\theta_1, \dots, \theta_d$, $\omega$ *can be chosen as generic linear forms.*

**Further discussion.** 1. For an integral simplicial $d$-polytope $P$ with $0 \in \operatorname{int} P$, there is a *toric variety* associated with $P$ denoted by $T(P)$. Let $R(P)$ be the face ring of $P$, and let $\theta_1, \dots, \theta_d$ be the system of parameters given by the rows of the coordinate matrix of the vertices. The cohomology ring $H^*$ of $T(P)$ is isomorphic to $\mathscr{H} = R(P)/\langle\theta_1, \dots, \theta_d\rangle$. (The grading is doubled.) This was proved in Danilov **[D]**. In particular, the $2i$th Betti number of $T(P)$ is $h_i(P)$. A direct link between shellings of $P$ and Morse functions of $T(P)$ which verifies this fact is given by Khovanskii **[Kho]**. The hard-Lefschetz theorem for $T(P)$ gives an action of $\mathrm{sl}_2(\mathbf{C})$ on $H^*$. This gives the necessary $\omega$ and in addition there is $\omega^* : \mathscr{H}^{i+1} \to \mathscr{H}^i$, $i \geq 0$, so that $U = \omega\omega^*$ restricted to $\mathscr{H}^k$ is equal to $(2k-d)I$. See **[St5]**.

2. Much effort was made by several researchers in order to find a more elementary proof for the necessity part of the $g$-theorem. There are two basic approaches to the problem.

I. *The generic approach.* Prove directly (perhaps, by some inductive argument based on a shelling or a related combinatorial operation) that a generic $\omega$ (for a generic system of parameters) obeys the $\omega$-hypothesis. This approach may apply for extending the $g$-theorem for larger classes of simplicial spheres.

II. *The geometric approach.* Identify geometrically the $\theta$'s and the $\omega$ (and perhaps even $\omega^*$) given by the hard-Lefschetz theorem in the face ring of boundary complexes of simplicial polytopes. This approach may be helpful for several extensions of the $g$-theorem for nonsimplicial polytopes **[St6]**. (As mentioned above, Danilov identified the $\theta$'s, and there is a guess made by many that $\omega$ should be $x_1 + x_2 + \cdots + x_n$. This guess, if true, shows that the hard-Lefschetz theorem reduces for 3-polytopes to the Dehn-Alexandrov infinitesimal-rigidity theorem. See, e.g., **[Le]**.)

## 6. Algebraic shifting

Let $\binom{\mathbf{N}}{k}$ denote the family of $k$-subsets of $\mathbf{N} = \{1, 2, 3, \dots\}$. Consider the partial ordering $<_p$ on $\binom{\mathbf{N}}{k}$ defined as follows: If $S = (s_1, s_2, \dots, s_k)_<$ and $T = (t_1, t_2, \dots, t_k)_<$ (the subscript $<$ indicates that the elements are listed in increasing order), then $S \leq_p T$ iff $s_i \leq t_i$ for every $i$.

A family $A \subset \binom{\mathbf{N}}{k}$ is *shifted* if $S \in A$ and $R \leq_p S$ imply $R \in A$.

For $A \in \binom{\mathbf{N}}{k}$ the shadow of $A$, $\partial A$, is defined by

$$\partial A = \left\{ R \in \binom{\mathbf{N}}{k-1} : \exists S \in A,\ R \subset S \right\}. \tag{6.1}$$

A shifting operation is a map which assigns to each family $A \in \binom{\mathbf{N}}{k}$ a family $\Delta(A)$ so that the following conditions hold:

(A1) $|A| = |\Delta(A)|$.
(A2) $\Delta(A)$ is shifted.
(A3) $\partial\Delta(A) \subset \Delta\partial(A)$.
(A4) If $B \subset A$, then $\Delta(B) \subset \Delta(A)$.

If $K$ is a simplicial complex, define $K_r = \{S \in K : |S| = r\}$ and $\Delta(K) = \bigcup \Delta(K_r)$. (A3) implies that $\Delta(K)$ is a simplicial complex. We can define a shifting operation on the class of simplicial complexes $K$ on the vertex set $\mathbf{N}$ by the following requirements:

(A1′) $f(K) = f(\Delta(K))$.
(A2′) $\Delta(K)$ is shifted (i.e., $K_r$ is shifted for every $r$).
(A3′) $\Delta(K)$ is a simplicial complex.
(A4′) If $L$ is a subcomplex of $K$, then $\Delta(L) \subset \Delta(K)$.

A combinatorial shifting operation introduced by Erdös, Ko, and Rado played a crucial rule in extremal combinatorics, see [**F**] and [**Bo**]. (Our notations are closer to Frankl's.) An algebraic shifting operation based on exterior algebra was introduced by the author [**Ka1**]. Here we need a variant based on the face ring.

For our purposes we need two additional properties of the shifting operation.

(B1) If $K$ is a Cohen-Macaulay complex, then $\Delta(K)$ is pure.

Definition. A $d$-subset $S$ of $[n]$ is *admissible* if $k \notin S$ implies $[k+1, d-k+2] \subset S$.

(B2) If $K$ is the boundary complex of a simplicial $d$-polytope, then every $(d-1)$-face of $\Delta(K)$ is admissible.

We describe now a shifting operation based on the face ring which satisfies conditions (A1)–(A4) and (B1), (B2). From that point on our arguments rely only on these conditions.

Let $\theta_1, \theta_2, \ldots, \theta_n$ be variables. We first need to define a lexicographic ordering on monomials. Let $m_1 = \theta_1^{k_1}\theta_2^{k_2}\cdots\theta_n^{k_2}$ and let $m_2 = \theta_1^{l_1}\theta_2^{l_2}\cdots\theta_n^{l_n}$. Define $m_1 <_L m_2$ if there exists a $j$ so that $k_u = l_u$ for $u < j$ and $k_j > l_j$. We also consider the majorization (partial) ordering $<_p$ on monomials defined by: For $m_1$ and $m_2$ as above, $m_1 \leq_p m_2$ if $\sum_{i=1}^t k_i \geq \sum_{i=1}^t l_i$ for every $t = 1, 2, \ldots, n$.

Now, let $R(K)$ be the face ring of $K$ (over $\mathbf{C}$.) Let $\theta_1, \theta_2, \ldots, \theta_n$ be $n$ linearly independent linear forms in $x_1, \ldots, x_n$. For an element $m$ in $\mathbf{C}[x_1, \ldots, x_n]$ denote by $\tilde{m}$ its image in $R(K)$. Let $M$ be the set of monomials in $\theta_1, \theta_2, \dots, \theta_n$. Define

(6.2) $$B(K) = \{m \in M : \tilde{m} \notin \operatorname{span}\{\tilde{l} : \deg l = \deg m, \, l <_L m\}\}.$$

(Here, for a set $A$ of monomials, span $A$ is the vector space spanned by $A$.)

LEMMA 6.1. *$B(K)$ is an order ideal of monomials, i.e., $m \in B(K)$ and $l \mid m$ implies $l \in B(K)$.*

PROOF. This is easy and well known **[St1]**. □

We assume now that $\theta_1, \theta_2, \dots, \theta_n$ are generic, i.e., $\theta_i = \sum a_{ij} x_j$, where the $a_{ij}$'s are transcendentals, algebraically independent over $\mathbf{Q}$.

LEMMA 6.2. *If $\theta_1, \theta_2, \dots, \theta_n$ are generic, then $B(K)$ is an initial set of monomials with respect to the majorization ordering.*

PROOF. The proof is completely analogous to the proof for algebraic shifting based on the exterior face-algebra, see **[BK1, Ka3]**. □

Define

$$\Delta_k = \{\{j_1, j_2, \dots, j_k\} : \theta_{j_1+k-1} \cdot \theta_{j_2+k-2} \cdot \dots \cdot \theta_{j_k} \in B(K)\},$$

and $\Delta(K) = \bigcup_{i \geq 0} \Delta_i$.

LEMMA 6.3. *Let $K$ be a simplicial complex of dimension $d-1$. Define*

$$U_r = \{m \in B(K) : \deg(m) = r, \;\; m \notin \langle \theta_1, \dots, \theta_{r-1} \rangle\}.$$

*Then the following conditions hold:*

(i) $|U_r| = f_{r-1}$.
(ii) *If $m \in U_r$ and $u$ is a monomial in the variables $\theta_1, \theta_2, \dots, \theta_r$, then $mu \in B(K)$.*

PROOF. First note that (ii) implies (i). From (ii) it follows that

$$\sum_{r=0}^{d} \binom{k-1}{r} |U_r| = |B(K)_k| = \sum_{r=0}^{d} f_{r-1} \binom{k-1}{r}.$$

(Here $B(K)_k = \{m \in B(K) : \deg(m) = k\}$.) The proof is by induction on $d$. The theorem is obvious for $d = 1$. Let $D$ be the $(d-2)$-skeleton of $K$. From the induction hypothesis and simple counting it follows that $|U_d| = f_{d-1}$. Assume that the theorem is violated and $mu \notin B(K)$ for $m$ and $u$ as above. From the induction hypothesis we may assume that $m \in U_d$, and since $B(K)$ is shifted we may assume that $u = \theta_d^r$ for some $r > 0$. We reach a contradiction by the following simple counting argument. Consider $m \gg d, r, n$. Then $|B(K)_m| = f_d \binom{m-1}{d} + o(m^d)$. On the other hand, the assumption $\theta^r m \notin B(K)$ implies that $|B(K)_m| < (f_d - 1)\binom{m-1}{d} + o(m^d)$. □

THEOREM 6.4. *The operation $K \to \Delta(K)$ satisfies properties* (A1′)–(A4′) *and* (B1), (B2).

PROOF. Properties (A1′), (A2′), and (A3′) follow respectively from Lemmas 6.3(i), 6.2, and 6.1. To prove (A4′) note that if $L$ is a subcomplex of $K$, then $R(L)$ is a quotient ring of $R(K)$. If $m_0 \notin B(K)$, then $(*)$ $\tilde{m}_0 = \sum\{a_m \tilde{m} : m <_L m_0\}$. This relation carries over to $R(L)$.

For a set $S$ of integers denote by $M(S)$ the set of monomials in the variables $\{\theta_i : i \in S\}$. To say that $K$ is Cohen-Macaulay and $\theta_1, \dots, \theta_d$ is a system of parameters for $R(K)$ is equivalent to the fact that if $m$ is a monomial in $\theta_1, \dots, \theta_d$ and $u$ is any monomial in $\theta_1, \dots, \theta_n$, then $mu \notin B(K)$ implies that $u \notin B(K)$.

Define

$$L(K) = B(K) \cap M([d+1, n]).$$

$L(K)$ is an order ideal of monomials of degrees at most $d$ in the variables $\theta_{d+1}, \dots, \theta_n$. If $K$ is Cohen-Macaulay, then $B(K)$ is determined from $L(K)$ by the relation

$$B(K) = \{m_1 m_2 : m_1 \in M([d]),\ m_2 \in L(K)\}.$$

It follows by an easy inspection that $\Delta(K)$ is the pure $(d-1)$-dimensional simplicial complex whose facets are given by

(6.3)

$$\Delta_d(K) = \bigcup\{[d-k] \cup \{i_1-(k-1), i_2-(k-2), \dots, i_k\} : \theta_{i_1}\theta_{i_2}\cdots\theta_{i_k} \in L(K)\}.$$

Let $K$ be a $(d-1)$-dimensional Cohen-Macaulay simplicial complex which satisfies the $\omega$-hypothesis. Regard $\theta_{d+1}$ as $\omega$. The $\omega$-hypothesis is equivalent to the fact that if $2k < d$, $m \in L(K)$, and $\deg m = d-k$, then $m = \theta_{d+1}^{d-2k} m'$, where $m' \in L(K)$. Define

$$U(K) = B(K) \cap M([d+2, n]).$$

The degree of a monomial $m$ in $U(K)$ is at most $\lfloor \frac{d}{2} \rfloor$. $L(K)$ is the set of all monomials of the form $\theta_{d+1}^u m$, where $m \in U(K)$ and $0 \le u \le d - 2\deg m$. The facets of $\Delta(K)$ which correspond to such monomials by (6.3) are admissible. Therefore $\Delta(K)$ satisfies (B2). □

Remark. If $K$ is a Cohen-Macaulay $(d-1)$-dimensional simplicial complex and $\theta_1, \dots, \theta_d$ is a system of parameters, then the monomials in $L(K)$ form a basis to $R(K)/\langle\theta_1, \dots, \theta_d\rangle$. The order ideal of monomials $L(K)$ was considered by Stanley in **[St1]**. Stanley deduced the fact that the $h$-vector of $K$ is an M-vector from the fact that the number of monomials in $L(K)$ of degree $r$ is $h_r$.

If the $\theta$'s are generic, then $L(K)$ is a shifted order ideal of monomials of degrees at most $d$ in the variables $\theta_{d+1}, \dots, \theta_n$. $\Delta(K)$ is determined by $L(K)$ by formula (6.3). Conversely, for every shifted order ideal of monomials $L$ of degrees at most $d$ in the variables $\theta_{d+1}, \dots, \theta_n$, formula (6.3) gives a pure shifted complex $K$, with $L(K) = L$.

If $K$ also satisfies the $\omega$-hypothesis, then $U(K)$ is a shifted order ideal of monomials with $h_k - h_{k-1}$ monomials of degree $k$, $k = 0, 1, \dots, \lfloor \frac{d}{2} \rfloor$, in the variables $\theta_{d+2}, \dots, \theta_n$. In **[Ka5]** there is a construction of a simplicial $(d-1)$-sphere $S(U)$ on $n$ vertices for every shifted order ideal of monomials

$U$ of degrees at most $\lfloor \frac{d}{2} \rfloor$ in the variables $\theta_{d+2}, \ldots, \theta_n$. It is plausible but yet to be proved that $U(S(U)) = U$.

**Further discussion.** 1. *Shifting by exterior algebra.* Let $A$ be a collection of subsets of $[n]$. Consider the exterior algebra over (say) the complex numbers $\bigwedge[x_1, x_2, \ldots, x_n]$. For a set $S = \{k_1, k_2, \ldots, k_s\}_< \subset [n]$ put $x_S = x_{k_1} \wedge x_{k_2} \wedge \cdots \wedge x_{k_s}$. Define $I(A) = \text{span}\{x_S : S \notin A\}$ and $\bigwedge(A) = \bigwedge[x_1, \ldots, x_n]/I(A)$. For an element $m$ in $\bigwedge[x_1, \ldots, x_n]$ denote by $\tilde{m}$ its image in $\bigwedge(A)$. Let $y_1, y_2, \ldots, y_n$ be generic linear forms. Define

$$\Delta^e(A) = \left\{S \in \binom{\mathbf{N}}{k} : \tilde{y}_S \notin \text{span}\{\tilde{y}_R : R <_L S\}\right\}.$$

The map $A \to \Delta^e(A)$ is an algebraic shifting operation which satisfies properties (A1)–(A4).

Properties (B1) and (B2) arose as conjectures for $\Delta^e$. (B1) was proved for shellable complexes in **[Ka4]** and recently a proof for arbitrary Cohen-Macaulay complexes was found **[Ka6]**. This can be regarded as an exterior algebra analogue of Reisner's theorem. A proof of (B2) for $\Delta^e$ is still unknown.

Algebraic shifting (both the symmetric and the exterior versions) can be carried over to other fields and the outcome depends on (and only on) the field characteristic. Properties (A1)–(A6), and (B1) are valid (for both versions) over arbitrary characteristic. Property (B2) is open for any nonzero characteristic.

2. *Further natural properties of algebraic shifting.* From this point until the end of this section we refer to the shifting operation based on the face-ring by $\Delta^s$. For $G = K_{3,3}$ the complete bipartite graph on 6 vertices, $\Delta^e(G) \neq \Delta^s(G)$.

All the shifting operations described here satisfy

(A5) If $A$ is shifted, then $\Delta(A) = A$.

This is obvious for the combinatorial shifting and proved in **[Ka3]** for $\Delta^e$. This proof can be easily modified to apply for $\Delta^s$.

A property which holds for the algebraic shiftings but not for the combinatorial shifting is:

(A6) If $B$ is combinatorially isomorphic to $A$ (i.e., $B$ is obtained from $A$ by changing the labels of the vertices), then $\Delta(A) = \Delta(B)$.

The combinatorial shifting operation preserves many combinatorial properties of the original family, see **[F; Bo, Chapter 17]**. Many of these properties are preserved also by $\Delta^e$ and $\Delta^s$. For example, if $A$ is intersecting, then so is $\Delta^e(A)$.

3. *Shifting and topology.* The main significance of algebraic shifting is that homological properties of $K$ are translated into combinatorial properties of $\Delta(K)$. In particular, there is the following simple interpretation for the $i$th Betti numbers of $K$.

(A7) $\beta_i(K) = |\{S \in \Delta^e(K) : |S| = i+1, \ \{1\} \cup S \notin \Delta\}|$.
((A7) probably carries over to $\Delta^s$.) This result yields a description of face numbers of acyclic complexes **[Ka2]** and more generally of complexes with a prescribed sequence of Betti numbers **[BK1]**. A survey of this approach is given in **[BK2]**. For simplicial manifolds other interpretations of the Betti numbers in terms of $\Delta^e$ can be given. The combinatorics of $\Delta^e(K)$ is described by certain iterated homology (and cohomology) groups based on successive applications of generic weighted boundary operators, see **[Ka1]**.

It is an interesting open problem to identify further combinatorial properties of $\Delta(K)$ which correspond to topological invariants of $K$.

## 7. Shifting and the $h$-vector

Let $D$ be a shifted $(d-1)$-dimensional simplicial complex. Define

$$U_k(D) = \{S \in D : |S| = k, \ S \cap [d-k+1] = \emptyset\}$$

and

$$W_k(D) = \{S \in D : |S| = d, \ [d-k] \subset S, \ (d-k+1) \notin S\}.$$

LEMMA 7.1. *Let $D$ be a pure shifted complex. Then*

$$h_k(D) = |U_k(D)| = |W_k(D)|. \tag{7.1}$$

PROOF. First note that if $S \in U_k$, then, since $D$ is pure, there exists $T \in D$, $|T| = d$, such that $S \subset T$. Clearly $(S \cup [d-k]) \leq_p T$ and therefore $(S \cup [d-k]) \in W_k(D)$. The map $\Phi : U_k(D) \to W_k(D)$, given by $\Phi(S) = S \cup [d-k]$, is a bijection. Therefore, $|U_k(D)| = |W_k(D)|$. To show that $h_k(D) = |U_k(D)|$ associate to each $S \in U_k(D)$, $0 \leq k \leq d$, the following family of $(r-1)$-faces. $F(A) = \{A : |A| = r, \ S \subset A \subset \Phi(S)\}$. Simple inspection shows that $|F(A)| = \binom{d-k}{r-k}$, $F(A) \cap F(B) = \emptyset$ for $A \neq B$, and $\bigcup F(A) = D_r$. The proof thus follows from the defining relations for the $h$-vector. □

(For an alternative proof note that every ordering of the facets of $D$ in which the facets in $W_i(D)$ precede the facets in $W_j(D)$ for $i < j$ is a shelling order and every facet in $W_j(D)$ has degree $j$ in this shelling.)

For a simplicial complex $K$ let $\Delta(K)$ be the algebraic shifting defined in the previous section. We rely on the properties (A1′)–(A4′) and (B1), (B2) of the shifting operation $K \to \Delta(K)$.

Let $K$ be a $(d-1)$-dimensional simplicial complex. Define

$$\tilde{h}_k(K) = |W_k(\Delta(K))|.$$

Define also $\tilde{f}_{r-1}(K) = \sum \tilde{h}_k(K)\binom{d-k}{r}$. Clearly,

$$f_{r-1}(K) = f_{r-1}(\Delta(K)) \geq \tilde{f}_{r-1}(K). \tag{7.2}$$

DEFINITION. A simplicial $(d-1)$-dimensional simplicial complex $K$ is P-*embedded* if it is a subcomplex of the boundary complex of some simplicial $d$-polytope.

THEOREM 7.2. *Let $K$ be a $(d-1)$-dimensional* P*-embedded simplicial complex. Then for every $k$, $k < \lfloor d/2 \rfloor$,*

$$\tilde{h}_k(K) \geq \tilde{h}_{d-k}(K). \tag{7.3}$$

PROOF. Put $W_j = W_j(\Delta(K))$, $j \geq 0$. From properties (B2) and (A3) we obtain that if $S \in W_{d-k}$, then $([k] \cup [k+2, d-k+1]) \subset S$. From this fact and the fact that $\Delta(K)$ is shifted we obtain that the map $\Psi : W_{d-k} \to W_k$ given by $\Psi(S) = (S \cup \{k+1\}) \backslash \{d-k+1\}$ is an injection and therefore $|W_k| \geq |W_{d-k}|$. □

COROLLARY 7.3. *Let $K$ be a $(d-1)$-dimensional Cohen-Macaulay* P*-embedded simplicial complex. Then $h_k(K) \geq h_{d-k}(K)$ if $k < \lfloor d/2 \rfloor$.*

REMARKS. 1. The fact that $\Delta(K)$ is shifted is used for Theorem 7.2 but Corollary 7.3 can be proved without using it. Just replace $\Psi$ by $\Phi : U_{d-k} \to U_k$ given by $\Phi(S) = S \backslash [k+2, d-k+1]$. The injectivity of $\Phi$ is equivalent to the fact that the map $\omega^{d-2k}$ is a surjection from $\mathscr{H}_k(K)$ to $\mathscr{H}_{d-k}(K)$.

2. (with T. Hibi and R. Stanley). Hibi asked the following: If $A$ is a Cohen-Macaulay ring and $B$ is a subring, is it true that $h_k(A) \geq h_k(B)$ for every $k$. It turns out that while this is not true in general it holds for the case that $A$ is the face ring of a simplicial complex and $B$ is the face ring of a subcomplex. This is an immediate corollary of Lemma 7.1 and (A4′).

## 8. The generalized lower bound inequalities for a large class of simplicial spheres

In this section we prove Theorem 3 (in a stronger form). A *Gorenstein complex* is a pure $d$-dimensional simplicial complex $K$ so that for every face $S$ of $K$

$$\tilde{H}_i(\mathrm{lk}(S, K)) = \begin{cases} 0 & \text{for } i < \dim \mathrm{lk}(S, K), \\ \mathbf{C} & \text{for } i = \dim \mathrm{lk}(S, K). \end{cases}$$

Clearly, every simplicial sphere is a Gorenstein complex.

THEOREM 8.1. *If $K$ is a* P*-embedded $(d-1)$-dimensional Gorenstein complex, then*

$$h_0(K) \leq h_1(K) \leq \cdots \leq h_{\lfloor d/2 \rfloor}(K).$$

PROOF. Assume that $K$ is embeddable in the boundary complex $B$ of a $(d+1)$-dimensional simplicial polytope $P$. From Alexander duality it follows that $K$ separates $B$ into two acyclic parts $B_1$ and $B_2$. The same fact holds for the link of each face in $K$, and from this we conclude that $B_1$ (and $B_2$) are Cohen-Macaulay complexes and $K = \partial B_1 = \partial B_2$. It is known (see, e.g., **[BL2]**) that

$$h_k(K) - h_{k-1}(K) = h_k(B_1) - h_{d-k}(B_1).$$

Corollary 7.3 asserts that $h_k(B_1) \geq h_{d-k}(B_1)$ and this completes the proof. □

REMARK. In our proof we pass from a simplicial $(d-1)$-sphere $K$ to a subcomplex $M$ of $K$ which is a simplicial $(d-2)$-sphere. We show that if $K$ satisfies the $\omega$-hypothesis, then $M$ obeys the generalized lower bound inequalities. If we could prove that $M$ itself satisfies the $\omega$-hypothesis, then this would yield a simple inductive proof of the $g$-theorem. (Starting with a simplex.) What we still miss is an answer to the following problem: Let $K$ be a simplicial $(d-1)$-ball and let $M = \partial K$. Is $\Delta(M)$ determined by $\Delta(K)$? If the answer is positive, how can $\Delta(M)$ be expressed in terms of $\Delta(K)$?

## 9. A proof of the generalized upper bound theorem

By relations (7.2) and (7.3) in order to give lower bounds for the number of $j$-faces in a P-embedded $d$-dimensional simplicial complex with $m$ facets we have to solve the following minimization problem:

$$(9.1)\quad \begin{array}{ll} \text{Minimize:} & f_{d-u}[h] = \sum_{k=0}^{d-u} h_k \binom{d-k}{u} \\ \text{subject to:} & \text{(i) } (h_0, h_1, \dots, h_{\lfloor d/2 \rfloor}) \text{ is an M-vector}, \\ & \text{(ii) } h_{d-i} \leq h_i, \ i < \lfloor d/2 \rfloor \\ & \text{(iii) } \sum_{i=0}^{d} h_i = m. \end{array}$$

We need the following definition.

DEFINITION. A *special* M-*vector* is a vector $h = (h_0, \dots, h_r)$ satisfying for some $j \leq r$ (denoted by $j(h)$):

(X1) $\partial^k(h_{k+1}) = h_k$ for $k \neq j$,

(X2) $\partial^k(h_{k+1} + 1) = h_k + 1$ for $k > j$,

(X3) $\partial^j(h_{j+1} + 1) = h_j$,

(X4) $\partial^j(h_{j+1}) = h_j - 1$.

The solution for the maximization problem is different for odd and even values of $d$. We start with the case that $d = 2r+1$ is odd.

LEMMA 9.1. *Given two integers $u, r > 0$ there is a unique way to express $u$ as a sum $u = \sum h_i$, where $(h_0, \dots, h_r)$ is a special* M-*vector.*

PROOF. Write the $r$-canonical representation of $u$ as

$$u = \binom{u_r + 1}{r} + \binom{u_{r-1} + 1}{r-1} + \cdots + \binom{u_i + 1}{i} + \vartheta,$$

where $u_k > u_{k-1} > \cdots > u_i \geq i \geq 1$, and $\vartheta < i$.

A representation of $u$ as $h_0 + h_1 + \cdots + h_r$ for a special M-vector $(h_0, \dots, h_r)$ is given by

$$h_{r-k} = \binom{u_r - k}{r-k} + \binom{u_{r-1} - k}{r-1-k} + \cdots + \binom{u_i - k}{i-k} + \varepsilon(u, k),$$

where $\varepsilon(u, k) = 1$ if $k \le \vartheta$. It is easy to verify that this is the unique representation required. □

Note that the representation for $u+1$ is obtained from the representation of $u$ by raising the value of one of the summands by 1.

THEOREM 9.2. *Let $d = 2r+1$ be an odd integer. The solution to the minimization problem* (9.1) *is given by the sequence $(h_1, h_2, \ldots, h_d)$ defined as follows: Put $n_1 = \lceil \frac{n}{2} \rceil$ and $n_2 = \lfloor \frac{n}{2} \rfloor$. Let $h' = (h'_0, \ldots, h'_r)$ and $h'' = (h''_0, \ldots, h''_r)$ be the special* M*-vectors with $\sum_{i=0}^r h'_i = n_1$ and $\sum_{i=0}^r h''_i = n_2$. Then $h = (h'_0, h'_1, \ldots, h'_r, h''_r, h''_{r-1}, \ldots, h''_0)$.*

PROOF OF THEOREM 9.2. Let $h = (h_0, h_1, \ldots, h_d)$ be an $h$-vector which minimizes $f_{d-u}[h]$.

CASE 1. For some $j < r$, $\partial^j(h_{j+1}) < h_j$. Choose $j$ minimal with this property. In this case we claim that (a) for every $l$, $r > l > j$, $\partial^l(h_{l+1}+1) = h_l+1$, (b) $h_i = h_{d-i}$ for every $i \ne j$, and (c) $h_{d-j} \ge h_j - 1$. Indeed, if (a) is violated, let $l$ be the smallest integer for which $\partial(h_{l+1}+1) \le h_l$. If for some $t$, $h_t > h_{d-t}$, let $t$ be the smallest integer for which $h_t > h_{d-t}$. If $t \ne j$ we can increase $f_{d-r}[h]$ by replacing $h_j$ by $h_j - 1$ and $h_{d-t}$ by $h_{d-t}+1$. If $t = j$, replace $h_j$ by $h_j - 1$ and $h_{l+1}$ by $h_{l+1}+1$. If $h_t = h_{d-t}$ for every $t$, then reduce the value of $h_j$ and $h_{d-j}$ by one and raise the values of $h_{l+1}$ and $h_{d-l-1}$ by one.

Thus property (a) is satisfied. If (b) is violated, then $h_t > h_{d-t}$ for some $t \ne j$. Replace $h_j$ by $h_j - 1$ and $h_{d-t}$ by $h_{d-t}+1$. If $h_j - h_{d-j} \ge 2$ replace $h_j$ by $h_j - 1$ and $h_{d-j}$ by $h_{d-j}+1$.

CASE 2. For every $j$, $\partial^j(h_j+1) = h_j$. In this case we claim that $h_t = h_{d-t}$ for $t \ne r$ and $h_r - h_{r+1} \le 1$. Indeed, if $h_t > h_{d-t}$ for some $t \ne r$, subtract one from $h_{r+1}$ and add one to $h_{d-t}$. If $h_r - h_{r+1} \ge 2$, subtract one from $h_r$ and add one to $h_{r+1}$.

In both cases we obtain that $h_i = h_{d-i}$ for every $i \le r$ with at most one exception for which $h_i = h_{d-i}+1$. By inspecting Cases 1 and 2 it is easily verified that $(h_0, \ldots, h_r)$ and $(h_d, \ldots, h_{r+1})$ both satisfy (X1)–(X4) and are thus special $h$-vectors. This completes the proof of Theorem 9.2. □

We describe now the case where $d = 2r$ is even.

THEOREM 9.3. *$n$ can be expressed as a sum of the form $h_r + 2\sum h_i + \varepsilon(n)$, where $(h_0, h_1, \ldots, h_r)$ is a special* M*-vector and $\varepsilon(n)$ is zero or one and is taken to be zero if it is possible. Also $\varepsilon(n) = 1$ implies $\varepsilon(n+1) = 0$. Given this representation, if $\varepsilon(n) = 0$ the optimal $h$-vector is given by $(h_0, h_1, \ldots, h_r, h_{r-1}, h_{r-2}, \ldots, h_0)$. Otherwise let $(h_0, h_1, \ldots, h_r)$ be the representation for $n+1$, and the optimal $h$-vector is $(h_0, h_1, \ldots, h_{r-1}, h_r - 1, h_{r-1}, \ldots, h_0)$.*

The proof of Theorem 9.3 is similar to the proof of Theorem 9.2 and is omitted.

In §11 we use the results of this section to estimate the magnifying constant of dual-to-neighborly polytopes. Here is a weaker but more convenient form of Theorems 9.2 and 9.3.

THEOREM 9.4. *Let $d$ be a fixed integer. Let $K$ be a* P*-embedded $(d-1)$-dimensional simplicial complex with $a$ facets. Then*:

(i) *For $d = 2r+1$ odd write $a$ in the form ($x > 0$ is real)*

$$a = 2\binom{x+r+1}{r}, \tag{9.2}$$

*then*

$$f_{d-u}(P) \geq \sum_{k=0}^{r} \binom{x+k}{k}\left(\binom{d-k}{u} + \binom{k}{u}\right). \tag{9.3}$$

(ii) *For $d = 2r$ even write $a$ in the form (again, $x > 0$ is real)*

$$a = \binom{x+r+1}{r} + 2\binom{x+r}{r-1}, \tag{9.4}$$

*then*

$$f_{d-u}(P) \geq \sum_{k=0}^{r-1} \binom{x+k}{k}\left(\binom{d-k}{u} + \binom{k}{u}\right) + \binom{x+r}{r}\binom{d}{r}. \tag{9.5}$$

Theorem 9.4 is easily derived from Theorems 9.2 and 9.3 using (3.1). Note that if $P$ is a cyclic $d$-polytope with $d+x+1$ vertices ($x$ is a positive integer), then equality holds in (9.2) and (9.3) for $d = 2r+1$ and in (9.4) and (9.5) for $d = 2r$.

We need the special case $u = 2$. The next theorem follows from Theorem 9.4 by direct computations.

THEOREM 9.5. *Let $K$ be a* P*-embedded $(d-1)$-dimensional simplicial complex with $a$ facets.*

(i) *For $d = 2r+1$, if $a = 2\binom{x+r+1}{r}$, then*

$$f_{d-3}(P) \geq S_d(a) = \tfrac{1}{2}a(r^2 + 2r/(x+1) + 2r^2/(x+1)x).$$

(ii) *For $d = 2r$, if $a = \binom{x+r+1}{r} + 2\binom{x+r}{r-1}$, then*

$$f_{d-3}(P) \geq S_d(a) = a\left(\binom{r}{2} + r/(x+2r-1) + 3(r-1)r/(x+2r-1)(x+1)\right.$$
$$\left. + 2(r-1)r(2r-1)/(x+2r-1)(x+1)x\right).$$

Equality holds for cyclic $d$-polytopes with $d+x+1$ vertices.

## 10. Toward a Kruskal-Katona type theorem for embedded complexes

PROBLEM 10.1. Let $K$ be a subcomplex of the boundary complex of a simplicial $d$-polytope. Given $f_j(K)$ find a sharp lower bound for $f_{j-1}(K)$.

Note that Theorems 9.2 and 9.3 give a fairly sharp lower bound on $f_j$ in terms of $f_{d-1}$. This bound is sharp whenever the solution of the maximizing problem satisfies the conditions of the $g$-theorems, and this occurs quite often (the gaps are of size at most $d$).

Denote by $C(d, n)$ the boundary complex of the cyclic $d$-polytope with $n$ vertices. Define $\Delta_d(n) = \Delta(C(d, n))$ and $\Delta_d = \bigcup \Delta_d(n)$. It follows from properties (B1) and (B2) that $\Delta_d(n)$ is the simplicial complex spanned by all admissible $d$-subsets of $[n]$. (Thus, it follows that $\Delta_d(n) \subset \Delta_d(n+1)$ and that $\Delta_d$ is the pure $(d-1)$-dimensional simplicial complex whose facets are the family of admissible sets in $\binom{\mathbb{N}}{d}$.)

From properties (A4), (B1), and (B2) of algebraic shifting it follows that if $K$ is a subcomplex of the boundary complex of a simplicial $d$-polytope, then

$$\Delta(K) \subset \Delta_d. \tag{10.1}$$

The problem of finding a sharp lower bound for the size of the shadow of a family of $m$ $(j+1)$-sets in $\Delta_d$ seems to be within reach. Such a bound would serve as a lower bound (which may be sharp) for $f_{j-1}(K)$ in terms of $f_j(K)$ for every subcomplex $K$ of the boundary complex of a simplicial $d$-polytope, and more generally for every simplicial complex $K$, for which (10.1) holds.

CONJECTURE 10.2. (i) Every simplicial complex $K$ that can be *topologically* embedded into $S^{d-1}$ satisfies relation (10.1). (See note added in proof.)

(ii) (Stronger than (i).) Let $E$ be a simplicial complex which satisfies relation (10.1). Then every simplicial complex $K$ that can be topologically embedded into $|E|$ satisfies relation (10.1). ($|E|$ is the topological space realized by $E$.)

(iii) Every Eulerian $(d-1)$-dimensional simplicial complex satisfies relation (10.1).

(iv) (Consequence of (i).) The generalized upper bound theorem applies to arbitrary simplicial complexes that can be topologically embedded into $S^{d-1}$.

(v) (Consequence of (iii).) The generalized upper bound theorem applies to arbitrary Eulerian $(d-1)$-dimensional simplicial complexes.

(vi) The generalized upper bound theorem applies to arbitrary (nonsimplicial) polytopes.

(vii) The generalized upper bound theorem applies to arbitrary Eulerian lattices.

Klee **[Kl2]** conjectured that the upper bound theorem applies to arbitrary Eulerian complexes and proves it in the special case where the number of vertices is large with respect to the dimension. Stanley conjectured that the upper bound theorem applies to arbitrary Eulerian lattices. For a recent paper on $f$-vectors of embedded simplicial complexes, see **[Sa]**.

## 11. Magnification via the generalized upper bound theorem

PROOF OF THEOREM 1. Let $P$ be a dual-to-neighborly $d$-polytope with $n$ facets. Put $x = n - d - 1$. Let $\{A, B\}$ be a partition of the vertices of $P$ into two sets with $a = |A|$ and $b = |B|$. Assume $a \leq b$. Since $P$ is dual-to-neighborly, $f_2(P) = S_d(a+b)$. Theorem 9.5 (dualized) asserts that there are at least $S_d(A)$ 2-faces incident to the vertices of $A$, and at least $S_d(b)$ 2-faces incident to the vertices of $B$. Assume that there are $u$ edges between $A$ and $B$. There are at most $u(d-1)$ 2-faces which are incident to both $A$ and $B$. It follows that

$$u(d-1) \geq S_d(a) + S_d(b) - S_d(a+b).$$

Let $d = 2r+1$ be an odd integer. Define $x_a$, $x_b$ by the relations $a = 2\binom{x_a+r+1}{r}$ and $b = 2\binom{x_b+r+1}{r}$. Note that $a + b = f_0(P) = 2\binom{x+r+1}{r}$. We obtain by Theorem 9.5:

$$\begin{aligned}(11.1)\quad ud &\geq \tfrac{1}{2}a(r^2 + 2r/(x+2) + 2r^2/(x+1)(x+2)) \\ &\quad + \tfrac{1}{2}b(r^2 + 2r/(x+2) + 2r^2/(x+1)(x+2)) \\ &\quad - \tfrac{1}{2}(a+b)(r^2 + 2r/(x+2) + 2r^2/(x+1)(x+2)) \\ &\geq ar(1/(x_a+2) - 1/(x+2)) + br(1/(x_b+2) - 1/(x+2)) \\ &\geq ar/(x^2).\end{aligned}$$

The case where $d$ is even is similar, and the computation is omitted. □

REMARK. From (11.1) a magnification function for $G(P)$ can be read which slightly improves the diameter estimate.

We mention now two geometric consequences of Theorem 1.

COROLLARY 11.1. *Let $P$ be a dual-to-neighborly $d$-polytope with $n$-facets. Let $H$ be a hyperplane which does not pass through vertices of $P$ so that there are at least $a$ vertices of $P$ in each of the two halfspaces determined by $H$. Then $H$ passes through at least $d(n-d)^{-2}a$ edges.*

COROLLARY 11.2. *Let $P$ be a dual-to-neighborly $d$-polytope with $n$-facets. If the (Euclidean) diameter of $P$ is at least* 1, *then there is an edge of $P$ whose length is at least $n^{-5}$.*

Note that the Hirsch conjecture implies the existence of an edge with length at least $\frac{1}{n-d}$ for every $d$-polytope with $n$ facets and Euclidean diameter 1.

The proof of Theorem 1 can be extended to simple $d$-polytopes with $n$ facets and close to $g(d, n)$ vertices.

THEOREM 11.3. *There are monotonic functions $w(a)$ and $u(a)$ which both tend to infinity with $a$, so that:*

(i) *The graph of every simple $d$-polytope with $n$ facets and $g(d, n) - n^a$ vertices is an $n^{-u(a)}$-magnifier (and thus for a fixed $a$ has a polynomial diameter).*

(ii) *There is a simple $d$-polytope with $n$ facets and $g(d, n) - n^a$ vertices whose graph is not an $n^{-w(a)}$-magnifier.*

PROOF (sketch). The proof of Theorem 1 applies to part (i). The example for part (ii) is obtained by gluing together two cyclic polytopes of appropriate sizes along a facet, and taking the dual polytope.

If the entire $f$-vector of $P$ is prescribed there are a few more cases where a magnification property is forced. Let $\mathscr{H}_d$ denote the class of $h$-vectors of simplicial $d$-polytopes. For a vector $a$ denote by $\|a\|$ the sum of its entries. Note that if a simplicial polytope $P$ is obtained by gluing two simplicial polytopes $P_1$ and $P_2$ along a facet, then $h(P) = h(P_1) + h(P_2)$. (In this case the magnifying constant of $G(P^*)$ can be extremely poor.) On the other hand, by similar reasoning to the proofs in this paper it can be proved that if it is impossible to represent $h = h(P)$ as a sum $h = h_1 + h_2 + u$, where $h_1, h_2 \in \mathscr{H}_d$, $\|h_1\| \leq \|h_2\|$, and $\|u\| < \lfloor n^{-c}\|h_1\|\rfloor$, then $G(P^*)$ has a magnifying property which implies that $\delta(P^*)$ is polynomial in the number of facets of $P^*$.

A larger class of $f$-vectors forces good magnification just for the case where the graph is partitioned into two parts which are roughly equal. For a simple $d$-polytope $P$ with $n$ facets, $E_2(P)$ denotes the average number of vertices of a 2-face of $P$. Note that $E_2(P) = \binom{d}{2} f_0(P)/f_2(P)$. Given $d$ and $n$, $E_2(P)$ is minimized when $P$ is dual-to-neighborly. Let $E_2(d, n)$ denote the minimal value of $E_2(P)$. Let $P$ be a simple $d$-polytope with $n$ facets. From the proof of Theorem 1 it follows that if $E_2(P) \geq (1 - 1/n^2)E_2(d, n)$, then for every set $A$ of half the vertices of $P$ there are at least $a/(dn^2)$ edges from $A$ to its complement.

REMARKS. 1. The proof of Theorem 1 is analogous to the proof of a magnification property of the solid $d$-dimensional ball via the isoperimetric inequality due to P. Levy **[Lev]**, see also the introduction in **[AM]**. Recall that the isoperimetric inequality asserts that a $d$-dimensional body with volume $V$ has minimal surface area $S(V)$, which is the surface area of a ball with volume $V$. If a $d$-ball $B$ of volume $V$ is separated by a $(d-1)$-dimensional set $U$ into two bodies $X$ and $Y$, then a lower bound on the $(d-1)$-volume of $U$ is given by $2V_{d-1}(U) \geq S(V_d(X)) + S(V_d(Y)) - S(V)$. Thus, the subadditivity of the function $S$ implies the good magnifying constant of the solid $d$-ball.

2. An improvement of Larman's bound for dual-to-neighborly polytopes can be obtained by the following elementary argument. If $P$ is a dual-to-neighborly $2d$-polytope, then every two $(\frac{3d}{4})$-faces of $P$ intersect, so $\delta(P) \leq 2\Delta(\frac{3d}{4}, n) \leq 2^{3d/4-1}n$.

## 12. Open problems on the magnification of graphs of polytopes, random walks, and linear programming

We start this section with some conjectures on the magnifying properties of graphs of simple polytopes. Since the magnifying property is a profound separation property of graphs, this is a good place to remark that separation properties of graphs of polytopes were studied extensively. A well-known result of Balinski asserts that the graph of a $d$-polytope is $d$-connected and Klee found the maximal possible number of connected components of a graph of a $d$-polytope after removing $n$ vertices, see **[Gr1, Chapter 11]**.

**Graphs of simple polytopes are not superb magnifiers.**

CONJECTURE 12.1. Let $d$ be fixed and put $r = \lfloor d/2 \rfloor$. The graph of every simple $d$-polytope with $n$ facets can be separated into two $O(n)$ components by $O(n^{(r-1)/r})$ vertices.

The case $d = 3$ is given by an important theorem of Lipton and Tarjan **[LT]** which asserts that the graph of any 3-polytope with $n$ vertices can be separated by $O(\sqrt{n})$ vertices into two components of size $O(n)$.

**Graphs of random simple polytopes are good magnifiers.**

CONJECTURE 12.2. A random simple $d$-polytope with $n$ facets is a $(1/(n-d))$-magnifier.

**The graph of the dual to the cyclic polytope.** Let $G(d, n)$ be the graph of the dual of the cyclic polytope $C(d, n)$. We can give a direct proof for the magnifying property of $G(d, n)$ using a canonical paths argument. This actually gives better magnifying functions than those obtained here by Theorem 1. It is possible that methods from extremal combinatorics can provide precise vertex- and edge-isoperimetric theorems for $G(d, n)$ (as in the case of the cube, see **[Bo, §16]**). It would be interesting to determine the spectrum of $G(d, n)$, and the precise rate of convergence of a random walk on $G(d, n)$ to a uniform distribution. The assertion of the Hirsch conjecture for $G(d, n)$ is well known, see **[KK1]**.

**The spectrum of graphs of polytopes and random walks.** Let $G$ be a graph of a dual-to-neighborly $d$-polytope with $n$ facets. Theorem 1 implies that the first positive eigenvalue of the Laplacian of $G$ is large (at least $n^{-9}$) and that a random walk on $G$ converges rapidly to a uniform distribution.

The spectrum of graphs of simple polytopes are of interest. Given the dimension $d$ and the number of facets $n$, how large can the first positive eigenvalue be? The best examples we know are Cartesian products of polygons. How small can the $t$th eigenvalue be? Note that for a polynomial bound on the diameter it is sufficient that the $t$th eigenvalue is larger than $\frac{1}{t}$ for $t = n^C$.

**Linear programming.** The presence of a magnification property suggests basing edge-following algorithms for linear programming on random walks. (The reader is referred to **[Kl3, KK1]** for a discussion on the relations between the diameter problem and linear programming.)

We describe now a probabilistic edge-following algorithm for linear programming. We consider only the phase in which a vertex $v$ of the feasible polyhedron $P$ is known and the aim is to reach the vertex which maximizes the objective function $\phi$.

Let $G(P, t)$ be the induced subgraph of $G(P)$ on the vertices $w$ which satisfy $\phi(w) \geq t$. The algorithm depends on the parameter $m$.

"HIGHJUMP $[m]$": Start from a vertex $v$ and set $BAR = \phi(v)$. Make a random walk from $v$ on $G(P, BAR)$, reaching a vertex $w$ after $m$ steps. Set $v = w$, and repeat.

Clearly, if $G(P, t)$ is an $n^{-C}$ magnifyer for every $t$, then we are in good shape (take $m = n^{C+1}$). We cannot expect this in general, but our (vague and tentative) conjectured picture of graphs of simple polytopes suggests that HIGHJUMP or a related algorithm based on random walks might have a (provable) good worst-case average performance.

## Notes added in proof (November 1990)

1. Recently the author proved that every simplicial complex $K$, that can be embedded into $S^{d-1}$ satisfies relation (10.1) $\Delta(K) \subset \Delta_d$, for the algebraic shifting based on exterior algebra.

The proof uses several new properties of algebraic shifting. An idea of K. S. Sarkaria **[Sa]** relating shifting with van Kampen obstructions for embeddability is central in the proof. And some of Sarkaria's results are used and refined.

Thus, all the results of this paper dealing with subcomplexes of simplicial polytopes apply to complexes that can be embedded in spheres.

2. Using relation (10.1) combined with the fact that if $K$ is Cohen-Macaulay then $\Delta(K)$ is pure, we can prove the $g$-theorem for arbitrary simplicial spheres!

3. Using elementary consideration the author recently proved the following subexponential bound for $\Delta(d, 2d)$. $\Delta(d, 2d) \leq 2^{2\sqrt{d \log d}}$. It follows that $\Delta(d, n) \leq 2^{\sqrt{(n-d)\log(n-d)}}$.

4. Expansion properties for simple zonotopes and their duals (or equivalently, simple arrangements of hyperplanes) are quite easy to prove by canonical path arguments. This matter will be discussed elsewhere. Mihail and Vasirani **[MV]** have recently studied expansion properties of graphs of polytopes all of whose vertices have 0-1 coordinates. The expansion of the graphs of dual-to-cyclic polytopes follows also from a recent theorem of Aldous and Babai (see **[B]**) asserting that vertex-transitive graphs with diameter $D$ are $(1/2D)$-magnifiers.

## References

[A] N. Alon, *Eigenvalues and expanders*, Combinatorics **6** (1986), 83–96.

[AM] N. Alon and V. Milman, $\lambda_1$, *isoperimetric inequalities for graphs and superconcentrators*, J. Combin. Theory Ser. B **38** (1985), 73–88.

[B] L. Babai, *Local expansion of vertex transitive graphs and random generation in finite groups*, preprint, 1990.

[Bi] F. Bien, *Constructions of telephone networks by groups representations*, Notices Amer. Math. Soc. **36** (1989), 5–22.

[BL1] L. J. Billera and C. W. Lee, *A proof of the sufficiency of McMullen's conditions for f-vectors of simplicial convex polytopes*, J. Combin. Theory Ser. A **31** (1981), 237–255.

[BL2] ——, *The numbers of faces of polytope pairs and unbounded polyhedra*, European J. Combin. **2** (1981), 307–322.

[Bj] A. Björner, *Face numbers of complexes and polytopes*, Proc. Internat. Congr. Math., Vol. II, Berkeley, 1986, pp. 1408–1416.

[BFS] A. Björner, P. Frankl, and R. Stanley, *The number of faces of balanced Cohen-Macaulay complexes and a generalized Macaulay theorem*, Combinatorica **7** (1987), 23–34.

[BK1] A. Björner and G. Kalai, *An extended Euler-Poincare formula*, Acta Math. **161** (1988), 279–303.

[BK2] ——, *On f-vectors and homology*, Combinatorial Mathematics (G. Bloom, R. L. Graham, and J. Malkevich, eds.), Ann. New York Acad. Sci., vol. 555, 1989, pp. 63–80.

[Bo] B. Bollobàs, *Combinatorics*, Cambridge Univ. Press, Cambridge, 1986.

[BrM] H. Bruggessor and P. Mani, *Shellable decompositions of cells and spheres*, Math. Scand. **29** (1972), 197–205.

[Br] A. Brønsted, *An introduction to convex polytopes*, Springer-Verlag, New York, Heidelberg, Berlin, 1983.

[C] F. R. K. Chung, *Diameters and eigenvalues*, J. Amer. Math. Soc. **2** (1989), 187–196.

[D] V. I. Danilov, *The geometry of toric varieties*, Russian Math. Surveys **33** (1978), 97–154.

[Da] G. B. Dantzig, *Linear programming and extensions*, Princeton Univ. Press, Princeton, NJ, 1963.

[F] P. Frankl, *The shifting technique in extremal set theory*, in Proc. British Comb. Colloq., Cambridge Univ. Press, London, 1987.

[Gr1] B. Grünbaum, *Convex polytopes*, Wiley-Interscience, New York, 1967.

[Gr2] ——, *Polytopal graphs*, in Studies in Graph Theory, Part II (D. R. Fulkerson, ed.), MAA Studies in Math., no. 12, Math. Assoc. Amer., Washington, D.C., 1975, pp. 201–224.

[Ha] L. Harper, *Optimal numberings and isoperimetric problems on graphs*, J. Combin. Theory **1** (1966), 385–393.

[Ho] M. Hochster, *Rings of invariants of tori, Cohen-Macaulay rings generated by monomials, and polytopes*, Ann. of Math. (2) **96** (1972), 318–337.

[Ka1] G. Kalai, *A characterization of f-vectors of families of convex sets in $R^d$, Part* I: *Necessity of Eckhoff's conditions*, Israel J. Math. **48** (1984), 175–195.

[Ka2] ——, *f-vectors of acyclic simplicial complexes*, Discrete Math. **55** (1985), 97–99.

[Ka3] ——, *Symmetric matroids*, J. Combin. Theory Ser. B **50** (1990), 54–64.

[Ka4] ——, *Algebraic shifting methods and iterated homology groups*, unpublished manuscript, 1985.

[Ka5] ——, *Many triangulated spheres*, Discrete Comput. Geom. **3** (1988), 1–14.

[Ka6] ——, *Cohen-Macaulayness is preserved under shifting* (in preparation).

[Kho] A. G. Khovanskii, *Hyperplanes section of polyhedra, toroidal manifolds, and discrete groups in Lobachevskii space*, Funktsional. Anal. i Prilozhen. **20** (1986), 50–61. (Russian)

[KiK] B. Kind and P. Kleinschmidt, *Schalbare Cohen-Macaulay Komplexe und ihre Parametrisierung*, Math. Z. **167** (1979), 173–179.

[Kl1] V. Klee, *Diameters of polyhedral graphs*, Canad. J. Math. **16** (1964), 602–614.

[Kl2] ——, The number of vertices of a convex polytope, Canad. J. Math. **16** (1964), 701–720.

[Kl3] ——, *Convex polyhedra and mathematical programming*, Proc. 1974 Internat. Congr. Math., Vancouver, 1975, pp. 51–56.

[KK1] V. Klee and P. Kleinschmidt, *The d-step conjecture and its relatives*, Math. Oper. Res. **12** (1987), 718–755.

[KK2] ——, *Convex polytopes and related complexes*, in Handbook in Combinatorics (to appear).

[KW] V. Klee and D. Walkup, *The d-step conjecture for polyhedra of dimension $d < 6$*, Acta Math. **133** (1967), 53–78.
[La] D. G. Larman, *Paths on polytopes*, Proc. London Math. Soc. (3) **20** (1970), 161–178.
[Le] C. W. Lee, *Some recent results on convex polytopes*, preprint, 1988.
[Lev] P. Levy, *Problemes concrets d'analyse functionelle*, Gauthier-Villars, Paris, 1951.
[LT] R. J. Lipton and R. E. Tarjan, *Applications of a planar separation theorem*, in Proc. 18th Ann. IEEE Sympos. Found. Comput. Sci., 1977, pp. 162–170.
[Lo] L. Lovàsz, *Combinatorial problems and exercises*, North-Holland, Amsterdam, 1979.
[LPS] A. Lubotzky, R. Phillips, and P. Sarnak, *Ramanujan graphs*, Combinatorica **8** (1988), 261–277.
[M] F. S. Macaulay, *Some properties of enumeration in the theory of modular systems*, Proc. London Math. Soc. **26** (1927), 531–555.
[Mc1] P. McMullen, *The maximum number of faces of a convex polytope*, Mathematika **17** (1970), 179–184.
[Mc2] ——, *The numbers of faces of simplicial polytopes*, Israel J. Math. **9** (1971), 559–570.
[MS] P. McMullen and G. C. Shephard, *Convex polytopes and the upper bound conjecture*, Cambridge Univ. Press, Cambridge, 1971.
[MW] P. McMullen and D. W. Walkup, *A generalized lower bound conjecture for simplicial polytopes*, Mathematika **18** (1971), 264–273.
[MV] M. Mihail and U. Vazirani, *On expansion of 0-1 polytopes*, preprint, 1990.
[PB] J. S. Provan and L. J. Billera, *Decompositions of simplicial complexes related to diameters and convex polyhedra*, Math. Oper. Res. **5** (1980), 576–594.
[R] G. A. Reisner, *Cohen-Macaulay quotients of polynomial rings*, Adv. Math. **21** (1976), 30–49.
[Sar] K. S. Sarkaria, *On shifting and embeddability*, preprint, 1989.
[Sh] I. Shemer, *Neighborly polytopes*, Israel J. Math. **43** (1982), 291–314.
[St1] R. P. Stanley, *The upper bound conjecture and Cohen-Macaulay rings*, Stud. Appl. Math. **54** (1975), 135–142.
[St2] ——, *Cohen-Macaulay complexes*, in Higher Combinatorics (M. Aigner, ed.), Reidel, Dordrecht, 1977, pp. 51–62.
[St3] ——, The number of faces of a simplicial convex polytope, Adv. in Math. **35** (1980), 236–238.
[St4] ——, *Combinatorics and commutative algebra*, Birkhäuser, Boston, 1983.
[St5] ——, *The number of faces of simplicial polytopes and spheres*, in Discrete Geometry and Convexity (J. E. Goodman and R. Pollack, eds.), Ann. New York, Acad. Sci., vol. 440, New York, 1985, pp. 212–223.
[St6] ——, *Generalized h-vectors, intersection cohomology of toric varieties, and related results*, Commutative Algebra and Combinatorics, Adv. Stud. Pure Math., no. 11, North-Holland, Amsterdam and New York, 1987, pp. 187–213.
[T] R. Tanner, *Explicit concentrators from generalized N-gons*, SIAM J. Algebraic Discrete Methods **5** (1984), 287–293.

Edmund Landau Center for Research in Mathematical Analysis, Institute of Mathematics, Hebrew University, Jerusalem 91904, Israel
*E-mail address*: kalai@humus huji.ac.il

DIMACS Series in Discrete Mathematics
and Theoretical Computer Science
Volume 4, 1991

Applied Geometry
and Discrete Mathematics
THE VICTOR KLEE FESTSCHRIFT

# Multiply Perspective Simplices, Desmic Triads and the Edelstein Theorems

L. M. KELLY

ABSTRACT. A finite collection of disjoint finite sets such that any line intersecting two of the set intersects at least one more is a generalized desmic configuration (GDC). C. Stephanos in 1893 observed that there is a triple of tetrahedra $\Delta_1, \Delta_2, \Delta_3$ in $E^3$ such that $\Delta_i$ and $\Delta_j$ are perspective in four different ways, the perspective centers being the vertices of $\Delta_k$; here $i, j, k$ is any permutation of 1, 2, 3. He called such a set a desmic system. It is a special instance of a GDC. The natural generalization to $n$ dimensions where the $\Delta_i$'s are simplices spanning a projective space $P^n$ we call a Stephanos triad. We prove, among other things, that Stephanos triads are confined to dimension 3 in any projective space as are multiply perspective simplices spanning $P^n$.

A survey of the history, development, and current status of the study of these concepts is included with emphasis on the work of M. Edelstein.

## 1. Introduction, historical background

A Sylvester-Gallai configuration (SGC) is a finite set of points in an affine or projective space such that the line joining any two of its points goes through at least one more point of the set. Interest in such sets dates from the discovery around 1730 by Maclaurin that the nine inflection points of a nonsingular cubic curve in the complex plane form such a set.

In 1893 Sylvester conjectured, and in 1933 Gallai proved, that SGCs in real space are confined to the line. The example above shows that this is not true in the complex plane but in 1966 J. P. Serre suggested [**16**] that SGCs in complex space might be confined to the plane. I recently confirmed this fact aided and abetted by the analysis of line arrangements in the complex plane by F. Hirzebruch. Interestingly, the key fact in that analysis stemmed from the now celebrated Miyaoka-Yau inequality.

The systematic study of line arrangements in the real plane carried out by Grünbaum and in the complex plane by Hirzebruch has fueled interest

1980 *Mathematics Subject Classification* (1985 *Revision*). Primary 51A20.
*Key words and phrases.* Configuration, desmic, perspectivity, tetrahedra, simplices.

in these configurational studies. Grünbaum's work was motivated by his wide ranging study of convex polytopes while Hirzebruch uses arrangements to produce surfaces satisfying (or nearly satisfying) the case of equality in the Miyaoka-Yau inequality $c_1^2 \leq 3c_2$ relating the first and second Chern numbers of a suitable surface.

A variety of generalizations of the Sylvester phenomenon have been proposed. In 1956 Grünbaum [**5**] suggested that "points" be replaced by "sets" which he took to be continua and he obtained the following.

THEOREM 1 (Grünbaum). *If a finite family of disjoint continua in $E^2$ is such that any line intersecting two of the sets intersects at least one more, then the sets are subsets of the same line.*

Edelstein, Herzog, and Kelly [**2, 3, 4**] pursued this type of generalization and their work culminated in the following remarkable theorem of Edelstein [**2**].

THEOREM 2 (Edelstein). *If $\{S_i\}$ is a finite family of disjoint compact sets in a real Hausdorff linear topological space such that any hyperplane intersecting two of the sets intersects at least one more then either* (1) $\bigcup S_i$ *is linear or* (2) $\bigcup S_i$ *is finite and spans a space of dimensions at most* 3.

A corollary of this is the following.

THEOREM 3. *If $\{S_i\}$ is a finite family of disjoint finite sets in an ordered projective or affine $n$-space such that any line intersecting two of the sets intersects at least one more then $\bigcup S_i$ is a subset of* 3*-space.*

One of the purposes of this paper is to explore some of the interesting implications of this theorem.

This brings us to a second classical configuration studied in considerable detail by early algebraic and projective geometers, namely the desmic configuration of Stephanos [**17**]. Cremona in 1877 seems to have been the first to explicitly observe that two disjoint tetrahedra in real space can be perspectively related in four different ways, i.e., they can be perspectively related from four different centers. But it was Stephanos who systematically explored the phenomenon and gave the configuration its name. The Greek word desmic can be translated as "related" or "pencil".

If $\Delta_1$ and $\Delta_2$ are vertices of two such tetrahedra their four centers of perspectivity are the vertices, $\Delta_3$, of a third tetrahedron and it is easy to see that $\Delta_i$ and $\Delta_j$ are in "four-fold" perspective from the four vertices $\Delta_k$, $i, j, k$ any permutation of 1, 2, 3. Stephanos refers to three such tetrahedra as a desmic system. To us it will be a specific instance of a generalized desmic configuration (GDC) which we define shortly.

First, however, we should exhibit the canonical configuration in $E^3$ which intrigued Stephanos. Take $\Delta_1$ to be one set of alternate vertices of a cube, $\Delta_2$ to be the other set, and $\Delta_3$ to be the center of the cube and the three ideal points on the edge lines. Of course, any projective image would do as well.

DEFINITION. A *generalized desmic configuration* (GDC) is a finite collection $\{S_i\}$ of disjoint finite sets such that any line intersecting two of the sets intersects at least one more. If $\{S_i\}$ consists of exactly three sets then the configuration is a *generalized desmic triad.*

Finally, a generalized desmic triad in projective $n$-space each set of which has cardinality $n+1$ and spans $n$-space is a *Stephanos triad* of dimension $n$.

Examples are given in §2. Of course, the desmic system of Stephanos described above and SGCs are special instances of GDCs.

Theorem 3 implies that in ordered projective spaces Stephanos triads are confined to dimension 3.

A very extensive analysis of multiply perspective triangles and tetrahedra appeared in a 90-page paper by E. Hess in 1887 [**7**].

While many other papers on the same subject have appeared from time to time, no move to extend these results to higher dimensions seems to have been made. We do that here.

To complete our historical survey we observe that the desmic system of Stephanos plays a major role in the early study of quartic surfaces. The opening paragraphs of R.W.H.T. Hudson's classic treatise on Kummer's quartic surface [**9**] are devoted to a description of a desmic system and its properties and the configuration plays a prominent part in the analysis of the 16 isolated singular points of the surface.

G. Humbert [**10**] in 1891 defined a desmic surface which is a quartic surface having 12 singular points forming a desmic system.

In §2 we extend Theorem 3 to other spaces and in §3 we explore the implications of these theorems. §4 is devoted to showing that multiply perspective $n$-dimensional simplices spanning $P^n$ can only occur if $n \le 3$.

## 2. Generalized desmic configurations (GDCs)

The principal contribution of this section is in the nature of an observation which extends Edelstein's theorem confining GDCs in ordered projective spaces to dimension 3, to confining GDCs in complex spaces to dimension 4.

THEOREM 4. *If in a class of projective or affine spaces SGCs are confined to dimension $k$, then GDCs are confined to dimension $k+2$.*

PROOF. The proof essentially parrots that of Edelstein for the real case, but since that argument is not very well known we present the details here.

Let $\Sigma$ be a GDC and suppose it spans a space $K$ of dimension $k+3$. Let $L$ be a line containing points $A$, $B$, and $C$ from three different sets of $\Sigma$. Now consider the family of planes $\pi = \{\pi_i\}$ containing $L$ and at least one additional point of $\Sigma$. Let $H$ be a hyperplane in $K$ intersecting

$L$ in a single point $P$. $J = \pi \cap H$ is a family of lines in $H$ through $P$ spanning $H$. Now consider a hyperplane $\overline{H}$ in $H - P$. $\overline{H}$ is a space of dimension $k+1$ in $H$ and hence by our assumption the set $\overline{H} \cap J$ which spans $\overline{H}$ is not an SGC, i.e., there is a pair of points $S$ and $T$ in $\overline{H} \cap J$ such that the line joining $S$ and $T$ contains no other points of $\overline{H} \cap J$. It follows that the two planes of $\pi$, one containing $S$ and the other $T$, span a 3-space containing no other planes of $\pi$. The intersection of these two planes with $\Sigma$ is a GDC spanning 3-space. But it is easy to see that a GDC spanning 3-space cannot lie wholly in two planes. This contradiction shows that $\Sigma$ cannot span a space of dimension $k+3$.

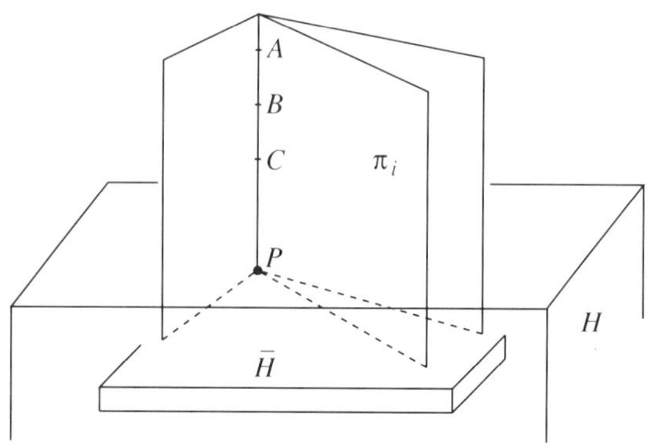


FIGURE 1

COROLLARY 4.1. *GDCs in $\mathbb{C}^n$ are confined to $\mathbb{C}^4$.*

PROOF. SGCs in $\mathbb{C}^n$ are confined to $\mathbb{C}^2$.

The next theorem is rather special but will be useful in §3. First we present a couple of lemmas, the obvious proofs of which will be omitted.

LEMMA ( i). *The minimum cardinality of an SGC is 7, and if the space satisfies the Fano axiom it is 9.*

LEMMA (ii). *The minimum cardinality of a GDC is 6.*

THEOREM 5. *The cardinality of a GDC spanning a Fano 4-space cannot be less than 30.*

PROOF. The proof is very similar to that of Theorem 4.

Suppose $\Sigma = \{S_i\}$ is a GDC spanning $P^4$ which satisfies the Fano axiom. Let $L$ be a line containing points $A$, $B$, $C$ from three different sets of $\Sigma$. Now consider the family of planes $\pi = \{\pi_i\}$ each containing $L$ and at least one additional point of $\Sigma$. Each $\pi_i \cap \Sigma$ is a GDC of cardinality at least 6. Hence, $\pi_i$ contains three points of $\Sigma$ not on $L$. If $\pi$ contains nine or more planes $\pi_i$ then the cardinality of $\Sigma$ is at least 30. So if the cardinality of $\Sigma$ is less than 30 then $\pi$ contains eight or fewer planes. Now if $H$ is a hyperplane in $P^4$ intersecting $L$ in a single point $P$ then $J = \pi \cap H$ is a family of eight or fewer lines in $H$ through $P$. Now consider a plane $\overline{H}$ in $H - P$ where $\overline{H} \cap J$ is a set of eight or fewer points spanning $\overline{H}$ and by Lemma (i) is not an SGC. There is a line intersecting $\overline{H} \cap J$ in exactly two points $S$ and $T$, one of which can be assumed to be in $\pi_1$ and the other in $\pi_2$. $\pi_1 \cup \pi_2$ spans a 3-space which contains no other planes in $\pi$. Hence $\Sigma \cap (\pi_1 \cup \pi_2)$ is a GDC spanning a 3-space which is a subset of two planes and this is clearly impossible.

Thus the cardinality of a GDC spanning a projective 4-space satisfying the Fano axiom cannot be less than 30. □

We now present some interesting and useful examples.

EXAMPLE 1. *The pappus configuration.*

$$\Delta_1 = \{A, A', Z\}, \quad \Delta_2 = \{C, C', X\}, \quad \Delta_3 = \{B, B', Y\}$$

$\Sigma = \{\Delta_1, \Delta_2, \Delta_3\}$ is a GDC, a generalized desmic triad, and if none of the sets $\Delta_i$ are linear it is a Stephanos triad.

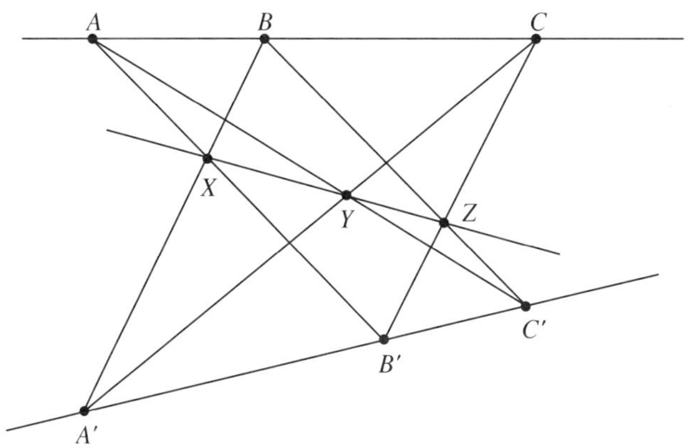


FIGURE 2

EXAMPLE 2. Let $\Delta_1$ be an alternate set of vertices of a regular polygon of $2n$ sides in $E^2$ and $\Delta_2$ the remaining vertices. Let $\Delta_3$ be the $n$ ideal points on the side lines of the polygon.

$\Sigma = \{\Delta_1, \Delta_2, \Delta_3\}$ is a generalized desmic triad but not, of course, a Stephanos triad. Figure 3 illustrates the case for $n = 3$.

EXAMPLE 3. As in Example 2 with $n$ odd and $\Delta_3$ containing the center of the polygon in addition to the ideal points. Here the sets in the generalized desmic triad do not have equal cardinality.

EXAMPLE 4. As in Example 2 let $\Delta_1$ be an alternate set of vertices of a regular polygon of $2n$ sides and $\Delta_2$ the remaining vertices. Let each of the ideal points on the side lines constitute a set $\Delta_i$, $i = 3, 4, \ldots, n+2$. Then $\Sigma = \{\Delta_i\}$, $i = 1, 2, \ldots, n+2$, is a GDC of $n+2$ sets, two of cardinality $n$

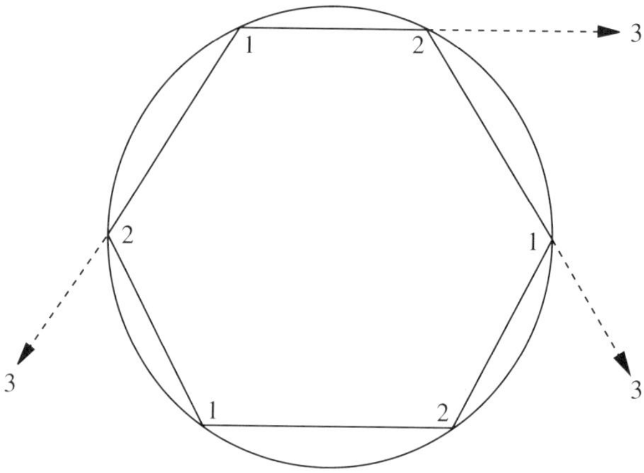


FIGURE 3

and $n$ of cardinality 1. By grouping the ideal points in varying combinations, more or less arbitrary cardinalities can be attained.

These examples and others suggest that characterizing GDCs in the real plane could be a very difficult problem. With this plethora of examples in mind it came as some surprise that our efforts to produce GDCs spanning $E^3$ were so fruitless. After rediscovering the desmic system of Stephanos we could find no others. We enlisted the aid of P. Borwein who proceeded to prove the following elegant theorem.

THEOREM 6 (P. BORWEIN). *The only generalized desmic triad spanning $E^3$ is the Stephanos triad.*

PROOF. See [**1**]. □

Observe that generalized desmic triads are not required to have equal cardinality and if they do it is not required to be 4. The force of the Borwein theorem is that for such triads spanning $E^3$ both these conditions prevail.

Edelstein and I have made determined but fruitless efforts to extend the Borwein result to GDCs spanning $E^3$ with more than three sets. Our conjecture that the only GDC spanning $E^3$ is the desmic system of Stephanos still stands.

## 3. Stephanos triads

The main objective of this section is to prove that Stephanos triads in any projective space $P^n$ are confined to dimension 3.

Before proving the more general theorem alluded to above we note that the important special cases of real and complex space follow at once from theorems already discussed.

THEOREM 7. *Stephanos triads in complex space are confined to dimension* 3.

PROOF. By Theorem 4 they are confined to dimension 4 since SGCs in complex space are confined to the plane. It remains to show that a Stephanos triad of dimension 4 is impossible. But such a triad contains exactly 15 points which span $\mathbb{C}^4$ and this, by Theorem 5, is impossible. □

Now we prove even more:

THEOREM 8. *Stephanos triads in any projective space are confined to three dimensions.*

PROOF. Let $\Sigma = \{\Delta_1, \Delta_2, \Delta_3\}$ be an $n$-dimensional Stephanos triad in $P^n$. The cardinality of $\Sigma$ is then $3(n+1)$. If $L$ is a line containing points. $A$, $B$, $C$, one from each $\Delta_i$, let $\pi = \{\pi_i\}$ be the family of planes such that each $\pi_i$ contains $L$ and at least one point of $\Sigma$ not on $L$. It is then easy to see that each $\pi_i$ contains at least three points of $\Sigma$ not on $L$. Hence the number of planes in $\pi$ is at most $n$.

Now let $H$ be a hyperplane in $P^n$ intersecting $L$ in exactly one point $P$. $J = \pi \cap H$ is a family of at most $n$ lines through $P$ spanning $H$. $H$, of course, is of dimension $n-1$. Suppose, now, that $\overline{H}$ is a hyperplane relative to $H$ not containing $P$. $K = \overline{H} \cap J$ is a set of at most $n$ points spanning $\overline{H}$. $\overline{H}$ is of dimension $n-2$ and any set of points spanning it must contain at least $n-1$ points. Hence, our set $K$ contains either $n$ or $n-1$ points. In either case it is transparent that the set is not an SGC and thus there is a line intersecting it in exactly two points. These two points lie in two planes of $\pi$ which span a 3-space $\Gamma$. This 3-space contains no other plane in $\pi$. Furthermore, $\Gamma \cap \Sigma$ is a GDC spanning a 3-space and contained wholly in two planes in that space. This is clearly impossible if $n-2>1$, i.e., if $n>3$.

A scrutiny of the diagram of Figure 4, which projective geometers might call a complete 6-plane. should suggest what can be carefully proved, to wit that a Stephanos triad can be realized in any projective space of dimension 3.

Thus, we claim that Stephanos triads exist in projective $n$-space iff $n \le 3$. □

REMARKS. We should note that the argument really only uses the properties of the triad $\Sigma$ that it spans $P^n$ and has cardinality $3(n+1)$. The reasoning, in fact, proves that there is no GDC spanning $P^n$ of cardinality $3(n+1)$ or less.

A key fact used in the proof (which we referred to as "transparent") is that a set of $n+1$ points spanning $P^n$ is not an SGC, i.e., there is at least one line intersecting the set in exactly two points. Motzkin and others call such lines "ordinary". In this case any line joining two points of the set is ordinary. It is also true that if the spanning set contains $n+2$ points it is not a SGC and while it may not be true that each line joining a pair of points is ordinary, it is true that if $n>2$ then each point is on at least one ordinary line. This fact will be needed in the next section.

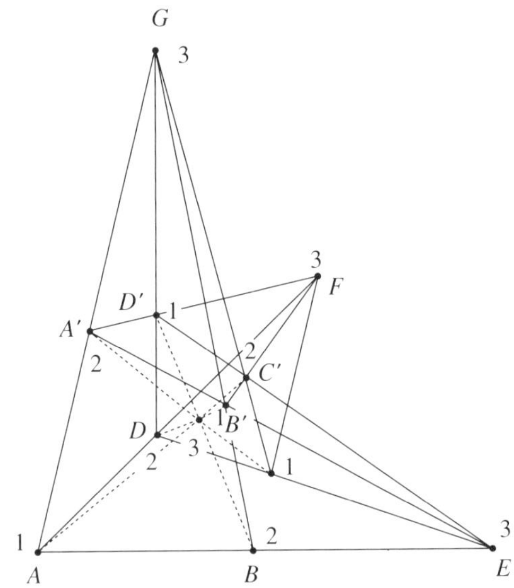


FIGURE 4

## 4. Multiply perspective simplices

DEFINITION. A 1-1 correspondence between two sets, $S_1$ and $S_2$, in $P^n$ such that there is a point $C$ with each pair of corresponding points on a line through $C$ is called a *perspectivity* of $S_1$ and $S_2$ and $C$ is called its *center.*

If $S_1 \cap S_2 = \varnothing$ (which will be the case in our discussion) this amounts to saying that the lines joining corresponding pairs are concurrent at the center.

While we have been viewing Stephanos triads as special cases of GDCs, it is also true, of course, that they are instances of multiply perspective $n$-dimensional simplices in $P^n$. That is, they are instances of a pair of disjoint $n$-dimensional simplices spanning $P^n$ which are perspective from $n+1$ different centers. Theorem 8 shows that this is only possible for $n \leq 3$.

We have already observed that the classical literature is replete with analyses of multiply perspective triangles in $E^2$ and $\mathbb{C}^2$ and tetrahedra in $E^3$ and $\mathbb{C}^3$. It was shown repeatedly that two triangles in the real plane can be perspective from at most four different points and that while this is maximal in $R^2$, in $\mathbb{C}^2$ six is possible and clearly maximal. In real 3-space tetrahedra can be four-fold perspective as the original desmic system of Stephanos shows but the situation cannot be improved in complex 3-space. While the situation in two and three dimensions was exhaustively analyzed we were unable to unearth any efforts to extend these interesting results to higher dimension. The following theorem may account for this.

THEOREM 9. *Two $n$-dimensional simplices spanning $P^n$ (or $A^n$) can be perspectively related from at most one center, if $n > 3$.*

More specifically, two disjoint sets, each of $n+1$ points and each spanning a projective space of $n$ dimensions, can be perspectively related from at most one center, if $n > 3$.

PROOF. The proof is only a slight modification of that of Theorem 8. We adopt that notation and diagram with obvious modifications.

Let $\Sigma = \{\Delta_1, \Delta_2\}$ be two disjoint $n$-dimensional simplices in $P^n$. The cardinality of $\Sigma$ is then $2(n+1)$. Now suppose that $\Delta_1$ and $\Delta_2$ are perspective from two different centers $C_1$ and $C_2$ and let $L$ be a line containing $C_1, A, B$; $A \in \Delta_1$, $B \in \Delta_2$.

We now consider the family of planes $\pi = \{\pi_i\}$, where each $\pi_i$ contains $L$ and at least one point of $\Sigma$ not on $L$. We claim that $C_2$ cannot be on $L$. Suppose it were and consider $\pi_i \cap \Sigma = \Gamma_i$. $\Delta_1 \cap \Gamma_i$ and $\Delta_2 \cap \Gamma_i$ are perspective from both $C_1$ and $C_2$ and a glance at the accompanying diagram (Figure 5) shows that $\pi_i$ must contain at least four points of $\Sigma$ not on $L$. Hence the number of planes in $\pi$ is at most $\frac{1}{4}[2(n+1) - 2] = \frac{n}{2}$. But $\frac{n}{2} < n - 1$ if $n > 2$ and hence $\pi$ cannot span $P^n$. This contradiction shows that $C_2$ cannot be on $L$.

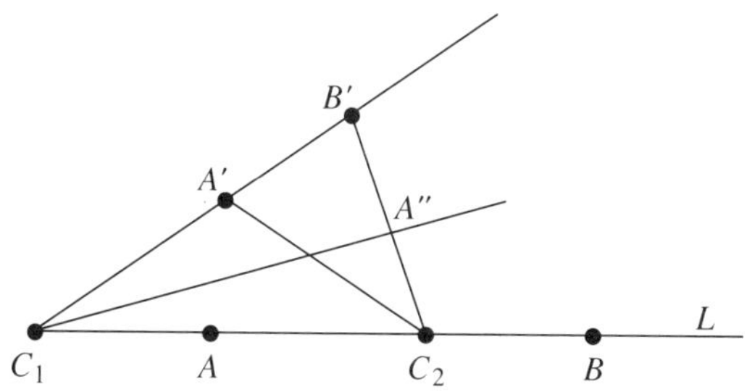


FIGURE 5

Note that $C_2$ must be on one of the planes $\pi_i$. As in Theorem 8 (and elsewhere) we consider a hyperplane, $H$, intersecting $L$ in exactly one point $P$ and the family of lines $J = \pi \cap H$ through $P$ spanning $H$. In this case we can only be sure that each $\pi_i$ contains at least two points of $\Sigma$ not on $L$ and hence that the number of planes in $\pi$ is at most $\frac{1}{2}[2(n+1) - 2] = n$. Since $\pi$ must span $P^n$, the number of planes in $\pi$ is at least $n - 1$ and hence the number of planes in $\pi$ is either $n - 1$ or $n$.

As in Theorem 8 we now suppose that $\overline{H}$ is a hyperplane relative to $H$ not containing $P$. $K = \overline{H} \cap J$ is a set of points spanning $\overline{H}$ and containing either $n - 1$ or $n$ points. Since $\overline{H}$ is of dimension $n - 2$ it follows from the remarks in §3 that there is an ordinary line through each point of $K$ if $n > 3$. Thus the plane $\pi_i$ containing $C_2$ and one additional plane $\pi_k$ span a 3-space in $P^n$ which contains no other planes of $\pi$. But this clearly contradicts the assumption that $C_2$ is a perspective center for $\Delta_1$ and $\Delta_2$ and it follows that $\Delta_1$ and $\Delta_2$ can have at most one perspective center. □

Theorem 8 can now be seen as a rather weak corollary of this theorem.

## REFERENCES

1. P. Borwein, *The desmic conjecture*, J. Combin. Theory Ser. A **35** (1983), 1–9.
2. M. Edelstein, *Hyperplanes and lines associated with families of compact sets in locally convex spaces*, Math. Scand. **25** (1969), 25–30.
3. M. Edelstein and L. M. Kelly, *Bisecants of finite collections of sets in linear spaces*, Canad. J. Math. **18** (1966), 375–380.
4. M. Edelstein, F. Herzog, and L. M. Kelly, *A further theorem of the Sylvester type*, Proc. Amer. Math. Soc. **14** (1963), 359–363.
5. B. Grünbaum, *A generalization of the problem of Sylvester*, Riveon Lematematiks **10** (1956), 46–48.
6. F. Herzog and L. M. Kelly, *A generalization of the theorem of Sylvester*, Proc. Amer. Math. Soc. **11** (1960), 327–331.
7. Edmund Hess, *Beiträge zur Theorie der mehrfach perspectiven Dreiecke und Tetraeder*, Math. Ann. **2** (1887), 167–260.
8. F. Hirzebruch, *Arrangements of lines and algebraic surfaces*, in Arithmetic and Geometry (Papers dedicated to I. R. Shafarevich), vol. 2, Birkhäuser, 1983, pp. 113–140.
9. R.W.H.T. Hudson, *Kummer's quartic surface*, Cambridge Univ. Press, 1905.
10. G. Humbert, *Sur la surface desmique du quatrième ordre*, J. Math. **7** (1891), 353–398.
11. L. M. Kelly and W. O. J. Moser, *On the number of ordinary lines determined by $n$ points*, Canad. J. Math. **10** (1958), 210–219.
12. L. M. Kelly, *A resolution of the Sylvester-Gallai problem of J.-P. Serre*, Discrete Comput. Geom. **1** (1986), 101–104.
13. L. M. Kelly and S. Nwankpa, *Affine embeddings of Sylvester-Gallai designs*, J. Comput. Theory **14** (1973), 422–438.
14. L. Klug, *Ueber mehrfach perspective Tetraeder*, Arch. Math. Phys. **6** (1888), 88–104.
15. T. Motzkin, *The lines and planes connecting the points of a finite set*, Trans. Amer. Math. Soc. **70** (1951), 451–469.
16. J.-P. Serre, *Problem* 5359, Amer. Math. Monthly **73** (1966), 89.
17. C. Stephanos, *Sur les systèmes desmiques de trois tetraèdres*, Bull. Sci. Math. Astronom. Ser II **3** (1879), 424–456.
18. J. J. Sylvester, *Mathematical question* 11851, Educational Times **59** (1893), 98.

DEPARTMENT OF MATHEMATICS, MICHIGAN STATE UNIVERSITY, EAST LANSING, MICHIGAN 48824

DIMACS Series in Discrete Mathematics
and Theoretical Computer Science
Volume 4, 1991

Applied Geometry
and Discrete Mathematics
THE VICTOR KLEE FESTSCHRIFT

# Submanifolds of the Cube

W. KÜHNEL AND CH. SCHULZ

ABSTRACT. In this paper we study subcomplexes of the $d$-dimensional cube which are $(2k)$- or $(2k+1)$-manifolds and which contain the full $k$-skeleton of the cube. This is well known for $k = 1$ where such 2-manifolds have been studied from various points of view. In particular, Danzer's $2^K$-construction provides a simple way for getting such even- and odd-dimensional examples for arbitrary $k$. For $K$ being the boundary complex of a cyclic polytope we determine the topological type of $2^K$, and we discuss related questions.

## Introduction and results

It is well known that there are subcomplexes $M$ of the 2-dimensional skeleton of the $d$-dimensional cube $C^d = [0, 1]^d$ with the following properties:

(i) $M$ is PL homeomorphic to a 2-dimensional manifold without boundary,

(ii) $M$ contains the complete 1-dimensional skeleton $\mathrm{Sk}_1(C^d)$ of this cube.

$M$ is also called 1-*Hamiltonian* if it satisfies (ii), generalizing the classical concept of (0-)Hamiltonian circuits.

*Orientable surfaces* of this type have been described by H. S. M. Coxeter [**Cox**] as examples of *regular maps on surfaces*: in his notation, $\{4, d|4^{[\frac{1}{2}d]-1}\}$. For $d = 3$ we get the boundary of $C^3$ itself or the hexahedron $\{4, 3\}$, and for $d = 4$ we get the toroidal map $\{4, 4\}_{4,0}$ in the boundary complex of $C^4$, consisting of 16 squares. Later G. Ringel [**Ri**] and Beineke and Harary [**BH**] rediscovered these examples, and determined the *genus* $g_d$ *of the* $d$-*cube* to be

$$g_d = 2^{d-3}(d-4) + 1,$$

which means that the edge graph of $C^d$ embeds into the surface of genus $g_d$ but not into a surface of lower genus. T. Banchoff [**Ba**] has used the

1980 *Mathematics Subject Classification* (1985 *Revision*). Primary 52A25; Secondary 57Q15.

same construction for getting examples of surfaces in euclidean space with the two-piece-property and with high essential codimension. All these papers seem to be independent of each other.

*Nonorientable surfaces* of this type have been studied by M. Jungerman [**J**] and independently by the second author [**Sch2**], and the nonorientable genus of the $d$-cube has been determined.

A geometric realization of such orientable surfaces embedded into $E^3$ has been given in [**MSW**]. Here it is somewhat surprising that for $d \geq 12$ the genus of this embedded surface exceeds the number $2^d$ of its vertices:

$$g_d = 2^{d-3}(d-4) + 1 \geq 2^d + 1 .$$

In the sequel we want to study a higher-dimensional generalization of this concept, as started already in [**Sch1**]. A *subcomplex of* $C^d$ is understood as a (combinatorial) subcomplex of the boundary complex $\partial C^d$ of $C^d$.

A subcomplex of $C^d$ (or of any other convex polytope) is called $k$-*Hamiltonian* if it contains the full $k$-dimensional skeleton $\mathrm{Sk}_k(C^d)$ of $C^d$ (or of the corresponding polytope). In particular, we are interested in $k$-Hamiltonian subcomplexes of $C^d$ whose underlying sets are PL $n$-manifolds without boundary (or with boundary, following [**RS, 1.8**]). We will refer to them as *$k$-Hamiltonian $n$-submanifolds of* $C^d$. Each vertex figure of such a manifold must be a combinatorial $(n-1)$-sphere with $d$ vertices which is $k$-neighborly. This implies the conditions $2k \leq n$ (see [**Gr, 11.1**]).

Recall that an $m$-dimensional simplicial polytope $P$ with $d$ vertices is called *$k$-neighborly* if every simplex spanned by an arbitrary $k$-tuple of vertices is a face of $P$ or, equivalently, if $f_{k-1}(P) = \binom{d}{k}$. $P$ is called *neighborly* if it is $k$-neighborly for each $k \leq m/2$. Particular examples are the so-called *cyclic polytopes* $C(d, m)$ which are defined as the convex hull of $d$ points on a moment curve in $E^m$ (see [**Gr, 4.7**]). A polytope $P$ is called *stacked* if there is a sequence $P_1, \dots, P_n$ of simplicial polytopes such that $P_1$ is a simplex, $P_n = P$, and each $P_k$ is constructed from $P_{k-1}$ by attaching a simplex along a facet of $P_{k-1}$.

A simple way of constructing submanifolds of $C^d$ is provided by *L. Danzer's $2^K$-construction* [**Da, Sch**]. Suppose $K$ is a simplicial complex of dimension $n-1$ with $d$ vertices $v_1, \dots, v_d$. For every simplex $\Delta = \langle v_{i_1}, \dots, v_{i_k} \rangle$ we define a collection $F(\Delta)$ of $2^{d-k}$ $k$-dimensional faces of $C^d$ by setting

$$A_j := \begin{cases} [0, 1] & \text{if } j \in \{i_1, \dots, i_k\}, \\ \{0, 1\} & \text{otherwise}, \end{cases}$$

and $F(\Delta) := A_1 \times A_2 \times \cdots \times A_d \leq C^d$. Then $2^K$ is defined as the collection of all $F(\Delta)$, where $\Delta$ ranges over all simplices of $K$. By construction $2^K$ contains the full 1-skeleton of $C^d$. Moreover, the vertex figure of each vertex in $2^K$ is combinatorially isomorphic to $K$, thus $2^K$ is a PL manifold if and only if $K$ is a combinatorial sphere, and in any case $\dim(2^K) = \dim K + 1 = n$.

The original intention in **[Da]** was to construct examples of regular incidence polytopes, i.e., those with a flag transitive automorphism group. In fact, if $K$ is regular then $2^K$ is regular and $|\operatorname{Aut}(2^K)| = 2^d \cdot |\operatorname{Aut}(K)|$.

EXAMPLES

$$2^{\{3\}} = \{4, 3\} \quad \text{(boundary of the 3-cube)},$$
$$2^{\{4\}} = \{4, 4\}_{4,0} = \{4\} \times \{4\} \quad \text{(toroidal map)},$$
$$2^{\{d\}} = \{4, d|4^{[\frac{1}{2}d]-1}\} \quad \text{(Coxeter's regular maps)},$$
$$2^{\{3^p\}} = \{4, 3^p\},$$
$$2^{\{3^{p-1}, 4\}} = \{4\}^{p+1}$$

If $\{\overline{d}\}$ denotes a triangulated planar $d$-gon without vertices in its interior, then $2^{\{\overline{d}\}}$ is a 3-manifold with boundary $2^{\{d\}}$. This is a 1-handlebody of genus $g_d$ (see Theorem 1(iii) below).

The following lemma is an almost immediate consequence of the definition of $2^K$, where $*$ denotes the join operator for simplicial complexes (see **[RS, Chapter 2]**).

LEMMA 1. *Let $K$ and $L$ be disjoint simplicial complexes, and let $v$ be an abstract vertex not in $K$. Then the following hold*:

(i) $2^{K\cup\{v\}} = 2^K \times \{0, 1\} \cup \operatorname{Vert}(2^K) \times [0, 1]$.
(ii) $2^{K*\{v\}} = 2^K \times [0, 1]$.
(iii) $2^{K*L} = 2^K \times 2^L$.
(iv) $2^{\operatorname{Sk}_{k-1}(\Delta^{d-1})} = \operatorname{Sk}_k(C^d)$.
(v) *$2^K$ is $k$-Hamiltonian iff $K$ is $k$-neighborly.*
(vi) *$2^{\partial K} = \partial(2^K)$ if $K$ is a combinatorial ball such that $\partial K$ contains all vertices of $K$.*

THEOREM 1. (i) *For every $d \geq 2k+1$ there exists a $k$-Hamiltonian $2k$-submanifold of $C^d$, and for $d \geq 2k+2$ there is a $k$-Hamiltonian $(2k+1)$-submanifold of $C^d$. These may be chosen as boundaries of $k$-Hamiltonian submanifolds of $C^d$ of dimensions $2k+1$ or $2k+2$, respectively. The $k$-Hamiltonian $2k$-submanifolds of $C^d$ are necessarily $(k-1)$-connected, and their Euler characteristic $\chi$ satisfies the equation*

$$\chi = 2 \cdot \chi(\operatorname{Sk}_k(C^{d-k-1})) = (-1)^k 2^{d-2k} \sum_{0\leq j\leq k/2} \binom{d-k-2j-3}{k-2j}.$$

(ii) *Particular examples for* (i) *are given by Danzer's $2^K$-construction applied to neighborly spheres. Especially, if $K$ is a suitable triangulation of* *the cyclic polytope* $C(d, 2k)$ *or* $C(d, 2k+1)$, *then* $2^K$ *is PL homeomorphic to a k-handlebody whose kth Betti number is just the kth Betti number of* $\mathrm{Sk}_k(C^{d-k-1})$ *or* $\mathrm{Sk}_k(C^{d-k-2})$. *Its boundary is PL homeomorphic to a connected sum of copies of* $S^k \times S^k$ *or* $S^k \times S^{k+1}$, *respectively.*

(iii) *Danzer's construction applied to a stacked n-polytope with d vertices leads to an orientable* 1*-handlebody of genus* $g_{d-n+2} = 2^{d-n-1}(d-n-2)+1$, *its boundary being PL homeomorphic to a connected sum of* $g_{d-n+2}$ *copies of* $S^1 \times S^{n-1}$.

Below we will give a direct combinatorial proof of Theorem 1, without using results from algebraic topology. By application of deeper results in topology we obtain the following theorem.

THEOREM 2. *Let P be an arbitrary neighborly 2k-polytope with d vertices. Then* $2^{\partial P}$ *is a k-Hamiltonian 2k-submanifold of* $C^d$ *which is homeomorphic (in fact PL homeomorphic if* $k \neq 2$*) to a connected sum of manifolds which are k-sphere bundles over* $S^k$. *For* $k = 1, 2, 3, 7$ *or* $k \equiv 6(8)$ *all these summands are products* $S^k \times S^k$.

CONJECTURE. Under the same assumptions, $2^{\partial P}$ is always PL homeomorphic to a connected sum of copies of $S^k \times S^k$.

QUESTION. Is it possible to construct $k$-Hamiltonian $2k$-submanifolds of $C^d$ which do not decompose into a connected sum as in Theorem 2, or where one of the summands is a nontrivial $k$-sphere bundle over $S^k$? Note that the latter is possible for $k = 1$ since the nonorientable examples necessarily contain summands which are nontrivial $S^1$-bundles over $S^1$ (Klein bottles).

REMARKS. (a) Theorem 2 applies to a large number of combinatorially distinct manifolds; compare **[Sh]** for the number of neighborly $2k$-polytopes.

(b) Any $k$-Hamiltonian $2k$-submanifold of $C^d$ is a tight submanifold of $E^d$ (see **[Kü, Oz]**).

(c) No topological uniqueness theorem can be expected in the case of $k$-Hamiltonian $(2k+1)$-submanifolds. If $P_1$ is a stacked 3-polytope with six vertices then $2^{\partial P_1}$ is the boundary of a 1-handlebody of genus 5 by Theorem 1(iii), and for $P_2$ = octahedron $2^{\partial P_2}$ is PL homeomorphic to the three-dimensional torus $S^1 \times S^1 \times S^1$ by Lemma 1(iii).

## Proofs of the theorems

PROOF OF THEOREM 1. We start with (iii). This will suggest the method for the proof of (ii), and the first part of (i) follows immediately from (ii).

We prove (iii) by induction on $d$. It is clear for $d = n+1$ because, for an $n$-simplex $\Delta$, $2^\Delta$ is a solid $d$-cube which may be regarded as a 1-handlebody of genus $g = 0$. Now let us assume that the assertion is true for every stacked $n$-polytope with $d$ vertices, and let $P = P' \cup \Delta$ be a stacked $n$-polytope with $d+1$ vertices, where $P'$ is one with vertices $v_1, \dots, v_d$, and where $\Delta$ is the simplex spanned by $v_{d-n+1}, \dots, v_d, v_{d+1}$. Lemma 1 implies

$$\begin{aligned}2^P &= 2^{P'\cup\Delta} = 2^{P'\cup\{v_{d+1}\}} \cup 2^{\{v_1, \dots, v_{d-n}\}\cup\Delta} \\ &= (2^{P'} \times \{0, 1\}) \cup (\{0, 1\}^{d-n} \times C^n \times [0, 1]).\end{aligned}$$

The two parts are glued together along $2^{d-n}$ disjoint $n$-cubes in each of the two components of $2^{P'} \times \{0, 1\}$. By assumption $2^{P'}$ is PL homeomorphic to an orientable 1-handlebody of genus $g_{d+2-n}$. It follows that $2^P$ is an orientable 1-handlebody of genus

$$\begin{aligned}2g_{d+2-n} + 2^{d-n} - 1 &= 2 \cdot 2^{d-n-1}(d-n-2) + 2^{d-n} + 1 \\ &= 2^{d-n}(d-n-1) + 1 = g_{d+3-n}.\end{aligned}$$

This establishes the recursion formula for the genus. Obviously $2^{\partial P} = \partial(2^P)$ is a connected sum of as many copies of $S^1 \times S^{n-1}$ as the genus tells us.

In the special case $n = 2$ we get the classical examples since every 2-polytope is stacked. Note that the genus of $2^P$ equals the genus of the two-dimensional surface $2^{\{d-n+2\}}$, and $\chi(2^P) = -2^{d-n-1}(d-n-2)$ equals the Euler characteristic of $\mathrm{Sk}_1(C^{d-n})$. In fact the same kind of induction holds for the construction of the 1-skeleton of $C^{d-n}$. This suggests we regard $2^P$ as a "thickened" 1-skeleton of $C^{d-n}$. We will use this idea essentially in the following proof of (ii).

If $K$ is an arbitrary neighborly $(2k-1)$-sphere with $d$ vertices then $2^K$ is a $2k$-submanifold of $C^d$ which is $k$-Hamiltonian by Lemma 1. Similarly, if $K$ is a neighborly $2k$-sphere then $2^K$ is a $k$-Hamiltonian $(2k+1)$-submanifold of $C^d$. The problem is to determine the topology of these examples. This seems to be difficult in general. Here we deal with the particular case of the cyclic polytope $C(d, m)$, where $m = 2k$ or $m = 2k + 1$ according to the two different cases.

By Gale's evenness condition (see **[Gr, 4.7]**) the simplices of $\partial C(d, m)$ are given by subsequent subsets $Y_1, X_1, \dots, X_t, Y_2$ of the set $\{v_1, \dots, v_d\}$ of vertices on the momentum curve, where $\#X_i$ is even for $i = 1, \dots, t$, and where $Y_1, Y_2$ are either empty or contain $v_1$ or $v_d$, respectively. We define two different triangulations of $C(d, m)$ as $m$-dimensional simplicial complexes. Let $C_1(d, m)$ be given by all $Y_1, X_1, \dots, X_t, Y_2$, where $\#X_i$ is even and $\#Y_2$ is odd, and let $C_2(d, m)$ be given by all $Y_1, X_1, \dots, X_t, Y_2$, where $\#X_i$ and $\#Y_2$ are even. By construction

$$\partial C_1(d, m) = \partial C(d, m) = \partial C_2(d, m).$$

$C_1$ and $C_2$ are combinatorially isomorphic for every even $m$, and $C_1(d, m) \cup C_2(d, m)$ and $\partial C(d, m+1)$ are combinatorially isomorphic.

The following recursion formula is easily verified, where $v_d$ denotes the last vertex in the subsequent labeling of the vertices:

$$C_1(d, m) = C_2(d-1, m-1) * \{v_d\},$$
$$C_2(d, m) = C_2(d-1, m) \cup C_1(d-1, m-1) * \{v_d\}.$$

By Lemma 1 this directly implies

$$2^{C_1(d, m)} = 2^{C_2(d-1, m-1)} \times [0, 1],$$
$$2^{C_2(d, m)} = (2^{C_2(d-1, m)} \times \{0, 1\}) \cup (2^{C_1(d-1, m-1)} \times [0, 1]).$$

The second equation shows a certain analogy with the following recursion formula for $\mathrm{Sk}_k(C^d)$:

$$\mathrm{Sk}_k(C^d) = (\mathrm{Sk}_k(C^{d-1}) \times \{0, 1\}) \cup (\mathrm{Sk}_{k-1}(C^{d-1}) \times [0, 1]).$$

This suggests we compare the topology of $2^{C_1}$ and $2^{C_2}$ with an appropriate $(m+1)$-dimensional "thickening" of the $k$-skeleton of a certain cube. First we have to define such a thickening. For arbitrary given numbers $d \geq k+1$ we define subsets $\mathrm{Sk}^d_{0,k} \subseteq \mathrm{Sk}^d_{1,k} \subseteq \cdots \subseteq \mathrm{Sk}^d_{k,k} \subseteq E^{2k+1}$ in the following way:

(i) for $d \leq 2k+1$ let $\mathrm{Sk}^d_{i,k} := \mathrm{Sk}_i(C^d) \subseteq E^d \subseteq E^{2k+1}$,

(ii) for given $d$ let $V$ be a nonzero vector of $E^{2k+1}$ in general position relative to $\mathrm{Sk}^d_{k,k}$, and let

$$\mathrm{Sk}^{d+1}_{i,k} := \mathrm{Sk}^d_{i,k} \cup \mathrm{Sk}^d_{i,k} + V \cup \{x + tV \mid 0 \leq t \leq 1,\ x \in \mathrm{Sk}^d_{i-1,k}\}.$$

We call this the *canonical embedding* of $\mathrm{Sk}_k(C^d)$ into $E^{2k+1}$, where each $\mathrm{Sk}^d_{i,k}$ corresponds to $\mathrm{Sk}_i(C^d)$. For $\varepsilon > 0$ being sufficiently small let $\overline{\mathrm{Sk}}^d_{i,k}$ denote the $\varepsilon$-*tube* around $\mathrm{Sk}^d_{i,k}$, i.e.,

$$\overline{\mathrm{Sk}}^d_{i,k} := \{x \in E^{2k+1} \mid \mathrm{dist}(x, \mathrm{Sk}^d_{i,k}) \leq \varepsilon\},$$

where the distance function dist is taken with respect to the maximum norm. Each face of $\mathrm{Sk}^d_{i,k}$ is the affine image of a cube of the corresponding dimension. Its tube is a $(2k+1)$-dimensional PL ball, and $\overline{\mathrm{Sk}}^d_{i,k}$ is a finite union of such balls, overlapping along smaller PL balls, all of dimension $2k+1$. Therefore $\overline{\mathrm{Sk}}^d_{i,k}$ is a $(2k+1)$-dimensional PL manifold with boundary, where the PL topological type is independent of the choice of $V$ and $\varepsilon$. (This is true also if we replace dist by the euclidean distance function; see **[Fe]**.)

We call $\overline{\mathrm{Sk}}^d_{i,k}$ the $(2k+1)$-dimensional *thickening* of $\mathrm{Sk}_i(C^d)$. By construction

$$\overline{\mathrm{Sk}}^d_{0,k} \subseteq \overline{\mathrm{Sk}}^d_{1,k} \subseteq \cdots \subseteq \overline{\mathrm{Sk}}^d_{k,k} \subseteq E^{2k+1}.$$

If $\phi_k$ denotes a suitable linear projection

$$\phi_k \colon E^{2k+1} \to E^{2k-1}$$

then $\phi_k(\mathrm{Sk}^d_{0,k}) \approx \mathrm{Sk}^d_{0,k-1}, \dots, \phi_k(\mathrm{Sk}^d_{k-1,k}) \approx \mathrm{Sk}^d_{k-1,k-1}$, and moreover

$$\overline{\mathrm{Sk}}^d_{i,k} \approx \overline{\mathrm{Sk}}^d_{i,k-1} \times [0,1] \times [0,1],$$

where "$\approx$" means "PL homeomorphic," and where this PL homeomorphism is compatible with the inclusions, in the following sense:

$$\begin{array}{ccc} \overline{\mathrm{Sk}}^d_{i,k} & \approx & \overline{\mathrm{Sk}}^d_{i,k-1} \times [0,1] \times [0,1] \\ \cup\!\!\uparrow & /\!\!/ & \cup\!\!\uparrow \\ \overline{\mathrm{Sk}}^d_{i-1,k} & \approx & \overline{\mathrm{Sk}}^d_{i-1,k-1} \times [0,1] \times [0,1] \end{array}$$

These homeomorphisms involve the choice of $\phi_k$ but their existence is independent of this choice.

LEMMA 2. *There exist the following PL homeomorphisms*:

(i) $2^{C_2(d,2k)} \approx \overline{\mathrm{Sk}}^{d-k-1}_{k,k}$,

(ii) $2^{C_1(d,2k+1)} \approx \overline{\mathrm{Sk}}^{d-k-2}_{k,k} \times [0,1]$;

*satisfying in addition the following compatibility*:

$$\begin{array}{ccccc} 2^{C_2(d,2k)} & \approx \overline{\mathrm{Sk}}^{d-k-1}_{k,k} & & \hookleftarrow & \\ \cup\!\!\uparrow & /\!\!/ & \overline{\mathrm{Sk}}^{d-k-1}_{k-1,k-1} \times [0,1] \times [0,1] & \approx & \overline{\mathrm{Sk}}^{d-k-1}_{k-1,k} \\ & & \cup\!\!\uparrow & & \\ 2^{C_1(d,2k-1)} & \approx & \overline{\mathrm{Sk}}^{d-k-1}_{k-1,k-1} \times [0,1] & & \end{array}$$

The proof of Lemma 2 uses induction on $d$. The assertion is easily verified for $d = 2k+1$ or $d = 2k+2$, respectively.

The induction step from $d-1$ to $d$ is easy for (ii):

$$2^{C_1(d,2k+1)} = 2^{C_2(d-1,2k)} \times [0,1] \approx \overline{\mathrm{Sk}}^{d-k-2}_{k,k} \times [0,1].$$

For (i) we have:

$$\begin{aligned} 2^{C_2(d,2k)} &= (2^{C_2(d-1,2k)} \times \{0,1\}) \cup (2^{C_1(d-1,2k-1)} \times [0,1]) \\ &\approx (\overline{\mathrm{Sk}}^{d-k-2}_{k,k} \times \{0,1\}) \cup (\overline{\mathrm{Sk}}^{d-k-2}_{k-1,k-1} \times [0,1] \times [0,1]) \\ &\approx \overline{\mathrm{Sk}}^{d-k-1}_{k,k}, \end{aligned}$$

where the last two steps involve the definition of $\overline{\mathrm{Sk}}$ and the compatibility.

The compatibility for $d$ follows from that for $d-1$ and from the induction step for $2^{C_2(d,2k)}$ and for $2^{C_2(d-1,2k-2)}$. This proves Lemma 2.

It seems to be a plausible conjecture that $2^{C_2(d,2k+1)}$ is a suitable thickening of $\mathrm{Sk}_{k+1}(C^{d-k-1})$ but this is not needed for the proof of our theorem.

From Lemma 2 it follows that $2^{C_1}$ and $2^{C_2}$ are both $k$-handlebodies because $\overline{\mathrm{Sk}}$ obviously has this property, and the $k$th Betti numbers are equal according to Lemma 2.

It remains to determine the topology of

$$2^{\partial C(d,2k)} \approx \partial\overline{\mathrm{Sk}}_{k,k}^{d-k-1}$$

and

$$2^{\partial C(d,2k+1)} \approx \partial(\overline{\mathrm{Sk}}_{k,k}^{d-k-2} \times [0,1]).$$

The $k$-dimensional skeleton of a cube is a union of a certain number of copies of $\partial C^{k+1} \approx S^k$ glued together along PL $k$-balls. Moreover in the embedding $\mathrm{Sk}_{k,k}^d \subseteq E^{2k+1}$ each such copy is embedded in the standard way, i.e., PL isotopic to $\partial C^{k+1} \subseteq E^{k+1} \subseteq E^{2k+1}$ or $C^k \subseteq E^k \subseteq E^{2k+1}$, respectively.

It follows that the thickening of each copy is PL homeomorphic to the product $\partial C^{k+1} \times C^{k+1}$, and that its boundary is PL homeomorphic to $\partial C^{k+1} \times \partial C^{k+1}$. Therefore $\partial\overline{\mathrm{Sk}}_{k,k}^d$ is PL homeomorphic to the connected sum of a certain number of sphere products $S^k \times S^k$.

Similarly, $\partial(\mathrm{Sk}_{k,k}^d \times [0,1])$ is PL homeomorphic to the connected sum of a certain number of sphere products $S^k \times S^{k+1}$. The number of these copies is just the $k$th Betti number of the $k$-handlebody. It can be calculated from the Euler characteristic of $\mathrm{Sk}_k^{d-k-1}$ or $\mathrm{Sk}_k^{d-k-2}$, respectively. This proves (ii).

To prove (i) we only have to show the last assertion since the examples in (ii) prove the first part. Let $M$ be an arbitrary $k$-Hamiltonian $2k$-submanifold of $C^d$. It is $(k-1)$-connected in the sense of homotopy theory, i.e., $\pi_1(M) = \pi_2(M) = \cdots = \pi_{k-1}(M) = 0$, because the $k$-skeleton of $C^d$ is homotopy equivalent to a wedge product of $k$-spheres. Furthermore the vertex figure of every vertex is a $k$-neighborly combinatorial sphere with $d$ vertices. The $f$-vector of such a triangulated sphere is known to coincide with the $f$-vector of $\partial C(d, 2k)$. Consequently the $f$-vector of $M$ coincides with that of $2^{\partial C(d,2k)}$, and by Lemma 2 it follows that $\chi(M) = 2\chi(\mathrm{Sk}_k(C^{d-k-1}))$. We prove the formula for the Euler characteristic in the following lemma:

LEMMA 3

$$\chi(\mathrm{Sk}_k(C^d)) = \sum_{i=0}^{k}(-1)^i 2^{d-i}\binom{d}{i} = (-1)^k 2^{d-k} \sum_{0\le j\le k/2} \binom{d-2j-2}{k-2j}.$$

PROOF. The first equation is just the definition because the $f$-vector of $C^d$ satisfies $f_i = 2^{d-i}\binom{d}{i}$.

The following recursion formula is obvious:

$$\chi(\mathrm{Sk}_k(C^{d+1})) = 2\chi(\mathrm{Sk}_k(C^d)) - \chi(\mathrm{Sk}_{k-1}(C^d)).$$

The same recursion formula holds for the expression on the right-hand side of Lemma 3. This follows from elementary properties of the Pascal triangle. This proves Lemma 3 by induction on $d$ since for $d = k+1$ the formula is trivial. This completes the proof of Theorem 1.

REMARK. If a combinatorial sphere $K$ is obtained from another one, $K'$, by one bistellar operation then $2^K$ is obtained from $2^{K'}$ by a number of "spherical modifications" or by "surgery," as far as the topology is concerned. However, it seems to be difficult to describe this procedure in a general formula.

PROOF OF THEOREM 2. By Theorem 1(i) the Euler characteristic and therefore the rank of the homology $H_k(2^{\partial P})$ of $2^{\partial P}$ coincide with the ones of $2^{\partial C(d,2k)}$. We will show that also the intersection forms on $H_k$ are the same. By the results of [**Br**] the $k$th homology of $2^{\partial P}$ is generated by the $(k-1)$st homology of all full subcomplexes of $\partial P$. By the neighborliness it is generated by the *missing k-simplices* in $\partial P$, i.e., the full subcomplexes combinatorially isomorphic to the boundary of a $k$-dimensional simplex. On the other hand such a missing $k$-simplex $\Delta$ is the boundary of a simplicial $k$-ball $B$ which is a subcomplex of $\partial P$. This follows easily by induction on the number of vertices. Thus every component of $2^{\partial B}$ is a boundary component of the $(k+1)$-submanifold $2^B$, and therefore it can be deformed isotopically into a disjoint position, and hence the self-intersection number of this homology class is zero. It follows that the free $\mathbb{Z}$-module $H_k(2^{\partial P};\mathbb{Z})$ is generated by elements with self-intersection zero, thus the intersection form has a matrix representation with only zeros in the main diagonal. For even $k$ it is therefore of type II with signature zero. In this case the results of [**MH**] say that this quadratic form is equivalent to a direct sum of matrices $\left(\begin{smallmatrix}0&1\\1&0\end{smallmatrix}\right)$. In any case, for $k \geq 3$ Lemma 9 in [**W**] says that the manifold is decomposable (as a combinatorial manifold) into summands of rank two whose intersection form is $\left(\begin{smallmatrix}0&1\\1&0\end{smallmatrix}\right)$ for even $k$ and $\left(\begin{smallmatrix}0&1\\-1&0\end{smallmatrix}\right)$ for odd $k$. It is also mentioned in [**W, pp. 185–186**] that the only possible indecomposable summands are $k$-sphere bundles over $S^k$ which are classified by $\pi_{k-1}(\mathrm{SO}_{k+1})$, the latter vanishes for $k=3, 7$ or $k \equiv 6(8)$. This proves the theorem for $k \geq 3$. All these arguments break down for $k=2$. In this case every PL manifold is smoothable, and the famous theorem by M. Freedman [**FR, Theorem 1.5**] says that our manifold is homeomorphic to a connected sum of copies of $S^2 \times S^2$. In the case $k=1$ the assertion is well known because $2^{\partial P}$ is an orientable surface.

## REFERENCES

[**Ba**] T. F. Banchoff, *Tightly embedded 2-dimensional polyhedral manifolds*, Amer. J. Math. **87** (1965), 462–472.

[**BH**] L. W. Beineke and F. Harary, *The genus of the n-cube*, Canad. J. Math. **17** (1965), 494–496.

[**Br**] U. Brehm, *The homology of the power-complex of a simplicial complex*, (in preparation).

[**Cox**] H. S. M. Coxeter, *Regular skew polyhedra in three and four dimensions, and their topological analogues*, Proc. London Math. Soc. (2) **43** (1937), 33–62; reprinted in *Twelve geometric essays*, Southern Illinois Univ. Press, 1968, pp. 76–105.

[**Da**] L. Danzer, *Regular incidence-complexes and dimensionally unbounded sequences of such*. I, Ann. Discrete Math. **20** (1984), 115–127: II (in preparation).

[**Fe**] S. Ferry, *When ε-boundaries are manifolds*, Fund. Math. **90** (1976), 199–210.

[Fr] M. H. Freedman, *The topology of four-dimensional manifolds*, J. Differential Geom. **17** (1982), 357–453.

[Gr] B. Grünbaum, *Convex polytopes*, Interscience, 1967.

[J] M. Jungerman, *The non-orientable genus of the $n$-cube*, Pacific J. Math. **76** (1978), 443–451.

[Kü] W. Kühnel, *Tight triangulations and tight polyhedral submanifolds of* $E^N$, preprint.

[MSW] P. McMullen, Ch. Schulz, and J. M. Wills, *Polyhedral 2-manifolds in* $E^3$ *with unusually large genus*, Israel J. Math. **46** (1983), 127–144.

[MH] J. Milnor and D. Husemoller, *Symmetric bilinear forms*, Springer-Verlag, 1973.

[Oz] T. Ozawa, *Relations between tightness and $k$-tightness for* PL-*manifolds in euclidean space*, preprint.

[Ri] G. Ringel, *Über drei Probleme am $n$-dimensionalen Würfel und Würfelgitter*, Abh. Math. Sem. Univ. Hamburg **20** (1955), 10–19.

[RS] C. P. Rourke and B. J. Sanderson, *Introduction to piecewise-linear topology*, Springer-Verlag, 1972.

[Sch] E. Schulte, *On a class of abstract polytopes constructed from binary codes*, Discr. Math. (to appear).

[Sch1] Ch. Schulz, *Mannigfaltigkeiten mit Zellzerlegung im Randkomplex eines konvexen Polytops und verallgemeinerte Hamilton-Kreise*, Dissertation Bochum, 1974.

[Sch2] ——, *Geschlossene Flächen im Rand des Würfels*, Abh. Math. Sem. Univ. Hamburg **50** (1980), 89–94.

[Sh] I. Shemer, *Neighborly polytopes*, Israel J. Math. **43** (1984), 291–314.

[W] C. T. C. Wall, *Classification of $(n-1)$-connected $2n$-manifolds*, Ann. of Math. (2) **75** (1962), 163–189.

FACHBEREICH MATHEMATIK, UNIVERSITÄT DUISBURG, LOTHARSTRASSE 65, 4100 DUISBURG 1, FEDERAL REPUBLIC OF GERMANY

FACHBEREICH INFORMATIK, FACHHOCHSCHULE WIESBADEN, KURT-SCHUMACHER-RING 18, 6200 WIESBADEN, FEDERAL REPUBLIC OF GERMANY

DIMACS Series in Discrete Mathematics
and Theoretical Computer Science
Volume 4, 1991

Applied Geometry
and Discrete Mathematics
THE VICTOR KLEE FESTSCHRIFT

# Finite Unions of Closed Subgroups of the $n$-Dimensional Torus

JIM LAWRENCE

ABSTRACT. Let $U$ be an open subset of the torus group $T^n$. We show that the set of maximal subgroups of $T^n$ which miss $U$ is of finite cardinality. This result is applied to show that the lattice of finite unions of closed subgroups of $T^n$ is a complete distributive lattice, and to show that, up to unimodular equivalence, there are only finitely many convex polytopes $P \subseteq \mathbb{R}^n$ having vertices in $\mathbb{Z}^n$ but no interior points in $\mathbb{Z}^n$ and such that each subgroup $G$ of the additive group $\mathbb{R}^n$ which properly contains $\mathbb{Z}^n$ does have points in common with the interior of $P$.

## 1. Introduction

Let $x = (x_1, \ldots, x_n)$ be an element of $\mathbb{R}^n$ and let $U \subseteq \mathbb{R}^n$ be an open neighborhood of 0. A classical theorem of Dirichlet asserts that there exist a positive integer $m$ and a point $z = (z_1, \ldots, z_n) \in \mathbb{Z}^n$ such that $mx - z \in U$. The numbers $x_1, \ldots, x_n$ and 1 are independent over the rational numbers if there is no $w \in \mathbb{Z}^n \sim \{0\}$ such that $\langle w, x\rangle \in \mathbb{Z}$. A classical theorem of Kronecker asserts that the numbers $x_1, \ldots, x_n$ and 1 are independent over the rational numbers if and only if for every open set $U \subseteq \mathbb{R}^n$ there exist a positive integer $m$ and $z \in \mathbb{Z}^n$ such that $mx - z \in U$. (These are Theorems 201 and 442 of [**4**]. See also Chapter VII of [**1**].)

In this paper we consider, for open sets $U \subseteq \mathbb{R}^n$, the nature of the sets $\tilde{\tau}(U) = \{x \in \mathbb{R}^n : \text{there exist } m \in \mathbb{Z} \text{ and } z \in \mathbb{Z}^n \text{ such that } mx - z \in U\}$. (Alternatively, $\tilde{\tau}(U) = \{x \in \mathbb{R}^n : \text{the (additive) group generated by } \{x\} \cup \mathbb{Z}^n \text{ intersects } U\}$.) We show $\mathbb{R}^n \sim \tilde{\tau}(U)$ is a finite union of closed subgroups of $\mathbb{R}^n$; and moreover, the set $M(\mathbb{R}^n, U)$ of maximal subgroups $G$ of $\mathbb{R}^n$ such that $G \cap U = \varnothing$ and $\mathbb{Z}^n \subseteq G$ is finite. This is Corollary 1.A, below.

As an example, let $n = 2$ and let $U$ be the triangular region $\{(x, y) \in \mathbb{R}^2 : 0 < x,\ 0 < y, \text{ and } x + y < 1\}$. Then the subgroups $H$ of $\mathbb{R}^2$ such that

---

1980 *Mathematics Subject Classification* (1985 *Revision*). Primary 10E05, 20K99.


1052-1798/91 $1.00 + $.25 per page

$\mathbb{Z}^2 \subseteq H$ and $H \cap U = \varnothing$ are precisely the subgroups of the following four groups:

$$H_1 = \{(x, y) \in \mathbb{R}^2 : x \in \mathbb{Z}\}, \qquad H_2 = \{(x, y) \in \mathbb{R}^2 : y \in \mathbb{Z}\},$$
$$H_3 = \{(x, y) \in \mathbb{R}^2 : x + y \in \mathbb{Z}\}, \qquad H_4 = \{(x, y) \in \mathbb{R}^2 : 2x \in \mathbb{Z} \text{ and } 2y \in \mathbb{Z}\}.$$

One of several interesting consequences of the general finiteness result concerns finite unions of closed subgroups of the $n$-dimensional torus group $T^n$. It is obvious that these subsets form a finitely distributive lattice under the operations of intersection and union. It follows from the finiteness result that they actually form a complete lattice: The intersection of an arbitrary family of finite unions of closed subgroups of $T^n$ is again a *finite* union of closed subgroups of $T^n$. (We will have occasion in this paper to use the word "lattice" in two different senses: We will use it as we have in this paragraph, to mean a partially ordered set with certain properties; we will use the phrase "point lattice" in its sense in the geometry of numbers, to mean a discrete, full-dimensional subgroup of $\mathbb{R}^n$.)

In §3 we present some consequences of these results concerning finiteness of certain sets of unimodular equivalence classes of polytopes with integer vertices. The results in this section are connected in spirit to those of Hensley **[5]** and Lagarias and Ziegler **[7]**.

This paper uses standard results concerning additive subgroups of $\mathbb{R}^n$. The best reference for this topic for our purposes is Chapter VII of **[1]**.

## 2. Preliminaries

We work with the lattice $\mathcal{G}$ of closed subgroups $G$ of $\mathbb{R}^n$ such that $\mathbb{Z}^n \subseteq G$. (We could equivalently work with closed subgroups of the torus group $T^n = \mathbb{R}^n/\mathbb{Z}^n$ in view of the bijective correspondence $G \to \pi(G)$ mapping the set of such subgroups to the set of subgroups of $T^n$, where $\pi: \mathbb{R}^n \to T^n$ is the canonical map. We remain in $\mathbb{R}^n$ in order to make easy use of results from the geometry of numbers.)

Any closed subgroup of $\mathbb{R}^n$ is linearly equivalent to $\mathbb{R}^a \times \mathbb{Z}^b$, for some nonnegative integers $a$ and $b$.

Let $\overline{\mathcal{G}}$ be the lattice of closed subgroups of $\mathbb{R}^n$. For $G \in \overline{\mathcal{G}}$, let $G^* = \{x \in \mathbb{R}^n : \langle x, u\rangle \in \mathbb{Z} \text{ for each } u \in G\}$. Then $G^*$ is also in $\overline{\mathcal{G}}$ and the map $G \to G^*$ is an anti-automorphism of $\overline{\mathcal{G}}$ (see **[1]**).

The lattice $\mathcal{G}$ satisfies the descending chain condition; that is, each nonempty subset of $\mathcal{G}$ possesses a minimal element. Equivalently, any chain $H_1 \supseteq H_2 \supseteq \cdots$ of distinct elements of $\mathcal{G}$ must be finite. To see this note that $H_1^* \subseteq H_2^* \cdots$ would be an ascending chain of subgroups of $(\mathbb{Z}^n)^* = \mathbb{Z}^n$, which satisfies the ascending chain condition, since it is a finitely generated abelian group.

For $S \subseteq \mathbb{R}^n$, let $\operatorname{pol}(S) = \{x \in \mathbb{R}^n : \langle x, y\rangle \leq 1 \text{ for each } y \in S\}$. Then $\operatorname{pol}(S)$ is a closed, convex set which contains the origin; $\operatorname{pol}(\operatorname{pol}(S))$ is the smallest closed, convex set which contains $S \cup \{0\}$; and pol is a dual automorphism of the partially ordered set of closed, convex sets containing the origin.

Our objective now is to establish a lemma (Lemma 3) which will be used in the proof in the next section of the main result.

LEMMA 1. *Suppose that $U$ is a convex subset of $\mathbb{R}^n$ and that $p \in U$. If $H$ is a subgroup of $\mathbb{R}^n$ such that $H \cap (1/n(U-p))$ contains a basis for $\mathbb{R}^n$, then $H + U = \mathbb{R}^n$.*

PROOF. Let $\{b_1, \dots, b_n\}$ be a basis for $\mathbb{R}^n$ contained in $H \cap (1/n(U - p))$. Let $P = \{\sum \alpha_i b_i : 0 \le \alpha_i \le 1 \text{ for } i = 1, \dots, n\}$. Then $P \subseteq \operatorname{conv}\{0, nb_1, \dots, nb_n\} \subseteq U - p$. Any $x \in \mathbb{R}^n$ can be expressed in terms of the basis $x = \sum \alpha_i b_i$, $i = 1, \dots, n$. We then have

$$x = \sum \lfloor \alpha_i \rfloor b_i + \sum (\alpha_i) b_i \in H + P,$$

so $H + P = \mathbb{R}^n$. (Here $\lfloor \alpha \rfloor$ denotes the greatest integer less than or equal to $\alpha$ and $(\alpha) = \alpha - \lfloor \alpha \rfloor$ is the fractional part of $\alpha$.) It follows that $H + (U - p) = \mathbb{R}^n$, i.e., $H + U = \mathbb{R}^n$. □

In the proof of Lemma 2 we will use a result of Mahler belonging to the theory of successive minima. Recall that for a point lattice $L \subseteq \mathbb{R}^n$ and a full-dimensional, compact, convex set $K$ symmetric about the origin (so that $K$ contains a neighborhood of the origin), the *successive minima* $\lambda_1, \dots, \lambda_n$ of $L$ with respect to $K$ are the smallest real numbers such that (for each $i$) $(\lambda_i K) \cap L$ contains a set of $i$ linearly independent points.

Let $\lambda_1, \dots, \lambda_n$ be the successive minima of $L$ with respect to $K$ (as above) and let $\lambda_1^*, \dots, \lambda_n^*$ be the successive minima of $L^*$ with respect to $\operatorname{pol}(K)$. Mahler's result is that (for each $i$) one has

$$1 \le \lambda_i \lambda_{n-i+1}^* \le n!.$$

(In Mahler's original result, the right-hand bound was $(n!)^2$. The statement as we have it is Theorem VI of Chapter VIII, §5, of **[2]**. The right-hand bound has been spectacularly improved by Lagarias, Lenstra, and Schnorr in **[6]**.)

LEMMA 2. *Let $K$ be a full-dimensional, convex, compact set with $K = -K$. Let $H$ be a closed subgroup of $\mathbb{R}^n$ such that $H \cap K$ does not contain a basis for $\mathbb{R}^n$. Then $H^* \cap (n! \operatorname{pol}(K))$ contains a nonzero element.*

PROOF. Suppose that there is a convex, full-dimensional, compact set $K$ symmetric about 0 and a closed subgroup $H$ such that $H \cap K$ contains no basis for $\mathbb{R}^n$ and $H^* \cap (n! \operatorname{pol}(K)) = \{0\}$. We may choose a basis $\{x_1, \dots, x_n\}$ for $\mathbb{R}^n$ such that

$$H = \left\{ \sum_{i=1}^n \alpha_i x_i : \alpha_i \in \mathbb{Z} \text{ for } i = a+1, \dots, b, \right.$$
$$\left. \text{and } \alpha_i = 0 \text{ for } i = b+1, \dots, n \right\}.$$

Let $L_m$ be the point lattice generated by $\{x_1/m, \dots, x_a/m, x_{a+1}, \dots, x_b, mx_{b+1}, \dots, mx_n\}$. It is clear that we may choose $m$ sufficiently large that $L_m \cap K$ contains no basis for $\mathbb{R}^n$, and $L_m^* \cap (n!\,\mathrm{pol}(K)) = \{0\}$. Let $\lambda_1, \dots, \lambda_n, \lambda_1^*, \dots, \lambda_n^*$ be the successive minima for $L_m$ with respect to $K$ and for $L_m^*$ with respect to $\mathrm{pol}(K)$, respectively. Since $L_m \cap K$ contains no basis for $\mathbb{R}^n$, we have $\lambda_n > 1$. Also, $L_m^* \cap (n!\,\mathrm{pol}(K)) = \{0\}$, so $\lambda_1^* > n!$. This contradicts Mahler's Theorem, since then $\lambda_n \lambda_1^* > n!$. □

LEMMA 3. *Let $G$ be a closed subgroup of $\mathbb{R}^n$. Let $U$ be a subset of $G$ which contains a nonempty relatively open set. Then there is a bounded set $X \subseteq \mathbb{R}^n$ such that if $H$ is a closed subgroup of $G$ for which $H + U \neq G$, then $H^* \cap X$ contains a point not in $G^*$.*

PROOF. It is clear that, if $\lambda: \mathbb{R}^n \to \mathbb{R}^n$ is a nonsingular linear transformation, then the statement holds for a given group $G$ and open set $U \subseteq G$ if and only if it holds for the images $\lambda(G)$ and $\lambda(U)$. We may therefore suppose that

$$G = \{(x_1, \dots, x_n) \in \mathbb{R}^n : x_{a+1}, \dots, x_b \in \mathbb{Z} \text{ and } x_{b+1} = \cdots = x_n = 0\},$$

where $a$ and $b$ are integers for which $0 \leq a \leq b \leq n$.

Let

$$A = \{(x_1, \dots, x_n) \in \mathbb{R}^n : x_i = 0 \text{ for } i > a\},$$
$$B = \{(x_1, \dots, x_n) \in \mathbb{R}^n : x_i = 0 \text{ for } i \leq a \text{ or } i > b\},$$

and

$$C = \{(x_1, \dots, x_n) \in \mathbb{R}^n : x_i = 0 \text{ for } i \leq b\};$$

and let $\alpha: \mathbb{R}^n \to A$, $\beta: \mathbb{R}^n \to B$, and $\gamma: \mathbb{R}^n \to C$ be the obvious projections. Then we may write

$$G = \{x \in \mathbb{R}^n : \beta(x) \in \mathbb{Z}^n \text{ and } \gamma(x) = 0\},$$
$$G^* = \{x \in \mathbb{R}^n : \alpha(x) = 0 \text{ and } \beta(x) \in \mathbb{Z}^n\}.$$

Let $P = \{x \in \mathbb{R}^n : \alpha(x) = \gamma(x) = 0 \text{ and } 0 \leq \beta(x) < 1\}$. Note that $P \cap G^* = \{0\}$ and $P$ is bounded. If $G$ is a discrete group, so that $a = 0$, then we may take $X = P$. Otherwise, let $W$ be the unit ball in $A$: $W = \{x \in A : \|x\| \leq 1\}$. Let $p \in U$ and choose $\varepsilon$ sufficiently small that $\varepsilon W \subseteq (U - p)/a$. Finally, take $X = (a!/\varepsilon)W + P$. Clearly, $X$ is bounded.

Suppose $H$ is a closed subgroup of $G$ such that $H + U \neq G$. We will show that $(H^* \cap X) \sim G^* \neq \varnothing$.

Suppose $\beta(H)$ is properly contained in $\beta(G) = \mathbb{Z}^n \cap B$. It follows that $H + A + C$ $(= \beta^{-1}(\beta(H)))$ is properly contained in $G + A + C$, so that $H^* \cap B = (H + A + C)^*$ properly contains $(G + A + C)^* = G^* \cap B = \mathbb{Z}^n \cap B$. Choose $x \in (H^* \cap B) \sim (G^* \cap B)$; say, $x = (0, \dots, 0, x_{a+1}, \dots, x_b, 0, \dots, 0)$. Then $\tilde{x} = (0, \dots, 0, \lfloor x_{a+1} \rfloor, \dots, \lfloor x_b \rfloor, 0, \dots, 0) \in \mathbb{Z}^n \cap B \subseteq H^* \cap B$, so

$x - \tilde{x}$ is a nonzero element of $P$ which is in $H^*$. Therefore, $x - \tilde{x} \in (H^* \cap X) \sim G^*$.

Finally, suppose $\beta(H) = \beta(G)$. If $a\varepsilon W + (H \cap A) = A$, then $a\varepsilon W + H = G$; so $U + H \supseteq (a\varepsilon W + p) + H = G$, contrary to our assumption. Therefore, $a\varepsilon W + (H \cap A) \neq A$, and we see by invoking Lemma 1 that $\varepsilon W \cap H$ contains no basis for $A$. By Lemma 2 applied to $A$ there is a nonzero vector in

$$(a!/\varepsilon)W \cap (H \cap A)^* = (a!/\varepsilon)W \cap (H^* + B + C),$$

i.e., we may find $x \in H^*$ such that $\alpha(x) \in (a!/\varepsilon)W$, $\alpha(x) \neq 0$. Suppose $x = (x_1, \dots, x_n)$. Then $\tilde{x} = (0, \dots, 0, \lfloor x_{a+1} \rfloor, \dots, \lfloor x_b \rfloor, x_{b+1}, \dots, x_n) \in H^*$ (since $H^*$ contains $G^*$), and $x - \tilde{x}$ is the required element of $(H^* \cap X) \sim G^*$. □

## 3. Results concerning subgroups of the torus

Let $G$ be a closed subgroup of $\mathbb{R}^n$. Suppose $U \subseteq G$. We shall call $U$ *full* if its intersection with each closed subgroup $H$ of $G$ is empty or contains a relatively open, nonempty subset of $H$. In particular, open sets are full.

THEOREM 1. *Suppose that $G$ is a closed subgroup of $\mathbb{R}^n$ and $U$ is a full subset of $G$. Let $M(G, U)$ be the set of subgroups $H \subseteq G$ maximal with respect to the property that $\mathbb{Z}^n \subseteq H$ and $H \cap U = \varnothing$. Then $M(G, U)$ is of finite cardinality.*

PROOF. Let $\Gamma$ denote the set of all closed subgroups $G$ of $\mathbb{R}^n$ containing $\mathbb{Z}^n$ for which there exists a full subset $U \subseteq G$ such that $M(G, U)$ is infinite. Suppose $G \in \Gamma$ and $U$ is a corresponding full subset. Clearly $U \neq \varnothing$ because $M(G, \varnothing) = \{\mathbb{R}^n\}$. By Lemma 3 there is a bounded set $X \subseteq \mathbb{R}^n$ such that if $H$ is a closed subgroup of $G$ such that $H + U \neq G$, then $H^* \cap X \not\subseteq G^*$. On the other hand, if $H \in M(G, U)$, then $H + U \neq G$ (since $0 \notin H + U$), so for such $H$ there is $b \in (H^* \cap X) \sim G^*$. It follows that

$$M(G, U) \subseteq \bigcup_b M(G_b, U_b),$$

where the union is taken over all $b \in (\mathbb{Z}^n \cap X) \sim G^*$, $G_b = \{x \in G\colon \langle x, b \rangle \in \mathbb{Z}\}$, and $U_b = U \cap G_b$. Notice that, for each such $b$, $G_b$ is a proper subgroup of $G$ (since $b \notin G^*$). Also, since $\mathbb{Z}^n \cap X$ is finite, it follows that $b$ runs over a finite set. It follows that $M(G_0, U)$ is of infinite cardinality for some $b \in (H^* \cap X) \sim G^*$, so that $G_b \in \Gamma$.

We have shown that $\Gamma$ has no minimal element. By the descending chain condition on $\mathcal{G}$, $\Gamma = \varnothing$. □

We present some corollaries of Theorem 1.

COROLLARY 1.A. *If $U$ is a full subset of $T^n$, then there are only finitely many maximal closed subgroups $H$ of $T^n$ such that $H \cap U = \varnothing$.*

COROLLARY 1.B. *Let $S$ be a closed subset of $T^n$ such that if $x \in S$ and*

*$m$ is a positive integer, then $mx \in S$. Then $S$ is a finite union of closed subgroups of $T^n$.*

We now consider an order relation on open subsets of the torus $T^n$. For open subsets $U$ and $V$ of $T^n$ we write $U \prec V$ if for each $x \in U$ there is a positive integer $m$ such that $mx \in V$. We write $U \approx V$ if $U \prec V$ and $V \prec U$. Then $\approx$ is an equivalence relation on the set of open subsets of $T^n$ and $\prec$ induces a partial ordering on the set $\mathscr{E}$ of equivalence classes. We wish to study this partially ordered set.

For open subsets $U$ of $T^n$ let $\tau(U)$ denote the complement of the union of the closed subgroups $G$ of $T^n$ such that $G \cap U = \varnothing$. We see from Theorem 1 that $\tau(U)$ is open. This fact is also a consequence of the following lemma.

LEMMA 4. $\tau(U) = \{x \in T^n$: *there is* $m \in \mathbb{Z}$, $m > 0$, *such that* $mx \in U\}$.

PROOF. Certainly if for $x \in T^n$ there is a positive integer $m$ such that $mx \in U$, then each subgroup $G \subseteq T^n$ such that $x \in G$ intersects $U$ nontrivially, so $x \in \tau(U)$. Suppose no such $m$ exists. The closure of the set $\{mx : m \in \mathbb{Z}, m > 0\}$ is then a closed subgroup $G$ of $T^n$ which misses $U$. Since $x \in G$, $x \notin \tau(U)$. $\square$

We see that $\tau$ is a closure operator on the collection of all open subsets of $T^n$: $U \subseteq \tau(U)$ for each open set $U$; if $U \subseteq V$, then $\tau(U) \subseteq \tau(V)$; and $\tau(\tau(U)) = \tau(U)$ for each open set $U$. Also from the lemma it is immediate that $\tau(U)$ is the largest open set such that $\tau(U) \prec U$. The following theorem, which is now immediate, characterizes the partial ordering of $\mathscr{E}$ induced by $\prec$.

THEOREM 2. *If $U$ and $V$ are open subsets of $T^n$, then $U \prec V$ if and only if $\tau(U) \subseteq \tau(V)$, and $U \approx V$ if and only if $\tau(U) = \tau(V)$. The partially ordered set $\mathscr{E}$ is dually isomorphic to the partially ordered set of finite unions of closed subgroups of $T^n$ (under inclusion). This partially ordered set is a finitely distributive complete lattice.*

Finally we wish to establish a chain condition for the partially ordered set of equivalence classes from Theorem 2.

THEOREM 3. *Let $U_1 \subseteq U_2 \subseteq \cdots$ be an ascending sequence of open subsets of $T^n$. Then there is an integer $M$ such that $\tau(U_M) = \tau(U_{M+1}) = \cdots$.*

PROOF. Let $\Gamma$ denote the set of closed subgroups $G$ of $T^n$ such that there exists an infinite ascending chain $\tau(U_1) \subseteq \tau(U_2) \subseteq \cdots$ of distinct $\tau$-closed open sets $\tau(U_i) \supseteq T^n \sim G$. We may write $\tau(U_1) = T^n \sim (G_1 \cup \cdots \cup G_m) = \bigcap_{j=1}^m (T^n \sim G_j)$ for some closed subgroups $G_1, \dots, G_m$. Then $\tau(U_i) = \tau(U_i) \cup \tau(U_1) = \bigcap_{j=1}^m (\tau(U_i) \cup (T^n \sim G_j))$. It is clear that for some $j$ the sequence of sets $\tau(U_i) \cup (T^n \sim G_j) \supseteq T^n \sim G_j$ must contain an infinite subsequence of distinct $\tau$-closed open sets. Since $G_j$ properly contains $G$, we see that $\Gamma$ contains no maximal element. By the chain condition on the closed subgroups of $T^n$, it follows that $\Gamma = \varnothing$. □

## 4. Some consequences concerning lattice points in polytopes

LEMMA 5. *Let* $U^n = \{(x_1, \dots, x_n) \in \mathbb{R}^n : x_i > 0$ *for* $i = 1, \dots, n$ *and* $x_1 + \cdots + x_n < 1\}$. *There is a number* $\chi < 1$ *such that if* $G$ *is a group for which* $\mathbb{Z}^n \subseteq G \subseteq \mathbb{R}^n$ *and* $G \cap U^n \neq \varnothing$, *then there is a point* $(x_1, \dots, x_n) \in G \cap U^n$ *for which* $x_1 + \cdots + x_n \leq \chi$.

PROOF. (Here $U^n$ is the standard open simplex.) Consider the sequence

$$\tau(\tfrac{1}{2}U^n) \subseteq \tau(\tfrac{2}{3}U^n) \subseteq \tau(\tfrac{3}{4}U^n) \subseteq \cdots.$$

By Theorem 3 there is an $m$ such that

$$\tau(m/(m+1)U^n) = \tau((m+1)/(m+2)U^n) = \cdots.$$

We may set $\chi = m/(m+1)$. □

In general it seems difficult to find a value for $\chi$. We know that for $n = 1$ we can take $\chi = 1/2$; for $n = 2$, $\chi = 5/6$. Any value for $n = 3$ must satisfy $\chi \geq 41/42$, but we do not know a value even in this case.

Let $\chi_n$ denote the least value for $\chi$ satisfying Lemma 5. It is easy to see that $\chi_n \leq \chi_{n+1}$ for $n = 1, 2, \dots,$ for if $G \subseteq \mathbb{R}^n$ is a group such that $G \cap U^n \neq \varnothing$ and $G \cap (\alpha U^n) = \varnothing$, then $G \times \mathbb{R}$ has the analogous properties in $\mathbb{R}^{n+1}$. For $S \subseteq \mathbb{R}^n$ denote by $S^\circ$ its interior. For a convex polytope $K \subseteq \mathbb{R}^n$ denote by $\operatorname{vert}(K)$ its vertex set.

LEMMA 6. *Let* $k = \lceil 1/(1 - \chi_{2n-2}) \rceil$. *Suppose the convex polytope* $K$, *having* $\operatorname{vert}(K) \subseteq \mathbb{Z}^n$, *contains at least* $(1+k)^n + 1$ *points of* $\mathbb{Z}^n$, *and* $K^\circ \cap \mathbb{Z}^n = \varnothing$. *Then there is a linear function* $A\colon \mathbb{R}^n \to \mathbb{R}^{n-1}$ *such that* $A(\mathbb{Z}^n) = \mathbb{Z}^{n-1}$ *and* $A(K)^\circ \cap \mathbb{Z}^{n-1} = \varnothing$.

PROOF. Clearly some pair of points of $K \cap \mathbb{Z}^n$ must be congruent modulo $1 + k$; the line $L$ containing these satisfies $|L \cap K \cap \mathbb{Z}^n| \geq k + 2$. Let $u, w \in \mathbb{Z}^n$ be such that $u, u + w, u + 2w, \dots,$ and $u + (k+1)w$ are consecutive points of $L \cap K \cap \mathbb{Z}^n$. We may choose a basis $\{w, b_2, b_3, \dots, b_n\}$ for $\mathbb{Z}^n$ which contains $w$. For $x = \alpha_1 w + \alpha_2 b_2 + \cdots + \alpha_n b_n \in \mathbb{Z}^n$, let $A(x) = (\alpha_2, \dots, \alpha_n) \in \mathbb{R}^{n-1}$. Then $A\colon \mathbb{R}^n \to \mathbb{R}^{n-1}$ is a linear function such that $A(\mathbb{Z}^n) = \mathbb{Z}^{n-1}$.

We will complete the proof by showing that if $A(K)^\circ \cap \mathbb{Z}^{n-1} \neq \varnothing$, then $K^\circ \cap \mathbb{Z}^n \neq \varnothing$. Suppose $p \in A(K)^\circ \cap \mathbb{Z}^{n-1}$. Then by a theorem of Steinitz **[8]** (see also Exercise 2.3.5 of **[3]**) we may choose a set of $m \leq 2(n-1)$ vertices of $A(K)$, say, $\{A(v_1), \dots, A(v_m)\}$, where $v_1, \dots, v_m$ are vertices of $K$, such that $p$ is in the interior of $\operatorname{conv}\{A(v_1), \dots, A(v_m)\}$. We may find $\alpha_1, \dots, \alpha_m$, and $\beta$, where $\alpha_i > 0$ $(i = 1, \dots, m)$, $\beta > 0$, $(\sum_{i=1}^m \alpha_i) + \beta = 1$, and $p = (\sum_{i=1}^m \alpha_i A(v_i)) + \beta A(u)$.

Let $G \subseteq \mathbb{R}^m$ be the subgroup

$$G = \left\{ (\nu_1, \ldots, \nu_m) \colon \sum_{i=1}^{m} \nu_i (A(v_i) - A(u)) \in \mathbb{Z}^{n-1} \right\}.$$

Clearly $G \supseteq \mathbb{Z}^m$, and $(\alpha_1, \ldots, \alpha_m) \in G$. By Lemma 5 it is possible to choose $(\tilde{\alpha}_1, \ldots, \tilde{\alpha}_m) \in G$ such that $\tilde{\alpha}_i > 0$ for $1 \le i \le m$ and $\tilde{\alpha}_1 + \cdots + \tilde{\alpha}_m \le \chi_m \le \chi_{2(n-1)}$. Let $\tilde{\beta} = 1 - \tilde{\alpha}_1 - \cdots - \tilde{\alpha}_m \ge 1 - \chi_{2(n-2)} > 1/(k+1)$. Consider $x = \sum_{i=1}^{m} \tilde{\alpha}_i v_i + \tilde{\beta} u$ and $y = \sum_{i=1}^{m} \tilde{\alpha}_i v_i + \tilde{\beta}(u + (k+1)w) = x + \tilde{\beta}(k+1)w$. Suppose $x = \tau_1 w + \tau_2 b_2 + \cdots \tau_n b_n$, so that $A(x) = (\tau_2, \ldots, \tau_n)$. Since $(\tilde{\alpha}_1, \ldots, \tilde{\alpha}_m) \in G$, $\tau_2, \ldots, \tau_n$ are integers. Since $\tilde{\beta}(k+1) > 1$, there is an integer $\tilde{\tau}_1$ such that $\tau_1 < \tilde{\tau}_1 < \tau_1 + \tilde{\beta}(k+1)$. Then $z = \tilde{\tau}_1 w + \tau_2 b_2 + \cdots + \tau_n b_n \in \mathbb{Z}^n$ is in the relative interior of the line segment connecting $x$ and $y$, so $z \in K^\circ \cap \mathbb{Z}^n$. □

THEOREM 4. *There are, up to unimodular equivalence, only finitely many convex polytopes $P$ satisfying*:

(i) $\operatorname{vert}(P) \subseteq \mathbb{Z}^n$;
(ii) $P^\circ \cap \mathbb{Z}^n = \varnothing$; *and*
(iii) $P^\circ \cap G \neq \varnothing$, *for each group $G \subseteq \mathbb{R}^n$ which properly contains $\mathbb{Z}^n$.*

PROOF. After Lemma 6, we need only show that there are only finitely many equivalence classes of such $P$ for which $|P \cap \mathbb{Z}^n| < m$ (where $m = (1+k)^n + 1$ as in Lemma 6). Indeed, if $|P \cap \mathbb{Z}^n| \ge m$ and if $P$ satisfies (i) and (ii), then $A^{-1}(\mathbb{Z}^{n-1})$ is a subgroup $G$ of $\mathbb{R}^n$ for which (iii) fails, where $A$ is the linear function guaranteed by the lemma.

Suppose that $P$ and $Q$ are convex polytopes, each satisfying conditions (i), (ii), and (iii), and neither having $m$ or more elements in common with $\mathbb{Z}^n$. Let

$$U^n = \{(x_1, \ldots, x_{m-1}) \in \mathbb{R}^{m-2} \colon x_i > 0 \text{ for } 1 \le i \le m-2 \text{ and } x_1 + \cdots + x_{m-2} < 1\}.$$

Let $B \colon \mathbb{R}^{m-2} \to \mathbb{R}^n$ and $C \colon \mathbb{R}^{m-2} \to \mathbb{R}^n$ be affine functions mapping $\operatorname{cl}(U^n)$ onto $P$ and $Q$ respectively and mapping $\mathbb{Z}^{m-2}$ onto $\mathbb{Z}^n$. The subgroups $G = B^{-1}(\mathbb{Z}^n)$ and $H = C^{-1}(\mathbb{Z}^n)$ then miss $U$, and are maximal such subgroups. If $G = H$, then there is an affine unimodular function $D \colon \mathbb{R}^n \to \mathbb{R}^n$ such that $B = DC$. In particular, $DQ = P$.

By Theorem 1, the number of maximal subgroups $G \supseteq \mathbb{Z}^{m-2}$ such that $G \cap U = \varnothing$ is finite. We see from the preceding paragraph that this number is an upper bound on the cardinality of any collection of unimodularly inequivalent convex polytopes $P$ satisfying (i), (ii), (iii), and $|\operatorname{vert}(P)| < m$. □

## 5. Unanswered questions

In this final section we present some problems and questions that seem natural but with which we have not dealt.

A. Is there a reasonable method for computing the finitely many groups of Theorem 1—say, when the dimension $n$ is small and the set $U$ is the interior of a convex polytope?

B. Compute $\chi_n$; or at least find numbers that can serve as the $\chi$'s of Lemma 5. (We know $\chi_1 = 1/2$, $\chi_2 = 5/6$, $\chi_3 \geq 41/42, \ldots$. In general, if $t_n$ is the $n$th term of the sequence having $t_1 = 2$ and $t_{n+1} = t_n^2 - t_n + 1$ for $n = 2, 3, \ldots$, then $\chi_n \geq 1 - 1/(t_{n+1} - 1)$, as is clear upon considering the group $G$ generated by $\mathbb{Z}^n$ and the point $(1/t_1, 1/t_2, \ldots, 1/t_n)$.)

C. Find the convex polytopes $P$ of Theorem 4, when (say) $n = 3$. (For $n = 1$, there is, up to unimodular equivalence, only the interval $[0, 1]$; for $n = 2$, only $\mathrm{conv}\{\left[\begin{smallmatrix}0\\0\end{smallmatrix}\right], \left[\begin{smallmatrix}2\\0\end{smallmatrix}\right], \left[\begin{smallmatrix}0\\2\end{smallmatrix}\right]\}$.

**Acknowledgment**

I am indebted to Mark Hartmann for a careful reading of and several improvements over a preliminary version of this paper.

## References

1. N. Bourbaki, *General topology*, Part 2, Hermann, 1966.
2. J. W. S. Cassels, *An introduction to the geometry of numbers*, Springer-Verlag, 1971.
3. B. Grunbaum, *Convex polytopes*, Interscience, 1967.
4. G. H. Hardy and E. M. Wright, *The theory of numbers*, Oxford Univ. Press, 1968.
5. D. Hensley, *Lattice vertex polytopes with few interior lattice points*, Pacific J. Math. **105** (1983), 183–191.
6. J. C. Lagarias, H. W. Lenstra, Jr., and C. P. Schnorr, *Korkine-Zolotarev bases and successive minima of a lattice and its reciprocal lattice*, Combinatorica, **10** (1990), 343–358.
7. J. C. Lagarias and G. Ziegler, *Bounds for lattice polytopes containing a fixed number of interior points in a sublattice*, to appear in Canad. J. Math.
8. E. Steinitz, *Bedingt Konvergente Reihen und Konvexe System*, J. Reine Angew. Math. **143** (1913), 128–175; **144** (1914), 1–40; **146** (1916), 1–52.

Center for Applied Mathematics, National Institute of Standards and Technology

Department of Mathematics, George Mason University, Fairfax, Virginia 22030
*E-mail address*: lawrence@gmuvax.gmu.edu

DIMACS Series in Discrete Mathematics
and Theoretical Computer Science
Volume **4**, 1991

Applied Geometry
and Discrete Mathematics
THE VICTOR KLEE FESTSCHRIFT

# Regular Triangulations of Convex Polytopes

CARL W. LEE

ABSTRACT. Let $V$ be a finite set of points in $\mathbf{R}^d$. We are interested in subdivisions of the convex hull of $V$ into polytopes whose vertices lie in $V$. We begin with a general introduction to lexicographic subdivisions, and refinements and subdivisions by pulling and placing, and present some implications for edge-path diameters of simple convex polytopes and linear programming. In particular, we show that if $P$ is any simple $d$-polytope with $n$ facets, then there exists a simple $(d+1)$-polytope $Q$ with $n+1$ facets, one of which is congruent to $P$, such that the edge-path diameter of $Q$ is at most $2(n-d)$. After reviewing the notions of Gale transforms and diagrams we discuss several equivalent definitions for the properly larger class of regular or Gale subdivisions. There is a simple characterization of such subdivisions, and shellings can be obtained from Gale transforms. If $V$ is a set of at most $n \le d+3$ points in $\mathbf{R}^d$ with $d$-dimensional convex hull, we prove that all subdivisions are regular and all triangulations are placeable. Examples show that these results break down when $n > d+3$. We conclude with some comments on the secondary polytope of all regular subdivisions of a given set $V$.

## 1. Faces, subdivisions and triangulations

Let $V = \{v_1, \ldots, v_n\}$ be a finite set of points in $\mathbf{R}^d$ such that $[V]$ is a $d$-dimensional convex polytope, where $[\cdot]$ denotes convex hull. We do not insist that each point of $V$ be a vertex of $[V]$ and in fact will not even require that they all be distinct. We say a subset $F$ of $V$ is a *face* of $V$ if $F$ is the intersection of $V$ with some supporting hyperplane of $[V]$. In addition, the empty set and $V$ itself will also be called faces. All faces except $V$ are *proper*. Note, in particular, that what we call a face depends not only on the geometrical structure of $[V]$ but also on the defining set $V$. The *dimension* of a face $F$, $\dim(F)$, equals the dimension of $[F]$. A $j$-dimensional face

1980 *Mathematics Subject Classification* (1985 *Revision*). Primary 52A25; Secondary 90C05.
Research supported in part by NSF grants DMS-8504050 and DMS-8802933, by NSA grant MDA904-89-H-2038, by an Alexander von Humboldt Foundation Fellowship, and by DIMACS (Center for Discrete Mathematics and Theoretical Computer Science), a National Science Foundation Science and Technology Center, NSF-STC88-09648.


1052-1798/91 \$1.00 + \$.25 per page

will be called a *j-face* for brevity. Faces of dimension 0, 1, and $d-1$ are *vertices, edges*, and *facets* of $V$, respectively.

A *subdivision* of $[V]$ is a collection $\mathcal{S} = \{S_1, \dots, S_m\}$ of subsets of $V$ such that (1) $\dim([S_i]) = d$ for all $i$, (2) $\bigcup_{i=1}^m [S_i] = [V]$, and (3) for $1 \le i < j \le m$ we have $[S_i] \cap [S_j] = [F]$ for some common proper face $F$ of $S_i$ and $S_j$. If in additional all $S_i$ have cardinality $d+1$ then the subdivision is called a *triangulation*. A subset $F$ is a face of $\mathcal{S}$ if it is a face of at least one of the $S_i$.

For two subdivisions $\mathcal{S} = \{S_1, \dots, S_m\}$ and $\mathcal{T} = \{T_1, \dots, T_p\}$ we call $\mathcal{T}$ a *refinement* of $\mathcal{S}$ if for each $S_i$, there exist $T_{i_1}, \dots, T_{i_k} \in \mathcal{T}$ such that $\{T_{i_1}, \dots, T_{i_k}\}$ is a subdivision of $S_i$. In this case we write $\mathcal{T} \le \mathcal{S}$, with strict inequality if the two subdivisions are different.

In §2 we give a general introduction to lexicographic subdivisions, and refinements and subdivisions by pulling and placing, and discuss some implications for edge-path diameters of simple convex polytopes and linear programming (Theorem 1). §3 reviews the notions of Gale transforms and diagrams. Several equivalent definitions for the properly larger class of regular or Gale subdivisions are presented in §4. Theorem 2 provides a simple characterization and Theorem 3 addresses shellability. In §5 we discuss subdivisions of small sets of points, proving that in such cases all subdivisions are regular and all triangulations are placeable (Theorem 4). The last section, §6, contains some examples of nonregular triangulations and some notes on the secondary polytope. Some of the results in this paper were announced in **[9]**.

## 2. Pulling, placing and lexicographic triangulations

Let $\mathcal{S}$ be a subdivision of $V$ and let $v \in V$. One refinement $p_v^-(\mathcal{S})$ of $\mathcal{S}$ can be obtained as follows. Consider each $S_i \in \mathcal{S}$.

1. If $v \notin S_i$ then $S_i \in p_v^-(\mathcal{S})$.
2. If $v \in S_i$ then $p_v^-(\mathcal{S})$ contains all sets of the form $F \cup \{v\}$, where $F$ is a facet of $S_i$ which does not contain $v$.

We define another refinement $p_v^+(\mathcal{S})$ of $\mathcal{S}$ by again considering each $S_i \in \mathcal{S}$.

1. If $v \notin S_i$ then $S_i \in p_v^+(\mathcal{S})$.
2. If $v \in S_i$ and $\dim([S_i \backslash \{v\}]) = d-1$ then $S_i \in p_v^+(\mathcal{S})$.
3. If $v \in S_i$ and $\dim([S_i \backslash \{v\}]) = d$, let $S_i' = S_i \backslash \{v\}$. Then $p_v^+(\mathcal{S})$ contains $S_i'$ together with all sets of the form $F \cup \{v\}$, where $F$ is a facet of $S_i'$ which $v$ is *beyond* (with respect to $[S_i']$ **[4, §5.2]**.

Let the points of $V$ be given in some order $v_1, \dots, v_n$, let $(\varepsilon_1, \dots, \varepsilon_n) \in \{\pm\}^n$, and consider any subdivision $\mathcal{S}$ of $V$. Then it is easy to check that $p_{v_n}^{\varepsilon_n} \cdots p_{v_1}^{\varepsilon_1}(\mathcal{S})$ is a triangulation of $V$. Triangulations that result from choosing $\mathcal{S}$ to be the trivial subdivision $\{V\}$ are *lexicographic* **[19]**.

If $\mathcal{S}$ is $\{V\}$ and $\varepsilon_i = -$ for all $i$, the resulting triangulation is said to be obtained by *pulling* the points of $V$ in the order $v_1, \dots, v_n$. On the other hand, if $\mathcal{S}$ is $\{V\}$ and $\varepsilon_i = +$ for all $i$, the resulting triangulation is said to be obtained by *placing* the points of $V$ in the order $v_n, \dots, v_1$ (although perhaps the term *pushing* would be more appropriate). In Figure 1, (a) is achievable by pulling but not by placing, (b) is achievable by placing but not by pulling, and (c) is lexicographic, but not achievable by either pulling or placing.

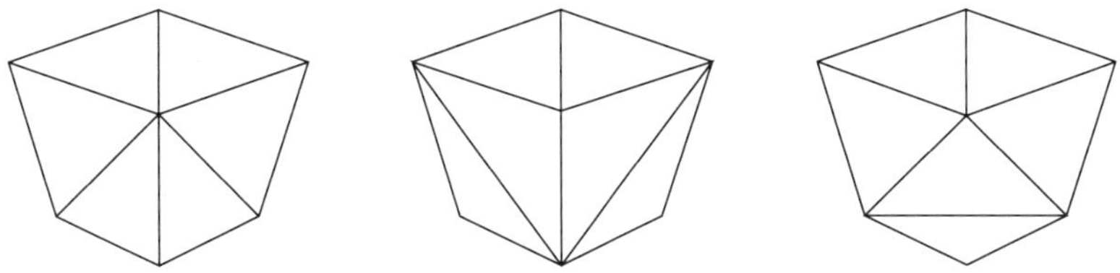

(a) (b) (c)

FIGURE 1. Some lexicographic triangulations

Pulling is described in Hudson [**7, Lemma 1.4**] and has been considered by others in several equivalent ways. Billera and Munson [**2**] defined triangulations by placing in the more general context of oriented matroids. Its designation is suggested by an equivalent formulation. One "places" the points of $V$ into position in the order $v_n, \dots, v_1$, successively updating the resulting convex hulls and their triangulations. This inductive method for constructing the convex hull of $V$ appears in Grünbaum [**4, §5.2**] and recent work of Seidel [**16**] shows that it can be implemented as a particularly simple and effective algorithm.

Now suppose that $\mathcal{T}$ is the placing triangulation $p^+_{v_n} \cdots p^+_{v_1}(\{V\})$. For $k = 1, \dots, d+1$, let $v_{i_k}$ be chosen recursively such that

$$i_k = \max\{j\colon \dim([\{v_{i_1}, \dots, v_{i_{k-1}}\} \cup \{v_j\}]) = k-1\}.$$

It is not difficult to show that if one reorders $v_1, \dots, v_n$ by bringing the points $v_{i_1}, \dots, v_{i_{d+1}}$ to the end of the list (without reordering any of the other points), then the resulting placing triangulation remains unchanged. So without loss of generality we may assume that $\dim([\{v_{n-d}, \dots, v_n\}]) = d$. In this case it is immediate that $\mathcal{T}$ is *weakly vertex decomposable* (see [**15**]) with decomposition order $v_1, \dots, v_n$. As a consequence, the diameter of $\mathcal{T}$ is no more than $2(n-d-1)$. That is, for any two $d$-simplices $S'$, $S''$ of $\mathcal{T}$ there is a chain of $d$-simplices $S' = S_0, S_1, \dots, S_{k-1}, S_k = S''$ such that $k \le 2(n-d-1)$ and $S_{i-1} \cap S_i$ has dimension $d-1$ for all $1 \le i \le k$.

This has a nice dual interpretation. Let $P \subset \mathbf{R}^d$ be a simple $d$-polytope containing the origin that has $n$ facets $F_1, \dots, F_n$. Let $V = \{v_1, \dots, v_n\}$

be the corresponding vertices of the simplicial polytope polar to $P$. Regard $P$ as sitting in $\mathbf{R}^{d+1}$ by appending a zero $(d+1)$st coordinate. Construct a pyramid $Q'$ over $P$ with apex $z = (\mathbf{O}, 1)$. There will be $n$ facets $F_1', \ldots, F_n'$ of $Q'$ containing $z$ such that $F_i' \cap P = F_i$. By "rotating" $F_1', \ldots, F_n'$ away from $z$ about the $F_i$ by successively smaller amounts, one obtains a simple $(d+1)$-polytope $Q$, one of whose facets is $P$. The vertices of $Q$ that do not lie in $P$ correspond to the $d$-simplices of $\mathcal{T} = p_{v_n}^+ \cdots p_{v_1}^+(\{V\})$. Similarly, the edges of $Q$ that do not lie in $P$ correspond to the $(d-1)$-simplices of $\mathcal{T}$. Each vertex in $P$ is joined by an edge to a unique vertex of $Q$ not in $P$. Hence we can conclude that the edge-path diameter of $Q$ is no more than $2(n-d-1)+2 = 2(n-d)$. Therefore we have proved the following theorem.

THEOREM 1. *Let $P$ be any simple $d$-polytope with $n$ facets. Then there exists a simple $(d+1)$-polytope $Q$ with $n+1$ facets, one of which is congruent to $P$, such that the edge-path diameter of $Q$ is at most $2(n-d)$.*

The significance of this result is due to the unresolved *Hirsch conjecture* that the edge-path diameter of any simple $d$-polytope with $n$ facets is not greater than $n-d$.

Suppose we start with the linear program

$$(1)\qquad \begin{gathered} \max c \cdot x, \\ a_i \cdot x \le b_i, \quad 1 \le i \le n \end{gathered}$$

where $x \in \mathbf{R}^d$. Assume that $b_i > 0$ for all $i$. From Theorem 1 and the accompanying discussion it is not hard to verify that if the feasible region is bounded (even if it is not simple) one can consider the augmented linear program

$$(2)\qquad \begin{gathered} \max c \cdot x + 0x_{d+1} \\ a_i \cdot x + (b_i - \varepsilon_i)x_{d+1} \le b_i, \ 1 \le i \le n, \\ x_{d+1} \ge 0. \end{gathered}$$

Here the $\varepsilon_i$ are indeterminates with lexicographic order $0 < \varepsilon_n \ll \cdots \ll \varepsilon_1 \ll 1$. The feasible region of (2) is obtained by making a pyramid over the feasible region of (1) with apex $(\mathbf{O}, 1)$, and then rotating the hyperplanes away from the apex in the order $i = 1, \ldots, n$. Because $\mathbf{O}$ is in the interior of the feasible region of (1) and the perturbations are small, one can conclude that $(x, 0)$ is optimal for (2) if and only if $x$ is optimal for (1). From any basic feasible solution to (2) there exists a sequence of at most $2(n-d)$ pivots to an optimal solution. So in principle the linear program can be solved in a linear number of pivots, if one only knew which pivots to make. Perhaps there is a way that the objective function can be used to suggest a favorable ordering of the constraints.

## 3. Gale transforms and diagrams

In this section we summarize some useful facts about Gale transforms and diagrams. More detailed expositions can be found in [**4**, §§**5.4**, **6.3**] and [**14**]. Let us return to a finite set of (not necessarily distinct) points $V$ in $\mathbf{R}^d$ such that $\dim([V]) = d$. List the points $v_i$ as columns of a matrix and append a row of ones, obtaining the $(d+1) \times n$ matrix

$$A = \begin{bmatrix} v_1 & \cdots & v_n \\ 1 & \cdots & 1 \end{bmatrix}.$$

This matrix has full row rank because the points of $V$ do not lie in a common hyperplane. The nullspace $\mathcal{N}(A)$ of $A$ is the space of all affine relations on $V$. Choose a basis for $\mathcal{N}(A)$ and list these vectors as the rows of an $(n-d-1) \times n$ matrix $\overline{A}$. The columns $\overline{v}_1, \ldots, \overline{v}_n$ of $\overline{A}$ are in a natural one-to-one correspondence with the original points $v_1, \ldots, v_n$ of $V$, which extends to a bijection between subsets $F$ of $V$ and subsets $\overline{F}$ of $\overline{V}$. The set $\overline{V} = \{\overline{v}_1, \ldots, \overline{v}_n\}$ is called a *Gale transform* of $V$. Note, in particular, that $\overline{v}_1 + \cdots + \overline{v}_n = \mathbf{O}$. The key property of Gale transforms is the following well-known characterization of faces.

PROPOSITION 1. *A subset* $F \subsetneq V$ *is a face of* $V$ *if and only if* $\mathbf{O} \in \operatorname{relint}([\overline{V} \backslash \overline{F}])$.

A set of $n$ points $\overline{V}'$ is a *Gale diagram* of $V$ if it is *isomorphic* to $\overline{V}$; i.e., $\mathbf{O} \in \operatorname{relint}([\overline{V} \backslash \overline{F}])$ if and only if $\mathbf{O} \in \operatorname{relint}([\overline{V}' \backslash \overline{F}])$ for all $F \subsetneq V$. For example, one can scale the points in a Gale transform independently by positive amounts—such a Gale diagram will be called a *scaled Gale transform*. If a Gale transform (respectively, diagram) is scaled so that the nonzero points lie on the unit $(n-d-2)$-sphere centered at $\mathbf{O}$, then we say we have a *normalized* Gale transform (respectively, diagram).

## 4. Regular subdivisions and triangulations

Again let $V$ be a finite set of (not necessarily distinct) points in $\mathbf{R}^d$ such that $\dim([V]) = d$. There are several equivalent ways to define the notion of a *regular* or *Gale* subdivision of $V$.

DEFINITION 1. Let $\overline{V}$ be a Gale transform of $V$ and $\overline{z} \in \mathbf{R}^{n-d-1}$. Put $\overline{V}' = \overline{V} \cup \{\overline{z}\}$. If $\mathbf{O} \in \operatorname{relint}([\overline{V}' \backslash \overline{S}])$ for $S \subseteq V$ and $S$ is maximal with respect to this property then $S$ is a $d$-face of the subdivision $\mathcal{S}$.

DEFINITION 2. Choose numbers $\lambda_1, \ldots, \lambda_n$ and let $W = \{(v_1, \lambda_1), \ldots, (v_n, \lambda_n)\} \subset \mathbf{R}^{d+1}$. If $S' = \{(v_{i_1}, \lambda_{i_1}), \ldots, (v_{i_k}, \lambda_{i_k})\}$ is a facet of $W$ in the *upper hull* of $W$ (i.e., the last component of the outward normal of its supporting hyperplane is positive) then $S = \{v_{i_1}, \ldots, v_{i_k}\}$ is a $d$-face of the subdivision $\mathcal{S}$. (If $\dim([W]) = d$ then the subdivision is taken to be the trivial one.)

DEFINITION 3. Choose numbers $\mu_1, \ldots, \mu_n$ and let $Q = \{(x, x_{d+1}) \in \mathbf{R}^{d+1}: x \cdot v_i + x_{d+1} \leq \mu_i,\ i = 1, \ldots, n\}$. Let $(w, w_{d+1})$ be a vertex of $Q$ and let $S = \{v_i: w \cdot v_i + w_{d+1} = \mu_i\}$. Then $S$ is a $d$-face of the subdivision $\mathscr{S}$.

Definition 2 appears, for example, in Gel′fand, Kapranov, and Zelevinskiĭ **[3]** who call such subdivisions regular because of their connection to solutions of systems of differential equations. McMullen **[14]** discusses Definition 1, and Definition 3 is due to Cottle and Hoffman **[6]**.

To show the equivalences, assume first that $\mathscr{S}$ is given via Definition 1. Scale $\overline{z}$ by a positive amount, if necessary, so that we can assume it lies in the interior of $[\overline{V}]$. Find numbers $\lambda_1, \ldots, \lambda_n$ such that $\overline{z} = \lambda_1\overline{v}_1 + \cdots + \lambda_n\overline{v}_n$, $\lambda_1 + \cdots + \lambda_n = 1$, and $0 < \lambda_i < 1$ for all $i$. One can show that $\overline{V}'$ is a Gale diagram of $V' = \{(1-\lambda_1)^{-1}(v_1, \lambda_1), \ldots, (1-\lambda_n)^{-1}(v_n, \lambda_n), (\mathbf{O}, -1)\}$. So the facets of $V'$ not containing $z = (\mathbf{O}, -1)$ correspond to the proposed elements of the subdivision of $V$. Noting that $(1-\lambda_i)^{-1}\lambda_i > 0$, apply the projective transformation $f(x, x_{d+1}) = (x, x_{d+1})/(x_{d+1} + 1)$ to $[V']$. This sends the point $z$ onto the hyperplane at infinity. The result is the unbounded polyhedron $[W] + [\mathbf{O}, -\infty)$, where $W = \{(v_1, \lambda_1), \ldots, (v_n, \lambda_n)\}$. The facets of $V'$ not containing $z$ correspond to the bounded facets of $[W]$, which in turn correspond to the facets in the upper hull of $W$. So $\mathscr{S}$ falls under Definition 2.

Conversely let $\mathscr{S}$ be given by Definition 2 and let $\overline{V}$ be a Gale transform of $V$. By translation and positive scaling of the last coordinate one can obtain $\lambda_1', \ldots, \lambda_n'$ that yield the same subdivision, such that $0 < \lambda_i' < 1$ and $\lambda_1' + \cdots + \lambda_n' = 1$. Set $\overline{z} = \lambda_1'\overline{v}_1 + \cdots + \lambda_n'\overline{v}_n$. Then with this $\overline{z}$ we get $\mathscr{S}$ using Definition 1.

The equivalence of Definitions 2 and 3 via $\lambda_i = -\mu_i$ is a straightforward consequence of duality.

EXAMPLE 1. The twisted triangle example of Figure 2(a) shows that there are regular triangulations that are not lexicographic.

EXAMPLE 2. Let $\lambda_k \neq 0$ and $\lambda_i = 0$ for $i \neq k$. Then the corresponding subdivision of $V$ is $p_{v_k}^{-sg(\lambda_k)}(\{V\})$. More generally, if $0 \ll \lambda_n' \ll \cdots \ll \lambda_1'$, $\varepsilon_i \in \{\pm\}$, and $\lambda_i = -\varepsilon_i\lambda_i'$ for $i = 1, \ldots, n$, then we get the lexicographic triangulation $p_{v_n}^{\varepsilon_n} \cdots p_{v_1}^{\varepsilon_1}(\{V\})$. In particular, there are points $\overline{z}$ and $-\overline{z}$ such that $\overline{z}$ corresponds to the triangulation by pulling the points in the order $v_1, \ldots, v_n$ and $-\overline{z}$ corresponds to placing the points in the order $v_n, \ldots, v_1$. So these two triangulations are in a sense opposite or *complementary.*

EXAMPLE 3. Let $\lambda_i = -\|v_i\|^2$ for $i = 1, \ldots, n$. Then we get the (nearest point) Delaunay subdivision of $V$. Of course this is invariant under rigid motion, and one can verify that $\overline{z} = \lambda_1\overline{v}_1 + \cdots + \lambda_n\overline{v}_n$ also remains unchanged when $V$ is moved rigidly. Similarly the furthest point Delaunay subdivision

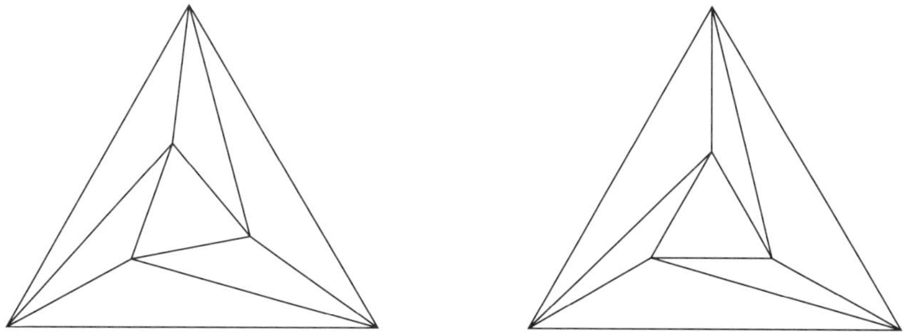

(a) (b)

FIGURE 2. A regular and a nonregular triangulation

is constructed by choosing $\lambda_i = \|v_i\|^2$, $i = 1, \ldots, n$. So these two subdivisions are also complementary. For more on complementary subdivisions see **[12]**.

Analogous to Shephard's theorem **[17]** we have the following test for regularity.

THEOREM 2. *Suppose $\mathcal{S} = \{S_1, \ldots, S_m\}$ is a subdivision of $V$. Let $\overline{V}$ be a Gale transform of $V$. Then $\mathcal{S}$ is a regular subdivision if and only if $\bigcap_{i=1}^m \operatorname{relint}(\operatorname{cone}(\overline{V}\backslash\overline{S}_i)) \neq \varnothing$.*

PROOF. The point $\overline{z}$ must be placed in $-\bigcap_{i=1}^m \operatorname{relint}(\operatorname{cone}(\overline{V}\backslash\overline{S}_i))$. □

The equivalence of Definitions 1 and 2 shows immediately that any regular subdivision is shellable. Let us make the connection more explicit when we have a regular triangulation. Suppose $\overline{V}$ is a Gale transform of $V$, where $\operatorname{card}(V) > d+1$. By applying an appropriate projective transformation to $V$, if necessary, we can assume that no hyperplane missing $\mathbf{O}$ contains an affinely dependent set of points of $\overline{V}$. Hence every set of $n-d-1$ linearly independent points of $\overline{V}$ determines a unique hyperplane. By translating $V$ we may also assume that $v_1 + \cdots + v_n = \mathbf{O}$. Let $\mathcal{T}$ be a regular triangulation induced by $\overline{z} = \lambda_1\overline{v}_1 + \cdots + \lambda_n\overline{v}_n$, where $\lambda_1 + \cdots + \lambda_n = 1$ and $0 < \lambda_i < 1$ for all $i$. It is easy to see that a set $S_i$ of cardinality $d+1$ is a $d$-simplex of $\mathcal{T}$ if and only if $[\overline{T}_i \cup \{\overline{z}\}]$ is an $(n-d-1)$-simplex containing $\mathbf{O}$ in its interior, where $\overline{T}_i = \overline{V}\backslash\overline{S}_i$. This happens if and only if the ray $-t\overline{z}$, $t > 0$, intersects the relative interior of $[\overline{T}_i]$ for some positive value $t_i$ of $t$. Because $\mathcal{T}$ is a triangulation, perturbing $\overline{z}$ by a small amount will not alter $\mathcal{T}$, so we may assume that the numbers $t_i$ are all distinct. Order such sets $\overline{T}_1, \ldots, \overline{T}_m$ in order of decreasing $t_i$. Consider the corresponding order $S_1, \ldots, S_m$ of the $d$-simplices of $\mathcal{T}$.

THEOREM 3. *The above ordering of the $d$-simplices of $\mathcal{T}$ constitutes a shelling order and in fact corresponds to the line shelling of the upper hull of $W = \{(v_1, \lambda_1), \ldots, (v_n, \lambda_n)\}$ induced by the ray $(\mathbf{O}, u)$, $u > 0$.*

PROOF. Straightforward. □

In [11] we discuss a connection between shellings, winding numbers in Gale transforms, and $f$-vectors of simplicial convex polytopes, and show how this can be used to understand some specific cases of the generalized lower-bound conjecture.

## 5. Subdivisions of sets with few points

Again let $V$ be a finite set of (not necessarily distinct) points in $\mathbf{R}^d$ such that $\dim([V]) = d$. If $\operatorname{card}(V) \leq d+3$ we say that $V$ has *few points.* The goal of this section is the following result.

THEOREM 4. *If $V$ has few points then every subdivision is regular and in fact every triangulation is placeable.*

PROOF. If $\operatorname{card}(V) = d+1$ then the only subdivision of $V$ is the trivial one. Suppose $\operatorname{card}(V) = d+2$. Consider any nontrivial subdivision $\mathscr{S}$ of $V$. Then $\mathscr{S}$ must be a triangulation. Choose any simplex in the triangulation, which is the convex hull of $d+1$ points, say, $v_1, \dots, v_{d+1}$. Any facet of this simplex that is not on the boundary of $V$ must be the base of another simplex whose apex is $v_{d+2}$. This implies that the triangulation is completely determined by the initially given simplex, and that it is $p^+_{v_{d+2}}(\{V\})$. Assume that the origin $\mathbf{O}$ is in the interior of $[V]$ in affinely general position with respect to the points in $V$. Then $V$ is the Gale diagram of some set of $d+2$ points in $\mathbf{R}^1$, not all identical. Since the convex hull of these points is one-dimensional it has two facets, implying that there are exactly two $d$-simplices containing $\mathbf{O}$ formed from the points in $V$. Hence $V$ admits exactly two triangulations.

Consider a Gale transform $\overline{V}$ of $V$. It will be one-dimensional, with $\overline{V}$ partitioned as $\overline{V} = \overline{W} \cup \overline{X} \cup \overline{Y}$, where $\overline{W}, \overline{X}, \overline{Y}$ are the points in $\overline{V}$ to the left of, at, and to the right of $\mathbf{O}$, respectively. It is well known and easy to see that if $S$ is a subset of $V$ that lies in no face of $V$ and is minimal with respect to this property, then either $S = W \cup X$ or $S = X \cup Y$. The two triangulations of $V$ consist of either (1) the simplices of the form $(W \cup X \cup Y)\backslash\{u\}$, where $u \in W$, or (2) the simplices of the form $(W \cup X \cup Y)\backslash\{u\}$, where $u \in Y$.

Now suppose $\operatorname{card}(V) = d+3$ and let $\mathscr{S}$ be any nontrivial subdivision of $V$. Assume first that $\mathscr{S}$ is not a triangulation. Then there must be some $S_k \in \mathscr{S}$ which consists of $d+2$ points, say, $v_1, \dots v_{d+2}$. Any facet of $S_k$ that is not on the boundary of $V$ must be the base of a pyramidal $d$-face whose apex is $v_{d+3}$. Hence $\mathscr{S} = p^+_{v_{d+3}}(\{V\})$.

Assume next that $\mathscr{S}$ is a triangulation. If we join the original boundary faces of $V$ to a new point $z$, then together with the faces of $\mathscr{S}$ the resulting complex is topologically a $d$-sphere with $d+4$ vertices. Hence by

Kleinschmidt [8] and Mani [13] this complex is realizable as the boundary complex of some $(d+1)$-polytope $P$. Consider a Gale transform $\overline{V}'$ of $P$ (which will be two-dimensional). Normalize $\overline{V}'$ by scaling the nonzero points of $\overline{V}'$ by positive amounts so that they all lie on the unit circle centered at $\mathbf{O}$. Note that $\overline{z} \neq \mathbf{O}$ since $\mathcal{S}$ is not trivial.

By appropriately "rotating diagonals" (see [4, §6.3] for the precise statement of the allowable operations), one can construct normalized Gale diagrams isomorphic to $\overline{V}'$. What we need to show is that there is a normalized Gale diagram $\overline{U}'$ such that (1) it is isomorphic to $\overline{V}'$, and (2) $\overline{U} = \overline{U}'\backslash\{\overline{z}\}$ is *strongly isomorphic* to a normalized Gale transform $\overline{V}$ of $V$. By strongly isomorphic we mean that the corresponding nonzero points of $\overline{U}$ and $\overline{V}$ occur in the same order around the perimeter of the circle, with $\overline{u}, \overline{v}$, coinciding in $\overline{U}$ if and only if $\overline{u}$, $\overline{v}$ coincide in $\overline{V}$. This will prove that $\mathcal{S}$ is regular.

Suppose there is no way to construct such a $\overline{U}'$ from $\overline{V}'$ by rotating diagonals. Then one of two types of obstacles must be encountered. The first possibility is that there are two nonzero points $\overline{u}$, $\overline{v}$ which coincide in $\overline{V}$ but cannot be merged in $\overline{V}'$ because of the presence of $-\overline{z}$ between them (see Figure 3). Then $V\backslash\{u, v\}$ is a $d$-simplex in $\mathcal{S}$ since $\mathbf{O}$ is in the interior of the triangle $\{\overline{u}, \overline{v}, \overline{z}\}$. On the other hand, if we begin with $\overline{V}$ (in which $\overline{u}$ and $\overline{v}$ coincide) and project orthogonally onto the linear span of $\{\overline{u}, \overline{v}\}$, then we obtain a one-dimensional Gale diagram for $V\backslash\{u, v\}$. Hence the dimension of $V\backslash\{u, v\}$ is $(d+1)-1-1 = d-1 < d$, which is a contradiction.

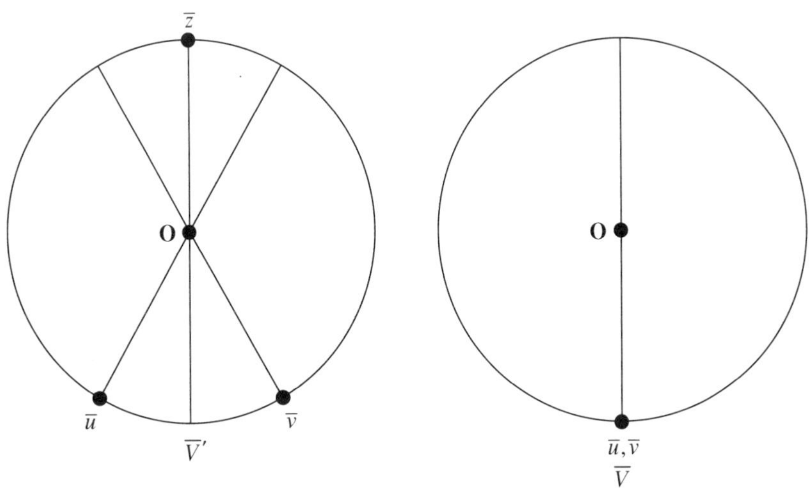


FIGURE 3. Obstacle 1

Assume the first obstacle does not occur. The second possibility is that everything can be ordered properly by rotating diagonals, except for a group of points which ought to appear in the order $\overline{v}_1, \ldots, \overline{v}_m$, but cannot be so ordered in $\overline{V}'$ because $-\overline{z}$ falls among them. This is depicted in Figure 4.

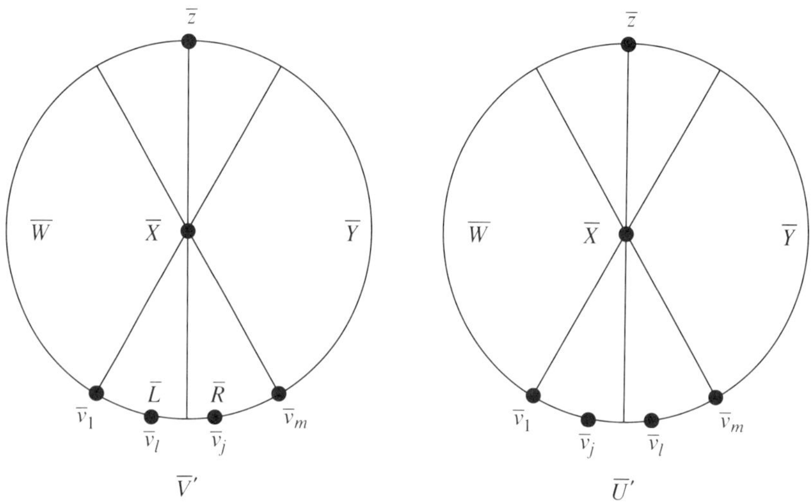


FIGURE 4. Obstacle 2

The only point in the upper closed arc of the circle in $\overline{z}$. There are nonempty sets $\overline{W}$, $\overline{Y}$ in the left and right open arcs, respectively. The (possibly empty) set of points at **O** is denoted $\overline{X}$. The points $\overline{v}_1, \ldots, \overline{v}_m$ fall in the lower closed arc. Of these, $\overline{L}$ and $\overline{R}$ are those points lying to the left and right of $-\overline{z}$, respectively. Assume that $\overline{v}_j$ is the point of smallest index in $\overline{R}$ and $\overline{v}_l$ is the point of largest index lying in $\overline{L}$. Let $\overline{v}_{j+1}, \ldots, \overline{v}_k$ be the points, if any, which coincide with $\overline{v}_j$ in $\overline{V}'$ (and $\overline{V}$ since we are assuming that the first obstacle does not occur). Our assumption implies that $j \leq k < l$.

Consider the set $S = X \cup Y \cup (R \backslash \{v_j\})$. First, it is a face of $\mathscr{S}$ since $[\overline{V}' \backslash \overline{S}]$ contains **O** in its interior. Second, it is not in the boundary of $\mathscr{S}$. If it were, then $S \cup \{z\}$ would be a face of $P$ and so $[\overline{V}' \backslash (\overline{S} \cup \{\overline{z}\})]$ should contain **O** in its interior, which is not the case. Third, $S \cup \{v_j\}$ is not a face of $\mathscr{S}$ because **O** is not in the interior of $[\overline{V}' \backslash (\overline{S} \cup \{\overline{v}_j\})]$. Therefore, $S$ cannot lie in any face of $V \backslash \{v_j\}$. But a Gale diagram for $V \backslash \{v_j\}$ is obtained by projecting $\overline{V}$ (in which the points $\overline{v}_1, \ldots, \overline{v}_m$ appear in the correct order) onto the orthogonal complement of the linear span of $\overline{v}_j$. The result is a one-dimensional Gale diagram. The sets projecting to the left of, on, and to the right of **O** are, respectively, $\overline{S}_1 = \overline{W} \cup \{\overline{v}_1, \ldots, \overline{v}_{j-1}\}$, $\overline{X} \cup \{\overline{v}_{j+1}, \ldots, \overline{v}_k\}$, and $\overline{S}_2 = \overline{Y} \cup \{\overline{v}_{k+1}, \ldots, \overline{v}_m\}$. Our contradiction now comes from the fact that $S$ contains neither $S_1 \cup X$ nor $S_2 \cup X$ (the minimal nonboundary sets of $V \backslash \{v_j\}$), since it does not contain $W$ or $v_l$. Therefore, $\mathcal{S}$ is regular and the appropriate $\overline{U}'$ can be constructed.

To show that $\mathcal{S}$ is placeable consider the point $\overline{v}$ nearest to $-\overline{z}$ in $\overline{U}'$ counterclockwise around the perimeter of the circle (see Figure 5). It will not coincide with $-\overline{z}$ because $\mathcal{S}$ is a triangulation. Let $\overline{W}$, $\overline{X}$, $\overline{Y}$ be the points to the left of, on, and to the right of the diagonal joining $\overline{z}$ and $-\overline{z}$, respectively. The triangles containing $\mathbf{O}$ which have $\overline{z}$ and $\overline{v}$ as two vertices are those of the form $[\overline{z}, \overline{v}, \overline{u}]$, where $\overline{u} \in \overline{W}$. So the $d$-simplices of $\mathcal{S}$ not containing $v$ are of the form $(W \cup X \cup Y) \backslash (\{u, v\})$, where $u \in W$. But by projecting $\overline{U}' \backslash \{\overline{z}\}$ onto the orthogonal complement of the linear span of $\overline{v}$ we see that these simplices are precisely those of one of the two triangulations of $V \backslash \{v\}$, which is vertex placeable. So $\mathcal{S}$ is vertex placeable. □

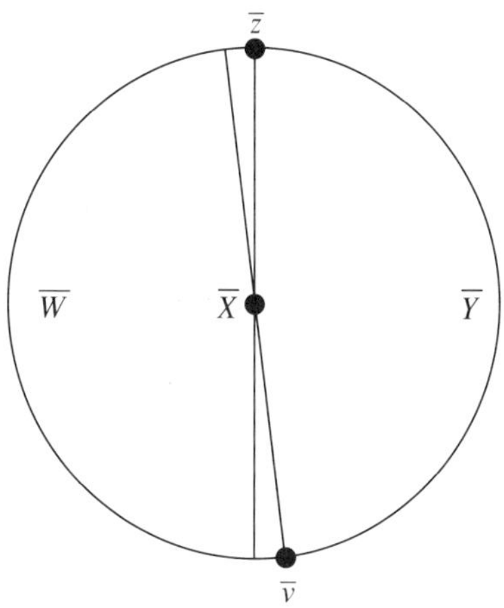


FIGURE 5. $\mathcal{S}$ is vertex placeable.

## 6. Subdivisions of larger sets

Consider again any subdivision $\mathcal{S}$ of a $d$-dimensional set of $n$ points $V$. If $d = 2$ and all the elements of $V$ are vertices of $V$ then it is easy to see that $\mathcal{S}$ is vertex placeable. However, if some points of $V$ are not vertices, the well-known example of Figure 2(b) (in which the three lines joining corresponding vertices of the outer and inner triangles meet at a common point when extended) shows that there can be nonregular subdivisions even when $n = 6$. We say that $\mathcal{S}$ is *weakly regular* if there exists a $d$-dimensional set $V'$ having a regular subdivision $\mathcal{S}'$ that is combinatorially isomorphic to $\mathcal{S}$. Equivalently, if one joins the original boundary faces of $V$ to a new point $z$, then together with the faces of $\mathcal{S}$ the result must be combinatorially equivalent to the boundary complex of some $(d+1)$-polytope. If $d = 2$ then $\mathcal{S}$ is always weakly regular by Steinitz's theorem.

In three dimensions a subdivision can be nonregular even if all of the points of $V$ are vertices of $V$. Consider the capped prism of Figure 6(a). The two triangulations {2368, 2568, 2347, 2378, 2456, 2457, 2578} and {2356, 2358, 2348, 2478, 2467, 2567, 2578} (using an abbreviated notation) are both nonregular, though they are weakly regular. On the other hand, if the upper capped triangle of the prism is twisted slightly the resulting polytope shown in Figure 6(b) has a triangulation {2356, 2358, 2348, 2478, 2467, 2567, 2578, 3568, 3478, 4567} that is not even weakly regular. For, joining the boundary faces to a new point 1 results in a triangulated 3-sphere that is dual to the Brückner sphere, which is not polytopal (see Grünbaum [**4, pp. 222–224**] where the points $A, \dots, H$ correspond to our points $1, \dots, 8$, respectively). This example was constructed with Peter Kleinschmidt.

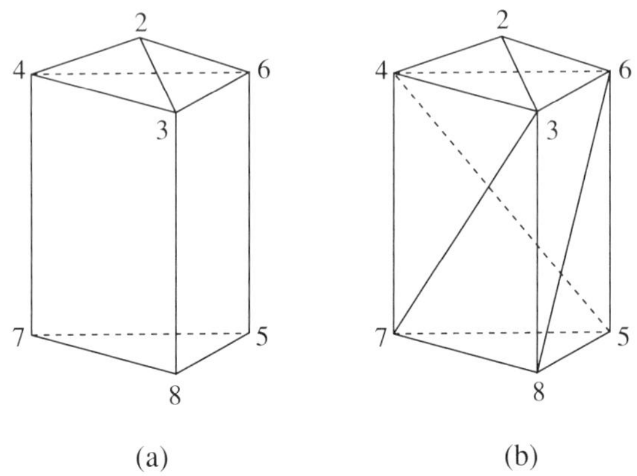


FIGURE 6. Two polytopes with nonregular triangulations

In general, one can take a Gale transform $\overline{V}$ of $V$ and consider the unit $(n-d-2)$-sphere centered at the origin. For every choice of a point $\overline{z}$ on this sphere there is an associated regular subdivision of $V$. The sphere can be partitioned into regions corresponding to identical subdivisions and in this manner one can construct a spherical complex of all the nontrivial regular subdivisions of $V$. The partial ordering of the faces of this complex by reverse inclusion coincides with the corresponding partial ordering of the regular subdivisions by refinement as defined in the first section. Gel′fand, Kapranov, and Zelevinskiĭ [**3**] proved that there is a convex $(n-d-1)$-polytope, known as the *secondary polytope*, that is dual to this spherical complex, and hence the complex itself is also polytopal. Billera, Filliman, and Sturmfels [**1**] present several proofs of this and they discuss the secondary polytope in greater detail.

When $V$ is the set of vertices of a two-dimensional convex $n$-gon the dual of the secondary polytope is known as the *associahedron* [**5, 10**]. Gel′fand,

Kapranov, and Zelevinskiĭ have pointed out that the secondary polytope of the $n$-gon already appears in work of Stasheff **[18]**.

We conclude with a sketch of the construction of the secondary polytope for a general set of points $V$. Assume that **O** is the centroid of $V$ and let $\overline{V}$ be a Gale transform of $V$. Then we may just as well regard $V$ as a Gale transform of $\overline{V}$. Let $z$ be any point in the interior of $[V]$, in affinely general position with respect to $V$. It is known that the translation $V - z$ is the Gale transform of some set of points $\overline{V}'$ which is a scaling of $\overline{V}$. Let $P(z)$ be the simplicial $(n-d-1)$-polytope $[\overline{V}']$. The facets of $P(z)$ are in one-to-one correspondence with the $d$-simplices in $V$ which contain $z$. We can put an equivalence relation on the regular triangulations of $V$ by saying $\mathcal{T}$ is equivalent to $\mathcal{T}'$ if the unique $d$-simplex containing $z$ is the same in both $\mathcal{T}$ and $\mathcal{T}'$. Then the facets of $P(z)$ correspond to the equivalence classes.

For any two different nontrivial regular triangulations $\mathcal{T}$ and $\mathcal{T}'$ there is always some point $z$ such that $\mathcal{T}$ and $\mathcal{T}'$ fall into two different equivalence classes relative to $z$. From here it is not too hard to see that the cones determined by the facets of the dual of the secondary polytope are precisely full-dimensional intersections of facet cones of the various possible $P(z)$ for different choices of $z$.

The set of all $d$-simplices in $V$ naturally induces a finite collection of open subsets $U_i$ of $[V]$, where $x$ and $y$ are in a common subset $U_i$ if and only if, for all $d$-simplices $S$, $x \in \text{int}([S]) \Leftrightarrow y \in \text{int}([S])$. Equivalently, $x$ and $y$ are in the same subset $U_i$ if and only if $P(x)$ and $P(y)$ are combinatorially equivalent simplicial polytopes. Finally (see **[1]**), one can show that the secondary polytope is combinatorially equivalent to the Minkowski sum of the polars of the $P(z_i)$, with one $z_i$ chosen from each subset $U_i$.

## References

1. L. J. Billera, P. Filliman, and B. Sturmfels, *Constructions and complexity of secondary polytopes*, Adv. in Math. (to appear).
2. L. J. Billera and B. S. Munson, *Triangulations of oriented matroids and convex polytopes*, SIAM J. Algebraic Discrete Methods **5** (1984), 515–525.
3. I. M. Gel′fand, A. V. Zelevinskiĭ, and M. M. Kapranov, *General theory of A-discriminants*, manuscript, 1990.
4. B. Grünbaum, *Convex polytopes*, Pure Appl. Math., vol. 16, Wiley-Interscience, New York, 1967.
5. M. Haiman, *Constructing the associahedron*, manuscript, 1984.
6. A. J. Hoffman, personal communication, 1984.
7. J. F. P. Hudson, *Piecewise linear topology*, Benjamin, New York and Amsterdam, 1969.
8. P. Kleinschmidt, *Sphären mit wenigen Ecken*, Geom. Dedicata **5** (1976), 307–320.
9. C. W. Lee, *Some notes on triangulating polytopes*, in Proceedings, 3. Kolloquium über Diskrete Geometrie, Institut für Mathematik, Universität Salzburg, May 1985, pp. 173–181.
10. ——, *The associahedron and triangulations of the n-gon*, European J. Combin. **10** (1989), 551–560.

11. ——, *Winding numbers and the generalized lower-bound conjecture*, manuscript, 1990.
12. C. W. Lee and B. Sturmfels, *Complementary triangulations* (in preparation).
13. P. Mani, *Spheres with few vertices*, J. Combin. Theory Ser. A **13** (1972), 346–352.
14. P. McMullen, *Transforms, diagrams and representations*, in Contributions to Geometry (Proc. Geom. Sympos., Siegen, 1978), Birkhäuser, Basel, 1979, pp. 92–130.
15. J. S. Provan and L. J. Billera, *Decompositions of simplicial complexes related to diameters of convex polyhedra*, Math. Oper. Res. **5** (1980), 576–594.
16. R. Seidel, *Linear programming and convex hulls made easy*, manuscript 1989.
17. G. C. Shephard, *Spherical complexes and radial projections of polytopes*, Israel J. Math. **9** (1971), 257–262.
18. J. D. Stasheff, *Homotopy associativity of H-spaces*. I, Trans. Amer. Math. Soc., **108** (1963), 275–292.
19. B. Sturmfels, *Gröbner bases of toric varieties*, Tôhoku Math. J. (submitted).

Department of Mathematics, University of Kentucky, Lexington, Kentucky 40506
*E-mail address*: lee@s.ms.uky.edu

DIMACS, Rutgers University, Piscataway, New Jersey 08903
*E-mail address*: lee@ukma.bitnet

DIMACS Series in Discrete Mathematics
and Theoretical Computer Science
Volume **4**, 1991

Applied Geometry
and Discrete Mathematics
THE VICTOR KLEE FESTSCHRIFT

# On the Number of Antipodal or Strictly Antipodal Pairs of Points in Finite Subsets of $\mathbb{R}^d$

E. MAKAI, JR. AND H. MARTINI

ABSTRACT. Let $X \subset \mathbb{R}^d$ be a set of $n$ points whose convex hull is a $d$-dimensional polytope $P$. Two elements of $X$ are called antipodal (strictly antipodal) if they lie in different parallel supporting hyperplanes of $P$ (if they are the intersections of $P$ and two such hyperplanes). In terms of $n$ and $d$, we shall give upper estimates on the number of antipodal (strictly antipodal) pairs in $X$. Additionally, known results on lower estimates are cited.

## 1. Introduction

We shall investigate the following problems, initiated in **[G1]**: Let $X = \{x_1, \dots, x_n\} \subset \mathbb{R}^d$ $(x_i \neq x_j$ for $i \neq j)$, and let $P = \text{co}\, X$ denote its convex hull. We always suppose that $X$ does not lie in a hyperplane. The points $x_i, x_j \in X$ are called antipodal (strictly antipodal) if there are different parallel supporting hyperplanes $H'$, $H''$ of $P$ with $x_i \in H'$, $x_j \in H''$ (with $\{x_i\} = P \cap H'$, $\{x_j\} = P \cap H''$). These concepts are due to V. Klee (cf. **[G2, p. 420]**) and B. Grünbaum **[G1]**, respectively. Evidently, $x_i$ and $x_j$ are antipodal if and only if $\text{co}\{x_i, x_j\}$ is a longest chord of $P$ among all parallel chords.

We denote by $\text{a}(X)$ $(\text{sa}(X))$ the number of antipodal (strictly antipodal) pairs $x_i, x_j$ of $X$. Obviously, $\text{sa}(X) \leq \text{a}(X)$. When investigating $\text{a}(X)$ $(\text{sa}(X))$, we may and will suppose that each $x_i$ lies on the boundary $\text{bd}\, P$ of $P$ (is a vertex of $P$).

In the present paper we shall give, under these assumptions, upper estimates for $\text{a}(X)$ $(\text{sa}(X))$, and also we shall estimate $\text{a}(X)$ under the stronger

---

1980 *Mathematics Subject Classification* (1985 *Revision*). Primary 52A25.

The first author's research was partially supported by the Hungarian National Foundation for Scientific Research, grant no. 1817.

condition that each $x_i$ is a vertex of $P$. Evidently,

$$\min\{\mathrm{a}(X) \mid |X| = n\},$$
$$\min\{\mathrm{sa}(X) \mid |X| = n\} \le \min\{\mathrm{a}(X) \mid |X| = n,\ X = \operatorname{vert} P\}$$

and

$$\max\{\mathrm{sa}(X) \mid |X| = n\} \le \max\{\mathrm{a}(X) \mid |X| = n,\ X = \operatorname{vert} P\}$$
$$\le \max\{\mathrm{a}(X) \mid |X| = n\}.$$

The quantities $\mathrm{a}(X)$ and $\mathrm{sa}(X)$ have a number of other interpretations. By **[G1]**, $\mathrm{sa}(X)$ is half the vertex number of the difference body $P + (-P)$ of $P$. Also $\mathrm{sa}(X)$ can be interpreted as a special case of a more general quantity. Namely, for $0 \le k \le d - 2$ we may consider the number of $k$-dimensional subcomplexes of $P$, which arise as shadow boundaries, i.e., the set of those points $p \in P$ for which for some (parallel) projection onto a $(k+1)$-subspace the image of $p$ lies on the boundary of the image of $P$ (cf. **[BES, p. 24]**). For $k = 0$ this is $\mathrm{sa}(X)$, while for $k = d - 2$ in **[M]** sharp upper and lower estimates on this quantity have been obtained (in terms of the number of nonoriented facet normals of $P$).

Another interpretation is the Minkowskian analogue of the Euclidean problem investigated in **[E]**, namely the question for the upper estimate on the number of Euclidean diameters of $X$. Let $\mathbb{R}^d$ be endowed with a Minkowski metric. We may then ask for the number of Minkowskian diameters, i.e., pairs $x_i$, $x_j$ whose Minkowski distance is maximal. This is $\le \mathrm{a}(X)$, and if $X$ is of constant Minkowski width (i.e., $P + (-P)$ is a Minkowskian ball), then equality holds here. Actually for our upper estimates we shall use the methods developed in **[E]**.

Antipodal (strictly antipodal) pairs are called extremal (exposed) diameters by Soltan and Nguen **[SN2]**, and they show that each $x_i \in \operatorname{vert} P$ belongs to a strictly antipodal pair, thus $\mathrm{sa}(X) \ge \lceil n/2 \rceil$ ($\lceil x \rceil$ denoting the smallest integer $\ge x$). For $n \ge 2d$ even, this estimate is sharp, since $\mathrm{sa}(X) = \frac{1}{2}|\operatorname{vert}(P + (-P))|$, and we may choose $P$ a centrally symmetric convex $d$-polytope. Soltan and Nguen **[SN2]** also showed $\mathrm{a}(X) \ge \binom{d+1}{2}$, with equality only for $n = d + 1$, and $\mathrm{sa}(X) \ge d$, with equality only for $X = \{\text{vertices of an affine cross-polytope}\}$. For $d = 3$, Nguen **[N]** determined the number $c(n) = \min\{\mathrm{sa}(X) \mid |X| = n\}$ for every $n$, namely

$$c(4) = c(5) = c(7) = c(9) = 6, \qquad (1)$$
$$c(n) = \lceil n/2 \rceil \quad \text{otherwise.}$$

Partially based on results of **[SN1]**, Nguen **[N]** derived the following for $d > 3$

with $c_d(n) = \min\{\mathrm{sa}(X) \mid |X| = n\}$:

$(2_1)$ $c_d(d+1) = d(d+1)/2$,

$(2_2)$ $c_d(n) \geq 3(d-1)$, $\quad d+2 \leq n \leq 2d-2$,

$(2_3)$ $c_d(2d-1) = 3(d-1)$,

$(2_4)$ $c_d(n) = 2d$, $\quad n$ odd with $2d+1 \leq n \leq 4d-1$,

$(2_5)$ $c_d(n) = \lceil n/2 \rceil$ otherwise.

Some of the above-mentioned results of **[N]** were independently rediscovered by the authors, before learning of **[N]**, in a way different from that of **[N]**, and were pointed out at the Oberwolfach conference on Combinatorial Convexity and Algebraic Geometry in 1989 (cf. the corresponding research report 35/1989). Details will be postponed to a sequel to this paper.

Continuing the citation of known results we again refer to **[SN1]**, where

$$\min\{\mathrm{sa}(X) \mid |X| = n,\ X = \mathrm{vert}\, P\} = n - 1 + \frac{d(d-1)}{2}$$

is proved. This implies that $\min\{\mathrm{a}(X) \mid |X| = n\}$ is the same quantity. Namely, any $X$ can be changed by an arbitrarily small amount to obtain a set $X'$ lying on a strictly convex surface, thus $X' = \mathrm{vert}\,\mathrm{co}\, X'$; further, for all sufficiently small perturbations $\mathrm{a}(X)$ does not increase (a chord $x_i x_j$ of Minkowski length $< 2$, measured with the unit ball $\frac{1}{2}(P + (-P))$, remains such after a sufficiently small perturbation).

For $d = 2$ **[G1]** shows $\max\{\mathrm{a}(X) \mid |X| = n,\ X = \mathrm{vert}\, P\} = \lfloor 3n/2 \rfloor$ (with $\lfloor x \rfloor$ denoting the largest integer $\leq x$) and $\max\{\mathrm{sa}(X) \mid |X| = n\} = n$ (this obviously follows from $\mathrm{sa}(X) = \frac{1}{2}|\mathrm{vert}(P + (-P))|$), which also shows that $\mathrm{sa}(X)$ can be any integer in $[\lceil n/2 \rceil, n]$ (cf. **[SN1, Theorem 1]**).

For $d = 3$, in **[G1, G2]** the inequalities

$$\max\{\mathrm{sa}(X) \mid |X| = n\} \geq \lfloor n^2/4 \rfloor + 2,$$
$$\max\{\mathrm{a}(X) \mid |X| = n,\ X = \mathrm{vert}\, P\} \geq \tfrac{1}{4}n^2 + \tfrac{3}{2}n + O(1)$$

are proved, and it is asked if

$$\max\{\mathrm{a}(X) \mid |X| = n,\ X = \mathrm{vert}\, P\} \sim \tfrac{1}{4}n^2,$$

respectively, if, more generally, for any $d \geq 3$ in $\mathbb{R}^d$ we have

$$\max\{\mathrm{a}(X) \mid |X| = n,\ X = \mathrm{vert}\, P\} \sim (\tfrac{1}{2} - 1/2^{d-1})n^2.$$

Some of our results (cf. §2) will answer this question in the negative for each $d \geq 3$.

As usual, we denote by co, aff, int, relint, and vert the convex hull, affine hull, interior, relative interior (in the affine hull) of a set, and the set of vertices of a convex polytope, in $\mathbb{R}^d$, respectively.

## 2. Results

THEOREM 1. *For $d = 2$ we have*

$$\left\lfloor \frac{n^2}{4} \right\rfloor \le \max\{\mathrm{a}(X) \,|\, |X| = n\} \le \frac{n^2}{4} + O(n).$$

In 3-space we obtain the inequalities of

THEOREM 2. *For $d = 3$ one has*

(I) $\lfloor n^2/3 \rfloor \le \max\{\mathrm{sa}(X) \,|\, |X| = n\} \le \frac{2}{5} n^2$.

(II) $\lfloor 3n^2/8 \rfloor + 4 \le \max\{\mathrm{a}(X) \,|\, |X| = n\} \le 7n^2/16$ (*the lower bound holding for* $n \ge 8$).

From Theorem 2 we also obtain the following

COROLLARY. *For $d = 3$*

$$\left\lfloor \frac{n^2}{3} \right\rfloor \le \max\{\mathrm{a}(X) \,|\, |X| = n, \ X = \operatorname{vert} P\} \le \frac{7n^2}{16}.$$

In the next theorem the proof of the lower estimate in (I) is based on a suggestion of J. Pach.

THEOREM 3. *For any fixed $d > 3$ we have*

(I)

$$\begin{aligned}
&\left(1 - \frac{\mathrm{const}}{1.0044\ldots^d}\right)\frac{n^2}{2} - O(1) \\
&= \left(1 - \frac{\mathrm{const}}{(1 + \frac{1}{36}\cdot(\frac{5}{6})^{10})^d}\right)\frac{n^2}{2} - O(1) \\
&\le \left(1 - \frac{1}{\lfloor \frac{1}{2}\varphi(d) \rfloor}\right)\frac{n^2}{2} - O(1) \\
&\le \left(1 - \frac{1}{\lfloor \frac{1}{2}\psi(d) \rfloor}\right)\frac{n^2}{2} - O(1) \\
&\le \max\{\mathrm{sa}(X) \,|\, |X| = n\} \le \left(1 - \frac{1}{2^d - 1}\right)\frac{n^2}{2},
\end{aligned}$$

*where $\varphi(d)$ (respectively $\psi(d)$) $= \max\{|Y| \,|\, Y \subset \mathbb{R}^d$, $\forall$ four different points $x_1, y_1, x_2, y_2$ of $Y$ spanning a 3-plane, there is a (not necessarily orthogonal) projection of $\mathbb{R}^d$ onto this 3-plane, carrying any $z \in Y$ different from $x_1, y_1, x_2, y_2$ into $\mathrm{co}\{x_1, y_1, x_2, y_2\}$, but not into $\mathrm{co}\{x_1, y_1\} \cup \mathrm{co}\{x_2, y_2\}$ (respectively carrying any such $z$ into the open slab in this 3-plane, bounded by two parallel 2-planes, one containing $\{x_1, y_1\}$, the other containing $\{x_2, y_2\}$)$\}$.*

(II)

$$\left(1-\frac{1}{3\cdot 2^{d-3}}\right)\frac{n^2}{2}-O(1)$$
$$\leq \max\{\mathrm{a}(X) \mid |X|=n,\ X=\operatorname{vert} P\} \leq \left(1-\frac{1}{2^d}\right)\frac{n^2}{2}.$$

(III)

$$\left(1-\frac{1}{2^{d-1}}\right)\frac{n^2}{2}-O(1) \leq \max\{\mathrm{a}(X) \mid |X|=n\} \leq \left(1-\frac{1}{2^d}\right)\frac{n^2}{2}.$$

Note that the Corollary and Theorem 3(II) answer the question of B. Grünbaum mentioned at the end of §1 in the negative. Observe further that in Theorem 3(II) and (III) the lower and upper estimates are rather close. It seems to be rather difficult to give an asymptotic estimate to $\max\{\mathrm{sa}(X) \mid |X|=n\}$ for $d>3$. We think that in Theorem 3(II) and in the Corollary the truth is closer to the lower estimate.

Conjecture. We have in $\mathbb{R}^d$, $d\geq 2$,

$$\max\{\mathrm{a}(X) \mid |X|=n\} \leq \left(1-\frac{1}{2^{d-1}}\right)\frac{n^2}{2},$$

and in $\mathbb{R}^3$

$$\max\{\mathrm{sa}(X) \mid |X|=n\} = \frac{n^2}{3}+O(1).$$

In the proof of Theorem 3 we use a statement which seems to be of independent interest, and whose proof we postpone to a sequel to this paper.

Proposition. *Let a simplex* $a_1\cdots a_{d-1}b_1b_2$ *in* $\mathbb{R}^d$, $d\geq 2$, *have two facets (say,* $a_1\cdots a_{d-1}b_1$ *and* $a_1\cdots a_{d-1}b_2$*), enclosing an angle* $\geq\varphi$, *where* $1/d\geq \cos\varphi\geq -1/(d^2-3d+1)$ *(the lower inequality holding for* $d\geq 3$*). Then the quotient of the lengths of the longest and shortest edges is*

$$\geq\sqrt{\frac{2-2\cos\varphi}{1+(d-2)(1-\cos\varphi)/(d-1)}}.$$

*Equality holds if and only if the angle of the mentioned two facets is* $\varphi$, *both facets are right pyramids over a regular* $(d-2)$*-simplex as base (namely over* $a_1\cdots a_{d-1}$*), and*

$$a_{i_1}a_{i_2}=b_1b_2=a_ib_j\cdot\sqrt{\frac{2-2\cos\varphi}{1+(d-2)(1-\cos\varphi)/(d-1)}}$$

*for each* $i, i_1, i_2\leq d-1$, $j\leq 2$.

An open question, connected with this, is the following: Let $\varphi> \arccos 1/d =$ angle of two facets of a regular simplex in $\mathbb{R}^d$, (respectively $\varphi<\arccos 1/d$), and suppose a simplex in $\mathbb{R}^d$ has two facets enclosing an angle $\geq \varphi$ (respectively $\leq \varphi$). Under this hypothesis determine for any fixed $\varphi$ the minimum of the quotient of lengths of the longest and shortest edges. The cases $\varphi = 0$, $\varphi = \pi$ are equivalent to a theorem of K. Schütte [**Sch**] (cf. also [**DW**]) about the minimal quotient of the maximal and minimal distances among $d+1$ points in $\mathbb{R}^{d-1}$ (thus the extremal simplices are in general different from the one given in the Proposition).

## 3. Proofs of the theorems

For the proofs of Theorems 1–3 we use graph-theoretic methods, developed in [**E**]. Let $K(l)$ denote a complete graph with $l$ vertices. More generally, let $K(t_1, \dots, t_l)$ denote a complete $l$-partite graph, with $t_i$ vertices in the $i$th class $(1 \leq i \leq l)$, i.e., the vertex set is the disjoint union of sets $V_1, \dots, V_l$ of cardinality $t_1, \dots, t_l$, respectively, and two vertices are connected by an edge if and only if they belong to different $V_i$'s.

The well-known theorem of P. Turán [**T**] implies the following: Let $G$ be a graph with $n$ vertices and with more than

$$\left(1 - \frac{1}{l-1}\right)\frac{n^2}{2}$$

edges. Then $G$ contains a subgraph $K(l)$. A generalization of this is the following (cf. [**Bo, Corollary 4.7**]): Let $G$ be a graph with $n$ vertices and with more than

$$\left(1 - \frac{1}{l-1}\right)\frac{n^2}{2} + \mathrm{const}_{t_1, \dots, t_l}\, n^{2-(1/t_1)}$$

edges. Then $G$ contains a subgraph $K(t_1, \dots, t_l)$.

In the proofs of our theorems, $G$ will be the graph whose edges correspond to the (strictly) antipodal pairs of $X$, and $K(l)$ and $K(t_1, \dots, t_l)$ will be suitably chosen.

PROOF OF THEOREM 1. The lower estimate follows from the following construction. Place $\lfloor n/2 \rfloor$, respectively $\lceil n/2 \rceil$, points on two different parallel lines.

For the upper estimate it suffices to show, in view of [**Bo, Corollary 4.7**], that the graph $G$ whose edges are the antipodal pairs of $X$ does not contain a subgraph $K(2, 2, 1)$. Note that if $x_i, x_j$ are (strictly) antipodal in $X$, then they are (strictly) antipodal in any $Y \subset X$ (taken in the affine hull aff $Y$ of $Y$), for which $x_i, x_j \in Y$.

Since $K(2, 2, 1)$ contains $K(2, 2)$ as a subgraph, we first investigate the $K(2, 2)$'s, i.e., quadrangles in $G$. IF $x_1, x_2, x_3, x_4 \in X$, then a graph-theoretic quadrangle $\{x_1, x_2, x_3, x_4\}$ forms a (possibly nonstrictly) convex quadrangle, which can (1) lie on a segment, or is not degenerate, and can be (2) a parallelogram, or (3) a trapezoid which is no parallelogram, or (4) a nontrapezoid. We investigate how the antipodal pairs of $\{x_1, x_2, x_3, x_4\}$ (with respect to $\{x_1, x_2, x_3, x_4\}$) may look. If the points $x_1, x_2, x_3, x_4$ lie on a segment, just one pair $x_i, x_j$, $1 \leq i, j \leq 4$, gives a maximal chord among all parallel chords, i.e., is antipodal. If we have a parallelogram, all pairs are antipodal. If we have a trapezoid, which is not a parallelogram, then each pair is antipodal, except the pair forming the shorter base. If we have a nontrapezoid, with order of vertices $x_1, x_2, x_3, x_4$, and, e.g.,

$$\sphericalangle x_3x_4x_1 + \sphericalangle x_4x_1x_2, \quad \sphericalangle x_4x_1x_2 + \sphericalangle x_1x_2x_3 < \pi$$

(this may be supposed after a cyclic permutation of the indices), then the antipodal pairs are $x_1x_2$, $x_1x_4$, $x_2x_4$, $x_1x_3$. Thus a quadrangle among the antipodal pairs of $\{x_1, x_2, x_3, x_4\}$ in the first case does not exist. In the second case it can be any quadrangle, i.e., geometrically four sides, or two opposite sides and two diagonals of the parallelogram. In the third case such a quadrangle consists of the two legs and of the two diagonals of the trapezoid. In the fourth case such a quadrangle does not exist.

First we investigate the case where the quadrangle consists of the four sides of the above parallelogram $P_0$. We have $P_0 \subset P = \operatorname{co} X$, but if $P_0$ is a proper subset of $P$, some side of $P_0$ is not a longest chord of $P$ among all parallel chords, i.e., does not give an antipodal pair of $X$, a contradiction. However, if $P_0 = P$, each point $x$ of $X$, other than $x_1, x_2, x_3, x_4$, lies in the interior of some side of $P_0$, thus not all $x_ix$ $(i = 1, 2, 3, 4)$ can be antipodal pairs of $X$. Hence $G$ does not contain a subgraph $K(2, 2, 1)$. Henceforward we may assume that any $K(2, 2)$ in $G$ consists of two parallel sides and two diagonals of a parallelogram or of the two legs and two diagonals of a trapezoid, which is not a parallelogram. Therefore the graph-theoretic diagonals of each quadrangle $K(2, 2)$ are geometrically two parallel segments, which are not collinear.

Now let us consider a subgraph $K(2, 2, 1)$ in $G$. Let the vertices of $K(2, 2, 1)$ in the first, second and third class be denoted by $x_1, x_2$, and $y_1, y_2$, and $z$, respectively. Thus $x_1x_2$, $y_1y_2$ are noncollinear, parallel segments. Let $a, b$ be any two of $x_1, x_2, y_1, y_2$. Then $z$ cannot lie on the open segment $ab$, since then $ab$ is longer than both $az$ and $bz$, thus $az$ and $bz$ are not longest chords of their direction. Also, $a$ cannot lie on the open segment $bz$, since then both $za$ and $zb$ are maximal chords of their direction, $zb$ being longer, a contradiction. Thus $z$ is not collinear with any two of $x_1, x_2, y_1, y_2$.

We may suppose that on the trapezoid $\operatorname{co}\{x_1, x_2, y_1, y_2\}$ the order of vertices is $x_1, x_2, y_1, y_2$. Recall that $x_1, x_2, y_1, y_2, z$ all lie on the boundary of $\operatorname{co} X$. Therefore, and also by the preceding paragraph, the point $z$ lies in one of the open domains outside our trapezoid which are bounded by a side of this trapezoid and the elongations of the neighbouring sides from the endpoints of the considered side (until their meeting point, if such exists). If this considered side is a base, $x_1y_2$ is not a maximal chord of $\operatorname{co}\{x_1, x_2, y_1, y_2, z\}$ among all parallel chords. Thus $x_1y_2$ is no antipodal pair. If the considered side is a leg, e.g., $x_2y_1$, then $zx_2$ is not a maximal chord of $\operatorname{co}\{x_1, x_2, y_1, y_2, z\}$ among all parallel chords, thus $zx_2$ is no antipodal pair. Hence, in any case a subgraph $K(2, 2, 1)$ cannot exist, finishing the proof of the theorem. □

PROOF OF THEOREM 2 AND ITS COROLLARY. The lower estimates follow from the following constructions. For (II) consider four parallel edges of a cube, and place on each $\lfloor n/4 \rfloor$ or $\lceil n/4 \rceil$ points, so that these points include both endpoints of the edges. For (I) consider three mutually skew edges of a cube. Any two of these have as convex hull a tetrahedron, and the centre $o$ of the cube is an edge midpoint of this tetrahedron. Now let us replace each of these three edges $E_i$ by circular arcs $I_i$ $(i = 1, 2, 3)$ of small curvature, lying in the plane spanned by the edge and $o$, passing through both endpoints of the corresponding edge, and turning with their concave sides towards $o$. Then any $x_i \in I_i$, $x_j \in I_j$ $(i \neq j)$ determine a strictly antipodal pair of points of $\operatorname{co}(I_i \cup I_j)$ (this being true even if $o$ is replaced in the above construction by any $o'$, lying in the open slab bounded by two parallel planes, one containing $E_i$, the other containing $E_j$). Note that the parallel distinct supporting planes of $\operatorname{co}(I_i \cup I_j)$ from the definition of strict antipodality are near the above-mentioned two parallel planes. Now let us place $\lfloor n/3 \rfloor$ or $\lceil n/3 \rceil$ points on each $I_1, I_2, I_3$, placing altogether $n$ points, and placing each of these points, say, in the middle third of the respective $I_i$. Then the above-mentioned pairs $x_i, x_j$, taken from these $n$ points, will remain strictly antipodal points of this set $X$ of $n$ points. The number of such pairs is however $\lfloor n^2/3 \rfloor$.

Now we turn to the upper estimates. By **[DG]** the set $X$ does not contain any nine-element subset $Y$, all of whose pairs are antipodal pairs of $Y$, taken in the affine hull of $Y$ (for such $Y$ with $\dim \operatorname{aff} Y = i$ we have $|Y| \le 2^i \le 8$). Thus $X$ does not contain any nine-element subset $Y$, all of whose pairs are antipodal pairs of $X$. By Turán's theorem $X$ cannot have more than $\frac{7}{16}n^2$ antipodal pairs. This also shows the Corollary.

Similarly, by **[G1]** the set $X$ does not contain any six-element subset $Y$ with $\dim \operatorname{aff} Y = 3$, all of whose paris are strictly antipodal pairs of $Y$ (and does not contain $Y$, $\dim \operatorname{aff} Y = i < 3$, with $|Y| = 3$ for $i = 1$, $|Y| = 5$ for $i = 2$, since if all pairs of $Y$ are antipodal in $Y$, taken in $\operatorname{aff} Y$, then $|Y| \le 2^i$). In the same way as above we obtain that $X$ cannot have more than $\frac{2}{5}n^2$ strictly antipodal pairs. □

For Theorem 3 we shall need a lemma. We take it from **[EF, proof of Theorem 3.6]**. Since it is formulated there in a way different from ours, we state it here in a bit sharper form.

LEMMA **[EF]**. *For any fixed* $\varepsilon \in (0, 1)$ *there is a set* $Y_d$ *in* $\mathbb{R}^d$, *with* $|Y_d| \ge \text{const} \cdot \sqrt{d}(1 + (\varepsilon^2/(1-\varepsilon)^{2-(2/\varepsilon)}))^d$, *such that*

$$\frac{\max x_i x_j}{\min x_i x_j} \le \frac{1}{\sqrt{1-\varepsilon}} \qquad (x_i, x_j \in Y_d).$$

PROOF. By [EF], there is a $Y_d$ in $\mathbb{R}^d$ with $\max x_i x_j / \min x_i x_j \leq 1/\sqrt{1-\varepsilon}$, such that

$$|Y_d| \geq \max_{1\leq k\leq d} \frac{\binom{d}{k}}{\binom{k}{x}\binom{d-x}{k-x}}, \quad \text{where } x = \lceil \varepsilon k \rceil .$$

We shall seek $k$ in the form $\lfloor \lambda\varepsilon d \rfloor$, $0 < \lambda < 1$. Then, by Stirling's formula,

$$\frac{\binom{d}{k}}{\binom{k}{x}\binom{d-x}{k-x}} \sim \text{const}\cdot\sqrt{d}(f(\lambda))^d ,$$

where $f(\lambda) = \lambda^{-\lambda\varepsilon^2}(1-\varepsilon)^{2(\lambda\varepsilon-\lambda\varepsilon^2)}\cdot(1-\lambda\varepsilon^2)^{\lambda\varepsilon^2-1}$ (the constant factor depending on $\varepsilon$, $\lambda$). We will determine $\max f(\lambda)$ for $0 < \lambda < 1$. We have $f(\lambda) > 0$ for $0 < \lambda < 1$, $f(\lambda) \to 1$ for $\lambda \to 0$, and

$$\begin{aligned} f(1) &= [(1-\varepsilon)^{(1-\varepsilon)/2}(1+\varepsilon)^{(1+\varepsilon)/2}]^{2(\varepsilon-1)} \\ &< \left[\frac{1-\varepsilon}{2}\cdot\frac{1}{1-\varepsilon} + \frac{1+\varepsilon}{2}\cdot\frac{1}{1+\varepsilon}\right]^{-2(\varepsilon-1)} = 1 \end{aligned}$$

by the weighted geometric-harmonic mean inequality. Also $\frac{d}{d\lambda}\log f(\lambda) \to \infty$ for $\lambda \to 0$, hence $\frac{d}{d\lambda}\log f(\lambda)$ has a root in $(0, 1)$. We have

$$\frac{d}{d\lambda}\log f(\lambda) = \varepsilon^2 \log\left(\frac{1}{\lambda} - \varepsilon^2\right) + 2(\varepsilon - \varepsilon^2)\log(1-\varepsilon),$$

which is strictly decreasing and hence has a unique root $\lambda_0 = [\varepsilon^2 + (1-\varepsilon)^{2-(2/\varepsilon)}]^{-1}$ in $(0, 1)$. Hence $\max f(\lambda) = f(\lambda_0) = 1 + (\varepsilon^2/(1-\varepsilon)^{2-(2/\varepsilon)})$. Then, choosing $k = \lfloor \lambda_0 \varepsilon d \rfloor$, we obtain the estimate for $|Y_d|$ given in the lemma. □

Erdös and Füredi [EF] asserted $|X_d| > (1 + 0.4\varepsilon^2)^d$, by the above proof, choosing $\lambda = \frac{1}{4}$. However, it seems that their method does not give the factor 0.4. In fact, for $\varepsilon = \frac{1}{6}$ (the only case we apply) this bound would be $1.0111\ldots^d$, which is in general greater than the bound obtained by optimization for all $\lambda \in (0, 1)$, namely $\text{const}\cdot\sqrt{d}\cdot 1.0044\ldots^d$. However, actually for $\varepsilon = \frac{1}{6}$ we have $f(\frac{1}{4}) = 1.0038\ldots(< 1.0044\ldots) < 1.0111\ldots$.

PROOF OF THEOREM 3. The lower estimates for (II) and (III) can be given analogously to Theorem 2. Namely, for (III) consider $2^{d-1}$ parallel edges of a cube, and place on each $\lfloor n/2^{d-1} \rfloor$ or $\lceil n/2^{d-1} \rceil$ points, altogether $n$ points (which may include both endpoints of the edges, if $n \geq 2^d$). For (II) consider the example $X_3$ of Theorem 2(I) with $|X_3| = n$, $\text{sa}(X_3) \geq \lfloor n^2/3 \rfloor$. In $\mathbb{R}^d$ define $X_d$ as the set of vertices of the convex polytope which is a $(d-3)$-fold prism over $\text{co}\, X_3$. This gives examples for $|X_d|$ a multiple of $2^{d-3}$. If we delete from these at most $2^{d-3} - 1$ points, we obtain examples $X_d$ with $|X_d|$ arbitrary.

The upper estimates have to be shown only for (I) and (III). Analogously to Theorem 2, for (III) observe that by **[DG]** $X$ does not contain any $(2^d+1)$-element subset $Y$, all of whose pairs are antipodal pairs of $Y$, taken in aff $Y$ (actually for such $Y$ with $\dim \operatorname{aff} Y = i$ we have $|Y| \le 2^i$), implying the upper bound for (III). For (I) suppose that $X$ contains a $2^d$-element subset $Y$, all of whose pairs are strictly antipodal pairs of $Y$, taken in aff $Y$. Then they are antipodal pairs of $Y$ as well, and by **[DG]** we have $\dim \operatorname{aff} Y = d$, and $Y$ consists of all vertices of a parallelotope. However, not all pairs of vertices of the parallelotope co $Y$ are strictly antipodal (a strictly antipodal pair $y_1, y_2 \in Y$ corresponds to a pair of opposite vertices of $\operatorname{co} Y + (-\operatorname{co} Y)$, i.e., $y_1$ and $y_2$ are opposite vertices of co $Y$), a contradiction. This implies the upper bound for (I).

It remains to show the three inequalities from below in (I). The example will be similar to that constructed in Theorem 2(I).

First we show the first inequality for which it suffices to verify

$$\varphi(d) \ge \text{const} \cdot \left(1 + \tfrac{1}{36}\left(\tfrac{5}{6}\right)^{10}\right)^d .$$

Let us consider the set $Y_d$ in $\mathbb{R}^d$, from Lemma **[EF]**, where $\varepsilon$ will be specified later. This will be the set $Y$ from the definition of $\varphi(d)$. We have to show that each four different points $x_1, y_1, x_2, y_2$ of $Y$ span a 3-plane. If they spanned an at most 2-dimensional plane, the quotient of the maximal and the minimal distances determined by $x_1, y_1, x_2, y_2$ would be $\ge \sqrt{2}$, (cf. **[Sch]**) (or, as is well known, these four points determine an angle $\ge 90^\circ$ **[DG]**). Therefore, for $\varepsilon < \frac{1}{2}$ the set $\operatorname{co}\{x_1, y_1, x_2, y_2\}$ is 3-dimensional.

Further we have to show the existence of a suitable (not necessarily orthogonal) projection on $\operatorname{aff}\{x_1, y_1, x_2, y_2\}$. We will choose the orthogonal projection and show that it satisfies the properties from the definition of $\varphi(d)$. In fact, let $z \in Y_d$, $z \ne x_1, y_1, x_2, y_2$. Similarly as above, if $\operatorname{co}\{x_1, y_1, x_2, y_2, z\}$ were at most 3-dimensional, by **[Sch]** the quotient of the maximal and the minimal distances determined by $x_1, y_1, x_2, y_2, z$ would be $\ge \sqrt{\frac{12}{7}}$, which is impossible if we choose $\varepsilon < \frac{5}{12}$. That means, in this case $\operatorname{co}\{x_1, y_1, x_2, y_2, z\}$ is a 4-simplex. If the orthogonal projection of $z$ onto $\operatorname{aff}\{x_1, y_1, x_2, y_2\}$ does not lie in $\operatorname{relint}\operatorname{co}\{x_1, y_1, x_2, y_2\}$, then $\operatorname{co}\{x_1, y_1, x_2, y_2, z\}$ has two facets (one being $\operatorname{co}\{x_1, y_1, x_2, y_2\}$) enclosing an angle $\ge \pi/2$. In this case by the Proposition the quotient of the maximal and minimal edge-lengths of $\operatorname{co}\{x_1, y_1, x_2, y_2, z\}$ is $\ge \sqrt{2/(1+\frac{2}{3})} = 1/\sqrt{1-\frac{1}{6}}$, with equality only if the orthogonal projection of $z$ onto $\operatorname{aff}\{x_1, y_1, x_2, y_2\}$ is the barycentre of a facet of $\operatorname{co}\{x_1, y_1, x_2, y_2\}$. The case of strict inequality is impossible for $\varepsilon = \frac{1}{6}$, thus for $\varepsilon = \frac{1}{6}$ we either have that the orthogonal projection of $z$ onto $\operatorname{aff}\{x_1, y_1, x_2, y_2\}$ lies in $\operatorname{relint}\operatorname{co}\{x_1, y_1, x_2, y_2\}$, or else it is the barycentre of a facet of $\operatorname{co}\{x_1, y_1, x_2, y_2\}$; hence anyway it lies in $\operatorname{co}\{x_1, y_1, x_2, y_2\}$ and does not lie in $\mathrm{co}\{x_1, y_1\} \cup \mathrm{co}\{x_2, y_2\}$. By Lemma [EF], applied for $\varepsilon = \frac{1}{6}$, we have

$$\varphi(d) \geq |Y| = |Y_d| \geq \mathrm{const} \cdot (1 + \tfrac{1}{36}(\tfrac{5}{6})^{10})^d .$$

The second inequality in (I) follows from the evident inequality $\varphi(d) \leq \psi(d)$.

Now we show the third inequality in (I). Let us consider the set $Y$ satisfying the requirements of the definition of $\psi(d)$, with $|Y| = \psi(d)$. Further, let us consider $\lfloor \frac{1}{2}\psi(d) \rfloor$ pairs $x_i, y_i$ of elements of $Y$, which are disjoint. Then let us consider circular arcs $I_i$ of small curvature, lying in the respective 2-planes spanned by $x_i, y_i$ and $o$ (= the barycentre of the point system $Y$), passing through $x_i, y_i$ and turning with their concave sides towards $o$. Then we claim that any $u_{i_1} \in I_{i_1}$, $u_{i_2} \in I_{i_2}$ $(i_1 \neq i_2)$ determine a strictly antipodal pair of points of $\mathrm{co}(I_{i_1} \cup I_{i_2})$. We may suppose $i_1 = 1$, $i_2 = 2$.

Let $o'$, respectively $I'_i$ $(i \leq \lfloor \frac{1}{2}\psi(d) \rfloor)$, denote the projection of $o$, resp. $I_i$, onto $\mathrm{aff}\{x_1, y_1, x_2, y_2\}$ by the projection from the definition of $\psi(d)$ (with the same notation). The point $o'$ lies on a segment, whose one endpoint is the barycentre of $\mathrm{co}(x_1, y_1, x_2, y_2\}$, and the other endpoint lies in the open slab from the definition of $\psi(d)$. Therefore $o'$ also lies in this open slab. In particular, for $i = 1, 2$ the set $\mathrm{aff}\{x_i, y_i, o'\}$ is 2-dimensional. For $i = 1, 2$, $I'_i$ is a convex arc and can be arbitrarily near to $\mathrm{co}\{x_i, y_i\}$ (provided $I_i$ has a sufficiently small curvature). Since $I_i$ lies in $\mathrm{aff}\{x_i, y_i, o\}$ and turns with its concave side towards $o$, then $I'_i$ also lies in $\mathrm{aff}\{x_i, y_i, o'\}$ and turns with its concave side towards $o'$. Now, as in the proof of Theorem 2(I), lower estimate, any $u'_1 \in I'_1$, $u'_2 \in I'_2$ determine a strictly antipodal pair of points of $\mathrm{co}(I'_1 \cup I'_2)$, considered in $\mathrm{aff}\{x_1, y_1, x_2, y_2\}$. (Moreover, the respective parallel supporting 2-planes are arbitrarily near to the two supporting 2-planes of $\mathrm{co}\{x_1, y_1, x_2, y_2\}$, parallel to $x_1y_1$, $x_2y_2$, i.e., the boundary 2-planes of the slab from the definition of $\psi(d)$.) This already implies any $u_1 \in I_1$, $u_2 \in I_2$ determine a strictly antipodal pair of points of $\mathrm{co}(I_1 \cup I_2)$. Namely, one obtains supporting planes in $\mathbb{R}^d$ showing the strict antipodality of $u_1, u_2$, as follows. Take the inverse images—by the projection $\mathbb{R}^d \to \mathrm{aff}\{x_1, y_1, x_2, y_2\}$ from the definition of $\psi(d)$—of the pair of parallel supporting 2-planes of $\mathrm{co}(I'_1 \cup I'_2)$ in $\mathrm{aff}\{x_1, y_1, x_2, y_2\}$, intersecting $\mathrm{co}(I'_1 \cup I'_2)$ in the respective projections of $u_1$, resp. $u_2$ (onto $\mathrm{aff}\{x_1, y_1, x_2, y_2\}$). Observe that each of these supporting planes in $\mathbb{R}^d$ is near to one $I_i$ only $(i = 1, 2)$, and the projection is one-to-one on each $I_i$, since $\mathrm{aff}\{x_i, y_i, o'\}$ is 2-dimensional. Thus each of these supporting planes has just one common point with the respective $I_i$ (and none with the other $I_i$), thus also has just one common point with $\mathrm{co}(I_1 \cup I_2)$, namely $u_i$. This proves our claim.

Now let us place $\lfloor n / \lfloor \frac{1}{2}\psi(d) \rfloor \rfloor$ or $\lceil n / \lfloor \frac{1}{2}\psi(d) \rfloor \rceil$ points $u_{ij}$ on each arc $I_i$, with altogether $n$ points. We assert any $u_{i_1 j_1}$, placed on $I_{i_1}$, and any

$u_{i_2 j_2}$, placed on $I_{i_2}$, determine strictly antipodal pairs of points of $\mathrm{co}(\bigcup I_i)$ (provided each $I_i$ has a sufficiently small curvature). We may suppose $i_1 = 1$, $i_2 = 2$. Then for any $i \neq 1, 2$ the projection of $I_i$ by the projection $\mathbb{R}^d \to \mathrm{aff}\{x_1, y_1, x_2, y_2\}$ from the definition of $\psi(d)$ lies very near to the segment joining the respective projections of $x_i$ and $y_i$. However, these projections all lie in the open slab from the definition of $\psi(d)$. Moreover, the supporting planes showing strict antipodality of $u_{1j_1}, u_{2j_2}$ in $\mathrm{co}(I_1 \cup I_2)$ were obtained as inverse images, also by the projection from the definition of $\psi(d)$, of planes very near the boundary 2-planes of our slab. These facts altogether imply that the supporting planes showing strict antipodality of $u_{1j_1}, u_{2j_2}$ in $\mathrm{co}(I_1 \cup I_2)$ also show strict antipodality of $u_{1j_1}, u_{2j_2}$ in $\mathrm{co}(\bigcup I_i)$. (Note that these planes have just one common point with $I_1$ or $I_2$, and none with the other $I_i$'s with all $I_i$'s lying on the same sides of these planes, thus these planes have just one common point with $\mathrm{co}(\bigcup I_i)$ as well.)

Hence, as asserted, for $i_1 \neq i_2$ the pairs $u_{i_1 j_1}, u_{i_2 j_2}$ are strictly antipodal with respect to $\mathrm{co}\{u_{ij}\}$. However, the number of such pairs is $(1 - 1/\lfloor \frac{1}{2}\psi(d) \rfloor)(n^2/2) - O(1)$. □

## 4. Concluding remarks

1. One can conjecture that, analogously to $\mathbb{R}^2$ (cf. the proof of Theorem 1), in $\mathbb{R}^d$ the graph of the antipodal pairs does not contain the complete $(2^{d-1}+1)$-partite graph $K(2, \dots, 2, 1)$, respectively that in $\mathbb{R}^3$ the graph of the strictly antipodal pairs does not contain the graph $K(2, 2, 2, 2)$, which would prove the conjecture stated after Theorem 3, up to an additive term $O(n)$, respectively $O(n^{3/2})$.

2. A sharpening of the (smallest) lower estimate of $\max\{\mathrm{sa}(X) \mid |X| = n\}$ in Theorem 3(I) is possible in the following way: Determine the minimum of the quotient $\frac{\text{maximal edge}}{\text{minimal edge}}$ of a 4-dimensional simplex, provided the foot-point of some altitude—which lies in a 3-plane spanned by four vertices $x_1, y_1, x_2, y_2$, say—does not lie in the open slab in $\mathrm{aff}\{x_1, y_1, x_2, y_2\}$ bounded by two parallel 2-planes, one containing $\{x_1, y_1\}$, the other containing $\{x_2, y_2\}$. If this minimum is $\geq \sqrt{1/(1-\varepsilon)}$ $(> \sqrt{\frac{6}{5}})$, where $\varepsilon < \frac{5}{12}$, the lower estimate in (I) can be replaced by

$$\left(1 - \frac{\text{const}}{(1 + (\varepsilon^2/(1-\varepsilon)^{2-(2/\varepsilon)}))^d}\right)\frac{n^2}{2} - O(1)$$

(by the same proof, only applying the largest lower estimate).

3. The definition of $\varphi(d)$ in Theorem 3 suggests we define (for $2 \leq i \leq$

$d+1$) $\varphi'_i(d)$, respectively $\varphi''_i(d)$, as

$$\max\{|Y| \mid Y \subset \mathbb{R}^d, \forall\ i\text{-element subset } Z \text{ of } Y \text{ spans an } (i-1)\text{-plane, and there is a (not necessarily orthogonal) projection of } \mathbb{R}^d \text{ onto this } (i-1)\text{-plane, carrying any } z \in Y,\ z \notin Z \text{ into } \operatorname{co} Z \text{ (respectively relint co } Z)\}.$$

This property of $Z$, for $i = 2$, is just the antipodality, respectively strict antipodality. We have (cf. **[DG]**)

$$2^d = \varphi'_2(d) \geq \varphi'_3(d) \geq \cdots, \quad \varphi''_2(d) \geq \varphi''_3(d) \geq \cdots, \quad \varphi'_i(d) \geq \varphi''_i(d).$$

For fixed $i$ and large $d$, both $\varphi'_i(d)$ and $\varphi''_i(d)$ are of exponential size in $d$. Namely, as in the proof of Theorem 3(I), lower bound, by **[Sch]** (stating that the quotient of the maximal and minimal distance for $i+1$ points in $\mathbb{R}^{i-1}$ is $\geq \sqrt{(i+1)/(i-1)}$) and by Lemma **[EF]** we see

$$\varphi''_i(d) \geq \text{const}_i \sqrt{d} \left(1 + \frac{1}{(2i-2)^2} \cdot \left(\frac{2i-3}{2i-2}\right)^{4i-6}\right)^d.$$

(Actually we obtain

$$\varphi''_i(d) \geq \text{const} \sqrt{d} \left(1 + \frac{1}{(2i-2)^2} \cdot \left(\frac{2i-3}{2i-2}\right)^{4i-6} - \delta\right)^d,$$

for arbitrary $\delta > 0$, but this already implies the given result.) One can ask for better bounds, and can also ask for bounds on the number of $i$-element subsets $Z$ of a set $X$, with the above properties.)

Another generalization of (strictly) antipodal pairs is the following. Let $X \subset \mathbb{R}^d$, $\operatorname{aff} X = \mathbb{R}^d$. We call two $i$-simplices spanned by points of $X$ (strictly) $i$-antipodal if they belong to two different parallel supporting planes (respectively are the intersections of $\operatorname{co} X$ and the supporting planes). One can ask about the maximal number of $i$-simplices, pairwise (strictly) $i$-antipodal, in some set $X$ (which is $\geq \lfloor \varphi'_{2i+2}(d)/(i+1) \rfloor$, respectively $\lfloor \varphi''_{2i+2}(d)/(i+1) \rfloor$, and is $\leq 2^d$); respectively, one can ask about bounds on the number of (strictly) $i$-antipodal pairs among $k$ $i$-simplices, spanned by some $i+1$ points of some set $X$. (We were informed that I. Bàrány and V. Soltan have also posed the first of these questions.)

It can be conjectured that for the $i$-antipodal case, $i \geq 1$, the answer to the first question, analogously to the Danzer-Grünbaum theorem **[DG]**, is $2^{d-i}$ (attained for simplices whose vertices lie on $2^{d-i}$ parallel $i$-faces of a cube), while the sharp upper bound for the second question for the $i$-antipodal case, $i \geq 1$, is $(1 - (1/2^{d-i}))(k^2/2) + O(1)$ (attained if the $i$-simplices lie on $2^{d-i}$ parallel $i$-faces of a cube, on each of these faces about $k/2^{d-i}$ ones), which would follow from the first stated conjecture.

4. As a partial answer to a special case of a question in Remark 3, we can prove the following proposition (which also shows that our proof of a lower estimate of $\max\{\mathrm{sa}(X) \mid |X| = n\}$ in Theorem 2(I) is in a sense optimal).

PROPOSITION $*$. *If* $\{E_i\}$ *is a system of pairwise skew edges of a convex polyhedron* $P$ *in* $\mathbb{R}^3$ *such that, for any* $i_1 \neq i_2$, $E_{i_1}$ *and* $E_{i_2}$ *are the intersections of the polyhedron* $P$ *and some different parallel supporting planes of* $P$, *then the number of the edges* $E_i$ *is at most* 3.

We postpone the proof to a sequel to this paper.

ACKNOWLEDGMENT. The authors are indebted to J. Pach for having proposed the construction of [EF] for the lower estimate in Theorem 3(I), and to Z. Füredi for stimulating discussions.

## REFERENCES

[Bá] I. Bàrány, *An extension of the Erdös-Szekeres theorem on large angles*, Combinatorica **7** (1987), 161–169.

[Bo] B. Bollobàs, *Extremal graph theory*, London Math. Soc. Monographs, vol. 11. Academic Press, London, New York, and San Franscisco, 1978.

[BES] H. Busemann, G. Ewald, and G. C. Shephard, *Convex bodies and convexity on Grassmann cones*. I–IV, Math. Ann. **151** (1963), 1–41.

[DG] L. Danzer and B. Grünbaum, *Über zwei Probleme bezüglich konvexer Körper von P. Erdös und V. L. Klee*, Math. Z. **79** (1962), 95–99.

[DW] B. V. Dekster and J. B. Wilker, *Edge lengths guaranteed to form a simplex*, Arch. Math. **49** (1987), 351–366.

[E] P. Erdös, *On sets of distances of* n *points in Euclidean space*, Magyar Tud. Akad. Mat. Kutató Int. Közl. **5** (1960), 165–169; also in P. Erdös, *The art of counting. Selected writings* (J. Spencer, ed.), MIT Press, Cambridge, Mass., and London, 1973, pp. 676–679.

[EF] P. Erdös and Z. Füredi, *The greatest angle among* n *points in the* d*-dimensional Euclidean space*, Combinatorial Math. (Proc. Internat. Colloq. Graph Theory Combin., Marseille-Luminy, 1981), North-Holland Math. Stud., vol. 75, North-Holland, Amsterdam and New York, 1983; Ann. Discrete Math. **17**.

[G1] B. Grünbaum, *Strictly antipodal sets*, Israel J. Math. **1** (1963), 5–10.

[G2] ——, *Convex polytopes*, Wiley-Interscience, London, New York, and Sidney, 1967.

[M] H. Martini, *Über scharfe Schattengrenzen und Schnitte konvexer Polytope*, Beiträge Algebra Geom. **19** (1985), 105–112.

[N] Manh Hung Nguen, *Lower estimates of the number of exposed diameters of a convex polytope with a small number of vertices*, manuscript (Kišinev). (Russian)

[Sch] K. Schütte, *Minimale Durchmesser endlicher Punktmengen mit vorgeschriebenem Mindestabstand*, Math. Ann. **150** (1963), 91–98.

[SN1] V. P. Soltan and Manh Hung Nguen, *Lower bounds for the numbers of extremal and exposed diameters of a convex polytope*, manuscript (Kišinev). (Russian)

[SN2] ——, *Lower bounds for the numbers of extremal and exposed diameters of a convex body*, Studia Sci. Math. Hungar. (to appear).

[T] P. Turan, *Eine Extremalaufgabe aus der Graphentheorie*, Mat. Fiz. Lapok **48** (1941), 436–452. (Hungarian, German Summary)

MATHEMATICAL INSTITUTE OF THE HUNGARIAN ACADEMY OF SCIENCES, PF. 127, H-1364 BUDAPEST, HUNGARY

SEKTION MATHEMATIK, PH DRESDEN, WIGARDSTRASSE 17, D-8060 DRESDEN, GERMANY

DIMACS Series in Discrete Mathematics
and Theoretical Computer Science
Volume 4, 1991

Applied Geometry
and Discrete Mathematics
THE VICTOR KLEE FESTSCHRIFT

# Multi-Order Convexity

JUAN-ENRIQUE MARTÍNEZ-LEGAZ AND IVAN SINGER

ABSTRACT. We introduce and study the concepts of segmental convexity and separational convexity in a multi-ordered set (i.e., a set endowed with a nonempty family of partial orders). We prove that, for a suitable family $\mathrm{T}$ of total orders on $R^n$, the segmentally convex sets and the separationally convex sets in the multi-ordered set $(R^n, \mathrm{T})$ coincide with the usual (vector) convex sets in $R^n$. Furthermore, we study two concepts of discrete convexity, namely, segmental and separational convexity in $(Z^n, \mathrm{T}')$, where $\mathrm{T}'$ is the family of restrictions to $Z^n$ of the orders belonging to $\mathrm{T}$. Also, we prove that the usual order convexity in a poset coincides with the segmental convexity and the separational convexity in some associated multi-ordered sets.

## 0. Introduction

The aim of the present paper is to introduce two natural general concepts of convexity of subsets $G$ of a set $S$, in terms of (partial) orders on $S$, encompassing, as particular cases, various known notions of (continuous and discrete) convexity. Our first concept is defined with the "*segmental*" (or "*inner*") approach, defining first the notion of a "segment" $\langle x, y\rangle$ in $S$ and then calling a set $G$ "convex" if the relations $x, y \in G$ imply $\langle x, y\rangle \subseteq G$. The second one of our concepts is defined with the "*separational*" (or "*outer*") approach, calling a set $G \subseteq S$ "convex" if every $x \in S\backslash G$ can be "separated" from $G$, in a certain sense (defined in terms of partial orders on $S$).

Concerning the well-known concept of "order convexity" of a subset $G$ of a poset $(S, \leq)$, we shall see in §2 that there exists no order relation $\leq$ on $R^n$ $(n \geq 2)$ for which the order convex subsets of $(R^n, \leq)$ are the usual (vector) convex subsets of $R^n$. Therefore we introduce, instead of posets, the following natural framework.

DEFINITION 0.1. A *multi-ordered set* is an ordered pair $(S, \mathscr{O})$, where $S$ is a set and $\mathscr{O}$ is a nonempty family of partial orders (i.e., reflexive antisymmetric transitive binary relations) on $S$.

1980 *Mathematics Subject Classification* (1985 *Revision*). Primary 52A01, 06A10.


1052-1798/91 $1.00 + $.25 per page

REMARK 0.1. (a) In the particular case when $\mathscr{O}$ is a singleton, say $\mathscr{O} = \{\leq\}$, the multi-ordered set $(S, \mathscr{O}) = (S, \{\leq\})$ can be identified with the poset $(S, \leq)$.

(b) Although not in the above generality, multi-ordered sets have been used, e.g., in **[4]** (in linear spaces) and **[13, 14]**; moreover, in **[14]** there was introduced a concept of convexity in multi-ordered topological spaces different from that introduced here. For other concepts of a set endowed with a family of binary relations, see also **[1]** and the references therein.

In §1 of the present paper we introduce and study $\mathscr{O}$-segments and $\mathscr{O}$-segmentally convex (or, briefly, $\mathscr{O}_{\text{seg}}$-convex) sets in a multi-ordered set $(S, \mathscr{O})$. We also consider the concept of an $\mathscr{O}_{\text{seg}}$-semispace in $(S, \mathscr{O})$, generalizing the notion of a semispace in a linear space, introduced by P. C. Hammer **[5]**; the first fundamental results about semispaces in linear spaces were obtained by P. C. Hammer **[5]** and V. Klee **[9]**.

In §2 we study segmental convexity in the multi-ordered set $(R^n, \mathrm{T})$, where $\mathrm{T}$ is the set of all total orders on $R^n$ which are compatible with the vector space structure of $R^n$. We prove that the $\mathrm{T}$-segments of $R^n$ coincide with the usual segments of $R^n$ and, hence, $\mathrm{T}_{\text{seg}}$-convexity in $R^n$ coincides with the usual convexity in $R^n$.

In §3 we show that segmental convexity in $(Z^n, \mathrm{T}')$, where $Z = \{\dots, -2, -1, 0, 1, 2, \dots\}$ and $\mathrm{T}'$ is the family of restrictions to the subset $Z^n$ of $R^n$ of the order relations belonging to $\mathrm{T}$ of §2, is equivalent to a concept of discrete convexity considered by L. Lupşa **[11, 12]**.

In §4 we prove that the usual order convex sets of a poset $(S, \leq)$ coincide with the segmentally convex sets of some associated multi-ordered sets, having the same "ground set" $S$.

In §5 we introduce and study $\mathscr{O}$-separationally convex (or, briefly, $\mathscr{O}_{\text{sep}}$-convex) sets in a multi-ordered set $(S, \mathscr{O})$, defined by means of a separation property by elements of $\mathscr{O}$. We show that every $\mathscr{O}_{\text{sep}}$-convex set is $\mathscr{O}_{\text{seg}}$-convex, but the converse is not true (in general). We also prove that, for any multi-ordered set $(S, \mathscr{O})$, the family of $\mathscr{O}_{\text{sep}}$-convex sets coincides with that of $\mathscr{T}_{\text{sep}}$-convex sets in an associated multi-ordered set $(S, \mathscr{T})$ (with the same $S$), such that $\mathscr{T}$ consists of total orders.

In §6 we prove that in $(R^n, \mathrm{T})$ (with $\mathrm{T}$ of §2), $\mathrm{T}_{\text{sep}}$-convexity, too, coincides with the usual convexity in $R^n$.

In §7 we give some characterizations of separational convexity in $(Z^n, \mathrm{T}')$ (with $\mathrm{T}'$ of §3), one of which shows that separational convexity in $(Z^n, \mathrm{T}')$ is equivalent to another concept of discrete convexity considered by L. Lupşa **[11, 12]**. Hence, we infer that there exist $\mathrm{T}'_{\text{seg}}$-convex sets which are not $\mathrm{T}'_{\text{sep}}$-convex.

Finally, in §8, we prove that the usual order convexity in a poset $(S, \leq)$ coincides with $\mathscr{O}_{\text{sep}}$-convexity in an associated multi-ordered set, having the same "ground set" $S$.

We emphasize that the concepts of segmental and separational convexity in multi-ordered sets, introduced in this paper, are motivated by the above-mentioned results of §§2 and 6, according to which the $\mathrm{T}_{\mathrm{seg}}$-convexity and $\mathrm{T}_{\mathrm{sep}}$-convexity in $(R^n, \mathrm{T})$ coincide with the usual convexity in $R^n$. The proofs of both results are based on a lexicographical separation theorem **[15, p. 258]**; see also **[16, Theorem 2.1]** closely related to a lexicographical separation theorem of V. Klee **[10, §2.4]**. Since the lexicographical order on $R^n$ will play an important role in the sequel, we recall that $x = \{\xi_i\}_1^n \in R^n$ is said to be "lexicographically less than" $y = \{\eta_i\}_1^n \in R^n$ (in symbols, $x <_L y$) if $x \neq y$ and if $\xi_k < \eta_k$ for $k = \min\{i \in \{1, \dots, n\} | \xi_i \neq \eta_i\}$. We write $x \leq_L y$ if $x <_L y$ or $x = y$. We denote by $(R^n)^*$ the conjugate space of $R^n$, identified with $R^n$ in the usual way, and by $O(R^n)$ the family of all linear isometries $v \colon R^n \to R^n$ (for the euclidean norm $\|\cdot\|$ on $R^n$).

## 1. Segmental multi-order convexity

DEFINITION 1.1. Let $(S, \mathscr{O})$ be a multi-ordered set.

(a) Given $x, y \in S$, we define *the $\mathscr{O}$-segment* $\langle x, y\rangle = \langle x, y\rangle_{\mathscr{O}}$ by

$$\langle x, y\rangle = \langle x, y\rangle_{\mathscr{O}} = \bigcap_{\rho\in\mathscr{O}} \langle x, y\rangle_\rho, \tag{1.1}$$

where

$$\langle x, y\rangle_\rho = \{z \in S | \text{either } x\rho z\rho y \text{ or } y\rho z\rho x\} \qquad (\rho \in \mathscr{O}). \tag{1.2}$$

(b) We call a set $G \subseteq S$ "*$\mathscr{O}$-segmentally convex*," or, briefly, *$\mathscr{O}_{\mathrm{seg}}$-convex*, if

$$\langle x, y\rangle \subseteq G \qquad (x, y \in G). \tag{1.3}$$

In the sequel we shall assume, without any special mention, that $(S, \mathscr{O})$ is a multi-ordered set.

PROPOSITION 1.1. *For any $x, y \in S$, the following statements are equivalent*:

1°. $\langle x, y\rangle \neq \varnothing$.
2°. $x, y \in \langle x, y\rangle$.
3°. *For each $\rho \in \mathscr{O}$, either $x\rho y$ or $y\rho x$.*

PROOF. 1° $\Rightarrow$ 3° by the transitivity of each $\rho \in \mathscr{O}$. Also, 3° $\Rightarrow$ 2° by the reflexivity of each $\rho \in \mathscr{O}$. Finally, the implication 2° $\Rightarrow$ 1° is obvious.

We recall that an order $\rho$ on $S$ is said to be *total* if for any $x, y \in S$ we have either $x\rho y$ or $y\rho x$.

COROLLARY 1.1. *The following statements are equivalent*:

1°. $\langle x, y\rangle_{\mathscr{O}} \neq \varnothing$ $(x, y \in S)$.
2°. $x, y \in \langle x, y\rangle_{\mathscr{O}}$ $(x, y \in S)$.
3°. *All $\rho \in \mathscr{O}$ are total.*

REMARK 1.1. (a) By Corollary 1.1, implication $1^\circ \Rightarrow 3^\circ$, if there exists $\rho \in \mathscr{O}$ which is not total, then there exist $x, y \in S$ such that $\langle x, y\rangle_{\mathscr{O}} = \emptyset$; thus, in this case, the axiom J1 for "join geometries" (see **[19, p. 209]**) is not satisfied. Moreover, let us also note that, by Proposition 1.1, equivalence $1^\circ \Leftrightarrow 3^\circ$, for $x, y \in S$ we have $\langle x, y\rangle_{\mathscr{O}} = \emptyset$ if and only if there exists $\rho = \rho(x, y) \in \mathscr{O}$ such that $x\overline{\rho}y$ *and* $y\overline{\rho}x$, where $\overline{\rho}$ denotes the negation of $\rho$.

(b) One may (and, sometimes, we shall) assume, without loss of generality, that for (some, or all) $\rho \in \mathscr{O}$ we have $\rho^{-1} \in \mathscr{O}$ (since the segments $\langle x, y\rangle$ remain the same), where $\rho^{-1}$ is the "*reverse order*" *to* $\rho$, defined by

$$x\rho^{-1}y \Leftrightarrow y\rho x \qquad (x, y \in S). \tag{1.4}$$

(c) We have

$$\langle x, y\rangle = \langle y, x\rangle \qquad (x, y \in S). \tag{1.5}$$

Clearly, the whole set $S$, the empty set $\emptyset$, and every singleton $\{x\}$ $(= \langle x, x\rangle)$, where $x \in S$, are $\mathscr{O}_{\text{seg}}$-convex. Furthermore, we have

PROPOSITION 1.2. *For any $x, y \in S$, the segment $\langle x, y\rangle$ is $\mathscr{O}_{\text{seg}}$-convex.*

PROOF. Let $z, t \in \langle x, y\rangle$, $u \in \langle z, t\rangle$, and $\rho \in \mathscr{O}$. Then, there are eight cases:

(i) If $z\rho u\rho t$, $x\rho z\rho y$, $x\rho t\rho y$, then $x\rho z\rho u\rho t\rho y$, whence $x\rho u\rho y$, so $u \in \langle x, y\rangle_\rho$.
(ii) If $z\rho u\rho t$, $x\rho z\rho y$, $y\rho t\rho x$, then $z\rho u\rho t\rho x\rho z\rho y\rho t$, whence $z = u = t = x = y$, so $u \in \langle x, y\rangle_\rho$.
(iii) If $z\rho u\rho t$, $y\rho z\rho x$, $x\rho t\rho y$, then $z\rho u\rho t\rho y\rho z\rho x\rho t$, whence $z = u = t = y = x$, so $u \in \langle x, y\rangle_\rho$.
(iv) If $z\rho u\rho t$, $y\rho z\rho x$, $y\rho t\rho x$, then $y\rho z\rho u\rho t\rho x$, whence $y\rho u\rho x$, so $u \in \langle x, y\rangle_\rho$.

Finally, the other four cases are obtained from (i)–(iv) by interchanging the roles of $z$ and $t$.

PROPOSITION 1.3. (a) *For any $\rho \in \mathscr{O}$, each set $M \subseteq S$ such that*

$$x \in M, \quad u \in S, \qquad x\rho u \Rightarrow u \in M, \tag{1.6}$$

*is $\mathscr{O}_{\text{seg}}$-convex.*

(b) *For any $\rho \in \mathscr{O}$, each set $N \subseteq S$ such that*

$$x \in N, \quad u \in S, \qquad u\rho x \Rightarrow u \in N, \tag{1.7}$$

*is $\mathscr{O}_{\text{seg}}$-convex.*

PROOF. (a) Assume that $\rho \in \mathscr{O}$ and $M \subseteq S$ satisfy (1.6), and let $z, t \in M$ and $u \in \langle z, t\rangle$. Then, for each $\rho' \in \mathscr{O}$, we have either $z\rho' u\rho' t$ or $t\rho' u\rho' z$. Let $\rho' = \rho$. If $z\rho u\rho t$, then, by $z \in M$ and (1.6), we have $u \in M$. On the other hand, if $t\rho u\rho z$, then by $t \in M$ and (1.6) we have, again, $u \in M$. Thus, $M$ is $\mathscr{O}_{\text{seg}}$-convex.

(b) The proof is similar to that of part (a). Alternatively, assuming that $\rho^{-1} \in \mathscr{O}$ (see Remark 1.1(b)), part (b) follows from part (a) applied to $\rho^{-1}$.

For each $a \in S$ and $\rho \in \mathscr{O}$, let us denote

$$M_{a\rho} = \{x \in S | x\bar{\rho}a\}, \qquad M'_{a\rho} = \{x \in S | a\bar{\rho}x\}, \tag{1.8}$$

$$C_{a\rho} = \{x \in S | a\rho x\}, \qquad C'_{a\rho} = \{x \in S | x\rho a\}; \tag{1.9}$$

clearly, $C_{a\rho} \subseteq M_{a\rho} \cup \{a\}$ and $C'_{a\rho} \subseteq M'_{a\rho} \cup \{a\}$. The sets $C_{a\rho}$ have been called "upper cones" in **[20]**, and the sets $C'_{a\rho}$ may be called "lower cones", in the poset $(S, \rho)$. The sets $M_{a\rho}$ and $M'_{a\rho}$ are useful in vector optimization.

DEFINITION 1.2. An $\mathscr{O}_{\text{seg}}$-*hemispace* is a set $M \subseteq S$ such that both $M$ and $S \backslash M$ are $\mathscr{O}_{\text{seg}}$-convex.

From Proposition 1.3 we obtain

COROLLARY 1.2. *For any $a \in S$ and $\rho \in \mathscr{O}$, we have the pairs of complementary $\mathscr{O}_{\text{seg}}$-hemispaces*

(1.10)

$$(M_{a\rho}, C'_{a\rho}), \quad (M_{a\rho} \cup \{a\}, C'_{a\rho} \backslash \{a\}), \quad (M'_{a\rho}, C_{a\rho}), \quad (M'_{a\rho} \cup \{a\}, C_{a\rho} \backslash \{a\}).$$

PROOF. $M_{a\rho}$, $M_{a\rho} \cup \{a\}$, $C_{a\rho}$, and $C_{a\rho} \backslash \{a\}$ satisfy (1.6), while $M'_{a\rho}$, $M'_{a\rho} \cup \{a\}$, $C'_{a\rho}$, and $C'_{a\rho} \backslash \{a\}$ satisfy (1.7).

DEFINITION 1.3. For $a \in S$, an $\mathscr{O}_{\text{seg}}$-*semispace at* $a$ (or, $\mathscr{O}_{\text{seg}}$-*copoint at* $a$) is a maximal (in the sense of inclusion) $\mathscr{O}_{\text{seg}}$-convex subset of $S \backslash \{a\}$.

COROLLARY 1.3. *If $a \in S$ is a $\rho$-minimal or a $\rho$-maximal element for some $\rho \in \mathscr{O}$, then $S \backslash \{a\}$ is the only $\mathscr{O}_{\text{seg}}$-semispace at $a$.*

PROOF. If $a \in S$ is $\rho$-minimal, i.e., if

$$M_{a\rho} = \{x \in S | x\bar{\rho}a\} = S \backslash \{a\}, \tag{1.11}$$

then, by Corollary 1.2, $S \backslash \{a\}$ is $\mathscr{O}_{\text{seg}}$-convex, and hence, clearly, a maximal $\mathscr{O}_{\text{seg}}$-convex subset of $S \backslash \{a\}$.

If $a \in S$ is $\rho$-maximal, i.e., if

$$M'_{a\rho} = \{x \in S | a\bar{\rho}x\} = S \backslash \{a\}, \tag{1.12}$$

the argument is similar.

PROPOSITION 1.4. *For any $\rho \in \mathscr{O}$, the family of all subsets $M$ (respectively, $N$) of $S$ which satisfy* (1.6) (*respectively,* (1.7)) *is closed under union and intersection. Hence, the union and the intersection of any family of subsets of $S$ all of whose members satisfy* (1.6), *or* (1.7), *is $\mathscr{O}_{\text{seg}}$-convex.*

We omit the easy proof.

COROLLARY 1.4. *For any set* $A \subseteq S$ *and any* $\rho \in \mathscr{O}$, *we have the following pairs of complementary* $\mathscr{O}_{\mathrm{seg}}$*-hemispaces*:

$$(1.13)\quad \left(\bigcup_{a\in A} M_{a\rho}, \bigcap_{a\in A} C'_{a\rho}\right), \qquad \left(\left(\bigcup_{a\in A} M_{a\rho}\right)\cup A, \bigcap_{a\in A}(C'_{a\rho}\backslash\{a\})\right),$$

$$(1.14)\quad \left(\bigcup_{a\in A} M'_{a\rho}, \bigcap_{a\in A} C_{a\rho}\right), \qquad \left(\left(\bigcup_{a\in A} M'_{a\rho}\right)\cup A, \bigcap_{a\in A}(C_{a\rho}\backslash\{a\})\right),$$

$$(1.15)\quad \left(\bigcap_{a\in A} M_{a\rho}, \bigcup_{a\in A} C'_{a\rho}\right), \qquad \left(\bigcap_{a\in A} M'_{a\rho}, \bigcup_{a\in A} C_{a\rho}\right).$$

PROOF. This follows from Proposition 1.4 and Corollary 1.2, observing also that the pairs in (1.13)–(1.15) are complementary. Indeed, e.g., for (1.13), we have

$$S\Big\backslash\left(\bigcup_{a\in A} M_{a\rho}\right) = \bigcap_{a\in A}(S\backslash M_{a\rho}) = \bigcap_{a\in A} C'_{a\rho},$$

$$S\Big\backslash\left(\left(\bigcup_{a\in A} M_{a\rho}\right)\cup A\right) = S\Big\backslash\left(\bigcup_{a\in A}(M_{a\rho}\cup\{a\})\right) = \bigcap_{a\in A}(C'_{a\rho}\backslash\{a\}).$$

COROLLARY 1.5. *If* $A \subseteq S$ *is an antichain* (*i.e., a set of pairwise incomparable elements*) *for some* $\rho \in \mathscr{O}$, *then the sets*

$$(1.16)\qquad \left(\bigcap_{a\in A} M_{a\rho}\right)\cup A, \qquad \left(\bigcap_{a\in A} M'_{a\rho}\right)\cup A$$

*are* $\mathscr{O}_{\mathrm{seg}}$*-hemispaces.*

PROOF. One can show that $(\bigcap_{a\in A} M_{a\rho})\cup A$ satisfies (1.6) and its complement satisfies (1.7). Hence, by Proposition 1.3, $(\bigcap_{a\in A} M_{a\rho})\cup A$ is an $\mathscr{O}_{\mathrm{seg}}$-hemispace.

Finally, the result for $(\bigcap_{a\in A} M'_{a\rho})\cup A$ follows by assuming that $\rho^{-1}\in\mathscr{O}$ (see Remark 1.1(b)) and applying the above to $\rho^{-1}$.

PROPOSITION 1.5. *Assume that each* $\rho \in \mathscr{O}$ *is a total order on* $S$ *and, for each* $a, x \in S$ *with* $x \neq a$, *let*

$$(1.17)\quad \mathscr{P}_{a,x} = \{\rho\in\mathscr{O}|x\rho a\}, \qquad \mathscr{P}'_{a,x} = \{\rho\in\mathscr{O}|a\rho x\} = \mathscr{O}\backslash\mathscr{P}_{a,x} = \mathscr{P}_{x,a}.$$

*Then, for any* $a, x \in S$ *with* $x \neq a$, *the sets*

$$(1.18)\quad D_{a,x} = \{y\in S\backslash\{a\}|\mathscr{P}_{a,y} = \mathscr{P}_{a,x}\}, \qquad D'_{a,x} = \{y\in S\backslash\{a\}|\mathscr{P}_{a,y} = \mathscr{P}'_{a,x}\}$$

*and* $D_{a,x}\cup\{a\}$, $D'_{a,x}\cup\{a\}$ *are* $\mathscr{O}_{\mathrm{seg}}$*-convex.*

PROOF. One can check that if $z, t\in D_{a,x}$ and $u\in\langle z, t\rangle$, then $u\in D_{a,x}$, and that if $z\in D_{a,x}$ and $u\in\langle z, a\rangle$, $u\neq a$, then $u\in D_{a,x}\subset D_{a,x}\cup\{a\}$. For the sets $D'_{a,x}$ and $D'_{a,x}\cup\{a\}$ the proofs are similar. We omit the details.

PROPOSITION 1.6. *The pair* $(S, \mathscr{C}_{\text{seg}})$, *where* $\mathscr{C}_{\text{seg}}$ *is the family of all* $\mathscr{O}_{\text{seg}}$-*convex subsets of* $S$, *is an aligned space in the sense of* [**8**], *i.e.*: A1. $\varnothing$ *and* $S$ *are* $\mathscr{O}_{\text{seg}}$-*convex.*

A2. *An arbitrary intersection of* $\mathscr{O}_{\text{seg}}$-*convex sets is* $\mathscr{O}_{\text{seg}}$-*convex.*

A3. *The union of any family of* $\mathscr{O}_{\text{seg}}$-*convex sets totally ordered by inclusion is* $\mathscr{O}_{\text{seg}}$-*convex.*

PROOF. A1 has been observed before Proposition 1.2. A2 and A3 hold for any "segmental convexity" (i.e., defined by(1.3), for some concept of "segment" $\langle x, y\rangle$).

By Proposition 1.6, one can apply to the pair $(S, \mathscr{C}_{\text{seg}})$ the theory of aligned spaces (see [**8**]). In particular, for any set $G \subseteq S$, the $\mathscr{O}_{\text{seg}}$-*convex hull* $\text{co}_{\mathscr{O}_{\text{seg}}} G$ of $G$ is defined as the smallest $\mathscr{O}_{\text{seg}}$-convex set containing $G$.

PROPOSITION 1.7. *We have*

$$\text{co}_{\mathscr{O}_{\text{seg}}}\{x, y\} = \langle x, y\rangle \qquad (x, y \in S). \tag{1.19}$$

PROOF. By Proposition 1.2, we have the inclusion $\subseteq$ in (1.19). The opposite inclusion $\supseteq$ follows from (1.3).

COROLLARY 1.6. *If there exists* $\rho \in \mathscr{O}$ *which is not total, then there exists no family* $\mathscr{T}$ *of total orders on* $S$, *such that* $\mathscr{T}_{\text{seg}}$-*convexity in* $S$ *coincides with* $\mathscr{O}_{\text{seg}}$-*convexity in* $S$.

PROOF. If $\mathscr{T}$ is any family of orders on $S$, such that $\mathscr{T}_{\text{seg}}$-convexity in $S$ coincides with $\mathscr{O}_{\text{seg}}$-convexity in $S$, then, for any subset $G$ of $S$, the $\mathscr{T}_{\text{seg}}$-convex hull of $G$ coincides with the $\mathscr{O}_{\text{seg}}$-convex hull of $G$, and hence, in particular, by (1.19),

$$\langle x, y\rangle_{\mathscr{T}} = \text{co}_{\mathscr{T}_{\text{seg}}}\{x, y\} = \text{co}_{\mathscr{O}_{\text{seg}}}\{x, y\} = \langle x, y\rangle_{\mathscr{O}} \qquad (x, y \in S). \tag{1.20}$$

But, since there exists an order $\rho \in \mathscr{O}$ which is not total, from Corollary 1.1 it follows that there exists no family $\mathscr{T}$ of total orders on $S$, satisfying (1.20).

## 2. Segmentally convex sets in $(R^n, \mathrm{T})$

Let us first show that the concept of order convexity is not sufficient to encompass, as a particular case, the usual (vector) convexity in $R^n$.

PROPOSITION 2.1. *There exists no order relation* $\leq$ *on* $R^n$ $(n \geq 2)$ *for which the order convex sets are the usual convex sets.*

PROOF. We recall (see, e.g., [**2**]) that, for any poset $(S, \leq)$, the *order segments* of $(S, \leq)$ are defined by

$$[x, y] = \{z \in S | x \leq z \leq y\} \qquad (x, y \in S), \tag{2.1}$$

and that a subset $G$ of $S$ is said to be *order convex* if

$$[x, y] \subseteq G \qquad (x, y \in G). \tag{2.2}$$

Clearly, for any $x \neq y$ in $S$ we have either $[x, y] = \varnothing$ or $[y, x] = \varnothing$, and hence $[x, y] \cup [y, x]$ is an order convex set containing $x$ and $y$; on the other hand, by (2.2), for any order convex set $G$ containing $x$ and $y$, we have $[x, y] \cup [y, x] \subseteq G$. Thus,

$$\mathrm{co}_{\leq}\{x, y\} = [x, y] \cup [y, x] \qquad (x, y \in S), \tag{2.3}$$

where $\mathrm{co}_{\leq}\{x, y\}$ denotes the order convex hull of $\{x, y\}$.

Assume now, a contrario, that there exists an order $\leq$ on $R^n$ such that the order convex sets are the usual convex sets. Then, by (2.3), we obtain, as in the above proof of Corollary 1.6, that

$$[x, y] \cup [y, x] = \mathrm{co}_{\leq}\{x, y\} = \mathrm{co}_{R^n}\{x, y\} \qquad (x, y \in R^n), \tag{2.4}$$

where $\mathrm{co}_{R^n}$ denotes the usual convex hull in $R^n$. Let $x, y, z, t \in R^n$ be any four distinct points with $z \in \mathrm{co}_{R^n}\{x, y\}$. We may assume that $\mathrm{co}_{R^n}\{x, y\} = [x, y]$, whence $z \in [x, y]$, i.e., $x \leq z \leq y$. Since $[z, t] \cup [t, z] = \mathrm{co}_{R^n}\{z, t\} \neq \varnothing$, we have either $z \leq t$ or $t \leq z$, whence either $x \leq z \leq t$, so $z \in [x, t] = \mathrm{co}_{R^n}\{x, t\}$, or $t \leq z \leq y$, so $z \in [t, y] = \mathrm{co}_{R^n}\{t, y\}$. Thus, in either case, $t$ belongs to the line determined by $x$, $y$, and $z$, contradicting the arbitrariness of $t \in R^n \backslash \{x, y, z\}$.

We recall that an order $\rho$ on $R^n$ is said to be *compatible with the vector space structure of* $R^n$, or, briefly, *compatible*, if

$$y_1 \rho y_1', \ y_2 \rho y_2' \Rightarrow y_1 + y_2 \rho y_1' + y_2', \tag{2.5}$$

$$y \rho y', \ \lambda \geq 0 \Rightarrow \lambda y \rho \lambda y'. \tag{2.6}$$

We shall consider the multi-ordered set $(R^n, \mathrm{T})$, where

$$\mathrm{T} = \text{the set of all compatible total orders on } R^n. \tag{2.7}$$

Let us recall the following result of [**18**].

THEOREM 2.1 [**18, Theorem 1.2**]. *For any* $\rho \in \mathrm{T}$ *there exists a unique* $v \in O(R^n)$ *such that*

$$y \rho y' \Leftrightarrow v(y) \leq_L v(y') \qquad (y, y' \in R^n). \tag{2.8}$$

*Conversely, given any* $v \in O(R^n)$, *if we define* $\rho = \rho_v$ *by* (2.8), *then* $\rho_v \in \mathrm{T}$. *Consequently, the mapping*

$$\delta : v \to \rho_v \tag{2.9}$$

*is a bijection of* $O(R^n)$ *onto* $\mathrm{T}$.

Now we can prove the main result of this section.

THEOREM 2.2. *In* $(R^n, \mathrm{T})$, *we have*

$$\langle x, y \rangle = \{(1-\lambda)x + \lambda y | \lambda \in [0, 1]\} \qquad (x, y \in R^n), \tag{2.10}$$

*i.e., the* T*-segments of* $R^n$ *coincide with the usual segments of* $R^n$. *Hence,* $\mathrm{T}_{\mathrm{seg}}$*-convexity in* $R^n$ *coincides with the usual convexity in* $R^n$.

PROOF. Let $z \in \langle x, y\rangle$ and assume, a contrario, that $z \notin \{(1-\lambda)x + \lambda y | \lambda \in [0, 1]\}$. Then, by the lexicographical separation theorem **[15, p. 258]**, there exists $v \in O(R^n)$ such that

$$v(z) <_L v((1-\lambda)x + \lambda y) \qquad (\lambda \in [0, 1]), \tag{2.11}$$

whence, in particular (for $\lambda = 0$ and $\lambda = 1$), $v(z) <_L v(x)$ and $v(z) <_L v(y)$, that is, $z\rho_v x$ and $z\rho_v y$, for $\rho_v \in \mathrm{T}$ of Theorem 2.1. On the other hand, since $z \in \langle x, y\rangle$, we have either $x\rho_v z\rho_v y$ or $y\rho_v z\rho_v x$, whence, by the antisymmetry of $\rho_v$, either $z = x$ or $z = y$, contradicting $z \notin \{(1-\lambda)x + \lambda y | \lambda \in [0, 1]\}$. Thus, we have the inclusion $\subseteq$ in (2.10).

Conversely, let $z = (1-\lambda)x + \lambda y$, where $\lambda \in [0, 1]$, and, for any $\rho \in \mathrm{T}$, let $v = \delta^{-1}(\rho) \in O(R^n)$ of Theorem 2.1. Then, since $\leq_L$ belongs to T, we have either $v(x) \leq_L v(y)$, or $v(y) \leq_L v(x)$. Hence, again since $\leq_L$ is in T, if $v(x) \leq_L v(y)$, then

$$v(x) \leq_L v((1-\lambda)x + \lambda y) = v(z) \leq_L v(y), \tag{2.12}$$

while if $v(y) \leq_L v(x)$, then

$$v(y) \leq_L v((1-\lambda)x + \lambda y) = v(z) \leq_L v(x). \tag{2.13}$$

Therefore (by (2.8)), we have either $x\rho z\rho y$ or $y\rho z\rho x$, whence, since $\rho \in \mathrm{T}$ has been arbitrary, $z \in \langle x, y\rangle$. Thus, we also have the inclusion $\supseteq$ in (2.10), and hence the equality.

REMARK 2.1. Since the hemispaces and semispaces in $R^n$ are known (see **[17, 5]**), so are, by Theorem 2.2, the $\mathrm{T}_{\mathrm{seg}}$-hemispaces and $\mathrm{T}_{\mathrm{seg}}$-semispaces. In the converse direction, from Corollary 1.2 and Theorem 2.1 it follows that for each $v \in O(R^n)$ and $a \in R^n$, the sets

$$M'_{a\rho_v} = \{x \in R^n | a\overline{\rho}_v x\} = \{x \in R^n | v(x) <_L v(a)\}, \tag{2.14}$$

$$C'_{a\rho_v} = \{x \in R^n | x\rho_v a\} = \{x \in R^n | v(x) \leq_L v(a)\} \tag{2.15}$$

are hemispaces in $R^n$, and, from **[17, Theorem 2.1]**, it follows that these exhaust all hemispaces " of rank $n$ ", i.e., all semispaces and all complements of semispaces in $R^n$. However, there are also other hemispaces in $R^n$, which might perhaps be encompassed by a more general theory of "*multiply preordered sets.*"

PROPOSITION 2.2. *In* $(R^n, \mathrm{T})$, *for any* $a \in R^n$ *and* $x \neq a$, *the sets* $D_{a,x}$ *and* $D'_{a,x}$ *of* (1.18) *are the open half-lines*

$$D_{a,x} = \{a + \lambda(x-a) | \lambda > 0\}, \qquad D'_{a,x} = \{a + \lambda(x-a) | \lambda < 0\}. \tag{2.16}$$

We omit the proof.

## 3. Segmentally convex sets in $(Z^n, \mathrm{T}')$

Now we shall consider the multi-ordered set $(Z^n, \mathrm{T}')$, where

$$\mathrm{T}' = \mathrm{T}|_{Z^n} = \{\rho|_{Z^n} | \rho \in \mathrm{T}\}, \tag{3.1}$$

with T of (2.7). By (3.1), the $\mathrm{T}'$-segments in $Z^n$ are

$$\langle x, y\rangle_{\mathrm{T}'} = Z^n \cap \langle x, y\rangle_{\mathrm{T}} \qquad (x, y \in Z^n). \tag{3.2}$$

Therefore, from Theorem 2.2 we obtain

THEOREM 3.1. *In* $(Z^n, \mathrm{T}')$ *we have*

$$\langle x, y\rangle_{\mathrm{T}'} = Z^n \cap \{(1-\lambda)x + \lambda y | \lambda \in [0, 1]\} \qquad (x, y \in Z^n). \tag{3.3}$$

*Hence, a subset* $G$ *of* $Z^n$ *is* $\mathrm{T}'_{\mathrm{seg}}$*-convex if and only if we have the implication*

$$x, y \in G, \ \lambda \in [0, 1], \quad (1-\lambda)x + \lambda y \in Z^n \Rightarrow (1-\lambda)x + \lambda y \in G. \tag{3.4}$$

REMARK 3.1. The sets $G \subseteq Z^n$ with property (3.4) have been called, in **[12]**, *2-convex sets.*

## 4. Segmentally convex sets in multi-ordered sets associated to posets

For any poset $(S, \leq)$, it is natural to consider the multi-ordered sets $(S, \{\leq\})$ and $(S, \{\leq, \geq\})$; note $\geq$ is nothing else than $\leq^{-1}$, in the sense of (1.4), so Remark 1.1(b) applies.

PROPOSITION 4.1. *For a poset* $(S, \leq)$, *the multi-ordered sets* $(S, \{\leq\})$ *and* $(S, \{\leq, \geq\})$, *and any* $x, y \in S$, *we have (with* $[x, y]$ *of* (2.1))

$$\begin{aligned} \langle x, y\rangle_{\{\leq\}} &= [x, y] \cup [y, x] \\ &= \left\{ \begin{array}{ll} [x, y], & \text{if } x \leq y \\ [y, x], & \text{if } y \leq x \\ \varnothing, & \text{if } x \not\leq y \text{ and } y \not\leq x \end{array} \right\} = \langle x, y\rangle_{\{\leq, \geq\}}. \end{aligned} \tag{4.1}$$

*Hence, for a subset* $G$ *of* $S$ *the following statements are equivalent*:

$1^\circ$. $G$ *is* $\{\leq\}_{\mathrm{seg}}$*-convex.*
$2^\circ$. $G$ *is* $\{\leq, \geq\}_{\mathrm{seg}}$*-convex.*
$3^\circ$. $G$ *is order convex.*

PROOF. Obvious from the definitions.

Some further remarks on posets and related multi-ordered sets are collected in

REMARK 4.1. (a) Given any multi-ordered set $(S, \mathscr{T})$, where $\mathscr{T}$ is a family of total orders on $S$, one can define on order $\leq_{\mathscr{T}}$ on $S$ by

$$x \leq_{\mathscr{T}} y \Leftrightarrow x \rho y \qquad (x, y \in S, \ \rho \in \mathscr{T}). \tag{4.2}$$

However, then segmental convexity in $(S, \mathcal{T})$ need not coincide with order convexity in $(S, \leq_{\mathcal{T}})$ and, in fact, $\leq_{\mathcal{T}}$ may be the "trivial order" on $S$, in which

$$(4.3) \qquad x \leq_{\mathcal{T}} y \Leftrightarrow x = y \qquad (x, y \in S);$$

indeed, in particular, if $\mathcal{T}$ contains $\rho$ and $\rho^{-1}$ (for example, if $(S, \mathcal{T}) = (R^n, \mathrm{T})$, with T of (2.7)), then we have (4.3), and hence every set in $(S, \leq_{\mathcal{T}})$ is order convex.

(b) R. Jamison-Waldner [7] observed that in a poset $(S, \leq)$, at each $a \in S$ there exist at most two semispaces; namely, if $a \in S$ is not maximal and not minimal, then $M_{a\leq}$ and $M'_{a\leq}$ of (1.8) above (with $\rho$ being $\leq$) are the only semispaces at $a$ (while if $a \in S$ is minimal or maximal, then Corollary 1.3 above applies). It is an unsolved problem to find the semispaces for segmental convexity in multi-ordered sets $(S, \mathcal{O})$.

(c) An aligned space $(S, \mathcal{L})$ (see Proposition 1.6 above) is said to be *determined by an order* [**6**] if there exists an order $\leq$ on $S$ such that $\mathcal{L}$ coincides with the family of all order convex sets in $(S, \leq)$. In [**6**], R. Jamison-Waldner gave necessary and sufficient conditions for an aligned space $(S, \mathcal{L})$ to be determined by an order (in terms of "natural" properties of aligned spaces). It is an unsolved problem to extend this result to an aligned space $(S, \mathcal{L})$ "determined by a family of orders" (replacing order convex sets by $\mathcal{O}_{\mathrm{seg}}$-convex sets, in the above definition).

In connection with Remark 4.1(b), let us give the following result on posets, which may have some interest for applications.

PROPOSITION 4.2. *In any poset* $(S, \leq)$, *the family*

$$(4.4) \qquad \mathcal{M} = \{M_{a\leq} | a \in S\} \cup \{M'_{a\leq} | a \in S\} \cup S$$

*is an intersectional basis for the family of all order convex sets, i.e., for every order convex set* $G \subseteq S$ *we have*

$$(4.5) \qquad G = \bigcap_{\substack{M \in \mathcal{M} \\ G \subseteq M}} M.$$

PROOF. Let $G \subseteq S$ be order convex. Since the inclusion $\subseteq$ in (4.5) is obvious, it will be enough to show that for each $a \notin G$ there exists $M \in \mathcal{M}$, namely, either $M = M_{a\leq}$ or $M = M'_{a\leq}$, such that $G \subseteq M$ (indeed, then, by $a \notin M$, $a$ does not belong to the right-hand side of (4.5)). If, a contrario, $G \not\subseteq M_{a\leq}$ and $G \not\subseteq M'_{a\leq}$, then there exist $g_1, g_2 \in G$ with $g_1 \leq a \leq g_2$, whence, since $G$ is order convex, it follows that $a \in G$, a contradiction.

REMARK 4.2. It is an unsolved problem to find an intersectional basis for the family of all segmentally convex sets in a multi-ordered set.

Finally, let us give a characterization of the family of all order segments in a poset, which will imply again Proposition 2.1, and may have some other applications.

PROPOSITION 4.3. *Let $S$ be a set and let $\{\langle a, b\rangle\}_{a,b\in S}$ be a family of subsets of $S$. The following conditions are equivalent:*

$1^\circ$. *There exists an order relation $\leq$ on $S$ such that*

$$[x, y] = \langle x, y\rangle \qquad (x, y \in S), \tag{4.6}$$

*where $[x, y]$ is the order segment* (2.1) *in $(S, \leq)$.*

$2^\circ$. *We have*

$$\langle x, x\rangle = \{x\} \qquad (x \in S), \tag{4.7}$$

$$\langle x, y\rangle = \{z \in S | \langle x, z\rangle \neq \emptyset, \langle z, y\rangle \neq \emptyset\} \qquad (x, y \in S). \tag{4.8}$$

*Moreover, in this case the order $\leq$ of $1^\circ$ is unique, namely, it is the order defined, for any $x, y \in S$, by*

$$x \leq y \Leftrightarrow \langle x, y\rangle \neq \emptyset. \tag{4.9}$$

PROOF. $1^\circ \Rightarrow 2^\circ$. Assume $1^\circ$. Then, by (4.6), for any $x \in S$ we have

$$z \in \langle x, x\rangle \Leftrightarrow x \leq z \leq x \Leftrightarrow z = x \Leftrightarrow z \in \{x\},$$

which proves (4.7). Now, let $x, y \in S$. Then, by (4.6),

$$z \in \langle x, y\rangle \Rightarrow x \leq z \leq y \Rightarrow x, z \in \langle x, z\rangle, \quad z, y \in \langle z, y\rangle,$$

which proves the inclusion $\subseteq$ in (4.8). Conversely, let $z \in S$ be such that $\langle x, z\rangle \neq \emptyset, \langle z, y\rangle \neq \emptyset$, and take any $t \in \langle x, z\rangle, s \in \langle z, y\rangle$. Then, by (4.6), we have $x \leq t \leq z \leq s \leq y$, whence, again by (4.6), $z \in \langle x, y\rangle$. This proves the inclusion $\supseteq$ in (4.8), and hence the equality.

$2^\circ \Rightarrow 1^\circ$. Assume $2^\circ$ and define a binary relation $\leq$ on $S$ by (4.9). Then, by (4.9), (4.7), and (4.8), $\leq$ is an order relation on $S$, satisfying (4.6).

Finally, in order to prove the uniqueness statement, let $\leq$ be any order on $S$ satisfying (4.6). Then, for any $x, y \in S$, we have

$$x \leq y \Rightarrow x, y \in [x, y] = \langle x, y\rangle \Rightarrow \langle x, y\rangle \neq \emptyset,$$

and, conversely,

$$\langle x, y\rangle \neq \emptyset \Leftrightarrow \exists z \in \langle x, y\rangle = [x, y] \Rightarrow x \leq y,$$

which proves (4.9).

REMARK 4.3. (a) Proposition 4.3 implies again Proposition 2.1, by (2.4) and since the family of all usual segments $\mathrm{co}_{R^n}\{x, y\}$ in $R^n$ does not satisfy (4.8).

(b) It is an unsolved problem to find a corresponding characterization of the family of all $\mathscr{O}$-segments in a multi-ordered set $(S, \mathscr{O})$.

## 5. Separational multi-order convexity

DEFINITION 5.1. Let $(S, \mathscr{O})$ be a multi-ordered set. We call a set $G \subseteq S$, "*$\mathscr{O}$-separationally convex,*" or, briefly, *$\mathscr{O}_{\mathrm{sep}}$-convex,* if for each $x_0 \in S \backslash G$

there exists an order $\rho = \rho(x_0) \in \mathscr{O}$ such that (where $\overline{\rho}$ denotes the negation of $\rho$)

$$x_0 \overline{\rho} g \qquad (g \in G). \tag{5.1}$$

REMARK 5.1. (a) In other words, $G$ is $\mathscr{O}_{\text{sep}}$-convex if every outside point $x_0$ can be "*separated from* $G$ *by some order* $\rho = \rho(x_0) \in \mathscr{O}$"; this motivates our terminology. Note also that this definition is of different type from that of $\mathscr{O}_{\text{seg}}$-convexity, which involved all $\rho \in \mathscr{O}$; also, $\mathscr{O}_{\text{sep}}$-convexity is a "one-sided" concept (see (5.1)), while $\mathscr{O}_{\text{seg}}$-convexity is a "symmetric" concept (see (1.5)).

(b) When all $\rho \in \mathscr{O}$ are total, a set $G \subseteq S$ is $\mathscr{O}_{\text{sep}}$-convex if and only if for each $x_0 \in S \backslash G$ there exists $\rho = \rho(x_0) \in \mathscr{O}$ such that

$$g \rho x_0 \qquad (g \in G). \tag{5.2}$$

(c) In contrast with the case of $\mathscr{O}_{\text{seg}}$-convexity (see Remark 1.1(b)), if $\mathscr{O}' = \mathscr{O} \cup \{\rho^{-1}\}$, for some $\rho \in \mathscr{O}$, then $\mathscr{O}'_{\text{sep}}$-convexity need not coincide with $\mathscr{O}_{\text{sep}}$-convexity (see Remark 5.2 and Proposition 5.2).

(d) In contrast with the case of $\mathscr{O}_{\text{seg}}$-convexity, a singleton $\{x\}$, where $x \in S$, need not be $\mathscr{O}_{\text{sep}}$-convex. Indeed, for example, when $\mathscr{O} = \{\rho\}$ (a singleton) and $x_0 \rho x$, $x_0 \neq x$, then $\{x\}$ is not $\mathscr{O}_{\text{sep}}$-convex. Also, an $\mathscr{O}$-segment $\langle x, y \rangle$ need not be $\mathscr{O}_{\text{sep}}$-convex. Indeed, for example, if $\mathscr{O} = \{\rho\}$ and $y \rho x$, $y \neq x$, then $\langle x, y \rangle = \{z \in S | y \rho z \rho x\}$. Hence, if $x_0 \rho y$, $x_0 \neq y$, then $x_0 \notin \langle x, y \rangle$, $x \in \langle x, y \rangle$, and $x_0 \rho y \rho x$, so (5.1) (with $G = \langle x, y \rangle$) is not satisfied, and therefore $\langle x, y \rangle$ is not $\mathscr{O}_{\text{sep}}$-convex. Thus, *$\mathscr{O}_{\text{seg}}$-convexity does not imply $\mathscr{O}_{\text{sep}}$-convexity.*

(e) Clearly, the whole set $S$ and the empty set $\varnothing$ are $\mathscr{O}_{\text{sep}}$-convex.

PROPOSITION 5.1. *For any $\rho \in \mathscr{O}$, each set $N \subseteq S$ satisfying* (1.7) *is $\mathscr{O}_{\text{sep}}$-convex.*

PROOF. Let $x_0 \notin N$. If (5.1) (with $G = N$) does not hold *for the given* $\rho$, then there exists $x' \in N$ such that $x_0 \rho x'$, whence, by (1.7), we obtain $x_0 \in N$, contradicting our assumption. Thus, for any $x_0 \notin N$ and for the given $\rho$ (which does not depend on $x_0$), we have (5.1) (with $G = N$).

COROLLARY 5.1. *For any set $A \subseteq S$ and any $\rho \in \mathscr{O}$, the sets $\bigcap_{a \in A} C'_{a\rho}$, $\bigcap_{a \in A}(C'_{a\rho} \backslash \{a\})$, $\bigcup_{a \in A} M'_{a\rho}$, $(\bigcup_{a \in A} M'_{a\rho}) \cup A$, $\bigcup_{a \in A} C'_{a\rho}$, and $\bigcap_{a \in A} M'_{a\rho}$ of* (1.13)–(1.15) *are $\mathscr{O}_{\text{sep}}$-convex.*

PROOF. By the arguments of §1, these sets satisfy (1.7), and hence, by Proposition 5.1, they are $\mathscr{O}_{\text{sep}}$-convex.

REMARK 5.2. The results corresponding to Proposition 1.3(a) and its consequences given in §1 do not hold, in general, for $\mathscr{O}_{\text{sep}}$-convexity. Moreover, let us note that if $\mathscr{O} = \{\rho\}$, where $\rho$ is total, then no set $M$ (different from $\varnothing$ and from $S$) satisfying (1.6) is $\mathscr{O}_{\text{sep}}$-convex; indeed, if $x_0 \notin M$

and (5.1) holds (with $G = M$), then, since $\rho$ is total, we have $x\rho x_0$ for all $x \in M$, whence, by (1.6), we obtain $x_0 \in M$, contradicting our assumption. However, for any multi-ordered set $(S, \mathscr{O})$, we have

PROPOSITION 5.2. *If for some* $\rho \in \mathscr{O}$ *we have* $\rho^{-1} \in \mathscr{O}$, *then each set* $M \subseteq S$ *satisfying* (1.6) *is* $\mathscr{O}_{\text{sep}}$*-convex.*

PROOF. Let $x_0 \notin M$. We shall show that

$$x_0 \overline{\rho^{-1}} x \qquad (x \in M), \tag{5.3}$$

which, since $\rho^{-1} \in \mathscr{O}$, will prove that $M$ is $\mathscr{O}_{\text{sep}}$-convex. If (5.3) does not hold, then there exists $x' \in M$ such that $x_0\rho^{-1}x'$, or, equivalently, $x'\rho x_0$. Hence, by (1.6), we obtain $x_0 \in M$, contradicting our assumption. Thus, $M$ is $\mathscr{O}_{\text{sep}}$-convex.

DEFINITION 5.2. An $\mathscr{O}_{\text{sep}}$*-hemispace* is a set $M \subseteq S$ such that both $M$ and $S \backslash M$ are $\mathscr{O}_{\text{sep}}$-convex.

COROLLARY 5.2. *If for some* $\rho \in \mathscr{O}$ *we have* $\rho^{-1} \in \mathscr{O}$, *then, for any set* $A \subseteq S$, (1.13)–(1.15) *are pairs of complementary* $\mathscr{O}_{\text{sep}}$*-hemispaces. If, in addition,* $A$ *is an antichain for* $\rho$, *then the sets* (1.16) *are* $\mathscr{O}_{\text{sep}}$*-hemispaces.*

PROOF. By the above proofs of Corollaries 1.4 and 1.5, these sets satisfy (1.6) or (1.7), so the result follows from Propositions 5.1 and 5.2.

PROPOSITION 5.3. *An arbitrary intersection of* $\mathscr{O}_{\text{sep}}$*-convex sets is* $\mathscr{O}_{\text{sep}}$*-convex.*

We omit the easy proof.

REMARK 5.3. We do not know under what conditions on $\mathscr{O}$ is the union of any family of $\mathscr{O}_{\text{sep}}$-convex sets, totally ordered by inclusion, again $\mathscr{O}_{\text{sep}}$-convex, i.e., under what conditions is the pair $(S, \mathscr{C}_{\text{sep}})$ an aligned space, where $\mathscr{C}_{\text{sep}}$ is the family of all $\mathscr{O}_{\text{sep}}$-convex subsets of $S$ (cf. Proposition 1.6). Nevertheless, Proposition 5.3 permits us to define the $\mathscr{O}_{\text{sep}}$-convex hull $\text{co}_{\mathscr{O}_{\text{sep}}} G$ of a set $G \subseteq S$ as the smallest $\mathscr{O}_{\text{sep}}$-convex set containing $G$.

Now we shall show that, as concerns total orders, the situation for $\mathscr{O}_{\text{sep}}$-convexity is better than the one for $\mathscr{O}_{\text{seg}}$-convexity, described in Corollary 1.6. To this end, we need some preparation.

Let us recall that an order $\rho_2$ on a set $S$ is said to be an *extension* of an order $\rho_1$ on $S$ if

$$y\rho_2 y' \qquad (y, y' \in S, y\rho_1 y'), \tag{5.4}$$

and that, by a classical theorem of Szpilrajn [**21**], every order relation $\rho$ on a set $S$ admits a total extension $\tau$ on $S$.

LEMMA 5.1. *Let* $\rho$ *be an order relation on a set* $S$ *and let* $G \subseteq S$ *and* $x_0 \in S \backslash G$ *satisfy* (5.1). *Then there exists a total extension* $\tau$ *of* $\rho$ *on* $S$, *such that*

$$g\tau x_0 \qquad (g \in G). \tag{5.5}$$

PROOF. Let

$$[G] = \{x \in S | \exists g \in G,\ x\rho g\}. \tag{5.6}$$

Then, by (5.1), $x_0 \in S \backslash [G]$. By Szpilrajn's theorem, there exist total order relations $\tau_1, \tau_2$ on $[G]$ and $S \backslash [G]$, respectively, extending the restrictions of $\rho$ to these sets. Let us define a binary relation $\tau$ on $S$ by saying that $x\tau y$ if one of the following conditions holds:

(a) $x, y \in [G]$ and $x\tau_1 y$;
(b) $x, y \in S\backslash[G]$ and $x\tau_2 y$;
(c) $x \in [G]$ and $y \in S\backslash[G]$.

Then, it is easy to see that $\tau$ is a total order on $S$ and $\tau$ is an extension of $\rho$, satisfying (5.5).

Now we can prove the result on total orders, announced above. Namely, in contrast with Corollary 1.6, we have

THEOREM 5.1. *For each multi-ordered set $(S, \mathscr{O})$ there exists a family $\mathscr{T}$ of total orders on $S$, such that $\mathscr{T}_{\text{sep}}$-convexity in $S$ coincides with $\mathscr{O}_{\text{sep}}$-convexity in $S$.*

PROOF. The family $\mathscr{T}$ defined by

$$\mathscr{T} = \bigcup_{\rho \in \mathscr{O}} \mathscr{T}_\rho, \tag{5.7}$$

where, for each $\rho \in \mathscr{O}$,

$$\mathscr{T}_\rho = \text{the family of all total extensions of } \rho, \tag{5.8}$$

has the required property. Indeed, each $\mathscr{T}_{\text{sep}}$-convex set $G$ is $\mathscr{O}_{\text{sep}}$-convex, since if $x_0 \notin G$ and $\tau \in \mathscr{T}$ satisfy $x_0 \bar{\tau} g$ $(g \in G)$, then for $\rho \in \mathscr{O}$ such that $\tau \in \mathscr{T}_\rho$, we have (5.1). Conversely, if $G$ is an $\mathscr{O}_{\text{sep}}$-convex set and if $x_0 \notin G$ and $\rho \in \mathscr{O}$ satisfy (5.1), then, by Lemma 5.1, there exists $\tau \in \mathscr{T}_\rho \subseteq \mathscr{T}$ satisfying (5.5), and hence, by Remark 5.1(b), $G$ is $\mathscr{T}_{\text{sep}}$-convex.

REMARK 5.4. By Theorem 5.1, in the study of $\mathscr{O}_{\text{sep}}$ convexity it is no restriction of generality to assume that each $\rho \in \mathscr{O}$ is total.

THEOREM 5.2. *$\mathscr{O}_{\text{sep}}$-convexity implies $\mathscr{O}_{\text{seg}}$-convexity.*

PROOF. Let $G$ be an $\mathscr{O}_{\text{sep}}$-convex subset of $S$, and assume, a contrario, that $G$ is not $\mathscr{O}_{\text{seg}}$-convex, i.e., that there exist $x, y \in G$ and $x_0 \in \langle x, y\rangle$, such that $x_0 \notin G$. Then, since $G$ is $\mathscr{O}_{\text{sep}}$-convex, there exists $\rho = \rho(x_0) \in \mathscr{O}$ satisfying (5.1), whence, in particular, $x_0 \bar{\rho} x$ and $x_0 \bar{\rho} y$ (since $x, y \in G$). But, by $x_0 \in \langle x, y\rangle$, for this $\rho$ we must have either $x\rho x_0 \rho y$, which contradicts $x_0 \bar{\rho} y$, or $y\rho x_0 \rho x$, which contradicts $x_0 \bar{\rho} x$.

REMARK 5.5. The converse implication need not hold, even when each $\rho \in \mathscr{O}$ is total, as shown by Remark 5.1(d). However, Theorems 2.2 and

6.1 and, respectively, Proposition 4.1 and Theorem 8.1, show two important cases when $\mathscr{O}_{\text{sep}}$-convexity coincides with $\mathscr{O}_{\text{seg}}$-convexity. In the general case, we do not know under what conditions on $\mathscr{O}$ is every $\mathscr{O}_{\text{seg}}$-convex set $\mathscr{O}_{\text{sep}}$-convex.

## 6. Separationally convex sets in $(R^n, \mathrm{T})$

Let us consider the multi-ordered set $(R^n, \mathrm{T})$, with $\mathrm{T}$ of (2.7) (so every $\rho \in \mathrm{T}$ is total).

THEOREM 6.1. *$\mathscr{O}_{\text{sep}}$-convexity in $R^n$ coincides with the usual convexity in $R^n$.*

PROOF. If $G \subseteq R^n$ is $\mathscr{O}_{\text{sep}}$-convex, then, by Theorem 5.1, $G$ is $\mathscr{O}_{\text{seg}}$-convex, and hence, by Theorem 2.2, $G$ is convex.

Conversely assume now that $G \subset R^n$ is convex, and let $x_0 \in R^n \backslash G$. Then, by the lexicographical separation theorem **[15, p. 256]**, there exists $v \in O(R^n)$ such that

$$v(g) <_L v(x_0) \qquad (g \in G). \tag{6.1}$$

Let $\rho = \rho_v \in \mathrm{T}$ of Theorem 2.1. Then, by (6.1) and (2.8), we have $g \rho_v x_0$ for all $g \in G$, and therefore, by Remark 5.1(b), $G$ is $\mathscr{O}_{\text{sep}}$-convex.

REMARK 6.1. Actually, Theorem 6.1 is equivalent to the lexicographical separation theorem.

## 7. Separationally convex sets in $(Z^n, \mathrm{T}')$

Let us consider now the multi-ordered set $(Z^n, \mathrm{T}')$ of §3 (so every $\rho \in \mathrm{T}'$ is total).

THEOREM 7.1. *For a subset $G$ of $Z^n$, the following statements are equivalent:*

$1^\circ$. *$G$ is $\mathrm{T}'_{\text{sep}}$-convex.*

$2^\circ$. *There exists a convex subset $C$ of $R^n$, such that*

$$G = Z^n \cap C. \tag{7.1}$$

$3^\circ$. *We have*

$$G = Z^n \cap \operatorname{co}_{R^n} G, \tag{7.2}$$

*where $\operatorname{co}_{R^n}$ denotes the (usual) convex hull in $R^n$.*

$4^\circ$. *For each $x_0 \in Z^n \backslash G$ there exists $v = v_{x_0} \in O(R^n)$ such that*

$$v(g) <_L v(x_0) \qquad (g \in G). \tag{7.3}$$

PROOF. $1^\circ \Rightarrow 4^\circ$. If $1^\circ$ holds and $x_0 \in Z^n \backslash G$, then, by Remark 5.1(b), there exists $\rho = \rho(x_0) \in \mathrm{T}'$ such that we have (5.2). By (3.1), let $\rho_0 \in \mathrm{T}$ be such that $\rho = \rho_0|_{Z^n}$ and, by Theorem 2.1, let $v = \delta^{-1}(\rho_0) \in O(R^n)$. Then, by (5.2), (2.8), and $x_0 \notin G$, we have (7.3).

$4^\circ \Rightarrow 3^\circ$. Since the inclusion $\subseteq$ in (7.2) is obvious, it is enough to show that $4^\circ$ implies the inclusion $\supseteq$ in (7.2). Assume $4^\circ$ and let $x_0 \in Z^n \cap \mathrm{co}_{R^n} G$. If $x_0 \notin G$, then by (7.3), we obtain $x_0 \notin \mathrm{co}_{R^n} G$, a contradiction.

The implication $3^\circ \Rightarrow 2^\circ$ is obvious.

$2^\circ \Rightarrow 1^\circ$. Assume $2^\circ$ and let $x_0 \in Z^n \backslash G$. Then, by (7.1), we have $x_0 \notin C$, and, by Theorem 6.1, $C$ is $\mathrm{T}_{\mathrm{sep}}$-convex. Hence, by Remark 5.1(b), there exists $\rho_0 = \rho_0(x_0) \in \mathrm{T}$ such that

$$x \rho_0 x_0 \qquad (x \in C). \tag{7.4}$$

Consequently, for $\rho = \rho_0|_{Z^n} \in \mathrm{T}'$ we have (5.2) (since $G \subset C$ by (7.1)), and thus, by Remark 5.1(b), $G$ is $\mathrm{T}'_{\mathrm{sep}}$-convex.

REMARK 7.1. (a) The sets $G \subseteq Z^n$ satisfying (7.2) have been called, in **[12]**, *strong convex sets.* Moreover, by **[12, Proposition 3.1]**, condition $2^\circ$ is equivalent to

$5^\circ$. *$G$ is $p$-convex, where $p = \dim G + 1$, i.e., we have the implication*

$$x_1, \dots, x_p \in G,\ \lambda_1, \dots, \lambda_p \geq 0,\quad \sum_{i=1}^p \lambda_i = 1,\ \sum_{i=1}^p \lambda_i x_i \in Z^n \Rightarrow \sum_{i=1}^p \lambda_i x_i \in G. \tag{7.5}$$

Also, by **[12, Example 1.3]**, the subset $G = \{(1, 3), (0, 2), (2, 1)\}$ of $Z^2$ is 2-convex, but not 3-convex; hence, by the above, *$G$ is $\mathrm{T}'_{\mathrm{seg}}$-convex, but not $\mathrm{T}'_{\mathrm{sep}}$-convex, in $Z^2$.* For other related results, in a more general framework, see **[12]**.

(b) The equivalence $1^\circ \Leftrightarrow 4^\circ$ means that a subset $G$ of $Z^n$ is $\mathrm{T}'_{\mathrm{sep}}$-convex if and only if each outside point $x_0 \in Z^n \backslash G$ "can be separated from $G$ by a linear isometry $v \in O(R^n)$." Note that this is similar to a concept of "convexity," due to Ky Fan **[3]**, in which one separates (strictly) $x_0$ and $G$ by real-valued functions (instead of operators $v: R^n \to R^n$).

## 8. Separationally convex sets in multi-ordered sets associated to posets

For a poset $(S, \leq)$, let us consider again the multi-ordered sets $(S, \{\leq\})$ and $(S, \{\leq, \geq\})$ (see §4).

THEOREM 8.1. *Let $(S, \leq)$ be a poset. Then, a subset $G$ of $S$ is $\{\leq, \geq\}_{\mathrm{sep}}$-convex if and only if it is order convex.*

PROOF. Assume that $G$ is $\{\leq, \geq\}$-convex, but not order convex, so there exist $g_1, g_2 \in G$ and $x_0 \in S \backslash G$, with $g_1 \leq x_0 \leq g_2$. Then, since $G$ is $\{\leq, \geq\}$-convex, we have (5.1) either for $\rho$ being $\leq$, or for $\rho$ being $\geq$, i.e., either $x_0 \nleq g \ \ (g \in G)$, which contradicts $x_0 \leq g_2$, or $x_0 \ngeq g \ \ (g \in G)$, which contradicts $g_1 \leq x_0$.

Conversely, assume now that $G$ is order convex, but not $\{\leq, \geq\}$-convex, so there exists $x_0 \in S \backslash G$ such that neither $x_0 \nleq g \ \ (g \in G)$, nor $x_0 \ngeq g$

$(g \in G)$. Then, there exist $g_1, g_2 \in G$ such that $g_1 \leq x_0 \leq g_2$. Hence, since $G$ is order convex, we obtain $x_0 \in G$, which contradicts our assumption on $x_0$.

REMARK 8.1. By Theorem 5.1 and Proposition 4.1, every $\{\leq\}_{\text{sep}}$-convex set is order convex. However, the converse is not true, as shown e.g. by Remark 5.1 (d).

## REFERENCES

1. E. Bannai and T. Ito, *Algebraic combinatorics*. I. *Association schemes*, Benjamin/Cummings, London, 1984.
2. G. Birkhoff, *Lattice theory*, revised ed., Amer. Math. Soc. Colloq. Publ., vol. 25, Providence, R. I., 1948.
3. K. Fan, *On the Krein-Milman theorem*, in Convexity (V. L. Klee, ed.), Proc. Sympos. Pure Math., no. 7, Amer. Math. Soc., Providence, R. I., 1963, pp. 211–220.
4. A. Ghika, *Convex polyhedroids and multiply ordered vector spaces*, Comun. Acad. R.P.R. **5** (1955), 311–313. (Romanian)
5. P. C. Hammer, *Maximal convex sets*, Duke Math. J. **22** (1955), 103–106.
6. R. Jamison-Waldner, *A convexity characterization of ordered sets*, Congr. Numer. **24** (1979), 529–540.
7. ——, *Copoints in antimatroids*, Congr. Numer. **29** (1980), 535–544.
8. ——, *A perspective on abstract convexity: Classifying alignments by varieties*, in Convexity and related combinatorial geometry (D. C. Kay and M. Breen, eds.), Marcel Dekker, New York, 1982, pp. 113–150.
9. V. Klee, *The structure of semispaces*, Math. Scand. **4** (1956), 54–64.
10. ——, *Separation and support properties of convex sets—a survey*, in Control theory and the calculus of variations (A. Balakrishnan, ed.), Academic Press, New York, 1969, pp. 235–303.
11. L. Lupşa, *Convexity with respect to a given set*, Univ. Babeş-Bolyai, Cluj, preprint no. 4, 1983.
12. ——, *Convexity with respect to a given set* (to appear).
13. H. van Maaren, *Generalized pivoting and coalitions*, in The computation and modelling of economic equilibria (Talman and van der Laan, eds.), North-Holland, Amsterdam, 1987, pp. 155–176.
14. ——, *Fixed point theory in multiply ordered spaces*, Delft Univ. of Technology, Report 89-74, 1989.
15. J.-E. Martínez-Legaz, *Exact quasiconvex conjugation*, Z. Oper. Res. Ser. A-B **27** (1983), 257–266.
16. J.-E. Martínez-Legaz and I. Singer, *Lexicographical separation in* $R^n$, Linear Algebra Appl. **90** (1987), 147–163.
17. ——, *The structure of hemispaces in* $R^n$, Linear Algebra Appl. **110** (1988), 117–179.
18. ——, *Compatible preorders and linear operators on* $R^n$, Linear Algebra Appl. (to appear).
19. W. Prenowitz and J. Jantosciak, *Join geometries: A theory of convex sets and linear geometry*, Springer-Verlag, New York, 1979.
20. I. Singer, *Infimal generators and dualities between complete lattices*, Ann. Mat. Pura Appl. (4) **148** (1987), 289–352.
21. E. Szpilrajn, *Sur l'extension de l'ordre partiel*, Fund. Math. **16** (1930), 386–389.

DEPARTMENT OF APPLIED MATHEMATICS AND ANALYSIS, UNIVERSITY OF BARCELONA, GRAN VÍA 585, 08007 BARCELONA, SPAIN

DEPARTMENT OF MATHEMATICS, INCREST, BD PĂCII 220, 79622 BUCUREŞTI, ROMANIA

DIMACS Series in Discrete Mathematics
and Theoretical Computer Science
Volume 4, 1991

Applied Geometry
and Discrete Mathematics
THE VICTOR KLEE FESTSCHRIFT

# Almost Orthogonal Lines in $E^d$

MOSHE ROSENFELD

ABSTRACT. It is proved that the maximum number of lines through the origin in $E^d$ such that among any three distinct lines there is at least one orthogonal pair is $2d$.

## Introduction

A set $\mathscr{L}$ of $n$ lines through the origin in $E^d$ is called almost orthogonal if among any three lines of $\mathscr{L}$ there is at least one orthogonal pair. We denote by $f(d)$ the maximum cardinality of a set of almost orthogonal lines in $E^d$. Note that if we choose a unit vector $\mathbf{u}_i$ along each line of $\mathscr{L}$, we obtain a set of $n$ unit vectors in $E^d$, such that among any three distinct vectors there is at least one orthogonal pair. Paul Erdös (private communication) asked what is the largest cardinality of a set of almost orthogonal vectors (lines) in $E^d$? Clearly, we can take two sets of unit vectors, each containing $d$ orthogonal vectors, and obtain a set of $2d$ almost orthogonal vectors (lines) in $E^d$. On the other hand, if $\mathscr{L}$ is almost orthogonal, it must contain a "large" subset of mutually orthogonal lines. Indeed, from the bound for the Ramsey number $R(3, n)$ **[S]** we get that if $n$ lines are almost orthogonal then they must contain a set of $\Omega(\sqrt{n \log(n)})$ mutually orthogonal lines, or $2d \le f(d) \le c \cdot d^2/\log(d)$. In this paper we prove that $f(d) = 2d$.

Erdös' question was motivated by results of Konyagin and Kashin. Answering a question of L. Lovàsz **[L]**, Konyagin **[K]** proved that

$$\tfrac{3}{2}n^{2/3} \ge \max \left\| \sum_{i=0}^{n} u_i \right\| \ge c \cdot n^{0.54},$$

(where the max is taken over all families of $n$ almost orthogonal unit vectors in $E^n$ and $||u||$ is the Euclidean norm of the vector $u$). In a subsequent paper, Kashin and Konyagin **[KaK]** improved these bounds to:

1980 *Mathematics Subject Classification* (1985 *Revision*). Primary 46C05.

$2^{1/3}n^{2/3} \geq \max ||\sum_{i=0}^{n} u_i|| \geq c \cdot n^{2/3}/\sqrt{\ln n}$. The lower bounds in both papers were obtained by constructing families of $n$ almost orthogonal vectors and estimating the norm of the sum. In both papers, the dimension of the subspace spanned by the almost orthogonal vectors used to obtain the lower bound is very close to $n$. We conclude this paper by noting the similarity between almost orthogonal lines and equiangular lines. (A family of lines in $E^d$ is equiangular if the angle between any pair of lines is the same, see **[LS]**).

## The main result

The following lemma will be used in our proof.

LEMMA. *Let* $\{a_1, \dots, a_n\}$ *be* $n$ *real numbers such that:*
(1) $a_i > -1$, $i = 1, \dots, n$;
(2) $\sum_{i=0}^{n} a_i = n + k$, *where* $k \geq 1$.
*Then* $\sum_{i=1}^{n} a_i^3 \geq n + 3k$.

PROOF. Assume first that all $a_i \geq 0$. If $a_i > a_j$ for some $i$ and $j$ then we define a new sequence as follows: $b_r = a_r$ if $r \neq i, j$ and $b_i = b_j = (a_i + a_j)/2$. Note that $b_i \geq 0$ and $\sum_{i=1}^{n} b_i = n+k$. Since for any non-negative numbers $a \neq b$, $a^3 + b^3 > 2((a+b)/2)^3$, it follows that $\sum_{i=1}^{n} a_i^3 > \sum_{i=1}^{n} b_i^3$. We may therefore assume that $b_i = (n+k)/n$ for all $i$ and hence:

$$\sum_{i=1}^{n} a_i^3 > \sum_{i=1}^{n} b_i^3 \geq n \left(\frac{n+k}{n}\right)^3 \geq n + 3k.$$

We now proceed by induction on $m$, the number of indices $i$ for which $a_i < 0$. For $m = 0$ the first part applies. If $m > 0$ we may assume that $a_1 = -\alpha < 0$. We replace $a_1$ by $0$ to obtain a set of $n$ real numbers $\{0, \dots, a_n\}$ that sums to $n + k + \alpha$. By the induction hypothesis we get (using the assumption that $\alpha < 1$):

$$\sum_{i=2}^{n} a_i^3 \geq n + 3(k+\alpha) \Longrightarrow \sum_{i=1}^{n} a_i^3 = \sum_{i=2}^{n} a_i^3 - \alpha^3 \geq n + 3(k+\alpha) - \alpha^3 \geq n + 3k.$$

We are now ready to prove the main result.

THEOREM. *If* $\{\mathbf{u_1}, \dots, \mathbf{u_n}\}$ *is an almost orthogonal set of unit vectors in* $E^d$, *then* $n \leq 2d$.

PROOF. Let $\{\mathbf{u_1}, \dots, \mathbf{u_n}\}$ be a maximal, almost orthogonal set of vectors in $E^d$. Clearly, the dimension of the subspace spanned by $\{\mathbf{u_1}, \dots, \mathbf{u_n}\}$ is $d$, otherwise we could add a vector orthogonal to all the vectors in the set and obtain $n+1$ almost orthogonal vectors.We now consider the Grammian matrix $\mathbf{A} = (\langle \mathbf{u_i}, \mathbf{u_j} \rangle)$ of these vectors. The matrix $\mathbf{A}$ is an $n \times n$ positive semidefinite matrix of rank $d$. The diagonal entries of $\mathbf{A}$ are all $1$. Since $n > d$, $0$ is an eigenvalue of $\mathbf{A}$ with multiplicity $n - d$. Since the vectors

$\{\mathbf{u}_1, \dots, \mathbf{u}_n\}$ form an almost orthogonal set, for all triples $i \neq j \neq k \neq i$ we have: $\mathbf{A}_{ij}\mathbf{A}_{jk}\mathbf{A}_{ki} = 0$.

It follows that the symmetric matrix $\mathbf{B} = \mathbf{A} - \mathbf{I}$ has the following properties:

(1) $\mathbf{B}_{ii} = 0$, $i = 1, \dots, n$.
(2) $-1$ is the smallest eigenvalue of $\mathbf{B}$; its multiplicity is $n - d$.
(3) $\mathbf{B}_{ij}\mathbf{B}_{jk}\mathbf{B}_{ki} = 0$ for all triples $i, j, k$.

Let $p(x) = \sum_{i=0}^{n} b_i x^{n-i}$ be the characteristic polynomial of $\mathbf{B}$, and let $\lambda_1, \dots, \lambda_n$ be the $n$ eigenvalues of $\mathbf{B}$. Let $\lambda_{d+1}, \dots, \lambda_n$ be the $n - d$ eigenvalues that are equal to $-1$. By (1) we have that the trace of $\mathbf{B}$ is 0 and hence:

$$\sum_{i=1}^{d} \lambda_i = n - d. \tag{1}$$

Since $p(x) = (x - \lambda_1) \cdots (x - \lambda_n)$,

$$-a_3 = \sum_{1 \le i \le j \le k \le n} \lambda_i \lambda_j \lambda_k. \tag{2}$$

On the other hand,

$$-a_3 = \sum_{i \neq j \neq k} \det \begin{pmatrix} 0 & \mathbf{B}_{ij} & \mathbf{B}_{ik} \\ \mathbf{B}_{ij} & 0 & \mathbf{B}_{jk} \\ \mathbf{B}_{ik} & \mathbf{B}_{jk} & 0 \end{pmatrix}. \tag{3}$$

From [3] it follows that at least one of the three off diagonal pairs in the matrix

$$\begin{pmatrix} 0 & \mathbf{B}_{ij} & \mathbf{B}_{jk} \\ \mathbf{B}_{ij} & 0 & \mathbf{B}_{jk} \\ \mathbf{B}_{ik} & \mathbf{B}_{jk} & 0 \end{pmatrix}$$

is $0$. It is simple to check that no matter which pair is $0$,

$$\det \begin{pmatrix} 0 & \mathbf{B}_{ij} & \mathbf{B}_{ik} \\ \mathbf{B}_{ij} & 0 & \mathbf{B}_{jk} \\ \mathbf{B}_{ik} & \mathbf{B}_{jk} & 0 \end{pmatrix} = 0,$$

and therefore $a_3 = 0$. From the identity

$$\begin{aligned} \left(\sum_{i=1}^{n} \lambda_i\right)^3 &= \sum_{i=1}^{n} \lambda_i^3 + 6\left(\sum_{1 \le i \le j \le k \le n} \lambda_i \lambda_j \lambda_k\right) + 3\left(\sum_{i \neq j} \lambda_i^2 \lambda_j\right) \\ &= \sum_{i=1}^{n} \lambda_i^3 + 6\left(\sum_{i \neq j \neq k} \lambda_i \lambda_j \lambda_k\right) \\ &\quad + 3\left(\sum_{i=1}^{n} \lambda_i\right)\left(\sum_{i \neq j} \lambda_i \lambda_j\right) - 9\left(\sum_{1 \le i \le j \le k \le n} \lambda_i \lambda_j \lambda_k\right), \end{aligned}$$

using $\sum_{i=1}^{n} \lambda_i = 0$ we get

$$\sum_{i=1}^{n} \lambda_i^3 = 3\left(\sum_{1\le i\le j\le k\le n} \lambda_i\lambda_j\lambda_k\right) = 0. \tag{4}$$

Assume now that $n > 2d$ by (1), the $d$ numbers $\lambda_1, \dots, \lambda_d$ satisfy the conditions of the lemma. Indeed $\lambda_i > -1$, $\sum_{i=1}^{d} \lambda_i = n - d = d + k$, and $k \ge 1$. From the lemma we obtain $\sum_{i=1}^{d} \lambda_i^3 \ge d + 3k$ and therefore $\sum_{i=1}^{n} \lambda_i^3 \ge d + 3k + d - n = 2k \ge 2$, contradicting (4). Hence $n \le 2d$.

## Concluding remarks

It is interesting to compare the Grammian matrices arising from almost orthogonal vectors to the Grammian matrices arising from equiangular lines. As was shown by Lemmens and J. J. Seidel [**LS**], in order to construct $n$ equiangular lines in $E^d$, one has to find a symmetric $n \times n$ matrix $\mathbf{A}$ such that $\mathbf{A}_{ii} = 0$, $\mathbf{A}_{ij} = \pm 1$, so that the smallest eigenvalue of $\mathbf{A}$ has multiplicity $n-d$. In [**LS**] it was proved that for large $d$ there exist such matrices of order $o(d^{3/2})$. Similarly, to construct $n$ almost orthogonal lines in $E^d$, we need to find matrices $\mathbf{B}$ such that $\mathbf{B}_{ii} = 0$, $\mathbf{B}_{ij}\mathbf{B}_{jk}\mathbf{B}_{ki} = 0$ for all triples $i, j, k$ so that the smallest eigenvalue of $\mathbf{B}$ has multiplicity $n-d$. While the matrices $\mathbf{A}$ appear to be more restrictive than the matrices $\mathbf{B}$, it seems surprising that they produce matrices having least eigenvalue with much greater multiplicity.

The author wishes to express his thanks to Micha Perles for reading the manuscript and suggesting changes that improved the paper. Micha also noted that the lemma remains valid under the weaker assumptions: $a_i \ge -1$ and $k \ge 0$. He gave a different, "more geometric" proof of the lemma.

## References

[**K**] S. V. Konyagin, *Systems of vectors in Euclidean space and an extremal problem for polynomials*, Mat. Zametki **29** (1981), 63–74.

[**KaK**] B. S. Kashin and S. V. Konyagin, *On systems of vectors in a Hilbert space*, Proc. Steklov Inst. Math. **3** (1983), 67–70.

[**L**] L. Lovàsz, *Combinatorial problems and exercises*, Akadémiai Kiado, Budapest, 1979.

[**LS**] P. W. H. Lemmens and J. J. Seidel, *Equiangular lines*, J. Algebra **24** (1973), 494–512.

[**S**] J. Spencer, *Asymptotic lower bounds for Ramsey functions*, Discrete Math. **20** (1979), 69-76.

Department of Mathematics and Computer Science, Pacific Lutheran University, Tacoma, Washington 98447

*E-mail address*: rosenfeld_m@plu.bitnet

DIMACS Series in Discrete Mathematics
and Theoretical Computer Science
Volume **4**, 1991

Applied Geometry
and Discrete Mathematics
THE VICTOR KLEE FESTSCHRIFT

# Chiral Polytopes

EGON SCHULTE AND ASIA IVIĆ WEISS

ABSTRACT. Abstract polytopes generalize the classical notion of a convex polytope to more general combinatorial structures with a distinctive geometrical and topological flavor. In recent years much work has been done on abstract regular polytopes, the class of abstract polytopes with maximal symmetry by reflection (flag-transitivity). By contrast, very little is known about chiral polytopes, the class of abstract polytopes with maximal symmetry by rotation. The paper discusses the basic theory of chiral polytopes.

## 1. Introduction

Following Sommerville [**28**] a *honeycomb* is a symmetrical face-to-face tessellation of a 3-dimensional manifold by polyhedral cells, all alike, such that each rotational symmetry of a cell is preserved as a symmetry of the tessellation. A honeycomb is called *twisted* if it is not symmetrical by reflection. For a detailed discussion of twisted honeycombs the reader is referred to Coxeter [**3**]. Many interesting twisted honeycombs arise from identifying points along right-handed or left-handed Petrie-polygons in hyperbolic tessellations. Accordingly, twisted honeycombs occur in right-handed or left-handed varieties. By contrast, if a honeycomb is symmetrical by reflection, then it is said to be *reflexible.*

In recent years the term "chiral" has been used for geometrical figures which are symmetrical by rotation but not by (hyperplane) reflection. In this sense twisted honeycombs are chiral (or irreflexible) honeycombs.

The purpose of this paper is to discuss the basic theory of abstract chiral polytopes. Traditionally a polytope is locally and globally of spherical type. In the 1970s Grünbaum [**13**] suggested studying abstract polytopes whose faces and vertex-figures are no longer necessarily spherical. The concept of incidence-polytopes provided a suitable setting for such abstract polytopes resembling the classical polytopes (cf. Danzer-Schulte [**9**]). For related notions see also McMullen [**15**], Dress [**10**], Buekenhout [**1**], and Tits [**31**]. In the past

1980 *Mathematics Subject Classification* (1985 *Revision*). Primary 51M20; Secondary 52A25.

Research of the second author was supported in part by NSERC grant No. A8857.


1052-1798/91 $1.00 + $.25 per page

ten years much work has been done on the construction and classification of abstract regular polytopes $\mathcal{P}$. Here *regularity* means that the automorphism group of $\mathcal{P}$ is transitive on the flags of $\mathcal{P}$. Abstract regular polytopes are symmetrical by reflection and thus may be called *reflexible.*

By contrast relatively little is known about chiral polytopes. The chiral polytopes of rank 3 are (essentially) in one-to-one correspondence with the irreflexible maps on surfaces in Coxeter-Moser [**5**]. Rank 4 twisted honeycombs provide one source of examples, but the list becomes even more sporadic if the 3-faces or vertex-figures are not spherical (cf. Colbourn-Weiss [**2**]). For higher ranks almost nothing is known.

## 2. Basic notions

Our notation will deviate slightly from [**9**], and will follow [**22**]. A *polytope* (incidence polytope) $\mathcal{P}$ of *rank* $n$, or simply an *n-polytope,* is defined by the following properties. $\mathcal{P}$ is a partially ordered set with a rank function $r$ with range $\{-1, 0, \dots, n\}$. The elements of rank $i$ are called the *i-faces* of $\mathcal{P}$. There exist a unique least face $F_{-1}$ and a unique largest face $F_n$. The *flags* (maximal totally ordered subsets) of $\mathcal{P}$ all contain exactly $n+2$ faces, including $F_{-1}$ and $F_n$. Two flags $\Phi$ and $\Psi$ of $\mathcal{P}$ are said to be *adjacent* if they differ in exactly one face; they are *i-adjacent* if this face is an $i$-face of $\mathcal{P}$.

We also assume that $\mathcal{P}$ is strongly *flag-connected* in the following sense: if $\Phi$ and $\Psi$ are two flags of $\mathcal{P}$ and $\Gamma := \Phi \cap \Psi$, then there exists a finite sequence of flags $\Phi = \Phi_0, \Phi_1, \cdots, \Phi_{k-1}, \Phi_k = \Psi$, all containing $\Gamma$, such that $\Phi_{j+1}$ and $\Phi_j$ are adjacent for each $j$. Finally, we require that for any $(i-1)$-face $F$ and any $(i+1)$-face $G$ with $F < G$, there are exactly two $i$-faces $H$ of $\mathcal{P}$ with $F < H < G$ $(i = 0, 1, \dots, n-1)$.

If $\Phi$ is a flag, we write $\Phi^i$ for the flag which is $i$-adjacent to $\Phi$. More generally, we define inductively $\Phi^{i_1, \dots, i_{k-1}, i_k} := (\Phi^{i_1, \dots, i_{k-1}})^{i_k}$ for $k \geq 2$ and $0 \leq i_1, \dots, i_k \leq n-1$.

If $F$ and $G$ are faces of $\mathcal{P}$ with $F \leq G$, we call $G/F := \{H | F \leq H \leq G\}$ a *section* of $\mathcal{P}$. There is little possibility of confusion if we identify a face $F$ with the section $F/F_{-1}$. The faces of ranks 0 and $n-1$ are also called the *vertices* and *facets* of $\mathcal{P}$, respectively. If $F$ is a face, then $F_n/F$ is said to be the *coface* of $F$, or the *vertex-figure* of $F$ if $F$ is a vertex. A *chain* of $\mathcal{P}$ is a totally ordered subset of $\mathcal{P}$. For a chain $\Gamma$ of $\mathcal{P}$ we call the set of ranks of its elements the *type* of $\Gamma$; the *cotype* of $\Gamma$ is the complement in $\{-1, 0, \dots, n\}$ of the type of $\Gamma$.

The dual $\mathcal{P}^*$ of a polytope $\mathcal{P}$ is obtained from $\mathcal{P}$ by reversing the partial order, while leaving the set of faces unchanged. We call $\mathcal{P}$ *self-dual* if $\mathcal{P}^*$ is isomorphic to $\mathcal{P}$.

An $n$-polytope $\mathcal{P}$ is *regular* if its (automorphism) group $A(\mathcal{P})$ is transitive on the flags. Let $\Phi := \{F_{-1}, F_0, \dots, F_n\}$ be a fixed flag, or *base flag,* of $\mathcal{P}$. For a regular polytope $\mathcal{P}$ its group $A(\mathcal{P})$ is generated by the involutions $\rho_0, \dots, \rho_{n-1}$, where $\rho_i$ is the unique automorphism which fixes all but the $i$-face of $\Phi$ $(i = 0, \dots, n-1)$. These *distinguished generators of* $A(\mathscr{P})$ satisfy the relations

$$(1) \qquad (\rho_i \rho_j)^{p_{ij}} = 1 \qquad (i, j = 0, \dots, n-1),$$

where $p_{ii} = 1$, $p_{ji} = p_{ij} =: p_i$ if $j = i-1$, and $p_{ij} = 2$ otherwise; here, the $p_i$'s are given by the *type* $\{p_1, \dots, p_{n-1}\}$ of $\mathscr{P}$. The group $A(\mathscr{P})$ and its generators $\rho_i$ satisfy the following *intersection property*:

$$(2) \qquad \langle \rho_i | i \in I \rangle \cap \langle \rho_i | i \in J \rangle = \langle \rho_i | i \in I \cap J \rangle \quad \text{for } I, J \subseteq \{0, \dots, n-1\}.$$

For a regular $n$-polytope $\mathscr{P}$, let the *rotation* $\sigma_i$ be defined by

$$(3) \qquad \sigma_i := \rho_{i-1}\rho_i \qquad (i = 1, \dots, n-1).$$

Then $\sigma_1, \dots, \sigma_{n-1}$ are the *distinguished generators* of the *rotation subgroup* $A^+(\mathscr{P})$ of $A(\mathscr{P})$. Note that $A^+(\mathscr{P})$ is of index at most 2 in $A(\mathscr{P})$ and can actually coincide with $A(\mathscr{P})$. The $\sigma_i$'s satisfy the relations

$$(4) \qquad \begin{cases} \sigma_i^{p_i} = 1 & \text{for } 1 \le i \le n-1, \\ (\sigma_i \sigma_{i+1} \cdot \dots \cdot \sigma_j)^2 = 1 & \text{for } 1 \le i < j \le n-1. \end{cases}$$

These will suffice for a presentation of $A^+(\mathscr{P})$ if and only if $\mathscr{P}$ equals the universal polytope $\{p_1, \dots, p_{n-1}\}$, with group the Coxeter group $[p_1, \dots, p_{n-1}]$ defined by the diagram $\bullet\overset{}{\underset{p_1}{—}}\bullet\cdots\bullet\underset{p_{n-1}}{—}\bullet$ (cf. **[24, 30]**). For a proof of this fact note that an involutary element $\rho_0$ (say) can be adjoined to the group abstractly defined by (4) to give a quotient of $[p_1, \dots, p_{n-1}]$ (which is indeed $[p_1, \dots, p_{n-1}]$ itself). The remaining generators $\rho_1, \dots, \rho_{n-1}$ are determined by (3); see also Theorem 1(c).

A regular $n$-polytope $\mathscr{P}$ is called *directly regular* if its rotation subgroup $A^+(\mathscr{P})$ is of index 2 in $A(\mathscr{P})$. Each 2-polytope is directly regular. A 3-polytope is directly regular if and only if it corresponds to a regular map on an orientable surface (cf. **[5]**). We will see in §4 that more generally a regular $n$-polytope $\mathscr{P}$ is directly regular if and only if its order complex $\Delta(\mathscr{P})$ is orientable. Recall that the order complex of a poset is the simplicial complex whose simplices are given by the totally ordered subsets of the poset (cf. Stanley **[29]**). Note that all classical regular polytopes are directly regular. In §§3 and 4 we shall see that directly regular polytopes and chiral polytopes have many properties in common.

Now let $\mathscr{P}$ be any polytope of rank $n \ge 3$, and again let $\Phi := \{F_{-1}, F_0, \dots, F_n\}$ be a base flag of $\mathscr{P}$. We call $\mathscr{P}$ *chiral* (or *irreflexible*) if $\mathscr{P}$ is not regular, but there still exist automorphisms $\sigma_1, \dots, \sigma_{n-1}$ of $\mathscr{P}$ such that $\sigma_i$ fixes all faces in $\Phi \backslash \{F_{i-1}, F_i\}$ and cyclically permutes consecutive $i$-faces of $\mathscr{P}$ in the (polygonal) rank 2 section $F_{i+1}/F_{i-2}$ of $\mathscr{P}$.

In Coxeter-Moser **[5]** the term "regular" has been used for two kinds of maps: first, for maps with a flag-transitive group (that is, reflexible maps);

and second, for maps which are chiral in the above sense (that is, irreflexible maps). Note that we use the term "regular" only for the flag-transitive case.

It is worth mentioning that the phenomenon of chirality cannot occur among convex polytopes. As we shall see in Proposition 3, any chiral polytope $\mathcal{P}$ has a Schläfli-symbol $\{p_1, \ldots, p_{n-1}\}$. But by a theorem of McMullen **[15, 16]** any convex polytope with a Schläfli-symbol must be isomorphic to a classical regular convex polytope (with the same symbol); for another proof see also Dress-Schulte **[11]**.

## 3. The groups of chiral polytopes

In this section we prove some basic facts about the group of a chiral polytope. We shall see that the group is generated by the "rotations" $\sigma_1, \ldots, \sigma_{n-1}$, so that it equals its rotation subgroup.

Throughout this section let $\mathcal{P}$ denote a chiral polytope of rank $n$, and let $\Phi$ and $\sigma_1, \ldots, \sigma_{n-1}$ be as in §2.

In dealing with chiral polytopes some care is necessary in handling the automorphisms $\sigma_i$. The definition of chirality leaves us with the possibility of replacing a $\sigma_i$ by its inverse if need be. For reasons which become clear below, we assume that the $\sigma_i$'s have been so chosen that the following property is satisfied. For $i = 0, \ldots, n-1$, denote by $F_i'$ the $i$-face of $\mathcal{P}$ with $F_{i-1} < F_i' < F_{i+1}$ and $F_i' \neq F_i$. Then we require that the "orientation" of the $\sigma_i$'s is such that $\sigma_i(F_i') = F_i$ for $i = 1, \ldots, n-1$; this implies $\sigma_i(F_{i-1}) = F_{i-1}'$. We call the corresponding set $\sigma_1, \ldots, \sigma_{n-1}$ of automorphisms the *distinguished generators* of $A(\mathcal{P})$. This notation is justified by the subsequent results.

For $1 \leq i \leq j \leq n-1$ define

$$\tau_{i,j} := \sigma_i \sigma_{i+1} \cdot \ldots \cdot \sigma_{j-1} \sigma_j. \tag{5}$$

Then $\tau_{i,i} = \sigma_i$ and $\tau_{i,i+1} = \sigma_i \sigma_{i+1}$ for all $i$. We also define $\tau_{0,i} := \tau_{i,n} := 1$ for $0 \leq i \leq n$.

LEMMA 1. $\tau_{i,j}(\Phi) = (\Phi^{i-1})^j =: \Phi^{i-1,j}$ *for all* $i, j$ *with* $1 \leq i \leq j \leq n-1$.

PROOF. The proof is by induction on $j - i$. If $j - i = 0$, then $\tau_{i,j}(\Phi) = \sigma_i(\Phi) = \Phi^{i-1,i}$ by our choice of the orientation of the $\sigma_i$'s. For the case $j - i = 1$ note first that the rank 3 section $F_{i+2}/F_{i-2}$ is (isomorphic to) a map on a surface. By our choice of the orientation of $\sigma_i$ and $\sigma_{i+1}$ the automorphism $\tau_{i,j} = \sigma_i \sigma_{i+1}$ is the half-turn about the center of the edge $F_i$ of $F_{i+2}/F_{i-2}$. Hence, $\tau_{i,i+1}(\Phi) = \Phi^{i-1,i+1}$, as required.

Let $j - i \geq 2$. Then $\tau_{i,j}(\Phi) = \sigma_i \tau_{i+1,j}(\Phi) = \sigma_i(\Phi^{i,j})$. Now, the $i$-face in $\Phi^{i,j}$ is $F_i'$, so that the $i$-face in $\tau_{i,j}(\Phi)$ is $F_i = \sigma_i(F_i')$. But then $\tau_{i,j}(\Phi) = \Phi^{i-1,j}$, since $\sigma_i$ fixes $F_k$ precisely for $k \neq i-1$, $i$ and also leaves invariant the $j$-face $F_j'$ of $\Phi^{i-1,j}$. This completes the proof. □

A flag $\Psi$ of $\mathcal{P}$ is called *even* (*with respect to* $\Phi$) if there exists a finite sequence of flags $\Phi = \Phi_0, \Phi_1, \ldots, \Phi_{2k-1}, \Phi_{2k} = \Psi$ where any two consecutive flags are adjacent. Note that the number of flags in the sequence distinct from $\Phi$ is even. In particular, $\Phi$ is even. A flag of $\mathcal{P}$ is *odd* (*with respect to* $\Phi$) if it is not even (with respect to $\Phi$).

PROPOSITION 1. *$A(\mathcal{P})$ has precisely two orbits on the flags. One is the set of even flags of $\mathcal{P}$, the other the set of odd flags. In particular, the flags of $\mathcal{P}$ adjacent to $\Phi$ are odd flags.*

PROOF. First we show that $A(\mathcal{P})$ is transitive on the even flags. Let $\Phi = \Phi_0, \Phi_1, \ldots, \Phi_{2k-1}, \Phi_{2k} = \Psi$ be a sequence of flags where $\Phi_m$ and $\Phi_{m+1}$ are adjacent for each $m$. Set $\Psi_l := \Phi_{2l}$ for $l = 0, \ldots, k$. Then $\Phi = \Psi_0, \Psi_1, \ldots, \Psi_{k-1}, \Psi_k = \Psi$ is a sequence of flags such that for each $l$ either $\Psi_{l+1} = \Psi_l$ or $\Psi_{l+1}$ and $\Psi_l$ differ in exactly two faces. We show by induction on $k$ that $\Psi = \varphi(\Phi)$ for some $\varphi \in \langle \sigma_1, \ldots, \sigma_{n-1} \rangle$. The case $k = 0$ is trivial. Let $k > 0$.

By the induction hypothesis there exists $\sigma \in \langle \sigma_1, \ldots, \sigma_{n-1} \rangle$ such that $\Psi_{k-1} = \sigma(\Phi)$. If $\Psi_k = \Psi_{k-1}$, we are done. If $\Psi_k \neq \Psi_{k-1}$, there exist distinct $i$ and $j$ such that

$$\Psi_k = \Psi_{k-1}^{i,j} = \sigma(\Phi)^{i,j} = \sigma(\Phi^{i,j}) = \sigma\rho(\Phi), \tag{6}$$

with $\rho = \tau_{i+1,j}$ if $i < j$, $\rho = \sigma_i^{-1}$ if $j = i - 1$, or $\rho = \tau_{j+1,i}$ if $j < i - 1$. Hence, $\varphi := \sigma\rho$ has the required properties and $A(\mathcal{P})$ is transitive on the even flags.

Since $\mathcal{P}$ is chiral and $A(\mathcal{P})$ is transitive on the even flags, odd flags must exist. Assume that $\Phi^i$ is even for some $i$. Then $\Phi^j$ is even for each $j$, since $\Phi^i, \Phi, \Phi^j$ is a sequence in which consecutive flags are adjacent. But then each flag of $\mathcal{P}$ must be even. If follows that $\Phi^i$ must be odd.

To see that $A(\mathcal{P})$ acts transitively on the odd flags, observe first that any odd flag can be mapped onto a suitable $\Phi^i$. But any two $\Phi^i$ are equivalent under $A(\mathcal{P})$; in fact, $\Phi^{i+1} = \sigma_{i+1}^{-1}(\Phi^i)$ and $\Phi^j = \tau_{i+1,j}(\Phi^i)$ for $j > i + 1$. Hence, each odd flag can be mapped onto $\Phi^0$ (or any other fixed flag $\Phi^i$).

Finally, an even flag cannot be mapped onto an odd flag, since otherwise $A(\mathcal{P})$ would be flag-transitive. This proves that $A(\mathcal{P})$ has two orbits, represented by the even and odd flags. □

The following proposition implies that faces of $\mathcal{P}$ of the same rank are mutually isomorphic. A similar remark applies to the cofaces.

PROPOSITION 2. *For each $i = 0, \ldots, n-1$ the group $A(\mathcal{P})$ is simply transitive on the chains of cotype $\{i\}$.*

PROOF. Extend the chains to flags of $\mathcal{P}$ and apply Proposition 1. For each chain, one of the two possible extensions is an even flag, the other an odd flag. If $\Gamma$ is a chain of cotype $\{i\}$ and $\varphi(\Gamma) = \Gamma$, then $\varphi = 1$, since otherwise $\varphi$ would map an even flag onto an odd flag. □

PROPOSITION 3. *We have* $A(\mathscr{P}) = \langle \sigma_1, \dots, \sigma_{n-1} \rangle$. *Also, the generators satisfy relations* (4), *with the* $p_i$*'s given by the (Schläfli-)type* $\{p_1, \dots, p_{n-1}\}$ *of* $\mathscr{P}$.

PROOF. By the arguments in the proofs of Propositions 1 and 2 the subgroup $\langle \sigma_1, \dots, \sigma_{n-1} \rangle$ is simply transitive on the chains of cotype $\{i\}$ for each $i$. Since the same is true for $A(\mathscr{P})$, the two groups must coincide. Furthermore, by the transitivity properties of $A(\mathscr{P})$ the polytope $\mathscr{P}$ has a Schläfli-symbol $\{p_1, \dots, p_{n-1}\}$. Then relations (4) follow from the defining properties of the $\sigma_i$'s and from Lemma 1. □

By Proposition 3 the group of a chiral polytope $\mathscr{P}$ equals its rotation subgroup. For convenience we also define

$$A^+(\mathscr{P}) := \langle \sigma_1, \dots, \sigma_{n-1} \rangle = A(\mathscr{P}).$$

PROPOSITION 4. (a) *The faces and cofaces of* $\mathscr{P}$ *are directly regular or chiral polytopes.*

(b) *For* $i = 0, \dots, n-1$ *we have* $A^+(F_i/F_{-1}) = \langle \sigma_1, \dots, \sigma_{i-1} \rangle$ *and* $A^+(F_n/F_i) = \langle \sigma_{i+2}, \dots, \sigma_{n-1} \rangle$.

PROOF. To prove (a) note that $\{\sigma_1, \dots, \sigma_{i-1}\}$ and $\{\sigma_{i+2}, \dots, \sigma_{n-1}\}$ give two sets of automorphisms which have the properties required for chirality of $F_i/F_{-1}$ and $F_n/F_i$, respectively. However, it can occur that $F_i/F_{-1}$ and/or $F_n/F_i$ are directly regular, in which case the $\sigma_j$'s play the role of the rotations corresponding to (3). Part (b) then follows. □

For a chain $\Gamma$ of $\mathscr{P}$ let $A_\Gamma(\mathscr{P})$ denote the stabilizer of $\Gamma$ in $A(\mathscr{P})$. If $\Gamma = \{F\}$ for some $F \in \mathscr{P}$, we write $A_F(\mathscr{P})$ instead of $A_{\{F\}}(\mathscr{P})$.

PROPOSITION 5. *For* $I \subseteq \{-1, 0, \dots, n\}$ *let* $\Gamma_I := \{F_i | i \notin I\}$. *Then*

$$A_{\Gamma_I}(\mathscr{P}) = \langle \tau_{r,s} | r \leq s \text{ and } r-1, s \in I \rangle.$$

PROOF. By Lemma 1 the group on the right side fixes $\Gamma_I$, hence is a subgroup of $A_{\Gamma_I}(\mathscr{P})$. Conversely, let $\varphi \in A_{\Gamma_I}(\mathscr{P})$ and $\Psi := \varphi(\Phi)$. Then $\Psi$ is an even flag of $\mathscr{P}$ with $\Gamma_I \subseteq \Psi$. By the flag-connectedness of $\mathscr{P}$ there exists a sequence of flags $\Phi = \Phi_0, \Phi_1, \dots, \Phi_{2k-1}, \Phi_{2k} = \Psi$, all containing $\Gamma_I$, such that any two consecutive flags are adjacent. (Note that $\Psi$ remains even if we consider it modulo $\Gamma_I$.) As in the proof of Proposition 1 we define $\Psi_l := \Phi_{2l}$ for $l = 0, \dots, k$ and consider the sequence $\Phi = \Psi_0, \Psi_1, \dots, \Psi_{k-1}, \Psi_k = \Psi$. The same arguments as in the proof of Proposition 1 show that $\Psi = \varphi(\Phi)$ with $\varphi \in \langle \tau_{r,s} | r \leq s \text{ and } r-1, s \in I \rangle =: B$. In fact, since $\Gamma_I \subseteq \Psi_{k-1}, \Psi_k$, we must have $i, j \in I$ in (6) and thus $\rho \in B$; note that $\sigma_i = \tau_{i,i}$. □

The following proposition is an immediate consequence of Proposition 5.

**Proposition** 6. *The stabilizers of the faces in* $\Phi$ *are given by*

$$A_{F_0}(\mathcal{P}) = \langle \sigma_2, \ldots, \sigma_{n-1} \rangle;$$
$$A_{F_i}(\mathcal{P}) = \langle \{\sigma_j | j \neq i, i+1\} \cup \{\sigma_i \sigma_{i+1}\} \rangle \quad \textit{for } i = 1, \ldots, n-2;$$
$$A_{F_{n-1}}(\mathcal{P}) = \langle \sigma_1, \ldots, \sigma_{n-2} \rangle.$$

**Proof.** Use Proposition 5 and observe that $\tau_{i,i} = \sigma_i$ and $\tau_{i,j}\tau_{j+1,k} = \tau_{i,k}$ if $i \leq j < k$. □

**Proposition** 7. *The following intersection property holds:*

$$(7) \qquad \begin{cases} \textit{If } I, J \subseteq \{-1, 0, \ldots, n\}, \textit{ then} \\ \langle \tau_{r,s} | r \leq s \textit{ and } r-1, s \in I \rangle \cap \langle \tau_{r,s} | r \leq s \textit{ and } r-1, s \in J \rangle \\ \qquad = \langle \tau_{r,s} | r \leq s \textit{ and } r-1, s \in I \cap J \rangle. \end{cases}$$

**Proof.** This follows from the fact that the subgroups are stabilizers of chains. But clearly $A_{\Gamma_I} \cap A_{\Gamma_J} = A_{\Gamma_I \cup \Gamma_J} = A_{\Gamma_{I \cap J}}$, so that Proposition 5 applies. □

The following proposition characterizes incidence in $\mathcal{P}$. Note that by the transitivity properties of $A(\mathcal{P})$ each $i$-face of $\mathcal{P}$ is of the form $\varphi(F_i)$ with $-1 \leq i \leq n$ and $\varphi \in A(\mathcal{P})$.

**Proposition** 8. *Let* $-1 \leq i, j \leq n$ *and* $\varphi, \tau \in A(\mathcal{P})$. *Then we have*

$$(8) \qquad \varphi(F_i) \leq \tau(F_j) \quad \textit{if and only if} \quad i \leq j \textit{ and } \varphi A_{F_i}(\mathcal{P}) \cap \tau A_{F_j}(\mathcal{P}) \neq \emptyset.$$

**Proof.** If $i \leq j$ and $\rho \in \varphi A_{F_i}(\mathcal{P}) \cap \tau A_{F_j}(\mathcal{P})$, then

$$\varphi(F_i) = \rho(F_i) \leq \rho(F_j) = \tau(F_j),$$

as required. Conversely, let $\varphi(F_i) \leq \tau(F_j)$. Clearly, $i \leq j$. Take $k \neq i, j$ and choose a chain $\Gamma$ of cotype $\{k\}$ such that $\varphi(F_i), \tau(F_j) \in \Gamma$. (Note that $n \geq 3$.) By Proposition 2 there exists $\rho \in A(\mathcal{P})$ with $\Gamma = \rho(\Phi \backslash \{F_k\})$. But then $\varphi(F_i) = \rho(F_i)$ and $\varphi(F_j) = \rho(F_j)$, so that $\rho \in \varphi A_{F_i}(\mathcal{P}) \cap \tau A_{F_j}(\mathcal{P})$. □

We proceed with a simple but quite surprising observation.

**Proposition** 9. *For a chiral n-polytope* $\mathcal{P}$ *the* $(n-2)$*-faces of* $\mathcal{P}$ *and the cofaces of edges of* $\mathcal{P}$ *are regular polytopes.*

**Proof.** This is trivial for $n = 3$ or 4. Let $n \geq 5$. Consider the $(n-2)$-face $F_{n-2}$ of $\mathcal{P}$. By Proposition 6 the automorphism $\sigma_{n-2}\sigma_{n-1}$ fixes $F_{n-2}$ and thus induces an automorphism of $F_{n-2}/F_{-1}$. By Lemma 1, $\sigma_{n-2}\sigma_{n-1}(\Phi) = \Phi^{n-3,n-1}$. Hence, considered as an automorphism of $F_{n-2}/F_{-1}$, $\sigma_{n-2}\sigma_{n-1}$ maps a flag (namely, $\Phi \backslash \{F_{n-1}, F_n\}$) onto an adjacent flag and thus cannot be in $A^+(F_{n-2}/F_{-1})$. It follows that $F_{n-2}/F_{-1}$ must be regular. A similar argument using $\sigma_1\sigma_2$ shows that $F_n/F_1$ must be regular too. □

Proposition 9 limits the possibilities for constructing chiral polytopes of rank $n \geq 5$. For example, the chiral toroidal maps $\{4, 4\}_{b,c}$ with $b, c \neq 0$

and $b \neq c$ cannot be the 3-faces of any chiral polytope of rank $n \geq 5$. However, this is not true for rank 4 (cf. **[27]**).

Concluding we briefly discuss self-dual chiral polytopes $\mathcal{P}$. Let $0 \leq i \leq n-1$ and $\varphi$ be any duality (incidence reversing bijection) of $\mathcal{P}$. By Proposition 2 there exists $\tau \in A(\mathcal{P})$ such that $\tau(\varphi(\Phi)) = \Phi$ or $\tau\varphi(\Phi) = \Phi^i$. Then $(\tau\varphi)^2(\Phi) = \Phi$ or

$$(\tau\varphi)^2(\Phi) = \tau\varphi(\Phi^i) = (\tau\varphi(\Phi))^{n-i-1} = \Phi^{i,n-i-1},$$

respectively. But $(\tau\varphi)^2 \in A(\mathcal{P})$, so that in the former case we have $(\tau\varphi)^2 = 1$. In the latter case we have

$$(\tau\varphi)^2 = \tau_{i+1,n-i-1},\ \sigma_{i+1},\ 1,\ \sigma_i^{-1},\ \text{or } \tau_{n-i,i}, \tag{9}$$

if $n > 2i+2$, $n = 2i+2$, $n = 2i+1$, $n = 2i$, or $n < 2i$, respectively.

Hence, if a duality (and thus all dualities) preserves the class of even flags and the class of odd flags, then by the first case $\tau\varphi$ is a polarity (duality of order 2) of $\mathcal{P}$. If a duality (and thus all dualities) interchanges the two classes, and if $n$ is odd, then by the second case with $i = (n-1)/2$ the duality $\tau\varphi$ is a polarity of $\mathcal{P}$. If the two classes are interchanged, and if for even $n$ and $i = n/2$ the entry $p_i$ in the Schläfli-symbol of $\mathcal{P}$ is odd, then by (9) the duality $(\tau\varphi)^{p_i}$ is a polarity. We do not know of an example where $\mathcal{P}$ is self-dual and nevertheless does not have a polarity.

If $\mathcal{P}$ admits a polarity $\omega$ (say), then the *extended group* $\mathcal{D}(\mathcal{P})$ of all automorphisms and dualities of $\mathcal{P}$ is given by $\mathcal{D}(\mathcal{P}) \simeq A(\mathcal{P}) \rtimes C_2$. If $\omega(\Phi) = \Phi$, then

$$\begin{aligned}\omega\sigma_i\omega(\Phi) &= \omega\sigma_i(\Phi) = \omega(\Phi^{i-1,i}) = \omega(\Phi)^{n-i,n-i-1}\\ &= \Phi^{n-i,n-i-1} = \sigma_{n-i}^{-1}(\Phi)\end{aligned}$$

and thus $\omega\sigma_i\omega = \sigma_{n-i}^{-1}$ for all $i$; it follows that there exists an involutory group automorphism $A(\mathcal{P}) \mapsto A(\mathcal{P})$ with $\sigma_i \mapsto \sigma_{n-i}^{-1}$ for all $i$.

Conversely, if such a group automorphism $\hat{\omega}$ (say) exists for a polytope $\mathcal{P}$, then Proposition 8 can be used to show that $\mathcal{P}$ is self-dual and admits a polarity $\omega$ with $\omega(\Phi) = \Phi$. In fact, we can define $\omega$ by $\omega(\varphi(F_i)) := \hat{\omega}(\varphi)(F_{n-i-1})$ for $\varphi \in A(\mathcal{P})$ and $i = -1, 0, \ldots, n$.

In §5 we shall characterize the groups of chiral polytopes. In doing so, we shall prove many further properties for the generators $\sigma_i$ and $\tau_{i,j}$. In particular, Lemmas 2–9 hold (with $A_{\Gamma_I}(\mathcal{P}) = A_I$ and $A_{F_i}(\mathcal{P}) = A^i$). A similar remark also applies to the groups of directly regular polytopes; see §4.

## 4. Directly regular polytopes

This section deals with the rotation subgroups $A^+(\mathcal{P})$ of directly regular $n$-polytopes $\mathcal{P}$. Again, let $\Phi := \{F_{-1}, F_0, \ldots, F_n\}$ be the base flag of $\mathcal{P}$ and $\rho_0, \ldots, \rho_{n-1}$ the corresponding generators of $A(\mathcal{P})$. Then the distinguished generators $\sigma_1, \dots, \sigma_{n-1}$ of $A^+(\mathcal{P})$ are given by (3); that is, $\sigma_i = \rho_{i-1}\rho_i$ for all $i$. We begin with the following obvious observation.

PROPOSITION 10. *The sections of directly regular polytopes are again directly regular.*

Directly regular polytopes and chiral polytopes share many properties. We summarize these in the proposition below. In the case of directly regular polytopes the proofs of the corresponding facts are generally easier, since the $\sigma_i$'s can be expressed in terms of the generators $\rho_0, \dots, \rho_{n-1}$. Note also that the orientation of the $\sigma_i$'s is consistent with the orientation of the generators in the chiral case.

Adopting the notation of §3, we again define

$$\tau_{i,j} := \sigma_i \sigma_{i+1} \cdot \dots \cdot \sigma_j = \rho_{i-1}\rho_j \quad \text{for } 1 \leq i \leq j \leq n-1.$$

It follows immediately that

$$\tau_{i,j}(\Phi) = \Phi^{i-1,j},$$

in accordance with Lemma 1. Also we write $A^+_\Gamma(\mathcal{P})$ and $A^+_F(\mathcal{P})$ for the stabilizer of a chain $\Gamma$ and a face $F$ of $\mathcal{P}$ in $A^+(\mathcal{P})$.

PROPOSITION 11. *Propositions* 1, 2, *and* 5–8 *carry over to directly regular polytopes* $\mathcal{P}$, *with* $A(\mathcal{P})$, $A_{\Gamma_I}(\mathcal{P})$, *and* $A_{F_i}(\mathcal{P})$ *replaced by* $A^+(\mathcal{P})$, $A^+_{\Gamma_I}(\mathcal{P})$, *and* $A^+_{F_i}(\mathcal{P})$, *respectively.*

PROOF. The proof is similar to, or the same as, the proofs for the corresponding results for chiral polytopes. Note that $\rho_j \notin A^+(\mathcal{P})$ for all $j$. □

Recall that every self-dual regular polytope $\mathcal{P}$ possesses a polarity which fixes the base-flag. It follows that the remarks on self-duality at the end of §3 extend also to directly regular polytopes.

Concluding this section, let us remark on the order complex $\Delta(\mathcal{P})$ of a chiral or regular polytope $\mathcal{P}$. With $\mathcal{P}$ we can associate the graph $\mathcal{G}(\mathcal{P})$ which is the dual 1-skeleton of $\Delta(\mathcal{P})$; that is, the vertices of $\mathcal{G}(\mathcal{P})$ correspond to the maximal simplices of $\Delta(\mathcal{P})$, and two distinct vertices are joined by an edge if and only if the corresponding simplices share a face of codimension 1. Then it is clear that $\Delta(\mathcal{P})$ is orientable if and only if $\mathcal{G}(\mathcal{P})$ is a bipartite graph. In terms of flags of $\mathcal{P}$ this means that $\Delta(\mathcal{P})$ is orientable if and only if $\mathcal{P}$ has (even and) odd flags.

It follows immediately that the order complexes of chiral or directly regular polytopes are orientable. Furthermore, the order complex of a regular polytope $(\mathcal{P})$ is orientable if and only if $\mathcal{P}$ is directly regular.

## 5. Chiral polytopes from groups

The groups of regular polytopes have been characterized as what are called $C$-groups (cf. **[18, 24]**). More precisely, a group is the group of a regular polytope of rank $n$ if and only if it is generated by involutions $\rho_0, \dots, \rho_{n-1}$ with properties (1) and (2).

Below we shall prove a similar result for chiral and directly regular polytopes. Their groups are characterized by (10) and (14). The proof follows a pattern which is familiar from the theory of regular polytopes, yet there are two essentially different approaches, namely either to assume or not assume the existence and properties of the universal polytopes $\{p_1, \dots, p_{n-1}\}$. In the latter case the proofs of Lemmas 4–9 and parts of Theorem 2 become considerably more complicated and tedious. However, as an advantage of this approach, Theorem 2 can be used to give a new construction of the universal polytopes $\{p_1, \dots, p_{n-1}\}$. For the sake of brevity we shall follow the former case and relate the construction to the universal polytopes $\{p_1, \dots, p_{n-1}\}$.

Let $n \geq 3$ and $p_1, \dots, p_{n-1} \geq 2$ be integers or $\infty$. Consider a group $A$ generated by elements $\sigma_1, \dots, \sigma_{n-1}$ which satisfy relations (4); that is,

$$(10) \qquad \begin{cases} \sigma^{p_i} = 1 & \text{if } 1 \leq i \leq n-1, \\ (\sigma_i \sigma_{i+1} \cdot \dots \cdot \sigma_j)^2 = 1 & \text{if } 1 \leq i < j \leq n-1. \end{cases}$$

Here we assume that $p_i$ is the exact order of $\sigma_i$ for each $i$. Note that $\sigma_{i+1}\sigma_i = \sigma_i^{-1}(\sigma_i\sigma_{i+1})\sigma_i$, so that we also have

$$(\sigma_{i+1}\sigma_i)^2 = 1 \quad \text{if } 1 \leq i \leq n-2.$$

We also define

$$\sigma_0 := \sigma_n := 1.$$

As in (5) we define

$$(11) \qquad \tau_{i,j} := \sigma_i \sigma_{i+1} \cdot \dots \cdot \sigma_j \quad \text{for } 1 \leq i \leq j \leq n-1.$$

Then $\tau_{i,i} = \sigma_i$ and $\tau_{i,j}^2 = 1$ if $i < j$; our assumption (14) below will imply that $\tau_{i,j} \neq 1$ for $1 \leq i \leq j \leq n-1$. We also define

$$\tau_{0,i} := \tau_{i,n} := 1 \quad \text{for } 0 \leq i \leq n.$$

Furthermore, set

$$(12) \qquad A_I := \langle \tau_{r,s} \mid r \leq s \text{ and } r-1, s \in I \rangle \quad \text{for } I \subseteq \{-1, 0, \dots, n\}.$$

Then

$$(13) \qquad \begin{cases} A_\emptyset = A_{\{i\}} = \{1\} & \text{for } i = -1, 0, \dots, n, \\ A_{\{-1,0,\dots,n\}} = A, \\ A_{\{i-1,\dots,j\}} = \langle \sigma_i, \dots, \sigma_j \rangle & \text{for } 0 \leq i \leq j \leq n. \end{cases}$$

From now on we shall assume that $A$ has the following *intersection property* (with respect to its generators $\sigma_1, \dots, \sigma_{n-1}$):

$$(14) \qquad A_I \cap A_J = A_{I \cap J} \quad \text{for } I, J \subseteq \{-1, 0, \dots, n\}.$$

Finally, set

$$(15) \qquad A^i := A_{\{-1,0,\dots,n\} \setminus \{i\}} = \langle \tau_{r,s} \mid r \leq s,\ r \neq i+1,\ s \neq i \rangle \quad \text{for } i = -1, \dots, n$$

and note that (as in Proposition 6) we have

$$(16)\qquad \begin{cases} A^0 = \langle \sigma_2, \ldots, \sigma_{n-1} \rangle, \\ A^i = \langle \{\sigma_j | j \neq i, i+1\} \cup \{\tau_{i,i+1}\} \rangle \quad \text{for } i = 1, \ldots, n-2, \\ A^{n-1} = \langle \sigma_1, \ldots, \sigma_{n-2} \rangle, \\ A^{-1} = A^n = A. \end{cases}$$

Before we construct a polytope from $A$ we deduce various properties of $A$ and its generators. Although these can be proved directly, we establish them by relating $A$ to the group $\widehat{A}$ abstractly defined by relations (10). As remarked in §2 this is the rotation subgroup of the group of the universal polytope $\widehat{\mathscr{P}} = \{p_1, \ldots, p_{n-1}\}$. By $\hat{\rho}_0, \ldots, \hat{\rho}_{n-1}$ and $\hat{\sigma}_1, \ldots, \hat{\sigma}_{n-1}$ we denote the distinguished generators of $A(\widehat{\mathscr{P}})$ and $A^+(\widehat{\mathscr{P}}) = \widehat{A} = \langle \hat{\sigma}_1, \ldots, \hat{\sigma}_{n-1} \rangle$, respectively. Clearly the mapping $\hat{\sigma}_i \mapsto \sigma_i$ $(i = 1, \ldots, n-1)$ defines a surjective homomorphism $f: \widehat{A} \mapsto A$. We also define such terms like $\hat{\tau}_{i,j}$, $\widehat{A}_I$, and $\widehat{A}^i$ (with respect to the $\hat{\sigma}_i$'s and $\widehat{A}$) in analogy to $\tau_{i,j}$, $A_I$, and $A^i$ (with respect to the $\sigma_i$'s and $A$). Note that $f(\hat{\tau}_{i,j}) = \tau_{i,j}$, $f(\widehat{A}_I) = A_I$, and $f(\widehat{A}^i) = A^i$.

It is a well-known fact that the rotation subgroup of a Coxeter group has index 2 (cf. [**5**]). This shows that $\widehat{\mathscr{P}}$ is directly regular, so that the results of §4 apply. In particular, $\widehat{A}$ has the intersection property (14) with respect to $\hat{\sigma}_1, \ldots, \hat{\sigma}_{n-1}$. We proceed with the proofs of several lemmas.

LEMMA 2. *For distinct index sets* $I \subseteq \{0, \ldots, n-1\}$ *with* $|I| \geq 2$, *the subgroups* $A_I$ *are distinct.*

PROOF. First note that

$$\langle \tau_{r,s} \rangle = A_{\{r-1,s\}} \quad \text{if } r \leq s.$$

But by (14) we have $A_{\{r-1,s\}} \cap A_K = \{1\}$ if $|\{r-1, s\} \cap K| \leq 1$. It follows that

$$\tau_{r,s} \in A_K \quad \text{if and only if} \quad r-1, s \in K.$$

Now, if $A_I = A_J$ for $I, J \subseteq \{0, \ldots, n-1\}$ with $|I|, |J| \geq 2$, then (14) implies $A_I = A_{I \cap J} = A_J$. To prove that $I = I \cap J = J$ assume that one of the inclusions $I \cap J \subseteq I, J$ is strict, say $I \cap J \subset I$ and $t \in I$ but $t \notin J$. Since $|I| \geq 2$, there exists $s \in I$ with $s \neq t$, say $t < s$. Then $\tau_{t+1,s} \in A_I$ but $\tau_{t+1,s} \notin A_{I \cap J}$, contradicting the equality. Hence $I = J$, as required. □

The proof of the following lemma is obvious.

LEMMA 3. *Let* $1 \leq i < j \leq n-1$. *With respect to its generators* $\sigma_i, \ldots, \sigma_j$, *the subgroup* $\langle \sigma_i, \ldots, \sigma_j \rangle$ $(= A_{\{i-1, \ldots, j\}})$ *of* $A$ *is itself a group with properties* (10) *and* (14).

LEMMA 4. *For* $1 \leq r < s-1 \leq n-2$ *we have*

$$(\tau_{r,s-1}\sigma_s^{-1})^2 = 1 = (\sigma_r^{-1}\tau_{r+1,s})^2.$$

PROOF. For the first equation, note that in $A(\widehat{\mathscr{P}})$ we have

$$\hat{\tau}_{r,s-1}\hat{\sigma}_s^{-1} = \hat{\rho}_{r-1}\hat{\rho}_{s-1}\hat{\rho}_s\hat{\rho}_{s-1} = \hat{\rho}_{s-1}(\hat{\rho}_{r-1}\hat{\rho}_s)\hat{\rho}_{s-1}$$

and thus $(\hat{\tau}_{r,s-1}\hat{\sigma}_s^{-1})^2 = 1$. Now apply $f$. The proof of the second equation is similar. □

LEMMA 5. $\sigma_i\sigma_j = \sigma_j\sigma_i$ *if* $|j-i| \geq 3$.

PROOF. In $A(\widehat{\mathscr{P}})$ we have

$$\hat{\sigma}_i\hat{\sigma}_j = \rho_{i-1}\rho_i\rho_{j-1}\rho_j = \rho_{j-1}\rho_j\rho_{i-1}\rho_i = \hat{\sigma}_j\hat{\sigma}_i.$$

Now apply $f$. □

LEMMA 6. *Let* $1 \leq i \leq n-2$ *and* $B^i := \langle \sigma_j | j \neq i, i+1\rangle$. *Then* $B^i$ *is the direct product of its subgroups* $\langle \sigma_j | j \leq i-1\rangle$ *and* $\langle \sigma_j | j \geq i+2\rangle$. *Also,* $A^i$ *is a semi-direct product of* $B^i$ *by* $\langle \sigma_i\sigma_{i+1}\rangle$ $(\simeq C_2)$. *In short,*

$$A^i \simeq (\langle \sigma_1, \dots, \sigma_{i-1}\rangle \times \langle \sigma_{i+2}, \dots, \sigma_{n-1}\rangle) \rtimes \langle \sigma_i\sigma_{i+1}\rangle.$$

PROOF. By Lemma 5 the two subgroups $\langle \sigma_j | j \leq i-1\rangle$ and $\langle \sigma_j | j \geq i+2\rangle$ of $B^i$ commute elementwise. Also, by (14),

$$\begin{aligned}&\langle \sigma_j | j \leq i-1\rangle \cap \langle \sigma_j | j \geq i+2\rangle \\ &\quad \subseteq A_{\{0,\dots,i-1\}} \cap A_{\{i+1,\dots,n-1\}} = A_{\varnothing} = \{1\}.\end{aligned}$$

This proves the first part of the lemma.

For the second part note that $\sigma_i\sigma_{i+1} \notin B^i$. In fact, if $\sigma_i\sigma_{i+1} \in B^i$, say $\sigma_i\sigma_{i+1} = \alpha\beta$ with $\alpha \in \langle \sigma_j | j \leq i-1\rangle$ and $\beta \in \langle \sigma_j | j \geq i+2\rangle$, then $\alpha^{-1}\sigma_i = \beta\sigma_{i+1}^{-1} \in \langle \sigma_j | j \leq i\rangle \cap \langle \sigma_j | j \geq i+1\rangle \subseteq A_{\{0,\dots,i\}} \cap A_{\{i,\dots,n-1\}} = A_{\{i\}} = \{1\}$, so that $\sigma_i = \alpha \in \langle \sigma_j | j \leq i-1\rangle$, contradicting Lemma 2.

To prove that $\sigma_i\sigma_{i+1}$ normalizes $B^i$ check the corresponding property for $\widehat{A}$ and apply $f$. Note that $\hat{\sigma}_i\hat{\sigma}_{i+1} = \hat{\rho}_{i-1}\hat{\rho}_{i+1}$ normalizes both $\langle \hat{\rho}_0, \dots, \hat{\rho}_{i-1}\rangle$ and $\langle \hat{\rho}_{i+1}, \dots, \hat{\rho}_{n-1}\rangle$. □

Proposition 8 suggests a way of constructing an $n$-polytope $\mathscr{P}$ from the group $A$. For $i = 0, \dots, n-1$, we take as $i$-faces of $\mathscr{P}$ the left cosets of $A^i$ in $A$; that is, the cosets $\varphi A^i$ with $\varphi \in A$. As improper faces (of rank $-1$ and $n$) we choose two copies of $A$, one denoted by $A^{-1}$, the other by $A^n$; then $\varphi A^{-1} = A^{-1}$ and $\varphi A^n = A^n$ for all $\varphi \in A$. We regard the two copies of $A$ as distinct faces.

On (the set of all faces of) $\mathscr{P}$ we define the partial order

(17) $\quad \varphi A^i \leq \tau A^j$ if and only if $\; -1 \leq i \leq j \leq n$ and $\varphi A^i \cap \tau A^j \neq \varnothing$.

Occasionally, if the dependence on $A$ and its generators $\sigma_1, \dots, \sigma_{n-1}$ is to be emphasized, we write $\mathscr{P}(A)$ or $\mathscr{P}(A; \sigma_1, \dots, \sigma_{n-1})$ for $\mathscr{P}$. We shall see below that (17) really defines a partial order on $\mathscr{P}$. Note that $\mathscr{P}$ is related to the universal $\widehat{\mathscr{P}}$ by the surjective homomorphism $\pi: \widehat{\mathscr{P}} \mapsto \mathscr{P}$ defined by

$$\hat{\varphi}\widehat{A}^i \mapsto f(\hat{\varphi})A^i \qquad (i = -1, 0, \dots, n; \hat{\varphi} \in \widehat{A}).$$

For the proof that $\widehat{\mathscr{P}}$ is indeed a polytope with the properties stated in Theorem 1 we need three further lemmas.

LEMMA 7. *Let $0 \le i, j \le n-1$, $i \ne j$, and let $\varphi \in A$ be such that $\varphi A^i \cap A^j \ne \emptyset$.*

(a) *If $i = 0$ and $j = 1$, then $\varphi A^i = \varphi A^0 = A^0$ or $\sigma_1\sigma_2 A^0$ $(= \sigma_1 A^0)$.*
(b) *If $i < j$ and $j \ge 2$, then there exists $\kappa \in \langle \sigma_1, \dots, \sigma_{j-1}\rangle$ such that $\varphi A^i = \kappa A^i$.*
(c) *If $j = n-2$ and $i = n-1$, then $\varphi A^i = \varphi A^{n-1} = A^{n-1}$ or $\sigma_{n-2}\sigma_{n-1}A^{n-1}$ $(= \sigma_{n-1}^{-1}A^{n-1})$.*
(d) *If $j \le n-3$ and $i > j$, then there exists $\kappa \in \langle \sigma_{j+2}, \dots, \sigma_{n-1}\rangle$ such that $\varphi A^i = \kappa A^i$.*

PROOF. To prove (b) let $i < j$ and $j \ge 2$. Since $\varphi A^i \cap A^j \ne \emptyset$, we have $\varphi\alpha = \beta$ with $\alpha \in A^i$ and $\beta \in A^j$. Choose $\hat{\varphi}, \hat{\alpha}, \hat{\beta} \in \widehat{A}$ such that $\hat{\alpha} \in \widehat{A}^i$, $\hat{\beta} \in \widehat{A}^j$, $f(\hat{\varphi}) = \varphi$, $f(\hat{\alpha}) = \alpha$, $f(\hat{\beta}) = \beta$, and $\hat{\varphi}\hat{\alpha} = \hat{\beta}$. (Note that $\hat{\varphi}$ is determined by $\hat{\alpha}$ and $\hat{\beta}$.) Then $\hat{\varphi}\widehat{A}^i \cap \widehat{A}^j \ne \emptyset$. But statement (b) is known to hold for $\widehat{\mathscr{P}}$. Hence, there exists $\hat{\kappa} \in \langle \hat{\sigma}_1, \dots, \hat{\sigma}_{j-1}\rangle$ such that $\hat{\varphi}\widehat{A}^i = \hat{\kappa}\widehat{A}^i$. Applying $f$ gives $\varphi A^i = \kappa A^i$ with $\kappa := f(\hat{\kappa}) \in \langle \sigma_1, \dots, \sigma_{j-1}\rangle$, as required. This method of "pulling back" works also for the remaining parts of the lemma. □

LEMMA 8. *Let $0 \le i \le j \le k \le n-1$ and let $\varphi, \tau \in A$ be such that $\varphi A^i \cap A^j \ne \emptyset$ and $A^j \cap \tau A^k \ne \emptyset$. Then $\varphi A^i \cap \tau A^k \ne \emptyset$.*

PROOF. As in the proof of Lemma 7 we pull back the faces. Then $\hat{\varphi}\widehat{A}^i \cap \widehat{A}^j \ne \emptyset$ and $\widehat{A}^j \cap \hat{\tau}\widehat{A}^k \ne \emptyset$, with $f(\hat{\varphi}) = \varphi$ and $f(\hat{\tau}) = \tau$. But now $\hat{\varphi}\widehat{A}^i \cap \hat{\tau}\widehat{A}^k \ne \emptyset$, implying $\varphi A^i \cap \tau A^k \ne \emptyset$. □

By construction the group $A$ acts on $\mathscr{P}$ as a group of automorphisms. It follows that Lemma 8 completes the proof that (17) really defines a partial order on $\mathscr{P}$.

LEMMA 9. *Let $-1 \le i-2 < k < j+1 \le n$ and let $\varphi \in A$ be such that $A^{i-2} \cap \varphi A^k \ne \emptyset$ and $\varphi A^k \cap A^{j+1} \ne \emptyset$.*

(a) *If $j = i-1$ (and thus $k = i-1$), then $\varphi A^k = A^{i-1}$ or $\sigma_i\sigma_{i+1}A^{i-1}$.*
(b) *If $j \ge i$, then there exists $\tau \in \langle \sigma_i, \dots, \sigma_j\rangle$ such that $\varphi A^k = \tau A^k$.*

PROOF. Lemma 7(d) applied to $A^{i-2} \cap \varphi A^k = \varnothing$ shows that we may assume $\varphi \in \langle \sigma_i, \dots, \sigma_{n-1} \rangle$. But also $\varphi A^k \cap A^{j+1} \neq \varnothing$, so that Lemma 6 implies $\varphi(\sigma_k\sigma_{k+1})^m \beta\alpha = (\sigma_{j+1}\sigma_{j+2})^l \gamma\delta$, with $m, l = 0$ or $1$, $\alpha \in \langle \sigma_1, \dots, \sigma_{k-1} \rangle$, $\beta \in \langle \sigma_{k+2}, \dots, \sigma_{n-1} \rangle$, $\gamma \in \langle \sigma_1, \dots, \sigma_j \rangle$, and $\delta \in \langle \sigma_{j+3}, \dots, \sigma_{n-1} \rangle$. Since $\alpha^{-1}\delta = \delta\alpha^{-1}$, we get $\varphi(\sigma_k\sigma_{k+1})^m \beta\delta^{-1} = (\sigma_{j+1}\sigma_{j+2})^l \gamma\alpha^{-1}$, with $\beta\delta^{-1} \in \langle \sigma_{k+2}, \dots, \sigma_{n-1} \rangle$ and $\gamma\alpha^{-1} \in \langle \sigma_1, \dots, \sigma_j \rangle$. Define $\kappa$ by $\kappa := (\sigma_k\sigma_{k+1})^m \beta\delta^{-1}(\sigma_k\sigma_{k+1})^m$, so that $\kappa \in \langle \sigma_{k+2}, \dots, \sigma_{n-1} \rangle$ by Lemma 6. Rewriting the equation gives

$$(\sigma_{j+1}\sigma_{j+2})^l \varphi\kappa = \gamma\alpha^{-1}(\sigma_k\sigma_{k+1})^m =: \chi.$$

Now, if $m = 0$, then by (14) we have

$$\begin{aligned} \chi \in \langle \sigma_i, \dots, \sigma_{n-1} \rangle \cap \langle \sigma_1, \dots, \sigma_j \rangle &\subseteq A_{\{i-1, \dots, n\} \cap \{-1, \dots, j\}} \\ &= A_{\{i-1, \dots, j\}} = \langle \sigma_i, \dots, \sigma_j \rangle. \end{aligned}$$

Similarly, if $m = 1$, then $\chi\sigma_{k+1}^{-1} \in \langle \sigma_i, \dots, \sigma_j \rangle$.

Let $j = i - 1$. Then $\chi = 1$ or $\chi = \sigma_{k+1}$ if $m = 0$ or $m = 1$, respectively. It follows that $\varphi\kappa = (\sigma_i\sigma_{i+1})^l$ or $\varphi\kappa = (\sigma_i\sigma_{i+1})^{l+1}\sigma_{i+1}^{-1}$. But $\sigma_{i+1} \in A^{i-1}$, so that

$$\varphi A^{i-1} = \varphi\kappa A^{i-1} = A^{i-1} \text{ or } \sigma_i\sigma_{i+1}A^{i-1},$$

as required.

Let $j \geq i$ and $\chi \in \langle \sigma_i, \dots, \sigma_j \rangle$. By Lemmas 3 and 6,

$$\hat{\chi} := (\sigma_{j+1}\sigma_{j+2})^l \chi (\sigma_{j+1}\sigma_{j+2})^l \in \langle \sigma_i, \dots, \sigma_j \rangle.$$

Then $\varphi A^k = \varphi\kappa A^k = \hat{\chi}(\sigma_{j+1}\sigma_{j+2})^l A^k$. Hence, if $l = 0$ or $k \leq j - 1$, then $\varphi A^k = \hat{\chi} A^k$, so that $\tau := \hat{\chi}$ has the properties required in the lemma. If $l = 1$ and $k = j$, then

$$\begin{aligned} \varphi A^k = \varphi A^j = \hat{\chi}\sigma_{j+1}\sigma_{j+2}A^j &= \hat{\chi}\sigma_{j+1}A^j \\ &= \hat{\chi}\sigma_{j+1}\sigma_{j+1}^{-1}\sigma_j^{-1}A^j = \hat{\chi}\sigma_j^{-1}A^j, \end{aligned}$$

so that $\tau := \hat{\chi}\sigma_j^{-1}$ has the required properties.

Let $j \geq i$ and $\chi\sigma_{k+1}^{-1} \in \langle \sigma_i, \dots, \sigma_j \rangle$. If $k \leq j - 1$, then $\chi \in \langle \sigma_i, \dots, \sigma_j \rangle$ and we are back in the previous case. Let $k = j$. If $l = 0$, then

$$\begin{aligned} \varphi A^k = \varphi A^j = \varphi\kappa A^j &= (\chi\sigma_{j+1}^{-1})\sigma_{j+1}A^j \\ &= (\chi\sigma_{j+1}^{-1})\sigma_{j+1}\sigma_{j+1}^{-1}\sigma_j^{-1}A^j = (\chi\sigma_{j+1}^{-1})\sigma_j^{-1}A^j, \end{aligned}$$

so that $\tau := (\chi\sigma_{j+1}^{-1})\sigma_j^{-1}$ has the required properties. Finally, if $l = 1$, then define $\hat{\chi} := \sigma_{j+1}\sigma_{j+2}(\chi\sigma_{j+1}^{-1})\sigma_{j+1}\sigma_{j+2}$, so that $\hat{\chi} \in \langle\sigma_i, \dots, \sigma_j\rangle$ by Lemmas 3 and 6. It follows that

$$\varphi A^k = \varphi A^j = \varphi\kappa A^j = \sigma_{j+1}\sigma_{j+2}(\chi\sigma_{j+1}^{-1})\sigma_{j+1}A^j = \hat{\chi}\sigma_{j+2}^{-1}A^j = \hat{\chi}A^j.$$

Hence we can choose $\tau := \hat{\chi}$. This completes the proof of the lemma. □

THEOREM 1. *Let $n \geq 3$, $2 \leq p_1, \dots, p_{n-1} \leq \infty$, and $A := \langle\sigma_1, \dots, \sigma_{n-1}\rangle$ be a group with properties* (10) *and* (14). *Let $\mathscr{P} = \mathscr{P}(A)$ be the poset corresponding to $A$. Then $\mathscr{P}$ has the following properties.*

(a) *$\mathscr{P}$ is a chiral or directly regular n-polytope such that $A^+(\mathscr{P}) = A$ and $\{\sigma_1, \dots, \sigma_{n-1}\}$ is the distinguished system of generators of $A^+(\mathscr{P})$ belonging to some base flag of $\mathscr{P}$.*

(b) *$\mathscr{P}$ is of type $\{p_1, \dots, p_{n-1}\}$. For $n \geq 4$ the facets and vertex-figures of $\mathscr{P}$ are isomorphic to the $(n-1)$-polytopes $\mathscr{P}(\langle\sigma_1, \dots, \sigma_{n-2}\rangle)$ and $\mathscr{P}(\langle\sigma_2, \dots, \sigma_{n-1}\rangle)$, respectively. More generally, if $n \geq 4$, $1 \leq i < j \leq n-1$, $F$ is an $(i-2)$-face, and $G$ is a $(j+1)$-face of $\mathscr{P}$ with $F \leq G$, then the section $G/F$ of $\mathscr{P}$ is isomorphic to the $(j-i+2)$-polytope $\mathscr{P}(\langle\sigma_i, \dots, \sigma_j\rangle)$.*

(c) *$\mathscr{P}$ is directly regular if and only if there exists an involutory group automorphism $\rho: A \mapsto A$ such that $\rho(\sigma_1) = \sigma_1^{-1}$, $\rho(\sigma_2) = \sigma_1^2\sigma_2$, and $\rho(\sigma_i) = \sigma_i$ for $i = 3, \dots, n-1$.*

PROOF. By construction $\mathscr{P}$ is a poset on which $A$ acts as a group of automorphisms. This action is faithful, since

$$\bigcap_{i=-1}^{n} A^i = A_\varnothing = \{1\}.$$

The rank function on $\mathscr{P}$ is given by $r(\varphi A^i) = i$. The $(-1)$-face and $n$-face are $A^{-1}$ and $A^n$, respectively. All flags of $\mathscr{P}$ contain exactly $n+2$ faces, that is, a face of each rank; this can be checked by pulling back faces to $\widehat{\mathscr{P}}$.

As the base flag of $\mathscr{P}$ we take $\Phi := \{A^{-1}, A^0, \dots, A^n\}$. By Lemma 7(a) the only flag of $\mathscr{P}$ which is 0-adjacent to $\Phi$ is

$$\Phi^0 := \{A^{-1}, \sigma_1\sigma_2A^0, A^1, \dots, A^n\}.$$

The group $A$ has precisely two orbits on the flags, one represented by $\Phi$ and the other by $\Phi^0$. To prove this, note first that $\Phi$ and $\Phi^0$ are not equivalent under $A$, since $\bigcap_{i\neq 0} A^i = A_{\{0\}} = \{1\}$. Then show by induction on $k$ that each flag of the form

$$\{A^{-1}, \varphi_0A^0, \dots, \varphi_kA^k, A^{k+1}, \dots, A^n\}$$

can be mapped onto $\Phi$ or $\Phi^0$; by Lemma 7(b) one may assume $\varphi_k \in \langle\sigma_1, \dots, \sigma_k\rangle$.

Next we consider the structure of the section $A^{j+1}/A^{i-2}$ of $\mathscr{P}$ for $-1 \leq i-2 < j \leq n-1$. If $j = i-1$, then Lemma 9(a) shows that $A^{j+1}/A^{i-2} = A^i/A^{i-2}$ has exactly two (distinct) $(i-1)$-faces, namely $A^{i-1}$ and $\sigma_i\sigma_{i+1}A^{i-1}$.

Let $j \geq i$. Then Lemma 9(b) implies that

$$A^{j+1}/A^{i-2} = \{\varphi A^k | i-2 \leq k \leq j+1, \varphi \in \langle \sigma_i, \dots, \sigma_j \rangle\}.$$

If $j = i$, then $\sigma_i A^{i-1} \leq A^i$ in $\mathscr{P}$, since $\sigma_i \sigma_{i+1} \in \sigma_i A^{i-1} \cap A^i \neq \emptyset$. Hence, if $p_i$ is finite, we have

$$\begin{aligned} A^{i-1} &\leq A^i \geq \sigma_i A^{i-1} \leq \sigma_i A^i \geq \sigma_i^2 A^{i-1} \leq \sigma_i^2 A^i \\ &\geq \cdots \geq \sigma_i^{p_i-1} A^{i-1} \leq \sigma_i^{p_i-1} A^i \geq \sigma_i^{p_i} A^{i-1} = A^{i-1} \end{aligned}$$

in $\mathscr{P}$. It follows that $A^{i+1}/A^{i-2}$ is isomorphic to a $p_i$-gon. If $p_i = \infty$, then there is a similar infinite sequence proving isomorphism with the apeirogon $\{\infty\}$.

Let $j \geq i+1$, so that $A^{j+1}/A^{i-2}$ is of rank at least 3. For $i-2 \leq k \leq j+1$ define the subgroups $D^k$ of $\langle \sigma_i, \dots, \sigma_j \rangle$ analogous to (15) or (16); that is,

$$\begin{aligned} D^k &:= A_{\{i-1, \dots, j\} \setminus \{k\}} \\ &= \langle \tau_{r,s} | r \leq s \text{ and } i \leq r, s \leq j \text{ and } r \neq k+1, s \neq k \rangle. \end{aligned}$$

Then

$$\begin{aligned} D^{i-2} &= D^{j+1} = \langle \sigma_i, \dots, \sigma_j \rangle, \\ D^{i-1} &= \langle \sigma_{i+1}, \dots, \sigma_j \rangle, \\ D^k &= \langle \{\sigma_l | i \leq l \leq j \text{ and } l \neq k, k+1\} \cup \{\sigma_k \sigma_{k+1}\} \rangle \quad \text{for } i \leq k \leq j-1, \\ D^j &= \langle \sigma_i, \dots, \sigma_{j-1} \rangle. \end{aligned}$$

The faces of $\mathscr{P}(\langle \sigma_i, \dots, \sigma_j \rangle)$ are precisely the cosets $\varphi D^k$ with $i-2 \leq k \leq j+1$ and $\varphi \in \langle \sigma_i, \dots, \sigma_j \rangle$.

Now consider the mapping

$$\begin{aligned} \eta \colon \mathscr{P}(\langle \sigma_i, \dots, \sigma_j \rangle) &\mapsto A^{j+1}/A^{i-2} \\ \varphi D^k &\mapsto \varphi A^k \qquad (i-2 \leq k \leq j+1; \varphi \in \langle \sigma_i, \dots, \sigma_j \rangle). \end{aligned}$$

We need to show that $\eta$ is a well-defined bijection which preserves incidence in both directions. Since $D^k \subseteq A^k$ for all $k$, the mapping $\eta$ is well defined and maps incident faces of $\mathscr{P}(\langle \sigma_i, \dots, \sigma_j \rangle)$ onto incident faces of $A^{j+1}/A^{i-2}$. By our remarks above, $\eta$ is also surjective.

For the proof of injectivity let $\varphi A^k = \tau A^k$ with $\varphi, \tau \in \langle \sigma_i, \dots, \sigma_j \rangle$. Then

$$\begin{aligned} \tau^{-1}\varphi \in A^k \cap \langle \sigma_i, \dots, \sigma_j \rangle &= A_{\{-1, 0, \dots, n\} \setminus \{k\}} \cap A_{\{i-1, \dots, j\}} \\ &= A_{\{i-1, \dots, j\} \setminus \{k\}} = D^k, \end{aligned}$$

so that $\varphi D^k = \tau D^k$.

Finally, let $\varphi A^k \leq \tau A^l$ with $k \leq l$ and $\varphi$, $\tau \in \langle \sigma_i, \dots, \sigma_j \rangle$. Then

$$\begin{aligned}&\emptyset \neq \varphi A^k \cap \tau A^l \cap \langle \sigma_i, \dots, \sigma_j \rangle \\ &= \varphi(A^k \cap \langle \sigma_i, \dots, \sigma_j \rangle) \cap \tau(A^l \cap \langle \sigma_i, \dots, \sigma_j \rangle) = \varphi D^k \cap \tau D^l.\end{aligned}$$

It follows that $\eta^{-1}(\varphi A^k) \leq \eta^{-1}(\tau A^l)$. This completes the proof that $\eta$ is an isomorphism.

For the proof of the connectivity properties of $\mathscr{P}$ it suffices to show that each section $\mathscr{P}$ (including $\mathscr{P}$ itself) is flag-connected; that is, any two flags can be joined by a sequence of flags where consecutive flags are adjacent. By our above considerations the sections have the same kind of structure as $\mathscr{P}$, so that it is sufficient to check this for $\mathscr{P}$ itself.

Let $\varphi \in A$. Pull back the flags $\Phi$ and $\varphi(\Phi)$ of $\mathscr{P}$ to $\widehat{\mathscr{P}}$; that is, take $\hat{\varphi} \in \widehat{A}$ with $f(\hat{\varphi}) = \varphi$ and consider in $\widehat{\mathscr{P}}$ the base flag $\hat{\Phi}$ and its transform $\hat{\varphi}(\hat{\Phi})$. By the flag-connectivity of $\widehat{\mathscr{P}}$ we can join $\hat{\Phi}$ and $\hat{\varphi}(\hat{\Phi})$ by a sequence of flags. Projecting down to $\mathscr{P}$ gives a sequence which connects $\Phi$ and $\varphi(\Phi)$.

It follows that $\mathscr{P}$ is an $n$-polytope with the properties of parts (a) and (b) of the theorem. It remains to prove (c).

Note first that the "orientation" of the generators is such that $\sigma_i(\Phi) = \Phi^{i-1,i}$; in fact, in the section $A^{i+1}/A^{i-2}$ we have $A^{i-1}$, $\sigma_i A^{i-1} \leq A^i$. Now assume that $\mathscr{P}$ is actually regular. Let $\rho_0, \dots, \rho_{n-1}$ be the distinguished generators of $A(\mathscr{P})$ with $\rho_i(\Phi) = \Phi^i$ for each $i$. Then $\sigma_i = \rho_{i-1}\rho_i$ for $i = 1, \dots, n-1$. Define the group automorphism $\rho \colon A \mapsto A$ by $\rho(\varphi) := \rho_0 \varphi \rho_0$. Then

$$\rho(\sigma_1) = \rho_0\sigma_1\rho_0 = \rho_1\rho_0 = \sigma_1^{-1},$$

$$\rho(\sigma_2) = \rho_0\sigma_2\rho_0 = \rho_0\rho_1\rho_2\rho_0 = \rho_0\rho_1\rho_0\rho_2 = (\rho_0\rho_1)^2\rho_1\rho_2 = \sigma_1^2\sigma_2,$$

and

$$\rho(\sigma_i) = \rho_0\sigma_i\rho_0 = \rho_0\rho_{i-1}\rho_i\rho_0 = \rho_{i-1}\rho_i = \sigma_i \quad \text{for } i \geq 3.$$

Conversely, assume there exists a group automorphism $\rho \colon A \mapsto A$ of the required type. Consider the semi-direct product

$$U = A \rtimes \langle \rho \rangle \quad (\sim A \rtimes C_2).$$

To simplify notation we assume that $\rho$ acts by conjugation; that is, $\rho\sigma_1\rho^{-1} = \sigma_1^{-1}$, and so on. Define

$$\begin{aligned}\rho_0 &:= \rho, \\ \rho_i &:= \rho\sigma_1\sigma_2 \cdot \ldots \cdot \sigma_i = \rho\tau_{1,i} \quad \text{for } i = 1, \dots, n-1.\end{aligned}$$

We show that $U$ is a $C$-group with distinguished generators $\rho_0, \dots, \rho_{n-1}$. Obviously these generators satisfy relations (1), so that it remains to check the intersection property (2).

First note that we have the semi-direct products

$$\langle \rho_i, \ldots, \rho_j \rangle = \langle \sigma_{i+1}, \ldots, \sigma_j \rangle \cdot \langle \rho_k \rangle \quad \text{for } 0 \le i \le k \le j \le n-1.$$

By our assumptions on $\rho$ the case $i = k = 0$ is obvious; the case $0 = i < k$ follows immediately from this case. For $i \ge 1$ we clearly have

$$\langle \rho_i, \ldots, \rho_j \rangle = \langle \rho_i, \sigma_{i+1}, \ldots, \sigma_j \rangle = \langle \{\rho_i\} \cup \{\tau_{i+1,l} | l \ge i+1\} \rangle.$$

But

$$\begin{aligned} \rho_i \tau_{i+1,l} \rho_i &= (\rho \sigma_1 \cdot \ldots \cdot \sigma_i \sigma_{i+1} \cdot \ldots \cdot \sigma_l \rho) \sigma_1 \cdot \ldots \cdot \sigma_i \\ &= (\tau_{1,i} \tau_{i+1,l}) \tau_{l,i} = \tau_{i+1,l}^{-1} \tau_{1,i}^{-1} \tau_{1,i} = \tau_{i+1,l}^{-1}, \end{aligned}$$

so that $\langle \rho_i, \ldots, \rho_j \rangle = \langle \sigma_{i+1}, \ldots, \sigma_j \rangle \cdot \langle \rho_i \rangle$. This proves the case $k = i$, and again the remaining cases follow from this case.

We shall show by induction on $j$ that (2) holds for the subgroups $\langle \rho_0, \ldots, \rho_j \rangle$ of $U$; if $j = n-1$ this proves (2) for $U$. The cases $j = 0$ or 1 are trivial. Assume that (2) holds for $\langle \rho_0, \ldots, \rho_j \rangle$ with $j \ge 1$. Then (2) holds for $\langle \rho_0, \ldots, \rho_{j+1} \rangle$ if we can show

$$\langle \rho_0, \ldots, \rho_j \rangle \cap \langle \rho_i, \ldots, \rho_{j+1} \rangle = \langle \rho_i, \ldots, \rho_j \rangle \quad \text{for } 1 \le i \le j$$

(cf. [**25**, p. 314]). But by (14)

$$\begin{aligned} &\langle \rho_0, \ldots, \rho_j \rangle \cap \langle \rho_i, \ldots, \rho_{j+1} \rangle \\ &\quad = (\langle \sigma_1, \ldots, \sigma_j \rangle \cdot \langle \rho_i \rangle) \cap (\langle \sigma_{i+1}, \ldots, \sigma_{j+1} \rangle \cdot \langle \rho_i \rangle) \\ &\quad = (\langle \sigma_1, \ldots, \sigma_j \rangle \cap \langle \sigma_{i+1}, \ldots, \sigma_{j+1} \rangle) \cdot \langle \rho_i \rangle \\ &\quad = (A_{\{0, \ldots, j\}} \cap A_{\{i, \ldots, j+1\}}) \cdot \langle \rho_i \rangle \\ &\quad = A_{\{i, \ldots, j\}} \cdot \langle \rho_i \rangle = \langle \sigma_{i+1}, \ldots, \sigma_j \rangle \cdot \langle \rho_i \rangle = \langle \rho_i, \ldots, \rho_j \rangle, \end{aligned}$$

as required. This completes the proof that $U$ is a $C$-group.

Since $U$ is a $C$-group, there exists a regular $n$-polytope $\mathcal{P}(U)$ with group $U$ and distinguished generators $\rho_0, \ldots, \rho_{n-1}$. But by construction, $A^+(\mathcal{P}(U)) = A$ and $|A(\mathcal{P}(U)) : A^+(\mathcal{P}(U))| = |U : A| = 2$. It follows that $\mathcal{P}(U)$ is directly regular. By Propositions 8 and 11 the construction of $\mathcal{P}(U)$ in terms of its rotation subgroup $A^+(\mathcal{P}(U))$ is equivalent to the construction of $\mathcal{P} = \mathcal{P}(A)$ from the group $A$. This establishes the isomorphism of $\mathcal{P}(U)$ and $\mathcal{P}$. In particular, $\mathcal{P}$ is directly regular. This completes the proof of the theorem. □

COROLLARY. *Let $\mathcal{P}$ be a chiral or directly regular polytope of rank $n$, and let $A^+(\mathcal{P}) = \langle \sigma_1, \ldots, \sigma_{n-1} \rangle$. Then $\mathcal{P}(A^+(\mathcal{P}))$ is isomorphic to $\mathcal{P}$.*

PROOF. By the results of §3 and §4, $A^+(\mathcal{P})$ has the required properties (10) and (14). Hence $\mathcal{P}(A^+(\mathcal{P}))$ is defined. But $A^i = A^+_{F_i}(\mathcal{P})$ for all $i$, so that by (17) and Propositions 8 and 11 the polytopes $\mathcal{P}(A^+(\mathcal{P}))$ and $\mathcal{P}$ are isomorphic. □

In actual applications the proof of the intersection property (14) for $A$ can often be reduced to the consideration of only a few cases. Here the following lemma is quite helpful. Its proof is lengthy and quite tedious, so we omit it here. The lemma itself is used in the proof of Theorem 2.

LEMMA 10. *Let* $n \geq 4$ *and* $A = \langle \sigma_1, \ldots, \sigma_{n-1} \rangle$ *be a group which satisfies* (10). *Assume that the subgroup* $A^{n-1} = \langle \sigma_1, \ldots, \sigma_{n-2} \rangle$ *of* $A$ *has the intersection property* (14) *with respect to its generators. Also, suppose that the following intersection conditions hold:*

(18)
$$\langle \sigma_1, \ldots, \sigma_{n-2} \rangle \cap \langle \sigma_i, \ldots, \sigma_{n-1} \rangle = \langle \sigma_i, \ldots, \sigma_{n-2} \rangle \quad \text{for } i = 2, \ldots, n-1.$$

*Then* $A$ *itself has the intersection property* (14).

The following lemma is an immediate consequence of Lemma 10 and covers the rank 4 case. There is also an easy direct proof which makes use of the fact that $A^1$ and $A^2$ are dihedral.

LEMMA 11. *Let* $A = \langle \sigma_1, \sigma_2, \sigma_3 \rangle$ *be a group which satisfies* (10) *with* $n = 4$. *Suppose that the following intersection conditions hold for* $A$:

$$(19) \qquad \begin{cases} \langle \sigma_1 \rangle \cap \langle \sigma_2 \rangle = \{1\} = \langle \sigma_2 \rangle \cap \langle \sigma_3 \rangle, \\ \langle \sigma_1, \sigma_2 \rangle \cap \langle \sigma_2, \sigma_3 \rangle = \langle \sigma_2 \rangle. \end{cases}$$

*Then* $A$ *has the intersection property* (14).

## 6. Universal polytopes

In this section we briefly discuss the problem of amalgamating two rank $n$ polytopes $\mathcal{P}_1$ and $\mathcal{P}_2$ so as to construct a chiral or directly regular $(n+1)$-polytope $\mathcal{P}$ with facets (isomorphic to) $\mathcal{P}_1$ and vertex-figures (isomorphic to) $\mathcal{P}_2$. By Propositions 4(a) and 10, $\mathcal{P}_1$ and $\mathcal{P}_2$ themselves have to be chiral or directly regular.

If $\mathcal{P}_1$ and $\mathcal{P}_2$ are regular $n$-polytopes, we denote by $\langle \mathcal{P}_1, \mathcal{P}_2 \rangle$ the class of all regular $(n+1)$-polytopes $\mathcal{P}$ with facets $\mathcal{P}_1$ and vertex-figures $\mathcal{P}_2$. From the theory of regular polytopes it is known that for any nonempty class $\langle \mathcal{P}_1, \mathcal{P}_2 \rangle$ there exists a universal polytope which covers all other polytopes in $\langle \mathcal{P}_1, \mathcal{P}_2 \rangle$. This is denoted by $\{\mathcal{P}_1, \mathcal{P}_2\}$ and is called the *universal polytope with facet type* $\mathcal{P}_1$ *and vertex-figure type* $\mathcal{P}_2$.

For chiral polytopes the definition of "class" is more complicated and involves the notion of an "oriented chiral (or directly regular) polytope"; for more details see our forthcoming paper "Chiral Groups and Projective Linear Groups". This refined notion takes care of the two ways how chiral $n$-polytopes can occur as the isomorphism types of the facets and vertex figures of a chiral $(n+1)$-polytope. If $\mathcal{P}_1$ and $\mathcal{P}_2$ are oriented chiral or directly regular polytopes, we write $\langle \mathcal{P}_1, \mathcal{P}_2 \rangle^{\text{ch}}$ for the class of all chiral $(n+1)$-polytopes $\mathcal{P}$ with (oriented) facets $\mathcal{P}_1$ and vertex figures $\mathcal{P}_2$. Below we will discuss existence of universal members of this class.

Our next proposition deals with the universal polytopes $\{\mathcal{P}_1, \mathcal{P}_2\}$ with directly regular facet type $\mathcal{P}_1$ and vertex-figure type $\mathcal{P}_2$. One of the implications of this proposition and Theorem 3 is that for directly regular $\mathcal{P}_1$ and $\mathcal{P}_2$ the class $\langle \mathcal{P}_1, \mathcal{P}_2 \rangle^{\mathrm{ch}}$ does not contain a "universal" member (since the natural candidate for this object, $\{\mathcal{P}_1, \mathcal{P}_2\}$, is directly regular). In the proof we make use of the following general fact (cf. [**5**]).

Let $\mathcal{P}$ be any regular $n$-polytope with group $A(\mathcal{P}) = \langle \rho_0, \dots, \rho_{n-1} \rangle$. By the *even subgroup* of $A(\mathcal{P})$ we mean the subgroup of $A(\mathcal{P})$ whose elements can be expressed as a product of an even number of generators $\rho_i$. But $\tau_{i,j} = \rho_{i-1}\rho_j$ for $i \leq j$, so that the even subgroup of $A(\mathcal{P})$ coincides with the rotation subgroup $A^+(\mathcal{P}) = \langle \sigma_1, \dots, \sigma_{n-1} \rangle$. It follows that $\mathcal{P}$ is directly regular if and only if for each (equivalently, any) presentation of $A(\mathcal{P})$ in terms of $\rho_0, \dots, \rho_{n-1}$ the number of generators in each defining relation is even.

PROPOSITION 12. *Let $\mathcal{P}_1$ and $\mathcal{P}_2$ be regular $n$-polytopes, and assume that the universal $\{\mathcal{P}_1, \mathcal{P}_2\}$ exists. Then $\{\mathcal{P}_1, \mathcal{P}_2\}$ is directly regular if and only if both $\mathcal{P}_1$ and $\mathcal{P}_2$ are directly regular.*

PROOF. Let $A(\{\mathcal{P}_1, \mathcal{P}_2\}) = \langle \rho_0, \dots, \rho_n \rangle$. Then a presentation of $A(\{\mathcal{P}_1, \mathcal{P}_2\})$ in terms of $\rho_0, \dots, \rho_n$ is obtained as the union of three sets of relations: first, a set of relations for $A(\mathcal{P}_1)$ in terms of $\rho_0, \dots, \rho_{n-1}$; second, a set of relations for $A(\mathcal{P}_2)$ in terms of $\rho_1, \dots, \rho_n$; third, the extra relation $(\rho_0\rho_n)^2 = 1$. Here, each relation involves an even number of generators if and only if this is true for each relation of $A(\mathcal{P}_1)$ and $A(\mathcal{P}_2)$. □

Now let $\mathcal{P}_1$ and $\mathcal{P}_2$ be (oriented) chiral or directly regular but not both directly regular. The construction of the universal member in $\langle \mathcal{P}_1, \mathcal{P}_2 \rangle^{\mathrm{ch}}$ follows the pattern of the corresponding construction for regular polytopes, except that now the groups are more complicated (cf. [**26**, §5]).

Assume that $A(\mathcal{P}_1) = \langle \sigma_1, \dots, \sigma_{n-1} \rangle$ and $A(\mathcal{P}_2) = \langle \sigma_2, \dots, \sigma_n \rangle$. Note that we have shifted the generators of $A(\mathcal{P}_2)$ by 1. Now, the search for the universal member in $\langle \mathcal{P}_1, \mathcal{P}_2 \rangle^{\mathrm{ch}}$ involves analysis of the group $A$ generated by $\sigma_1, \dots, \sigma_n$ subject to the relations dictated by $A(\mathcal{P}_1)$ (for $\sigma_1, \dots, \sigma_{n-1}$) and by $A(\mathcal{P}_2)$ (for $\sigma_2, \dots, \sigma_n$) together with the extra relation

$$(\sigma_1\sigma_2 \cdot \dots \cdot \sigma_n)^2 = 1. \tag{20}$$

This group clearly satisfies (10). The crucial question is whether the intersection property (14) holds for $A$ or not. If (14) holds too, then Theorem 1 applies and $A$ is the group of a chiral $(n+1)$-polytope. By construction, this must be universal.

Although the following plausible existence theorem could be deduced from more general results on coverings of chamber complexes in the sense of Tits [**31**] (see also Ronan [**23**]), we give here a simple proof in terms of free products with amalgamation (cf. [**14**, **26**]).

THEOREM 2. *Let* $\mathcal{P}_1$ *and* $\mathcal{P}_2$ *be (oriented) chiral or directly regular but not both directly regular. Assume that* $\langle \mathcal{P}_1, \mathcal{P}_2\rangle^{\mathrm{ch}} \neq \emptyset$. *Then there exists a chiral* $(n+1)$*-polytope in* $\langle \mathcal{P}_1, \mathcal{P}_2\rangle^{\mathrm{ch}}$ *such that any other* $\mathcal{P}$ *in* $\langle \mathcal{P}_1, \mathcal{P}_2\rangle^{\mathrm{ch}}$ *is obtained from it by suitable identifications. This polytope is denoted* $\{\mathcal{P}_1, \mathcal{P}_2\}^{\mathrm{ch}}$ *and is called the universal chiral* $(n+1)$*-polytope with facet type* $\mathcal{P}_1$ *and vertex-figure type* $\mathcal{P}_2$.

PROOF. Consider the group $\widehat{A} = \langle \hat{\sigma}_1, \dots, \hat{\sigma}_n\rangle$ which is the free product of $A(\mathcal{P}_1) = \langle \hat{\sigma}_1, \dots, \hat{\sigma}_{n-1}\rangle$ and $A(\mathcal{P}_2) = \langle \hat{\sigma}_2, \dots, \hat{\sigma}_n\rangle$ with amalgamation along their joint subgroup $\langle \hat{\sigma}_2, \dots, \hat{\sigma}_{n-1}\rangle$; this subgroup is the rotation subgroup of the vertex-figures of $\mathcal{P}_1$, or, equivalently, of the facets of $\mathcal{P}_2$. Here we have slightly changed notation in order to distinguish the generators of $A$ (as defined above) and $\widehat{A}$.

Note that the group $A$ is a homomorphic image of $\widehat{A}$ under the projection $\hat{\pi}: \widehat{A} \mapsto A$ given by $\hat{\pi}(\hat{\sigma}_i) = \sigma_i$ for $i = 1, \dots, n$. Also the groups $A(\mathcal{P}_1)$ and $A(\mathcal{P}_2)$ are canonically embedded into $\widehat{A}$, but a priori we do not know this for $A$.

By assumption we have $\langle \mathcal{P}_1, \mathcal{P}_2\rangle^{\mathrm{ch}} \neq \emptyset$, say $\mathcal{P} \in \langle \mathcal{P}_1, \mathcal{P}_2\rangle^{\mathrm{ch}}$. Let $A(\mathcal{P}) = \langle \overline{\sigma}_1, \dots, \overline{\sigma}_n\rangle$, with $\overline{\sigma}_1, \dots, \overline{\sigma}_n$ the distinguished generators of $A(\mathcal{P})$. By the construction of $A$ the mapping $\overline{\pi}: A \mapsto A(\mathcal{P})$ defined by $\overline{\pi}(\sigma_i) = \overline{\sigma}_i$ is a surjective homomorphism. This gives the surjective homomorphism $\overline{\pi}\hat{\pi}: \widehat{A} \mapsto A(\mathcal{P})$ mapping $\hat{\sigma}_i$ to $\overline{\sigma}_i$. Now, when restricted to the subgroups $\langle \hat{\sigma}_1, \dots, \hat{\sigma}_{n-1}\rangle$ and $\langle \hat{\sigma}_2, \dots, \hat{\sigma}_n\rangle$ of $\widehat{A}$, this homomorphism must be injective, since $\mathcal{P}$ has facets $\mathcal{P}_1$ and vertex-figures $\mathcal{P}_2$. It follows that the restrictions of $\overline{\pi}$ to the subgroups $\langle \sigma_1, \dots, \sigma_{n-1}\rangle$ and $\langle \sigma_2, \dots, \sigma_n\rangle$ of $A$ are injective too. Hence, $A(\mathcal{P}_1)$ and $A(\mathcal{P}_2)$ are canonically embedded into $A$.

To check (14) for $A$ we apply Lemma 10 to $A$. As we have just proved, $\langle \sigma_1, \dots, \sigma_{n-1}\rangle$ is isomorphic to $A(\mathcal{P}_1)$ and thus has the intersection property (14). It remains to verify (18).

Let $1 \leq i \leq n$ and $\varphi \in \langle \sigma_1, \dots, \sigma_{n-1}\rangle \cap \langle \sigma_i, \dots, \sigma_n\rangle$. Then, by the intersection property (14) for $A(\mathcal{P})$, we have

$$\overline{\pi}(\varphi) \in \langle \overline{\sigma}_1, \dots, \overline{\sigma}_{n-1}\rangle \cap \langle \overline{\sigma}_i, \dots, \overline{\sigma}_n\rangle = \langle \overline{\sigma}_i, \dots, \overline{\sigma}_{n-1}\rangle \subseteq \langle \overline{\sigma}_1, \dots, \overline{\sigma}_{n-1}\rangle.$$

But $\overline{\pi}$ restricted to $\langle \sigma_1, \dots, \sigma_{n-1}\rangle$ is injective. It follows that $\varphi \in \langle \sigma_i, \dots, \sigma_{n-1}\rangle$. This proves one inclusion of (18); the other holds trivially.

Our considerations imply that $A$ is the group of a chiral $n$-polytope with facets $\mathcal{P}_1$ and vertex-figures $\mathcal{P}_2$. We denote it by $\{\mathcal{P}_1, \mathcal{P}_2\}^{\mathrm{ch}}$. To prove universality of $\{\mathcal{P}_1, \mathcal{P}_2\}^{\mathrm{ch}}$ observe that $\overline{\pi}: A \mapsto A(\mathcal{P})$ induces the surjective homomorphism

$$\begin{aligned} \{\mathcal{P}_1, \mathcal{P}_2\}^{\mathrm{ch}} &\mapsto \mathcal{P} \\ \varphi A^i &\mapsto \overline{\pi}(\varphi) A_{F_i}(\mathcal{P}) \qquad (-1 \leq i \leq n+1;\ \varphi \in A). \end{aligned}$$

Here, $\{F_{-1}, F_0, \ldots, F_{n+1}\}$ is the base flag of $\mathcal{P}$. By Proposition 8 this mapping preserves incidence. This completes the proof of Theorem 2. □

Concluding, we briefly return to the case when $\mathcal{P}_1$ and $\mathcal{P}_2$ are directly regular. Our next theorem shows that $\{\mathcal{P}_1, \mathcal{P}_2\}$ covers indeed not only the regular but also the chiral $(n+1)$-polytopes with facets $\mathcal{P}_1$ and vertex-figures $\mathcal{P}_2$.

THEOREM 3. *Let $\mathcal{P}_1$ and $\mathcal{P}_2$ be directly regular n-polytopes. Then the universal (regular) $\{\mathcal{P}_1, \mathcal{P}_2\}$ exists if $\langle\mathcal{P}_1, \mathcal{P}_2\rangle^{\text{ch}} \neq \emptyset$ or $\langle\mathcal{P}_1, \mathcal{P}_2\rangle \neq \emptyset$. If $\{\mathcal{P}_1, \mathcal{P}_2\}$ exists, then each $(n+1)$-polytope in $\langle\mathcal{P}_1, \mathcal{P}_2\rangle^{\text{ch}}$ or $\langle\mathcal{P}_1, \mathcal{P}_2\rangle$ is obtained from $\{\mathcal{P}_1, \mathcal{P}_2\}$ by making identifications.*

PROOF. We know that $\{\mathcal{P}_1, \mathcal{P}_2\}$ exists if $\langle\mathcal{P}_1, \mathcal{P}_2\rangle \neq \emptyset$. Now let $\langle\mathcal{P}_1, \mathcal{P}_2\rangle^{\text{ch}} \neq \emptyset$. We proceed as in the proof of Theorem 2 and construct a group $A$. The construction of $A$ remains basically the same except that now we replace $A(\mathcal{P}_1)$ and $A(\mathcal{P}_2)$ by $A^+(\mathcal{P}_1) = \langle\sigma_1, \ldots, \sigma_{n-1}\rangle$ and $A^+(\mathcal{P}_2) = \langle\sigma_2, \ldots, \sigma_n\rangle$, respectively. The arguments of the proof of Theorem 2 carry over and prove that $A$ is the rotation group of a chiral or directly regular polytope $\mathcal{P}(A)$.

To prove that $\mathcal{P}(A)$ is indeed directly regular we apply Theorem 1(c). First we construct an involutory group automorphism $\alpha: A \mapsto A$ which would correspond to conjugation with $\rho_1$ in the group of $\mathcal{P}(A)$.

We define $\alpha$ on the generator of $A$ by $\alpha(\sigma_1) = \sigma_1^{-1}$, $\alpha(\sigma_2) = \sigma_2^{-1}$, $\alpha(\sigma_3) = \sigma_2^2\sigma_3$, and $\alpha(\sigma_i) = \sigma_i$ for $i = 4, \ldots, n$. To see that $\alpha$ defines a homomorphism note first that $\mathcal{P}_1$ and $\mathcal{P}_2$ are directly regular. Hence, when restricted to the sets $\sigma_1, \ldots, \sigma_{n-1}$ and $\sigma_2, \ldots, \sigma_n$ of generators, $\alpha$ defines automorphisms of $\langle\sigma_1, \ldots, \sigma_{n-1}\rangle$ and $\langle\sigma_2, \ldots, \sigma_n\rangle$, respectively. It follows that these isomorphisms extend to the free product $\widehat{A}$ of these groups along their joint subgroup $\langle\sigma_2, \ldots, \sigma_{n-1}\rangle$; note that this subgroup is mapped onto itself. But the definition of $\alpha$ is also compatible with the relation (20) defining $A$. In fact, by Lemma 4 we have

$$(\alpha(\sigma_1)\alpha(\sigma_2)\cdot\cdots\cdot\alpha(\sigma_n))^2 = (\sigma_1^{-1}\sigma_2\sigma_3\cdot\cdots\cdot\sigma_n)^2 = 1.$$

It follows that $\alpha$ indeed defines a homomorphism of $A$. Since $\alpha^2 = 1$, this is an isomorphism.

Now, to apply Theorem 1(c) define $\rho: A \mapsto A$ by $\rho := \alpha \circ \beta$, the composition of $\alpha$ and the inner automorphism $\beta: A \mapsto A$ given by $\beta(\varphi) := \sigma_1^{-1}\varphi\sigma_1$. Then

$$\rho(\sigma_1) = \sigma_1^{-1}, \qquad \rho(\sigma_2) = \sigma_1\sigma_2^{-1}\sigma_1^{-1} = \sigma_1^2\sigma_2,$$
$$\begin{aligned}\rho(\sigma_3) &= \sigma_1\sigma_2^2\sigma_3\sigma_1^{-1} = \sigma_1\sigma_2(\sigma_2\sigma_3)\sigma_1^{-1}\\ &= (\sigma_1\sigma_2\sigma_3^{-1})\sigma_2^{-1}\sigma_1^{-1} = (\sigma_1\sigma_2\sigma_3^{-1})\sigma_1\sigma_2 = \sigma_3,\end{aligned}$$

and $\rho(\sigma_i) = \sigma_i$ for $i \geq 4$; note that to find $\rho(\sigma_3)$ we have used Lemma 4. Hence Theorem 1(c) applies and proves that $\mathcal{P}(A)$ is directly regular.

Finally, to show that $\mathcal{P}(A)$ is indeed the universal $\{\mathcal{P}_1, \mathcal{P}_2\}$ let $\mathcal{P} \in \langle\mathcal{P}_1, \mathcal{P}_2\rangle^{\text{ch}}$ or $\mathcal{P} \in \langle\mathcal{P}_1, \mathcal{P}_2\rangle$. By construction of $A$ we have a surjective homomorphism $\overline{\pi}: A \mapsto A^+(\mathcal{P})$. As in the proof of Theorem 2 this induces the homomorphism

$$\begin{aligned} &\mathcal{P}(A) \mapsto \mathcal{P} \\ &\varphi A^i \mapsto \overline{\pi}(\varphi)A^+_{F_i}(\mathcal{P}) \qquad (-1 \leq i \leq n+1\,;\ \varphi \in A). \end{aligned}$$

It follows that $\mathcal{P}(A)$ covers all members of $\langle\mathcal{P}_1, \mathcal{P}_2\rangle^{\text{ch}}$ and $\langle\mathcal{P}_1, \mathcal{P}_2\rangle$. In particular, $\mathcal{P}(A) = \{\mathcal{P}_1, \mathcal{P}_2\}$. This completes the proof. □

In conclusion we wish to thank H. S. M. Coxeter and A. W. M. Dress for helpful discussions.

## REFERENCES


1. F. Buekenhout, *Diagrams for geometries and groups*, J. Combin. Theory Ser. A **27** (1979), 121–151.
2. C. J. Colbourn and A. I. Weiss, *A census of regular* 3*-polystroma arising from honeycombs*, Discrete Math. **50** (1984), 29–36.
3. H. S. M. Coxeter, *Twisted honeycombs*, Regional Conf. Ser. in Math., Amer. Math. Soc., Providence, RI, 1970.
4. ——, *Regular polytopes*, 3rd ed., Dover, New York, 1973.
5. H. S. M. Coxeter and W. O. J. Moser, *Generators and relations for discrete groups*, 4th ed., Springer, Berlin, 1980.
6. H. S. M. Coxeter and G. C. Shephard, *Regular* 3*-complexes with toroidal cells*, J. Combin. Theory Ser. B **22** (1977), 131–138.
7. H. S. M. Coxeter and A. I. Weiss, *Twisted honeycombs* $\{3, 5, 3\}_t$ *and their groups*, Geom. Dedicata **17** (1984), 169–179.
8. L. Danzer, *Regular incidence-complexes and dimensionally unbounded sequences of such*. I, Ann. Discrete Math. **20** (1984), 115–127.
9. L. Danzer and E. Schulte, *Reguläre Inzidenzkomplexe*. I, Geom. Dedicata **13** (1982), 295–208.
10. A. W. M. Dress, *Regular polytopes and equivariant tessellations from a combinatorial point of view*, Algebraic Topology (Göttingen 1984), Lecture Notes in Math., vol. 1172, Springer, 1985, pp. 56–72.
11. A. W. M. Dress and E. Schulte, *On a theorem of McMullen about combinatorially regular polytopes*, Simon Stevin **61** (1987), 265–273.
12. B. Grünbaum, *Convex polytopes*, Wiley, London, 1967.
13. ——, *Regularity of graphs, complexes and designs*, in Problèmes Combinatoire et Théorie des Graphes, Coll. Internat. CNRS, No. 260, Orsay, 1977, pp. 191–197.
14. R. C. Lyndon and P. E. Schupp, *Combinatorial group theory*, Springer, Berlin, 1977.
15. P. McMullen, *Combinatorially regular polytopes*, Mathematika **14** (1967), 142–150.
16. ——, *On the combinatorial structure of convex polytopes*, Ph.D. Thesis, Birmingham, England, 1968.
17. ——, *Realizations of regular polytopes*, Aequationes Math. **37** (1989), 38–56.
18. P. McMullen and E. Schulte, *Regular polytopes from twisted Coxeter groups*, Math. Z. **201** (1989), 209–226.
19. ——, *Regular polytopes from twisted Coxeter and unitary reflexion groups*, Adv. in Math. **82** (1990), 35–87.
20. ——, *Hermitian forms and locally toroidal regular polytopes*, Adv. in Math. **82** (1990), 88–125.
21. ——, *Regular polytopes of type* $\{4, 4, 3\}$ *and* $\{4, 4, 4\}$, Combinatorica (to appear).
22. ——, *Abstract regular polytopes*, monograph (in preparation).

23. M. A. Ronan, *Coverings and automorphisms of chamber systems*, European J. Combin. **1** (1980), 259–269.
24. E. Schulte, *Reguläre Inzidenzkomplexe*. II, Geom. Dedicata **14** (1983), 33–56.
25. ——, *Regular incidence-polytopes with Euclidean or toroidal faces and vertex-figures*, J. Combin. Theory Ser. A **40** (1985), 305–330.
26. ——, *Amalgamations of regular incidence-polytopes*, Proc. London Math. Soc. (3) **56** (1988), 303–328.
27. E. Schulte and A. I. Weiss, *Chiral polytopes from projective linear groups* (in preparation).
28. D. M. Y. Sommerville, *An introduction to the geometry of n dimensions*, Methuen, London, 1929.
29. R. P. Stanley, *Enumerative combinatorics*, Vol. I, Wadsworth & Brooks/Cole, Belmont, California, 1986.
30. J. Tits, *Groupes et géométries de Coxeter*, Notes polycopiées, I.H.E.S., Paris, 1961.
31. ——, *A local approach to buildings*, in The Geometric Vein (The Coxeter-Festschrift) (Ch. Davis, B. Grünbaum, and F. A. Sherk, eds.), Springer, Berlin, 1981, pp. 519–547.
32. A. I. Weiss, *Incidence-polytopes with toroidal cells*, Discrete Comput. Geom. **4** (1989), 55–73.

Department of Mathematics, Northeastern University, Boston, Massachusetts 02115

Department of Mathematics, York University, North York, Ontario M3J 1P3, Canada

DIMACS Series in Discrete Mathematics
and Theoretical Computer Science
Volume 4, 1991

Applied Geometry
and Discrete Mathematics
THE VICTOR KLEE FESTSCHRIFT

# Exact Upper Bounds for the Number of Faces in $d$-Dimensional Voronoi Diagrams

RAIMUND SEIDEL

ABSTRACT. In 1980 Victor Klee presented asymptotically tight upper bounds for the number of faces in higher-dimensional Voronoi diagrams. In this paper we give *exact* upper bounds for the number of faces in such diagrams along with exact upper bounds for the number of faces in so-called *furthest point Voronoi diagrams*.

## 1. Introduction

Let $T$ be a finite set in $\mathbf{R}^d$ and let $d(x, y)$ denote the Euclidean distance between $x, y \in \mathbf{R}^d$. We define two functions $R_c$ and $R_f$ that map $\mathbf{R}^d$ into the powerset of $T$ as follows:

$$R_c(x) = \{p \in T \mid d(x, p) \leq d(x, q) \text{ for all } q \in T\},$$
$$R_f(x) = \{p \in T \mid d(x, p) \geq d(x, q) \text{ for all } q \in T\}.$$

The function $R_c$ defines an equivalence relation $\rho_c$ on $\mathbf{R}^d$, where $x\rho_c y$ iff $R_c(x) = R_c(y)$. The partition of $\mathbf{R}^d$ induced by $\rho_c$ is called the (*closest point*) *Voronoi diagram of* $T$, for short $\mathrm{VOD}_c(T)$. In a similar way $R_f$ defines an equivalence relation $\rho_f$ on $\mathbf{R}^d$ and the partition of $\mathbf{R}^d$ induced by that relation is called the *furthest point Voronoi diagram of* $T$, for short $\mathrm{VOD}_f(T)$.

The equivalence classes of a Voronoi diagram are called *faces* and we denote them by $V_S = R_c^{-1}(S)$ and $W_S = R_f^{-1}(S)$ for appropriate subsets $S$ of $T$. For example, if $S = \{p\}$ is a singleton set, $V_S$ is the set of all points in $\mathbf{R}^d$ being at least as close to $p$ as to any point in $T$. It is well known that the topological closure of a face in a Voronoi diagram is a polyhedral set, i.e., it is representable as the intersection of a finite number of closed half-spaces

1980 *Mathematics Subject Classification* (1985 *Revision*). Primary 51M20, 52A25, 52A40.

Supported by National Science Foundation grant CCR-8809040. A major portion of this research was carried out while the author was with the IBM Almaden Research Center.


1052-1798/91 $1.00 + $.25 per page

[4]. Thus one can assign in a natural way to each face in a Voronoi diagram a *dimension*.

Let $\mu_i(T)$ denote the number of $i$-dimensional faces in $\mathrm{VOD}_c(T)$ and define

$$\mu_i(n, d) = \max\{\mu_i(T) \mid T \subset \mathbf{R}^d, \ |T| = n\}.$$

Similarly, let $M_i(T)$ denote the number of $i$-dimensional faces in $\mathrm{VOD}_f(T)$ and define

$$M_i(n, d) = \max\{M_i(T) \mid T \subset \mathbf{R}^d, \ |T| = n\}.$$

We are interested in upper bounds for these functions.

Consider first the case of closest point Voronoi diagrams: Trivially, $\mu_d(n, d) = n$ for all $n$ and $d$. For small $n$, namely $n \le d+1$, we have $\mu_i(n, d) = \binom{n}{d+1-i}$, which is realized by any set of $n$ points in $\mathbf{R}^d$ in nondegenerate position. Trivially, $\mu_0(n, 1) = n-1$ holds. A routine application of Euler's relation yields $\mu_0(n, 2) = 2n-5$ and $\mu_1(n, 2) = 3n-6$, provided $n > 2$. Preparata [9] showed that $\mu_i(n, 3) = \Theta(n^2)$ for $i = 0, 1, 2$. Dewdney and Vranch [3] showed that $\mu_2(n, 3) = \binom{n}{2}$. Victor Klee [5] proved that $\mu_{d-1}(n, d) = \binom{n}{2}$ for all $n > d > 3$; furthermore, $\mu_i(n, d) = \binom{n}{d+1-i}$ for $\lceil d/2 \rceil < i \le d$; most importantly, Klee gave an asymptotically sharp bound that states that for any fixed $d$

$$\mu_0(n, d) = \Theta(n^{\lceil d/2 \rceil}).$$

Finally, Paschinger [8] and independently Seidel [10, 11] gave exact values for $\mu_i(n, d)$ for all $n > d+1 > 1$, namely:

$$\mu_i(n, d) = \begin{cases} C_{d-i}(n, d+1) - 1, & i = 0, \\ C_{d-i}(n, d+1), & 0 < i \le d. \end{cases}$$

Here $C_j(n, d+1)$ denotes the maximal number of $j$-dimensional faces of a $(d+1)$-polytope with $n$ vertices and is known to be

$$C_j(n, d+1) = \begin{cases} \displaystyle\sum_{i=1}^{s} \frac{n}{i} \binom{n-i-1}{i-1} \binom{i}{j-i+1}, & d+1 = 2s, \\ \displaystyle\sum_{i=1}^{s} \frac{j+2}{i+1} \binom{n-i-1}{i} \binom{i+1}{j-i+1}, & d+1 = 2s+1. \end{cases}$$

Much less work has been done on furthest point Voronoi diagrams. The bounds $M_1(n, 1) = 2$ and $M_0(n, 1) = 1$, provided $n > 1$, and $M_d(n, d) = n$, provided $d > 1$, are trivial. Again for small $n \le d+1$ the bound $M_i(n, d) = \binom{n}{d+1-i}$ is realized by any set of $n$ points in $\mathbf{R}^d$ in nondegenerate position. Shamos [12] proved $M_0(n, 2) = n-2$ and $M_1(n, 2) = 2n-3$, provided $n > 1$. For $d > 2$ only the asymptotic bound

$$M_0(n, d) = \Theta(n^{\lceil d/2 \rceil})$$

was known [10, 11].

It is the purpose of this paper to give *exact* values for $M_i(n, d)$ for all $i$ and $d$ and $n > d+1$. Specifically, we show

$$M_i(n, d) = \begin{cases} C_{d-i}(n, d+1) - n + d, & i = 0, \\ C_{d-i}(n, d+1) - n + d + 1, & i = 1, \\ C_{d-i}(n, d+1), & 2 \le i \le d. \end{cases}$$

§2 of this paper shows how $d$-dimensional Voronoi diagrams relate to $(d+1)$-dimensional convex polytopes. For the sake of completeness and also because of the present lack of ready accessibility we include the proof of the exact upper bound for the face numbers of closest point Voronoi diagrams in §3. The somewhat more involved proof of the corresponding upper bound for furthest point Voronoi diagrams is presented in §4.

## 2. Voronoi diagrams, Delaunay triangulations, and polytopes

In this section we show how the problem of maximizing the number of faces in a Voronoi diagram in $\mathbf{R}^d$ can be reduced to a problem of maximizing certain faces of a $(d+1)$-dimensional polytope. For this purpose define

$$\mathrm{DT}_c(T) = \{\mathrm{conv}\, S \mid S \subset T,\ V_S \text{ not empty}\},$$
$$\mathrm{DT}_f(T) = \{\mathrm{conv}\, S \mid S \subset T,\ W_S \text{ not empty}\},$$

where $\mathrm{conv}\, S$ denotes the convex hull of $S$. $\mathrm{DT}_c(T)$ and $\mathrm{DT}_f(T)$ are polyhedral complexes and are known as *closest point* and *furthest point Delaunay triangulation of* $T$, respectively. The term triangulation is used in this context because in the nondegenerate case, when no $d+2$ points of $T$ lie on a common sphere, both $\mathrm{DT}_c(T)$ and $\mathrm{DT}_f(T)$ are simplicial complexes that "fill" $\mathrm{conv}\, T$. For the maximization questions addressed in this paper such degeneracies turn out to be irrelevant. Thus we assume from now on that we only deal with point sets $T$ that are nondegenerate in the sense stated above. In this case $\mathrm{DT}_c(T)$ and $\mathrm{DT}_f(T)$ are simplicial complexes, and $V_S$ (resp. $W_S$) is the $i$-dimensional face in $\mathrm{VOD}_c(T)$ (resp. $\mathrm{VOD}_f(T)$) iff $|S| = d+1-i$ and $\mathrm{conv}\, S$ is a $(d-i)$-simplex in $\mathrm{DT}_c(T)$ (resp. $\mathrm{DT}_f(T)$). It follows that determining $\mu_i(n, d)$ or $M_i(n, d)$ amounts to determining the maximum number of $(d-i)$-simplices in the closest point or furthest point Delaunay triangulation of $n$ points in $\mathbf{R}^d$. We attack the latter problem by mapping it into a polytope problem by a method suggested by K. Q. Brown [2].

Let $K$ be a $d$-sphere in $\mathbf{R}^{d+1}$, let $T'$ be a set of $n$ points on $K$ such that $P = \mathrm{conv}\, T'$ is a simplicial $(d+1)$-polytope, and let $x$ be a point on $K$ that does not lie in any of the hyperplanes spanned by the facets of $P$. We say a facet $F$ of $P$ is *visible* from $x$ iff $x$ and $P$ do not lie in the same half-space bounded by the hyperplane that is spanned by $F$. Call $F$ *invisible* from $x$ otherwise. Now consider the inversion of $\mathbf{R}^{d+1}$ with center $x$ given by $I_x(z) = (z-x)/\langle z-x, z-x\rangle + x$. This function is an involutary 1-1 mapping of $\mathbf{R}^{d+1} - \{x\}$ into itself and maps the sphere $K$ into a hyperplane $H$ that contains the set $T = I_x(T')$. The hyperplane $H$ can be identified with $\mathbf{R}^d$ and thus we can form the Delaunay triangulations $\mathrm{DT}_c(T)$ and $\mathrm{DT}_f(T)$. The following lemma is essentially due to Brown [**2**]. For a more complete proof see [**11**].

LEMMA 1. *Let $S' \subset T'$ and let $S = I_x(S')$. Then the following assertions are true:*

(i) $\operatorname{conv} S$ *is an $i$-simplex in* $\mathrm{DT}_c(T)$ *iff* $\operatorname{conv} S'$ *is an $i$-face of $P$ that is contained in some facet of $P$ that is invisible from $x$.*
(ii) $\operatorname{conv} S$ *is an $i$-simplex in* $\mathrm{DT}_f(T)$ *iff* $\operatorname{conv} S'$ *is an $i$-face of $P$ that is contained in some facet of $P$ that is visible from $x$.*

Every finite set $T$ in $\mathbf{R}^d \cong H$ can be realized as the image of some $T' \subset K$ under $I_x$. Thus this lemma implies that $\mu_i(n, d)$ as well as $M_i(n, d)$ are bounded from above by $C_{d-i}(n, d+1)$, the maximal possible number of $(d-i)$-faces of a $(d+1)$-polytope with $n$ vertices. Furthermore, since every facet of $P$ is either visible from $x$ or invisible, the following holds:

LEMMA 2. *For any $T \subset \mathbf{R}^d$ with $|T| = n$*

$$\mu_0(T) + M_0(T) \le C_d(n, d+1).$$

Lemma 2 implies that a lower bound for $M_0(T)$ yields an improved upper bound for $\mu_0(n, d)$, and similarly a lower bound for $\mu_0(T)$ yields an improved upper bound for $M_0(n, d)$. How few maximal simplices (i.e., $d$-simplices) can a triangulation of $k$ points in $\mathbf{R}^d$ have? It is not hard to see that there must be at least $k - d$ of them. Since, as is well known, every one of the $n$ points in $T$ must be a vertex of some maximal simplex of $\mathrm{DT}_c(T)$ it follows that $\mu_0(T) \ge n - d$. On the other hand, it is known (see [**12, 11**]) that exactly the extreme points of $T$ are vertices of maximal simplices in the furthest point Delaunay triangulation $\mathrm{DT}_f(T)$. Thus one can only guarantee $M_0(T) \ge 1$. These observations together with the fact that every $(d-1)$-face of a $(d+1)$-polytope $P$ is contained in exactly two facets of $P$ yields the following theorems:

THEOREM 3. *For $n > d + 1 > 1$*

$$\mu_i(n, d) \le \begin{cases} C_{d-i}(n, d+1) - 1, & i = 0, \\ C_{d-i}(n, d+1), & 0 < i \le d. \end{cases}$$

THEOREM 4. *For $n > d + 1 > 1$*

$$M_i(n, d) \le \begin{cases} C_{d-i}(n, d+1) - n + d, & i = 0, \\ C_{d-i}(n, d+1) - n + d + 1, & i = 1, \\ C_{d-i}(n, d+1), & 2 \le i \le d. \end{cases}$$

In the following sections we argue that these bounds can be realized for all $n$ and $d$ with $n > d+1$.

## 3. The "closest-point" case

The purpose of this section is to prove the following:

THEOREM 5. *For* $n > d+1 > 1$

$$\mu_i(n, d) = \begin{cases} C_{d-i}(n, d+1) - 1, & i = 0, \\ C_{d-i}(n, d+1), & 0 < i \le d. \end{cases}$$

Theorem 3 already established the indicated values as upper bounds for $\mu_i(n, d)$. Thus we only need to show that these bounds can be achieved. By Lemma 1 and because of the fact that every proper face $F$ of a polytope $P$ is contained in more than one facet of $P$ (unless $F$ itself is a facet), it suffices to solve the following construction problem for every $n > d+1 > 1$.

CONSTRUCTION 1. Find a set $T'$ of $n$ points on a $d$-sphere $K$ and a point $x$ on $K$ with the following properties:

(i) The $(d+1)$-polytope $P = \operatorname{conv} T'$ has a maximal number of faces.

(ii) The point $x$ is not in the affine span of any facet of $P$. Moreover, exactly one facet of $P$ is visible from $x$.

The first condition is relatively easy to satisfy since there is a generic way of constructing polytopes with a maximal number of faces [7]: Call a curve $\gamma$ in $\mathbf{R}^{d+1}$ a $(d+1)$-*order algebraic curve* iff every hyperplane intersects $\gamma$ in at most $d+1$ points. Choose any set $T$ of $n$ points on such a curve. Then $\operatorname{conv} T$ will have $C_i(n, d+1)$ $i$-faces for all $i$.

Thus in order to satisfy condition (i) above, it suffices to exhibit a $(d+1)$-order curve $\gamma$ that lies on a $d$-sphere $K$. For this purpose we need to specify the sphere $K$. The most convenient way turns out to be as follows: As sphere $K$ we choose the image of the hyperplane

$$H = \{(1, x_d, x_{d-1}, \dots, x_1) \mid x_j \in \mathbf{R} \quad \text{for all } j\}$$

under the inversion $I = I_0$ of $\mathbf{R}^{d+1}$. In other words,

$$K = \left\{ \frac{1}{1 + \sum_{i=1}^{d} x_i^2}(1, x_d, x_{d-1}, \dots, x_1) \mid x_j \in \mathbf{R} \quad \text{for all } j \right\}.$$

As curve on $K$ we choose

$$\gamma(t) = I((t, t^2, \dots, t^d)), \qquad t > 0.$$

Why is $\gamma$ a $(d+1)$-order algebraic curve? Consider an arbitrary hyperplane $\{z \mid \langle z, a\rangle + c = 0\} \subset \mathbf{R}^{d+1}$, where $a \in \mathbf{R}^{d+1}$ and $c \in \mathbf{R}$. The number of intersections of this hyperplane with the curve $\gamma$ is given by the number of positive roots of $\langle \gamma(t), a\rangle + c = 0$. Simple transformations yield that the number of these roots is bounded from above by the number of positive roots of a polynomial

$$ct^{2d} + ct^{2d-2} + ct^{2d-4} + \dots + ct^{2s} + b_d t^d + b_{d-1} t^{d-1} + \dots + b_1 t + b_0 = 0,$$

where $s = \lceil (d+1)/2 \rceil$ and the coefficients $b_j$ depend on $a$ and $c$ only.

The *sign rule of Descartes* states that the number of positive roots of a polynomial cannot exceed the number of sign changes in the degree ordered enumeration of the coefficients of that polynomial. As in our case the coefficients for terms of degree greater than $d$ are all the same, it follows that the number of sign changes, and hence the number of positive roots, is at most $d+1$, which is what we wanted.

In order to deal with condition (ii) of Construction 1 we need a good characterization of the invisibility of facets of a polytope. The following condition is fairly obvious. We omit its proof.

LEMMA 6. *Let $P \subset \mathbf{R}^d$ be a $d$-polytope and let $x \in \mathbf{R}^d$ be a point not in $P$ and not in the affine span of any facet of $P$. A facet $F$ of $P$ is invisible from $x$ if for some $z \in F$ the line segment connecting $z$ and $x$ intersects the relative interior of some other facet of $P$.*

We now have all the tools ready to show how a set $T'$ and a point $x$ satisfying conditions (i) and (ii) can be constructed. Choose $d+2$ distinct points $p_1, \dots, p_{d+2}$ on curve $\gamma$. Let $G$ be the $d$-simplex formed by the convex hull of $\{p_1, \dots, p_{d+1}\}$ and let $u$ be a point in the relative interior of $G$. As $\gamma$ is a $(d+1)$-order algebraic curve, the point $p_{d+2} = \gamma(t^*)$ does not lie in the affine span of $G$ and hence the straight line $L$ through the points $p_{d+2}$ and $u$ intersects $G$ only in $u$. The line $L$ intersects the sphere $K$ in two points. One of them is $p_{d+2}$. Call the other $x$.

Clearly there is some $\varepsilon > 0$ such that for all $t$ with $|t-t^*| < \varepsilon$ the straight line segment joining $\gamma(t)$ and $x$ intersects the relative interior of $G$. Choose $n-d-2$ distinct values $t_i$, with $t_i \neq t^*$ and $|t_i - t^*| < \varepsilon$ for $d+2 < i \leq n$, and let $p_i = \gamma(t_i)$.

We claim that the set $T' = \{p_1, \dots, p_n\}$ together with point $x$ satisfies the desired conditions (i) and (ii). Condition (i) holds since all points of $T'$ lie on the $(d+1)$-order algebraic curve $\gamma$. Condition (ii) is satisfied since by virtue of our construction and because of Lemma 6 the only facet of $\operatorname{conv} T'$ that is visible from $x$ is $G$.

## 4. The "furthest-point" case

In this section we want to prove the following:

THEOREM 7. *For $n > d+1 > 1$*

$$M_i(n, d) = \begin{cases} C_{d-i}(n, d+1) - n + d, & i = 0, \\ C_{d-i}(n, d+1) - n + d + 1, & i = 1, \\ C_{d-i}(n, d+1), & 2 \leq i \leq d. \end{cases}$$

From Theorem 4 we know that the indicated values are indeed upper bounds for $M_i(n, d)$. Lemma 1 again implies that in order to establish equality it suffices to solve the following construction problem for every $n > d+1 > 1$:

CONSTRUCTION 2. Find a set $T$ on a $d$-sphere $K$ and a point $x$ on $K$ with the following properties:

(i) The $(d+1)$-polytope $P = \operatorname{conv} T$ has a maximal number of faces.
(ii) The point $x$ is not in the affine span of any facet of $P$. Moreover, exactly $n-d$ facets of $P$ are invisible from $x$, and for exactly $n-d-1$ ridges (i.e., $(d-1)$-faces) of $P$ both containing facets are invisible from $x$.

Condition (i) of Construction 2 is again easy to satisfy by choosing $n$ points on a $(d+1)$-order algebraic curve $\gamma$ that lies on $K$. However, for technical reasons our choice of the sphere $K$ and curve $\gamma$ will be slightly different from the one in the last section.

In particular, as sphere $K$ we choose the image of the hyperplane

$$H = \{(x_d, 1, x_{d-1}, \dots, x_1) \mid x_j \in \mathbf{R} \quad \text{for all } j\}$$

under the inversion $I = I_0$ of $\mathbf{R}^{d+1}$. In other words,

$$K = \left\{ \frac{1}{1+\sum_{i=1}^d x_i^2}(x_d, 1, x_{d-1}, \dots, x_1) \mid x_j \in \mathbf{R} \quad \text{for all } j \right\}.$$

As curve on $K$ we choose

$$\gamma(t) = I((t, 1, t^{-1}, t^{-2}, \dots, t^{1-d})), \qquad t > 0.$$

To see that $\gamma$ is indeed a $(d+1)$-order algebraic curve, consider again an arbitrary hyperplane $\{z \mid \langle z, a\rangle + c = 0\} \subset \mathbf{R}^{d+1}$, where $a \in \mathbf{R}^{d+1}$ and $c \in \mathbf{R}$. The number of intersections of this hyperplane with the curve $\gamma$ is given by the number of positive roots of $\langle \gamma(t), a\rangle + c = 0$. Simple transformations yield that the number of these roots is bounded from above by the number of positive roots of a polynomial

$$ct^{2d} + b_1 t^{2d-1} + b_2 t^{2d-2} + \cdots + b_{d+1} t^{d-1} + ct^{2s} + ct^{2s-2} + \cdots + ct^2 + c = 0,$$

where $s = \lfloor (d-2)/2 \rfloor$ and the coefficients $b_j$ depend on $a$ and $c$ only. When $d$ is odd the *sign rule of Descartes* implies that a polynomial with such a sequence of coefficients can have at most $d+1$ positive zeros, which is what we wanted.

For even $d$ this polynomial can have $d+2$ positive zeros. Thus the construction about to be described is only applicable to odd dimensions $d$. However, it turns out that there is a relatively straightforward way of "lifting" this construction to even dimensions, which we will describe at the end of the section.

We have now an easy way of satisfying condition (i) stated at the beginning of this section: For our point set $T$ we can choose any $n$ points from the curve $\gamma$. What about condition (ii)? It says that exactly $n-d$ facets of $\operatorname{conv} T$ must be invisible from some point $x$ on $K$. How do we find such an $x$? How do we have to choose $T$ to make sure such an $x$ exists?

Let us pose a slightly different construction problem whose solution is clearly also a solution for Construction 2:

CONSTRUCTION 3. Choose a set $T$ of $n > d+1$ points on curve $\gamma$ and a point $x$ on $K$ such that the polytope $P' = \operatorname{conv}(T \cup \{x\})$ is a simplicial $(d+1)$-polytope with $n+1$ vertices that has exactly $n-d$ facets that do not contain $x$.

It is convenient to think of $P'$ as "glued together" out of $n-d$ simplices of dimension $d+1$, where each simplex has $x$ as one vertex and the remaining $d+1$ vertices are drawn from $T$. The particular collection of $(d+1)$-simplices that we have in mind is as follows: Let $p_1 = \gamma(t_1), \dots, p_n = \gamma(t_n)$ be the points in $T$ in order along the curve $\gamma$, i.e., $0 < t_1 < \cdots < t_n$. For each $k$ with $d \le k \le n-1$ let $S_k$ be the simplex $\operatorname{conv}\{p_n, p_k, p_{k-1}, x, p_{d-2}, p_{d-3}, \dots, p_1\}$.

Note that for each $k$ the simplices $S_k$ and $S_{k+1}$ intersect in the common facet $\operatorname{conv}\{p_n, p_k, x, p_{d-2}, p_{d-3}, \dots, p_1\}$ and every other pair of simplices intersects in the $(d-1)$-simplex $\operatorname{conv}\{p_n, x, p_{d-2}, p_{d-3}, \dots, p_1\}$. Thus the union of all these $(d+1)$-simplices forms a "polytope" with the desired property, namely exactly $n-d$ facets do not contain $x$ (for each $S_k$ the facet $\operatorname{conv}\{p_n, p_k, p_{k-1}, p_{d-2}, p_{d-3}, \dots, p_1\}$). Now it just remains to show that it is possible to choose $T$ and $x$ such that this polytope is convex.

What are the conditions for this to happen? Suppose for some $l > d$ the points $x, p_n, p_1, p_2, \dots, p_{l-1}$ have been chosen already so that the union of the simplices $S_d, \dots, S_{l-1}$ forms their convex hull. Where is $p_l$ to be placed? It must be in a location so that it can "see" only one facet of that polytope, namely the facet $\operatorname{conv}\{p_n, p_{l-1}, x, p_{d-2}, \dots, p_1\}$ of $S_{l-1}$. It is not difficult to convince oneself that this is equivalent to saying for every $k$ with $d \le k < l$ the only facet of the simplex $S_k$ that is visible from $p_l$ is $\operatorname{conv}\{p_n, p_k, x, p_{d-2}, \dots, p-1\}$.

We now need to give rigorous definitions for the intuitive terms "see" and "visible." For $q_i \in \mathbf{R}^{d+1}$ let $\Delta(q_1, q_2, \dots, q_{d+2})$ denote the $(d+2) \times (d+2)$ determinant

$$\begin{vmatrix} 1 & q_1 \\ 1 & q_2 \\ \vdots & \vdots \\ 1 & q_{d+2} \end{vmatrix}.$$

Let $\Delta(q_1, \dots, q_{d+2})[p \backslash q_i]$, where $1 \le i \le d+2$, denote $\Delta(q_1, \dots, q_{i-1}, p, q_{i+1}, \dots, q_{d+2})$. A point $p \in \mathbf{R}^{d+1}$ "sees" the facet of a $(d+1)$-simplex $\operatorname{conv}\{q_1, \dots, q_{d+2}\}$ that does not contain vertex $q_i$ iff the sign of $\Delta(q_1, \dots, q_{d+2})$ is opposite to the sign of $\Delta(q_1, \dots, q_{d+2})[p \backslash q_i]$ ($p$ does "not see" that facet iff the signs agree). With these definitions we can formulate our problem as follows: Find $n$ points $p_1, \dots, p_n$ on the curve $\gamma$ and find a point $x$ on the sphere $K$ such that for $d \le k < n$ and $k < l < n$ the following conditions hold $(i = 1, \ldots, d-2, k, n)$:

$$\Delta(p_n, p_k, p_{k-1}, x, p_{d-2}, p_{d-3}, \ldots, p_1) > 0,$$
$$\Delta(p_n, p_k, p_{k-1}, x, p_{d-2}, p_{d-3}, \ldots, p_1)[p_l \backslash p_{k-1}] < 0,$$
$$\Delta(p_n, p_k, p_{k-1}, x, p_{d-2}, p_{d-3}, \ldots, p_1)[p_l \backslash p_i] > 0,$$
$$\Delta(p_n, p_k, p_{k-1}, x, p_{d-2}, p_{d-3}, \ldots, p_1)[p_l \backslash x] > 0.$$

Using the fact that $p_i = \gamma(t_i)$, parametrizing $x$ as $I((x_d, 1, x_{d-1}, -x_{d-2}, -x_{d-3}, \ldots, -x_1))$, and performing appropriate row multiplications on the matrices that underly $\Delta$, this problem can be reformulated as follows: Let $A_k$ denote the $(d+2)\times(d+2)$ matrix

$$\begin{pmatrix}
\sum_{j=0}^{d} t_n^{2j} & t_n^{2d-1} & t_n^{2d-2} & t_n^{2d-3} & t_n^{2d-4} & t_n^{2d-5} & \cdots & t_n^{d} & t_n^{d-1} \\
\sum_{j=0}^{d} t_k^{2j} & t_k^{2d-1} & t_k^{2d-2} & t_k^{2d-3} & t_k^{2d-4} & t_k^{2d-5} & \cdots & t_k^{d} & t_k^{d-1} \\
\sum_{j=0}^{d} t_{k-1}^{2j} & t_{k-1}^{2d-1} & t_{k-1}^{2d-2} & t_{k-1}^{2d-3} & t_{k-1}^{2d-4} & t_{k-1}^{2d-5} & \cdots & t_{k-1}^{d} & t_{k-1}^{d-1} \\
1+\sum_{j=1}^{d} x_j^2 & x_d & 1 & x_{d-1} & -x_{d-2} & -x_{d-3} & \cdots & -x_2 & -x_1 \\
\sum_{j=0}^{d} t_{d-2}^{2j} & t_{d-2}^{2d-1} & t_{d-2}^{2d-2} & t_{d-2}^{2d-3} & t_{d-2}^{2d-4} & t_{d-2}^{2d-5} & \cdots & t_{d-2}^{d} & t_{d-2}^{d-1} \\
\sum_{j=0}^{d} t_{d-3}^{2j} & t_{d-3}^{2d-1} & t_{d-3}^{2d-2} & t_{d-3}^{2d-3} & t_{d-3}^{2d-4} & t_{d-3}^{2d-5} & \cdots & t_{d-3}^{d} & t_{d-3}^{d-1} \\
\vdots & \vdots & \vdots & \vdots & \vdots & \vdots & \ddots & \vdots & \vdots \\
\sum_{j=0}^{d} t_2^{2j} & t_2^{2d-1} & t_2^{2d-2} & t_2^{2d-3} & t_2^{2d-4} & t_2^{2d-5} & \cdots & t_2^{d} & t_2^{d-1} \\
\sum_{j=0}^{d} t_1^{2j} & t_1^{2d-1} & t_1^{2d-2} & t_1^{2d-3} & t_1^{2d-4} & t_1^{2d-5} & \cdots & t_1^{d} & t_1^{d-1}
\end{pmatrix}.$$

We want to find $n$ positive real numbers $0 < t_1 < \cdots < t_n$ and $d$ reals $x_1, \ldots, x_d$ such that for $d \leq k < n$ and $k < l < n$ the following conditions are satisfied:

the determinant of $A_k$ is positive;
if the third row of $A_k$ is replaced by the row

$$R_l = \left(\sum_{j=0}^{d} t_l^{2j} \; t_l^{2d-1} \; t_l^{2d-2} \; t_l^{2d-3} \; t_l^{2d-4} \; t_l^{2d-5} \cdots t_l^{d} \; t_l^{d-1}\right)$$

then the determinant of the newly obtained matrix is negative, whereas if any other row is replaced, the determinant stays positive.

Thus, we have $n-d+(d+2)\binom{n-d}{2}$ matrices expressed in terms of $n+d$ variables and we want to find values for these variables so that the determinants of these matrices have certain prescribed signs. This problem looks quite messy and unwieldy. Surprisingly, there is a rather simple and relatively straightforward way of showing that in our case a solution must exist.

The key idea is to view each determinant as a multivariate polynomial in the variables $t_1, \ldots, t_n$ and $x_1, \ldots, x_d$. Our problem can then be expressed as follows: Given a set of $n-d+(d+2)\binom{n-d}{2}$ multivariate polynomials in $n+d$ variables, find values for these variables so that all polynomials evaluate to prescribed signs. In the following we give a sufficient condition for this to be possible.

First we give some notation. Let $Z_1, \dots, Z_m$ be variables. For $I = (i_1, \dots, i_m) \in \mathbf{N}^m$ let $Z^I$ denote the monomial $Z_1^{i_1} Z_2^{i_2} \cdots Z_m^{i_m}$, where $\mathbf{N}$ stands for the nonnegative integers. We write a multivariate polynomial $P$ as

$$P(Z_1, \dots, Z_m) = \sum_{I \in \mathbf{N}^m} c_I Z^I ,$$

where only a finite number of the (real) coefficients are nonzero. We say $Z^I$ is the *leading monomial* and $c_I$ is the *leading coefficient* of $P$ iff $I$ is the lexicographically largest $m$-tuple with nonzero $c_I$. The following lemma can be easily proved by induction on the number of variables [**1**]:

LEMMA 8. *Let $P_1, \dots, P_r$ be multivariate polynomials in $Z_1, \dots, Z_m$ with leading coefficients $c_1, \dots, c_r$, respectively. Then there exist positive real numbers $T_1 > T_2 > \cdots > T_m > 0$ such that*

$$\operatorname{sign}(P_j(T_1, \dots, T_m)) = \operatorname{sign}(c_j) \quad \textit{for } 1 \le j \le r .$$

In order to apply this lemma to our problem we have to impose an ordering on the variables so that one can talk about lexicographically largest monomials. The following ordering turns out to have the right properties:

$$t_n \succ t_{n-1} \succ \cdots \succ t_d \succ t_{d-1} \succ x_{d-1} \succ t_{d-2} \succ x_{d-2} \succ t_{d-3} \succ \cdots \\ \succ x_2 \succ t_1 \succ x_1 \succ x_d .$$

For instance, under this ordering the leading monomial of the determinant of the matrix $A_k$ is given by the main diagonal of $A_k$ (with $t_n^{2d}$ taken from the sum in the upper left-hand corner). The coefficient of this monomial is $+1$, and we indeed want the determinant of $A_k$ to evaluate to a positive value for every $k$.

What about the other sign conditions? If the first row of $A_k$ is replaced by $R_l$, then the same argument applies, since $k < l$; the main diagonal contributes the leading monomial, and it has coefficient $+1$ as desired. The same holds if the second row of $A_k$ is replaced by $R_l$.

If the third row of $A_k$ is replaced by $R_l$ and rows 2 and 3 of the resulting matrix are exchanged, then the leading monomial of the determinant of that matrix is again contributed by the main diagonal with coefficient $+1$. We indeed want that determinant to have positive sign, as then, because of the row swap, the determinant of $A_k$ with the third row replaced by $R_l$ must have negative sign, as desired.

The sign of the determinant of $A_k$ with the fourth row replaced by $R_l$ is the same as the sign of the determinant of $A_k$ with the fourth row deleted and $R_l$ inserted as the second row. The leading monomial of the latter determinant is again contributed by the main diagonal and its coefficient is $+1$, as desired.

The interesting case is when the $i$th row of $A_k$, counting from the bottom, is replaced by $R_l$. To deal with this matrix do the following: swap $R_l$ with the fourth row (the " $x$-row") and then move $R_l$ two rows up, which results in the following matrix $A_{k,l}^{(i)}$:

$$\begin{pmatrix}
\sum_{j=0}^{d} t_n^{2j} & t_n^{2d-1} & t_n^{2d-2} & t_n^{2d-3} & t_n^{2d-4} & \cdots & t_n^{d+i-1} & t_n^{d+i-2} & t_n^{d+i-3} & \cdots & t_n^{d-1} \\
\sum_{j=0}^{d} t_l^{2j} & t_l^{2d-1} & t_l^{2d-2} & t_l^{2d-3} & t_l^{2d-4} & \cdots & t_l^{d+i-1} & t_l^{d+i-2} & t_l^{d+i-3} & \cdots & t_l^{d-1} \\
\sum_{j=0}^{d} t_k^{2j} & t_k^{2d-1} & t_k^{2d-2} & t_k^{2d-3} & t_k^{2d-4} & \cdots & t_k^{d+i-1} & t_k^{d+i-2} & t_k^{d+i-3} & \cdots & t_k^{d-1} \\
\sum_{j=0}^{d} t_{k-1}^{2j} & t_{k-1}^{2d-1} & t_{k-1}^{2d-2} & t_{k-1}^{2d-3} & t_{k-1}^{2d-4} & \cdots & t_{k-1}^{d+i-1} & t_{k-1}^{d+i-2} & t_{k-1}^{d+i-3} & \cdots & t_{k-1}^{d-1} \\
\sum_{j=0}^{d} t_{d-2}^{2j} & t_{d-2}^{2d-1} & t_{d-2}^{2d-2} & t_{d-2}^{2d-3} & t_{d-2}^{2d-4} & \cdots & t_{d-2}^{d+i-1} & t_{d-2}^{d+i-2} & t_{d-2}^{d+i-3} & \cdots & t_{d-2}^{d-1} \\
\vdots & \vdots & \vdots & \vdots & \vdots & \ddots & \vdots & \vdots & \vdots & & \vdots \\
\sum_{j=0}^{d} t_{i+1}^{2j} & t_{i+1}^{2d-1} & t_{i+1}^{2d-2} & t_{i+1}^{2d-3} & t_{i+1}^{2d-4} & \cdots & t_{i+1}^{d+i-1} & t_{i+1}^{d+i-2} & t_{i+1}^{d+i-3} & \cdots & t_{i+1}^{d-1} \\
1+\sum_{j=1}^{d} x_j^2 & x_d & 1 & x_{d-1} & -x_{d-2} & \cdots & -x_{i+1} & -x_i & -x_{i-1} & \cdots & -x_1 \\
\sum_{j=0}^{d} t_{i-1}^{2j} & t_{i-1}^{2d-1} & t_{i-1}^{2d-2} & t_{i-1}^{2d-3} & t_{i-1}^{2d-4} & \cdots & t_{i-1}^{d+i-1} & t_{i-1}^{d+i-2} & t_{i-1}^{d+i-3} & \cdots & t_{i-1}^{d-1} \\
\vdots & \vdots & \vdots & \vdots & \vdots & & \vdots & \vdots & \vdots & \ddots & \vdots \\
\sum_{j=0}^{d} t_1^{2j} & t_1^{2d-1} & t_1^{2d-2} & t_1^{2d-3} & t_1^{2d-4} & \cdots & t_1^{d+i-1} & t_1^{d+i-2} & t_1^{d+i-3} & \cdots & t_1^{d-1}
\end{pmatrix}$$

The determinant of $A_k$ with the $i$th row from the bottom replaced by $R_l$ is the negative of $A_{k,l}^{(i)}$. Because of the ordering that we imposed on the variables, in particular $t_{i+1} \succ x_{i+1}$ and $x_i \succ t_{i-1}$, the leading monomial of the polynomial represented by the determinant of $A_{k,l}^{(i)}$ is again contributed by the diagonal. Its coefficient is $-1$, as desired.

Thus, we managed to fulfill all sign conditions, and therefore, by Lemma 8, for every $n > d+1$ the set $T$ on $\gamma$ and the point $x$ on $K$ satisfying the conditions of Construction 3 and hence Construction 2 exist. Thus we have proved that for odd $d > 2$ and for every $n > d+1$ indeed $M_0(n, d) = C_d(n, d+1) - n + d$, as claimed.

Our construction also shows that $M_1(n, d) = C_{d-1}(n, d+1) - n + d + 1$, since the only $(d-1)$-faces of our polytope $P$ that are contained only in facets that are invisible from $x$ are $\operatorname{conv}\{p_n, p_k, p_{d-2}, p_{d-3}, \ldots, p_1\}$ for $d \le k < n-1$. Obviously, our construction also realizes the bound $M_i(n, d) = C_{d-i}(n, d+1)$ for $2 \le i \le d$, and thus Theorem 4 is proven for odd $d > 0$. (Obviously, it holds for $d = 1$.)

What about even $d$? The following "lifting construction" shows that the same bound applies for even dimensions as well. But first a definition and a lemma whose routine proof is omitted.

Let $P$ be an $(n-1)$-vertex $d$-polytope lying on a hyperplane $H$ of $\mathbf{R}^{d+1}$, and let $u$ and $v$ be two points in $\mathbf{R}^{d+1}$ whose connecting line segment intersects $H$ in vertex $p$ of $P$. We call the $n$-vertex $(d+1)$-polytope $P' = \operatorname{conv}(P \cup \{u, v\})$ a *pseudo-bipyramid* with *base* $P$, *corner* $p$, and *apices* $u$ and $v$.

LEMMA 9. *The faces of a pseudo-bipyramid $P'$ with base $P$, corner $p$, and apices $u$ and $v$ are characterized as follows:*

- *If $F$ is an $i$-face of $P$ with $p \in F$, then $F' = \operatorname{conv}(F \cup \{u, v\})$ is an $(i+1)$-face of $P'$ (and $F$ is* not *a face of $P'$).*
- *If $F$ is an $i$-face of $P$ with $p \notin F$, then $F' = \operatorname{conv}(F \cup \{u\})$ and $F'' = \operatorname{conv}(F \cup \{v\})$ are $(i+1)$-faces of $P'$, and $F$ is an $i$-face of $P'$.*

COROLLARY 10. *A pseudo-bipyramid $P'$ with an even-dimensional $(n-1)$-vertex simplicial neighborly $d$-polytope as base is a simplicial neighborly $(d+1)$-polytope with $n$ vertices, and hence has $C_i(n, d+1)$ $i$-faces.*

We now perform Construction 2 for even $d$ as follows: Let $K$ be a $d$-sphere in $\mathbf{R}^{d+1}$ that is intersected by a hyperplane $H$ in a $(d-1)$-sphere $K'$. Perform Construction 2 as described above in $H$ to yield a set $T$ of $n-1$ points on $K$ and an additional point $x$ on $K$ so that $(n-1)-(d-1) = n-d$ facets of $P = \operatorname{conv} T$ are invisible from $x$. Point $p_{n-1}$ of this construction is contained in all the invisible facets. Perturb $p_{n-1}$ slightly into the open ball bounded by $K$ so that it stays on $H$ and so that the face structure of $P$ and the set of facets visible from $x$ remain unchanged. Since $P$ is simplicial, this is possible. Let $L$ be a straight line in $\mathbf{R}^{d+1}$ that intersects $H$ in the perturbed point $p_{n-1}$. The line $L$ intersects the sphere $K$ in two points, call them $u$ and $v$. The convex hull of the $n$-point set $T' = T \cup \{u, v\} - \{p_{n-1}\} \subset K$ is a pseudo-bipyramid $P'$ with base $P$, corner $p_{n-1}$, and apices $u$ and $v$. Since $P$ is an even-dimensional simplicial neighborly polytope, it follows from Corollary 10 that $P'$ is simplicial and neighborly and therefore has $C_i(n, d+1)$ $i$-faces. Finally, it is easy to check that the only facets of $P'$ that are invisible from $x$ are the facets of the form $\operatorname{conv}(F \cup \{u, v\})$, where $F$ is a facet of $P$ that is invisible from $x$. But there are exactly $n-d$ such facets of $P$. Thus, $P'$ is a $(d+1)$-polytope with $n$ vertices, all lying on sphere $K$, that has $C_d(n, d+1)$ facets of which exactly $n-d$ are not visible from point $x \in K$. Therefore we have successfully completed Construction 2 for even $d$, and Theorem 7 holds for all $d$.

## 5. Conclusion

Although the existence proofs presented in this paper are "constructive" in the mathematical sense, they are unsatisfactory from the computational point of view. The inductive proof of the crucial Lemma 8 shows how one can actually construct the numbers $T_i$, whose existence is asserted. But it is unclear whether in our application these numbers can be computed in subexponential time. Moreover it is not clear how large these numbers must get. It remains to be seen whether the results of this paper can be proven without these pitfalls.

ACKNOWLEDGMENTS. I would like to thank the referee for pointing out some rather consequential typos in the original manuscript.

## REFERENCES

1. L. J. Billera and C. W. Lee, *A proof of the sufficiency of McMullen's conditions for f-vectors of simplicial convex polytopes*, J. Combin. Theory Ser. A **31** (1981), 237–255.
2. K. Q. Brown, *Voronoi diagram from convex hulls*, Inform. Process. Lett. **9** (1979), 223–228.
3. A. K. Dewdney and J. K. Vranch, *A convex partition of* $\mathbf{R}^3$ *with applications to Crum's problem and Knuth's post-office problem*, Utilitas Math. **12** (1977), 193–199.
4. H. Edelsbrunner, *Algorithms in combinatorial geometry*, Springer-Verlag, 1987.
5. V. Klee, *On the complexity of d-dimensional Voronoi diagrams*, Arch. der Math. **34** (1980), 75–80.
6. J. Linhart, *Dirichletsche Zellenkomplexe mit maximaler Eckenzahl*, Geom. Dedicata **11** (1981), 363–367.
7. P. McMullen and G. C. Shepard, *Convex polytopes and the upper bound conjecture*, London Math. Soc. Lecture Notes Ser., vol. 3, Cambridge Univ. Press, 1971.
8. I. Paschinger, *Konvexe Polytope und Dirichletsche Zellenkomplexe*, Ph.D. Thesis, Math. Inst., Univ. Salzburg, 1982.
9. F. P. Preparata, *A nearest-point Voronoi polyhedron of n points may have* $O(n^2)$ *vertices*, in Steps Into Computational Geometry, Tech. Rep., Coordinated Science Lab., Univ. of Illinois, Urbana, 1977, pp. 23–24.
10. R. Seidel, *The complexity of Voronoi diagrams in higher dimensions*, Proc. 20th Allerton Conf. on Communication, Control, and Computing, Sponsored by the Department of Electrical Engineering and the Coordinated Science Laboratory of the University of Illinois at Urbana-Champaign, 1982.
11. ——, *Voronoi diagrams in higher dimensions*, Diplomarbeit, Tech. Univ. Graz, Austria, 1982.
12. M. I. Shamos, *Computational geometry*, Ph.D. Thesis, Yale University, 1978.

COMPUTER SCIENCE DIVISION, UNIVERSITY OF CALIFORNIA, BERKELEY, CALIFORNIA 94720
*E-mail address*: seidel@lyra.berkeley.edu

DIMACS Series in Discrete Mathematics
and Theoretical Computer Science
Volume 4, 1991

Applied Geometry
and Discrete Mathematics
THE VICTOR KLEE FESTSCHRIFT

# Stretchability of Pseudolines is NP-Hard

PETER W. SHOR

ABSTRACT. We prove that the problem of determining whether a pseudoline arrangement is stretchable is NP-hard. We also use our techniques to find a symmetrical pseudoline arrangement that is stretchable but not stretchable to a symmetrical line arrangement. The NP-hardness result can also be obtained from a paper by Mnëv (Lecture Notes in Math., vol. 1346, Springer, 1988, pp. 527–544) which implies the stronger result that determining stretchability is equivalent to the existential theory of the reals. We give a short explanation of Mnëv's proof, viewed from a complexity theory point of view, which may be more comprehensible than the original paper to readers who do not know much topology.

## 1. Introduction

A *line arrangement* is the partition of the plane induced by a set of lines in the plane. A *pseudoline* is a simple curve in the plane that goes to infinity in two directions. (In other words, a pseudoline is the image of a line under a homeomorphism of the plane.) A *collection of pseudolines* is a set of pseudolines such that any two members of the set intersect at most once, and cross if they intersect. A *pseudoline arrangement* is the partition of the plane induced by a collection of pseudolines. A pseudoline arrangement is *stretchable* (or *realizable*) if there is an arrangement of lines with the same combinatorial structure. A line or pseudoline arrangement is *uniform* if no three lines intersect in a point and no two lines are parallel.

The stretchability of pseudoline arrangements has a long history. In particular, given a pseudoline arrangement, the question of finding a realization of it has been studied extensively [**BS**]. We show that this problem is NP-hard. We also answer a question of Bokowski and Sturmfels [**BS**, p. 80] by showing that there exists a symmetrical pseudoline arrangement which is stretchable, but which is not stretchable to a symmetrical line arrangement.

The *realization space* of a pseudoline arrangement is the set of line arrangements realizing this pseudoline arrangement. Mnëv [**Mn**] has shown

1980 *Mathematics Subject Classification* (1985 *Revision*). Primary 05B35.

that the topology of the realization space of a pseudoline arrangement can be the same as the topology of any semialgebraic variety (a *semialgebraic variety* is the solution space of a set of polynomial inequalities and equations over the reals). Mnëv's result also implies that determining the stretchability of a pseudoline arrangement is equivalent to the existential theory of the reals. This is stronger than our result.

In §2 of this paper we give our proof of the NP-hardness of determining if a pseudoline arrangement is stretchable. In §3, we use our techniques to show that there exists a symmetric pseudoline arrangement which is stretchable, but not stretchable to a symmetric line arrangement. In §4, we give Mnëv's proof that determining the stretchability of a pseudoline arrangement is equivalent to the existential theory of the reals, with one argument in his proof simplified so as to require less topology than he uses. In §2 and 3, we will be working with pseudoline arrangements—our basic objects will be lines. In §4, we will (as Mnëv does) work with "pseudo-point" arrangements, i.e., a configuration of points for which we know the orientation of all triples. This is an equivalent problem, as pseudo-point arrangements are the projective dual of pseudoline arrangements.

## 2. Proof of NP-hardness

Our proof is based on incidence theorems of projective geometry, namely Pappus' and Desargues' theorems. More specifically, we use the nonrealizable arrangements of pseudolines that can be obtained from these two theorems. The Pappus and Desargues configurations are shown in Figures 1 and 2. The Pappus configuration contains nine lines, each incident with three of the points, and nine points, each on three of the lines. In the Desargues configuration, there are ten lines and ten points. (Note that we do not draw the Desargues configuration in the standard manner.) Pappus' theorem is that in the Pappus configuration, if any eight of these triples of lines are concurrent (or eight of the triples of points are collinear), the last triple must also be concurrent (collinear). Similarly, Desargues' theorem is that if nine of these triples of lines (points) are concurrent (collinear), the last triple must also be. From either of these configurations, we can obtain a nonrealizable arrangement of pseudolines. We do this by slightly bending each of the lines to "go around" the points; that is, we replace each of the points in the configuration by a small triangle. By bending all lines in the right way (see Figure 3 and 4), we obtain nonrealizable uniform configurations [**Gr**].

We prove NP-completeness by reducing a variant of the NP-complete 3-SAT problem to the stretchability problem. The 3-SAT problem is: Given a Boolean expression in conjunctive normal form containing only three variables in each clause, is there an assignment of the variables making the expression true [**GJ**]? It is easy to show that this problem is still NP-complete if we require that in each clause either only nonnegated variables or only negated variables appear. We use this variant of 3-SAT, called *monotone* 3-SAT

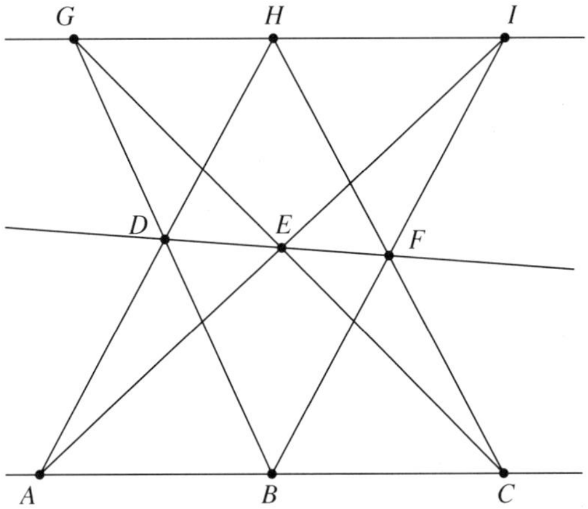

FIGURE 1. A Pappus configuration.

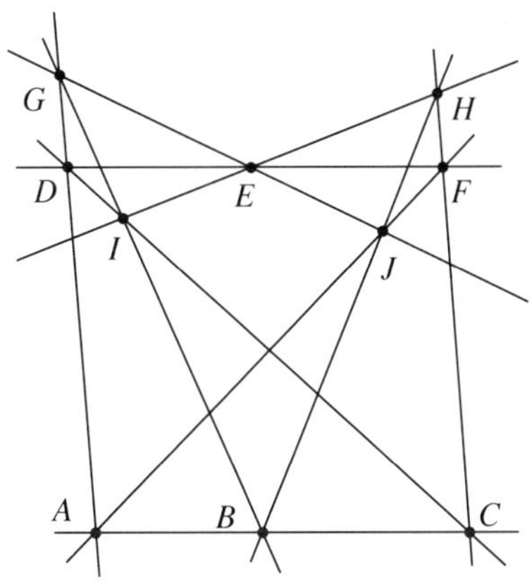

FIGURE 2. A Desargues configuration.

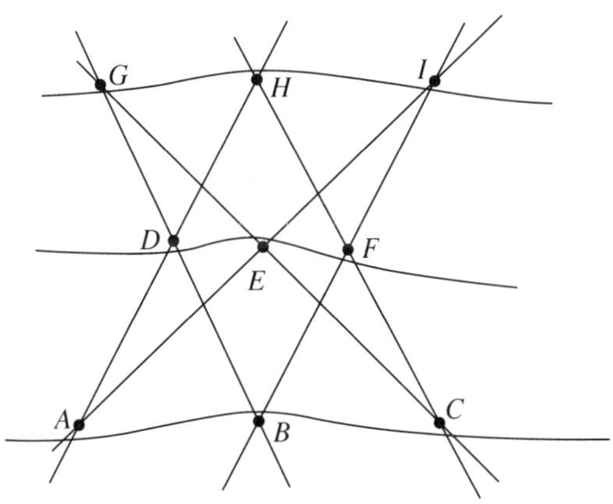

FIGURE 3. A nonrealizable Pappus configuration.

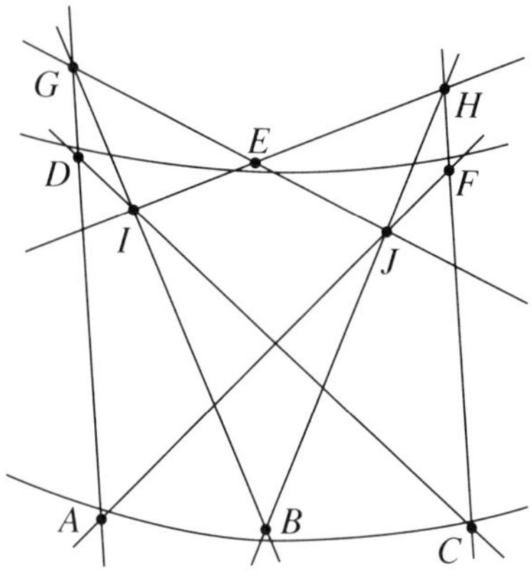


FIGURE 4. A nonrealizable Desargues configuration.

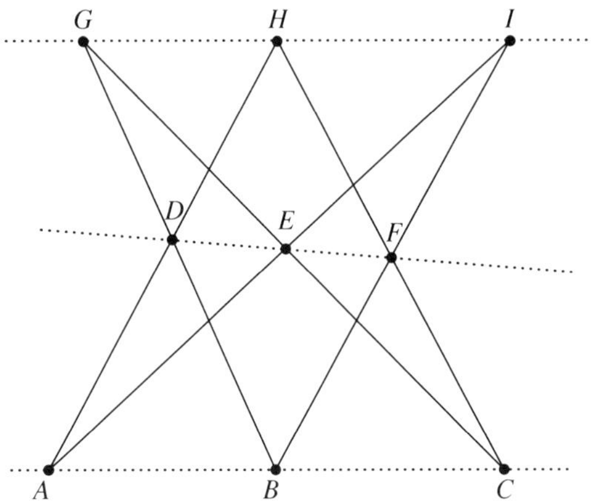


FIGURE 5. A Pappus configuration with three "imaginary" lines.

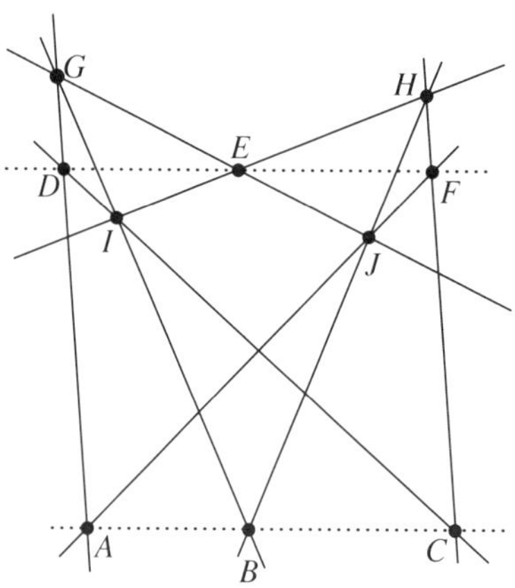


FIGURE 6. A Desargues configuration with two "imaginary" lines.

[GJ, p. 259]. We call a clause *positive* if it contains only nonnegated variables and *negative* if it contains only negated variables. To show monotone 3-SAT is NP-hard, we can reduce 3-SAT to it by adding a separate variable $y_i$ for each variable $x_i$, and setting $x_i = \overline{y_i}$ by adding the clauses $x_i \vee y_i \vee 0$ and $\overline{x_i} \vee \overline{y_i} \vee \overline{1}$. We then replace $\overline{x_i}$ by $y_i$ where needed so as to make all clauses either positive or negative.

Given a monotone 3-SAT formula, we will construct a pseudoline arrangement which is stretchable if and only if the formula is satisfiable. In the pseudoline arrangement, clauses will correspond to modified Pappus configurations, variables will correspond to certain triples of points, and variables will be linked to clauses by modified Desargues configurations.

We now look at the Pappus configuration more closely. We let three of the lines be "imaginary" (see Figure 5), i.e., the lines do not appear in our final pseudoline arrangement, but are used implicitly in the proof. We call the three points on each of these "imaginary" lines a *triple.* In our construction, each Pappus configuration corresponds to a clause of the Boolean formula and each of the three triples in the Pappus configuration corresponds to one of the variables in the clause. Furthermore, the position of the points in the triples corresponds to the truth (or falsehood) of the variables. Specifically, let $PQR$ be the triple corresponding to the variable $x_i$, with point $Q$ between points $P$ and $R$. We will put point $Q$ above (or on) the line $PR$ if $x_i$ is true and below if $x_i$ is false. Unless all three triples of points are collinear, the Pappus configuration is realizable if and only if not all the variables are the same; i.e., if in at least one of three triples $ABC$, $DEF$, and $GHI$ the middle point is above the segment between the end points of that triple, and in at least one of these triples the middle point is below the segment.

In our pseudoline arrangement, we will have many different Pappus configurations, one corresponding to each clause. This gives rise to many triples all corresponding to the same variable. We must connect these triples somehow so as to ensure that they all give the same value of $x_i$. To do this, we introduce three new points for the variable $x_i$, and make the position of these three points correspond to the value of $x_i$. We will call these three points the *master triple* for $x_i$. We then hook this master triple up to all the *slave triples* that correspond to $x_i$ in the Pappus configurations. We could do this by using more Pappus configurations, but for reasons which we will explain later we use Desargues configurations instead. In a Desargues configuration, if we take out the two lines $ABC$ and $DEF$ (see Figure 6), the resulting configuration is realizable only if point $B$ is above line $AC$ and point $E$ is below line $DF$ or if point $B$ is below and point $E$ is above. Assume we are dealing with a positive clause. We wish to hook up one of the slave triples in the clause (corresponding to variable $x_i$) with the master triple for $x_i$. To do this we introduce a new Desargues configuration with the bottom triple $ABC$ being the master triple and the top triple $DEF$ being the slave triple. We now define true and false for a master triple to the opposite of true and false for a slave triple; i.e., in a master triple, for a false variable, the middle point is above the other two points, and for a true variable, the middle point is below the other two points. This ensures that if the points in the master triple are false, then the points in the slave triple are false (and vice versa). Our actual construction will use a modified Desargues configuration having only the property that if the points in the master triple are false, then the points in the slave triple are. However, this is exactly what we want: in a Pappus configuration corresponding to a positive clause, the configuration is nonrealizable if and only if the three corresponding master triples are in the false position. For Pappus configurations corresponding to a negative clause, we will turn the Desargues configuration upside-down, so the Pappus configuration is nonrealizable if and only if the three master triples are in the true position.

The astute reader may have noticed that we are using both Desargues and Pappus configurations, where intuitively it seems that only one of these configurations would suffice. We could probably obtain a proof using just one of these configurations, but each has properties which make the proof simpler if we use them both. The advantage of the Desargues configuration is that it has more degrees of freedom than the Pappus configuration; this makes it easier to show that a pseudoline configuration corresponding to a satisfiable formula is stretchable. The advantage of the Pappus configuration is that there are three disjoint lines contained in a Pappus configuration, one for each of the variables in a clause, whereas a Desargues configuration contains only two disjoint lines.

We are now ready to give the construction. First, we describe how to construct the pseudoline arrangement corresponding to a given Boolean formula. This arrangement will be stretchable if and only if the formula is satisfiable. During most of the construction of our pseudoline arrangement, we will actually be constructing a line arrangement. It is only in the last step of our construction that we will perturb this line arrangement to get a pseudoline arrangement that may not be stretchable.

To start constructing the arrangement, for each variable we place a triple of three points (the master for this variable) on a line. All these variables will be placed near some horizontal line, say $y = 0$, but in "general position"; i.e., the only relations between the points are that the three points in a triple lie on a line. This can be accomplished by putting all the triples down on the line $y = 0$ and then perturbing them slightly (see Figure 7). We will later show how to accomplish this in polynomial time.

We must include the values 1 and 0 among our variables. There are two ways of doing this: we can either add an extra line in the master triples corresponding to 0 and 1, which forces the middle point to be above (or below) the line through the end points, or we can add a set of clauses which forces one variable to be 0 and another to be 1.

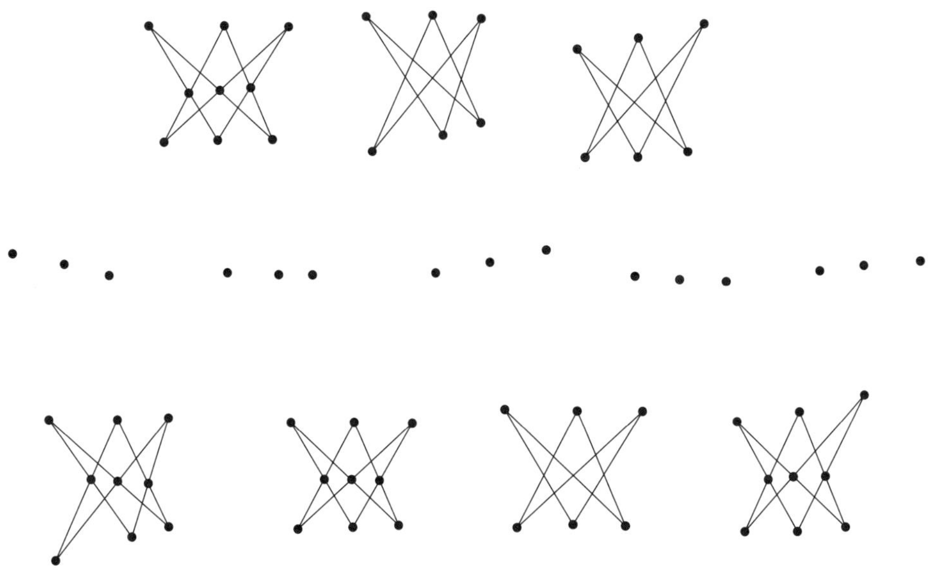

FIGURE 7. The general layout of our construction, with Pappus configurations corresponding to positive clauses on top, Pappus configurations corresponding to negative clauses on bottom, and master triples in the middle.

Next, for each positive clause, we place a modified Pappus configuration above the horizontal line $y = 0$ which all the master triples lie near, and for each negative clause we place a modified Pappus configuration below this horizontal line (see Figure 7). Here, by modified Pappus configurations, we mean a Pappus configuration missing the three horizontal lines as in Figure 5. Again, although all the triples in these modified Pappus configurations are nearly horizontal, all these configurations must lie in general position, so the only relations between points are those implied by the fact that they are all modified Pappus configurations. The important part of this placement is that the top side of each of the master triples "sees" the bottom of the triples in positive Pappus configurations, and the bottom of each of the master triples "sees" the top of the triples in negative Pappus configurations.

Next, with a "contracted" Desargues configuration (see Figure 8(a)), we will connect each master variable to the Pappus configurations corresponding to clauses containing it (see Figure 9(a)). A *contracted Desargues configuration* has had points $D$, $G$, and $I$ identified and points $F$, $H$, and $J$ identified, so that it contains only six points, as in Figure 8(a).

We have now placed down essentially all the lines that we need in our arrangement. We will obtain the final arrangement by slightly perturbing the arrangement that we already have. In our perturbation, we will replace certain lines by pairs of lines which differ by a small angle. Also, in the neighborhood of certain points which have many lines passing through them, we will perturb the lines so that they no longer are all concurrent. We can perturb the lines to make only the desired changes because everything is in general position.

We perturb the configuration by replacing each of the contracted Desargues configurations by a new, more complete Desargues configuration. Specifically, we replace the configuration in Figure 8(a), first by the configuration in Figure 8(b), and then by the one in Figure 8(c). Thus, we perturb the two lines $b_1$ (originally $BD$) and $b_2$ (originally $BF$) slightly inward, and add the new points $G = b_1 \cap AD$, $H = b_2 \cap CF$, $I = b_1 \cap CD$, and $J = b_2 \cap AF$. Next, line $DEF$ is replaced by the two lines $GEJ$ and $IEH$. This gives the configuration in Figure 8(b). Figure 9(b) shows this configuration connected to a Pappus configuration as it would appear in our construction.

As the points (and Pappus configurations) were placed in general position, the only triple intersections containing lines $b_1$, $b_2$ or $DF$ in the original Desargues configuration are $B$, $D$, $E$, and $F$. If the amount we perturb the lines $b_1$ and $b_2$ by is sufficiently small, the only place that the arrangement changes is in the neighborhood of $D$, $E$, and $F$. It is easy enough to see what happens at $D$ and $F$ (see blown-up neighborhood of $D$ in Figure 10); all lines previously passing through $E$ still pass through it.

So far, it is possible to make all our perturbations using not just pseudoline arrangements but actual line arrangements, as in Figure 9(b). The arrangement can still be realized with every triple of points lying on a line and with all our incomplete Pappus and Desargues configurations being actual Pappus and Desargues configurations and not perturbed versions of them. This all changes in the next step.

The last step in our construction is to replace the points $I$ and $J$ by very small triangles $I_1 I_2 I_3$ and $J_1 J_2 J_3$, as shown in Figures 8(c) and 9(c). This is easy to do with pseudolines. In a line arrangement, by Desargues' theorem, replacing $I$ and $J$ by triangles in this way either forces point $E$ or point $B$ to move down. This makes it no longer possible to realize the pseudoline arrangement while having every triple of points lie on a line, and thus forces decisions about how to perturb these triples. We will show that it is possible to find a consistent set of such decisions if and only if the Boolean formula is satisfiable.

We will show that the above procedure finds, given a monotone 3-SAT Boolean formula, a pseudoline arrangement which is equivalent. That is, the pseudoline arrangement is realizable if and only if the expression is satisfiable. We still have not quite shown how to produce the pseudoline arrangement in polynomial time—we must give an algorithm for our initial step of putting the Pappus configurations and the sets of three points in a row down in "general position." As one might expect, this step is not hard. We also have not yet obtained a *uniform* pseudoline configuration, which we must do to show that the problem is NP-complete in the uniform case. There is a trick discovered by both Mnëv and Sturmfels and White [**Mn, SW**] which we can apply to turn the above configuration into a uniform one. We will discuss these steps in more detail later.

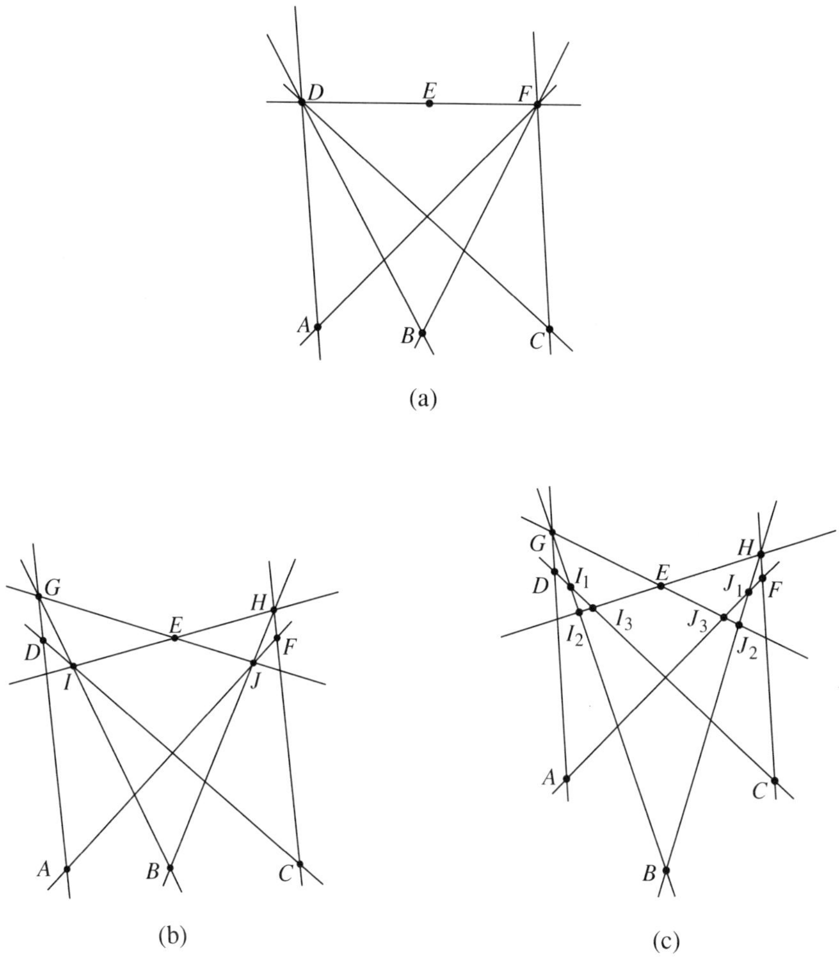


FIGURE 8(a) A "contracted" Desargues configuration. (b) The intermediate step in expanding a contracted Desargues configuration. (c) The final step: replacing points $I$ and $J$ with triangles.

What we do now is to show that the above pseudoline configuration is stretchable if and only if the expression generating it is satisfiable. We first do the easy direction: if the configuration is stretchable, then the expression is satisfiable. We later do the hard direction.

Suppose the arrangement is stretchable. Consider a realization of the arrangement. Let the value of a variable $x_i$ be true if, in the master triple corresponding to $x_i$, the middle point is below the line segment joining the other two points. Let the variable be false if the middle point is above this segment. If the point is on the segment, we may let the variable be either false or true. Now if this truth assignment makes the Boolean expression

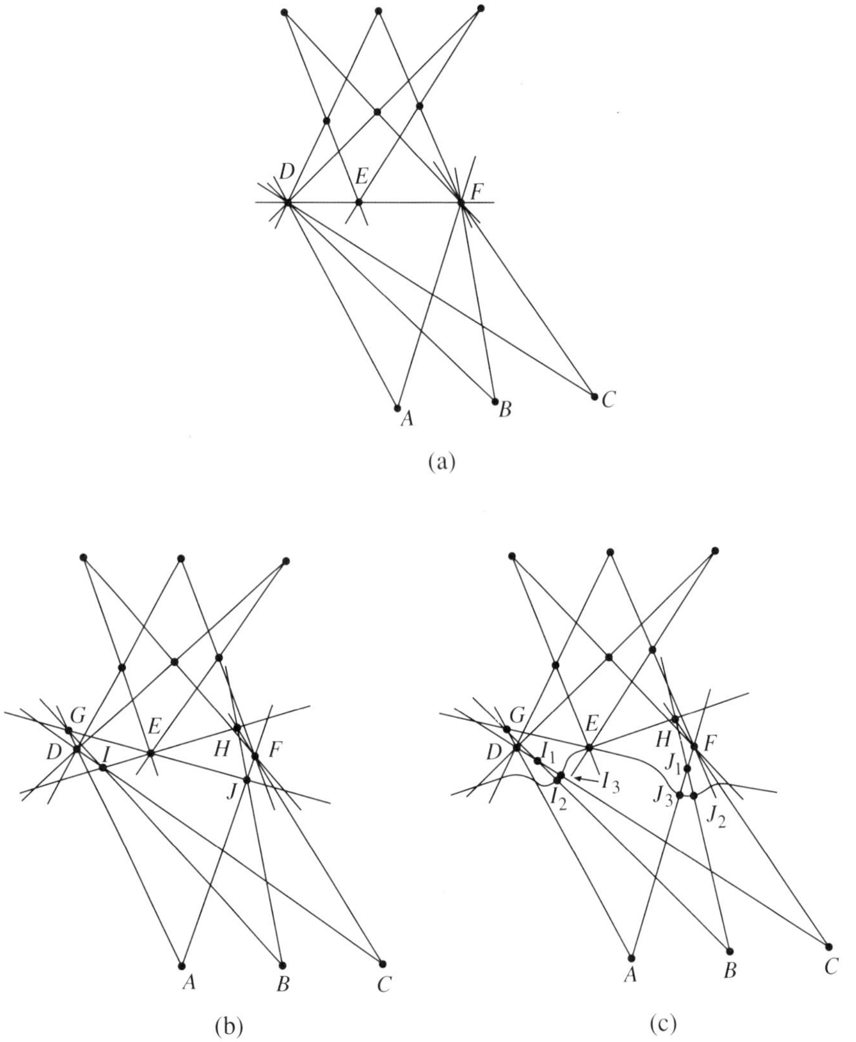


FIGURE 9(a) A contracted Desargues configuration in place: $ABC$ is the master triple and $DEF$ the corresponding slave triple in the Pappus configuration at top. (b) The intermediate step in expanding the contracted Desargues configuration in (a). (c) Replacing points $I$ and $J$ in (b) with triangles.

false, there must be some clause which is false. If this clause is a positive clause, say $x_{i_1} \vee x_{i_2} \vee x_{i_2}$, then all the variables in it are false. This means that, in the master triple corresponding to each of these variables, the middle point is above (or on) the line segment joining the other two. Since the points in this master triple are connected by a modified Desargues configuration to the points in the slave triples in the Pappus configuration corresponding to this clause, the middle point lies strictly below the line segment joining the other two points in all three triples in this Pappus configuration. However,

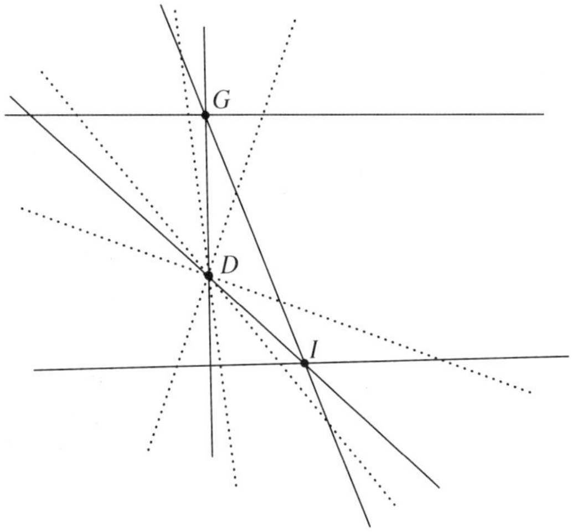


FIGURE 10. The neighborhood of point $D$ before point $I$ is replaced by a triangle.

this gives a nonstretchable version of the Pappus configuration. Similarly, a false negative clause also gives a nonstretchable configuration.

The other direction is harder. Assume that we are given a satisfying assignment of the Boolean formula. We must show that the corresponding pseudoline arrangement is stretchable. We start by putting the Pappus configuration and the master triples down in general position, with the three points in all triples collinear. We will then show that we can move the middle points of the master triples up or down, depending on whether the corresponding variable is false or true, perturb the Pappus configurations by a very small amount, and add the Desargues configurations linking the slave triples to the "master" triples to give a realization of the desired pseudoline configuration. We will do this by proving a series of lemmas showing that we can realize the configuration by perturbing the original configuration in a certain way.

Before we can proceed with the lemmas, we must give some definitions. For each line in our configuration, we will call two or three points on it "anchor points." We will consider the line to be "fixed" to these points, so that when these points are perturbed, the line is perturbed. For the Desargues configurations, the lines, listed by their two anchor points, are $AD$, $AF$, $CD$, $CF$, $BG$, $BH$, $EG$, $EH$. For the Pappus configurations, the anchor points will be $ADH$, $AEI$, $BDG$, $BFI$, $CEG$, and $CFH$; when we move anchor points in a line having three anchor points, we must make sure that they remain collinear. We use anchor points to bound the effects of small perturbations on the overall configuration.

LEMMA 1. *There exists an $\varepsilon$ such that if all anchor points are moved by at most $\varepsilon$, any three lines not originally concurrent retain their relative orientations.*

Proof. This is clear for any specific three lines. To obtain an $\varepsilon$ that works for all lines, simply choose $\varepsilon$ to be the minimum over all sets of three lines of the $\varepsilon$'s for each set. □

For our next lemma, when we talk about the line arrangement of a Desargues configuration, we also include all lines passing through points in the configuration, although these lines may technically not be in the configuration.

Using this lemma, we will show that there is some $\varepsilon_1$ and a corresponding $\varepsilon_2$ such that if we move all the middle points of the master triples either up or down by $\varepsilon_1$, depending on whether the corresponding variable in the satisfying assignment is true or false, then the points of the Pappus configurations can be moved by a distance $\varepsilon_2$ so as to realize the arrangement. We first show the existence of these $\varepsilon$'s for each Desargues configuration separately, and then show that we can choose these $\varepsilon$'s to be valid for the whole arrangement. Showing that these $\varepsilon$'s exist for each Desargues configuration is the substance of Lemma 2.

Lemma 2. *For each Desargues configuration connecting a master triple to a positive clause, the following assertions hold*:

(1) *For every* $\varepsilon_1$, $0 < \varepsilon_1 \leq \varepsilon/2$, *there exists an* $\varepsilon_2 > 0$ *such that if point* $B$ *is moved up by* $\varepsilon_1$ *to* $B'$, *if points* $D$, $E$, *and* $F$ *are moved by less than* $\varepsilon_2$ *to* $D'$, $E'$, *and* $F'$, *and if point* $E'$ *is below line* $D'F'$, *then the points* $G$, $H$, $I_1$, $I_2$, $I_3$, $J_1$, $J_2$, *and* $J_3$ *can be added so as to make the resulting line arrangement of the Desargues configuration be combinatorially the desired arrangement, even if all the anchor points not in this Desargues configuration are moved by at most* $\varepsilon/2$.

(2) *For every* $\varepsilon_1$, $0 < \varepsilon_1 \leq \varepsilon/2$, *there exists an* $\varepsilon_2 > 0$ *such that if point* $B$ *is moved down by* $\varepsilon_1$ *to* $B'$, *if points* $D$, $E$, *and* $F$ *are moved by less than* $\varepsilon_2$ *to* $D'$, $E'$, *and* $F'$, *and if point* $E'$ *is above line* $D'F'$, *then the points* $G$, $H$, $I_1$, $I_2$, $I_3$, $J_1$, $J_2$, *and* $J_3$ *can be added so as to make the resulting line arrangement of the Desargues configuration be combinatorially the desired arrangement, even if all the anchor points not in this Desargues configuration are moved by at most* $\varepsilon/2$.

(3) *There exists an* $\varepsilon_1 > 0$ *such that if point* $B$ *is moved down by at most* $\varepsilon_1$ *to* $B'$, *then the points* $G$, $H$, $I_1$, $I_2$, $I_3$, $J_1$, $J_2$, $J_3$ *can be added so as to make the resulting line arrangement of the Desargues configuration be combinatorially the desired arrangement, even if all the anchor points not in this Desargues configuration are moved by at most* $\varepsilon/2$.

Proof. (1) We will first show (1), assuming that $I_1$, $I_2$, and $I_3$ are all identical to point $I$, and similarly the $J_i$'s are identical to $J$, as in Figure 8(b). Let us consider points $D'$ and $F'$ to be fixed. Let us also consider all anchor points outside the Desargues configuration to be fixed (except on lines with three anchor points, one of them a point in the Desargues configuration, in which case we only fix one of the two outside anchor points). As $B$ was moved upward to produce $B'$, by Desargues' theorem no matter where we place the lines $BG$ and $BH$, $E'$ will be below $D'F'$. However, we can make $E'$ approach segment $D'F'$ by moving line $BG$ towards $D'$ and $BH$ towards $F'$. By moving only $BG$ towards $D'$, we move $E'$ to the right, and by moving only $BH$ towards $F'$, we move $E'$ to the left. We can therefore move these two lines simultaneously so as to make $E'$ approach any point on segment $D'F'$. Thus, in some sufficiently small neighborhood of $E$, we can put $E'$ anywhere below $D'F'$. This involves moving points $G$, $H$, $I$, and $J$ by some amount that we can make arbitrarily small by making $\varepsilon_2$ arbitrarily small. It is easy to see that by making $\varepsilon_2$ sufficiently small, the combinatorial structure of the arrangement near points $B$, $D$, $E$, and $F$ is unaffected. By Lemma 1, the combinatorial structure cannot be affected anywhere else. Finally, we can perturb the lines $GE$ and $HE$ by a tiny amount to make triangles at $I$ and $J$ without affecting the combinatorial structure of the arrangement elsewhere.

We have now proved (1) with points $D'$, $F'$, and anchor points outside the Desargues configuration all fixed. However, by compactness, this implies (1) even when these points are not fixed: the set of all positions for these points at most $\varepsilon/2$ away from their original positions is a compact set, and as the maximum possible $\varepsilon_2$ is a continuous function of these positions, $\varepsilon_2$ as a function of the positions of these points must be bounded away from 0 on this set.

(2) This proof is very similar to the proof of (1), and will be omitted.

(3) We first place points $G$ and $H$ on lines $AD$ and $CF$, sufficiently close to points $D$ and $F$, respectively, so that the combinatorial configuration is as in Figure 8(b), even if all the other anchor points are moved by at most $\varepsilon/2$. Now, for the time being, let us fix all these other anchor points. Suppose $B'$ is below line $AC$. Let $B''$ be the intersection of lines $AC$ and $B'E$. Since $B''$ is on line $AC$ and $E$ is on line $DF$, by moving point $G$ closer to $D$ or point $H$ closer to $F$, we can keep point $E$ fixed and move point $B$ to $B''$. Now, note that if we move $B''$ downward along line $B'E$ to point $B'$, keeping points $G$ and $H$ fixed, we open small triangles at points $I$ and $J$, producing the desired configuration of Figure 8(c). The amount that we can move $B''$ downward is constrained by what happens to other lines through $D$ and $F$ (see Figure 11). For example, if we move $B''$ down too far, we may put $I_2$ on the wrong side of a different line through point $D$ (to another master triple). However, if we keep points $B''$ and $B'$ close enough to $B$ so as to move lines $BG$ and $BH$ through a sufficiently small angle, we do not change the combinatorial configuration by moving lines $BG$ or $BH$. Thus, if we make $\varepsilon_1$ small enough, the angle $B'GB$ becomes arbitrarily small, and we obtain the results in (3). Again, by compactness, the proof with anchor points fixed implies the result with nonfixed anchor points. □

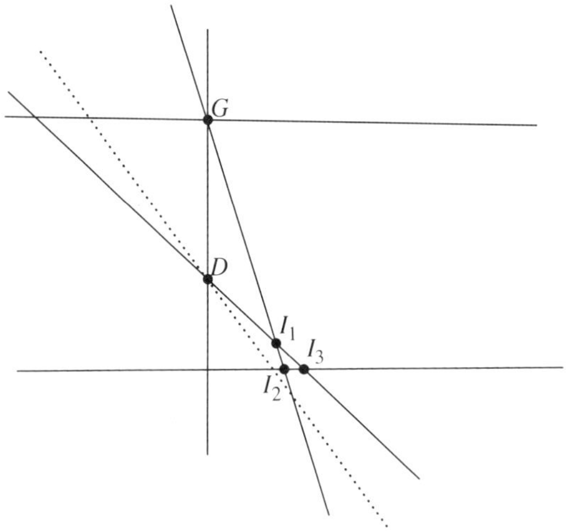


FIGURE 11. The dotted line must pass below point $I_2$.

Now, we use Lemma 2 to prove the following:

LEMMA 3. *In our construction, if the Boolean formula is satisfiable then the corresponding pseudoline arrangement is realizable.*

PROOF. To show the pseudoline arrangement is realizable, we use Lemma 2. Assume we are given a satisfying assignment for the Boolean formula. Consider a Pappus configuration in our construction. This configuration corresponds to a clause of the Boolean formula. Assume that this is a positive clause, so the Pappus configuration is above the line $y = 0$. (If it corresponds to a negative clause, the argument is symmetric.) In the satisfying assignment, this clause must have either one, two, or three variables set to true, and the rest set to false. If all three variables are set to true, we do not move any points of the Pappus configuration. If two variables are true and one false or if two variables are false and one true, we move all the points of the Pappus configuration by at most $\varepsilon_2$ (which is a quantity that will be determined later) so that for the true slave variables, the middle point is above the other two, and for the false slave variables, the middle point is below the other two. (This can easily be done.) Now, we go back to Lemma 2. We first choose an $\varepsilon_1$ smaller than all the $\varepsilon_1$'s of Lemma 2(3) for the Desargues configurations in our construction. We then move all middle points of the master triples up or down by exactly $\varepsilon_1$, depending on whether the variable is false or true. We then choose $\varepsilon_2$ so it is smaller than all the $\varepsilon_2$'s of Lemma 2(1), (2) for the Desargues configurations in our construction. Now, by Lemma 2, each Desargues configuration can be perturbed appropriately, and by Lemmas 1 and 2 none of these perturbations interferes with another (all the anchor points have been moved by less than $\varepsilon/2$), so we have a realization of our pseudoline arrangement. □

To conclude the proof that determining stretchability of pseudolines is NP-hard, we need to show how to construct the pseudoline arrangement in polynomial time. To do this, we need to show how to lay out the master triples and the Pappus configurations in general position. One way to do this is to first lay out all the configurations, but not necessarily in general position, and then perturb them to obtain a configuration in general position. We do not need to obtain actual coordinates for the perturbed lines; we merely need the combinatorial structure of the resulting line arrangement. It thus suffices to perturb the lines symbolically. To perturb a Pappus configuration, we first symbolically translate the configuration by an "infinitesimal" amount. This eliminates all degeneracies between lines in the Pappus configuration and lines outside it, except for those at points at infinity. To remove these, we symbolically rotate the Pappus configuration. A similar procedure works to put the master triples into general position.

We have now shown that determining the stretchability of a pseudoline arrangement is NP-hard. The arrangements we have constructed, though, are not uniform. To show that it is still NP-hard with a uniform arrangement, we use a lemma proved both by Mnëv and by Strumfels and White on *constructible* pseudoline arrangements.

A pseudoline arrangement is *constructible* if we can produce it by adding the pseudolines one at a time, while never placing a pseudoline through more than two points defined by intersections of previously placed pseudolines.

LEMMA 4 [**Mn, SW**]. *Given a constructible pseudoline arrangement $\mathscr{L}$ one can find (in polynomial time) a uniform pseudoline arrangement $\mathscr{L}'$ which is stretchable if and only if $\mathscr{L}$ is stretchable.*

PROOF. To produce $\mathscr{L}'$, consider the order of placing the pseudolines that shows the arrangement $\mathscr{L}$ is constructible. We will examine the pseudolines in the reverse order, and replace pseudolines that passed through one or two points previously defined by intersections by either two or four new pseudolines. When processing a pseudoline $L$ that passes through one point $P$ defined by intersections of pseudolines, replace it by two pseudolines as in Figure 12. When processing a pseudoline $L$ that passes through two points defined by intersections of pseudolines, replace it by four pseudolines as in Figure 13. This process will produce a uniform pseudoline arrangement $\mathscr{L}'$. Clearly, if $\mathscr{L}$ is stretchable, $\mathscr{L}'$ is. If $\mathscr{L}'$ is stretchable, one can produce a stretching of $\mathscr{L}$ by starting with the realization of $\mathscr{L}'$ and undoing the process we used to construct $\mathscr{L}'$, as shown in more detail in [**Mn, SW**]. □

Now, to show the NP-hardness result applies to uniform arrangements, we need to show that the arrangement we produced is constructible. If we first put down the master triples and the Pappus configurations, and then add

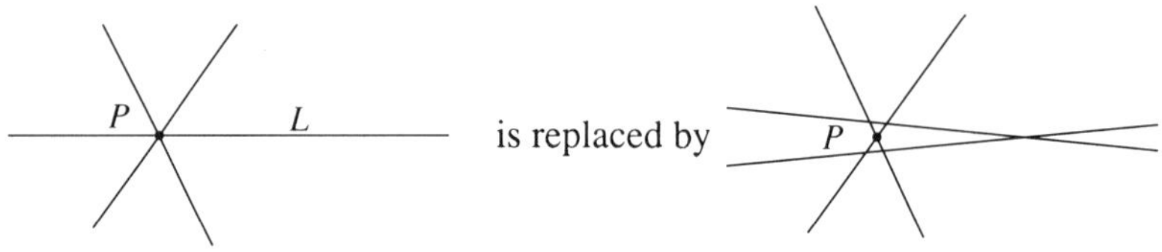


FIGURE 12. Making a constructible arrangement uniform: replacing a line through one point.

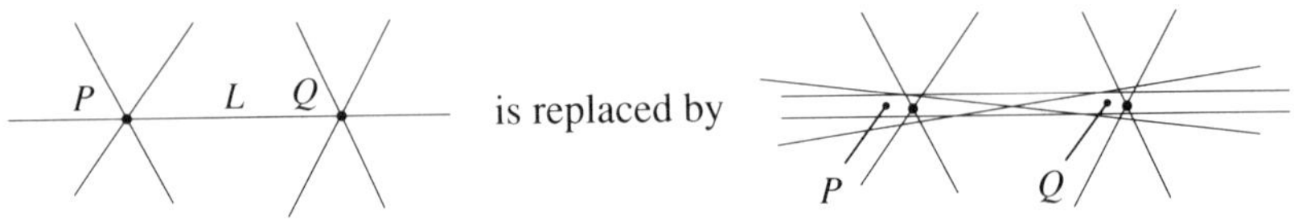


FIGURE 13. Making a constructible arrangement uniform: replacing a line through two points.

the lines in each of the Desargues configurations in the order $ADG$, $CFH$, $EH$, $EG$, $CD$, $AF$, $BG$, $BH$, we never put a line through more than two previously defined points. Our arrangement is thus constructible, so by Lemma 4 realizability of uniform pseudoline arrangements is also NP-hard.

## 3. A symmetric pseudoline arrangement not symmetrically stretchable

By using the tools introduced in the previous section, it is easy to produce a symmetrical pseudoline arrangement that is not stretchable to a symmetrical line arrangement. We will construct this in much the same manner that we constructed the pseudoline arrangement from an arbitrary 3-SAT formula in the previous section. This arrangement will correspond to the formula $(x \vee y) \wedge (\bar{x} \vee \bar{y})$. The only satisfying assignments for this formula are, clearly, $x = 1$, $y = 0$, and $x = 0$, $y = 1$. We will construct the symmetric pseudoline arrangement with a vertical axis of symmetry. To do this, we put down on the left of the vertical axis, in general position, a master triple for $x$, and on the right of the vertical axis, symmetrical to the triple for $x$, a master triple for $y$. On the left of the axis of symmetry, we put in general position a Pappus configuration both above and below the master triples. We then put these Pappus configurations symmetrically to the right of the line (see Figure 14). Since these Pappus configurations correspond to clauses containing two variables, we add an extra line through one of the three triples (say, through points $DEF$ in Figure 5). Now, using Desargues configurations as in the previous section, we connect the master triple for $x$ to points $ABC$ in the Pappus configurations on the left and to points $GHI$

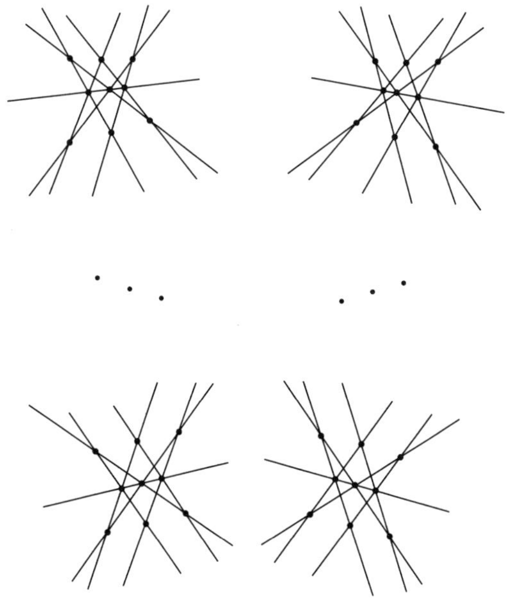

FIGURE 14. The master triples and Pappus configurations for a symmetrical pseudoline arrangement not symmetrically stretchable.

in the Pappus configurations on the right. Similarly, we connect the master triple for $y$ to points $GHI$ in the Pappus configurations on the left and to points $ABC$ in the Pappus configurations on the right. By the arguments in the previous section, the resulting pseudoline configuration is realizable only by configurations in which the middle point of the master triple for $x$ is above the line through the other two points in this triple and the middle point of the master triple for $y$ is below the line through the other two points, or vice versa. Again, by the arguments in the previous section, this configuration can indeed be realized. Thus, we have a symmetric pseudoline arrangement which is not symmetrically stretchable.

By using Lemma 4, we can turn this pseudoline arrangement into a uniform one. All we need to show is that we can apply Lemma 4, and retain the symmetry. We can do this because in this arrangement the intersection of a symmetric pair of lines never lies on more than these two lines.

## 4. A computer scientist's view of Mnëv's universality theorem

Mnëv [**Mn**] proved an even stronger theorem. When translated into complexity theory terms, his theorem implies that determining the stretchability of pseudoline arrangements is equivalent to the existential theory of the reals. What he actually shows is the following stronger theorem: given any primary semialgebraic variety (i.e., the solution space of a set of equations and strict inequalities over the reals) there exists a pseudoline arrangement whose realization space has the same topology (more precisely, the semialgebraic variety and the realization space are stably equivalent). For more details, see [**Mn**]. If the semialgebraic variety is empty (i.e., the equations and inequalities have no solution) the pseudoline arrangement will not be stretchable. To show that this implies the complexity result, all that is necessary is to show how to find this pseudoline arrangement in polynomial time. This can be done by carefully following Mnëv's proof [**Po**]. We will show how to find such a pseudoline arrangement using a variant of his proof which is not as topological, and is thus easier for computer scientists to understand.

Mnëv uses an intermediate step to reduce stretchability of pseudoline arrangements to the existential theory of the reals. That is, he reduces the existential theory of the reals to an intermediate problem, which he then reduces to the stretchability of pseudoline arrangements. We will call the intermediate problem "the existential theory of totally ordered real variables." The problem is:

> Given a set of variables $x_1, x_2, x_3, \ldots, x_n$, a set of equations on these variables of the forms
>
> $$x_i + x_j = x_k, \qquad x_i \times x_j = x_k,$$
>
> and the inequalities
>
> $$1 = x_1 < x_2 < x_3 < x_4 < \cdots < x_n,$$
>
> does there exist a set of real numbers satisfying these equations and inequalities?

Instead of proving directly that this problem is equivalent to the stretchability of pseudoline arrangements, we will show that it is equivalent to the realizability of point configurations, as Mnëv does. A *point configuration* is a set of points together with an orientation on every triple of points (i.e., we know whether each ordered triple is clockwise, counterclockwise, or collinear). Because line arrangements and point configurations are related by projective duality, stretchability of line arrangements is equivalent to realizability of point configurations. We produce a point configuration which is realizable if and only if there is a vector of real numbers satisfying our equations and inequalities.

We first show how to reduce the existential theory of totally ordered real variables to realizability of point configurations and then we show how to reduce the existential theory of the reals to the existential theory of totally ordered real variables.

For the first reduction (to realizability of point configurations) we use Mnëv's proof. We first place three lines in the plane. We can assume without loss of generality that these are the $x$-axis, the $y$-axis, and the line at infinity. (Otherwise we apply a projective transformation to obtain these.)

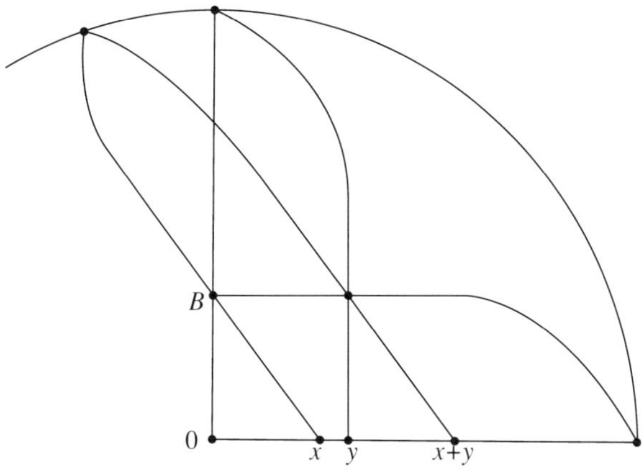


FIGURE 15. Addition in Mnëv's construction. The large circle is the line at infinity.

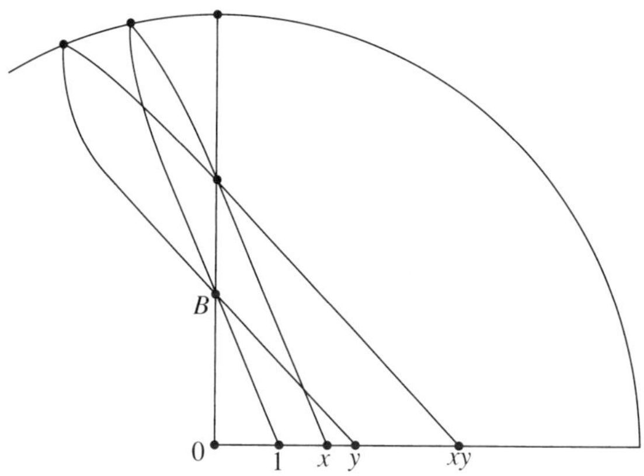


FIGURE 16. Multiplication in Mnëv's construction. The large circle is the line at infinity.

We next place points $P_{x_1}, P_{x_2}, \ldots, P_{x_n}$, corresponding to our variables, on the $x$-axis. The $x$-coordinate of point $P_{x_i}$ will be the value of the corresponding variable $x_i$. To perform an addition we introduce the set of points in Figure 15, and to perform a multiplication we introduce the set of points in Figure 16. The $y$-coordinate of point $B$ in Figures 15 and 16 will be different for each equation; we will denote these points by $B_1, B_2, \ldots,$ in the order that we place these equations in the point configuration. Performing additions and multiplications in a similar manner is an old technique; Mnëv's contribution was to realize that if the multiplications and additions are done in this manner, and point $B_i$ in Figures 15 and 16 is placed sufficiently closer to $y_\infty$ than points $B_1, B_2, \ldots, B_{i-1}$, then the resulting line arrangement has a unique combinatorial structure, and thus the realizability of the point configuration is equivalent to the solvability of these equations and inequalities.

The next step is to reduce the existential theory of the reals to the existential theory of totally ordered real variables. We do this in three steps. The first is by reducing the existential theory of the reals to the problem: Given a set of equations and inequalities in the forms

$$x_i + x_j = x_k, \qquad x_i \times x_j = x_k, \qquad x_i < x_j,$$

is there a solution?

One can take any polynomial equation and reduce it to an equation of this form just by introducing variables for all the intermediate steps and building the polynomial term by term. For example, the equation $x^5 + y^2 = 2$ reduces to the following set of equations (where new variables are introduced by $V$'s, with subscripts standing for the expression which the $V$'s are supposed to represent):

$$V_{x^5} + V_{y^2} = 2, \qquad V_{x^5} = V_{x^4} \times x,$$
$$V_{x^4} = V_{x^2} \times V_{x^2}, \qquad V_{x^2} = x \times x, \qquad V_{y^2} = y \times y.$$

Constants can be built by starting with 1 and adding and multiplying to obtain integers, dividing to obtain rationals, and solving polynomial equations to obtain algebraic numbers. The number of such equations is clearly no more than a constant factor times the size of the input. Thus, we only require the basic constant 1. For complexity purposes, an inequality of the form $x \geq y$ can be taken care of by replacing it with the equation $x = y + V^2$, where $V$ is a new variable. For the topological equivalence to hold, it seems that we need to restrict ourselves to strict inequalities (i.e., *primary* semialgebraic varieties).

We now reduce this problem to the same problem of determining if a set of equations of the above form has a solution, but with the additional restriction that all the variables be greater than 1. To do this, we replace each variable $x_i$ with a variable $V_{x_i+a}$, which will be assumed to have the value $x_i+a$ for some $a$, where $a$ can be arbitrarily large. We also must introduce the variables $V_a$, $V_{a^2}$, and $V_{a+a^2}$, and the relations $V_a \times V_a = V_{a^2}$ and $V_a + V_{a^2} = V_{a+a^2}$. We now show that by introducing a few extra equations, we can add and multiply using only these new variables.

Comparisons are easy, since $x_i < x_j$ is equivalent to $V_{x_i+a} < V_{x_j+a}$. To add, we replace $x_i + x_j = x_k$ with

$$V_{x_i+a} + V_{x_j+a} = V_{x_i+x_j+2a}, \qquad V_{x_i+x_j+2a} = V_{x_k+a} + V_a.$$

It is easy to see that if we let $x_i = V_{x_i+a} - V_a$, and similarly obtain $x_j$, $x_k$, then the above equations imply that $x_i + x_j = x_k$.

Multiplying is somewhat more complicated. Instead of $x_i \times x_j = x_k$ we use

$$\begin{aligned} V_{x_i+a} \times V_{x_j+a} &= V_{x_ix_j+ax_i+ax_j+a^2}, \\ V_a \times V_{x_i+a} &= V_{ax_i+a^2}, \\ V_a \times V_{x_j+a} &= V_{ax_j+a^2}, \\ V_{ax_i+a^2} + V_{ax_j+a^2} &= V_{ax_i+ax_j+2a^2}, \\ V_{x_ix_j+ax_i+ax_j+a^2} + V_{a+a^2} &= V_{x_ix_j+ax_i+ax_j+a+2a^2}, \\ V_{ax_i+ax_j+2a^2} + V_{x_k+a} &= V_{x_ix_j+ax_i+ax_j+a+2a^2}. \end{aligned}$$

It is again easy to see that these equations force the desired relation $x_k = x_ix_j$ to hold.

Thus, by replacing variables $x_j$ with variables $V_{x_j+a}$, and replacing equations and inequalities as described above, we obtain an equivalent set of equations and inequalities. It is easy to see that they are equivalent, because any solution for the $x_i$ can be turned into a solution for $V$'s satisfying $V > 1$ for all $V$'s by simply choosing $a > |x_i| + 1$ for all $x_i$. Similarly, one can obtain the $x_i$ from a solution for the $V$'s by letting $x_i = V_{x_i+a} - V_a$.

Finally, we show how to obtain an ordered set of variables. The idea is essentially the same as the previous one, but somewhat more complicated. Again, we introduce a new variable, this time $b$. We make $b$ larger than any of the previous variables (these were previously denoted by $V_j$ but will now be denoted by $x_1, x_2, \dots, x_n$). Now, we will work with a variable $V_{x_i+b^i} = x_i + b^i$ instead of with the original $x_i$'s. Note that we are now using a different power of $b$ for each variable. To obtain the powers of $b$, we introduce the equations $V_{b^{i+1}} = V_{b^i} \times V_b$ for $1 \le i \le r$, where we will determine the value of $r$ later. Since $b > x_i$ for all $x_i$, we know the ordering of the variables $V_{x_i+b^i}$.

To add, multiply, and compare the $V_{x_i+b^i}$'s, we again choose different powers of $b$, at most three for each equation or inequality. The idea is to choose several unused powers of $b$, say $b^\alpha$, for each equation, and work with $V_{x_i+b^\alpha}$ instead of $V_{x_i+b^i}$. The easiest case is again the inequality $x_i < x_j$. For this, we choose an unused power of $b$, say $\alpha > n$, and introduce the equations and inequalities

$$\begin{aligned} V_{b^\alpha} &= V_{b^i} + V_{b^\alpha - b^i}, & V_{b^\alpha} &= V_{b^j} + V_{b^\alpha - b^j}, \\ V_{x_i+b^i} + V_{b^\alpha - b^i} &= V_{x_i+b^\alpha}, & V_{x_j+b^j} + V_{b^\alpha - b^j} &= V_{x_j+b^\alpha}, \end{aligned}$$
$$V_{x_i+b^\alpha} < V_{x_j+b^\alpha}.$$

The last inequality will not be in our final set of equations because it is implicit in the ordering of the variables.

We need to check that we know the ordering of all the new variables we just introduced. This is true: if $i < j$, then

$$V_{b^\alpha - b^j} < V_{b^\alpha - b^i} < V_{b^\alpha} < V_{x_i + b^\alpha} < V_{x_j + b^\alpha}.$$

If $j < i$, the order of $V_{b^\alpha - b^i}$ and $V_{b^\alpha - b^j}$ is reversed. Because these variables are the only ones with a $V_{b^\alpha}$ term in them, we know their relative ordering with respect to the other variables in our problem.

To perform addition, suppose we have the equation $x_i + x_j = x_k$. We first choose three unused powers of $b$, say $b^\alpha$, $b^\beta$, and $b^\gamma$, with $n < \alpha < \beta < \gamma$. We then introduce the equations

$$\begin{aligned} V_{b^\alpha} &= V_{b^i} + V_{b^\alpha - b^i}, \\ V_{b^\beta} &= V_{b^j} + V_{b^\beta - b^j}, \\ V_{b^\gamma} &= V_{b^k} + V_{b^\gamma - b^k}, \end{aligned}$$

$$\begin{aligned} V_{x_i + b^i} + V_{b^\alpha - b^i} &= V_{x_i + b^\alpha}, \\ V_{x_j + b^j} + V_{b^\beta - b^j} &= V_{x_j + b^\beta}, \\ V_{x_k + b^k} + V_{b^\gamma - b^k} &= V_{x_k + b^\gamma}; \end{aligned}$$

$$\begin{aligned} V_{b^\gamma} &= V_{b^\alpha} + V_{b^\gamma - b^\alpha}, \\ V_{b^\gamma - \alpha} &= V_{b^\beta} + V_{b^\gamma - b^\alpha - b^\beta}, \\ V_{x_i + b^\alpha} + V_{x_j + b^\beta} &= V_{x_i + x_j + b^\alpha + b^\beta}, \\ V_{x_i + x_j + b^\alpha + b^\beta} + V_{b^\gamma - b^\alpha - b^\beta} &= V_{x_k + b^\gamma}. \end{aligned}$$

The first six equations produce variables of the form $V_{x_i + b^\alpha}$ instead of the form $V_{x_i + b^i}$. The remaining four do the work of adding $x_i$ and $x_j$. We know the ordering of the variables, because

$$\begin{aligned} V_{b^\alpha - b^i} < V_{b^\alpha} &< V_{x_i + b^\alpha}, \\ V_{b^\beta - b^j} < V_{b^\beta} &< V_{x_j + b^\beta} < V_{x_i + x_j + b^\beta + b^\alpha}, \\ V_{b^\gamma - b^\alpha - b^\beta} < V_{b^\gamma - b^\alpha} &< V_{b^\gamma - b^k} < V_{b^\gamma} < V_{x_k + b^\gamma}. \end{aligned}$$

Multiplication is, as usual, the most complicated operation. As before, we will choose three unused powers of $b$, $b^\alpha$, $b^\beta$, and $b^\gamma$, and $n < \alpha < \beta < \gamma$, but this time we also require that $\gamma = \alpha + \beta$. We now introduce the same first six equations as for the case of addition, as well as the equations

$$V_{b^\gamma} + V_{b^\gamma} = V_{2b^\gamma},$$
$$V_{x_i+b^\alpha} \times V_{x_j+b^\beta} = V_{x_ix_j+b^\beta x_i+b^\alpha x_j+b^\gamma},$$
$$V_{x_ix_j+b^\beta x_i+b^\alpha x_j+b^\gamma} + V_{2b^\gamma} = V_{x_ix_j+b^\beta x_i+b^\alpha x_j+3b^\gamma},$$
$$V_{x_i+b^\alpha} \times V_{b^\beta} = V_{b^\beta x_i+b^\gamma},$$
$$V_{x_j+b^\beta} \times V_{b^\alpha} = V_{b^\alpha x_j+b^\gamma},$$
$$V_{b^\alpha x_j+b^\gamma} + V_{b^\beta x_i+b^\gamma} = V_{b^\alpha x_j+b^\beta x_i+2b^\gamma},$$
$$V_{x_ix_j+b^\beta x_i+b^\alpha x_j+3b^\gamma} = V_{b^\alpha x_j+b^\beta x_i+2b^\gamma} + V_{x_k+b^\gamma}.$$

Again, it is easily checked that we know the ordering of the variables.

By going through the above proof with more care, we can show that if the semialgebraic variety was defined only by inequalities, then we can produce an equivalent *uniform* point configuration. We do this by showing that the point configuration we obtain is constructible, and then applying Lemma 4 (this time in the dual version for points and not lines, as it is used in [**Mn, SW**]).

The above proof also gives the original version of Mnëv's theorem. We have to show that the solution space of our set of equations on ordered variables is topologically equivalent to the solution space of the semialgebraic variety. This follows in the first (or second) reduction if, given a solution with some value of $a$ (or $b$), we can always increase $a$ (or $b$) and still have a solution. This can easily seen to be the case for $a$. For $b$, this is not clear; however, we can show this holds for $b$ if we introduce some extra variables and equations. We introduce the extra variables $V_{b^i+1}$ and $V_{b^i+b}$ for $1 < i < n$, along with the corresponding equations that force them to have the appropriate values, thus ensuring that $V_{b^i+1} < V_{x_i+b^i} < V_{b^i+b}$, so $1 < x_i < b$. We also introduce another equation ensuring that $b > 6$. Now, if we choose any $b > \max(6, x_1, x_2, \ldots, x_n)$ we can show that for any set of $x_i$, given a solution, this $b$ leads to a set of $V_{f(x_i, b)}$ which is a solution of the equations with ordered variables.

**Acknowledgments**

I would like to thank Steve Fortune, Ricky Pollack, and Bernd Sturmfels for useful discussions.

## References

[**BS**] J. Bokowski and B. Sturmfels, *Computational synthetic geometry*, Lecture Notes in Math., vol. 1355, Springer, Berlin, 1989.

[**GJ**] M. R. Garey and D. S. Johnson, *Computers and intractability: A guide to the theory of NP-completeness*, Freeman, New York, 1979.

[**Gr**] B. Grünbaum, *Arrangements and spreads*, Regional Conf. Ser. in Math., Amer. Math. Soc., Providence, RI, 1972.

[**Mn**] N. E. Mnëv, *The universality theorems on the classification problem of configuration varieties and convex polytopes varieties*, in Topology and Geometry—Rohlin Seminar (O. Y. Viro, ed.), Lecture Notes in Math., vol. 1346, Springer, Berlin, 1988, pp. 527–544.

[**Po**] R. Pollack, Private communication.

[**SW**] B. Sturmfels and N. White, *Constructing uniform oriented matroids without the isotopy property*, IMA Preprint Ser. **32** (1988).

AT&T Bell Laboratories, Room 2D-149, 600 Mountain Avenue, Murray Hill, New Jersey 07974

*E-mail address*: shor@research.att.com

DIMACS Series in Discrete Mathematics
and Theoretical Computer Science
Volume 4, 1991

Applied Geometry
and Discrete Mathematics
THE VICTOR KLEE FESTSCHRIFT

# A Zonotope Associated with Graphical Degree Sequences

RICHARD P. STANLEY

ABSTRACT. Let $D_n$ denote the convex hull in $\mathbf{R}^n$ of all (ordered) degree sequences of simple $n$-vertex graphs. Using the fact that $D_n$ is a zonotope, an explicit generating function is found for the number of these degree sequences. The $f$-vector of $D_n$ is found using Zaslavsky's theory of signed graph colorings. Finally we give a generalization based on a result of Fulkerson, Hoffman, and MacAndrew.

## 1. Introduction

Let $G$ be a simple graph (i.e., no loops or multiple edges) with vertex set $[n] := \{1, 2, \ldots, n\}$. Let $\deg(i)$ denote the degree of vertex $i$. We call the sequence $d(G) := (\deg(1), \ldots, \deg(n))$ the *degree sequence* of $G$. Note that our degree sequences are *ordered*; we do not require, as is often done, that the degrees are listed in descending order. Regard the vector $d(G)$ as a point in the vector space $\mathbf{R}^n$, and define

$$D_n = \operatorname{conv}\{d(G) : G \text{ is a simple graph with vertex set } [n]\},$$

where conv denotes convex hull. Thus $D_n$ is a convex polytope, first considered by Koren [6] (suggested to him by M. Perles) and called the *polytope of degree sequences* (of length $n$). Koren determined the vertices of $D_n$ and gave a system of (redundant) linear inequalities which determine $D_n$. Subsequently Peled and Srinivasan [8] obtained considerable further information about $D_n$, including a description of its facets and edges.

It follows from the Erdös-Gallai inequalities (explained in more detail below) that an integer point $(d_1, \ldots, d_n) \in D_n \cap \mathbf{Z}^n$ is the degree sequence of some graph $G$ if and only if $d_1 + \cdots + d_n$ is even. Peled suggested to this writer that the number of distinct (ordered) degree sequences of $n$-vertex graphs might be closely approximated by half the volume of $D_n$, and

1980 *Mathematics Subject Classification* (1985 *Revision*). Primary 05C30; Secondary 52A25.
Partially supported by NSF grant #DMS-8401376.

asked whether it was possible to compute or estimate the volume of $D_n$. It turns out that $D_n$ is a *zonotope*, a special kind of polytope (defined below) with a well-developed theory for computing the number of integer points, the volume, the $f$-vector, etc. Moreover, there is a slight modification of $D_n$, which we denote by $\widetilde{D}_n$, which is a zonotope combinatorially equivalent to $D_n$, but whose integer points correspond *exactly* to the degree sequences of graphs. Using the theory of zonotopes we are able to determine exactly the number $f(n)$ of degree sequences of $n$-vertex graphs, viz.,

$$\sum_{n\geq 0} f(n)\frac{x^n}{n!} = \frac{1}{2}\left[\left(1+2\sum_{n\geq 1} n^n\frac{x^n}{n!}\right)^{1/2} \times\left(1-\sum_{n\geq 1}(n-1)^{n-1}\frac{x^n}{n!}\right)+1\right] e^{\sum_{n\geq 1} n^{n-2}x^n/n!}.$$

Moreover, using the connection between the zonotope $\widetilde{D}_n$ and Zaslavsky's theory of signed coloring [**14**], we are able to compute exactly the $f$-vector of $\widetilde{D}_n$ (or of $D_n$).

## 2. The Ehrhart polynomial of an integer zonotope

A *zonotope* is by definition a (Minkowski) sum of closed line segments. In other words, if $L_1, \dots, L_r$ is a set of closed line segments in $\mathbf{R}^n$, then the corresponding zonotope is given by

$$(1) \qquad Z(L_1, \dots, L_r) = \{\alpha_1 + \cdots + \alpha_r : \alpha_1 \in L_1, \dots, \alpha_r \in L_r\}.$$

If $L_i$ joins the origin with the vector $\beta_i \in \mathbf{R}^n$, then we also write $Z(\beta_1, \dots, \beta_r)$ for $Z(L_1, \dots, L_r)$.

For any positive integer $q$ and any convex polytope $\mathscr{P}$ with integer vertices, let $i(\mathscr{P}, q)$ denote the number of points $\alpha \in \mathscr{P}$ satisfying $q\alpha \in \mathbf{Z}^n$. Then $i(\mathscr{P}, q)$ is a polynomial function of $q$ of degree $\dim\mathscr{P}$, called the *Ehrhart polynomial* of $\mathscr{P}$. Thus $i(\mathscr{P}, 1)$ is equal to the number of integer points in $\mathscr{P}$. Moreover, if $\mathscr{P} \subset \mathbf{R}^n$, then the coefficient of $q^n$ in $i(\mathscr{P}, q)$ is equal to the $n$-dimensional volume of $\mathscr{P}$. For these and other basic facts concerning Ehrhart polynomials, see, e.g., [**11**, pp. 235–241]. The following lemma is a refinement of a result of Shephard [**9**, Theorem 54]. Define a *half-open cube* $C$ to be a Minkowski sum of linearly independent half-open line segments $L_1, \dots, L_r$. In other words, if $\alpha_1, \dots, \alpha_s, \beta_1, \dots, \beta_s \in \mathbf{R}^n$ and $\beta_1 - \alpha_1, \dots, \beta_s - \alpha_s$ are linearly independent, then $C$ has the form

$$C = \{a_1\alpha_1+(1-a_1)\beta_1+a_2\alpha_2+(1-a_2)\beta_2+\cdots+a_s\alpha_s+(1-a_s)\beta_s : 0 \leq a_i < 1\}.$$

We call the vectors $\gamma_i := \beta_i - \alpha_i$ the *generating vectors* of $C$. (If $s = 0$ then $C$ is a single point, whose set of generating vectors is empty.)

2.1. LEMMA. *Let $L_i$ be the line segment connecting the origin to $\gamma_i \in \mathbf{R}^n$, $1 \le i \le r$. Then the zonotope $Z = Z(L_1, \dots, L_r) = Z(\gamma_1, \dots, \gamma_r)$ is a disjoint union $Z = \biguplus_X C_X$ of half-open cubes $C_X$, where* (a) *$X$ ranges over all linearly independent subsets of $\{\gamma_1, \dots, \gamma_r\}$, and* (b) *if $X = \{\delta_1, \dots, \delta_s\}$ then there are signs $e_i \in \{-1, 1\}$ such that $C_X$ is generated by $e_1\delta_1, \dots, e_s\delta_s$. Moreover, if each $\gamma_i \in \mathbf{Z}^n$ then the vertices of each $C_X$ lie in $\mathbf{Z}^n$.*

SKETCH OF PROOF (suggested by G. Ziegler). The proof is a "Tutte-Grothendieck argument", in the sense of **[2]**. We proceed by induction on $r$. The lemma is clear for $r = 0$, when $Z$ consists of a single point. Assume now the case for $r - 1$. Let $Z' = Z(L_1, \dots, L_{r-1})$, so $Z'$ has a desired decomposition $Z' = \biguplus C_X$, where $X$ ranges over all linearly independent subsets of $\{\gamma_1, \dots, \gamma_{r-1}\}$. Now orthogonally project $Z'$ to a hyperplane orthogonal to $L_r$. We obtain a zonotope $\overline{Z}$ generated by the projections $\overline{L}_1, \dots, \overline{L}_{r-1}$ of the line segments $L_1, \dots, L_{r-1}$. By induction decompose $\overline{Z} = \biguplus_{\overline{Y}} C_{\overline{Y}}$, where the sets $\overline{Y}$ are projections of subsets $Y$ of $\{\gamma_1, \dots, \gamma_{r-1}\}$ for which $Y \cup \{\gamma_r\}$ is linearly independent. Now "lift" this decomposition back up to $\mathbf{R}^n$. Each half-open cube $C_{\overline{Y}}$ is lifted to a product $(O, \gamma_r] \times C_Y$, where $(O, \gamma_r]$ denotes the half-open interval from the origin $O$ to $\gamma_r$ which excludes $O$ and includes $\gamma_r$. Then

$$Z = \left(\bigcup_X C_X\right) \cup \left(\bigcup_Y (O, \gamma_r] \times C_Y\right)$$

gives a desired decomposition of $Z$. $\square$

The next theorem is a basic result about the Ehrhart polynomial of an integer zonotope. It was stated without proof in [**10**, Example 3.1; **11**, Exercise 4.31]. The special case of the volume of a zonotope $Z \subset \mathbf{R}^n$ (the coefficient of $q^n$ in $i(Z, q)$) is due to McMullen (see [**9**, (57)]) and also appears in [**7**].

2.2. THEOREM. *Let $\beta_1, \dots, \beta_r \in \mathbf{Z}^n$. Let $Z$ denote the zonotope $Z(\beta_1, \dots, \beta_r)$. Then*

$$i(Z, q) = \sum_X h(X) q^{|X|},$$

*where $X$ ranges over all linearly independent subsets of $\{\beta_1, \dots, \beta_r\}$, and where $h(X)$ denotes the greatest common divisor of all minors of size $|X|$ of the matrix whose rows are the elements of $X$.*

PROOF. By the preceding lemma, we have $Z = \biguplus C_X$, where $X$ runs over all linearly independent subsets $\{\delta_1, \dots, \delta_s\}$ of $\{\beta_1, \dots, \beta_r\}$, and where $C_X$ is a half-open cube with integer vertices generated by $\pm\delta_1, \dots, \pm\delta_s$ for some choice of signs. Suppose $C$ is any half-open cube with integer vertices generated by vectors $\zeta_1, \ldots, \zeta_s \in \mathbf{Z}^n$. Let $Y = \{\zeta_1, \ldots, \zeta_s\}$. Let $\Lambda$ be the (additive) abelian group of integer vectors in the subspace of $\mathbf{R}^n$ spanned by $Y$, so $\Lambda \cong \mathbf{Z}^s$. Let $\Gamma$ be the subgroup of $\Lambda$ generated by $Y$. By standard properties of determinants we have $h(Y) = [\Lambda : \Gamma]$, the index of $\Gamma$ in $\Lambda$. On the other hand, it is easy to see that $C \cap \Lambda$ is a set of coset representatives for $\Gamma$ in $\Lambda$. Hence $i(C, 1) = \#(C \cap \Lambda) = h(Y)$. By a simple scaling argument it follows that $i(C, q) = h(Y)q^{|Y|}$ for any integer $q \geq 1$. Hence

$$i(Z, q) = \sum_X i(C_X, q) = \sum_X h(X)q^{|X|}. \qquad \square$$

## 3. Counting degree sequences

We will now apply Theorem 2.2 to a modification $\widetilde{D}_n$ of the polytope $D_n$. For a graph $G$ on the vertex set $[n]$ with degree sequence $d(G) = (d_1, \ldots, d_n)$, define the *extended degree sequence*

$$\tilde{d}(G) = (d_1, \ldots, d_n, \tfrac{1}{2}(d_1 + \cdots + d_n)).$$

Then define the *polytope of extended degree sequences*,

$$\widetilde{D}_n = \operatorname{conv}\{\tilde{d}(G) : G \text{ is a simple graph on the vertex set } [n]\} \subset \mathbf{R}^{n+1}.$$

Since the last component of vectors in $\widetilde{D}_n$ is a fixed linear combination of the first $n$ components, it follows that $D_n$ and $\widetilde{D}_n$ are linearly equivalent and hence combinatorially equivalent. However (and this is the point of defining $\widetilde{D}_n$), they do not have the same Ehrhart polynomial.

Let us recall the Erdös-Gallai conditions (see any graph theory text) characterizing the degree sequences of simple graphs: A vector $(d_1, \ldots, d_n)$ of positive integers $d_i$ satisfying $d_1 \geq \cdots \geq d_n$ is the degree sequence of some simple graph $G$ if and only if the following two conditions hold:

$$\text{(EG1)} \qquad \sum_{i=1}^{j} d_i - j(j-1) \leq \sum_{i=j+1}^{n} \min(j, d_i), \qquad 1 \leq j \leq n.$$

$$\text{(EG2)} \qquad d_1 + \cdots + d_n \text{ is even.}$$

From these conditions it is easy to deduce (see [**6**]) that the condition for an arbitrary vector $(d_1, \ldots, d_n)$ of positive integers (i.e., not necessarily satisfying $d_1 \geq \cdots \geq d_n$) to be the degree sequence of a simple graph is given by a system of linear inequalities, together with (EG2). (These conditions in fact follow from an earlier result of Tutte [**12**] and are also obtained in a more general context in [**4**, Theorem 2.1].) It follows that $(d_1, \ldots, d_n) \in \mathbf{Z}^n$ is a degree sequence if and only if $(d_1, \ldots, d_n) \in D_n$ and $d_1 + \cdots + d_n$ is even. Hence from the definition of $\widetilde{D}_n$ we conclude:

3.1. PROPOSITION. *There is a one-to-one correspondence between degree sequences $d(G)$ of simple graphs $G$ on the vertex set $[n]$ and integer points in $\widetilde{D}_n$, given by $d(G) \leftrightarrow \tilde{d}(G)$.*

More generally, an analogue of the Erdös-Gallai condition for multigraphs (graphs with repeated edges, but no loops) with edge multiplicity bounded by $q$ yields the following result (which is generalized further in Proposition 5.2).

3.2. PROPOSITION. *Let $q$ be a positive integer. Then $i(\widetilde{D}_n, q)$ is equal to the number of distinct degree sequences of multigraphs on the vertex set $[n]$ such that no edge multiplicity can exceed $q$.*

Now let $e_i$ denote the $i$th unit coordinate vector in $\mathbf{R}^n$, and write $e_{ij} = e_i + e_j$. Let $\tilde{e}_{ij}$ be the vector in $\mathbf{R}^{n+1}$ obtained by adjoining to $e_{ij}$ an $(n+1)$st coordinate equal to 1. Clearly for a graph $G$ on the vertex set $[n]$,

$$d(G) = \sum e_{ij}, \qquad \tilde{d}(G) = \sum \tilde{e}_{ij},$$

where both sums range over all edges $\{i, j\}$ of $G$. From this it follows easily that $D_n$ and $\widetilde{D}_n$ are zonotopes, given explicitly by

$$\begin{aligned} D_n &= Z(e_{ij} : 1 \le i < j \le n), \\ \widetilde{D}_n &= Z(\tilde{e}_{ij} : 1 \le i < j \le n). \end{aligned} \tag{2}$$

(Of course, since $D_n$ and $\widetilde{D}_n$ are linearly equivalent, it follows that one is a zonotope if and only if the other is.)

Now define a *quasitree* to be a connected graph which is either a tree or has exactly one cycle, which is of odd length. A *quasiforest* is a graph whose components are all quasitrees. We have come to the main result of this section.

3.3. THEOREM. *The Ehrhart polynomial $i(\tilde{D}_n, q)$ of $\tilde{D}_n$ is given by*

$$i(\widetilde{D}_n, q) = a_n(n)q^n + a_n(n-1)q^{n-1} + \cdots + a_n(0),$$

*where*

$$a_n(i) = \sum_X \max\{1, 2^{c(X)-1}\};$$

*here $X$ ranges over all quasiforests with $i$ edges on the vertex set $[n]$, and $c(X)$ denotes the number of (odd) cycles of $X$.*

PROOF. By Theorem 2.2 and (2), we have

$$a_n(i) = \sum_X h(X),$$

where $X$ ranges over all $i$-element linearly independent subsets of $\widetilde{E}_n := \{\tilde{e}_{ij} : 1 \le i < j \le n\}$, and $h(X)$ has the meaning of Theorem 2.2. Let us identify a subset $X$ of $\widetilde{E}_n$ with the graph on the vertex set $[n]$ which has an edge $\{i, j\}$ if and only if $\tilde{e}_{ij} \in X$. When is $X$ linearly independent? We could appeal to Zaslavsky's extensive theory of signed graphs **[13]** to answer this question, but it is also easy to proceed directly. If $X$ is linearly independent, then every component $Y$ of $X$ with $i$ vertices may contain at most $i$ edges, since any $i+1$ vectors in $\mathbf{R}^i$ are dependent. Thus every component $Y$ of $X$ is either a tree or contains one cycle. It is easily checked that a connected graph $Y$ with at most one cycle is linearly independent if and only if $Y$ is a tree or the unique cycle of $Y$ has odd length. Thus we have that $X$ is a quasiforest.

Assume then that $X$ is a quasiforest, and set $m = |X|$. We need to compute $h(X)$, the gcd of the $m \times m$ minors of the matrix $M$ whose rows are the vectors $\tilde{e}_{ij} \in X$. Thus the rows of $M$ are indexed by the edges of $X$ and the columns by the vertices, together with a last column of 1s. An $m \times m$ minor of $M$ corresponds to choosing either $m$ vertices of $X$ (i.e., $m$ numbers from $[n]$) or $m-1$ vertices of $X$ and a final column of 1s. In the first case, in order for the columns to be linearly independent we must choose all vertices from any component of $X$ with an odd cycle, and all but one vertex from any component of $X$ which is a tree. Each odd cycle contributes to a factor $\pm 2$ to the minor, and we obtain a value $\pm 2^{c(X)}$.

Now consider the submatrix $N$ obtained by choosing $m-1$ vertices of $X$ and a final column of 1s. Again, in order for the columns to be linearly independent we must choose our $m-1$ vertices in one of the following ways: (a) choose all but one vertex from every component $Y$ which is a tree, choose all but one vertex from some component $T_0$ which has an odd cycle, and choose every vertex from the remaining components $T$ with one odd cycle, or (b) choose all but two adjacent vertices $i$ and $j$ from some component which is a tree, choose all but one vertex from the remaining components which are trees, and choose every vertex from the components with one odd cycle.

In case (a), in the submatrix $N$ subtract one-half of every column except the last from the last column. Every entry of the last column is now 0 or 1/2. Factor out 1/2 from the last column. The resulting matrix $N'$ is the incidence matrix of a graph $X'$ for which every component has exactly one cycle, which is of odd length. Moreover, $X'$ has the same number of odd cycles as $X$. Hence $\det N' = \pm 2^{c(X)}$, so $\det N = \pm 2^{c(X)-1}$.

In case (b), there is a row which is all 0s except a 1 in the last column. Hence we may delete this row and the last column without affecting the value of the minor (except possibly for sign). Let $X'$ denote the graph $X$ with edge $\{i, j\}$ deleted (but retain all vertices). We now have a matrix obtained from the incidence matrix of $X'$ by deleting a column (vertex) from every component which is a tree. Since $X'$ still has $c(X)$ odd cycles, it follows that the value of the minor is $\pm 2^{c(X)}$.

Thus we have seen that all nonzero minors of $M$ are equal to $\pm 2^{c(X)}$ or $\pm 2^{c(X)-1}$. The latter value is possible if and only if $c(X) \geq 1$. Hence the gcd of the minors of $M$ is given by $\max\{1, 2^{c(X)-1}\}$, and the proof follows. □

3.4. Corollary. *The number $f(n)$ of distinct degree sequences of simple graphs on the vertex set $[n]$ is given by*

$$f(n) = \sum_X \max\{1, 2^{c(X)-1}\}, \tag{3}$$

*where $X$ ranges over all quasiforests on the vertex set $[n]$, and $c(X)$ denotes the number of (odd) cycles of $X$.*

Proof. Put $q = 1$ in Theorem 3.3. □

It would be interesting to find a direct combinatorial proof of (3). An analogous problem to determining $f(n)$ is to find the number $g(n)$ of distinct outdegree sequences of orientations of the complete graph $K_n$. Using the theory of zonotopes as we have done here, it was shown in [**10**, Example 3.1] (see also [**11**, Exercise 4.32]) that $g(n)$ is equal to the number of (labelled) forests on $n$ vertices. A combinatorial proof of this result was subsequently given by Kleitman and Winston [**5**]. Perhaps their techniques can be used to establish (3). However, while the enumeration of outdegree sequences can be extended to orientations of *any* undirected graph (as discussed in the previous three references), an analogous generalization for degree sequences of subgraphs of a given graph $G$ seems only to hold for a special class of graphs $G$ (see §5 for further details).

It is a fairly routine exercise in enumerative combinatorics to convert Theorem 3.3 into a generating function for $i(\widetilde{D}_n, q)$.

3.5. Proposition. *We have*

$$\sum_{n\geq 0} i(\widetilde{D}_n, q)\frac{x^n}{n!} = \frac{1}{2}\left[\left(1 + 2\sum_{n\geq 1} n^n \frac{q^n x^n}{n!}\right)^{1/2} \times \left(1 - \sum_{n\geq 1}(n-1)^{n-1}\frac{q^n x^n}{n!}\right) + 1\right] e^{\sum_{n\geq 1} n^{n-2}q^{n-1}x^n/n!}.$$

*Here and below we set $0^0 = 1$ (which arises when $n = 1$ in the second sum on the right-hand side).*

Proof. We assume knowledge of the theory of exponential generating functions, in particular the exponential formula, as expounded for instance in [**3**, Theorem 4.3]. The number of rooted trees on $n$ vertices is $n^{n-1}$, with exponential generating function

$$R(x) = \sum_{n\geq 1} n^{n-1}\frac{x^n}{n!}.$$

Hence the exponential generating function for $k$-tuples of rooted trees is $R(x)^k$ and so for undirected $k$-cycles of rooted trees (i.e., graphs with exactly one cycle, which is of length $k \geq 3$) is $R(x)^k/2k$.

Let $h(j, n)$ be the number of graphs on the vertex set $[n]$ such that every component has exactly one cycle, which is of odd length $\geq 3$, and with $j$

cycles. (Such graphs have exactly $n$ edges.) Then by the exponential formula we have

$$\begin{aligned}\sum_{j,n\geq 0} h(j,n)\frac{t^j x^n}{n!} &= \exp\sum_{k\geq 1}\frac{t}{2(2k+1)}R(x)^{2k+1}\\ &= \exp\frac{t}{2}\left[\frac{1}{2}(\log(1-R(x))^{-1}-\log(1+R(x))^{-1})-R(x)\right]\\ &= \left(\frac{1+R(x)}{1-R(x)}\right)^{t/4} e^{-tR(x)/2}.\end{aligned}$$

Thus,

$$\begin{aligned}1+\sum_{j,n\geq 1} 2^{j-1}h(j,n)\frac{x^n}{n!} &= 1+\frac{1}{2}\left[\left(\frac{1+R(x)}{1-R(x)}\right)^{1/2} e^{-R(x)}-1\right]\\ &= \frac{1}{2}\left[\left(-1+\frac{2}{1-R(x)}\right)^{1/2} e^{-R(x)}+1\right].\end{aligned}$$

It is well known (and easily deduced from $R(x)=xe^{R(x)}$) that

$$\frac{1}{1-R(x)}=\sum_{n\geq 0} n^n\frac{x^n}{n!}, \qquad e^{-R(x)}=1-\sum_{n\geq 1}(n-1)^{n-1}\frac{x^n}{n!},$$

so we get

$$\begin{aligned}&1+\sum_{j,n\geq 1} 2^{j-1}h(j,n)\frac{x^n}{n!}\\ &= \frac{1}{2}\left[\left(1+2\sum_{n\geq 1} n^n\frac{x^n}{n!}\right)^{1/2}\left(1-\sum_{n\geq 1}(n-1)^{n-1}\frac{x^n}{n!}\right)+1\right].\end{aligned} \tag{4}$$

There are $n^{n-2}$ free (i.e., unrooted) trees with $n$ labelled vertices, each with $n-1$ edges. Hence if $r(i,n)$ denotes the number of forests with $i$ edges on the vertex set $[n]$, then

$$\sum_{i,n\geq 0} r(i,n)\frac{q^i x^n}{n!}=\exp\sum_{n\geq 1} n^{n-2}\frac{q^{n-1}x^n}{n!}.$$

Since all graphs enumerated by $h(n,j)$ have exactly $n$ edges, there follows

$$\sum_{n\geq 0} i(\widetilde{D}_n,q)\frac{x^n}{n!}=\left[1+\sum_{j,n\geq 1} 2^{j-1}h(j,n)\frac{(qx)^n}{n!}\right]\cdot\exp\sum_{n\geq 1} n^{n-2}\frac{q^{n-1}x^n}{n!}.$$

Comparing with (4) completes the proof. □

Putting $q=1$ in Proposition 3.5 yields:

3.6. COROLLARY. *We have*

$$\sum_{n\geq 0} f(n)\frac{x^n}{n!} = \frac{1}{2}\left[\left(1+2\sum_{n\geq 1} n^n\frac{x^n}{n!}\right)^{1/2} \times\left(1-\sum_{n\geq 1}(n-1)^{n-1}\frac{x^n}{n!}\right)+1\right] e^{\sum_{n\geq 1} n^{n-2}x^n/n!}.$$

Some small values of $i_n := i(\widetilde{D}_n, q)$ are given by:

$$\begin{aligned}
i_1 &= 1\\
i_2 &= 1+q\\
i_3 &= 1+3q+3q^2+q^3\\
i_4 &= 1+6q+15q^2+20q^3+12q^4\\
i_5 &= 1+10q+45q^2+120q^3+195q^4+162q^5\\
i_6 &= 1+15q+105q^2+455q^3+1320q^4+2508q^5+2540q^6\\
i_7 &= 1+21q+210q^2+1330q^3+5880q^4+18564q^5+39809q^6\\
&\quad +46035q^7\\
i_8 &= 1+28q+378q^2+3276q^3+20265q^4+93240q^5+317800q^6\\
&\quad +749200q^7+951552q^8.
\end{aligned}$$

Moreover,

$$\begin{aligned}
f(1) &= 1\\
f(2) &= 2\\
f(3) &= 8\\
f(4) &= 54\\
f(5) &= 533\\
f(6) &= 6944\\
f(7) &= 111850\\
f(8) &= 2135740\\
f(9) &= 47003045\\
f(10) &= 1168832808.
\end{aligned}$$

The coefficient of $q^n$ in $i(\widetilde{D}_n, q)$ is the $n$-dimensional *relative volume* $V(\widetilde{D}_n)$ of $\widetilde{D}_n$ (see, e.g., [**10**, p. 335] or [**11**, pp. 238–239] for the definition). Putting $1/q$ for $q$ and $qx$ for $x$ in Proposition 3.5 and then setting $q = 0$ yields:

3.7. Corollary. *We have*

$$\sum_{n\geq 0} V(\widetilde{D}_n)\frac{x^n}{n!} = \frac{1}{2}\left[\left(1+2\sum_{n\geq 1} n^n\frac{x^n}{n!}\right)^{1/2}\left(1-\sum_{n\geq 1}(n-1)^{n-1}\frac{x^n}{n!}\right)+1\right].$$

We also have that

$$V(\widetilde{D}_n) = \sum_X 2^{c(X)-1}, \tag{5}$$

where $X$ ranges over all graphs on the vertex set $[n]$ for which every component has exactly one cycle, which is of odd length, and where $c(X)$ denotes the number of (odd) cycles of $X$. Moreover, in regard to Peled's question about the volume of $D_n$, we have by similar reasoning that

$$V(D_n) = \sum_X 2^{c(X)},$$

summed over the same range as (5), so $V(D_n) = 2V(\widetilde{D}_n)$. Moreover, $V(D_n)$ is the ordinary volume of $D_n$ for $n \geq 3$.

## 4. The $f$-vector of $D_n$ and $\widetilde{D}_n$

We mentioned in §1 that Koren, Peled, and Srinivasan obtained a description of the vertices, edges, and facets of $D_n$ (or $\widetilde{D}_n$). Beissinger and Peled [**1**] used the description of the vertices to obtain a formula for the number of vertices (see equation (7)). Here we extend this result to a determination of the entire $f$-vector $f(D_n) = (f_0(D_n), \ldots, f_n(D_n)) = (f_0, \ldots, f_n)$, where $D_n$ has $f_i$ $i$-dimensional faces. (Normally the $f$-vector of a $d$-polytope only includes $f_i$ for $0 \leq i \leq d-1$, but it will be convenient for us to allow $0 \leq i \leq n$.) Thus, $f(D_1) = (1, 0)$, $f(D_2) = (2, 1, 0)$, and $f(D_3) = (8, 12, 6, 1)$, the latter since $D_3$ is a 3-cube.

Our derivation will be based on a result of Zaslavsky concerning the coloring of signed graphs. Let $-K_n$ denote the *negative complete signed graph*, i.e., the complete graph $K_n$ with every edge labelled with a minus sign [**13**, §7D]. The matroid $M = M(-K_n)$ which Zaslavsky associates with $-K_n$ is just the linear matroid $\{e_{ij} : 1 \leq i < j \leq n\} \subset \mathbf{R}^n$. Hence by (2), Zaslavsky's zonotope $Z[-K_n]$ (defined in [**14**, p. 226]) is just $D_n$.

In [**14**, pp. 217–218], Zaslavsky defines a *signed coloring* of $-K_n$ (more generally, Zaslavsky deals with an arbitrary signed graph $\Sigma$) in $\mu \geq 0$ colors to be a function $k : N \to [-\mu, \mu] = \{-\mu, -\mu+1, \ldots, 0, \ldots, \mu-1, \mu\}$, where $N = [n]$ denotes the vertex set of $-K_n$. The *set of impropriety* $I(k)$ of $k$ consists of all edges $\{i, j\}$ of $-K_n$ for which $k(i) = -k(j)$ [**14**, p. 218]. Call such edges *improper*. The *rank* of $I(k)$ is defined by $\operatorname{rk} I(k) = n - b(I(k))$, where $b(I(k))$ is the number of bipartite components of $I(k)$ (considered as a graph on the vertex set $[n]$) [**14**, p. 216]. The *Whitney* *polynomial* of $-K_n$ is defined by

$$w_{-K_n}(x, 2\mu+1) = \sum_k x^{\operatorname{rk} I(k)},$$

summed over all signed colorings in $\mu$ colors.

We can now state the result of Zaslavsky [**14**, Corollary 4.1″] (specialized to $\Sigma = -K_n$ and $Z[\Sigma] = D_n$), which will be our main tool for evaluating the $f$-vector $f(D_n)$.

4.1. Proposition. *We have*

$$\sum_{i=0}^{n} f_i(D_n)x^i = (-1)^n w_{-K_n}(-x, -1).$$

The main result of this section is the following:

4.2. Theorem. *We have*

$$\sum_{n\geq 0}\sum_{i=0}^{n} f_i(D_n)\frac{x^i t^n}{n!} = \frac{e^{xt} - (1+x)t - x(1+x)\frac{t^2}{2}}{1 + 2(e^{-t}-1) + \frac{2}{x}(e^{xt}-1) - \frac{1}{x}(e^{2xt}-1)}.$$

Proof. We may choose a signed coloring $k : N \to [-t, t]$ of $-K_n$ as follows. First choose the sets $A_i = k^{-1}\{-i, i\}$. Thus $(A_0, A_1, \ldots, A_\mu)$ is a *weak ordered partition* of $[n]$ into $\mu+1$ blocks, i.e., $\bigcup A_i = [n]$ and $A_i \cap A_j = \varnothing$ if $i \neq j$. ("Weak" means that we allow $A_i = \varnothing$.) Let $a_i = |A_i|$. Now for each block $A_i$ with $1 \leq i \leq \mu$, choose a subset $B_i = k^{-1}(i)$. For two of these choices (when $A_i \neq \varnothing$), namely $B_i = A_i$ and $B_i = \varnothing$, there will be no improper edges incident to vertices in $A_i$, so $A_i$ forms $a_i$ bipartite components of $I(k)$, each consisting of a single vertex. For the remaining $2^{a_i} - 2$ choices of $B_i$ (when $a_i \geq 1$), all edges from $B_i$ to $A_i - B_i$ are improper; these edges form a single bipartite component of $I(k)$. There remains the case of $A_0$. All edges between vertices in $A_0$ are improper, so we get a component of $I(k)$ isomorphic to the complete graph $K_{a_0}$. This component will be bipartite if and only if $a_0 = 0$, 1, or 2. It follows that

$$(6)\quad w_{-K_n}(x, 2\mu+1) = x^n \sum_{(A_0, \ldots, A_\mu)} v(a_0)\prod_{i=1}^{\mu}(2x^{a_i} + (2^{a_i}-2)\chi(a_i \neq 0)x^{-1}),$$

where: (a) the sum ranges over all weak ordered partitions of $[n]$ into $\mu+1$ blocks, (b) $\chi(a_i \neq 0) = 1$ if $a_i \neq 0$ and $= 0$ if $a_i = 0$, and (c) $v(a_0)$ is defined by

$$v(a) = \begin{cases} 1 & \text{if } a = 0 \text{ or } a \geq 3, \\ x^{-1} & \text{if } a = 1 \text{ or } a = 2. \end{cases}$$

Now suppose $g_0, \ldots, g_\mu$ are any functions defined on $\mathbf{N}$, and define (using the notation of (6))

$$h(n) = \sum_{(A_0, \ldots, A_\mu)} g_0(a_0)\cdots g_\mu(a_\mu).$$

It then follows from standard properties of exponential generating functions that

$$h(n) = \left[\frac{t^n}{n!}\right] \prod_{i=0}^{\mu} \left( \sum_{j \geq 0} g_i(j) \frac{t^j}{j!} \right),$$

where $[\frac{t^n}{n!}]F(t)$ denotes the coefficient of $t^n/n!$ in the generating function $F(t)$. Hence from (6) we have

$$\begin{aligned} W_{K_n}&(x, 2\mu+1) \\ &= x^n \left[\frac{t^n}{n!}\right] \left( x^{-1}t + x^{-1}\frac{t^2}{2} + \sum_{j \geq 3} \frac{t^j}{j!} \right) \left( 1 + \sum_{j \geq 1} (2x^{-j} + (2^j - 2)x^{-1}) \frac{t^j}{j!} \right)^{\mu} \\ &= \left[\frac{t^n}{n!}\right] \left( 1 + \frac{xt^2}{2} + \sum_{j \geq 3} \frac{(xt)^j}{j!} \right) \left( 1 + \sum_{j \geq 1} (2x^{-j} + (2^j - 2)x^{-1}) \frac{(xj)^j}{j!} \right)^{\mu} \\ &= \left[\frac{t^n}{n!}\right] \left( (1-x)t + x(1-x)\frac{t^2}{2} + e^{xt} \right) \\ &\quad \times \left( 1 + 2(e^t - 1) - \frac{2}{x}(e^{xt} - 1) + \frac{1}{x}(e^{2xt} - 1) \right)^{\mu}. \end{aligned}$$

Substituting $-x$ for $x$, $-t$ for $t$, and $-1$ for $\mu$ yields

$$\begin{aligned} (-1)^n w_{-K_n}(-x, -1) = \left[\frac{t^n}{n!}\right] &\left( e^{xt} - (1+x)t - x(1+x)\frac{t^2}{2} \right) \\ &\times \left( 1 + 2(e^{-t} - 1) + \frac{2}{x}(e^{xt} - 1) - \frac{1}{x}(e^{2xt} - 1) \right)^{-1}. \end{aligned}$$

The proof follows from Proposition 4.1. □

From Theorem 4.2 one can compute:

$f(D_1) = (1, 0)$
$f(D_2) = (2, 1, 0)$
$f(D_3) = (8, 12, 6, 1)$
$f(D_4) = (46, 108, 84, 22, 1)$
$f(D_5) = (332, 1020, 1080, 450, 60, 1)$
$f(D_6) = (2874, 10830, 14880, 9160, 2460, 224, 1)$
$f(D_7) = (29874, 129486, 220920, 182770, 75670, 14238, 882, 1)$
$f(D_8) = (334982, 1726648, 3529344, 3679872, 2074660,$
$610288, 81144, 3322, 1).$

Putting $x = 0$ in Theorem 4.2 yields

$$\sum_{n \geq 0} f_0(D_n) \frac{t^n}{n!} = \frac{1-t}{-1+2e^{-t}} = \frac{(1-t)e^t}{2-e^t}, \tag{7}$$

agreeing with a result of Beissinger and Peled [**1**, p. 216] mentioned at the beginning of this section.

## 5. A generalization

We wish to generalize Theorem 3.3 by considering degree sequences of graphs which are contained in a fixed graph $G$. More precisely, let $G$ be a multigraph (i.e., allowing multiple edges, but not loops) on the vertex set $[n]$. Define a convex polytope $\widetilde{D}(G) \subset \mathbf{R}^{n+1}$ by

$$\widetilde{D}(G) = \operatorname{conv}\{\tilde{d}(H) : H \text{ is a spanning subgraph of } G\}.$$

Thus $\widetilde{D}_n = \widetilde{D}(K_n)$, where $K_n$ denotes the complete graph on $[n]$. It is easily seen that

$$\widetilde{D}_n = Z(\tilde{e}_{ij} : \{i, j\} \text{ is an edge of } G).$$

(Here we take $\tilde{e}_{ij}$ a total of $m$ times if there are $m$ edges between $i$ and $j$.) The following generalization of Theorem 3.3 is proved in exactly the same way as Theorem 3.3.

5.1. Theorem. *The Ehrhart polynomial $i(\widetilde{D}(G), q)$ of $\widetilde{D}(G)$ is given by*

$$i(\widetilde{D}(G), q) = a_G(n)q^n + a_G(n-1)q^{n-1} + \cdots + a_G(0),$$

*where*

$$a_G(i) = \sum_X \max\{1, 2^{c(X)-1}\};$$

*here $X$ ranges over all spanning quasiforests of $G$ with $i$ edges, and $c(X)$ denotes the number of (odd) cycles of $X$.*

We would like to interpret $i(\widetilde{D}(G), q)$ in terms of degree sequences analogously to Proposition 3.2. We therefore need to know when Proposition 3.1 extends to $\widetilde{D}(G)$. Call $G$ an *FHM-graph* (named after the authors of [**4**]) if the following property holds:

(FHM). No induced subgraph of $G$ consists of two vertex-disjoint odd cycles (with no other edges). (Equivalently, every induced subgraph of $G$ has at most one nonbipartite component.)

Given a multigraph $G$ on the vertex set $[n]$, let

$$\Delta(G) = \{d(H) : H \text{ is a spanning subgraph of } G\}.$$

Define a map $\varphi_G : \Delta(G) \to \mathbf{Z}^{n+1}$ by $\varphi_G(d(H)) = \tilde{d}(H)$. Clearly $\varphi_G$ is one-to-one. It is immediate from the definition of $\widetilde{D}(G)$ that

$$\varphi_G(\Delta(G)) \subseteq \widetilde{D}(G) \cap \mathbf{Z}^{n+1}. \tag{8}$$

5.2. Proposition. *We have $\varphi_G(\Delta(G)) = \widetilde{D}(G) \cap \mathbf{Z}^{n+1}$ if and only if $G$ is an FHM-graph.*

Proof. The "if" part follows from [**4**, Theorem 2.1], while the "only if" part (which is easy) follows from the last paragraph on p. 170 of [**4**]. □[1]

[1] I am grateful to L. Lovász for bringing the paper [**4**] to my attention.

Now given any multigraph $G$ on $[n]$ and any integer $q \geq 1$, define $G^{(q)}$ to be the multigraph obtained from $G$ by replacing each edge by $q$ edges, and let $f(G^{(q)})$ denote the number of distinct degree sequences of spanning subgraphs of $G^{(q)}$.

5.3. THEOREM. *For any multigraph $G$ on $[n]$ and any integer $q \geq 1$, we have*

$$f(G^{(q)}) \leq i(\widetilde{D}(G), q).$$

*Equality holds if and only if $G$ is an FHM-graph.*

PROOF. From the definition of FHM-graph we see that for any integer $q \geq 1$, $G$ is an FHM-graph if and only if $G^{(q)}$ is. Moreover, by the definition of $\widetilde{D}(G)$ we have $\widetilde{D}(G^{(q)}) = q\widetilde{D}(G)$, where for any polytope $\mathscr{P}$ we define

$$q\mathscr{P} = \{q\alpha : \alpha \in \mathscr{P}\}.$$

Since

$$i(\widetilde{D}(G), q) = \#(q\widetilde{D}(G) \cap \mathbf{Z}^{n+1}) = \#(\widetilde{D}(G^{(q)}) \cap \mathbf{Z}^{n+1}),$$

the proof follows from (8) and Proposition 5.2 (and the fact that $\varphi_{G^{(q)}}$ is one-to-one). □

A special case of FHM-graphs is bipartite graphs. Theorems 5.1 and 5.3 yield for these graphs the following result.

5.4. COROLLARY. *Let $G$ be a bipartite (multi)graph on the vertex set $[n]$. Then*

$$f(G^{(q)}) = b_G(n-1)q^{n-1} + b_G(n-2)q^{n-2} + \cdots + b_G(0),$$

*where $b_G(i)$ is the number of spanning forests of $G$ with $i$ edges. In particular, $f(G)$ is equal to the number of spanning forests of $G$.*

We mentioned after Corollary 3.4 that the number $g(G)$ of outdegree sequences of orientations of any graph (or multigraph) $G$ is equal to the number of spanning forests of $G$, and that this result has a combinatorial proof. Hence by the preceding corollary we have $f(G) = g(G)$ when $G$ is bipartite. It is easy to give a combinatorial proof of this fact. Namely, if $G$ has vertex bipartition $(A, B)$ and if $D$ is an orientation of $G$, then let $\sigma(D)$ be the spanning subgraph of $G$ consisting of those edges which are oriented from $A$ to $B$ in $D$. Then $D$ and $D'$ have the same outdegree sequence if and only if $\sigma(D)$ and $\sigma(D')$ have the same degree sequence, yielding the desired bijection. Hence the combinatorial proof **[5]** yields a combinatorial proof of Corollary 5.4 when $q = 1$, and this leads easily to a combinatorial proof for any $q$.

Combining Theorems 5.1 and 5.3 when $q = 1$ yields that for an FHM-graph $G$,

$$f(G) = \sum_X \max\{1, 2^{c(X)-1}\}, \tag{9}$$

where $X$ ranges over all spanning quasiforests of $G$. It would be interesting to find a generalization of (9) valid for all graphs $G$. While we have been unable to do this, there is a more general class of graphs than FHM-graphs which we can handle. Recall that a *closed trail* $T$ of length $m$ in a graph $G$ is a sequence $v_0e_1v_1e_2v_2\cdots e_mv_m$ such that each $v_i$ is a vertex, each $e_i$ is an edge, $v_0 = v_m$, any two adjacent terms are incident in $G$, and finally all the $e_i$'s are distinct. Call the closed trail $T$ *even* if $m$ is even.

5.5. PROPOSITION. *Let $e$ be an edge of the graph $G$. Then $f(G) = 2f(G-e)$ if and only if $e$ does not belong to an even closed trail of $G$.*

PROOF. Note that $f(G) = 2f(G-e)$ if and only if there do not exist spanning subgraphs $H$, $H'$ of $G$ such that $d(H) = d(H')$, $e$ is an edge of $H$, and $e$ is not an edge of $H'$. Suppose $e$ is an edge of the even closed trail $v_0e_1v_1\cdots e_{2r}v_{2r}$. Let $H$ have edges $e_1, e_3, \ldots, e_{2r-1}$, and $H'$ have edges $e_2, e_4, \ldots, e_{2r}$. Then $d(H) = d(H')$, and $e$ is an edge of exactly one of $H$ and $H'$. Thus $f(G) \neq 2f(G-e)$.

Conversely, suppose $H$ and $H'$ are spanning subgraphs of $G$ such that $d(H) = d(H')$, and let $e$ be an edge of $H$ but not of $H'$. Let $B$ be the set of edges of $H$ which are not edges of $H'$, and let $R$ be the set of edges of $H'$ which are not edges of $H$. Regard the edges in $B$ as "blue" and in $R$ as "red". Consider the graph $G'$ whose edges are $B \cup R$. Every vertex $G'$ is incident to the same number of blue edges as red edges. It follows easily that there is a closed trail $T$ containing $e$ whose edge alternate blue and red. Hence $T$ is even, as desired. □

The point of Proposition 5.5 is that even though $G$ may not be an FHM-graph, after we remove all edges not belonging to even closed trails the resulting graph may then be an FHM-graph. Thus equation (9) and Proposition 5.5 gives us a formula for $f(G)$. We also may assume $G$ is connected, since clearly $f(G+H) = f(G)f(H)$, where $G+H$ denotes the disjoint union of $G$ and $H$.

## REFERENCES


1. J. S. Beissinger and U. N. Peled, *Enumeration of labelled threshold graphs and a theorem of Frobenius involving Eulerian polynomials*, Graphs Combin. **3** (1987), 213–219.
2. T. Brylawski, *A decomposition for combinatorial geometries*, Trans. Amer. Math. Soc. **171** (1972), 235–282.
3. D. Foata, *La série génératrice exponentielle dans les problèmes d'énumération*, Sém. Math. Sup., Presses Univ. Montréal, 1974.
4. D. R. Fulkerson, A. J. Hoffman, and M. H. McAndrew, *Some properties of graphs with multiple edges*, Canad. J. Math. **17** (1965), 166–177.
5. D. Kleitman and K. Winston, *Forests and score vectors*, Combinatorica **1** (1981), 49–54.
6. M. Koren, *Extreme degree sequences of simple graphs*, J. Combin. Theory Ser. B **15** (1973), 213–224.
7. H. L. Montgomery and J. D. Vaaler, E3301, American Math. Monthly **96** (1989), 54.
8. U. N. Peled and M. K. Srinivasan, *The polytope of degree sequences*, Linear Algebra Appl. **114/115** (1989), 349–377.

9. G. C. Shephard, *Combinatorial properties of associated zonotopes*, Canad. J. Math. **26** (1974), 302–321.
10. R. Stanley, *Decompositions of rational convex polytopes*, Ann. Discrete Math. **6** (1980), 333–342.
11. ——, *Enumerative combinatorics*, vol. 1, Wadsworth & Brooks/Cole, Pacific Grove, Cal., 1986.
12. W. T. Tutte, *The factors of graphs*, Canad. J. Math. **4** (1952), 314–329.
13. T. Zaslavsky, *Signed graphs*, Discrete Appl. Math. **4** (1982), 47–74.
14. ——, *Signed graph coloring*, Discrete Math. **39** (1982), 215–228.

Department of Mathematics, Massachusetts Institute of Technology, Cambridge, Massachusetts 02139

*E-mail address*: rstan@math.mit.edu

DIMACS Series in Discrete Mathematics
and Theoretical Computer Science
Volume **4**, 1991

Applied Geometry
and Discrete Mathematics
THE VICTOR KLEE FESTSCHRIFT

# Geometry of Spaces of Homogeneous Polynomials on Banach Lattices

K. SUNDARESAN

ABSTRACT. An important subspace of the Banach space of $n$-homogeneous polynomials on a Banach lattice $E$ is isolated, and a concrete representation of the space is provided when $E$ is $L_p(l_p)$, $1 \leq p < \infty$. Geometric and topological properties of these spaces are discussed.

Polynomials on Banach spaces are very useful in function theory and differential analysis on these spaces. Motivated by certain problems that naturally arise in extending well-known theorems of classical function theory and approximation theory on finite-dimensional domains to domains in Banach spaces, Aron and his collaborators Globevenik, Hervés, Llavona, Prolla Schottenloher and Valdivia [**2–6**], Bombal and Llavona [**7**], Dineen [**9**], and Nemirovskii and Semenov [**16**] have made important contributions to the theory of homogeneous polynomials on Banach spaces in the past decade. In spite of these important contributions very little is known about the geometry of these spaces. This paper is concerned with the study of the geometry of the important subspace of the space of continuous homogeneous polynomials on the classical Banach lattices $L_p(l_p)$, $1 \leq p < \infty$, consisting of orthogonally additive polynomials.

In this introductory section the Banach space of continuous $n$-homogeneous orthogonally additive polynomials on a Banach lattice $E$ is defined. In §1 a concrete representation of the space when $E$ is $L_p(l_p)$, $1 \leq p < \infty$, is obtained. In §2 applications of the representation theorem are illustrated by obtaining geometric and topological properties of these spaces.

Before proceeding to the main results in the paper, basic definitions and some known results are recalled. Let $E, F$ be two real Banach spaces. If $n$ is an integer, $n \geq 1$, $L^n(E, F)$ $(L^n_S(E, F))$ denotes the space of all

1980 *Mathematics Subject Classification* (1985 *Revision*). Primary 46E15, 46B30, 46B20.

This research was supported in part by a research grant awarded by Cleveland State University.

continuous (continuous and symmetric) $n$-linear mappings on $E$ into $F$. A continuous $n$-homogeneous polynomial $P$ on $E$ into $F$ is a mapping of the form $P = A \circ \Delta_n$ for some $A \in L_S^n(E, F)$, where $\Delta_n$ is the diagonal map on $E \to E^n$. It is verified that the correspondence between $P$ and $A$ as described here is 1-1. In fact, while $P$ associated with a given $A$ is determined as above, $A$ can be recovered from $P$ by the following well-known polarization formula:

$$A(z_1, z_2, \dots, z_n) = \frac{1}{2^n n!} \sum_{\substack{\varepsilon_i = \pm 1 \\ 1 \le i \le n}} \varepsilon_1 \varepsilon_2 \cdots \varepsilon_n P(\varepsilon_1 z_1 + \cdots + \varepsilon_n z_n).$$

The space of continuous $n$-homogeneous polynomials on $E$ into $F$ is noted here by $P({}^nE; F)$, and simply by $P({}^nE)$ if $F$ is the real line $R$. $P({}^nE; F)$ is verified to be a vector space, and with the norm defined by

$$\|P\| = \sup_{x \ne 0} \frac{\|p(x)\|}{\|x\|^n} = \sup\{\|P(x)\| \mid \|x\| \le 1\},$$

it is a Banach space. The norm on $L_S^n(E, F)$ defined by

$$\begin{aligned}\|A\| &= \sup_{x_1, x_2, \dots, x_n \ne 0} \frac{\|A(x_1, x_2, \dots, x_n)\|}{\|x_1\| \, \|x_2\| \cdots \|x_n\|} \\ &= \sup\{\|A(x_1, x_2, \dots, x_n)\| \mid \|x_i\| \le 1, \ i = 1, 2, \dots, n\}\end{aligned}$$

renders $L_S^n(E, F)$ into a Banach space, and it is verified from the polarization formula that $P$ is continuous iff $A$ is continuous, and

$$\|A\| \le \frac{n^n}{n!} \|P\|.$$

For a comprehensive account of the spaces of homogeneous polynomials with applications to approximation theory, see the recent monograph of Llavona on approximation of continuously differentiable functions [**13**].

As in the case of continuous linear operators on $E$ it is verified that every polynomial $P \in P({}^nE; F)$ is Lipschitzian (in particular, uniformly continuous) on bounded subsets of $E$ by using the multilinearity of the symmetric form $A$ determining $P$.

Let $E$ be a Banach lattice. Two vectors $x, y$ in $E$ are said to be orthogonal (mutually disjoint), in symbols $x \perp y$, if $|x| \wedge |y| = 0$. A real-valued function $f$ on $E$ is said to be orthogonally additive if $f(x+y) = f(x) + f(y)$ whenever $x \perp y$. If $X$ is the space of all orthogonally additive real-valued continuous $n$-homogeneous polynomials on $E$, it is verified that $X$ is a closed subspace of the Banach space $P({}^nE)$. In the rest of the paper the Banach space $X$ is noted by $P_0({}^nE)$.

$L_p$, $1 \le p < \infty$, is the usual Banach space of $p$-summable functions on the unit interval $I = [0, 1]$ with respect to the Lebesgue measure $\mu$, with the norm $\|f\|_p = [\int_I |f|^p \, d\mu]^{1/p}$. For $1 \le p < \infty$, $l_p$ is the Banach space of $p$-summable real sequences $a = \{a_i\}_{i \ge 1}$, equipped with the norm

$\|a\|_p = [\sum_{i\geq 1} |a_i|^p]^{1/p}$. Similarly $L_\infty$ ($l_\infty$) is the Banach space of essentially bounded measurable functions (bounded sequences) with the usual norms.

For several properties from the geometry of Banach spaces, such as super reflexivity and uniform convexity etc., the reader is referred to Day [**8**] and Sundaresan [**18**]. For the terms and symbols from Banach lattices used in this paper, see Lacy [**11**] and Lindenstrauss and Tzafriri [**12**]. For some recent work on orthogonally additive functionals and their importance in the study of convex functions on Banach lattices, see Sundaresan [**19, 20**].

## 1. Orthogonally additive homogeneous polynomials

Before stating the main results in this section a theorem of Mizel [**14**] on the representation of a class of orthogonally additive functionals on the Banach spaces $L_p$, $1 \leq p < \infty$, is recalled.

If $R$ is the real line and $I = [0, 1]$ is equipped with the Lebesgue measure $\mu$, then a real-valued function $\varphi$ on $R \times I$ is said to be Carathéodory if (i) $\varphi(\cdot, t)\colon I \to R$ is continuous for $\mu$ a.e. $t$, and (ii) $\varphi(c, \cdot)\colon I \to R$ is Lebesgue measurable for all $c \in R$. If $f$ is a measurable function on $I$ and $\varphi$ is a Carathéodory function, then it is verified that the function $\varphi \circ f$ defined on $I$ by setting $\varphi \circ f(t) = \varphi(f(t), t)$ is also measurable. A Carathéodory function $\varphi$ on $R \times I$ is said to be in the Carathéodory $p$-class ($\mathrm{Car}^p(I)$) if for all $f$ in $L_p$, $\varphi \circ f \in L_1$. With the above terminology the following representation theorem in a slightly more general setting has been obtained by Mizel [**14**, Theorem 2]. For certain extensions of the representation theorem, see Mizel and Sundaresan [**15**].

THEOREM 1. *Let $p$ be a fixed number, $1 \leq p < \infty$, and $F$ be a continuous orthogonally additive functional on $L_p$ such that*

(*) *$F$ is uniformly continuous relative to the $L_\infty$ norm on each bounded subset of $L_\infty$.*

*Then there exists a $\varphi \in \mathrm{Car}^p(I)$ such that*

$$F(f) = \int_I \varphi \circ f \, d\mu \quad \text{for all } f \in L_p. \tag{$**$}$$

*Moreover, $\varphi$ may be chosen so that $\varphi(0, \ ) = 0$ a.e. and is then unique up to sets of the form $R \times N$, where $N$ is a $\mu$ null set.*

*Conversely, for every $\varphi \in \mathrm{Car}^p(I)$ such that $\varphi(0) = 0$ a.e. the formula* (**) *defines a unique continuous orthogonally additive function $F$ verifying condition* (*).

The next theorem provides a concrete representation of orthogonally additive continuous $n$-homogeneous polynomials on $L_p$ spaces.

THEOREM 2. *Let $p$ be a fixed number, $1 \leq p < \infty$, and $P$ be an orthogonally additive continuous $n$-homogeneous polynomial, $1 \leq n < p$. Then there* *exists a (unique) function* $\xi \in L_{p/(p-n)}$, *such that for all* $z \in L_p$,

$$P(z) = \int_I \xi z^n \, d\mu .$$

*Conversely, if* $\xi \in L_{p/(p-n)}$ *the above formula determines an orthogonally additive n-homogeneous polynomial on* $L_p$. *If* $p$ *is an integer and* $n = p$ *then* $P \in P_0({}^nL_p)$ *iff there exists a (unique) function* $\xi \in L_\infty$ *such that*

$$P(z) = \int_I \xi z^n \, d\mu \quad \text{for all } z \in L_p .$$

PROOF. Since $P$ is an $n$-homogeneous polynomial on $L_p$, as noted in the introduction, $P$ is uniformly continuous on bounded subsets of $L_p$. Thus, if $B$ is a bounded subset of $L_\infty$, clearly $B$ is a bounded subset of $L_p$, and $P$ is uniformly continuous on $B$ w.r.t. the $L_\infty$ norm since $\|z\|_p \le \|z\|_\infty$ for all $z \in L_\infty$. Thus it follows from Theorem 1 that there is a Carathéodory function $\varphi \in \mathrm{Car}^p I$ with $\varphi(0, t) = 0$ a.e. $t$, such that

$$P(z) = \int_I \varphi(z(t), t)\, d\mu(t) .$$

In particular, if $z$ is the constant function 1 on $I$ taking the value 1 on $I$ then

$$P(1) = \int_I \varphi(1, t)\, d\mu(t) . \tag{1}$$

Let $\varphi(1, t) = \xi(t)$. From equation (1) it follows that $\xi \in L_1$. Now let $s$ be a $\mu$-simple function, $s = \sum_{i=1}^m a_i \chi_{A_i}$, where $\{A_i\}_{i=1}^m$ is a finite measurable partition of $I$, and $\chi_T$ is the characteristic function of the set $T$. Then, since $P$ is orthogonally additive,

$$\begin{aligned} P(s) &= \sum_{i=1}^m P(a_i \chi_{A_i}) = \sum_{i=1}^m a_i^n P(\chi_{A_i}) \\ &= \sum_{i=1}^m \int_I a_i^n \varphi(\chi_{A_i}(t), t)\, d\mu(t) \\ &= \sum_{i=1}^m \int_{A_i} a_i^n \xi(t)\, d\mu(t) = \int_I \xi(t) s^n(t)\, d\mu(t), \end{aligned} \tag{2}$$

for all simple functions $s$.

It is now verified that $P(z) = \int_I \xi z^n \, d\mu$ as follows, for all $z \in L_p$. Let $z$ be a nonnegative function with support of $z \subset$ support of $\xi^+$, where $\xi = \xi^+ - \xi^-$. Consider a sequence of nonnegative simple functions $\{s_m\}_{m \ge 1}$ such that $s_m \uparrow z$. Since $\|s_m - z\|_p \to 0$, it follows that $\lim_{m\to\infty} P(s_m) = P(z)$. It is noted that, by our choice of $z$, $0 \le \xi s_m^n \uparrow \xi z^n$. Hence, using (2)

$$\begin{aligned} P(z) &= \lim_{m\to\infty} P(s_m) = \lim_{m\to\infty} \int_I \xi s_m^n \, d\mu \\ &= \int_I \xi z^n \, d\mu, \quad \text{if support } z \subset \text{support } \xi^+ . \end{aligned} \tag{3}$$

Similarly it is verified that if $z$ is a nonnegative function in $L_p$, and support of $z \subset$ support of $\xi^-$ (support of $z \subset I \sim$ support $\xi$), then

$$P(z) = \int_I \xi z^n \, d\mu \qquad (P(z) = 0). \tag{4}$$

If $z \in L_p$ is an arbitrary nonnegative function, and $A^+$, $A^-$, and $A_0$ are respectively the supports of $\xi^+$, $\xi^-$, and $I \sim (A^+ \cup A^-)$, then since $z = z\chi_{A^+} + z\chi_{A^-} + z\chi_{A_0}$ and $z\chi_{A^+}$, $z\chi_{A^-}$, $z\chi_{A_0}$ are pairwise orthogonal, it follows from the orthogonal additivity of $P$ that

$$P(z) = P(z\chi_{A^+}) + P(z\chi_{A^-}) + P(z\chi_{A_0}).$$

Thus, from (3) and (4) above it is verified that

$$P(z) = \int_{A^+} \xi z^n \, d\mu + \int_{A^-} \xi z^n \, d\mu = \int_I \xi z^n \, d\mu \tag{5}$$

for all nonnegative $z$ in $L_p$.

If $z = z^+ - z^-$ is an arbitrary function in $L_p$, then

$$P(z) = P(z^+) + P(-z^-), \quad \text{since } z^+ \perp z^-.$$

Hence, $P(z) = P(z^+) + P(z^-)$ or $P(z) = P(z^+) - P(z^-)$ according as $n$ is even or odd. Using the above two formulas and formula (5) it is verified that

$$P(z) = \int_I \xi z^n \, d\mu \quad \text{for all } z \in L_p, \tag{6}$$

noting that $z^n = (z^+)^n + (z^-)^n$ if $n$ is even, and $z^n = (z^+)^n - (z^-)^n$ if $n$ is odd.

Let $n < p$. In order to verify that the function $\xi$ representing $P$ given by formula (6) above is in $L_{p/(p-n)}$ it is enough to show that $\int_I \xi x \, d\mu < \infty$ for all $x \in L_{p/n}$, since $n/p + (p-n)/p = 1$. Let $x$ be a nonnegative function in $L_{p/n}$. Then $x^{1/n} \in L_p$, and using (6) it follows that $P(x^{1/n}) = \int_I \xi(t)x(t)\, d\mu(t) < \infty$. If $x$ is an arbitrary function in $L_{p/n}$, let $x = x^+ - x^-$. Thus, it follows that

$$\begin{aligned}\int_I \xi(t)x(t)\, d\mu(t) &= \int_I \xi(t)x^+(t)\, d\mu(t) - \int_I \xi(t)x^-(t)\, d\mu(t)\\ &= P((x^+)^{1/n}) - P((x^-)^{1/n}) < \infty.\end{aligned}$$

Hence $\xi \in L_{p/(p-n)}$.

Conversely, if $\xi \in L_{p/(p-n)}$ and $z \in L_p$, then

$$\begin{aligned}\left|\int_I \xi z^n \, d\mu\right| &\le \left[\int_I |\xi|^{p/(p-n)} \, d\mu\right]^{(p-n)/p} \left[\int_I |z^n|^{p/n} \, d\mu\right]^{n/p}\\ &= \|\xi\|_{p/(p-n)} \|z\|_p^n, \quad \text{using Hölder's inequality}.\end{aligned} \tag{7}$$

Hence, $P(z) = \int_I \xi z^n \, d\mu < \infty$ for all $z \in L_p$.

Further, if $A(z_1, z_2, \dots, z_n) = \int_I \xi z_1 z_2 \cdots z_n \, d\mu$ for $z_i \in L_p$, $1 \le i \le n$, then $A$ is a symmetric $n$-linear form and it is verified by using the generalized Hölder inequality that

$$|A(z_1, z_2, \dots, z_n)| \le \|\xi\|_{P/(p-n)} \|z_1\|_p \|z_2\|_p \cdots \|z_n\|_p .$$

Hence, $A$ is continuous. Since clearly $P = A \circ \Delta_n$, it follows that $P$ is a continuous $n$-homogeneous form. It is verified that $P \in P_0({}^n L_p)$, completing the proof of the part of the theorem stated for $n < p$.

To prove the representation when $P \in P_0({}^n L_p)$, $p = n$, once again applying Theorem 1, there is a measurable function $\xi$ such that

$$P(z) = \int_I \xi z^n \, d\mu \quad \text{for all } z \in L_p = L_n .$$

Now let $x$ be a nonnegative function in $L_1$. Then $x^{1/n} \in L_n$ and

$$P(x^{1/n}) = \int_I \xi x \, d\mu < \infty .$$

If $x$ is an arbitrary function in $L_1$, $x = x^+ - x^-$, then applying the preceding observation it follows that

$$\begin{aligned} \int_I \xi x \, d\mu &= \int_I \xi x^+ \, d\mu - \int_I \xi x^- \, d\mu \\ &= P((x^+)^{1/n}) - P((x^-)^{1/n}) < \infty . \end{aligned}$$

Hence, $\xi \in L_\infty$.

Conversely, if $\xi \in L_\infty$, it is verified as in the case $1 \le n < p$ that the function $P$ on $L_p = L_n$ defined by $P(z) = \int_I \xi z^n \, d\mu$ is an $n$-homogeneous polynomial, which is continuous since

$$(7') \qquad |P(z)| \le \|\xi\|_\infty \int_I |z|^n \, d\mu = \|\xi\|_\infty \|z\|_p^n .$$

Finally the uniqueness of the function $\xi$ representing $P \in P_0({}^n L_p)$, $1 \le n \le p$, is readily verified by noting that for any measurable set $A \subset I$, $P(\chi_A) = \int_A \xi \, d\mu$. □

REMARK 3. Let $1 \le n \le p$ and let $\sigma$ be the mapping on $P_0({}^n L_p)$ defined by

$$\sigma(P) = \xi , \quad \text{where } P(z) = \int_I \xi z^n \, d\mu \text{ for all } z \in L_p ,$$

$\xi$ being the function in $L_{p/(p-n)}(L_\infty)$ according as $n < p$ $(n = p)$, guaranteed by the preceding theorem. It is verified that $\sigma$ is a linear 1-1 surjection. Further, since $\|P\| \le \|\xi\|_{p/(p-n)}(\|\xi\|_\infty)$ if $n < p$ $(n = p)$ from inequalities (7) and (7′), the mapping $\sigma$ is a linear continuous 1-1 surjection. Hence, $\sigma$ is an isomorphism on $P_0({}^n L_p)$ onto $L_{p/(p-n)}(L_\infty)$ according as $n < p$ $(n = p)$.

Before determining the Banach spaces $P_0({}^nL_p)$ modulo a congruence it is noted that the norm $\| \, \|_p$ on $L_p$, $1 \le p \le \infty$, defined by

$$|||f|||_p = \max[\|f^+\|_p, \|f^-\|_p],$$

where $f = f^+ - f^-$, is equivalent to the usual norm $\| \, \|_p$ on $L_p$ if $1 \le p < \infty$, and coincides with $\| \, \|_\infty$ if $p = \infty$. For, if $1 \le p < \infty$, since $\| \, \|_p^p$ is orthogonally additive on $L_p$, it follows that

$$\begin{aligned} |||f|||_p &= \max[\|f^+\|_p, \|f^-\|_p] \\ &\le [\|f^+\|_p^p + \|f^-\|_p^p]^{1/p} = \|f\|_p \\ &\le 2^{1/p} \max[\|f^+\|_p, \|f^-\|_p] = 2^{1/p}|||f|||_p. \end{aligned}$$

From the definition of $||| \, |||_\infty$ it follows that $||| \, |||_\infty = \| \, \|_\infty$.

In the rest of the paper if $1 \le n < p$, for convenience, the numbers $p/(p-n)$ and $p/n$ are noted as $r$ and $r'$ respectively. It is verified that $1/r + 1/r' = 1$.

THEOREM 4. *The Banach space* $P_0({}^nL_p)$ *is isometrically isomorphic with* $L_r$ $(L_r, ||| \cdot |||_r)$ *if* $n$ *is odd (even), under the mapping* $\sigma$ *of the preceding remark, where* $r = p(p-n)$ *if* $1 \le n < p$, *and* $r = \infty$ *if* $n = p$.

PROOF. As already noted in Remark 3, $\sigma$ is an isomorphism on $P_0({}^nL_p)$ onto $L_r$ $(L_\infty)$, according as $n < p$ or $n = p$. Now it is proved that $\sigma$ is an isometry as described in the theorem. It is enough to show that $\|\sigma(P)\| = \|P\|$ for $P \ne 0$. Let $P \ne 0$ and $\sigma(P) = \xi$. Hence, $\xi \ne 0$. As a preliminary case let $\sigma(P) = \xi$ be nonnegative. Thus

$$\|P\| = \sup_{\|z\|_p \le 1} \left| \int_I \xi z^n \, d\mu \right| \le \sup_{\|z\|_p \le 1} \int_I \xi |z|^n \, d\mu.$$

Since $z$ and $|z|$ are of the same norm, it follows that

$$\|P\| = \sup_{\|z\|_p \le 1} \int_I \xi |z|^n \, d\mu \le \|\xi\|_r, \tag{8}$$

using the observations in Remark 3. It is verified now that $\|P\| = \|\xi\|_r$. Since $\xi \ne 0$, $\xi \ge 0$, the function $z = \xi^{(r-1)/n}/\|\xi\|_r^{(r-1)/n}$ is well defined and

$$\|z\|_p^p = \int_I \frac{\xi^{p(r-1)/n}}{\|\xi\|^{p(r-1)/n}} \, d\mu = \int_I \frac{\xi^r}{\|\xi\|_r^r} \, d\mu = 1.$$

With this choice of $z$,

$$P(z) = \int_I \xi z^n \, d\mu = \int_I \frac{\xi^r}{\|\xi\|_r^{(r-1)}} \, d\mu = \|\xi\|_r.$$

The above equation and (8) imply that $\|P\| = \|\xi\|_r$. Of course, since $\xi \ge 0$, in this case $|||\xi|||_r = \|\xi\|_r$ trivially.

Now let the representing function $\xi$, $\xi = \xi^+ - \xi^-$, be a nonzero arbitrary function in $L_r$ with $\xi^- \neq 0$, and $n$ be an odd integer, $1 \leq n < p$. Note that $\|\xi\|_r^r = \|\xi^+\|_r^r + \|\xi^-\|_r^r$. Let $z$ be the function on $I$ defined by

$$z = \frac{(\xi^+)^{(r-1)/n} - (\xi^-)^{(r-1)/n}}{\|\xi\|_r^{(r-1)/n}}.$$

Since $\xi^+ \perp \xi^-$,

$$\|z\|_p^p = \int_I \frac{(\xi^+)^r + (\xi^-)^r}{\|\xi\|_r^r}\, d\mu = 1,$$

using the fact $p(r-1)/n = r$. Since $n$ is odd and $\xi^+ \perp \xi^-$,

$$z^n = \frac{(\xi^+)^{(r-1)} - (\xi^-)^{(r-1)}}{\|\xi\|_r^{(r-1)}}$$

and

$$\begin{aligned} P(z) &= \int_I \xi z^n\, d\mu \\ &= \int_I (\xi^+ - \xi^-)\left[\frac{(\xi^+)^{(r-1)} - (\xi^-)^{(r-1)}}{\|\xi\|_r^{(r-1)}}\right] d\mu \\ &= \int_I \frac{(\xi^+) + (\xi^-)^r}{\|\xi\|_r^{(r-1)}}\, d\mu = \frac{\|\xi\|_r^r}{\|\xi\|_r^{(r-1)}} = \|\xi\|_r. \end{aligned}$$

Since $\|z\|_p = 1$, using the fact $\|P\| \leq \|\xi\|_r$, noted in Remark 3, it follows that $\|P\| = \|\xi\|_r$, completing the proof of the isometric representation claim when $n$ is an odd integer.

Now let $n$ be even, $1 \leq n < p$, and $P \in P_0(^nL_p)$. If $\sigma(P) = \xi \geq 0$, it has already been shown that $\|P\| = \|\xi\|_r = |||\xi|||_r$. Now let $\xi = \xi^+ - \xi^-$ with $\xi^- \neq 0$. Since $n$ is an even integer,

$$\begin{aligned} \|P\| &= \sup_{\|z\|_p \leq 1} |P(z)| = \sup_{\|z\|_p \leq 1} \left|\int_I (\xi^+ - \xi^-)|z|^n\, d\mu\right| \\ &\leq \max\left[\sup_{\|z\|_p \leq 1} \int_I \xi^+|z|^n\, d\mu,\ \sup_{\|z\|_p \leq 1} \int_I \xi^-|z|^n\, d\mu\right] \\ &= \max[\|\xi^+\|_r, \|\xi^-\|_r], \end{aligned}$$

since $\xi^+, \xi^-$ are nonnegative functions. Now in the case $\|\xi^+\|_r \geq \|\xi^-\|_r$ choose $z = (\xi^+)^{(r-1)/n}/\|\xi^+\|_r^{(r-1)/n}$. It is verified that $\|z\|_p = 1$. Since $\xi^+ \perp \xi^-$,

$$\begin{aligned} P(z) &= \int_I (\xi^+ - \xi^-)\frac{(\xi^+)^{(r-1)}}{\|\xi^+\|_r^{(r-1)}}\, d\mu \\ &= \int_I \frac{(\xi^+)^r}{\|\xi^+\|_r^{(r-1)}}\, d\mu = \|\xi^+\|_r. \end{aligned}$$

In the case $\|\xi^-\|_r \geq \|\xi^+\|_r$, choosing $z = (\xi^-)^{(r-1)/n}/\|\xi^-\|_r^{(r-1)/n}$, it is verified that $\|z\|_p = 1$ and $|P(z)| = \|\xi^-\|_r$. Hence if $n$ is an integer $< p$ and $P \in P_0(^n L_p)$, it follows that $\|P\| = \||\xi|\|_r$, where $r = p/(p-n)$.

Finally, if $n = p$, then it is enough to consider the case $n > 1$. For if $n = p = 1$, then $P_0(^n L_p) = L_1^*$, the dual of $L_1$, which by a well-known theorem is isometrically isomorphic with $L_\infty$, under the mapping $\sigma$ of Remark 3.

Let $n = p > 1$. Let $P \in P_0(^n L_p)$. Then $\sigma(P) = \xi \in L_\infty$ from Theorem 1. Let $\xi = \xi^+ - \xi^-$. Then $\|\xi\|_\infty = \max[\|\xi^+\|_\infty, \|\xi^-\|_\infty]$. Let $\|\xi^+\|_\infty \geq \|\xi^-\|_\infty$. Since $\xi^+ \in L_\infty$, and $L_\infty$ is isometrically isomorphic with $L_1^*$ under the mapping $\sigma$, it follows that there exists a sequence $\{x_m\}_{m\geq 1}$ in $L_1$ of positive functions of unit norm with $(*)$ the support of $x_m \subset$ support of $\xi^+$ for all $m \geq 1$ such that

$$\lim_{m\to\infty} \int_I \xi^+ x_m \, d\mu = \|\xi^+\|_\infty. \tag{9}$$

Since each $x_m \geq 0$, it is verified that the functions $z_m = x_m^{1/n}$ are in $L_p$, and $\|x_m\|_1 = \|z_m\|_p = 1$ for all $m \geq 1$. Now since $\sigma(P) = \xi$, it follows from property $(*)$ and the orthogonality of $\xi^+$ and $\xi^-$ that

$$\lim_{m\to\infty} P(z_m) = \lim_{m\to\infty} \int_I \xi x_m \, d\mu = \lim_{m\to\infty} \int_I \xi^+ x_m \, d\mu. \tag{10}$$

Since $\|P\| \leq \|\xi\|_\infty = \|\xi^+\|_\infty$, it follows from (9) and (10) that $\|P\| = \|\xi\|_\infty$. In the case $\|\xi^-\|_\infty \geq \|\xi^+\|_\infty$, a similar argument shows that $\|P\| = \|\xi\|_\infty$. This completes the proof of the theorem. □

REMARK 5. The case $n > p$ has not been discussed in Theorems 2 and 4. For, in this case there are no nontrivial $P \in P_0(^n L_p)$. For, if $P \in P_0(^n L_p)$, then as shown in the proof of Theorem 2 there is a measurable function $\xi$ such that

$$P(z) = \int_I \xi z^n \, d\mu \quad \text{for all } z \in L_p.$$

If possible let there be a positive number $a$ such that $\mu\{t|\xi(t) \geq a\} > 0$. Let $A = \{t|\xi(t) \geq a\}$. Since $n > p$, and $\mu$ is finite and nonatomic there is a positive function $x$ such that $\int_A x^p \, d\mu < \infty$ and $x\chi_A \notin L_n$. Since $P(x\chi_A) = \int_A \xi x^n \, d\mu \geq a \int_A x^n \, d\mu$, $P$ is not defined on $L_p$. Using a similar argument it is shown that if $\mu\{t|\xi(t) \leq -a\} > 0$ for some positive number $a$ then $P$ is not defined. Thus, if $n > p$ and $P \in P_0(^n L_p)$, then $\sigma(P) = \xi = 0$ $\mu$-a.e. Hence, $P$ is trivial as remarked.

In the remaining part of this section the representation theorem for the polynomials in $P_0(^n l_p)$ is stated. The results and the proofs for the case $1 \leq n < p$ are similar to the case of polynomials on $L_p$, and the proofs are not supplied. However, if $n > p$, unlike the case of $L_p$ spaces there are plenty of nontrivial polynomials in $P_0(^n l_p)$. For this reason and since the statement and proof of the representation theorem are the same for the cases $n > p$ and $n = p$, the theorem dealing with the case $n \geq p$ is stated separately.

The following notations are adopted in the statement of the next theorem. $e^i$ is the sequence of real numbers with 1 in the $i$th place and 0 elsewhere. If $P \in P_0({}^n l_p)$ let $\xi$ be the sequence $\{\xi_i\}_{i\geq 1}$ such that $\xi_i = P(e^i)$. Since $P$ is a continuous orthogonally additive $n$-homogeneous polynomial if $z = \{z_i\} \in l_p$,

$$P(z) = \lim_{m\to\infty} \sum_{i=1}^{m} P(z_i e^i) = \lim_{m\to\infty} \sum_{i=1}^{m} \xi_i z_i^n = \sum_{i=1}^{\infty} \xi_i z_i^n,$$

noting that $z = \sum_{i\geq 1} z_i e^i$ and $e^i \perp e^j$ if $i \neq j$.

THEOREM 6. *If $1 \leq n < p$, then the Banach space $P_0({}^n l_p)$ is linearly isomorphic with the Banach space $l_r$, where $r = p/(p-n)$, under the mapping $\sigma$ such that $\sigma(P) = \xi$ if and only if*

$$P(z) = \sum_{i\geq 1} \xi_i z_i^n \quad \textit{for all } z \in l_p.$$

*If $n$ is odd, then the mapping $\sigma\colon P_0({}^n l_p) \to l_r$ is a linear isometric surjection. If $n$ is even, then the mapping $\sigma$ is a linear isometric surjection if $l_r$ is equipped with the equivalent norm $|||\cdot|||_r$ defined by $|||z|||_r = \max[\|z^+\|_r, \|z^-\|_r]$, where $z = z^+ - z^-$.*

THEOREM 7. *Let $n$ be a positive integer $\geq p$. Then the space $P_0({}^n l_p)$ is isometrically isomorphic with $l_\infty$ under the mapping $\sigma$ where for $P \in P_0({}^n l_p)$, $\sigma(P) = \xi$ is such that*

$$P(z) = \sum_{i\geq 1} \xi_i z_i^n \quad \textit{for all } z = \{z_i\}_{i\geq 1} \in l_p.$$

PROOF. Let $P \in P_0({}^n l_p)$, where $n \geq p \geq 1$. Let $P(e^i) = \xi_i$ for all $i \geq 1$. Since $\|e^i\|_p = 1$ for all $i \geq 1$, it follows at once that $|\xi_i| \leq \|P\|$ for all $i \geq 1$, and $\|\xi\|_\infty \leq \|P\|$. Further from the properties of $P$ it is verified that

$$\text{(A)} \qquad P(z) = \sum_{i\geq 1} \xi_i z_i^n \quad \text{for all } z \in l_p.$$

Hence, if $\sigma(P) = \xi$, $\sigma$ is a continuous linear 1-1 mapping on $P_0({}^n l_p)$ into $l_\infty$ such that $\|\sigma(P)\|_\infty \leq \|P\|$. Further, if $a = \{a_i\}_{i\geq 1} \in l_\infty$ and $P(z) = \sum_{i\geq 1} a_i z_i^n$, then $P$ is a function on $l_p \to R$ for

$$\text{(B)} \qquad |P(z)| \leq \sum_{i\geq 1} |a_i|\,|z_i|^n \leq \|a\|_\infty \sum_{i\geq 1} |z_i|^n \leq \|a\|_\infty \|z\|_p^n$$

since $n \geq p$. Further, $P = A \circ \Delta_n$ if $A\colon l_p^n \to R$ is defined by

$$A(z^1, z^2, \dots, z^n) = \sum a_i z_i^1 z_i^2 \cdots z_i^n$$

for $z^j \in l_p$, $1 \leq j \leq n$, and $\Delta_n$ is the diagonal map on $l_p$ into $l_p^n$. The function $A$ on $l_p^n \to R$ is clearly a symmetric $n$-linear mapping, and it is verified that $A$ is continuous by applying the generalized Hölder inequality. Thus, $P$ is a continuous $n$-homogeneous polynomial. It is verified from formula (A) that $P$ is orthogonally additive. Thus, $P \in P_0({}^n l_p)$ and $\sigma(P) = a$. Hence, $\sigma$ is an isomorphism on $P_0({}^n l_p)$ onto $l_\infty$.

It is verified now that $\sigma$ is an isometry. It has been noted at the outset of this proof that $\|\xi\|_\infty = \|\sigma(P)\| \leq \|P\|$. Further, if $\sigma(P) = a = \{a_i\}_{i \geq 1}$, then from inequality (B) it is verified that

$$\|P\| = \sup_{\|z\|_p \leq 1} |P(z)| \leq \|a\|_\infty = \|\sigma(P)\|.$$

Hence,

$$\|\sigma(P)\| = \|P\|$$

and $\sigma$ is an isometric isomorphism as claimed. □

## 2. Geometry and related properties

By way of illustrating the importance of the representation theorems obtained in §1, geometric properties, reflexivity, and separability of the spaces $P_0({}^n E)$ and $P({}^n E)$ when $E = L_p$ or $l_p$ are discussed. Before proceeding to discuss these applications some known results which are useful in the proofs are recalled. It is known that if $p \geq 2$ there is a subspace of $L_p$ isomorphic with $l_2$. The result stated in the next theorem is due to Kadec and Pelczyński (See Theorem 3 in **[10]**).

THEOREM 8. *Let $p > 2$ and $E$ be an infinite-dimensional subspace of $L_p$. Then the following two conditions are equivalent:*

(i) *$E$ is isomorphic with the space $l_2$, and*
(ii) *the norms $\| \ \|_p$ and $\| \ \|_2$ are equivalent on $E$, i.e., there is a constant $C_E$ such that for all functions $x \in E$, $\|x\|_p \geq \|x\|_2 \geq C_E \|x\|_p$, where $\| \ \|_2$ is the $L_2$ norm of the function $x$.*

Before discussing the applications it is noted that any continuous linear functional on a Banach space $E$ is weakly continuous. However there are $n$-homogeneous polynomials, $n \geq 2$, which are not weakly continuous. For example, if $E$ is the Hilbert space $l_2$, consider the polynomial $P(x) = \sum_{i \geq 1} x_i^2$. $P \in P({}^2 l_2)$, however it is verified that $P$ is not weakly sequentially continuous; in particular, it is not weakly continuous. In function theory and approximation theory on Banach spaces, it is important to know whether an $n$-homogeneous polynomial is weakly continuous on the unit ball or not (see **[13]**, **[1]**).

THEOREM 9. *If $P$ is a nonzero polynomial in $P_0({}^n L_p)$, $1 \leq p < \infty$, $n \geq 2$, then $P$ is not weakly sequentially continuous on the unit ball of $L_p$.*

PROOF. It follows from Remark 5 that the only case to be considered is when $2 \leq n \leq p$.

Consider the case when $n = 2$. If $p = 2$ and $P \in P_0(^2L_2)$, $P \neq 0$, then from the representation Theorem 3, it follows that there is a function $\xi \in L_\infty$, $\xi \neq 0$, such that

$$P(z) = \int_I \xi z^2 \, d\mu .$$

Since $\xi \neq 0$, there is a positive number $a$ such that either the set $A_1 = \{t | \xi(t) \geq a\}$ or the set $A_2 = \{t | \xi(t) \leq -a\}$ is of positive $\mu$-measure. Let $\mu(A_1) > 0$. Consider the measure $\mu_{A_1}$, the contraction of $\mu$ to the measurable subset $A_1$ of $I$. Consider the closed subspace $E$ of $L_2$ consisting of functions in $L_p$ vanishing outside $A_1$. $E$ is a separable Hilbert space, and let $\{e^i\}_{i\geq 1}$ be a complete orthonormal system of functions in $E$. Then $e^i \to 0$ weakly in $L_2$. However,

$$P(e^i) = \int_{A_1} \xi (e^i)^2 \, d\mu \geq a \int_{A_1} (e^i)^2 \, d\mu = a .$$

Hence, $P(e^i) \nrightarrow 0$, and $P$ is not weakly sequentially continuous on the unit ball of $L_2$. The case when $\mu(A_2) > 0$ and $\mu(A_1) = 0$ is similarly dealt with.

Next let $n = 2$ and $p > 2$. From the representation theorem it follows that there is a $\xi \in L_r$, where $r = p/(p-2)$, such that

$$P(z) = \int_I \xi z^2 \, d\mu \quad \text{for all } z \in L_p .$$

Since $P \neq 0$, as noted in the preceding part of the proof, there is a positive number $a$ such that for the sets $A_1$ and $A_2$ as defined there, either $\mu(A_1) > 0$ or $\mu(A_2) > 0$. As in the earlier part of the proof it is enough to consider the case $\mu(A_1) > 0$. Consider as before the contraction $\mu_{A_1}$ and the space $L_p(\mu_{A_1})$. Clearly $L_p(\mu_{A_1})$ is a subspace of $L_p$ isomorphic with $L_p$. Hence, from Theorem 8 (due to Kadec and Pelczyński), since $p > 2$, it follows that there is a subspace $E$ of $L_p(\mu_A)$ isomorphic with $l_2$ and a constant $C$ such that for all $z \in E$

$$(*) \qquad \|z\|_2 \leq \|z\|_p \leq C\|z\|_2 ,$$

where $\| \; \|_2$ is the norm on $L_2(\mu_{A_1})$. Let $T$ be an isomorphism on $l_2$ onto $E$, and $\{e^i\}_{i\geq 1}$ be the sequence of unit vectors in $l_2$. Let $z_i = Te^i$, $i \geq 1$. Since $T$ is an isomorphism, there are positive numbers $m$, $M$, such that $m \leq \|z_i\|_p \leq M$. Thus it follows from $(*)$ above that $\|z_i\|_2 \geq m/C$ for all $i \geq 1$. Since $e^i \to 0$ weakly in $l_2$, it follows that $Te^i = z_i \to 0$ weakly in $L_p$ as $i \to \infty$. Thus if $x_i = z_i/\|z_i\|_p$, then $\{x_i\}_{i\geq 1}$ is a sequence of vectors of unit norm in $L_p$ weakly converging to $0$. However,

$$\begin{aligned} P(x_i) &= P\left(\frac{z_i}{\|z_i\|_p}\right) \geq \frac{1}{M^n} P(z_i) \\ &= \frac{1}{M^n} \int_{A_1} \xi z_i^2 \, d\mu \geq \frac{a}{M^n} \left(\frac{m}{C}\right)^2 . \end{aligned}$$

Hence, $P(x_i) \nrightarrow 0$ and $P$ is not weakly sequentially continuous on the unit ball of $L_p$. Next let $2 < n \leq p$, and $P \in P_0({}^n L_p)$, $P \neq 0$. If possible, let $P$ be weakly sequentially continuous. Then applying the representation Theorem 2 it follows that there is a function $\xi \in L_r$, where $r = p/(p-n)$ if $n < p$, and $r = \infty$ if $n = p$, such that

$$P(z) = \int_I \xi z^n \, d\mu \quad \text{for all } z \in L_p .$$

It is clear that the $n$-linear symmetric form $A$ of $L_p$ determining the polynomial $P$ is given by

$$A(z_1, z_2, \dots, z_n) = \int_1 \xi z_1 z_2 \cdots z_n \quad \text{for all } z_i \in L_p .$$

Since $P$ is weakly sequentially continuous, it is verified from the polarization formula that $A$ is weakly sequentially continuous, i.e., if $\{z_j^i\}_{i \geq 1}$, $1 \leq j \leq n$, are $n$ sequences in $L_p$ such that $z_j^i \to z_j$ in the weak topology for $1 \leq j \leq n$, then $A(z_1^i, z_2^i, \dots, z_n^i) \to A(z_1, z_2, \dots, z_n)$. In particular, if each of the sequences $\{z_j^i\}$, $1 \leq j \leq n-2$, is the constant sequence with range $\{1\}$, $1$ being the function taking the value $1$ on $I$, then the symmetric linear two-form $B$ on $L_p \times L_p$, defined by

$$B(y_1, y_2) = A(\underbrace{1, 1, \dots, 1}_{n-2}, y_1, y_2) = \int_I \xi y_1 y_2 ,$$

is weakly sequentially continuous. Hence the 2-homogeneous polynomial $Q$ determined by $B$, which is given by

$$Q(z) = \int_I \xi z^2 \, d\mu ,$$

is weakly sequentially continuous. Since $Q \in P_0({}^2 L_p)$, $Q \neq 0$, this contradicts the result proved earlier in the proof, completing the proof. □

In the next theorem the weak sequential continuity of the polynomials $P \in P_0({}^n l_p)$ is discussed. If $p = 1$, then a sequence $\{x^{(k)}\}_{k \geq 1}$ weakly converges to $z$ in $l_1$ iff $z^{(k)} \to z$ strongly by the well-known theorem of Schur. Thus every polynomial $P \in P({}^n l_1)$ is weakly sequentially continuous. If $p \geq 1$ and $n < p$, then it is known that every polynomial $P \in P({}^n l_p)$ is weakly sequentially continuous (see 4.2 in Pelczyński **[17]**). Thus, the only case that concerns us here arises when $n \geq p > 1$.

THEOREM 10. *Let $n$ be a positive integer and $p$ be a fixed number $> 1$. If $n \geq p$, then a polynomial $P \in P_0({}^n l_p)$ is weakly sequentially continuous if and only if $\sigma(P)$ is a null sequence, where $\sigma$ is the mapping described in Theorem 7.*

PROOF. Let $P \in P_0({}^n l_p)$ and $\sigma(P) = \xi$. Let $\xi$ be a null sequence. Let $\{z^{(k)}\}_{k\geq 1}$ be a sequence in $l_p$ weakly converging to $0$. Thus, the sequence $\{\|z^{(k)}\|_p\}$ is bounded and, further, since $n \geq p$, $z^{(k)} \in l_n$ for all $k \geq 1$ and $\|z^{(k)}\|_n^n \leq \|z^{(k)}\|_p^{n/p}$. Now it is verified that $P(z^{(k)}) \to 0$ as $k \to \infty$ as follows. Let $\varepsilon > 0$. Since $\xi_i \to 0$, there is a positive integer $m$ such that $|\xi_i| < \varepsilon$ if $i > m$. Let $M$ be an upper bound of the sequence $\{\|z^{(k)}\|_p^n\}_{k\geq 1}$. It is verified that

$$
\begin{aligned}
|P(z^{(k)})| &\leq \sum_{i\geq 1} |\xi_i|\,|z_i^{(k)}|^n \leq \sum_{i=1}^{m} |\xi_i|\,|z_i^{(k)}| + \varepsilon \sum_{i\geq m+1} |z_i^{(k)}|^n \\
&\leq \sum_{i=1}^{m} |\xi_i|\,|z_i^{(k)}|^n + \varepsilon M .
\end{aligned}
$$

Since $z^{(k)} \to 0$ weakly in the space $l_p$, $z_i^{(k)} \to 0$ as $k \to \infty$ for every $i \geq 1$. Thus, the finite sum $\sum_{i=1}^m |\xi_i|\,|z_i^{(k)}|^n \to 0$ as $k \to \infty$. Hence, it follows that $P(z^{(k)}) \to 0$ and $P$ is weakly sequentially continuous at the origin. In case $z^{(k)} \to z$ weakly in $l_p$ and $z \neq 0$, by a slight modification of the above argument it is verified that $P(z^{(k)}) \to P(z)$.

If $\sigma(P) = \xi$ is not a null sequence, then there is a positive number $\delta$ and a subsequence $\{\xi_{i_j}\}_{j\geq 1}$ of $\{\xi_i\}$ such that $|\xi_{i_j}| \geq \delta$ for all $j \geq 1$. For $j \geq 1$ let $e^{(i_j)}$ be the sequence $\{a_t\}$ in $l_p$ with $a_t = 0$ if $t \neq i_j$, and $a_t = 1$ if $t = i_j$. Thus,

$$
|P(e^{(i_j)})| = \left| \sum_{t\geq 1} \xi_t a_t^n \right| = |\xi_{i_j}| \quad \text{for all } j \geq 1 .
$$

Since $e^{(i_j)} \to 0$ weakly, it follows that $P$ is not weakly sequentially continuous at $0$, since $|\xi_{i_j}| \geq \delta$ for all $j \geq 1$, completing the proof of the theorem. □

THEOREM 11. *The Banach spaces $P_0({}^n L_p)$, $1 \leq n < p$, are superreflexive. In case $n$ is an odd integer these spaces are uniformly convex. If $n = p$ the spaces $P_0({}^n L_p)$ are nonseparable and nonreflexive.*

PROOF. Let the number $p$ and integer $n$ be fixed, $1 \leq n < p$. It follows from Theorem 4 that the space $P_0({}^n L_p)$ is isomorphic with $L_r$, where $r = p/(p-n)$. Since $r > 1$, the Banach space $L_r$ is uniformly convex and the space $P_0({}^n L_p)$ is superreflexive. If, further, $n$ is odd, the same theorem implies that $P_0({}^n L_p)$ is isometrically isomorphic with $L_r$, with $r$ as above. Hence $P_0({}^nL_p)$ is uniformly convex. In case $n = p$, since $P_0({}^nL_p)$ is isomorphic with $L_\infty$, as shown in Theorem 4, it is nonseparable and nonreflexive. □

Theorem 11 remains true if $L_p$ and $n = p$ in the statement of the theorem are replaced by $l_p$ and $n \geq p$, respectively. The proof is an easy application of Theorems 6 and 7 in this paper.

Before concluding the paper, it is noted that since the homogeneous forms associated with the $n$th derivative of the norm of an $L_p$ space, $1 \leq n < p$, are orthogonally additive, orthogonally additive homogeneous polynomials on a $L_p$ space are of interest in the geometry of these spaces. Further the representation Theorem 2 in this paper answers the natural problem of providing the precise description of the spaces $L_{p/(p-n)}$, $1 \leq n < p$, in terms of the space $L_p$, thus extending the well-known theorem identifying the space $L_{p/(p-1)}$ as the dual of the space $L_p$, $1 < p < \infty$.

Representation theorems for the Banach spaces $P_0({}^nE, F)$ when $E$ is one or the other of the classical Banach lattices and $F$ is an arbitrary Banach space, together with applications to symmetrically monotone nonlinear operators (see **[19]**) on these Banach lattices, will be presented elsewhere.

## References

1. Raymundo Alencar, R. M. Aron, and S. Dineen, *A reflexive space of holomorphic functions in infinitely many variables*, Proc. Amer. Math. Soc. **90** (1984), 407–411.
2. R. M. Aron and J. Globevenik, *Analytic functions on* $c_0$, preprint.
3. R. M. Aron, C. Hervés, and M. Valdivia, *Weakly continuous mappings on Banach spaces*, J. Funct. Anal. **52** (1983), 189–203.
4. R. M. Aron and J. Llavona, *Composition of weakly uniformly continuous functions*, preprint.
5. R. M. Aron and J. B. Prolla, *Polynomial approximation of differentiable functions on a Banach space*, J. Reine Angew. Math. **313** (1980), 195–216.
6. R. M. Aron and R. M. Schottenloher, *Compact holomorphic mappings on a Banach space and approximation property*, J. Funct. Anal. **21** (1976), 7–30.
7. F. Bombal and J. Llavona, *La propiedad de approximación en espacios de funciones differenciables*, Rev. Real Acad. Cienc. Exact. Fís. Natur. Madrid **70** (1976), 337–346.
8. M. M. Day, *Normed linear spaces*, 3rd ed., Springer-Verlag, New York, 1973.
9. S. Dineen, *Entire functions of* $c_0$, J. Funct. Anal. **52** (1983), 205–218.
10. M. I. Kadec and A. Pelczyński, *Bases, lacunary sequences, and complemented subspaces in the spaces* $L_p$, Studia Math. **21** (1962), 161–176.
11. H. E. Lacey, *The isometric theory of classical Banach spaces*, Springer-Verlag, New York, 1974.
12. J. Lindenstrauss and L. Tzafriri, *Classical Banach spaces*, Vol. 2, Springer-Verlag, New York, 1979.
13. J. Llavona, *Approximation of continuously differentiable functions*, Math. Studies, no. 130, North-Holland, 1983.
14. V. Mizel, *Characterization of nonlinear transformations possessing kernels*, Canad. J. Math. **22** (1970), 449–471.
15. V. Mizel and K. Sundaresan, *Characterization of nonlinear vector valued functions*, Trans. Amer. Math. Soc. **159** (1971), 111–127.

16. A. M. Nemirovskii and S. M. Semenov, *On polynomial approximation of functions on a Hilbert space*, Mat. Sb. USSR **21** (1973), 255–277.
17. A. Pelczyński, *A property of multilinear operations*, Studia Math. **16** (1957), 173–182.
18. K. Sundaresan, *Geometry and nonlinear analysis in Banach spaces*, Pacific J. Math. **103** (1982), 487–498.
19. ——, *Convex functions on Banach lattices*, Contemp. Math., no. 54, Amer. Math. Soc., Providence, R.I., 1986, pp. 79–94.
20. ——, *Monotone gradients on Banach lattices*, Proc. Amer. Math. Soc. **98** (1986), 448–454.

Department of Mathematics, Cleveland State University, Cleveland, Ohio 44115

DIMACS Series in Discrete Mathematics
and Theoretical Computer Science
Volume 4, 1991

Applied Geometry
and Discrete Mathematics
THE VICTOR KLEE FESTSCHRIFT

# The Combinatorics of Bivariate Splines

WALTER WHITELEY

ABSTRACT. For a triangulation $\Delta$ of a plane polygonal disk, we examine the vector space $S_d^r(\Delta)$ of $C^r$ functions over $\Delta$ which are polynomial, of degree at most $d$, when restricted to the triangles of $\Delta$. We prove that projective transformations of the plane induce isomorphisms of these spaces. Using a combinatorial induction on the rows of a corresponding matrix, we show that $S_d^r(\Delta)$ contains only trivial splines (with the same polynomial over all triangles), for $d \leq (3r+1)/2$ and generic positions of the vertices. For $S_2^1(\Delta)$ and $S_3^1(\Delta)$, we also characterize the effects of singular vertices (where the edges lie on two lines).

## 1. Introduction

Consider a plane polygonal domain $D$ (Figure 1(a)) subdivided into a set $\Delta$ of polygonal faces (Figure 1(b)). A *bivariate spline over* $\Delta$ is a function over $D$ which is polynomial piecewise over the faces and has a specified continuity over the entire region $D$. Specifically, if the precise polynomials have degree at most $d$, and the overall function has continuous $r$th derivatives, then it is a *bivariate* $C_d^r$*-spline.* (Figure 1(c) shows a $C_1^0$-spline.) The set of all $C_d^r$-splines over this subdivision forms a vector space $S_d^r(\Delta)$.

For the past several decades these spaces have been studied using numerical and algebraic methods [**1–4, 12–14, 21–24**], and more recently using algebraic and homological methods [**8–10**]. For the special case of $S_1^0(\Delta)$, the piecewise linear, globally continuous functions are also called "polyhedral pictures". These polyhedral pictures have been studied as a separate field of "picture recognition" or "scene analysis" by a variety of distinctive algebraic, geometric, and combinatorial methods [**25, 28, 32**].

Here, we study the bivariate splines $S_d^r(\Delta)$ for $d \leq (3r+1)/2$, as well as $S_3^1(\Delta)$, using a combined matrix and graph theory approach. The methods draw on a century long dialogue among the geometric theories of

---

1980 *Mathematics Subject Classification* (1985 *Revision*). Primary 41A15, 51N15, 65D07; Secondary 05C50, 52A37, 68U05.

Work supported in part by grants from FCAR (Québec) and NSERC (Canada).

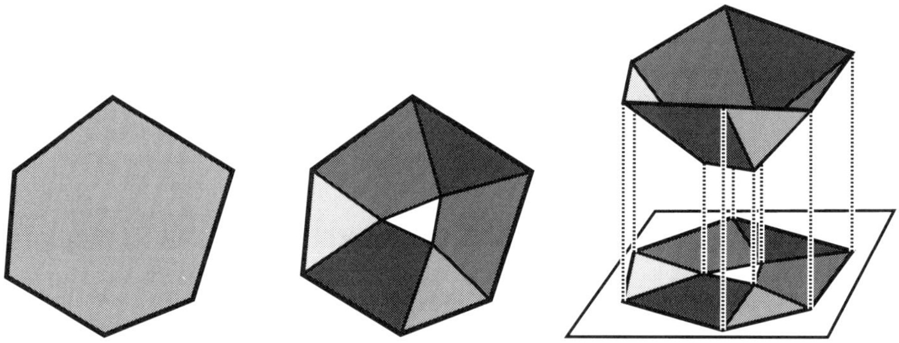

(a) (b) (c)

FIGURE 1

polyhedral pictures, of statics of bar frameworks, and of mechanics of rigid panel structures [**20, 15, 16, 6, 27, 28, 31**].

It has been observed that an affine transformation of the underlying subdivision induces a corresponding isomorphism of the space of splines [**2**]. In §2 we show that any nonsingular projective transformation which takes a subdivision $\Delta$ to a subdivision $\Delta'$ induces a simple isomorphism of the spline spaces. Thus the dimension and the pattern of the spline space is *projectively invariant*, extending classical results in scene analysis and statics. This observation has essential implications for the analysis of trivariate splines by projection to bivariate splines [**5**].

In §3 we introduce a matrix over the graph of interior edges of a disk subdivision and prove the connection with splines. In many "general" cases, with $d \geq 3r+1$, the known values for $\dim(S_d^r(\Delta))$ correspond to the independence of the columns of this matrix. However for $d \leq \frac{3}{2}(r+1)$, we show that the rows are independent for generic vertices and arbitrary faces inside a triangular boundary (§4), proving there are only trivial splines. This case extends our recent result for $S_2^1(\Delta)$ [**29, 31**].

For $S_2^1(\Delta)$ and $S_3^1(\Delta)$, we also extend the previous results to include "singular" vertices with all edges lying in two lines (§5). The singular vertices each drop the rank of the matrix by exactly 1, confirming that, for generic positions of the nonsingular vertices, $S_d^1(\Delta)$ achieves the lower bound of [**22, 23**].

In the conclusion, §6, we highlight an important set of still unsolved cases for the dimension of $S_d^r(\Delta)$, $(3r+1)/2 < d < 3r+1$.

In a companion paper [**33**], we extend this approach to multiple boundaries, etc. and examine the geometric "special positions" which yield a larger dimension for some $S_2^1(\Delta)$. Related extensions for other $S_d^r(\Delta)$ will appear in further papers. Many of these results (such as projective invariance) also extend to splines in higher dimensions. We have chosen to emphasize bivariate splines for simplicity and for space.

Within the history of mathematics, the study of multivariate splines is quite new. The fields of rigidity and scene analysis draw on a much longer tradition. In exposing the deep similarities among the fields, we hope to encourage the transfer of techniques and results among the fields. Such a transfer has already begun in [**30, 31**] leading to recent results in both fields [**8, 17**]. To aid this transfer, we have included some details and examples to make the presentation more self-contained. We ask the indulgence of workers on splines, who may find some discussions elementary.

There are important differences among the questions asked in the different applications of the common geometry and combinatorics. The methods used here work best for small-dimensional spaces of splines. In this spirit, the specific techniques introduced here may be more relevant to splines as objects in a geometric design than to splines as a tool of numerical or geometric interpolation.

**Acknowledgments**

We thank Louis Billera for introducing us to these problems—and for ongoing conversations on these spaces. We also thank Peter Alfeld for patiently explaining the known results and the point of view of workers in approximation theory. Both have contributed to the evolution of the results presented here. In the background of this work lies a 15-year exchange of combinatorial and geometric ideas within the Structural Topology Research Group at the Université de Montréal. We also thank the referees for helpful comments on the presentation.

## 2. Projective invariance of spaces of splines

For simplicity, we will focus on plane disks—regions without holes, or pinch points. This assumption is not essential to the results (see [**33**]), but it does simplify the presentation.

A *face* $P^i$ is a simple polygon in $\mathbb{R}^2$ and its interior (it need not be convex). A *plane disk subdivision* is a collection $\Delta = \{\dots, P^i, \dots\}$ of faces in $\mathbb{R}^2$, such that

(i) the union of the faces is a simple plane polygon and its interior (Figure 2(a));

(ii) the intersection of any two faces is either empty, or a union of vertices and edges shared by the faces (Figure 2(b)).

From the definition, we have a strong connectedness:

(iii) any two faces $P^0$ and $P^n$ can be joined by a *face-edge path*:

$$P^0, \dots, P^i, P^{i+1}, \dots, P^n,$$

such that any adjacent faces $P^i$, $P^{i+1}$ of the path have a common edge.

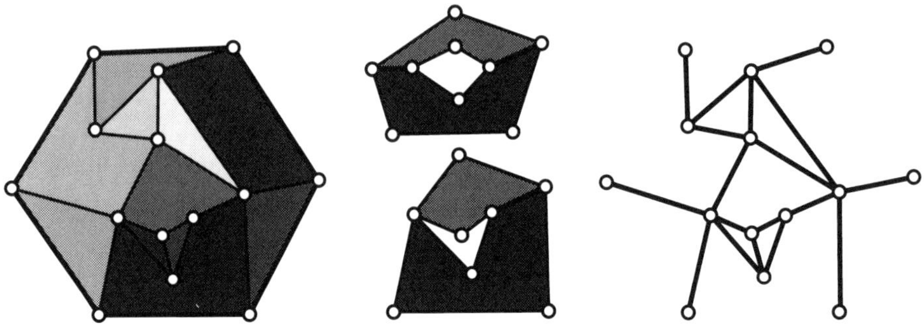

(a) (b) (c)

FIGURE 2

The disk subdivision $\Delta$ defines an *interior graph* $G_\Delta^\circ = (V, E^\circ)$ of all vertices and all *interior edges* which separate two faces (Figure 2(c)).

A $C_d^r$*-spline* over the disk subdivision $\Delta$ is an assignment of a polynomial $S^j$ of degree at most $d$ to each face $P^j$ such that the overall function has continuous $r$th derivatives. Since this set is closed under addition and multiplication by a scalar, the $C_d^r$-splines over a fixed disk subdivision $\Delta$ form a vector space $S_d^r(\Delta)$.

There is a simple algebraic criterion for $C^r$ continuity. Consider the example of piecewise linear, globally continuous splines. Two nonvertical planes

$$S^j = A^j x + B^j y + C^j = z \quad \text{and} \quad S^k = A^k x + B^k y + C^k = z$$

meet in a spatial line above the plane line $A^{jk}x + B^{jk}y + C^{jk} = 0$ if and only if

$$A^j x + B^j y + C^j - (A^k x + B^k y + C^k) = \beta^{jk}(A^{jk}x + B^{jk}y + C^{jk})$$

for some scalar $\beta^{jk}$, called a "smoothing cofactor" in [**13, 14**], and an "angular velocity" in [**27**]. This idea generalizes to an algebraic criterion for general splines [**10, 12–14**].

THEOREM (Wang). *Two polynomials $S^j$ and $S^k$ meet with continuous $r$th derivatives over the line $A^{jk}x + B^{jk}y + C^{jk} = 0$ if and only if*

$$S^k - S^j = \beta^{jk}(A^{jk}x + B^{jk}y + C^{jk})^{r+1}$$

*for some polynomial $\beta^{jk}$ of degree at most $d - r - 1$.*

For convenience, we rewrite the plane points $\mathbf{q}_i = (x_i, y_i)$ with affine coordinates $\mathbf{p}_i = (x_i, y_i, 1)$. An interior edge separating two faces $P^j$, $P^k$

is defined by two distinct vertices $\mathbf{p}_h$, $\mathbf{p}_i$. The line of this edge then has the equation

$$|[\mathbf{p}_h \mathbf{p}_i \mathbf{X}]| = 0,$$

where $\mathbf{X} = (x, y, 1)$ gives the affine coordinates of a variable point, $[abc]$ is the $3 \times 3$ matrix with these columns, and $|\ |$ is the determinant.

We now show that a projective transformation on the disk subdivision induces an isomorphism of the space of splines. For the projective plane we use homogeneous coordinates $(u_i, v_i, w_i)$, where the finite point $(x_i, y_i)$ is expressed by any of the triples $(w_i x_i, w_i y_i, w_i)$ for a nonzero $w_i$. (The coordinates $(u_i, v_i, 0)$ represent points at infinity.) A *projective transformation* of the projective plane is expressed as a nonsingular linear transformation of the projective coordinates $T(u_i, v_i, w_i) = (u_i^*, v_i^*, w_i^*)$. Such a transformation takes points to points and lines to lines.

For a disk subdivision $\Delta$, we are interested in projective transformations $T$ which produce a new disk subdivision $T(\Delta)$. We say a projective transformation is *admissible for* $\Delta$ if $T(u_i, v_i, w_i)$ is finite for all points in the faces of $\Delta$. The reader can check that an admissible transformation for $\Delta$ takes $\Delta$ to a disk subdivision $T(\Delta)$.

1. THEOREM. *Given an admissible projective transformation $T$ of a plane disk subdivision $\Delta$, there is an induced isomorphism between $S_d^r(\Delta)$ and $S_d^r(T(\Delta))$.*

PROOF. A spline in $S_d^r(\Delta)$ is defined by a choice of the polynomial function $S^j(\mathbf{X})$ for each face $P^j$, such that for each interior edge there is a polynomial $\beta^{jk}(\mathbf{X})$:

$$S^j - S^k = \beta^{jk}(\mathbf{X}) |[\mathbf{p}_h \mathbf{p}_i \mathbf{X}]|^{r+1}.$$

To respect the projective setting, we "homogenize" the variable point $\mathbf{X}$ and the polynomials $S^j$, $S^k$, and $\beta^{jk}(\mathbf{X})$. We change the variable point to $\mathbf{Z} = (x, y, z)$. We alter the polynomials accordingly—by substituting $(x/z)$ for $x$ and $(y/z)$ for $y$, and then multiplying by $z^d$, clearing fractions. This gives homogeneous polynomials $\beta_*^{jk}(\mathbf{Z})$, $S_*^j(\mathbf{Z})$, and $[\mathbf{p}_h \mathbf{p}_i \mathbf{Z}]$ satisfying

$$(1) \qquad S_*^j(\mathbf{Z}) - S_*^k(\mathbf{Z}) = \beta_*^{jk}(\mathbf{Z}) |[\mathbf{p}_h \mathbf{p}_i \mathbf{Z}]|^{r+1}.$$

The identity remains true on substitution and on multiplication by $z^d$, $z \neq 0$. Therefore, it is also true for $z = 0$.

Assume that an admissible projective transformation for $\Delta$ is expressed by the nonsingular matrix $[T]$. We can replace the variables $\mathbf{Z}$ by the variables $[T]^{-1}(\mathbf{Z})$ without altering the polynomial identity:

$$(2) \quad S_*^j([T]^{-1}(\mathbf{Z})) - S_*^k([T]^{-1}(\mathbf{Z})) = \beta_*^{jk}([T]^{-1}(\mathbf{Z})) |[\mathbf{p}_h \mathbf{p}_i [T]^{-1}(\mathbf{Z})]|^{r+1}.$$

If we expand the $S_*^j([T]^{-1}(\mathbf{Z}))$ and $\beta_*^{jk}([T]^{-1}(\mathbf{Z}))$ into the original variables, we obtain the *induced polynomials* $T^*(S^j(\mathbf{Z}))$, $T^*(\beta^{jk}(\mathbf{Z}))$. Next, we multiply and divide the right-hand side of equation (2) by the nonzero constant $|[T]|^{r+1} = |T|^{r+1}$. This gives the equation

$$T^*(S^j(\mathbf{Z})) - T^*(S^k(\mathbf{Z})) = |T|^{-(r+1)}T^*(\beta^{jk}(\mathbf{Z}))|[T][\mathbf{p}_h\mathbf{p}_i[T]^{-1}(\mathbf{Z})]|^{r+1}$$
$$= |T|^{-(r+1)}T^*(\beta^{jk}(\mathbf{Z}))|[T(\mathbf{p}_h)T(\mathbf{p}_i)(\mathbf{Z})]|^{r+1}.$$

Finally, for each vertex with affine coordinates $\mathbf{p}_i = (x_i, y_i, 1)$, the projective image under $T$ will have affine coordinates $\mathbf{p}_i^* = 1/w_i^* T(\mathbf{p}_i)$. The equation is now written as

$$(3) \quad T^*(S^j(\mathbf{Z})) - T^*(S^k(\mathbf{Z})) = \frac{1}{(w_h^* w_i^* |T|)^{r+1}} T^*(\beta^{jk}(\mathbf{Z}))|[\mathbf{p}_h^*\mathbf{p}_i^*\mathbf{Z}]|^{r+1} = 0.$$

Since the equations $|[\mathbf{p}_i^*\mathbf{p}_i^*\mathbf{Z}]| = 0$ now refer to the edges of $T(\Delta)$, we have the required spline condition for the new disk subdivision. □

For those familiar with splines in higher dimensions, or over more general regions, it will be clear that Theorem 1, and the proof, both extend immediately. This projective invariance, and the projective thinking which flows from this invariance, will apply to all multivariate splines. For those familiar with Billera's module of $C^r$-splines, we note that $T^*(S_1 \cdot S_2) = T^*(S_1) \cdot T^*(S_2)$, and we have a module isomorphism.

## 3. Splines and matrices on graphs

For each spline in $S_d^r(\Delta)$ we now have an assignment $(\ldots, \beta^{jk}, \ldots)$ of polynomials of degree at most $d-r-1$ for the interior edges $(\ldots, P^jP^k, \ldots)$ as pairs of faces. In particular, for any *trivial spline*, in which all faces are assigned the same fixed polynomial, we obtain the *trivial assignment* $(\ldots, 0, \ldots)$. If we add a fixed polynomial, of degree at most $d$, to each face polynomial then the spline changes but the assignment does not change.

It is possible to characterize these assignments by a local vertex condition (see **[13]** or **[15, 27]** for such characterizations). Instead, we will switch to polynomials on the interior edges as pairs of vertices. This change may seem minor—but it is central to our approach throughout this series of papers.

First we order all edges lexicographically and define:

$$L_{hi}^{r+1} = \begin{cases} [\mathbf{p}_h\mathbf{p}_i\mathbf{X}]^{r+1} & \text{if } h < i, \\ -[\mathbf{p}_h\mathbf{p}_i\mathbf{X}]^{r+1} & \text{if } i < h. \end{cases}$$

Note that this means that $L_{hi}^{r+1} = -L_{ih}^{r+1}$ for all edges and all $r$.

Next, we orient each face clockwise (Figure 3(a)). Consider an interior edge $\mathbf{p}_h, \mathbf{p}_i$ separating polgons $P^j, P^k$. If the vertices $\mathbf{p}_h, \mathbf{p}_i$ appear in that order in the polygon $P^j$ (Figure 3(b)), then we define $\beta_{hi}$ by the equation

$$\beta_{hi}L_{hi}^{r+1} = S^j - S^k.$$

By our conventions, this implies

$$\beta_{ih}L_{ih}^{r+1} = S^k - S^j = -(S^j - S^k) = -(\beta_{hi}L_{hi}^{r+1}) = \beta_{hi}(-L_{hi}^{r+1}).$$

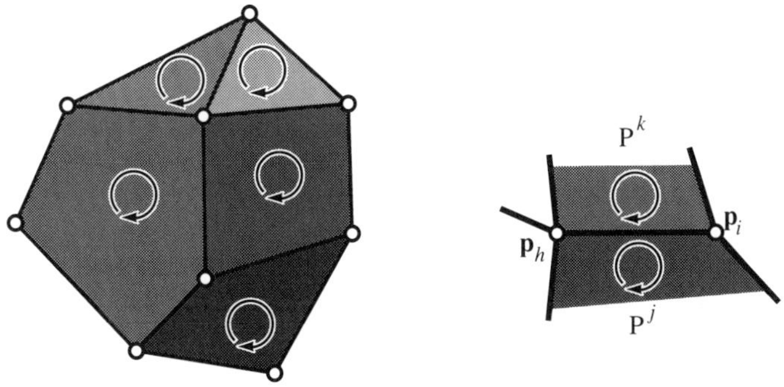


(a) (b)

FIGURE 3

Thus, $\beta_{ih} = \beta_{hi}$ for all interior edges.

Consider any closed face-edge path or *cycle*, $N$, in the subdivision $\Delta$:

$$P^0, \ldots, P^i, P^{i+1}, \ldots, P^0.$$

Around this cycle we have

$$\mathbf{0} = \sum_{\{i,i+1\}\in N} S^{i+1} - S^i = \sum_{\{h,i\}\in N} \beta_{hi} L_{hi}^{r+1}.$$

In particular, this holds for the cycle of faces and edges around any interior vertex of the subdivision.

Conversely, an assignment $(\ldots, \beta_{hi}, \ldots)$ of polynomials of degree at most $d-r-1$ to the edges of the interior graph of a disk subdivision $\Delta$ is called an $S_d^r$*-dependence* if, for the edges around any interior vertex $h$,

$$(*) \qquad \sum_{\{i|\{h,i\}\in E^\circ\}} \beta_{hi} L_{hi}^{r+1} = \mathbf{0}.$$

Since these are linear equations in the coefficients of the $\beta_{hi}$, the $S_d^r$-dependencies form a vector space $B_d^r(\Delta)$. We call this equation the *balance condition.* It is the extension of the "equilibrium condition" for statics (and $C_1^0$-splines) [**15, 16, 27**] and an analogue of the "conformality condition" of [**13, 14**].

2. THEOREM. *For any disk subdivision* $\Delta$, *and any chosen initial face* $P^0$, *there is an isomorphism between the subspace of* $C_d^r$*-splines which are* $\mathbf{0}$ *on* $P^0$ *and the space* $B_d^r(\Delta)$ *of* $S_d^r$*-dependencies.*

PROOF. We have already shown that each spline in $S_d^r(\Delta)$ leads to an $S_d^r$-dependence. It remains to show that each $S_d^r$-dependence comes from some spline in $S_d^r(\Delta)$ with $S^0 = \mathbf{0}$.

Consider an arbitrary $S_d^r$-dependence in $B_d^r(\Delta)$, $\beta = (\dots, \beta_{hi}, \dots)$. For each face $P^k$, we use a face-edge path $N$ from $P^0$ to $P^k$ to define $S^k$, as follows:

$$S^k = \mathbf{0} + \sum_{\{g,h\}\in N} \beta_{gh} L_{gh}^{r+1},$$

where $\{g, h\}$ has that order in its second face along the path. This polynomial $S^k$ is well defined. Each $S^k$ is a polynomial of degree at most $d$ and we will show that this polynomial does not depend on the path chosen. If we have two distinct paths $N$ and $N'$, then $N, -N'$ forms a closed cycle $M$. Now any closed cycle on a disk breaks down into the sum of cycles around a set of interior vertices "inside" the cycle. Specifically,

$$\sum_{\{g,h\}\in M} \beta_{gh} L_{gh}^{r+1} = \sum_f \sum_{\{i|\{f,i\}\in E^\circ\}} \beta_{fi} L_{fi}^{r+1} = \sum_f \mathbf{0} = \mathbf{0}.$$

Thus our "balance condition" applies to all closed cycles.

In particular, for the cycle $M$ defined above we have

$$\sum_{\{g,h\}\in N} \beta_{gh} L_{gh}^{r+1} - \sum_{\{g',h'\}\in N'} \beta_{g'h'} L_{g'h'}^{r+1} = \mathbf{0}$$

or

$$\sum_{\{g,h\}\in N} \beta_{gh} L_{gh}^{r+1} = \sum_{\{g',h'\}\in N'} \beta_{g'h'} L_{g'h'}^{r+1}.$$

These polynomials $S^j$ have the required continuity for an $S_d^r$-spline. For any two faces $P^j$ and $P^k$ sharing an edge $\{h, i\}$, a path $N$ from $P^0$ to $P^j$, with the added edge $\{h, i\}$, is a defining path for $P^k$:

$$S^k = \sum_{\{f,g\}\in N} \beta_{fg} L_{fg}^{r+1} + \beta_{ih} L_{ih}^{r+1} = S^j + \beta_{ih} L_{ih}^{r+1},$$

as required. This completes the proof that a unique spline with $S^0 = \mathbf{0}$ corresponds to each $S_d^r$-dependence.

Clearly the image of a sum of splines is the sum of the image dependencies, and a scalar times a spline is the same scalar times the image dependence. We have the required vector space isomorphism. □

REMARK. In the study of splines, it is not traditional to extract all these topological details of cycles, etc. However, in the theory of plane stresses, it is common to go far beyond $C_1^0$-splines to the projections of oriented polyhedra [**16, 27**]. The topological details become crucial. In the full theory of $S_d^r$-dependencies [**33**], we also work with more general objects than the interior graph of a disk, so the topological details bear an explicit treatment.

3. COROLLARY. *A disk subdivision* $\Delta$ *has a nontrivial spline in* $S_d^r(\Delta)$ *if and only if the disk subdivision has a nontrivial* $S_d^r$*-dependence.*

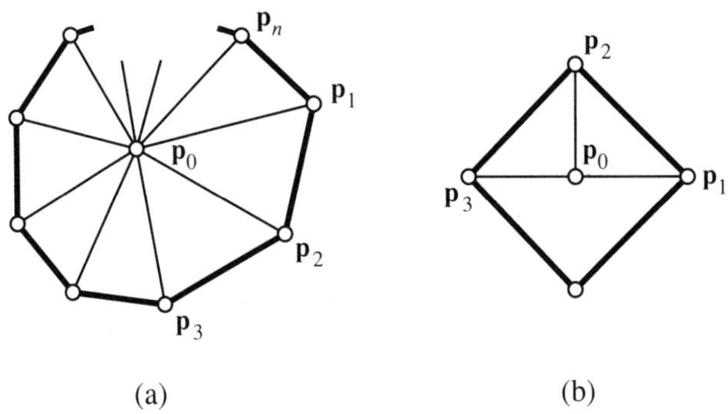


FIGURE 4

1. EXAMPLE. Consider $B_2^1(\Delta)$ for the simple disk subdivision of Figure 4(a). This is called a "cell" by people in splines and a "wheel" by people in graph theory. This selects $n$ scalars $\beta_{01}, \beta_{02}, \ldots, \beta_{0n}$ for the $n$ interior edges, which must satisfy the one polynomial equation

$$\beta_{01}|[\mathbf{p}_0\mathbf{p}_1\mathbf{X}]|^2 + \cdots + \beta_{0n}|[\mathbf{p}_0\mathbf{p}_n\mathbf{X}]|^2 = \mathbf{0}.$$

On the surface, this represents six linear equations on the scalars, for the six monomials $x^2, xy, y^2, x, y, 1$. However, if we choose the origin $(0, 0)$ as $\mathbf{p}_0$, it is clear that all lines have the form $Ax + By = 0$, and produce nonzero coefficients only for the three polynomials $x^2, xy, y^2$. Are the three remaining constraints actually independent?

Consider the three points $\mathbf{p}_1 = (0, 1)$, $\mathbf{p}_2 = (1, 0)$, $\mathbf{p}_3 = (-1, -1)$. These translate to the constraints

$$\beta_{01}(x^2) + \beta_{01}(y^2) + \beta_{03}(x^2 + 2xy + y^2) = \mathbf{0}.$$

Clearly there is only the trivial solution, and the three remaining constraints are independent. Any general set of four points, no three collinear, is projectively equivalent to this set—so the constraints remain independent as long as the lines at the vertex cover at least three directions (Figure 4(a)). This yields an $(n-3)$-dimensional space of $S_2^1$-dependencies.

If the edges use only two lines (Figure 4(b)), this reduces the number of independent constraints, giving an $(n-2)$-dimensional space of $S_2^1$-dependencies. (In the extreme case of two edges on one line, we have a $(n-1)$-dimensional space of $S_2^1$-dependencies.)

For a general point $\mathbf{p}_0$, the reduction from six constraints to three constraints at the vertex can also be expressed by the fact that all the polynomials $|[\mathbf{p}_0\mathbf{p}_i\mathbf{X}]|^2$ already satisfy the following three equations:

$$|[\mathbf{p}_0\mathbf{p}_i\mathbf{p}_0]|^2 = 0, \quad \frac{\partial}{\partial x}|[\mathbf{p}_0\mathbf{p}_i\mathbf{X}]|^2|_{\mathbf{p}_0} = 0, \quad \frac{\partial}{\partial y}|\mathbf{p}_0\mathbf{p}_i\mathbf{X}]|^2|_{\mathbf{p}_0} = 0.$$

More generally, consider $B_d^r(\Delta)$ for the disk subdivision of Figure 4. This selects $n$ polynomials $\beta_{01}, \beta_{02}, \dots, \beta_{0n}$ for the $n$ interior edges. The single polynomial equation

$$\beta_{01}L_{01}^{r+1} + \cdots + \beta_{0n}L_{0n}^{r+1} = \mathbf{0}$$

appears to be $\binom{d+2}{2}$ linear equations for the monomials $x^d, x^{d-1}y, \dots, y, 1$ in the $n\binom{r+2}{2}$ coefficients of the $\beta_{0i}$. Choosing the origin $(0, 0)$ as $\mathbf{p}_0$ makes $L_{0i}^{r+1}$ zero for the monomials of degree $\leq r$. Therefore we really have $\binom{d+2}{2} - \binom{r+2}{2}$ linear equations. If there are "enough" directions in the edges ($\geq r+2$), these equations are independent.

For a general point $\mathbf{p}_0$, the reduction from $\binom{d+2}{2}$ linear equations to $\binom{d+2}{2} - \binom{r+2}{2}$ linear equations at the vertex $\mathbf{p}_0$ can also be expressed by the fact that the polynomials $\beta_{01}\|[\mathbf{p}_0\mathbf{p}_i\mathbf{X}]\|^{r+1}$ already satisfy the $\binom{r+2}{2}$ equations:

$$\frac{\partial^k}{\partial x^j \partial y^{k-j}} \beta_{01}\|[\mathbf{p}_0\mathbf{p}_i\mathbf{X}]\|^{r+1}|_{\mathbf{p}_0} = 0 \quad \text{for } 0 \leq k \leq r+1.$$

These equations guarantee that for any vertex, we can derive the lowest coefficients, for $x^k y^m 0 \leq k+m \leq r$, of the polynomial $\beta_{hi}L_{hi}^{r+1}$ from the "truncation" $T_r$ which only retains the coefficients for $r < k+m$.

This analysis holds at each interior vertex for any $S_d^r$-dependence. Therefore, we can truncate the vector (polynomial) equations

$$(*) \qquad \sum_{\{i|\{h,i\}\in E^\circ\}} \beta_{hi}L_{hi}^{r+1} = \mathbf{0},$$

eliminating all powers $\leq r$. The defining balance equations for a $S_d^r$-spline are now

$$(**) \qquad \sum_{\{i|\{h,i\}\in E^\circ\}} T_r(\beta_{hi}L_{hi}^{r+1}) = \mathbf{0}.$$

We write these linear equations in the coefficients of the $\beta_{hi}$ as $\beta M_d^r(\Delta) = \mathbf{0}$, where $M_d^r(\Delta)$ is a $\binom{d-r+1}{2}|E^\circ| \times (\binom{d+2}{2} - \binom{r+2}{2})|V^\circ|$ matrix.

For example, for $S_{r+1}^r(\Delta)$, the truncation applies directly to $L_{hi}^{r+1}$, producing the polynomial $T_r(L_{hi}^{r+1}) = D_{hi}^{r+1}$. If the disk subdivision has interior edges $(1, 2)$, $(13)$, $\dots$, $(h, i)$, $\dots$, the matrix $M_2^1(\Delta)$ has the form:

$$M_2^1(\Delta) = \begin{bmatrix} D_{12}^2 & -D_{12}^2 & 0\,0\,0 & \dots & 0\,0\,0 & \dots & 0\,0\,0 & \dots & 0\,0\,0 \\ D_{13}^2 & 0\,0\,0 & -D_{13}^2 & \dots & 0\,0\,0 & \dots & 0\,0\,0 & \dots & 0\,0\,0 \\ \vdots & \vdots & \vdots & \ddots & \vdots & \ddots & \vdots & \ddots & \vdots \\ 0\,0\,0 & 0\,0\,0 & 0\,0\,0 & \dots & D_{hi}^2 & \dots & -D_{hi}^2 & \dots & 0\,0\,0 \\ \vdots & \vdots & \vdots & \ddots & \vdots & \ddots & \vdots & \ddots & \vdots \end{bmatrix}.$$

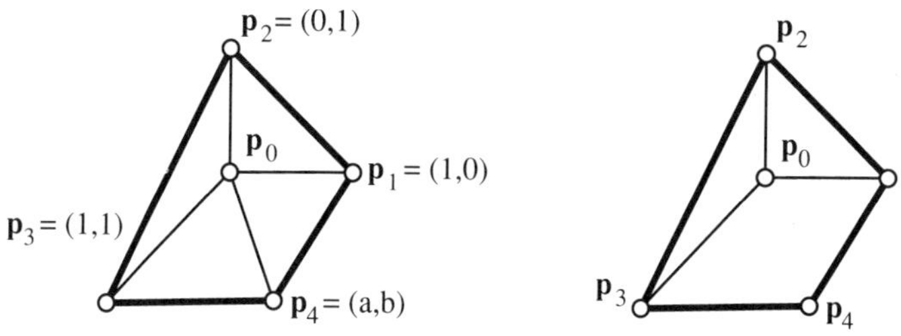


(a) (b)

FIGURE 5

This matrix is a full analogue of the rigidity matrix of a graph [**31**]. This similarity in form is the basis for the strong transfer of techniques between rigidity theory and the theory of splines.

2. EXAMPLE. Consider a quadrilateral disk with a single 4-valent interior vertex (Figure 5). With the labelled vertices $\mathbf{p}_0 = (0, 0)$, $\mathbf{p}_1 = (1, 0)$, $\mathbf{p}_2 = (0, 1)$, $\mathbf{p}_3 = (-1, -1)$, $\mathbf{p}_4 = (a, b)$ the edges give equations $D^2_{01} = (1, 0, 0)$, $D^2_{02} = (0, 0, 1)$, $D^2_{03} = (1, -2, 1)$, $D^2_{04} = (b^2, -2ab, a^2)$. This projectively general configuration gives the matrix

$$M_2^1(\Delta) = \begin{bmatrix} 0 & 0 & 1 \\ 1 & 0 & 0 \\ 1 & -2 & 1 \\ b^2 & -2ab & a^2 \end{bmatrix}.$$

These rows are always dependent, but any three are independent. The columns are independent and remain so if we delete one edge.

Consider $M_4^2(\Delta)$ for the same disk subdivision. We have $D^3_{01} = (1, 0, 0, 0)$, $D^3_{02} = (0, 0, 0, 1)$, $D^3_{03} = (1, -3, 3, -1)$, and $D^3_{04} = (b^3, -3b^2a, 3ba^2, -a^3)$. This gives:

$$M_4^2(\Delta) = \begin{bmatrix} 0 & 0 & 0 & 0 & 0 & 0 & 0 & 0 & 1 \\ 0 & 0 & 0 & 1 & 0 & 0 & 0 & 0 & 0 \\ 0 & 0 & 0 & 0 & 1 & 0 & 0 & 0 & 0 \\ 0 & 0 & 0 & 0 & 0 & 1 & 0 & 0 & 0 \\ 1 & 0 & 0 & 0 & 0 & 0 & 0 & 0 & 0 \\ 0 & 1 & 0 & 0 & 0 & 0 & 0 & 0 & 0 \\ 0 & 0 & 0 & 0 & 0 & 1 & -3 & 3 & -1 \\ 1 & -3 & 3 & -1 & 0 & 0 & 0 & 0 & 0 \\ 0 & 1 & -3 & 3 & -1 & 0 & 0 & 0 & 0 \\ 0 & 0 & 0 & 0 & 0 & b^3 & -3b^2a & 3ba^2 & -a^3 \\ b^3 & -3b^2a & 3ba^2 & -a^3 & 0 & 0 & 0 & 0 & 0 \\ 0 & b^3 & -3b^2a & 3ba^2 & -a^3 & 0 & 0 & 0 & 0 \end{bmatrix}.$$

The rows are dependent—even if we remove one edge and its three rows. The columns of this matrix are independent, but become dependent if we remove one edge.

If the rows of $M_d^r(\Delta)$ are independent, then we have only the zero $S_d^r$-dependence—and only trivial splines. On the other hand, if the columns of $M_d^r(\Delta)$ are independent, then we have a simple count for the dimension of the space of $S_d^r$-dependencies—and the space of $C_d^r$-splines. What is already known about these cases?

We write $|E^\circ| = e^\circ$ and $|V^\circ| = v^\circ$ and subtract the number of equations from the number of variables. This gives a simple form of the standard lower bound:

$$\dim(B_d^r(\Delta)) \geq \binom{d-r+1}{2} e^\circ - \left(\binom{d+2}{2} - \binom{r+2}{2}\right) v^\circ .$$

For simplicity, consider a disk subdivision with triangular faces and a boundary of $v^b$ vertices. By Euler's formula for a disk, we have

$$v^\circ - e^\circ + f = 1 \quad \text{and} \quad 3f = 2e^\circ + v^b \quad \text{implies} \quad e^\circ = 3v^\circ + (v^b - 3).$$

If we substitute and simplify, we have

$$\dim(B_d^r(\Delta)) \geq (d-r)(d-2r)v^\circ + \binom{d-r+1}{2}(v^b - 3).$$

This (or zero, if it is negative) is the best available lower bound, provided that there are no *defective vertices*: interior vertices which involve fewer than $r+2$ directions in their edges [**22, 23, 1**]. The bound is known to be exact for $r = 0$, for $d \geq 3r+2$ [**17**], and for $r = 1, d = 4$ [**3**], for all disk subdivisions without singular or defective vertices. In these cases the columns of the matrix are independent. (Notice that for $r \geq 4$, a triangulated triangle must contain vertices of valence $\leq 5$—which are topologically defective.)

For a triangular boundary, this proposes $\dim(B_d^r(\Delta)) = 0$ for $d \leq 2r$. This bound is not exact for some "special" triangulated disk subdivisions (see [**21, 2, 33**]). How do we separate a "special" disk subdivision from the "general" disk subdivision?

The combinatorial pattern of this matrix $M_d^r(\Delta)$ is determined by the combinatorics of the disk subdivision, not the geometric details of the positions of the vertices. The final disk subdivision $\Delta$, and the rank of $M_d^r(\Delta)$, depend on both the interior graph $G_\Delta = (V; E^\circ)$ and the positions $P_\Delta$ for the vertices. We can think of the matrix as $M_d^r(G_\Delta, P_\Delta)$.

The rank of a matrix is determined by the largest nonzero minor. If we use variables for the vertices, we quickly see that the rank is governed by a finite set of polynomials—which are either identically zero, or almost always nonzero. For example, if we choose algebraically independent transcendentals for the coordinates of the points, the largest possible set of polynomials will be nonzero, and the matrix will have maximum rank and maximal independent sets. This is the *generic* or *combinatorial* behavior for the graph $G_\Delta$ and therefore for the disk subdivision $\Delta$ for $C_d^r$-splines.

If we have any disk subdivision, the set of positions $P$ near $P_\Delta$ in $\mathbb{R}^{2|V|}$ which make all possible minors nonzero forms an open dense subset of the neighborhood. These *generic points* near $P_\Delta$ also correspond to disk subdivisions, although a random realization of the graph will not. If the matrix $M_d^r(G_\Delta, P)$ has independent rows at a generic point $P$, we say that $\Delta$ is *combinatorially independent.* Of course this rank may drop due to the value of $P$—causing a *geometric dependence.* Similarly, we can speak of the *combinatorial rank* of the disk subdivisions, and even the *combinatorial dimension* of $S_d^r(\Delta)$.

Recent results show that $\dim(B_2^1(\Delta)) = 0$ and $\dim(B_3^1(\Delta)) = 2v^\circ + 3(v^b - 3)$ for generic disk subdivisions [**8, 29, 31**]. It is also known that the bound is exact for generic disk subdivisions without singular or defective vertices for $d = 3r + 1$ [**4**]. We will show that 0 is the exact value for generic disk subdivisions, for $d \leq (3r+1)/2$, even if there are topologically defective vertices.

REMARK. For disk subdivisions with some nontriangular faces, these bounds can also fail to be exact for topological reasons. Figure 6 shows two disk subdivisions for which the estimated and correct lower bounds for $\dim(S_1^0(\Delta))$ at a generic point are the following:

$$\text{Estimated(A)} = 3, \qquad \text{Correct(A)} = 3,$$
$$\text{Estimated(B)} = 3, \qquad \text{Correct(B)} = 4.$$

The "failure" of $B$ is topological, not geometric. (The reader can see that the vertex $\mathbf{p}_0$ can be lifted from the plane, creating a nontrivial $C_1^0$-spline.) The results of plane statics can be used to generate the correct lower bounds for all of these cases, using standard methods of matroid theory [**27, 31**]. Similar counter examples with nontriangular faces exist for $S_{2r+1}^r(\Delta)$ for all $r$.

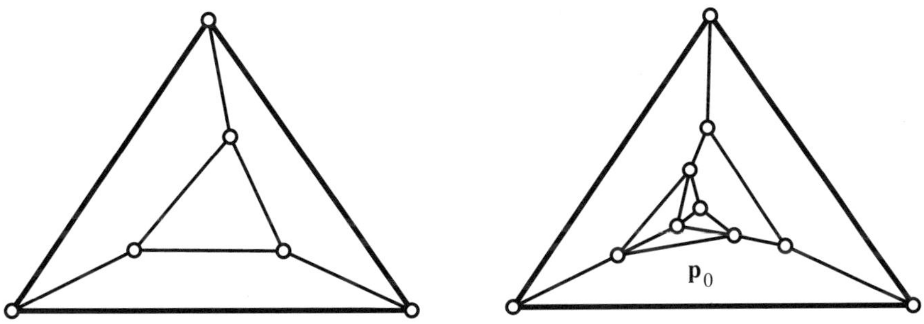


(a) (b)

FIGURE 6

## 4. Combinatorial behavior of $S_d^r(\Delta)$ for $d \leq (3r+1)/2$

Consider a triangulated disk $\Delta$ with a triangle boundary. In the previous section, we have seen that for $d \leq 2r$ there is a chance that the rows of $M_d^r(\Delta)$

are independent. Our approach is to analyze the matrix by an induction on the number of interior vertices. Consider the simplest cases.

3. EXAMPLE. Consider a single interior vertex $(0, 0)$ connected by three interior edges to the boundary vertices $(1, 0)$, $(0, 1)$, and $(-1, -1)$. For $S_3^2(\Delta)$ this creates the matrix

$$M_3^2(\Delta) = \begin{bmatrix} 0 & 0 & 0 & 1 \\ 1 & 0 & 0 & 0 \\ 1 & -3 & 3 & -1 \end{bmatrix}.$$

These rows are independent—and we can hope to proceed with an induction.

Consider $S_4^2(\Delta)$ for the same disk subdivision (recall Example 2). $M_4^2(\Delta)$ has dependent rows since the six rows added to $M_3^2(\Delta)$ are only nonzero in the five added columns—hence are themselves dependent. Neither the rows nor the columns of this matrix are independent, since this vertex is topologically defective. Here, we cannot use an induction which includes 3-valent vertices.

These examples indicate a basic limit to "independent rows". For $M_d^r(\Delta)$ on a single 3-valent interior vertex, we need the last added $3(d-r)$ rows to be no more than the last added $(d+1)$ columns. That is, we need $3(d-r) \leq (d+1)$ or $d \leq (3r+1)/2$. (For example, $r = 1$, $d = 2$.)

4. LEMMA. *For a disk subdivision $\Delta$ with a single 3-valent interior vertex using three distinct directions for the edges and for $d \leq (3r+1)/2$, the rows of $M_d^r(\Delta)$ are independent.*

PROOF. By a projective transformation, we assume that the lines are $x = 0$, $y = 0$, and $x - y = 0$. For $d = r + 1$, $r \geq 1$, we have

$$M_{r+1}^r(\Delta) = \begin{bmatrix} 0 & 0 & \cdots & 0 & 1 \\ 1 & 0 & \cdots & 0 & 0 \\ 1 & -(r+1) & \cdots & (-1)^r(r+1) & (-1)^{r+1} \end{bmatrix}.$$

These three rows are clearly independent.

Assume that the rows were independent for $M_k^r(\Delta)$, and $k+1 \leq (3r+1)/2$. As we pass from $k$ to $k+1$ we add the specific block:

$$C = \begin{bmatrix} 1 & 0 & 0 & \cdots & 0 & 0 & 0 \\ 0 & 1 & 0 & \cdots & 0 & 0 & 0 \\ \vdots & \vdots & \vdots & \ddots & \vdots & \vdots & \vdots \\ \vdots & \vdots & \vdots & \ddots & \vdots & \vdots & \vdots \\ 0 & 0 & 0 & \cdots & 0 & 1 & 0 \\ 0 & 0 & 0 & \cdots & 0 & 0 & 1 \\ 1 & -(r+1) & \binom{r+1}{2} & \cdots & 0 & 0 & 0 \\ 0 & 1 & -(r+1) & \cdots & 0 & 0 & 0 \\ \vdots & \vdots & \vdots & \ddots & \vdots & \vdots & \vdots \\ 0 & 0 & 0 & \cdots & (-1)^r(r+1) & (-1)^{r+1} & 0 \\ 0 & 0 & 0 & \cdots & (-1)^{r-1}\binom{r+1}{2} & (-1)^r(r+1) & (-1)^{r+1} \end{bmatrix}.$$

It is clear that the upper rows for the first two edges are independent. We leave it as a direct exercise with binomial coefficients to show that the entire matrix is independent provided that $k+1 \leq (3r+1)/2$. □

For the induction, we need one further result about triangulated disks. Given a triangulated triangle $\Delta$ with at least two interior vertices, we have two adjacent interior vertices $\mathbf{p}_0, \mathbf{p}_1$, joined by an interior edge. We wish to *shrink* this edge, identifying the two vertices, dropping the two triangles $\mathbf{p}_0, \mathbf{p}_1, \mathbf{p}_2$ and $\mathbf{p}_0, \mathbf{p}_1, \mathbf{p}_3$, and identifying the edges $\mathbf{p}_0\mathbf{p}_2$ and $\mathbf{p}_1, \mathbf{p}_2$, and the edges $\mathbf{p}_0\mathbf{p}_3$ and $\mathbf{p}_1, \mathbf{p}_3$ (Figure 7(a)). If this creates a smaller triangular disk, we call this a *shrinkable edge.* (Figure 7(b) shows a nonshrinkable edge.) We adapt a lemma of [7].

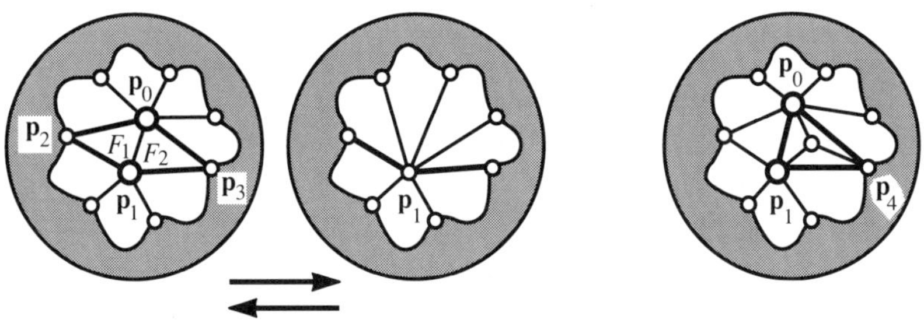

(a) (b)

FIGURE 7

5. LEMMA. *Every triangulated disk $\Delta$ with at least two interior vertices contains a shrinkable edge joining two interior vertices.*

PROOF. Take an edge joining two interior vertices $\mathbf{p}_0, \mathbf{p}_1$ in $\Delta$. If this edge is not shrinkable, then the identification fails because there is a (nonfacial) triangle $\mathbf{p}_0, \mathbf{p}_1, \mathbf{p}_4$ in the original disk subdivision.

In this case, the problem triangle and its interior is a smaller triangulated disk $\Delta$ (Figure 7(b)), and we show by induction that there is a shrinkable edge inside this disk (possibly going to the boundary of $\Delta^*$ but not of $\Delta$). If the smaller disk contains only one interior vertex, then all interior edges of it are shrinkable. One of these edges also connects to a selected vertex $\mathbf{p}_0$, which is an interior vertex of $\Delta$.

Assume there is a shrinkable edge connected to vertex $\mathbf{p}_0$ in all triangles with $k$ interior vertices. Take any interior edge from $\mathbf{p}_0$ in a triangle with $k+1$ vertices. If this is shrinkable we are done. Otherwise, a shrinking identifies a smaller triangle sharing the vertex, which, by induction, also contains a shrinkable edge. We conclude that every triangulated interior triangle contains a shrinkable edge to $\mathbf{p}_0$. This is an edge between two interior vertices of the original disk $\Delta$. □

6. THEOREM. *For a disk subdivision* $\Delta$ *with a triangular boundary, the rows of* $M_d^r(\Delta)$ *are combinatorially independent if* $d \leq (3r+1)/2$.

PROOF. We first prove the theorem for a disk subdivision with all faces triangular by induction on the number of interior vertices. The theorem is clearly true for a single triangle with no interior vertices.

(1) The smallest nontrivial disk subdivision has one interior vertex and three concurrent edges. We have seen, in Lemma 4, that the rows are independent.

(2) Consider a triangulated triangle $\Delta$ with $n+1$ interior vertices. Shrinking along any shrinkable edge creates a smaller triangulated triangle $\Delta'$ which is combinatorially independent, by the induction hypothesis. We will reverse this shrinking to create the original pattern—a process called *vertex splitting* which adds one vertex, two triangles, and three edges. Following the labeling of Figure 7(a), we split the vertex $\mathbf{p}_1$ into $\mathbf{p}_0$ and $\mathbf{p}_1$, adding triangles $F^1$ and $F^2$.

We choose, as a limiting initial case, to add $\mathbf{p}_0$ at $\mathbf{p}_1 = (0, 0)$, with an arbitrary "direction" for the "line" $\mathbf{p}_0, \mathbf{p}_1$ and hence the vector $D_{01}^{r+1}$. Assume there is a dependence $(\dots, \beta_{hi}, \dots)$ on the rows of the matrix of this "degenerate" $\Delta$. This will reduce to a dependence $(\dots, \beta'_{hi}, \dots)$ satisfying $\sum \beta'_{hi} D_{hi}^{r+1} = \mathbf{0}$ on the matrix of the smaller triangulation $\Delta'$, if we set $\beta'_{12} = \beta_{12} + \beta_{02}$, $\beta'_{13} = \beta_{03} + \beta_{13}$, and drop $\beta_{01}$. (The balance of $\beta'$ at $\mathbf{p}_1$ follows by the sum of the equations for $\beta$ at $\mathbf{p}_0$ and $\mathbf{p}_1$, and the cancellation of the term for $\beta_{01}$.)

By our induction hypothesis, all $\beta'_{hi} = \mathbf{0}$ on a generic disk subdivision $\Delta'$. Thus, on the new subdivision $\Delta$ the only possible nonzero values are

$$\beta_{02} = -\beta_{12}, \quad \beta_{03} = -\beta_{13}, \quad \text{and} \quad \beta_{01}.$$

Around $\mathbf{p}_0$ in the larger triangulation $\Delta$, we have

$$\beta_{01} D_{01}^{r+1} + \beta_{02} D_{02}^{r+1} + \beta_{03} D_{03}^{r+1} = \mathbf{0}.$$

However, in the proof of Lemma 4 we saw that this requires $\beta_{01} = \beta_{02} = \beta_{03} = \mathbf{0}$ if no two lines are collinear. Thus, for generic choices of $\Delta'$ and of a direction for $D_{01}^2$, the rows of the matrix are independent.

A small change in the position of $\mathbf{p}_0$ leads to nearby nondegenerate disk subdivision $\Delta$ for which the matrix is also independent. By induction, the theorem holds for all triangulated disks with a triangular boundary and vertices in generic position.

(3) If we have nontriangular faces in the disk subdivision, then triangulating these faces would only increase the possible assignments $(\dots, \beta_{hi}, \dots)$. Since the triangulated structure produced independent rows, the original disk subdivision produced independent rows. □

REMARK. This result is an analogue of a classical rigidity theorem of Cauchy [**11, 18, 30, 31**]. Cauchy's Theorem proves the infinitesimal rigidity of convex triangulated spheres in 3-space by the independence of the rows of the corresponding rigidity matrix. For $S_2^1$-dependencies, this is a direct proof of a result which Billera **[8]** derived from a related technical theorem of Whiteley **[29]**.

7. COROLLARY. *For $d \leq (3r+1)/2$, a disk subdivision $\Delta$ with a triangular boundary and vertices in generic position has only trivial $C_d^r$-splines with a global degree $d$ polynomial.*

What happens if we enlarge the boundary beyond a triangle? We have already seen the lower bound:

$$\dim(B_2^1(\Delta)) \geq e^\circ - 3v^\circ = (v^b - 3).$$

Since this is positive for larger boundaries, we cannot hope to prove the rows of $M_2^1(\Delta)$ are independent. It can easily be shown that the theorem does extend to $S_d^r$ with $d \leq (3r+1)/2$ and a $k$-gon boundary with $k(d-r) \leq d+1$ (i.e., $k \leq (d+1)/(d-r)$). For example, $S_3^2$ with a quadrilateral boundary is covered.

For the special case of $M_2^1(\Delta)$ we have a further observation which will be used in the next section. For a triangulated triangle, $e^\circ = 3v^\circ$, and the matrix is square. Since the matrix has independent rows, the columns are also independent. This independence is retained if we delete a boundary edge and its face, enlarging the polygon.

8. COROLLARY. *For a triangular disk subdivision $\Delta$ with an arbitrary $n$-gon boundary, the columns of $M_2^1(\Delta)$ are combinatorially independent.*

As we noted above, the previous results actually imply the combinatorial independence of the columns of $M_d^1(\Delta)$ for all $d$ on all $n$-gons. This result can also be proven directly by a matrix induction illustrated in **[31]**.

## 5. Singular vertices in $S_r^1(\Delta)$

It is well known that each singular vertex, with only two directions (Figure 8(a)), increases the dimension of $S_d^1(\Delta)$ (and of $B_d^1(\Delta)$). In our terminology, the standard lower bounds of **[22, 23]** are

$$\dim(B_d^1(\Delta)) \geq (d-1)(d-2)v^\circ + \binom{d}{2}(v^b - 3) + v^s.$$

It is also known that this bound is exact, for all disk subdivisions with $d \geq 4$.

If we reexamine these singular vertices, we see that each such vertex induces an automatic column dependence in $M_d^1(\Delta)$ in the columns of this vertex. We can adapt our techniques to verify this refined bound is exact for $d = 2$ and $d = 3$, provided that the nonsingular vertices remain in "generic" position.

9. LEMMA. *In a triangulated disk subdivision $\Delta$, no two singular vertices are adjacent.*

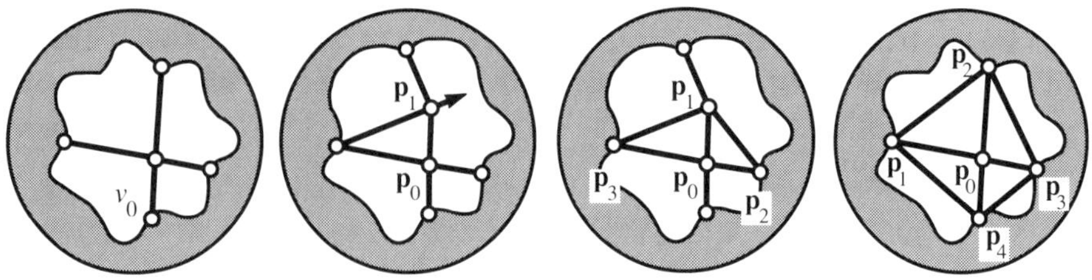


(a) (b) (c) (d)

FIGURE 8

PROOF. Assume that $\mathbf{p}_1$ is a singular vertex adjacent to $\mathbf{p}_1$ (Figure 8(b)). Since we have a triangulated disk, the vertex $\mathbf{p}_1$ must share two triangles with $\mathbf{p}_0\mathbf{p}_2$ and $\mathbf{p}_0\mathbf{p}_3$ (Figure 8(c)). For a nondegenerate disk subdivision, the three lines $\mathbf{p}_1\mathbf{p}_1$, $\mathbf{p}_1\mathbf{p}_2$, and $\mathbf{p}_1\mathbf{p}_3$ are distinct. Therefore, $\mathbf{p}_1$ is not singular. □

The next induction lemma allows us to trace the rank of the matrix $M_2^1(\Delta)$ and $M_3^1$ as we add singular vertices.

10. LEMMA. *If a disk subdivision $\Delta$ has independent columns in $M_2^1(\Delta)$, or $M_3^1(\Delta)$, for all nonsingular vertices, then inserting a singular vertex inside an interior edge between nonsingular vertices adds exactly one dependent column among the columns for this vertex.*

PROOF. We add the singular vertex $v_0$ to the matrix $M_d^1(\Delta^*)$ in two stages. Assume that the edge $\mathbf{p}_1\mathbf{p}_3$ is already present in the disk subdivision $\Delta$ (Figure 9(a)). First we add the rows for an "edge" $\mathbf{p}_2\mathbf{p}_4$ to the matrix creating $M^*$, which does not directly correspond to any disk subdivision (Figure 9(b)). Since the columns of $M_d^1(\Delta^*)$ corresponding to the nonsingular vertices are independent, this clearly does not increase the rank of the matrix.

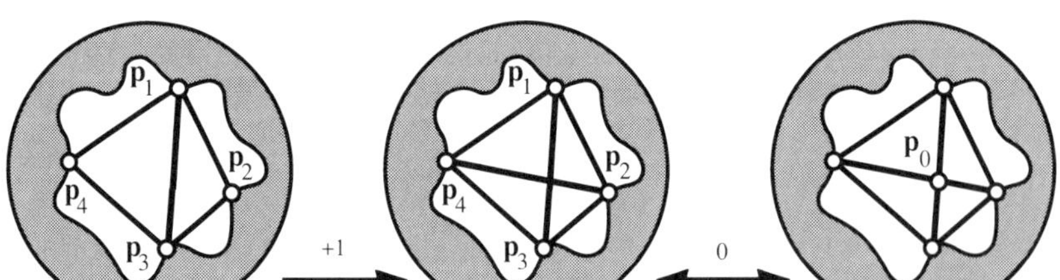


(a) (b) (c)

FIGURE 9

We now "insert" a new singular vertex $\mathbf{p}_0$ at the crossing point of the lines $\mathbf{p}_1\mathbf{p}_3$ and $\mathbf{p}_2\mathbf{p}_4$, splitting these edges but preserving the lines (Figure 9(c)).

This gives us the graph of the disk subdivision $\Delta$ and the matrix $M_d^1(\Delta)$. We must account for the independent columns of this modified matrix.

Adding the crossing point can be described in two steps. First we add the vertex with two edges $\mathbf{p}_0\mathbf{p}_1$ and $\mathbf{p}_0\mathbf{p}_2$. This adds $\frac{1}{2}(d-1)(d+4)$ columns and $d(d-1)$ rows in the pattern:

$$M' = \begin{bmatrix} 0 & M_d^1(\Delta^*) \\ B & C \end{bmatrix}.$$

Without loss of generality, the two lines are, projectively, the lines $x = 0$ and $y = 0$. For $d = 2$, the added block $B$ has a simple form:

$$B = \begin{bmatrix} 0 & 0 & 1 \\ 1 & 0 & 0 \end{bmatrix}.$$

These give two independent columns, leaving one dependent column in the middle.

For $d = 3$, the added columns and rows take the form

$$B'' = \begin{bmatrix} 0 & 0 & 1 & 0 & 0 & 0 & 0 \\ 0 & 0 & 0 & 1 & 0 & 0 & 0 \\ 0 & 0 & 0 & 0 & 0 & 0 & 1 \\ 1 & 0 & 0 & 0 & 0 & 0 & 0 \\ 0 & 1 & 0 & 0 & 0 & 0 & 0 \\ 0 & 0 & 0 & 0 & 1 & 0 & 0 \end{bmatrix}.$$

This makes all the columns independent—except one column inherited from $d = 2$. The net result is that all but one of the added columns are independent, as required.

A simple row reduction, which preserves the column independence and dependence, exchanges the rows for $\mathbf{p}_0\mathbf{p}_1$ and $\mathbf{p}_1\mathbf{p}_3$ for the parallel rows for $\mathbf{p}_0\mathbf{p}_1$ and $\mathbf{p}_0\mathbf{p}_3$, and the rows of $\mathbf{p}_0\mathbf{p}_2$ and $\mathbf{p}_2\mathbf{p}_4$ for the parallel rows for $\mathbf{p}_0\mathbf{p}_2$ and $\mathbf{p}_0\mathbf{p}_4$. This takes us to the matrix $M_d^1(\Delta)$ with one additional column dependence.

Wc have assumed, implicitly, that the vertices $\mathbf{p}_1$, $\mathbf{p}_3$ were interior vertices. Even if all four of the vertices $\mathbf{p}_1$, $\mathbf{p}_2$, $\mathbf{p}_3$, $\mathbf{p}_4$ are boundary vertices, the above process actually works—even though the added rows for $\mathbf{p}_2\mathbf{p}_4$ were zero rows.

If we take a vertex off the origin, and other directions for the two lines, the column dependence will "spread out" across the columns for the vertex, but remain restricted to this set of columns. □

For $d \geq 4$, the same lemma follows from the known results for $\dim(S_d^1(\Delta))$.

11. PROPOSITION. *For a triangulated disk subdivision $\Delta$ with $v^s$ singular vertices and all nonsingular vertices in generic position*

$$\dim(B_d^1(\Delta)) = (d-1)(d-2)v^\circ + \binom{d}{2}(v^b - 3) + v^s.$$

PROOF. We create a disk subdivision $\Delta'$ with the nonsingular vertices and edges, and one extended edge in place of each singular vertex. With its vertices in generic position, the columns of the matrix $M_d^1(\Delta')$ are independent.

If we now insert the singular vertices at the crossing points, we have, by Lemma 10,

$$\operatorname{rank}(M_d^1(\Delta)) = \frac{1}{2}\left(\binom{d+2}{2} - 3\right)v^\circ - v^s.$$

Since we have $\binom{d}{2}e^\circ$ rows and $e^\circ = 3v^\circ = (v^b - 3)$, we find that

$$\dim(B_d^1(\Delta)) = (d-1)(d-2)v^\circ + \binom{d}{2}(v^b - 3) + v^s. \quad \square$$

12. COROLLARY. *For a disk subdivision $\Delta$ with $v^\circ$ interior vertices, $v^s$ singular vertices, and $v^b$ boundary vertices and all nonsingular vertices in generic position, we have*

$$\dim(S_d^1(\Delta)) = \binom{d+2}{2} + (d-1)(d-2)v^\circ + \binom{d}{2}(v^b - 3) + v^s.$$

This confirms that Schumaker's lower bound is exact for $d = 2$ and $d = 3$, if the nonsingular vertices are in generic position. Again this bound was proven exact for all positions of the vertices, for $S_d^1$ with $d \geq 4$ **[3, 21]**. The examples of **[21, 2]**, etc. show that this is not exact for some "special" $S_2^1(\Delta)$.

We *conjecture* that this bound, $2v^\circ + 3v^b + v^s - 3$, is also exact for $S_3^1(\Delta)$ for *all* triangulated disk subdivisions. We note that a substantial unsolved problem in approximation theory asks whether this space of splines permits the arbitrary assignment of $v^\circ + v^b$ heights to the vertices. The space is clearly large enough, but are these heights independent parameters in the space?

## 6. Conclusions

These matrix and graph methods have given the combinatorial dimension of $S_d^r(\Delta)$ for a number of "small" values: $r = 1$, $d = 2, 3$, or $d \leq (3r+1)/2$. Other methods, based on the Bézier-Bernstein nets of a spline, have provided an explicit construction for a basis of $S_d^r(\Delta)$ for larger values with $d \geq 3r+2$ and any triangulated disk subdivision $\Delta$ **[3, 19]**, and for $d = 3r + 1$ and generic vertices **[4]**.

There remains a significant gap between these two sets of results. As Table 1 indicates there is an important wedge of unsettled cases (shaded).

For simplicity in the table, we assumed a triangular disk subdivision with triangular boundary and generic positions of the vertices. Further, in entries indicated by $^*$, we must add a correction factor $v_3^s$ for the 3-valent vertices (which are "singular", even in generic position). In the entries indicated by †, we must add a correction factor $2v_3^s + v_4^s$ to the conjectured and proven values, for the 3-valent vertices $v_3^s$, and the 4-valent vertices $v_4^s$. In the

| r \ d | 1 | 2 | 3 | 4 | 5 | 6 | 7 | 8 | 9 | 10 | $k = d - r$ |
|---|---|---|---|---|---|---|---|---|---|---|---|
| 0 | $v$ | $4v$ | $9v$ | $16v$ | $25v$ | $36v$ | $49v$ | $64v$ | $81v$ | $100v$ | $k^2v$ |
| 1 | | 0 | $2v$ | $6v$ | $12v$ | $20v$ | $30v$ | $42v$ | $56v$ | $72v$ | $k(k-1)v$ |
| 2 | | | 0 | 0* | $3v$* | $8v$* | $15v$* | $24v$* | $35v$* | $48v$* | $k(k-3)v$* |
| 3 | | | | 0 | 0 | 0† | $4v$† | $10v$† | $16v$† | $28v$† | $k(k-4)v$† |
| 4 | | | | | 0 | 0 | 0? | 0+ | $5v$+ | $12v$+ | $k(k-5)v$+ |

TABLE 1. Dimensions of $B_d^r(\Delta)$ for generic positions for the vertices

entries indicated by +, we must add the correction factor $4v_3^s + 2v_4^s + v_5^s$ for the vertices of valence 3, 4, and 5.

The smallest case without a proven value, even for generic vertices, is $S_4^2(\Delta)$. We conjecture that, for triangulated triangles with all vertices at least 4-valent, $\dim(B_4^2(\Delta)) = 0$. We view this wedge, which includes cases such as $S_4^2(\Delta)$, $S_5^3(\Delta)$, and $S_6^3(\Delta)$, as an essential challenge to all the available techniques. In the context of Billera's modules of splines **[9, 10]**, a detailed analysis of the borderline cases, $S_{2r}^r(\Delta)$, remains an essential step for developing an algebraic theory of splines.

## BIBLIOGRAPHY


1. P. Alfeld, *On the dimension of multivariate piecewise polynomials*, in Proceedings of the Biennial Dundee Conference on Numerical Analysis, Pitman, Harlow, Essex, 1985.
2. ——, *A case study of multivariate piecewise polynomials*, in Geometric Modeling: Algorithms and New Trends (G. Farin, ed.), SIAM, Philadelphia, Pa., 1987, pp. 149–160.
3. P. Alfeld, B. Piper, and L. L. Schumaker, *An explicit basis for $C^1$ quartic bivariate splines*, SIAM J. Numer. Anal. **24** (1987), 891–911.
4. ——, *On the dimension of bivariate spline spaces of smoothness $r$ and degree $d = 3r+1$*, Numer. Math. **57** (1990), 651–661.
5. P. Alfeld, L. L. Schumaker and W. Whiteley, *Splines by projection*, (to appear).
6. P. Ash, E. Bolker, H. Crapo, and W. Whiteley, *Convex polyhedra, Dirichlet tesselations and spider webs*, in Shaping Space, a Polyhedral Approach (M. Senechal and G. Fleck, eds.), Birkhauser, Boston, 1988, pp. 231–250.
7. D. Barnette, *Generating triangulations of the projective plane*, J. Combin. Theory Ser. B **33** (1982), 222–230.
8. L. Billera, *Homology theory of smooth splines: generic triangulations and a conjecture of Strang*, Trans. Amer. Math. Soc. **310** (1988), 325–340.
9. ——, *The algebra of continuous piecewise polynomials*, Adv. in Math. **76** (1989), 170–183.
10. L. Billera and L. Rose, *A dimension series for multivariate splines*, preprint, Dept. of Mathematics, Cornell University, Ithaca, New York, 1989.
11. A. Cauchy, *Deuxiéme memoire sur les polygons et les polyédres*, J. Ecole Polytech. **16** (1831), 87–98.
12. C. K. Chui, *Multivariate splines*, SIAM, Philadelphia, Pa., 1988.
13. C. K. Chui and R. H. Wang, *Multivariate spline spaces*, J. Math. Anal. Appl. **94** (1983), 197–221.
14. ——, *On smooth multivariate spline functions*, Math. Comp. **41** (1983), 131–142.

15. H. Crapo and W. Whiteley, *Stresses on frameworks and motions of panel structures: a projective geometric introduction*, Structural Topology **6** (1982), 42–82.
16. ——, *Plane stresses and projected polyhedra*, preprint Champlain Regional College, St. Lambert, Québec, 1978, 1985.
17. A. Fogelsanger, *The generic rigidity of minimal cycles*, Ph.D. Thesis, Dept. of Mathematics, Cornell University, Ithaca, New York, June 1988.
18. H. Gluck, *Almost all simply connected surfaces are rigid*, in Geometric Topology, Lecture Notes in Math., vol. 438, Springer-Verlag, Berlin and New York, 1975, pp. 225–239.
19. D. Hong, *Spaces of multivariate spline functions over triangulations*, preprint, 1988.
20. J. C. Maxwell, *On reciprocal diagrams and diagrams of forces*, Philosophical Mag. (4) **27** (1964), 250–261.
21. J. Morgan and R. Scott, *A nodal basis for $C^1$ piecewise polynomials of degree $n \geq 5$*, Math. Comp. **29** (1975), 736–740.
22. L. L. Schumaker, *On the dimension of spaces of piecewise polynomials in two variables*, in Multivariate Approximation Theory (W. Schemp and K. Zeller, eds.), Birkhauser, Basel, 1979, pp. 396–412.
23. ——, *Bounds on the dimension of spaces of multivariate piecewise polynomials*, Rocky Mountain J. Math. **14** (1984), 251–264.
24. G. Strang, *The dimension of piecewise polynomial spaces and onesided approximations*, Proc. Conf. Numerical Solution of Differential Equations (Dundee, 1973), Lecture Notes in Math., vol. 365, Springer-Verlag, New York, 1974, pp. 144–152.
25. K. Sugihara, *Machine interpretation of line drawings*, MIT Press, 1986.
26. R. H. Wang, *The structural characterization and interpolation for multivariate splines*, Acta Math. Sinaca **18** (1975), 91–106. (English transl., ibid. **18** (1975), 10–39.)
27. W. Whiteley, *Motions, stresses and projected polyhedra*, Structural Topology **7** (1983), 13–38.
28. ——, *Two algorithms for polyhedral pictures*, Proc. 2nd Annual ACM Symposium on Computational Geometry, Assoc. Computing Machinery, New York, 1986, pp. 142–149.
29. ——, *A matrix for splines*, in *Progress in approximation theory* (Nevai and Pinkus, eds.) Academic Press, Boston, 1991.
30. ——, *Vertex splitting in isostatic frameworks*, Structural Topology (to appear).
31. ——, *Matroids and rigid structures*, Applications of Matroid Theory (N. White, ed.), Cambridge Univ. Press (to appear).
32. ——, *A matroid for hypergraphs, with applications in scene analysis and geometry*, Discrete Comput. Geom. **4** (1989), 75–95.
33. ——, *The geometry of bivariate $C_2^1$-splines*, preprint, Champlain Regional College, St. Lambert, Québec, 1990.

Department of Mathematics, Champlain Regional College, St. Lambert, Québec, J4P-3P2

*E-mail address*: Whiteley@CC.UMontreal.CA